AF688286

Clinical Embryology

Robert Carachi • Sameh Helmi Edward Doss
Editors

Clinical Embryology

An Atlas of Congenital Malformations

Springer

Editors
Robert Carachi
Yorkhill NHS Trust
University of Glasgow Surgical Paediatrics
Glasgow
UK

Sameh Helmi Edward Doss
Anatomy & Embryology
Kasr Al Ainy-Cairo University
Cairo
Egypt

Associate Editors
Sharon F. Sneddon
School of Medicine, Dentistry and Nursing
University of Glasgow
Glasgow
UK

Suzanne V. McMahon
Department of Paediatric Surgery
Royal Hospital for Children
Glasgow
UK

ISBN 978-3-319-26156-0 ISBN 978-3-319-26158-4 (eBook)
https://doi.org/10.1007/978-3-319-26158-4

Library of Congress Control Number: 2019930718

Cover illustration: Cartoon by Leonardo Da Vinci(1452-1519) Depicting Studies of the foetus in the womb.

This Springer imprint is published by the registered company Springer Nature Switzerland AG
The registered company address is: Gewerbestrasse 11, 6330 Cham, Switzerland

Cartoon by Leonardo Da Vinci (1452–1519) Depicting Studies of the foetus in the womb. Permission From The Royal Windsor Collection of Her Majesty the Queen

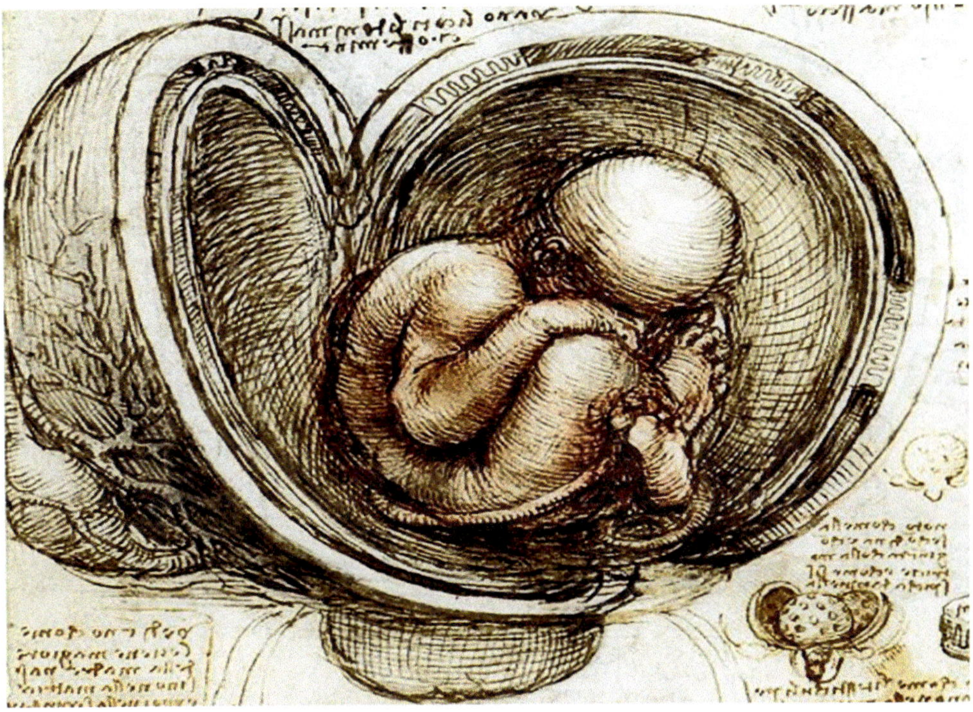

Preface

This book is a unique compilation of notes first documented and written in 1996 by Dr. S. Doss, one of the co-editors who is an anatomist and embryologist at Cairo University. The clinical photographs are a unique compilation of three generations of paediatric surgeons, a collection of over 40,000 images, the lifetime work of the late Professor D. G. Young, University of Glasgow. This book is dedicated to him.

Clinical embryology is defined as the study of the embryo and its development into the human form, first as a foetus until birth. It explores the relationship to clinical malformations that present at birth.

Congenital malformations affect an estimated 3.2 million birth defect-related disabilities each year worldwide (WHO statistics 2014). Three children out of every hundred born are affected by a congenital malformation and an estimated 270,000 newborn die within the first month each year. The ones who survive from the medical and surgical care they receive often result in long-term disabilities. These have significant impact on the patients, their families and carers, the health systems and society as a whole. The most common severe congenital malformations affect the heart, the nervous system, spine and Downs syndrome. In 2010 the World Health Assembly adopted a resolution calling all member states to promote primary prevention and the health of children with congenital malformations by:

1. Developing and strengthening registration and surveillance systems
2. Developing expertise and building capacity
3. Strengthening research and studies on aetiology, diagnosis and prevention
4. Promoting international co-operation

The problem with teaching embryology as an undergraduate subject is that all too often it is done in isolation as a science and part of the anatomy teaching. Since anatomy teaching has been dropped by many medical schools worldwide, the teaching of embryology has also been neglected. A recent survey of the whole of the medical undergraduate school of one of the largest universities in the UK revealed that only 17% of students were confident in their embryology knowledge and that 80% declared that it should definitely be included in the medical school curriculum. Eighty per cent found it a difficult subject to understand and only 36% were satisfied with the teaching of embryology. As a subject embryology is difficult to conceptualise, complex to understand and difficult to teach. The clinical abnormalities encountered in most conditions can be explained by a failure of this developmental process.

This book is unique because it combines the embryological development of the human and the malformations encountered in clinical practice.

Undergraduate medical students as well as nurses in training will find it helpful in their studies. Postgraduate trainees in paediatric surgery, paediatrics and neonatology will find it easy to follow, and it enables them to explain the mystery of congenital malformations to relatives and patients. General practitioners who will encounter more and more of these patients with their disabilities will be able to understand how these developed. It is hoped that people who read this book will find it helpful in their work.

We would like to thank Ms. Suzanne McMahon and Dr. Sharon F. Sneddon as subeditors for all of their hard work in seeing this book to completion.

I would like to thank my wife Annette for all the work she has helped me over the years to see this to completion.

Glasgow, UK — Robert Carachi
Cairo, Egypt — Sameh Helmi Edward Doss

You created my inmost self,
knit me together in my mother's womb.
For so many marvels I thank you;
A wonder am I, and all your works are wonders.
You knew me through and through

Psalm 139 verses 13, 14 Jerusalem Bible

ولقد خلقنا الإنسان من سلالة من طين ١٢
ثم جعلناه نطفة في قرار مكين ١٣
ثم خلقنا النطفة علقة فخلقنا العلقة مضغة فخلقنا
المضغة عظاما فكسونا العظام لحما ثم أنشأناه خلقا آخر
فتبارك الله أحسن الخالقين ١٤

These verses appear in Surah Al Mu'minūn.

A simple translation of these verses (12–14) is as follows:

We created man from an essence of clay (12),

then We placed him as a drop of fluid (nutfah) in a safe place (13),

then We made that drop into a clinging form (alaqah), and We made that form into a lump of flesh (mudghah), and We made that lump into bones (idhaam), and We clothed those bones with flesh (lahm), and later We made him into other forms (nash'ah)—glory be to God, the best of creators (14)!

Contents

1 **A Brief History of Embryology: Historical Vignettes in Embryology** 1
Basith Amjad

2 **General Embryology** 11
Sameh H. Doss and Sharon F. Sneddon

3 **Embryology of the Foetal Membranes and Placenta** 31
Sharon F. Sneddon

4 **Foetal Circulation** 39
Maria Ilina

5 **Teratogenesis and Infection** 45
Suzie Wills

6 **Embryological Basis of Congenital Anomalies** 55
Srinivas Annavarapu

7 **Antenatal Screening/Prenatal Diagnosis** 69
Nicola Brindley

8 **Antenatal Counselling and Genetics** 73
Nicola Brindley

9 **The Embryological Basis of Behavioural and Psychiatric Conditions** 77
Laxmi Kathuria

10 **Animal Models** 85
Piotr Hajduk

11 **Embryology Education for Nurses and Midwives** 91
Gian Battista Parigi and Gloria Pelizzo

12 **Oral and Dental Malformations** 101
Eiling Wu, Catherine Wicks, Tom W. M. Walker, and Robert Carachi

13 **Development of Cleft Lip and Palate** 111
Ambika Chadha and Alistair R. M. Cobb

14 **Facial Clefts** 119
Tom W. M. Walker, Ben C. Green, Caroline Mills, and Peter Ayliffe

15 **TMJ and Mandibular Congenital Malformations** 127
Nabeela Ahmed and N. Shaun Matthews

16 **Craniofacial Syndromes** 133
Elizabeth Anne Gruber and Michael Stephen Dover

17 **External, Middle, and Inner Ear** 143
Frank G. Garritano and Vito C. Quatela

18 **Nose and Paranasal Sinuses** 149
Catherine Lau and Steve Goudy

19 **Congenital Abnormalities of the Thyroid and Parathyroid** 157
John D. Collin and Ceri Hughes

20 **Embryology of the Salivary Glands** 165
Catherine Lau and Mark McGurk

21 **Embryology of the Branchial Arches** 169
Mark Wilson and Margaret Coyle

22 **Congenital Malformations of the Larynx and Trachea** 177
Christopher Barringer, Ramanathan Kasivisvanathan, Mark Catolico, and Steve Goudy

23 **Vascular Anomalies of the Head and Neck** 185
Ross Elledge, Kevin McMillan, Andrew Monaghan, and Rhodri Williams

24 **Eyes** 191
Aaron Jamison and Gerard McGowan

25 **Skin and Soft Tissues** 201
Mairi Steven

26 **The Brain and Central Nervous System** 213
Roddy O'Kane and Thomas Begg

27 **Chest Wall Abnormalities** 237
James Andrews

28 **Cardiac Abnormalities** 243
Maria Ilina and Stuart Lilley

29 **Embryology of the Great Vessels** 275
Maria Ilina and Stuart Lilley

30 **Pulmonary and Airways** 289
Peter Carachi

31 **The Diaphragm** 297
Rania Kronfli

32 **Abdominal Wall Defects** 303
Lynne A. Mcintosh

33 **Hernia and Hydrocele** 311
Aileen Rooney

34 **Gastrointestinal Tract 1: Foregut** 317
Gregor M. Walker

35 **Gastrointestinal Tract II: Midgut** 325
Marie Steven

36 **Gastrointestinal Tract III: Hindgut** 333
Tim J. Bradnock

37 **The Umbilicus** 347
Paul Cullis

38 **The Liver and Gallbladder** 353
Mark Davenport

39 The Pancreas 365
Suzanne McMahon

40 Spleen 371
Mohamed Abdel-Latif and Mohamed Sameh Shalaby

41 Anorectal Malformations and Cloacal Anomalies 379
Constantinos A. Hajivassiliou

42 The Kidneys and Ureters 391
Boma Lee and Martyn Flett

43 The Bladder and Urachus 409
Emily Broadis

44 Normal Development of the Penis and Urethra 417
Ahmed T. Hadidi

45 Disorders of Sex Development (DSD) 427
John Hutson

46 The Testis 437
John Hutson

47 Congenital Abnormalities of the Testis and Epididymis 443
Prabhu Sekaran

48 The Limbs 449
Laura Burton and Robert Carachi

49 Embryological Basis of Congenital Tumours 463
Philip Hammond and Srinivas Annavarapu

50 Conjoined Twins 475
Mohamed Abdel-Latif, Mohamed Sameh Shalaby, and Sameh Abd Elhay

Index 481

A Brief History of Embryology: Historical Vignettes in Embryology

1

Basith Amjad

What is embryology? It is perhaps an arbitrary term to describe a vast and fascinating science. Most embryologists consider their discipline as the study of "developmental biology", essentially encompassing biological development from a fertilized egg to a multicellular organism.

Human embryology considers the developmental aspects of life as a whole and not just the first 8 weeks of the embryo. The appreciation of human embryology is important, not just to give us an understanding of how we come into our adult form but also to explain the biological, chemical and physical forces, which shape the growing embryo. These very forces then continue to define and sustain the individual throughout their lifetime.

Within the last couple of decades, the science of embryology has taken tremendous strides providing us with the knowledge and understanding to bring about significant improvement in both maternal and child health. We are now equipped with techniques for improving fertility, diagnosing and managing prenatal conditions, sustaining the pregnancy and the foetus and preventing birth anomalies.

These breakthroughs in modern medicine have made a vast difference in human history and continue to do so. Thus embryology has brought about long-term social, cultural, religious, political and financial shifts.

Apart from the obvious better birth outcome, as an estimated 276,000 babies still die within 4 weeks of birth every year worldwide, the science of applied embryology has long-term effects on the physical and mental wellbeing of an individual.

The study of embryology, which is the formation and development of the embryo and the foetus, can be traced back to ancient Egypt. Around 1400 BC, the ancient Egyptians made references to the placenta and its importance as the seat of the external soul. However they did not consider the embryo to be alive until the baby was born. They also discovered that chicken eggs could be incubated in ovens, a finding later used for studying embryos during different periods of development (Fig. 1.1).

The science of embryology as an entity first appears in works by early Greek philosophers such as Empedocles, Anaxagoras and Diogenes. They were interested in the study of reproduction, development, differentiation and regeneration of body parts. They believed that new organisms could arise through sexual or asexual reproduction or spontaneous generation. A strongly held view was of a fire inside the embryo, which set the parts in order as the foetus developed.

Joseph Needham, one of the great historians of embryology, called Hippocrates (c460 BC–c370 BC) the first true embryologist. Hippocrates believed in what came to be known as preformationism, a concept that all organisms were fully formed in miniature within the womb before birth. He also felt that the embryo derived its blood supply from the placenta and developed by extracting moisture and breath from the mother. He then went on to identify what he called

Fig. 1.1 An Egyptian sculpture shows a pharaoh, and attendants in a ceremonial procession carry the pharaoh's soul, his placenta with the umbilical cord

B. Amjad (✉)
Department of Surgical Paediatrics, The Royal Hospital for Children, Glasgow, UK
e-mail: Syed.Amjad@ggc.scot.nhs.uk; Basith.Amjad@ggc.scot.nhs.uk

R. Carachi, S. H. E. Doss (eds.), *Clinical Embryology*, https://doi.org/10.1007/978-3-319-26158-4_1

Fig. 1.2 A bust of Hippocrates: the first true embryologist

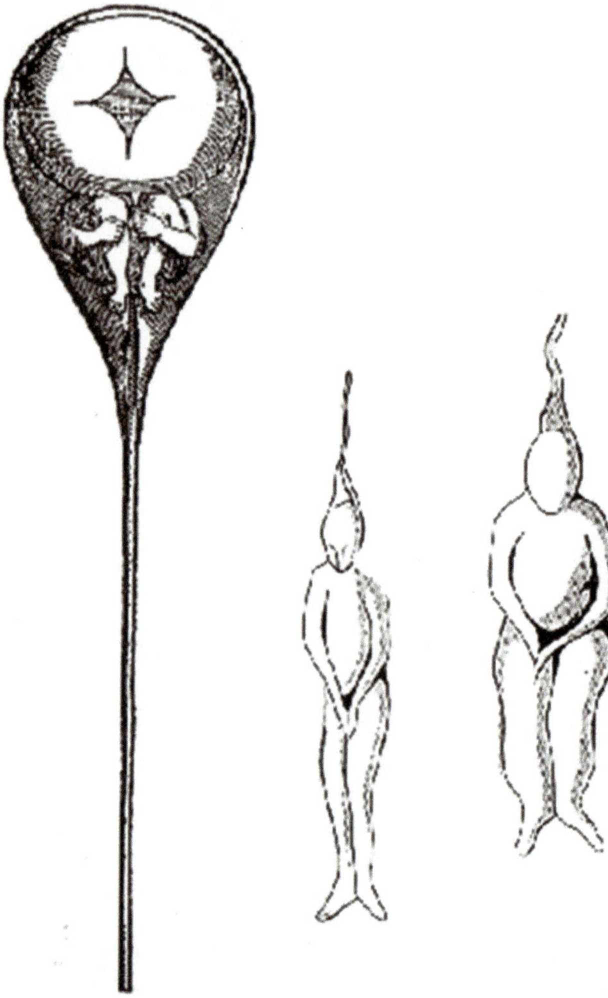

Fig. 1.3 "Preformation" drawn by Nicolaas Hartsoeker in *Essai de Dioptrique*, 1694

a number of condensations and fires that led to the formation and development of bones, flesh, gut and circulation within the embryo and foetus (Fig. 1.2).

Even though Plato deals with natural phenomenon in Timaeus (Dialogues 360 BC), the next major progress in embryology came about under his pupil Aristotle (384 BC–322 BC). His main embryological book was titled *On the Generation of Animals*, but most of his well-known observational science on embryology is found in the four compendiums, the *History of Animals*, the *Parts of Animals*, *On Respiration* and *On the Motion of Animals*. His writings confirm that he studied embryos of different organisms by not only opening up bird eggs at different stages of development but also dissecting mammalian embryos. He may have also observed human embryos, an almost impossible task in antiquity. Aristotle believed that the semen supplied a substance to give form to the embryo and then the mothers supplied another substance, which aided development (Fig. 1.3). He also believed that the menstrual blood had some part to play in the formation of the embryo. Aristotle also believed that all young embryos were similar with universal characteristics, but as the embryo grew, it started to show differentiation (Fig. 1.4).

Following the death of Aristotle, there was not much progress in the science of embryology till the time of Galen of Pergamon (129 AD–216 AD). Even though Galen did not give much attention to embryology, he wrote and propagated the strong belief that the umbilical cord was necessary for foetal respiration. Ali ibn Sahl Al-Tabari of Baghdad (808–864) wrote a seven-part system of medicine called *Firdous al-Hikmah* in which an entire part was devoted to embryology, a mixture of Greek and Arabic thinking at that time. The great Avicenna or to give him his proper name Abu Ali al-Hasan ibn Sina (980–1037) devoted certain chapters of his *Canon Medicinae* to the development of the foetus but added nothing new to the science past Galen (Fig. 1.5).

Fig. 1.4 Plato and Aristotle: *The School of Athens* by Raphael

Fig. 1.5 Avicenna at work on the *Canon of Medicine*

Again quoting Joseph Needham, the credit for helping to bring embryology back into the scientific realm belongs to Albertus Magnus of Cologne (c1200 AD–1280 AD). Albertus read extensively and then interpreted almost all the works of Aristotle and the Arabic commentaries that accompanied them and was therefore in the position of replacing speculative and theological ideas with observational techniques and attention to detail. Albertus believed that women had seeds and that female seeds congealed after coming into contact with male seeds, and once this egg came into contact with menstrual blood, there was nutrition available for the egg to develop into an embryo and then a foetus. A large portion of Albertus's knowledge came from studying chick and fish embryos.

Leonardo da Vinci (15 April 1452–2 May 1519) has been described by art historian Helen Gardner as "an individual of unquenchable curiosity and feverishly inventive imagination, the very epitome of renaissance humanist ideals". He was a polymath whose interests included sculpture, painting, architecture, music, engineering, cartography, literature, botany, anatomy and embryology. Among the artists of the Renaissance, he was not alone in his interest in human anatomy. But unlike Michelangelo, Raphael and Durer among others who undertook human dissection to increase their understanding of the human body, Leonardo was actually interested in biology. Leonardo's embryology is contained in the third volume of his notebooks called Quaderni d Anatomia. His drawings show the dissection of the pregnant uterus along with the amniotic and chorionic membranes. He also undertook dissection of the human embryo at various stages and produced quantitative measurements of the growth of the embryo. He was therefore the first to show that embryos can be measured chronologically and that they change in size, shape and weight as they grew. Leonardo's era also saw the emergence of midwifery and gynaecology as a field of science and practice, and this had direct implications on the emergence of modern embryology (Fig. 1.6a–c).

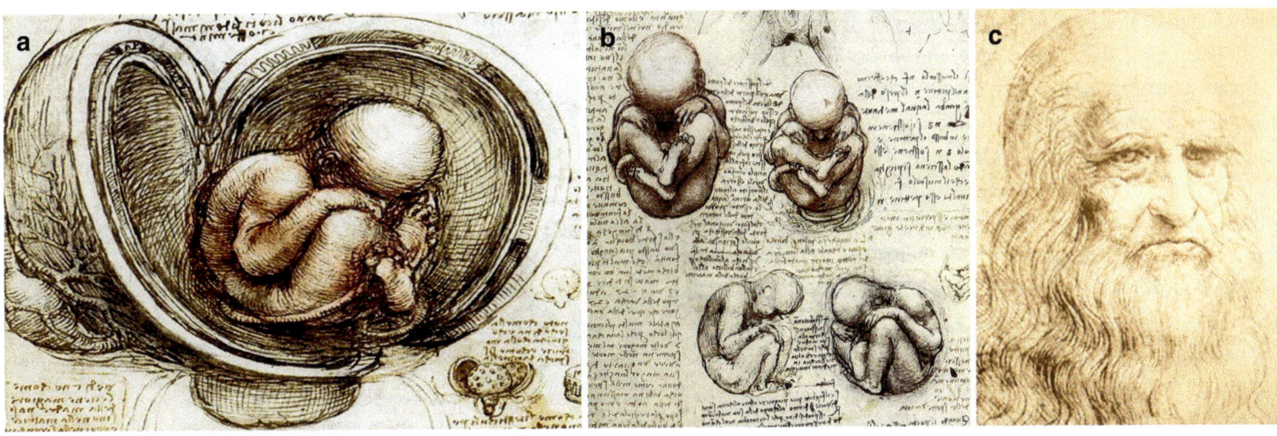

Fig. 1.6 (**a**) From *Studies of the Foetus in the Womb* by Leonardo da Vinci. (**b**) Views of a foetus in the womb by Leonardo da Vinci. (**c**) Self Portrait: Leonardo da Vinci

Fig. 1.7 First edition of Fabricius's treatise on human embryology Padua1600 AD

Hieronymus Fabricius (20 May 1537–21 May 1619) of Latium was a student of Falloppio and succeeded him as the professor of surgery and anatomy at the University of Padua. He is considered by many to be the Father of Embryology. In reality his work in comparative anatomy is where his legacy lies as he set up the first permanent theatre for public dissection. He dissected embryos of man, rabbit, guinea pig, dog, cat, sheep, ox, deer and viper, among others, a feat never before accomplished. This allowed him to investigate the formation of the foetus, the oesophagus, the stomach and the intestines. He also studied the function of the eye, the ear and the larynx (Fig. 1.7). The drawings and illustrations of Fabricius' anatomical works are incredibly accurate and beautiful and a testament to his genius. Unfortunately his works in embryology which included the belief that the chalaza found inside reptile and bird eggs and now known to suspend the yolk was the precursor of the brain, heart and liver and that the heart and other organs of the foetus had no proper function were erroneous and subsequently forced William Harvey to spend a considerable amount of his time challenging and refuting them (Fig. 1.8).

Fig. 1.8 Robert Hannah: William Harvey 1848

American historian Arthur M. Schlesinger Jr. has placed William Harvey (1 April 1578–3 June 1657) among the ten most influential people of the second millennium. A physician by training, Harvey's most defining contribution in human physiology and anatomy was to describe in complete detail the circulation of blood. Published in 1628, his work called *On the Motion of the Heart and Blood* gave a clear and detailed account of the function of the heart and the movement of blood around the body in a circuit. Less well known are Harvey's contributions to the study of embryology. Using low-powered lenses, Harvey had carried out extensive dissection work in deer and chicken embryos, described the blastoderm as the site of origin of the embryonic body, explained the importance of the amniotic fluid and, having shown that even the lowest organisms arise from eggs, finally laid the theory of spontaneous generation to rest.

Another unique individual who added to the advancement of embryology in the seventeenth century was the Italian physician Marcello Malpighi (10 March 1628–29 November 1694). Malpighi had an illustrious career as an academic professor of medicine and anatomy at Pisa but decided to retire from academic pursuit and return home to Bologna and dedicate his life to anatomical studies. Although some of his work involved gross anatomy, his most significant work appears to be based on the use of the microscope. Because of this many microscopic structures are named after him including the Malpighi layer in the skin, Malpighi corpuscles in the spleen and kidneys, etc. Needham credits him as the person responsible for the doctrine of preformation, metamorphosis and the development of the embryo as a simple unfolding of an already miniature adult organism.

Two rival schools of thought dominated the historical narrative of embryology in the eighteenth century. The preformationists, steeped in the writings of Malpighi, Swammerdam

Fig. 1.9 Albrecht von Haller: Swiss physiologist and naturalist

Fig. 1.10 Caspar Friedrich Wolff: German physiologist and embryologist

and Bonnet, believed that the embryo pre-existed in some form in either the maternal egg or the male sperm. They also advocated that all embryos had been formed by God at creation and encased within one another to await their future appointed time of development. Epigeneticists on the other hand, influenced by the legacy of Aristotle and Harvey, argued that each egg was newly produced through progressive development from unorganized material and proposed various theories to explain how this gradual formation occurred.

This discourse and dispute between the preformationists and the epigeneticists continued throughout the age of enlightenment and is best showcased by the debate that took place between Albrecht von Haller (Fig. 1.9) (16 October 1708–12 December 1777), a Swiss anatomist, physiologist and naturalist, and Caspar Friedrich Wolff (18 January 1733–22 February 1794), a German physiologist who is considered one of the founders of modern embryology (Fig. 1.10). Haller had been an ardent supporter of preformation since 1758, a year before the publication of Wolff's dissertation, which endorsed epigenesis. The tussle between the two men lasted over a decade and came to define and symbolize the key questions faced by biological sciences in that era, the idea of God in relation to the biology of creation, the question of spontaneous generation, the role of mechanism in developmental embryology, the issue of regeneration and the dilemma of "monstrous births". Haller, a "Newtonian mechanist" and a deeply religious man, held beliefs about the nature of the world and scientific views that were fundamentally very different to those of Wolff, whose own world views and scientific outlook derived largely from the tradition of German rationalism. Wolff's research work covered both the fields of anatomy and microscopic embryology. In a series of ground-breaking scientific papers, he laid the foundation of modern embryology. He was the first to describe the primitive kidneys or mesonephros or the "Wolffian bodies" and its excretory ducts as laid out in his dissertation entitled "Theoria Generationis". In 1768 he published *De formatione intestinorum* in which he explained the development of the intestine and foreshadows the idea of germ layers in the embryo. He demonstrated that the chick intestine was formed by the folding of tissue that detaches from the embryo's ventral surface. The folds eventually transform into a closed tube. He then argued that this observation proved that the intestine was not preformed and that the organs appeared gradually. Wolff's observations are now recognized as the most fundamental conception in structural embryology. Wolff also examined and dissected the so-called embryonic monsters and assessed correctly that they were

Fig. 1.11 Herman Boerhaave: Dutch pioneer of chemical embryology

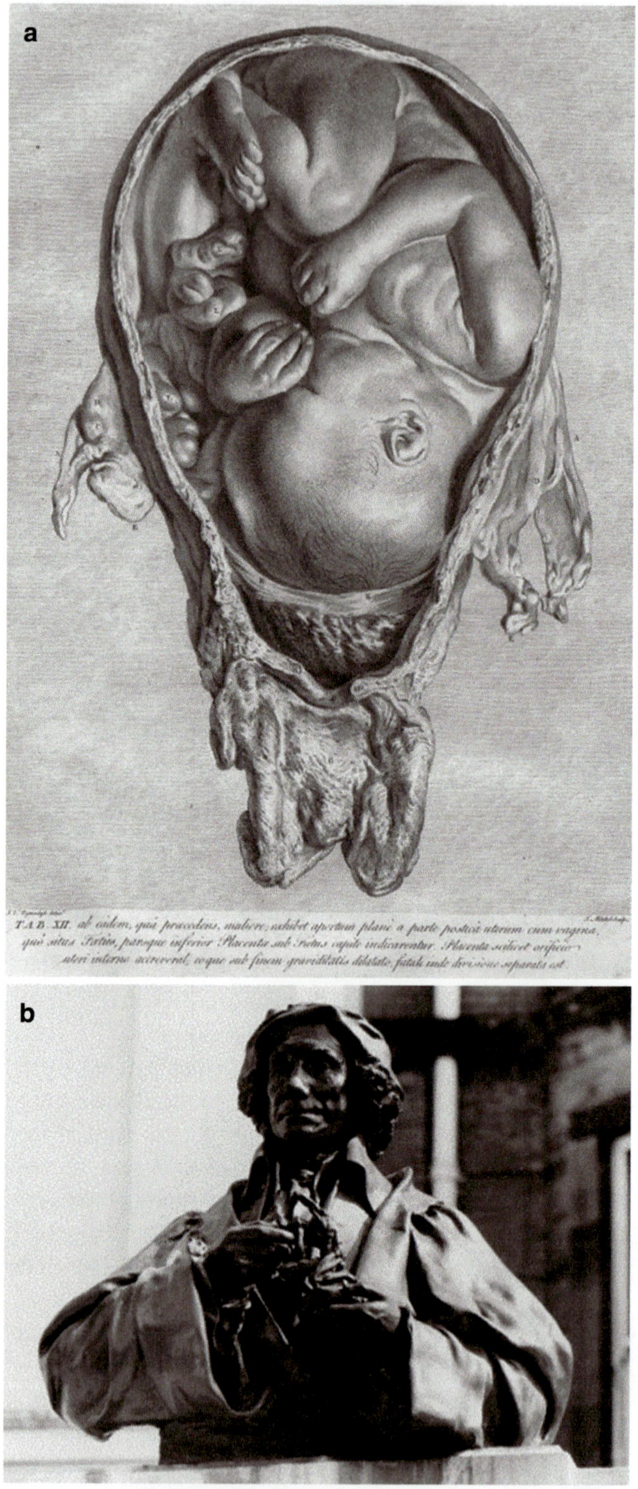

Fig. 1.12 (**a**) Page from the *Anatomy of the Human Gravid Uterus* by William Hunter. (**b**) Statue of John Hunter outside St George's Hospital in London

formed by the mechanics of nature and thus were examples of epigenesis rather than preformationism. In spite of all this work by Wolff, Haller's reputation was such that his assertions continued to cast a powerful influence within the scientific community even though ultimately Wolff was vindicated by posterity.

Another embryological puzzlement that the embryologist wrestled with was the issue of foetal nutrition. The ideas put forward regarding the source of nutrition included amniotic fluid ingested by the foetus, a wholesome fluid made available to the foetus called uterine milk, nutrition circulating within the menstrual blood and passing via the umbilical cord. Without clear evidence backed up by experimental techniques, these ideas remained within the sphere of theory and conjecture.

Another remarkable individual was the Dutch physician and anatomist Herman Boerhaave (31 December 1668–23 September 1738). Boerhaave separated egg white from the yolk and conducted various chemical and physical experimentations including adding various acids and bases and shaking, heating and boiling the components to see the effects produced. He thereafter published his results in a first detailed account of chemical embryology. This work in turn led to the science of experimental work in the field of biology (Fig. 1.11).

The brothers William (23 May 1718–30 March 1783) and John Hunter (13 February 1728–16 October 1793), anatomists and surgeons, are both giants of modern medicine who have greatly advanced the scientific method in medicine. One of their most significant discoveries, published in their work on the anatomy of the gravid uterus, was to show clearly that maternal and foetal circulations were two distinct physiological systems (Fig. 1.12a, b).

The nineteenth century saw one of the great advancements in modern biology when cell theory came to be

Fig. 1.13 Matthias Schleiden: German botanist and proponent of cell theory

Fig. 1.14 Theodor Schwann: German biologist and physiologist

accepted. It was the work of two remarkable German scientists: Matthias Schleiden (Fig. 1.13) (5 April 1804–23 June 1881) and Theodor Schwann (Fig. 1.14) (7 December 1810–11 January 1882). While holding the chair of Botany at the University of Jena, Schleiden studied plant structure under the microscope and authored *Contributions to Phytogenesis* in which he stated that each and every part of a plant is made up of cells. He also mentions the cell nucleus and its role in cell division. Theodor Schwann working at the University of Berlin in Physiology was undertaking research in animal tissue. A meeting with Schleiden, where they talked about plant cells, made him realize that he had observed similar cells in animal notochord. This similarity was confirmed by both scientists working together, and the results appeared in Schwann's famous "Microscopic Investigations on the Accordance in the Structure and Growth of Plants and Animals", in which he declared that "All living things are composed of cells and cell products". This has now come to be known as cell theory or cell doctrine. In the course of this work, he went on to prove the cellular origin of highly differentiated tissues including nails and tooth enamel. Vitally he also studied the ovum and established that it is a single cell that eventually develops into a complete organism, thus confirming a basic principle of embryology. Schwann is also remembered for the discovery of Schwann cells in the peripheral nervous system and his discovery and study of pepsin.

One of the founders of modern embryology, Karl Ernst von Baer (17 February 1792–16 November 1876), also belonged to the nineteenth century and hailed from Estonia. He spent most of his productive years at the St. Petersburg Academy of Sciences, studying the embryonic development of animals (Fig. 1.15). His many achievements included the discovery of the mammalian ovum, the blastula stage of development and the notochord. Building on the work of Wolff, he described the germ layer theory of development and established that mammals developed from eggs. His book *Uber Entwickelungsgeschichte der Thiere* established the foundation of comparative embryology and laid down what have come to be known as Baer's laws of embryology. Another leading light in nineteenth-century embryology was the German zoologist and experimental embryologist Wilhelm Roux (9 June 1850–15 September 1924) who worked mostly with chicken embryos and frog's eggs to study developmental embryology. Maintaining his embryonic material in warm saline, he was the first to establish the idea of tissue culture and helps establish the mosaic theory of epigenesis.

By the middle of the last century, a sound body of basic knowledge was finally established which could describe the events of development within the embryo. Thereafter two themes have dominated the progress and development

Fig. 1.15 Karl Ernst von Baer: established the foundations of modern embryology

Fig. 1.16 Hans Spemann: German embryologist and Nobel laureate in Medicine

of embryology as a science. One of these themes, chemical embryology, coming into its own between the 1940s and the 1970s and using experimental techniques, has tried to explain the nature of embryonic induction or embryogenesis. This process of induction helps to define and then direct the development of a group of cells into particular tissues and organisms. The disciples of this field have tried to do this by seeking out and characterizing these inducing signals. A concurrent theme has been modern molecular embryology, which beginning in the 1980s has led to the current revolution in biological sciences. It seeks to explain at the genomic level the differentiation of cells into specific tissues and structure within the same organisms. Thus embryology is now concerned with the development of the organism from the telescope of activation and transcription of the DNA, thus allowing us to understand the genetic mechanisms of development and its consequences.

Hans Spemann (27 June 1869–9 September 1941), a German embryologist and Nobel laureate in Medicine in 1935, spent a lifetime of work in embryonic induction (Fig. 1.16). After qualifying as a doctor of medicine, he was initially intent on a career in medicine, but fate had other plans for him. He contracted tuberculosis and was confined to a sanatorium in 1896 and there, while recuperating, read *The Germ-Plasm: A Theory of Heredity* by August Weismann. Weismann, considered one of the great biologists of all times, is today best remembered for his germ plasm theory which states that in a multicellular organism, inheritance takes place only by means of the germ cells, that is, the egg cells and the sperm cells, and other cells of the body, that is, the somatic cells, do not play any part in this process. After reading Weismann, Spemann switched to the field of zoology and embryology and embarked on a career that eventually led to his appointment at the Institute of Biology in Berlin. Working with a protégé named Hilde Mangold, he used microsurgical techniques to transplant a specific group of cells (which he called the primitive knot and now named Spemann's organizer) from one embryo to another. This organizer upon transplantation was then able to induce secondary embryonic primordial regardless of its location. He then went on to show how different parts of this organizer would produce different parts of the embryo. Thereafter the work of Johannes Holtfreter (9 January 1901–13 November 1992), Joseph Needham (9 December 1900–24 March 1995) and Conrad Waddington (8 November 1905–26 September 1975) showed that even if these foci of cells were killed, by either fixing, boiling or freezing them, they would continue to cause induction within the embryo. Their conclusion that these were inert molecules was better understood by the turn of the century when there was an appreciation of signalling within cells and the embryo.

Fig. 1.17 Gerald Edelman: American biologist and Nobel laureate

One of the giants of molecular embryology would surely be Gerald Edelman (1 July 1929–17 May 2014), an American biologist who won the Nobel Prize in Medicine for his work on the molecular structure of antibodies (Fig. 1.17).

He was initially intent on a career as a concert violinist but decided that he did not have the inner drive and willpower to succeed in that enterprise and switched, by a gift of providence, to medical research. In his most seminal work called *Topobiology: An Introduction to Molecular Embryology*, Edelman presented a theory that "morphogenesis is driven by differential adhesive interactions among heterogeneous cell populations and it explains how a single cell can give rise to a complex multicellular organism". Topobiology therefore as proposed by Edelman is a biological process that creates and maintains differentiated tissues and is acquired by segregation of cells through heterologous cellular interactions.

The advent of the twenty-first century saw the mapping of the human genome, and it appears likely that this will be the epoch of biological sciences and that it will play a key role in tackling global challenges. Embryology has not yet exhausted all its fascinating and mysterious possibilities.

Suggested Reading

A history of embryology by Joseph Needham. 2nd edition revised by Arthur Hughes. Cambridge: Cambridge University Press; 1959.

Gilbert S, editor. Developmental biology Volume 7: a conceptual history of modern embryology. New York: Plenum Press; 1991.

Matter, life and generation. Eighteenth-century embryology and the Haller-Wolff debate by Shirley A Roe. Cambridge: Cambridge University Press; 1981.

Murillo-Gonzalez J. Evolution of embryology: a synthesis of classical, experimental and molecular perspectives. Clin Anat. 2001;14(2):158–63.

Sadler TW. Langman's medical embryology. 10th ed. Philadelphia: Lippincott Williams & Wilkins; 2006.

Sebastian A. A dictionary of the history of medicine. Nashville: The Parthenon Publishing Group; 1999.

Wellner K. On line review of a history of embryoilogy by Joseph Needham. In: The Embryo Project Encyclopedia, Embryo Project, US National Science Foundation, Washington DC; Arizona State University, Tempe, AZ; 2010.

General Embryology

2

Sameh H. Doss and Sharon F. Sneddon

Female Reproductive Tract

Ovary

The female reproductive system shows a monthly cycle of growth and development throughout adult life. In the female, the reproductive tract comprises of a gland producing both gametes and steroid hormones and a duct system for transport of the gametes. The ducts develop from two sets of precursors found in both sexes. In the female, it is the paramesonephric ducts that are maintained and develop into the uterine tubes and uterus while the mesonephric ducts (of the male tract) degenerate. The anatomy of the female reproductive tract and structure of the ovary containing ovarian follicles at various stages of development can be seen in Fig. 2.1.

Genital Ducts

Body: the main part of the uterus.

Ovaries (Fig. 2.2)

The ovary is composed of an outer cortex (the functional zone of the organ), which function to produce gametes and endocrine secretion (oestradiol and progesterone), and an inner medulla where vessels and nerves predominate. The mesovarium, a fold of peritoneum carrying blood vessels, nerves and lymphatics to the ovary, is attached to the ovary at the hilum. The stroma of the cortex is richly cellular, and is looser than that of the medulla, to accommodate the growth of blood vessels that occurs when the ovary is active. Usually, only one of the two ovaries is active in each cycle. A cohort of follicles begin to develop together, but only the secondary oocyte at the metaphase of the second meiotic division from the dominant Graafian follicle is ovulated.

After ovulation, the follicle collapses and forms the corpus luteum. The basement membrane between the granulosa and thecal layers breaks down, and blood vessels from the theca interna can invade the granulosa. Granulosa cells develop into lutein cells, seen in cordlike associations separated by blood vessels. This sort of structural arrangement is typical of endocrine glands: each secreting cell is close to a blood vessel. Lutein cells are large and contain lipid inclusions and produce oestrogens and progesterone. Unless there is a pregnancy, the corpus luteum will degenerate after a fortnight. If there is a pregnancy, the embryo is able to signal its presence and the corpus luteum is saved, developing into a corpus luteum of pregnancy which maintains the pregnancy until placental hormones take over.

Uterine Tubes

The uterine tube is a duct for the passage of the oocyte. It has no direct continuity with the ovary, opening instead into the peritoneal cavity. One of the problems with the female system is that this provides a potential route of infection from the external environment to the peritoneal cavity (vagina to uterus to uterine tube to peritoneal cavity). The uterine tube has four constituent parts:

S. H. Doss (✉)
Anatomy Department, Cairo University, Cairo, Egypt
e-mail: Sameh.doss@bue.edu.eg

S. F. Sneddon
School of Medicine, Dentistry and Nursing,
University of Glasgow, Glasgow, UK
e-mail: Sharon.sneddon@glasgow.ac.uk

R. Carachi, S. H. E. Doss (eds.), *Clinical Embryology*, https://doi.org/10.1007/978-3-319-26158-4_2

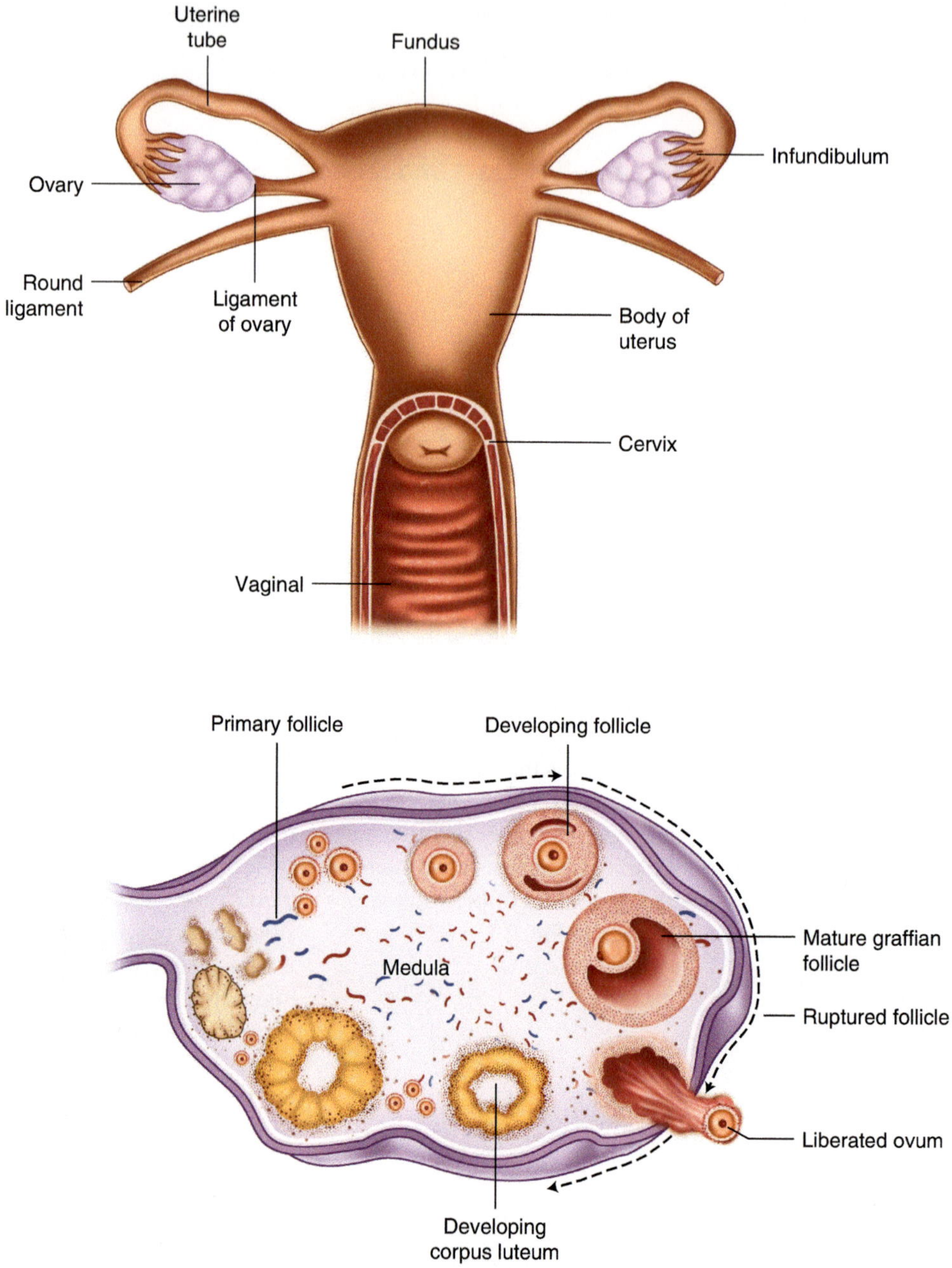

Fig. 2.1 Anatomy of the female reproductive tract and structure of the ovary containing ovarian follicles at various stages of development

- Interstitial part—through the wall of the uterus.
- Isthmus—medial third of tube, next to interstitial part, narrow.
- Ampulla—occupies more than half the length of the tube; fertilisation occurs here.
- Infundibulum—funnel-shaped end, wafts over ovary at ovulation and has folds on free margin—fimbriae.

The mucosa of the fallopian tube is composed of a simple columnar epithelium and a supporting connective tissue lamina propria. The epithelium has both ciliated and secretory cells. These may be variations of a single cell type as the ciliated cells are oestrogen-dependent. After the menopause, the epithelium becomes low cuboidal, but on administration of oestrogens, cells become ciliated. Cilia mainly beat towards the uterus – aiding the passage of the oocyte and zygote and beating back infection. The lamina propria has reticular fibres and fusiform cells, similar to cells of the uterine stroma. External to the mucosa is a smooth muscle coat, the muscularis, whose waves of contraction aid the rapid transport of the oocyte – which takes only 10 min after ovulation to reach the ampulla. Movement of the zygote, which must be slower to allow the uterus time to prepare, depends more on ciliary beating. The uterine tube has an outer covering of peritoneum, the serosa.

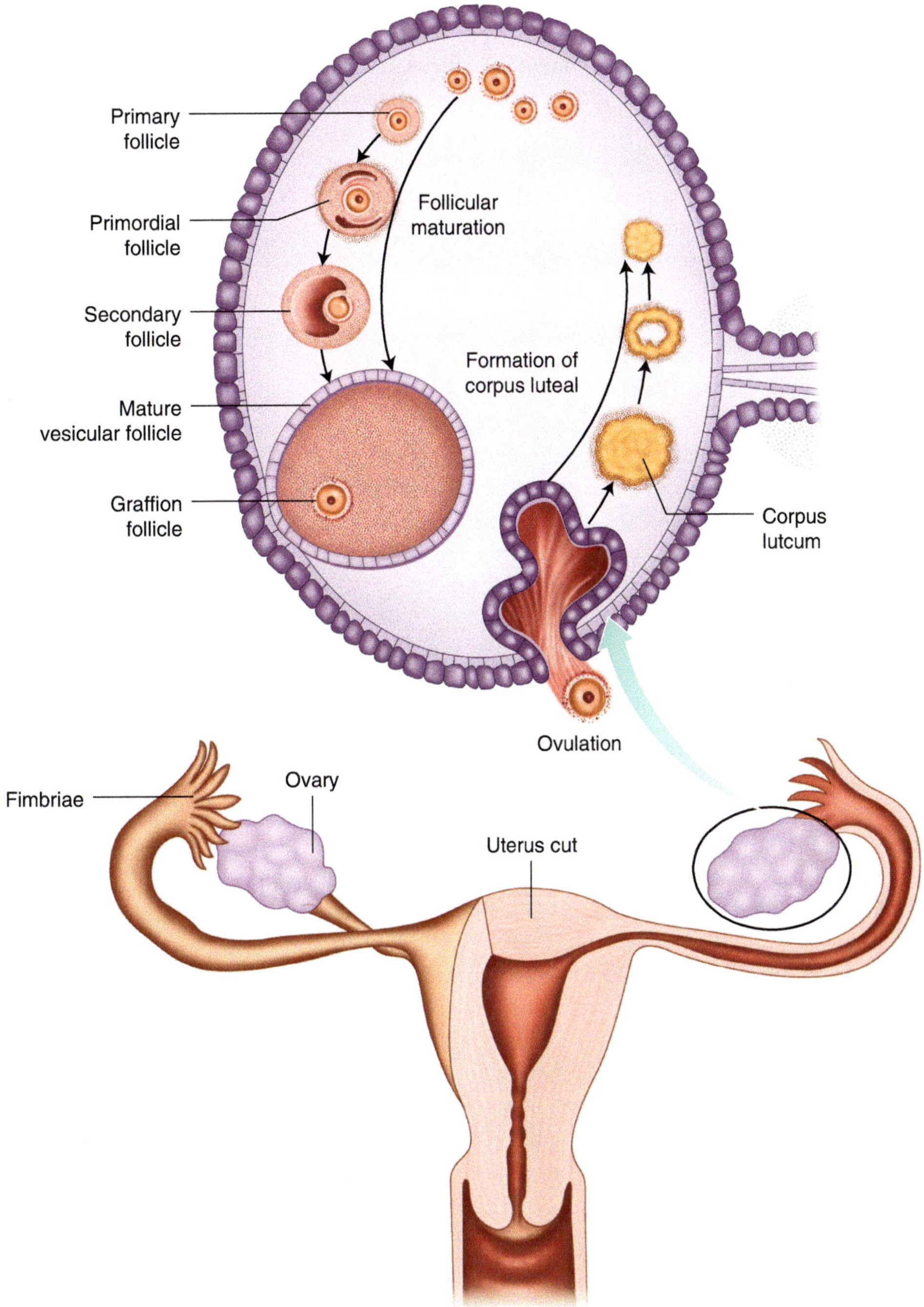

Fig. 2.2 Follicular maturation and ovulation within the ovary

Uterus

The uterus is composed of:

- *A body (main part)*
- *Fundus (part that extends above level of uterine tube entry)*
- *Cervix (neck)*

The uterine wall consists of:

- *A mucosa (secretory layer), known as the endometrium*
- *A muscle layer or myometrium*
- *An outer serosa or perimetrium*

Like the ovary, the uterus undergoes cyclical changes, which are seen in the endometrium. This can be divided into an upper functional layer and a lower basal layer, which is adjacent to the myometrium. The basal layer contains the bases of the endometrial glands, which are simple tubular glands. The functional layer can itself be subdivided into an upper stratum compactum, where the necks of the glands are, and a stratum spongiosum beneath it. The stratum spongiosum is named for its appearance in the latter half of the cycle, when the glands are coiled and distended with secretion. The myometrium has smooth muscle arranged in three layers.

Male Reproductive Tract

In the male, the reproductive tract comprises a gland producing both gametes and steroid hormones and a duct system for transport of the gametes. In the male, each mesonephric duct develops into the ductus (or 'vas') deferens, whereas the paramesonephric ducts of the female tract degenerate.

Testes

Testes are found in the scrotum, partially covered by a serous membrane derived from the peritoneum, the tunica vaginalis. Deep to this is a connective tissue capsule, the tunica albuginea, which sends septa into the gland dividing it into 200–300 lobules. Each lobule of the testis contains from 1 to 4 seminiferous tubules; each tubule is 30–70 cm long and coiled on itself, the two ends being joined together where they open into the straight tubules. Straight tubules connect to an anastomosing network, the rete testis. Fifteen to twenty efferent ductules leave the rete. Up to this point, movement of sperm is brought about by the flow of luminal fluid produced by the Sertoli cells; from now on, muscle is responsible.

Epididymis

The ductules join to the epididymis, a single convoluted tube about 4–6 m in length. The tube is so convoluted that gross anatomical parts of the epididymis can be distinguished: the head or caput, body or corpus and the tail or cauda. The lining epithelium of the epididymis is pseudostratified columnar epithelium bearing long (15 μm) microvilli, termed stereocilia. In the caput region, the epithelium is absorptive, resorbing fluid from the testis and probably modifying the composition of the seminal fluid. Although the functions of the epididymis remain to be fully elucidated, it is thought to be very important, since it is while in the epididymis that the spermatozoa mature; this process is androgen-dependent.

Vas (Ductus) Deferens

The epididymis is continuous with the vas or ductus deferens, the main duct of the testis. This tube is found in the spermatic cord, a collection of structures, blood vessels, nerves, ductus and striated cremaster muscle, running from the anterior abdominal wall to the scrotum. The ductus has a very thick wall. Its mucosa forms longitudinal folds; the pseudostratified columnar epithelium bears stereocilia and is supported by a lamina propria rich in elastic fibres. The muscularis is very well developed and moves sperm by peristalsis. At its distal end, the ductus is dilated to form the ampulla – here the mucosa is folded to provide recesses for sperm storage, though most sperm are stored in the epididymis. On each side, the seminal vesicle joins the end of the ampulla to form an ejaculatory duct, which opens into the prostatic urethra.

Accessory Glands

Accessory glands associated with the male reproductive tract include the seminal vesicles and the prostate gland. Each seminal vesicle is a tortuous tube coiled upon itself. The folded mucosa is composed of a pseudostratified columnar epithelium whose cells show ultrastructural characteristics of protein-secreting cells. The alkaline secretion which accumulates forms the main bulk of the fluid which is ejaculated during copulation. It contains fructose as an energy source for sperm, globulin and vitamin C for nutrition and motility of sperm. Functional activity of the seminal vesicles is testosterone-dependent.

The prostate gland surrounds the first part of the urethra, into which its ducts empty. It is composed of 30–50 branched tubulo-alveolar glands surrounded by a fibroelastic capsule rich in smooth muscle and supported by a fibromuscular stroma. Glands fall into three types: mucosal, submucosal and main glands. The latter contribute most to the prostatic secretion which contains acid phosphatase, citric acid and fibrinolysin, which has a role in the liquefaction of semen. The lumen of the glands often contains pink circular or oval structures – prostatic concretions or corpora amylacea.

Penis

The penis contains three masses of erectile tissue: there is a corpus cavernosum on each side, each containing a deep artery, and below them is the corpus spongiosum bearing the penile urethra. The erectile tissue is composed of endothelial-lined vascular spaces with fibrous tissue between the vascular spaces. It can be inflated with blood.

The finer mesh of the fibrous tissue in the corpus spongiosum does not prevent distension of the urethra when the erectile tissue is filled with blood. In the male, the urethra forms the final duct for both the reproductive and the urinary systems.

Gametogenesis

Gametogenesis is the process of production of mature gametes in the testis and ovary. The process results in a reduction in the halving of chromosome content by meiotic division to allow fertilisation to occur.

In the female, by the time of birth, oogonia have all differentiated into primary oocytes, arrested in the dictyotene stage in the prophase of the first meiotic division. This stock will be steadily depleted throughout the woman's reproductive life. Spontaneous degeneration of oocytes (atresia) occurs leading to the theory that dictyotene is unstable. There are approximately 6.8 million oocytes at the peak in the 5th month of intrauterine life; 2 million remain at birth and only 40,000 survive until puberty. However, a total of only about 480 will be ovulated in a woman's reproductive life (1 a month for 12 months × 40 years of reproductive life).

Oogenesis

The production of a mature ovum ready for fertilisation by sperm.

Oogenesis occurs in the cortex of the ovary. The aim of the process is:

1. A reduction in the number of chromosomes from the diploid number to the haploid number
2. Increase in the size of the ovum from 30 μm to 120 μm

Oogenesis occurs during the fertile period of the female, starting at puberty (11–14 years) and ending at menopause (45–55 years). During this fertile period, one mature ovum develops in the ovary every cycle.

The oogonia (primitive germ cells) lie in the cortex of the ovary. At puberty, each ovary contains about 40,000 oogonia. Each oogonium is surrounded by a layer of flat epithelial cells called follicular cells to become the primary oocyte.

Oogenesis includes two processes:

1. Maturation of the primary oocyte
   (a) To become a mature ovum containing the haploid number of chromosomes
2. Maturation of the follicular cells
   (a) Around the oocyte to become a mature follicle for protection of the ovum and production of hormones

The female gamete is the major source of cytoplasm for the new individual after fertilisation; meiosis in the female is geared towards producing a gamete with a lot of cytoplasm. Thus cytoplasmic organelles are inherited through the maternal line; of particular significance, mitochondria and hence mitochondrial genes are inherited via the mother.

The oogonium contains a diploid number of chromosomes (44 + X). It grows and becomes surrounded by a single layer of follicular cells to become the primary oocytes. The primary oocytes and the follicular cells are together known as the primary follicle (Fig. 2.3).

The primary oocytes undergo meiosis to give:

(a) The secondary oocytes— a large cell containing 23 chromosomes (22 + X)
(b) The first polar body— a small cell with a very small amount of cytoplasm (22 + X)

The secondary oocytes undergo mitosis to give rise to the mature ovum and a second polar body. The first polar body divides into two by mitosis, and then all the polar bodies disappear.

The simple flat epithelial cells which surround the primary oocytes enlarge and become cuboidal and then columnar which divide to form many layers around the oocyte (Fig. 2.4).

Development of the Oogonium (Oogenesis)

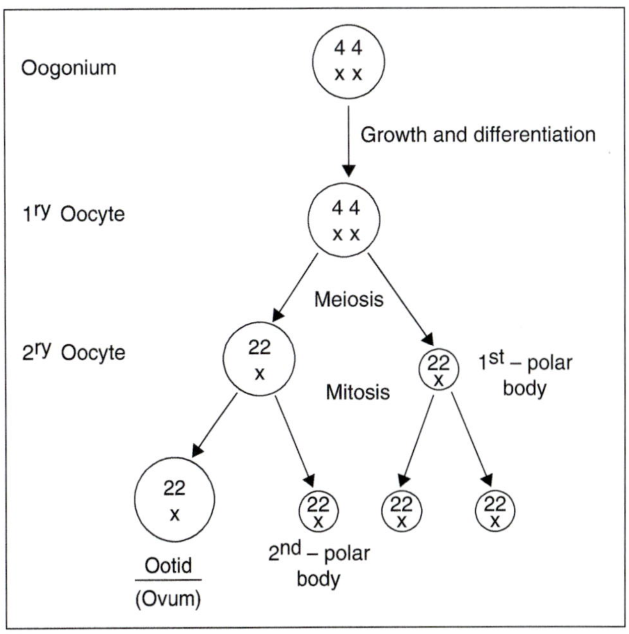

Fig. 2.3 Development of the oogonium during oogenesis

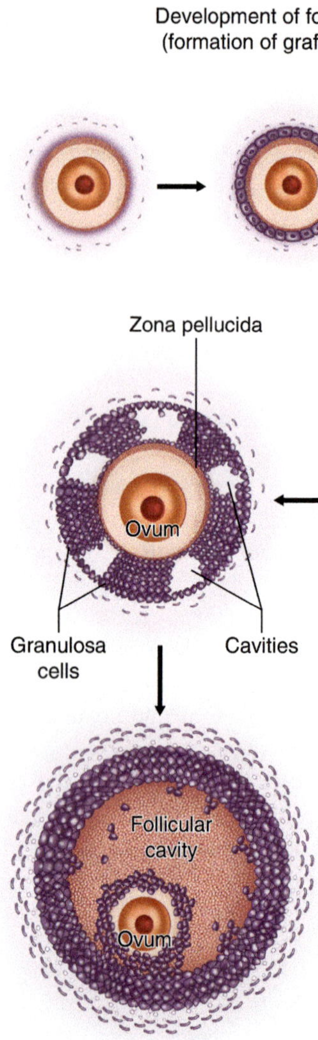

Fig. 2.4 Development of the follicular cells

The follicular cells deposit a glycoprotein coat around the oocyte known as the zona pellucida, and the follicular cells are now called granulosa cells.

Formation of the Graafian Follicle

Small irregular spaces appear between the granulosa cells and later join each other to form one large follicular cavity. This fills with follicular fluid from the granulosa cells.

The follicular wall is formed of two layers, an outer fibrous layer, the theca externa, and an inner vascular theca interna (Fig. 2.5).

The ruptured follicle in the ovary undergoes morphological change to become the corpus luteum, the fate of which depends on whether fertilisation takes place. If no fertilisa-

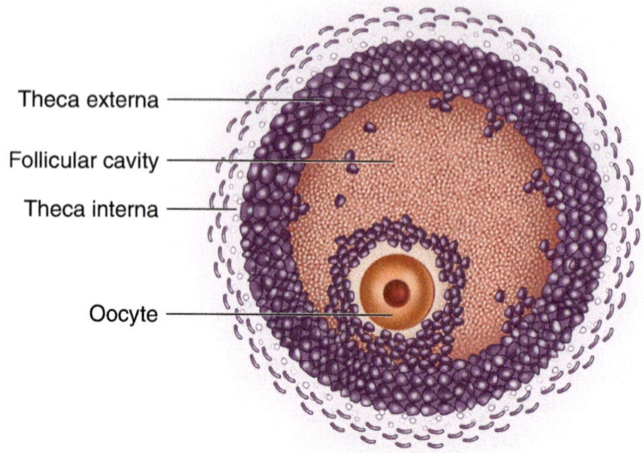

Fig. 2.5 Formation of the follicular wall

tion occurs, the oocyte will degenerate after 24–36 h, and the corpus luteum will collapse and progesterone levels will fall (Fig. 2.6).

Fate of the Graafian Follicle and the Ovum

The mature Graafian follicle ruptures at the time of ovulation, releasing the mature ovum. This enters the uterine tube where it awaits fertilisation. Within the follicle, the oocyte is surounded by follicular cells known as cumulus oophorus; at ovulation, some of these follicular cells will be released along with the ovum, now termed the corona radiata.

If no fertilisation occurs, the ovum degenerates after 24–36 h. The ruptured Graafian follicle is transformed into the corpus luteum, a yellow-bodied structure which produces hormones.

Spermatogenesis and Spermiogenesis

Spermatogenesis is the process of sperm formation in the seminiferous tubules of the testis (shown in Fig 2.7). Its aim is to

1. Reduce the number of chromosomes from diploid (46) to haploid (23) by meiosis.
2. Change the shape of the male germ cells to produce a highly motile sperm ready for fertilisation of the ovum.
3. Increase the number of cells.

Spermatogenesis occurs continuously in the seminiferous tubules of the testis from puberty and continues throughout life.

The process begins with the differentiation of spermatogonia into spermatids.

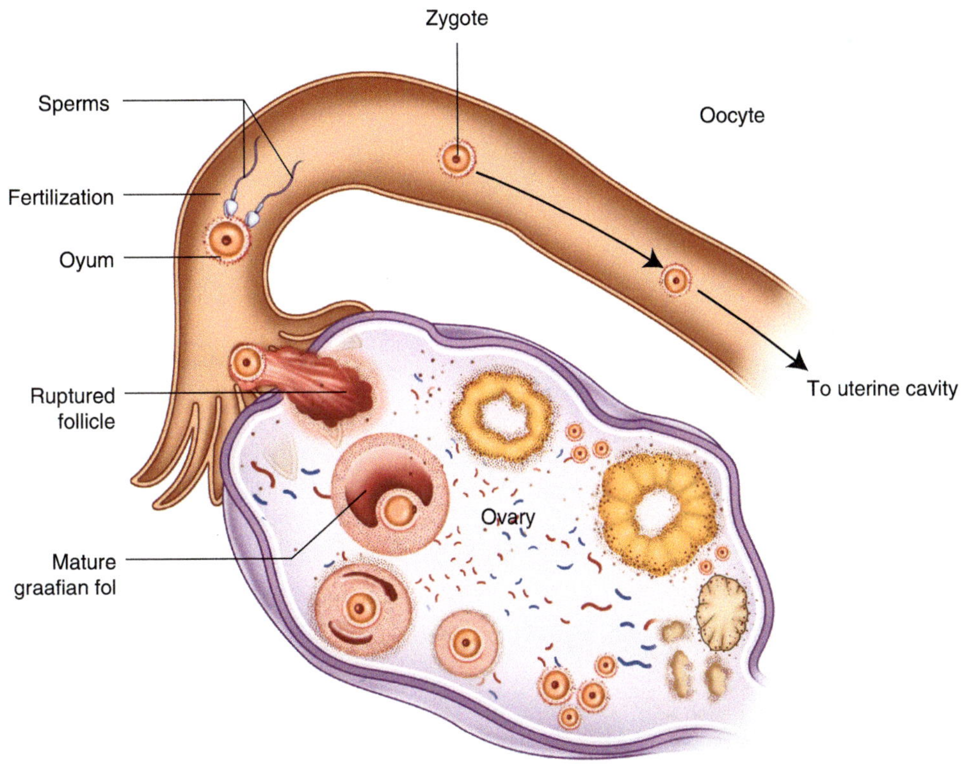

Fig. 2.6 Ovulation of the mature ovum from the Graffian follicle

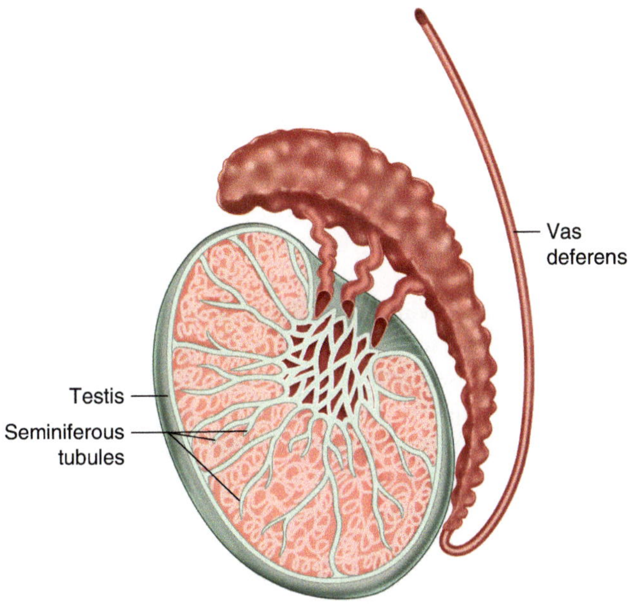

Fig. 2.7 Structure of the testis and vas deferens

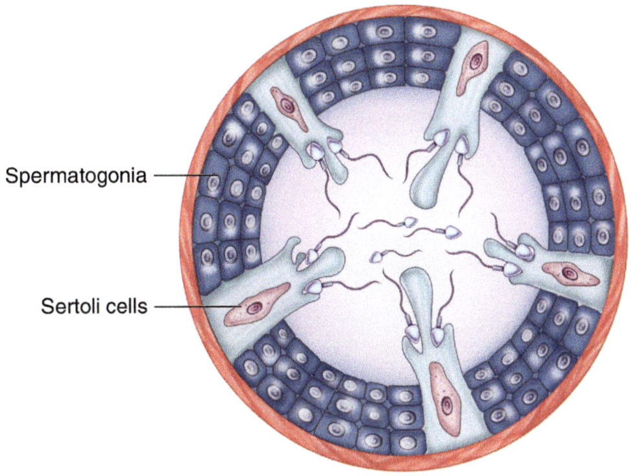

Fig. 2.8 Spermatogenesis

Spermatogonia are the most primitive male germ cells. They contain a diploid number of chromosomes (44 autosomes and 2 sex chromosomes). They lie on the wall of the seminiferous tubules and are supported by Sertoli cells.

Each spermatogonium undergoes mitotic division to give two daughter cells. Each of these grows to produce a primary spermatocyte (Fig. 2.8).

Primary spermatocytes undergo meiotic division, giving rise to two secondary spermatocytes, each of which contains a haploid number of chromosomes

(22 + X or 22 + Y).

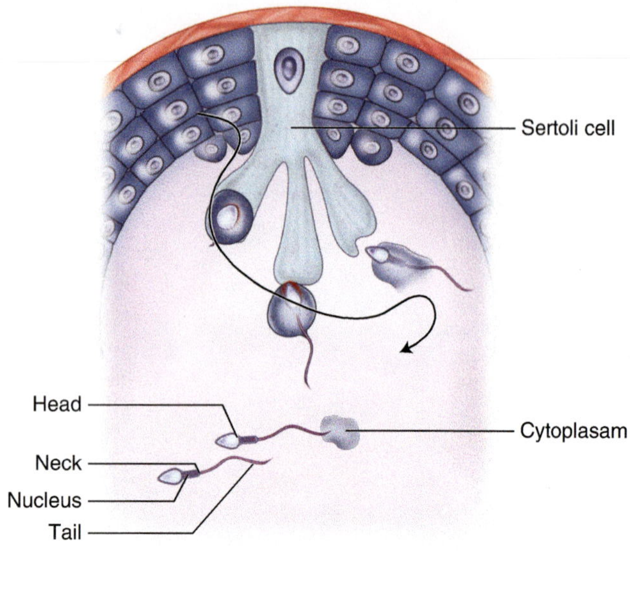

Fig. 2.9 Arrangement of cells within the seminiferous tubule

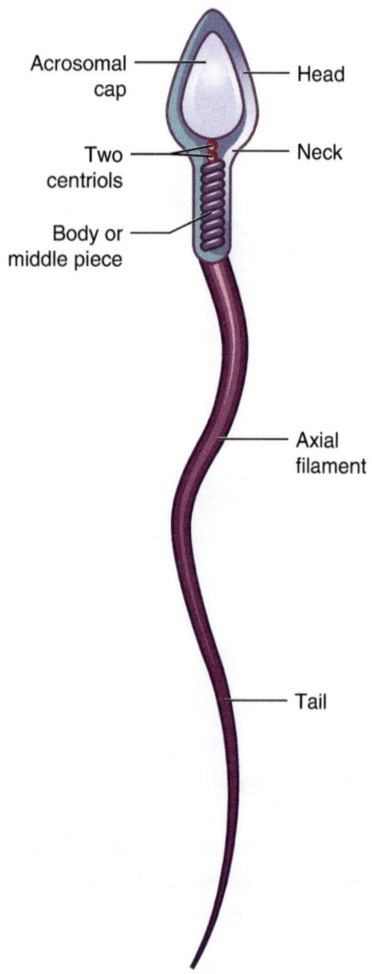

Fig. 2.10 The structure of the mature spermatozoa

Each secondary spermatocyte undergoes meiosis to give two spermatids.

Spermiogenesis

The spermatid which is rounded undergoes morphological and structural changes to become transformed into a mature sperm as follows (Figs. 2.9 and 2.10):

1. The nucleus becomes condensed and forms most of the sperm head.
2. Golgi apparatus forms the head cap covering the anterior half of the nucleus.
3. The centriole elongates to form the axial filament which traverses the neck, mid piece and tail of the sperm.
4. The mitochondria form a spiral sheath around the mid piece.
5. The remainder of the cytoplasm is shed.

The mature sperm is 60 μm in length and is formed from the following parts:

Head: 5 μm, formed mainly by the nucleus of the spermatid and carries the genetic information. It is partly covered by the acrosomal cap which contains enzymes to aid in penetration of the ovum.

Neck: Very short, contains two centrioles.

Mid piece: 5 μm, formed by the axial filaments surrounded by a mitochondrial sheath and concerned with energy production.

Tail: 50 μm, formed of axial filaments covered by a thin protoplasmic membrane and is concerned with motility.

Each spermatogonium divides by mitosis to produce two different daughter cells—one replaces the stem cell population, one divides again by mitosis and its progeny becomes the primary spermatocytes. These divide by meiosis to produce secondary spermatocytes which are very short lived and quickly enter the second meiotic division to produce spermatids which then undergo a maturation process, spermiogenesis, to form spermatozoa. In the human, the cycle of development from stem cell to sperm takes 64 days.

The mature male gamete or sperm is essentially a nucleus with a flagellum for motility; most of its cytoplasm has been shed during spermiogenesis when the mature spermatozoon develops from the spermatid, which itself is the end product of meiosis.

In the male, testosterone, produced by the Leydig cells in the testis, is essential for spermatogenesis because of its local effect on the Sertoli cells. The Leydig cells in turn depend on the secretion of LH by the gonadotrophs of the anterior pituitary. FSH is also essential for spermatogenesis because of its direct action on the Sertoli cells. Prolactin, from the anterior pituitary, potentiates the effects of testosterone and LH. Secretion of GnRH from the hypothalamus and LH and FSH from the anterior pituitary is controlled by negative feedback, as in the female. The normal functioning of the feedback control is essential for fertility. Testosterone in the plasma reduces the frequency of the pulses of GnRH released and also the quantity of LH released per pulse by the anterior pituitary. In females progesterone acts this way, rather than testosterone. Inhibin, secreted by the Sertoli cells, suppresses the secretion of FSH.

Fertilisation

Human development begins at fertilisation with the fusion of sperm with a newly ovulated Graafian follicle released from the ovary. Ovulation occurs at mid-cycle under the influence of LH released from the anterior pituitary at mid-cycle. The follicle ruptures releasing the secondary oocyte which is ovulated at metaphase II of the second meiotic division. The movement of the fimbriae of the uterine tube pulls the oocyte towards the uterine tube where fertilisation can occur.

The site of fertilisation is the ampullary region of the uterine tube, the oocyte moves to the site of fertilisation by ciliary action, and sperm reach the oocyte aided by contraction of tubal musculature. Fertilisation is a multistep process occurring 12–24 h after ovulation:

1. Capacitation of sperm. This takes place once the sperm enters the female reproductive tract. Taking approximately 7 h, this process involved the removal of the glycoprotein coat and seminal plasma proteins.
2. Penetration of the corona radiata and cumulus oophorus. At ovulation, the oocyte is surrounded by a cellular layer known as the corona radiata. The cellular layer is composed of protein and hyaluronic acid, and cumulus oophorus cells released by the follicle at ovulation may also surround the oocyte complex penetration driven by enzymatic degradation of the cellular layer by hyaluronidase produced in the sperm head. Active swimming of the sperm also plays a role in penetration of the corona radiata.
3. Penetration of the zona pellucida. 13um thick in the human, the zona pellucida is composed of four glycoproteins, ZP 1–4.
4. Acrosome reaction. The acrosome is derived from the Golgi apparatus and contains its own membrane. Fusion occurs between the acrosome membrane and the sperm membrane.
5. Sperm contacts the oocyte cell membrane. At this point, polyspermy must be prevented.
   (a) Fast block to polyspermy – rapid electrical depolarisation of oocyte plasma membrane.
   (b) Slow block to polyspermy – wave of Ca2+ ions from the site of sperm fusion acts on the cortical granules. These fuse with the plasma membrane, and this releases enzymes which break down the sperm receptors in the zona.

Fusion of the oocyte and the sperm occurs when the nuclear membranes around the male and female pronuclei break down nucleus allowing fusion of chromosomes resulting in formation of a single-celled diploid zygote.

Once fertilisation is complete, this leads to resumption of the oocytes' second meiotic division. In the adult human, there are 10^{14} cells, taking 45 generations of mitoses to produce these cells from a single fertilised egg, and cell division, growth, morphogenetic movement, differentiation and cell death are all required to shape the developing human.

Results of Fertilisation

Fertilisation results in the formation of the zygote which then divides to form a blastocyst, as seen in Fig. 2.11.

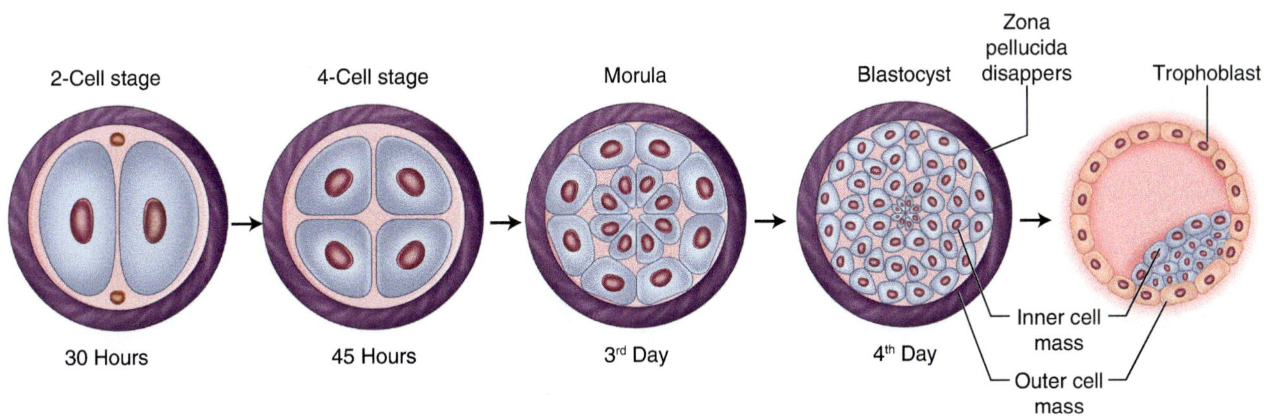

Fig. 2.11 Division of the zygote to produce a blastocyst stage embryo

1. Diploid number of chromosomes is restored.
2. Sex of embryo is determined.
3. Activation of the zygote occurs structural and molecular.
4. Cleavage commences.

The fertilised egg passes along the uterine tube, undergoing a series of mitotic divisions. There is no increase in size, so that component cells become smaller with each division. This is known as cleavage and allows the conceptus to pass through the narrow tubal isthmus. Early cleavage-stage embryos are totipotent. The zona pellucida is still intact preventing premature implantation in ectopic site.

The two-cell stage appears around 30 h after fertilisation, and by 45 h post fertilisation, the four-cell stage should have been reached. By day 3, the developing embryo has reached the 16-cell stage and is termed a morula, and the process of compaction occurs. During compaction, individual cells flatten against each other and form cell junctions. Cells of the morula show the first sign of cellular differentiation, cells flatten against each other, form junctions and two populations of different cells arranged into an inner cell mass population and an outer layer of cells, the trophectoderm, in the periphery. Post compaction, the blastomeres continue to divide, and fluid enters between spaces in the cells forming a blastocoel cavity marking the formation of the blastocyst which enters the uterine cavity by the 5th day.

Allocation of cells to an inner cell mass which will form the embryo or trophectoderm lineage depends on relevant positions of cells during cleavage. Segregation of cells to different positions is guided by cell-cell interactions. This may begin as early as the two-cell stage. Control of differentiation is exerted by micro environmental factors and development of polarity within the cells.

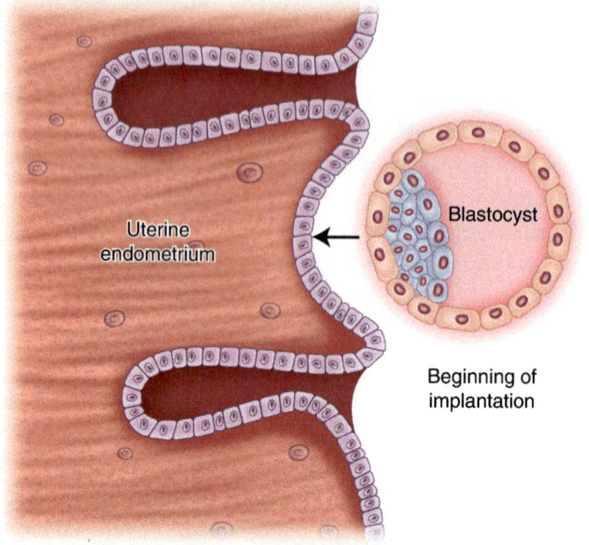

Fig. 2.12 Beginning of implantation

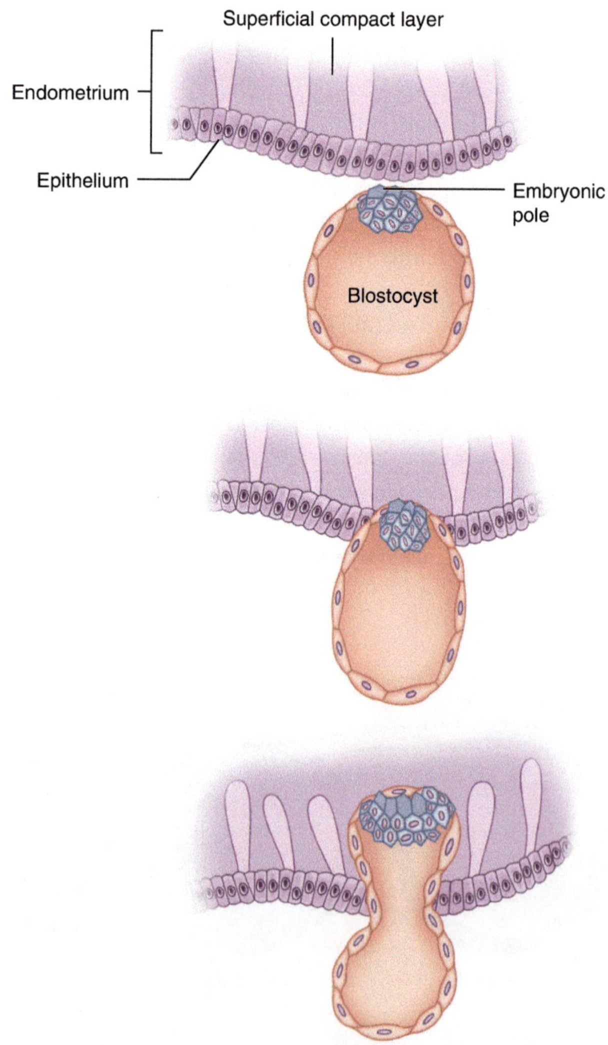

Fig. 2.13 Attachment of the blastocyst to the endometrium followed by enzymatic invasion

Implantation

Once in the uterine cavity, the blastocyst penetrates the superficial compact layer of the uterine endometrium. This begins around day 6 or 7 and is complete by around day 12 after fertilisation. The normal site of implantation is the endometrium of the posterior wall of the fundus of the uterus and can be seen in Fig. 2.12.

Implantation Is a Multistep Process and can be summarised as shown in (Fig. 2.13).

1. The blastocyst becomes attached to the endometrium.
2. The trophoblast cells lying over the inner cell mass begin to erode the endometrium by enzymatic action.
3. The blastocyst burrows into the endometrium.
4. After complete embedding of the blastocyst, the endometrium wall is closed by a fibrin clot. Implantation is completed by the growth of epithelium to cover the entry site.

Week 2 of Pregnancy

During week 2, the major events result in the development of two trophoblast layers, two embryonic layers and two cavities.

Trophoblast

During week 2, the trophoblast, seen in Fig. 2.14 shows a rapid rate of growth compared to the growth of the embryonic disc.

Entry to the uterine cavity occurs 4–5 days after fertilisation. The blastocyst can remain free in the uterine lumen for several days before implanting, but demand for space and nutrients is a driver for implantation. Pressure in the blastocyst cavity allows the cells of the blastocyst to 'hatch' from the zona pellucida, and the trophoblast immediately over the inner cell mass, and invade the maternal tissue allowing attachment in the posterior wall of the uterus, near the midsagittal plane. Cells push between the uterine epithelial cells and through their basement membrane. Stromal cells of the uterine stroma undergo decidualisation, filling with glycogen and lipids which act as a source of nourishment for the invading embryo, as well as creating an immunologically privileged site for the embryo. Implantation is complete by day 12 in the human. It is estimated that up to 50% of all pregnancies are aborted around this time by a natural screening process for chromosomal abnormalities.

The trophoblast over the inner cell mass is the first to implant and differentiates as it erodes the endometrial stroma, forming:

1. An inner mitotic layer, the cytotrophoblast
2. An outer syncytial (multinucleate) layer, the syncytiotrophoblast

Spaces called lacunae begin to appear in the syncytiotrophoblast to allow exchange of foetal and maternal blood.

Embryonic Layers and Cavities

At day 8, the inner cell mass rearranges to give two distinct cell layers as seen in Fig. 2.15:

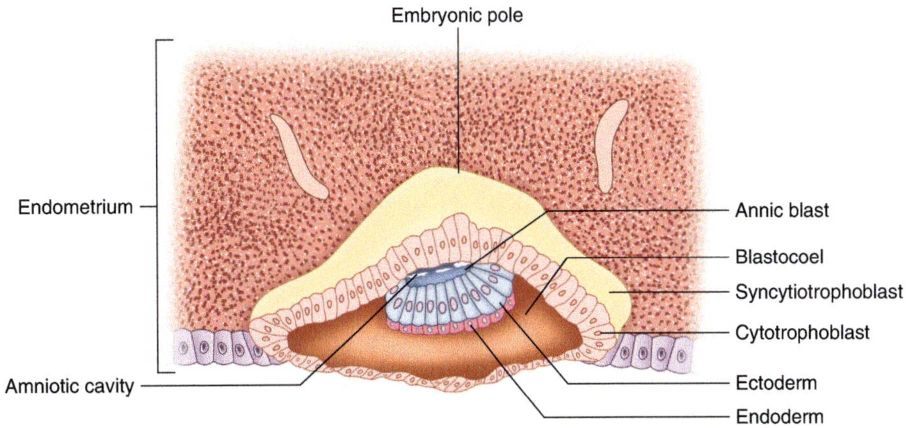

Fig. 2.14 Early stages of implantation

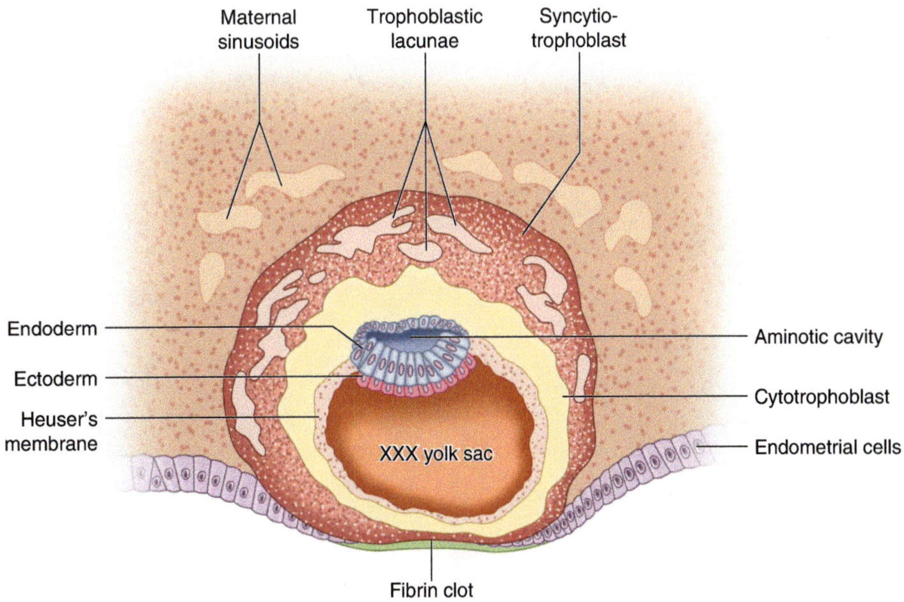

Fig. 2.15 Establishment of the embryonic layers and extra-embryonic membranes and cavities

1. Epiblast – high columnar cells adjacent to the cytotrophoblast layer
2. Hypoblast – small cuboidal cells nearest to the blastocyst cavity

The embryo is now in the form of a bilaminar germ disc.

The epiblast layer will go on to form the three germ layers of the embryo. The hypoblast is concerned with the derivation of the extraembryonic membranes. Each of these layers becomes continuous at its margins with an extraembryonic membrane which develops in association – the epiblast with the amnion and the hypoblast with the yolk sac. Each layer associated with a membrane forms a vesicle enclosing a cavity: the amniotic cavity and the yolk sac cavity.

Amniotic Cavity

Small clefts begin to appear between the ectodermal cells and the trophoblast. These clefts join each other forming the amniotic cavity. The cytotrophoblast develops a layer of flat cells called amnioblasts which form the roof of the amniotic cavity while its floor is formed by the ectodermal layer.

Yolk Sac

About 12 days after fertilisation, extraembryonic mesoderm appears outside these two vesicles in the blastocyst cavity. It is of great significance in the development of the placenta. Almost as soon as it appears, a cavity, the extraembryonic or chorionic cavity, forms within it. The trophoblast and its lining of extraembryonic mesoderm are referred to as the chorion and can be seen in Fig. 2.16.

The yolk sac forms on the ventral aspect of the embryonic disc. Cells of the endoderm layer grow and line the inner surface of the cytotrophoblast forming Heuser's membrane. The yolk sac replaces the cavity of the blastocyst.

Both cavities are fluid filled which facilitate diffusion of nutrients at this early stage before the placenta and its circulation are established. The amniotic cavity persists throughout development allowing symmetrical growth and protection for the foetus.

At the end of week 2, cavities are formed inside the extraembryonic mesoderm. These cavities then fuse together forming the extraembryonic coelom. This coelom divides the mesoderm incompletely into somatopleuric mesoderm which lines the cytotrophoblast and covers the amniotic cavity and the splanchnopleuric mesoderm which covers the yolk sac. The roof of the amniotic cavity is connected to the trophoblast by the body or connecting stalk.

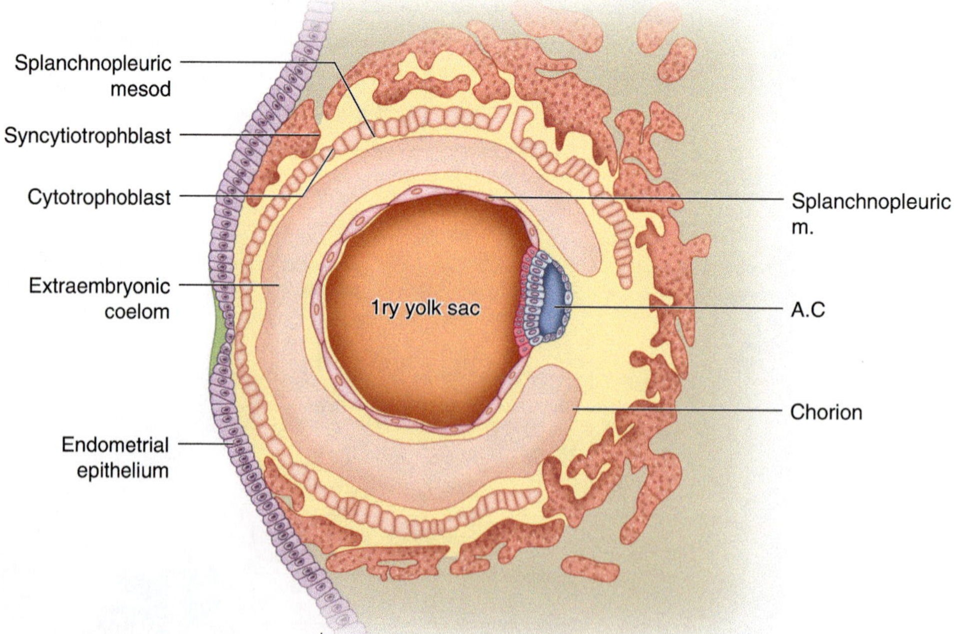

Fig. 2.16 Development of the embryonic coelom and chorionic vesicle at the end of week 2 of development

Week 3

In the third week of life, the three definitive germ layers, ectoderm, mesoderm and endoderm, are formed, transforming the bilaminar germ disc into a trilaminar structure (Fig. 2.17).

The three germ layers arise from the epiblast by a process of cell proliferation and migration known as gastrulation. Three important structures are involved in gastrulation, all of which make an appearance in week 3, the primitive streak, the notochord and the neural plate.

Ectodermal cells in the caudal part of the bilaminar disc migrate to the midline forming the primitive streak. Cells of the primitive streak proliferate and invaginate through the streak. A population of cells replace the hypoblast cells, which are pushed out laterally into the yolk sac to form the endoderm. Others form a new middle layer between the epiblast and hypoblast/endoderm called the mesoderm. At the end of gastrulation, the remaining epiblast is termed the ectoderm.

At the head end of the streak, a thickened area of ectoderm appears; this primitive node consists of a small central depression called the primitive pit. Cells of the primitive node proliferate to form a solid rod of cells, the notochord, a mesodermal structure which grows in a cephalic direction along the midline between the ectoderm and endoderm. This is an important structure as it forms a primitive skeletal axis

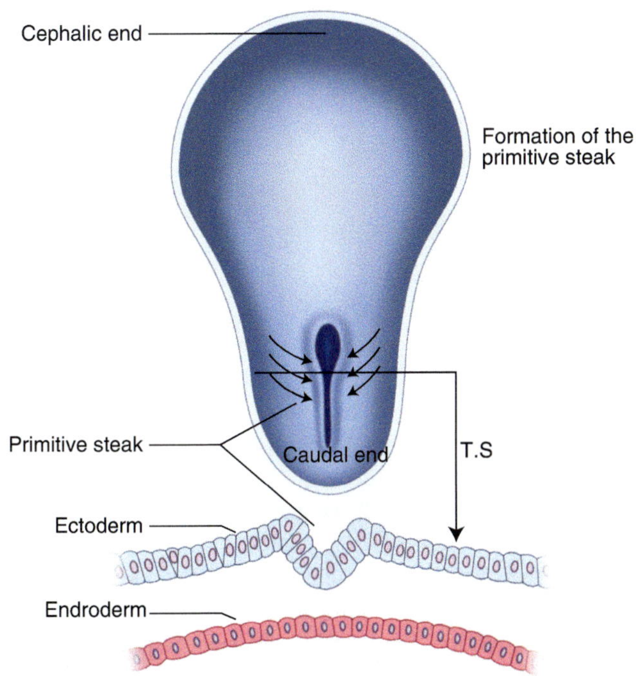

Fig. 2.17 Formation of the primitive streak signalling the start of gastrulation

and plays an important role in the development of the central nervous system.

The notochord induces changes in the overlying ectoderm, which thickens to form the neural plate. The edges of the plate fold and eventually meet in the midline to form the neural tube, the forerunner of the central nervous system which sinks inwards as the gap in the surface ectoderm is repaired (Fig. 2.18).

A population of tissue at the summit of the neural fold is excluded from the tube and pinched off separately to lie alongside the tube on each side. This is the neural crest tissue. Neural crest cells are highly migratory; they give rise to many derivatives including cells of the spinal ganglia, autonomic ganglia, pigmented cells, Schwann cells, mesoderm in the head, adrenal medulla and leptomeninges.

Closure of the neural tube begins in the future cervical region of the spinal cord; for a while openings are present at the head and tail ends – the anterior and posterior neuropores. These will eventually close in week 4. Failure of their closure leads to defects such as anencephaly and spina bifida.

Two areas are present which do not contain mesodermal cells, just ectoderm and endoderm cells. The fate of these areas is to break down and form the two openings to the gut, the oral cavity and the anus.

Differentiation of the germ layers during the embryonic period (Table 2.1).

Ectoderm

At first, the ectoderm forms the dorsal layer of the embryonic disc and makes the floor of the amniotic cavity. After the embryo begins the process of folding, the ectoderm becomes the outer layer of the body of the embryo. The ectoderm differentiates into the following:

1. Epidermis of the skin, including hair, nails and skin glands
2. Neural tube, which will form the nervous system
3. Sensory epithelium of the sense organs
4. Pituitary gland

The Neural Tube

Development of the neural tube in week 3 of development happens with the following steps: (Fig. 2.19a)

1. Formation of the neural plate

   At the beginning of week 3, the ectoderm over the notochord thickens, forming a median band called the neural plate. This is neuroectoderm in origin and extends

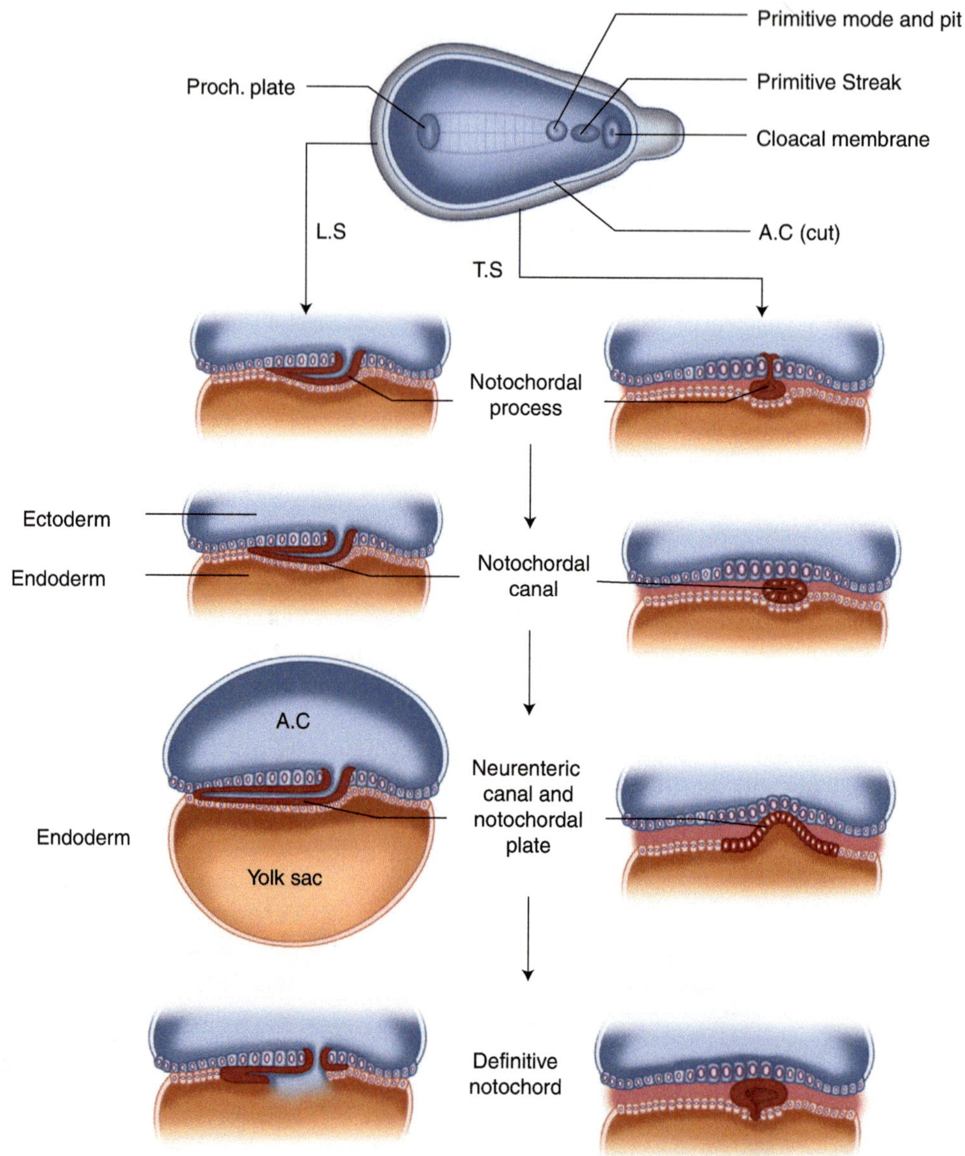

Fig. 2.18 Formation of the notochord

Table 2.1 The significance of the germ layers and their derivatives

Germ Layer	Derivative
Ectoderm	Forms epidermis of skin and nervous tissue (through the neural tube and neural crest)
Mesoderm	Forms cardiovascular system, urogenital system, muscle and connective tissue
Endoderm	Forms epithelial linings of the gut and respiratory systems

from the primitive node to the buccopharyngeal membrane (Fig. 2.19b).

2. Formation of the neural groove

   The edges of the neural plate become elevated forming right and left neural folds, and as a result, the neural plate is transformed into the neural groove.

3. Formation of the neural tube

   As the neural groove deepens, the right and left neural folds approach each other in the midline and fuse at the region of the fourth somite. The fusion of the two folds then proceeds in both a cranial and caudal direction, transforming the neural groove into a tube structure which is buried under the surface ectoderm.

4. Closure of the anterior and posterior neuropores

   As the closure of the tube proceeds in a cranial and caudal direction, the anterior and posterior ends remain open and are connected to the amniotic cavity. These are the anterior and posterior neuropores. The anterior neuropore closes at the 20 somite stage, while the posterior neuropore closes at the 25 somite stage.

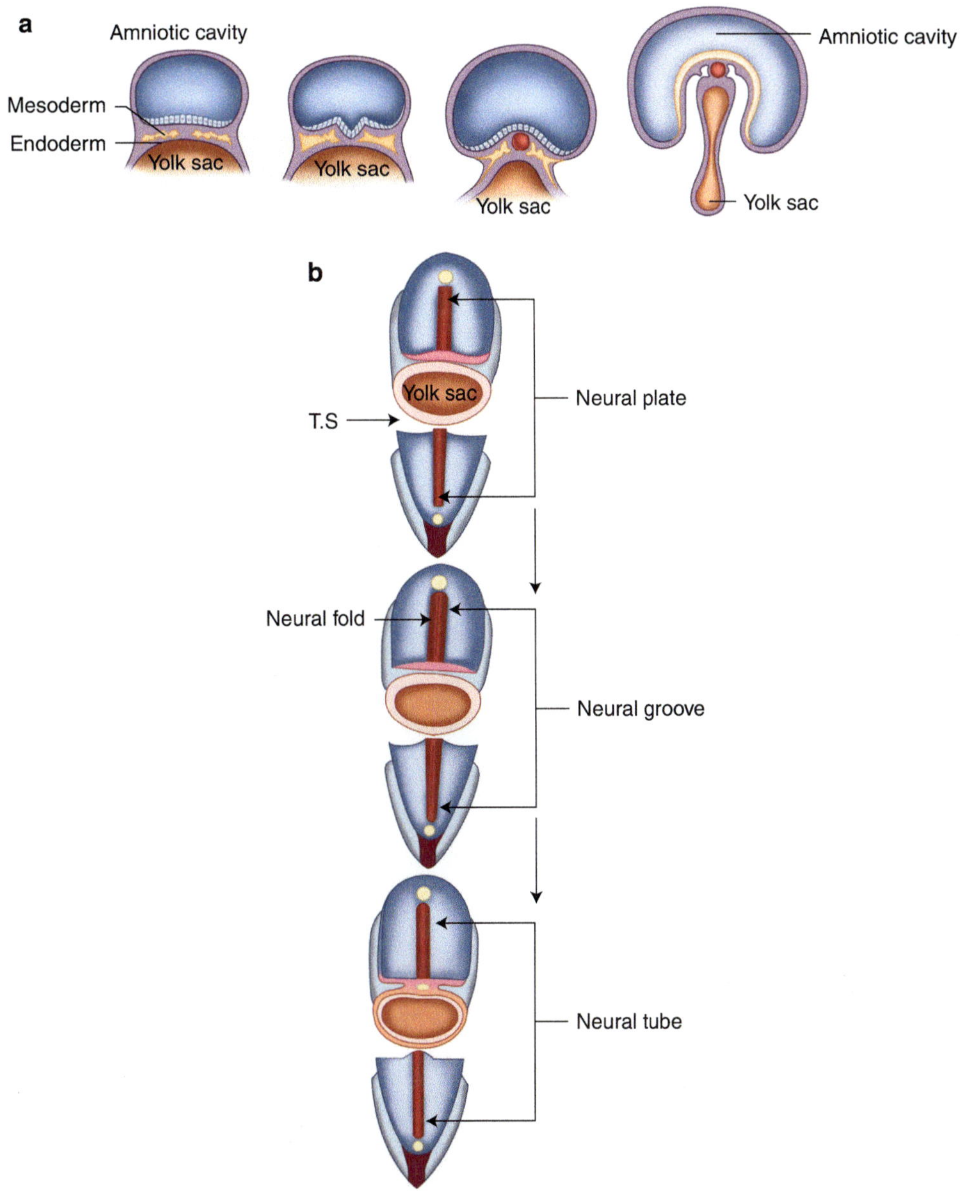

Fig. 2.19 (**a**) Establishment of the ectoderm layer from the embryonic disc. (**b**) Establishment of the neural plate from the ectoderm layer

Differentiation of the Neural Tube

1. The broad cranial part of the neural tube will become the brain, while the narrower caudal portion of the tube will form the spinal cord (Fig. 2.20).

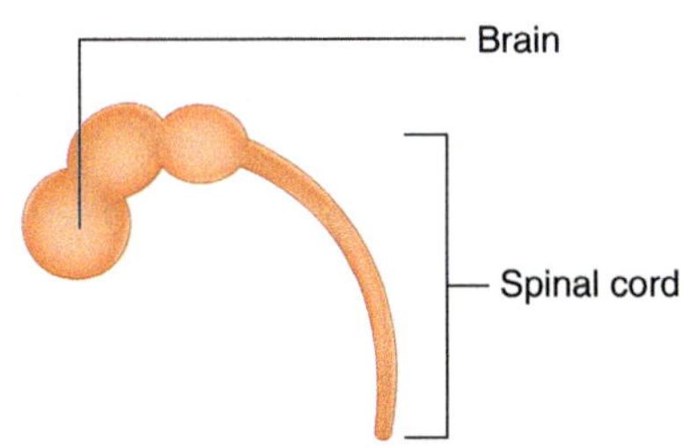

Fig. 2.20 Differentiation of the neural tube

Mesoderm (Fig. 2.21)

Initially, the mesoderm is a sheet of loose tissue between the ectoderm and the endoderm on either side of the notochord (Fig. 2.19b).

As development proceeds, two longitudinal grooves appear in the mesoderm on either side of the notochord dividing it into three parts:

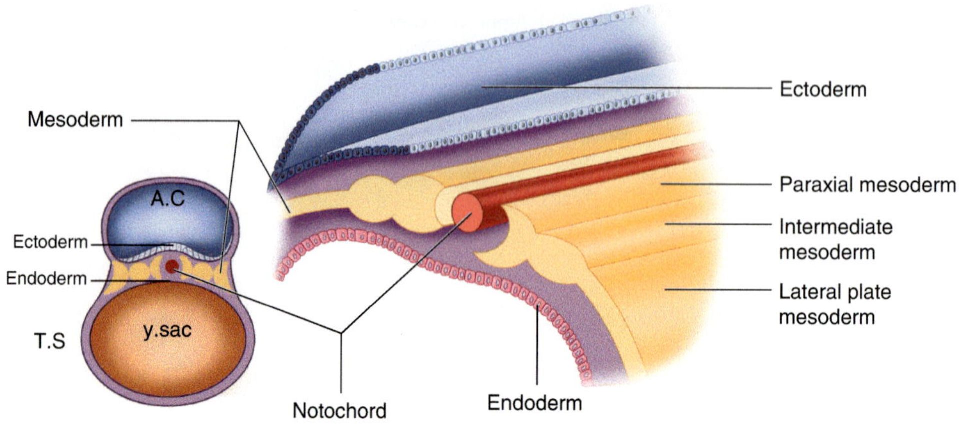

Fig. 2.21 Formation of the mesoderm layer

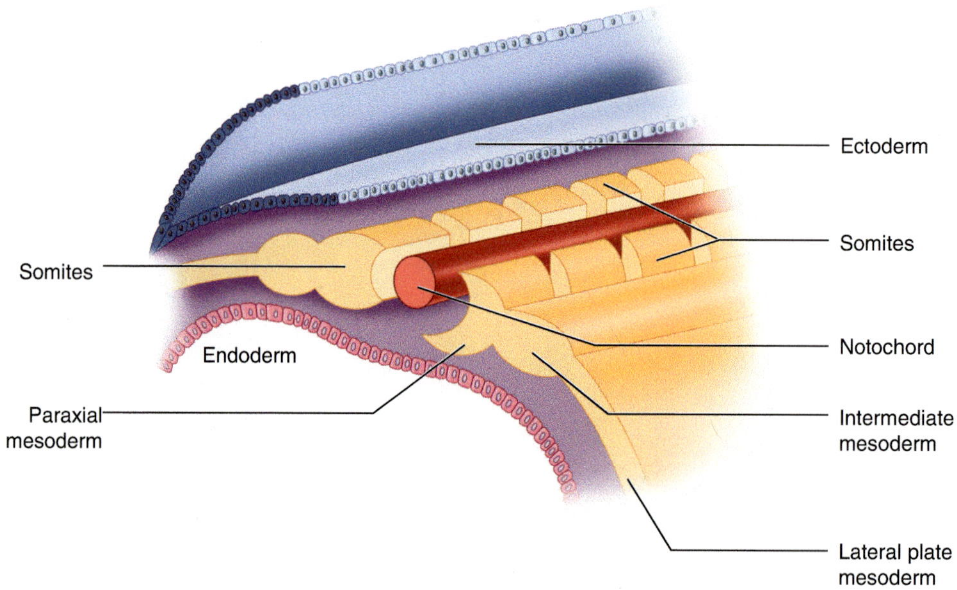

Fig. 2.22 Segmentation of the paraxial mesoderm into somites

1. Paraxial mesoderm
2. Intermediate mesoderm
3. Lateral plate mesoderm

Paraxial mesoderm consists of two thick longitudinal bands, one on either side of the notochord. At day 20, transverse grooves appear in the paraxial mesoderm dividing it into small mesodermal blocks called somites (Fig. 2.22).

Somites appear in a cranio-caudal manner at a rate of approximately three pairs per day. By the end of the first month, around 30 pairs of somites will have been formed; between days 30 and 40, the rate of somite formation slows, until 42–44 pairs are present. The somites are arranged as follows: 4 occipital, 8 cervical, 12 thoracic, 5 lumbar, 5 sacral and 8–10 coccygeal (Fig. 2.23).

By the end of the 4th week, the somites start to differentiate, again in a cranio-caudal direction. Each somite divides into two parts, a ventromedial part called the sclerotome and a dorsolateral part called the dermomyotome (Fig. 2.24).

The sclerotome differentiates into mesenchyme which will give rise to the connective tissue, cartilage and cells of the axial skeleton. The dermomyotome divides further, a lateral part – the dermatome – which spreads underneath the ectoderm to form the dermis of the skin and a medial part, the myotome, which develops into myoblasts that later will form the skeletal muscles of the body.

During the period of somite formation, the age of the embryo can be roughly estimated by counting the number of somite pairs as shown in Table 2.2.

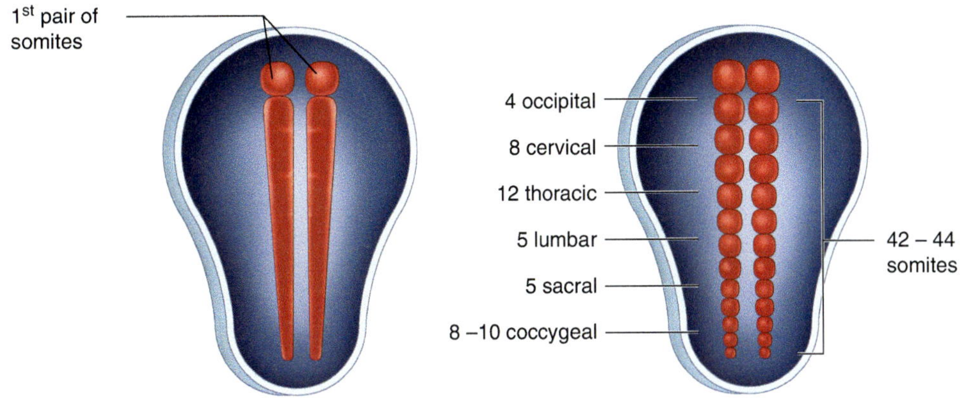

Fig. 2.23 Arrangement of the somites

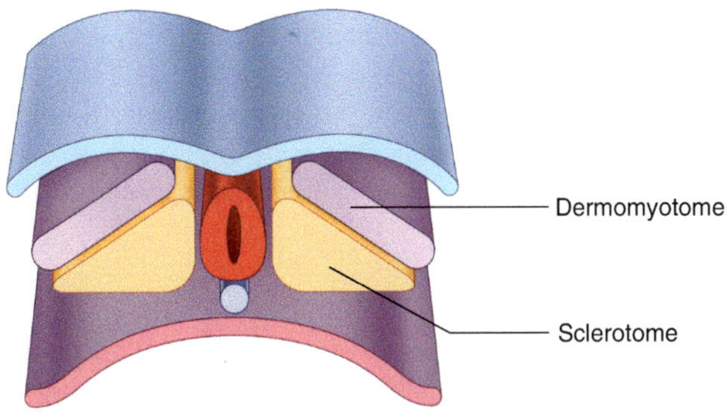

Fig. 2.24 Differentiation of the somites

Table 2.2 Age determination during somite development

Number of somite	1	4	7	10	13	16	19	22	25	28	31
Age in days	20	21	22	23	24	25	26	27	28	29	30

Intermediate Mesoderm

This is found between the paraxial mesoderm and the lateral plate mesoderm. It gives rise to the cortex of the adrenal gland, the nephrons of the kidney and the gonads.

Lateral Plate Mesoderm

This lies lateral to the intermediate mesoderm near the edges of the embryonic disc and extends in a cephalic direction to the prechordal plate. Cavities appear inside the lateral plate mesoderm and unite together forming the intraembryonic coelom which divides the lateral plate mesoderm into two layers:

Somatic or parietal layer becomes adherent to the ectoderm and forms the muscles, connective tissue and supportive elements of the body wall.

Splanchnic or visceral layer becomes adherent to the endoderm and gives rise to:

1. The serous membranes: pleura, pericardium and peritoneum
2. The smooth muscles, connective tissue of the gastrointestinal and respiratory tracts
3. The cardiovascular system

Endoderm

Initially, the endoderm forms the ventral layer of the embryonic disc and lines the yolk sac. As a result of folding, the upper part of the endodermal-lined yolk sac becomes incorporated into the body of the embryo forming the primitive git. This gives rise to the following endodermal derivatives:

1. Epithelial lining of the digestive tract, respiratory tract, the middle ear and Eustachian tube and most of the lining of the bladder and urethra
2. Parenchyma of the tonsil, thyroid, parathyroid and thymus glands, the liver and the pancreas (Figs. 2.25, 2.26 and 2.27)

In week 4 of life, the embryonic body changes in shape from a flat disc to a rolled-up cylinder, curved at the head and tail ends. The flat embryonic disc becomes folded ventrally and bulges into the amniotic cavity.

The embryonic disc folds in two directions simultaneously, in a cranio-caudal direction, which forms the head and the tail folds, and in a lateral direction, forming two lateral folds. The folding is caused by differential growth in the embryonic disc; the central portion of the embryonic disc grows more rapidly than the peripheral parts.

As this happens, the gut tube is formed from the bending of the endoderm. The embryonic body becomes more constricted from the yolk sac which remains connected to the midgut and balloons up into the amniotic cavity. For a period, the body of the embryo is too small to accommodate the rapidly lengthening gut tube, and so the midgut is herniated out into the umbilical cord from weeks 5 to 10.

As a result of the formation of the head and tail folds, the primitive gut becomes divided into three parts, the foregut, within the head fold; the hindgut, which lies in the tail fold; and the midgut which lies in between the foregut and midgut.

The cranial end of the newly folded embryo has a number of distinct features. The forebrain swelling is visible, caused by the growing forebrain. A pericardial swelling is also apparent due to the growing heart structures.

The caudal end of the embryo contains the cloacal membrane which occupies the most caudal end and the allantois and connecting stalk lie cranial to this membrane.

Following the formation of the three germ layers, the foetus undergoes differentiation and the organ systems develop as shown in table 2.3. The features of the developing foetus up until the point of birth can be observed in (Figs. 2.28).

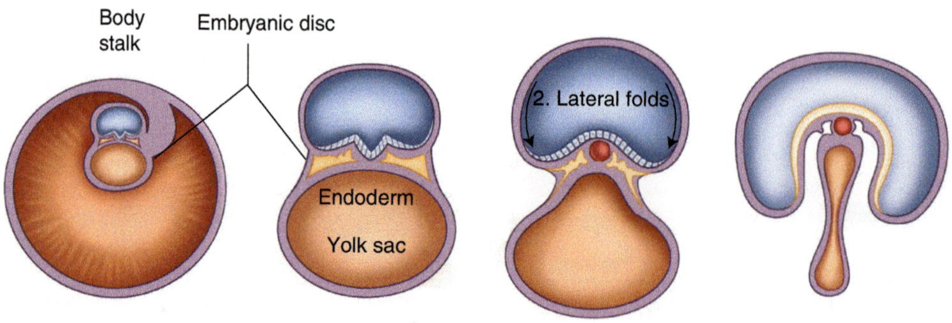

Fig. 2.25 Embryonic folding during week 4 of development

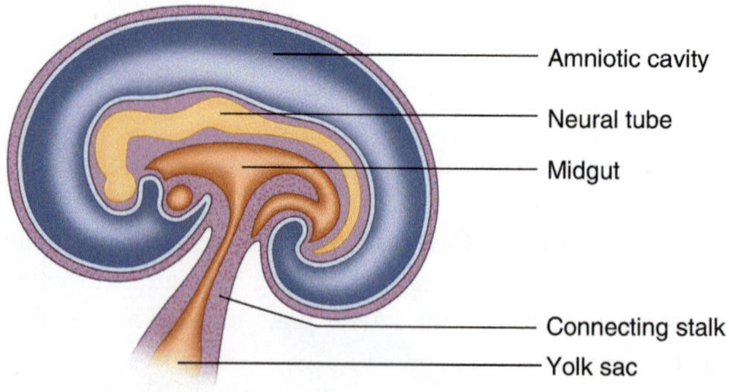

Fig. 2.26 Result of embryonic folding

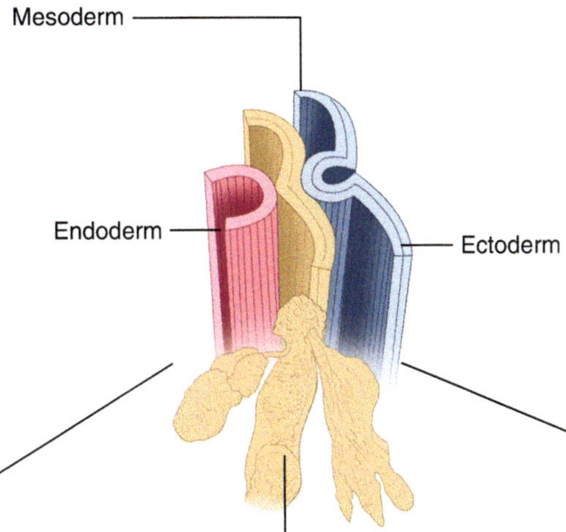

Endoderm	Mesoderm	Ectoderm
I-Epithelial lining of: (1) the alimentary canal except: –ant.part of mouth –lower part of anal canal (2) the respiratory tract (3) the urinary bladder (except the trigone). (4) most of urethra (5)Ear: (a)pharyngotympanic tube (b)middle ear cavity (c) inner layer of eardrum II- parenchyma of: (1) Tonsil (2) Thyroid &parathyroid glands (3) Thymus gland	(1) All types of connective tissue, cartilage, bone & joints (2) All types of muscles: Skeletal, smooth and cardiac muscle (3) All cardiovascular system: the heart, blood vessels (4) Bone marrow & blood cells (5) Lymph vessels & all lymph tissue: lymph nodes & spleen (6) Serous membranes: pleura, pericardium & peritoneum (7) Dura matter & microglia of the nervous system (8) Dermis of the skin (9) Urogenital system except most of the urinary bladder & urethra (10) Cortec of the suprareral gland.	I-Nervous system: –brain & spinal cord –pia & arachnoid mater –sensory epithelium of the sense organs (2) Skin: epidermis, hair & nails (3) Glands: pituitary gland & the medulla of suprarenal gland (4) Digestive system: lining of: –ant.part of the oral cavity –lower part of anal canal (5) Respiratory system: Epithelium of the nose (6) Ear: External auditory meatus & outer layer of eardrum

Fig. 2.27 Derivatives of the three germ layers

Table 2.3 Overview of first 16 weeks of life

Overview of first 16 weeks of life	
Weeks 1–2 – pre-embryo	Fertilisation, cleavage, compaction, blastocyst formation, implantation, bilaminar germ disc (see below)
Weeks 3–8 – embryonic period (organogenesis)	
Week 3	Gastrulation, formation of trilaminar germ disc (with *three* definitive germ layers); neural and heart tubes are forming (see above)
Week 4	Neural folds fuse, body folds, heart starts to pump and limb buds appear
Weeks 4–10	The face develops
Weeks 5–7	Septation of the heart – partitions develop dividing the heart tube into four chambers. Conducting system first appears in the middle of week 6
Week 6	Bronchopulmonary segment primordia appear; major and minor calyces of kidney form. Sertoli cells appear in the male gonad, the first sign of sexual differentiation
Weeks 6–7	The clavicle (first bone to ossify) begins to ossify
Weeks 6–16	Pseudoglandular stage of lung development (tracheobronchial tree is formed as far as terminal bronchioles)
End of week 8	All components of the upper and lower limbs are distinct
Week 12	Kidney produces dilute urine

Fig. 2.28 External features of the foetus

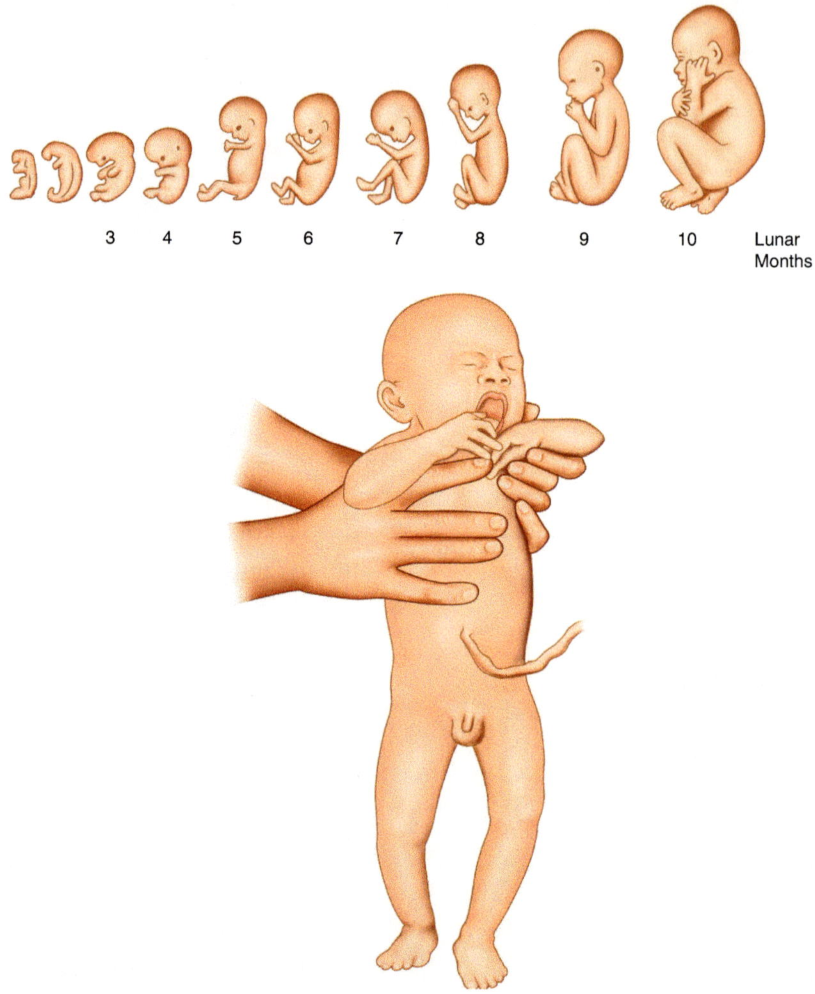

Embryology of the Foetal Membranes and Placenta

3

Sharon F. Sneddon

The foetal membranes include all the extraembryonic structures which are derived from the primitive blastomeres and do not enter into the formation of the embryo itself. The membranes are the amnion, the yolk sac and the chorion.

The Amnion

Continuous with the ectodermal germ layer, the amnion is the membrane which bounds the amniotic cavity. It starts to appear on day 7 of development as small intercellular clefts between the ectodermal cells and the trophoblast. The clefts unite together to form a small space between the trophoblast and the ectoderm. This is the amniotic cavity. The cavity enlarges in size and becomes roofed by a layer of flattened amnioblasts, which develop from the inner surface of the trophoblast. After the formation of the extraembryonic mesoderm and the development of the extraembryonic coelom the roof of the amniotic cavity becomes separated from the trophoblast by a mass of extraembryonic mesoderm known as the connecting stalk, or body stalk. The amniotic cavity is now bounded by a membrane formed of amnioblasts and extraembryonic mesoderm and is termed the amnion (Fig. 3.1).

As pregnancy proceeds, the amniotic cavity enlarges rapidly and by the third month surrounds the embryo almost completely. The amnion eventually comes in contact with the chorion, thus obliterating the extraembryonic coelom (Fig. 3.2).

Secreted by the amnioblasts, the amniotic cavity is filled with amniotic fluid. It also receives urine secreted by the foetal kidney. By the end of pregnancy, the cavity will contain approximately 1.5 L of fluid. The amniotic fluid acts as a watery protective cushion around the embryo, protecting against trauma and external pressure. It prevents adhesions between the embryo and the amnion and helps maintain a constant temperature around all parts of the body of the embryo. The amniotic cavity and fluid provide a space for the embryo to freely move, which encourages muscle development. It also acts as a reservoir for the accumulation of foetal urine and meconium. The foetus will swallow the amniotic fluid, which helps develop the suckling reflex.

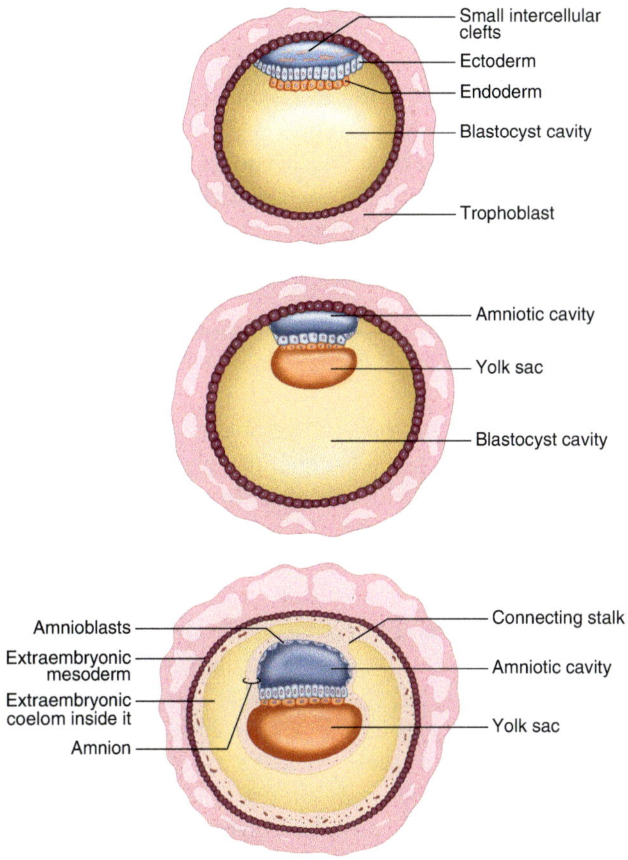

Fig. 3.1 Formation of the amnion and amniotic cavity

S. F. Sneddon (✉)
School of Medicine, Dentistry and Nursing,
University of Glasgow, Glasgow, UK
e-mail: Sharon.sneddon@glasgow.ac.uk

R. Carachi, S. H. E. Doss (eds.), *Clinical Embryology*, https://doi.org/10.1007/978-3-319-26158-4_3

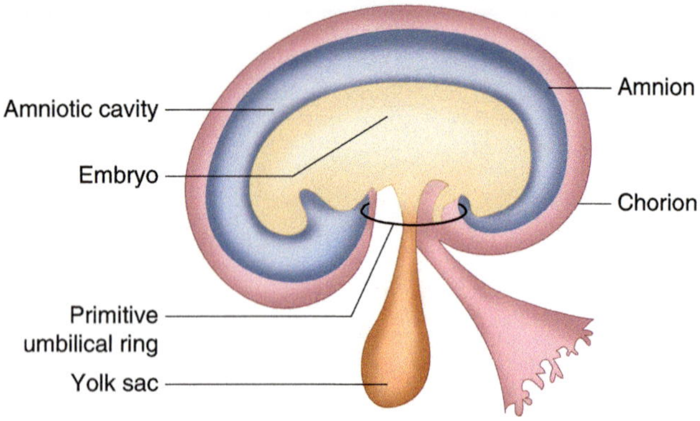

Fig. 3.2 Enlargement of the amniotic cavity and obliteration of the extra embryonic coelom

Oligohydramnios is the condition arising when the volume of amniotic fluid is less than half a litre. This may lead to adhesions between the embryo and the amnion. The opposite condition, polyhydramnios, when the volume of fluid is more than 2 L may lead to premature rupture of the amnion.

The Yolk Sac

Development of the yolk sac begins at around day 9 when the endoderm of the embryonic disc grows down on the inner surface of the trophoblast forming a membrane called Heuser's membrane. The primary yolk sac is roofed by the endodermal germ layer, while the rest of its wall is formed of Heuser's membrane with its covering extraembryonic mesoderm. The primary yolk sac later separates from the trophoblast by the development of the extraembryonic coelom.

The secondary yolk sac begins development at the end of the second week when the terminal end of the primary yolk sac becomes cut-off. The secondary yolk sac is covered by the splanchnic layer of the extraembryonic mesoderm, and a network of vitelline vessels develop in the mesoderm covering the secondary yolk sac (Fig. 3.3).

As a result of folding of the embryo, the roof of the secondary yolk sac becomes enclosed inside the body of the embryo. The secondary yolk sac becomes divided into three parts:

1. The primitive gut: the part inside the body of the embryo. This will eventually subdivide to form the foregut, midgut and hindgut.

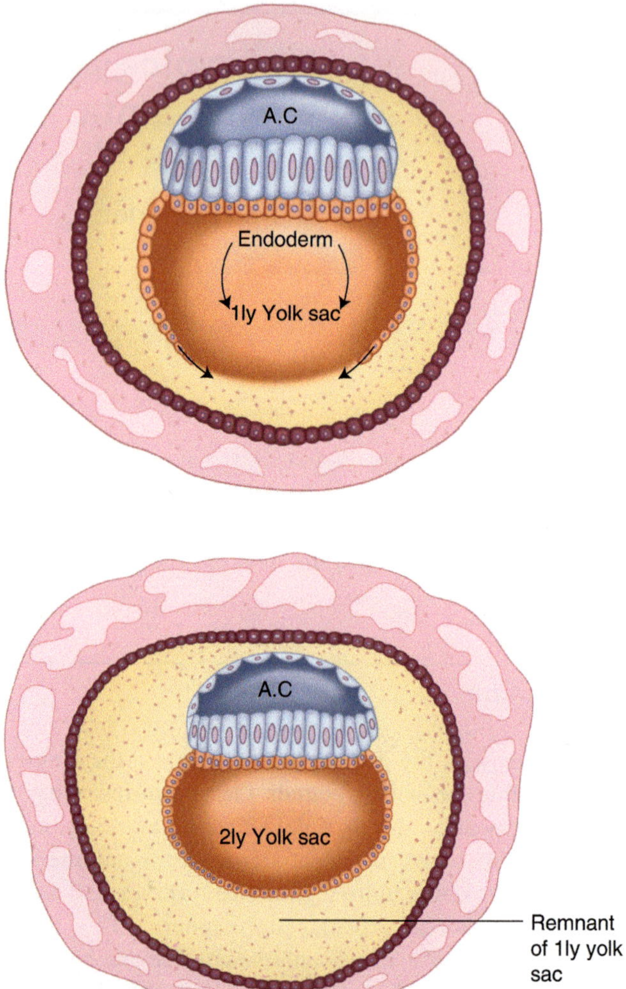

Fig. 3.3 Development of the primary and secondary yolk sac

2. The definitive yolk sac: the part of the yolk sac lying outside the embryo but inside the umbilical cord. This grows very slowly, never exceeding 0.5 cm in diameter, and eventually shrinks to form a small body within the cord.
3. The yolk sac stalk or vitello-intestinal duct: this connects the primitive gut with the definitive yolk sac. This disappears later in development (Fig. 3.4).

During week 3, a tubular invagination from the caudal part of the secondary yolk sac which extends into the connecting stalk develops into the allantois. When the hindgut is formed, the allantois becomes connected to the ventral aspect of the cloaca. The allantois has two parts, an intra-embryonic part which forms the urachus, connecting the urinary bladder to the umbilicus, and an extraembryonic part inside the umbilical cord which becomes obliterated.

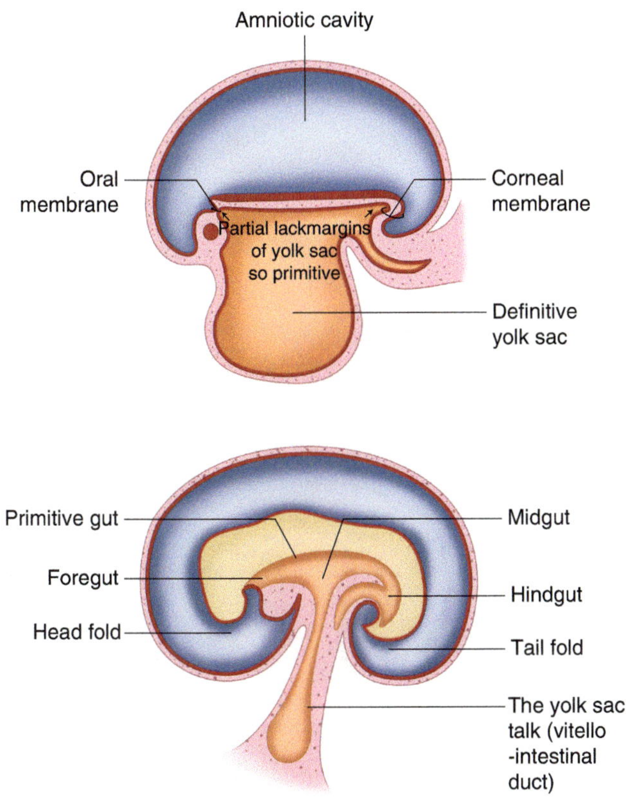

Fig. 3.4 Development of the definitive yolk sac

The yolk sac has no nutritive function in the human. As well as a role in the development of the primitive gut and the allantois, it has several other important functions. The primordial germ cells arise from the caudal wall of the yolk sac and then migrate to enter the developing gonads of the embryo. Blood is formed in the mesoderm of the wall of the yolk sac between weeks 4 and 6 of development. The vitelline vessels of the yolk sac will go on to form some of the embryonic vessels.

The Chorion

The chorion forms from the trophoblast after the formation of the extraembryonic mesoderm from its inner surface. At implantation, the trophoblast differentiates into an outer syncytiotrophoblast layer and an inner cytotrophoblast. The cytotrophoblast develops a layer of extraembryonic mesoderm on its inner surface which later splits into somatopleuric and splanchnopleuric layers. Both the trophoblast and the somatopleuric mesoderm are called the chorion, and the blastocyst is now termed the chorionic vesicle.

At the end of the second week of development, formation of the placenta proceeds with the appearance of the primary chorionic villi. These first appear at the embryonic pole of the chorionic vesicle and increase in number at the beginning of week 3. Each primary villus is formed of a central core of cytotrophoblast covered by a layer of syncytiotrophoblast (Fig. 3.5).

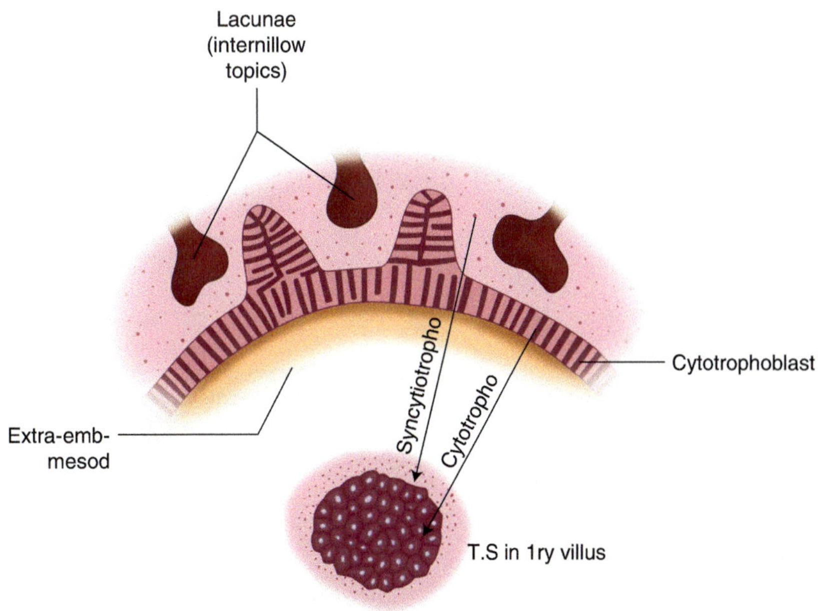

Fig. 3.5 Primary villus with central core of cytotrophoblast covered with syncytiotrophoblast during week 3 of development

Cells from the extraembryonic mesoderm which line the cytotrophoblast start to penetrate the primary villi to form the secondary chorionic villi. Each villus now consists of a central core of extraembryonic mesoderm, a middle zone of cytotrophoblast and an outer layer of syncytiotrophoblast (Fig. 3.6).

By the end of the third week, a loop of an afferent and an efferent capillary appears in the mesodermal core of the secondary villi, transforming them into tertiary villi. The afferent capillary loop is connected to the umbilical artery, while the efferent is connected to the umbilical vein. The tertiary villi branch in the intervillous spaces and oxygen and nutrients diffuse from the maternal blood in the intervillous space to the capillary loop in the tertiary villus (Fig. 3.7).

By the end of the third week, the embryo and embryonic structures appear as show in (Fig. 3.8).

The tertiary villi branch extensively to form a villous tree. The majority of these branching villi are free and surrounded by maternal blood. A few villi penetrate into the decidua basalis and fix the chorionic vesicle to the wall of the uterus; these are the anchoring villi (Fig. 3.9).

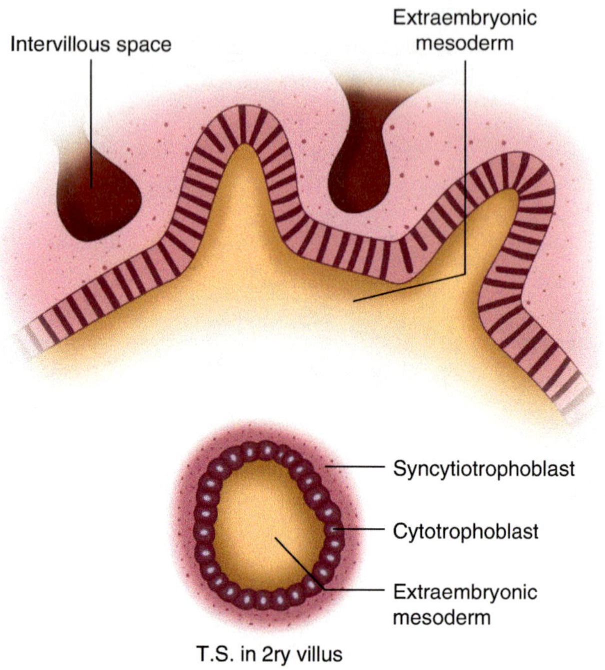

Fig. 3.6 Secondary villus with the development of a central core of mesoderm

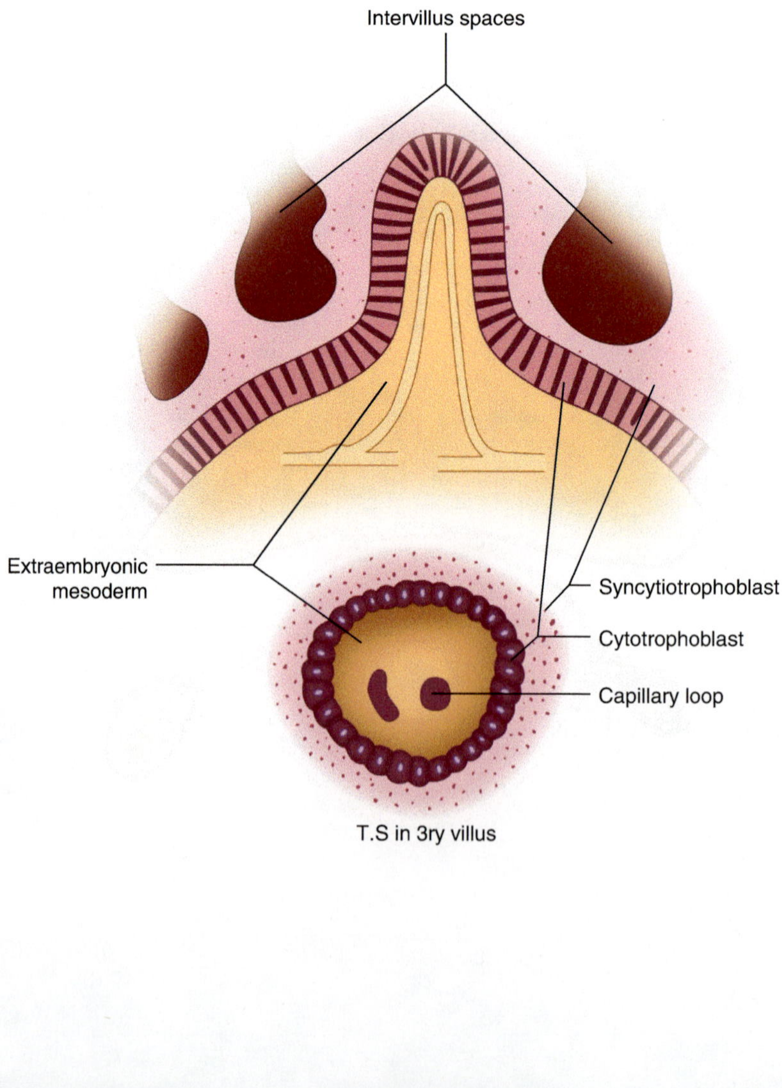

Fig. 3.7 Formation of the tertiary villi, the functional villi which allows diffusion of nutrients and oxygen from the maternal blood

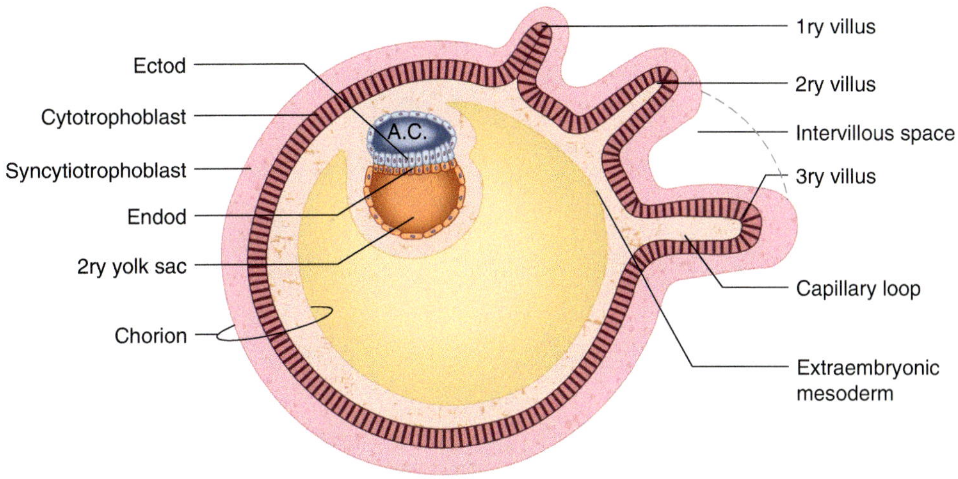

Fig. 3.8 The fully formed chorionic villi at week 3 of development

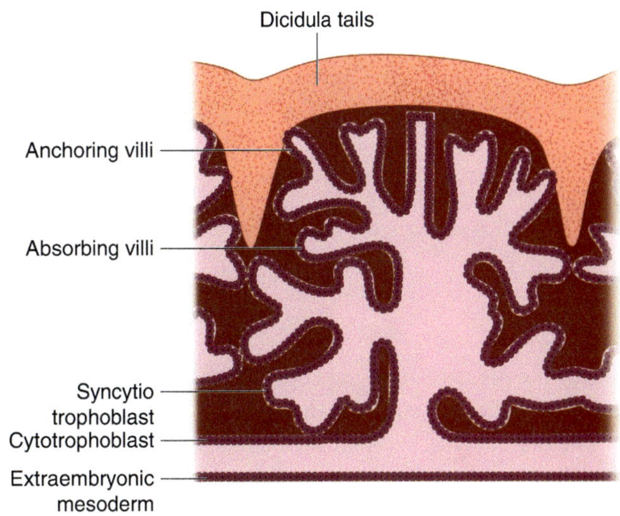

Fig. 3.9 The villous tree—the main structure of the placenta which anchors the placenta to the wall of the uterus

In the early weeks of pregnancy, the chorionic villi cover the whole surface of the chorionic vesicle. As pregnancy advances, the following changes occur: The chorionic villi which lie over the embryonic pole become more numerous and well developed giving this part of the chorion a leaflike appearance. For this reason, this part of the chorion is termed the chorion frondosum and is the only functioning part of the chorion. The villi that lie over the remaining part of the chorionic vesicle begin to degenerate by the end of the third month. This part of the chorion becomes smooth, having no villi, and is termed the chorion laeve.

The Placenta

The placenta is a vital organ of connection between the foetus and mother allowing physiological exchange between foetal and maternal circulations. It acts as a nutritive, respiratory, excretory and endocrine organ for the foetus during intrauterine life. The placenta is formed of two parts, a foetal part, the chorion frondosum, described above, and the decidua basalis, the maternal part (Fig. 3.10).

After implantation, the endometrium is termed the decidua. According to its relation to the chorionic vesicle, it can be divided into three parts. The decidua basalis is the part which lies over the embryonic pole of the chorionic vesicle, i.e. the part facing the chorion frondosum. The thin layer of decidua which covers the abembryonic pole of the chorionic vesicle is termed the decidua capsularis, as it is covering the chorion laeve forming a thin capsule. The rest of the lining of the uterine cavity is termed the decidua parietalis (Fig. 3.11).

At first, the different parts of the decidua are similar in structure. As the embryo grows, the decidua basalis develops and remains functional to form the maternal

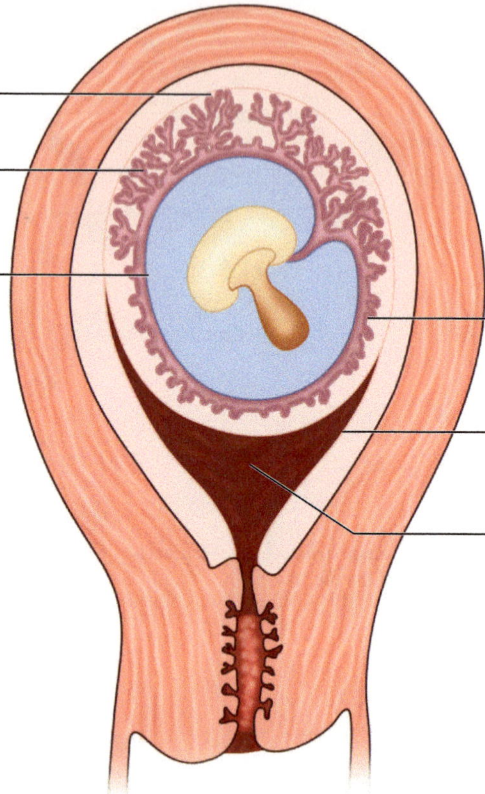

Fig. 3.10 The foetal and maternal contributions to the placenta

part of the placenta, and the capsularis and parietalis fuse together and eventually degenerate.

The placenta has several components:

1. The chorionic villi

   As described above, the tertiary villi contain a central core of extraembryonic mesoderm and blood vessels. The villi are connected to the chorion at the chorionic plate by the anchoring villi.
2. Intervillous space

   The space between the basal plate (forming its roof) and the chorionic plate (forming its floor) is filled with maternal blood coming from the maternal arteries (Fig. 3.12).
3. Cytotrophoblastic shell

   Lying in the roof of the intervillous space, this is formed at the basal plate by fusion of the cytotrophoblast of the adjacent anchoring villi to form a continuous sheet (Fig. 3.13).
4. Placental septa and cotyledons

   The septa are fingerlike processes which project from the roof of the intervillous space but do not reach the chorionic plate. The bases of the septa correspond to grooves on the outer surface of the decidual basalis bounding elevated masses of tissue called cotyledons.

The placental barrier consists of the layers of the villous wall which separates the foetal blood in the capillary loop of the floating villus from the maternal blood in the intervillous space. In early pregnancy, the barrier is formed of four layers, the endothelial lining of the capillary loop in the villus, the extraembryonic core of mesoderm in the villus, a layer of cytotrophoblast and a layer of syncytiotrophoblast. After the fourth month, the placental barrier becomes reduced in thickness to allow easier exchange of gases and nutrient substances and is composed of only the endothelial layer of the capillary loop and the syncytiotrophoblast layer.

The placenta is fully formed by the third month of pregnancy. After this, it grows in size by elongation of the villi and widening of the intervillous spaces which increase the thickness of the placenta. Its diameter increases secondary to the growth of the uterine wall. During the final month of pregnancy, the placenta undergoes degeneration which is manifested by fibrosis of the villi resulting in reduction of

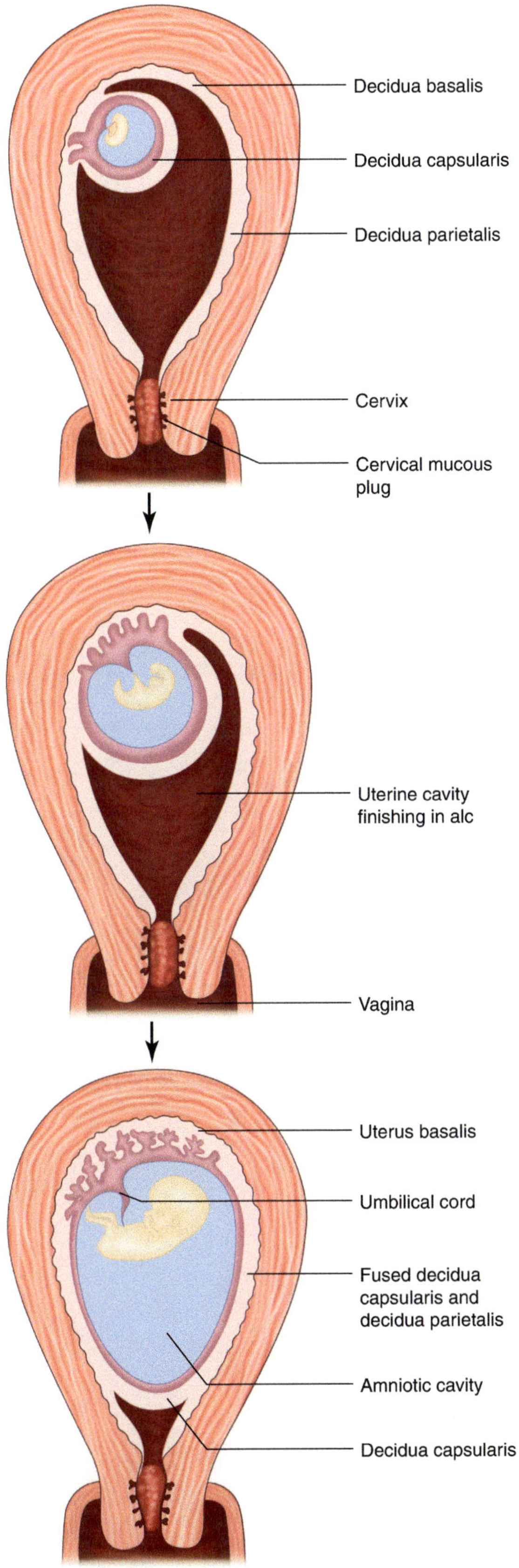

Fig. 3.11 Decidual components of the forming placenta

placental function, and the placenta begins to separate from the wall of the uterus.

Maternal blood (oxygenated) passes from the spiral arterioles of the decidua basalis to the intervillous spaces between the villi and then leaves via numerous thin-walled decidual veins. The deoxygenated blood from the foetus reaches the placenta via the branches of the two umbilical arteries. Blood flows through the arterioles where exchange of gases takes place. Oxygenated blood returns to the foetus via venules and veins which drain into the umbilical vein (Fig. 3.14).

Abnormalities of the Placenta

Size and Shape Anomalies

Placenta membranacea or diffuse placenta is seen when the placenta lines the greater part of the uterine cavity. It is due to persistence of the chorionic villi of the chorion laeve. Placenta succenturiata is when the placenta has one or more accessory lobes which are completely separate from the main placenta. Placenta accreta is the condition where there is abnormal fixation of the placenta to the wall of the uterus. This is caused by extensive invasion of the stem villi to the myometrium and can cause an increased risk of haemorrhage at the time of placental delivery.

Abnormalities of Position

Normally, implantation occurs in the posterior wall of the fundus, and the placenta develops in the upper part of the uterus. If implantation occurs in the lower part of the uterus, the placenta will develop in the lower uterine segment; this is known as placenta previa. Depending on the relation of the placenta to the internal os of the cervix, placenta previa is classified into three types:

1. Placenta previa lateralis—where the placenta encroaches on the lower uterine segment but does not reach the internal os
2. Placenta previa marginalis—where the margin of the placenta overlies the internal os of the cervix
3. Placenta previa centralis—where the centre of the placenta overlies the internal os of the cervix

Placenta previa is a dangerous abnormality leading to premature separation of the placenta from the uterine wall before labour. This can result in haemorrhage.

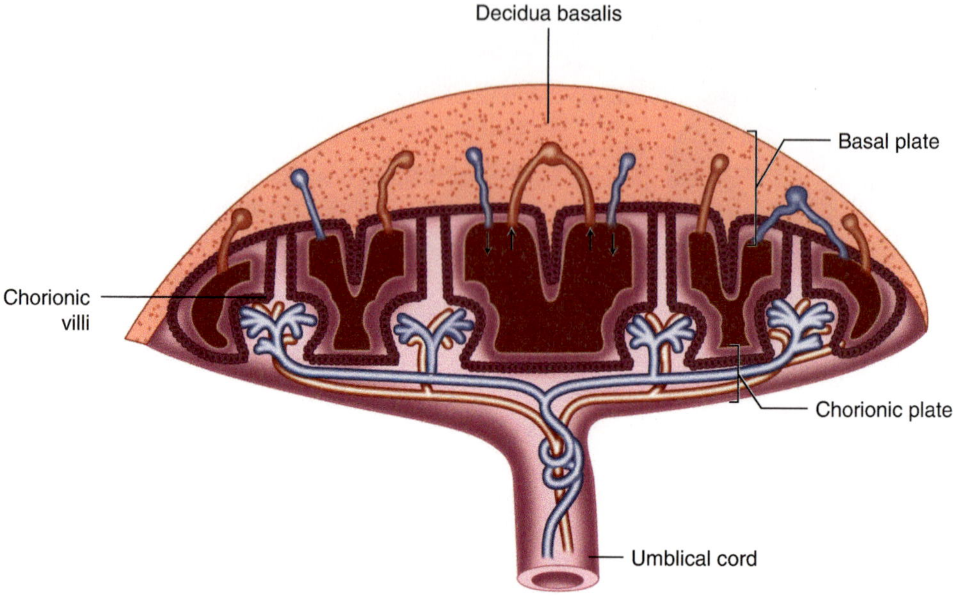

Fig. 3.12 Arrangement of the chorionic villi within the placenta

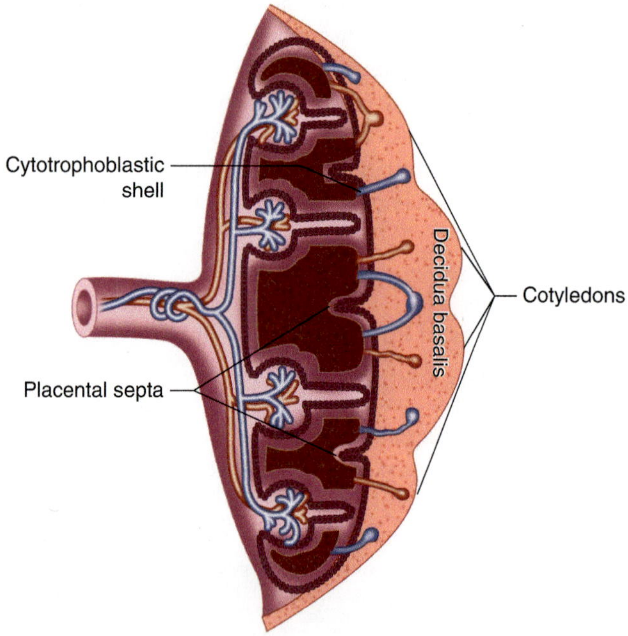

Fig. 3.13 Placental septa and cotyledons

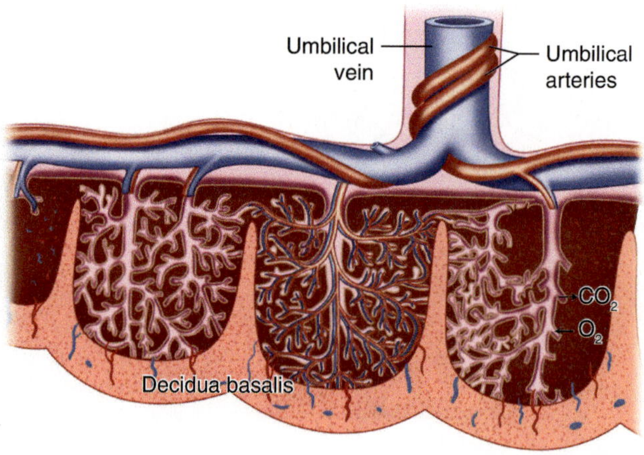

Fig. 3.14 Maternal and foetal circulation in the fully formed placenta

Foetal Circulation

4

Maria Ilina

I. **The oxygenated blood coming from the placenta:**
   - **The placenta** acts as a lung for oxygenation of the foetal blood.
   - The oxygenated blood is carried from the placenta to the foetus via the left umbilical vein which passes to the liver.
   - **In the liver:** Most of the oxygenated blood passes through the ductus venosus to reach the inferior vena cava (IVC). Little amount of blood passes through the liver sinusoids and then reaches the IVC.
   - **The I.V.C.:** carries the oxygenated blood (from the placenta) which mixes with little amounts of deoxygenated blood reaching the IVC from the lower ½ of the body. The IVC finally opens into the right atrium.
   - **In the Rt. right atrium:** most of the blood of the IVC is directed, through the foramen ovale, to the left atrium because:
   1. The opening of the IVC faces the foramen ovale.
   2. The valve of the IVC, or Eustachian valve, directs the blood towards foramen ovale and away from the tricuspid valve.
   3. The pressure in the left atrium is lower than that of the right atrium.
   - **From the left Lt. atrium** the blood passes to the left ventricle and the aorta where it is distributed mainly to the heart, head and neck and upper limbs.

II. **The deoxygenated blood carried by the superior vena cava S.V.C:**
   - Reaches the right atrium where it passes directly to the right ventricle because:
   1. The opening of the S.V.C faces the tricuspid orifice.
   2. The lower border of the septum secundum prevents the blood from entering the foramen ovale.
   - On reaching the right ventricle, the deoxygenated blood passes to the pulmonary trunk.
   - From the pulmonary trunk: little amount of blood goes to the lungs (collapsed), while the majority of blood escapes through the *ductus arteriosus* to reach the distal part of the arch of aorta where it mixes with the oxygenated blood.

III. **The Dorsal Descending aorta:**
   - Carries partially oxygenated blood which is distributed to the abdomen, lower limbs and finally passes through the two umbilical arteries to the placenta to be oxygenated and returned to the embryo again via the umbilical V.

M. Ilina (✉)
Department of Paediatric Cardiology, Royal Hospital for Children, Glasgow, UK
e-mail: Maria.Ilina@ggc.scot.nhs.uk

R. Carachi, S. H. E. Doss (eds.), *Clinical Embryology*, https://doi.org/10.1007/978-3-319-26158-4_4

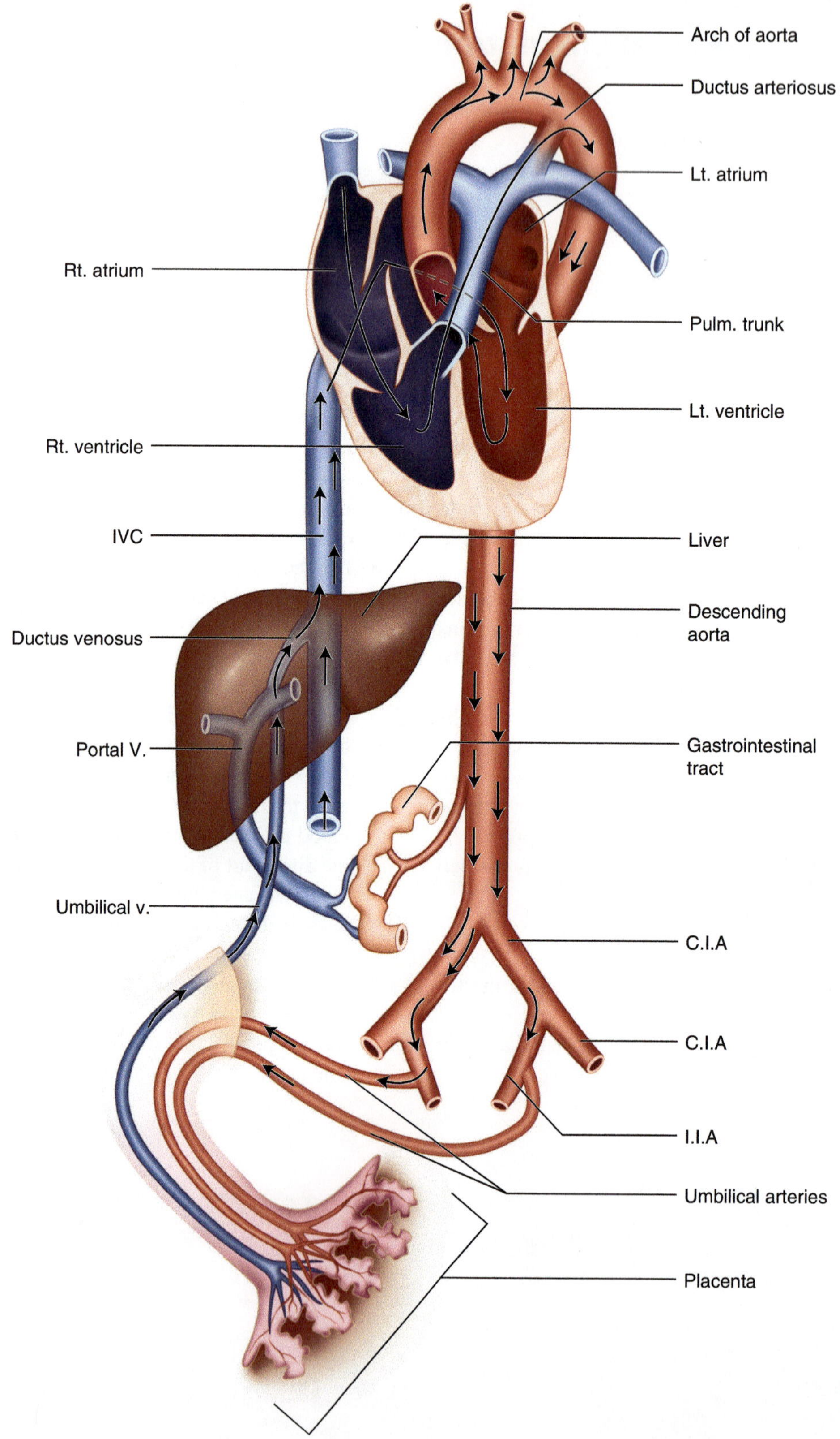
Arch of aorta
Ductus arteriosus
Lt. atrium
Rt. atrium
Pulm. trunk
Lt. ventricle
Rt. ventricle
IVC
Liver
Descending aorta
Ductus venosus
Portal V.
Gastrointestinal tract
Umbilical v.
C.I.A
C.I.A
I.I.A
Umbilical arteries
Placenta

Changes in the Circulation After Birth

A. **Immediate changes:**

1. **Establishment of pulmonary circulation:**

– Immediately after birth, the lungs expand due to the anoxia resulting from stoppage of the placental blood flow (stimulates the respiratory centres of the foetus).

– The expansion of the lungs creates negative intra-thoracic pressure leading to suction of blood into the lungs and establishment of the pulmonary circulation.

2. **Functional closure of the foramen ovale:**

– *The increased pressure inside the Lt. left atrium* (due to the establishment of the pulmonary circulation) and *the decreased pressure inside the Rt. right atrium* (due to stoppage of the placental blood flow) cause firm *apposition of the septum primum to the septum secondum* leading to closure of foramen ovale.

3. **Functional closure of ductus arteriosus:**

– The ductus arteriosus becomes functionally closed immediately after birth by contraction of its thick muscular wall leading to:

(a) Cutting short the shunt between the left pulmonary artery and the arch of aorta.

(b) Passage of all the blood of the pulmonary trunk to the lungs.

B. **Late fibrotic changes:** During the first year of postnatal life some of the vessels become fibrosed and change into ligaments as follows:

1. **The Lt. left umbilical vein:** becomes the ligamentum teres of the liver which extends from the umbilicus to the left branch of the portal vein.
2. **Ductus venosus:** becomes the ligamentum venosum of the liver which extends from the left branch of the portal vein to the IVC.
3. **Ductus arteriosus:** becomes the ligamentum arteriosum connecting the left pulmonary artery to the arch of the aorta.
4. **The umbilical arteries:** become the lateral umbilical ligaments. The proximal part of each umbilical artery, however, remains patent and gives the superior vesical artery which supplies the urinary bladder.

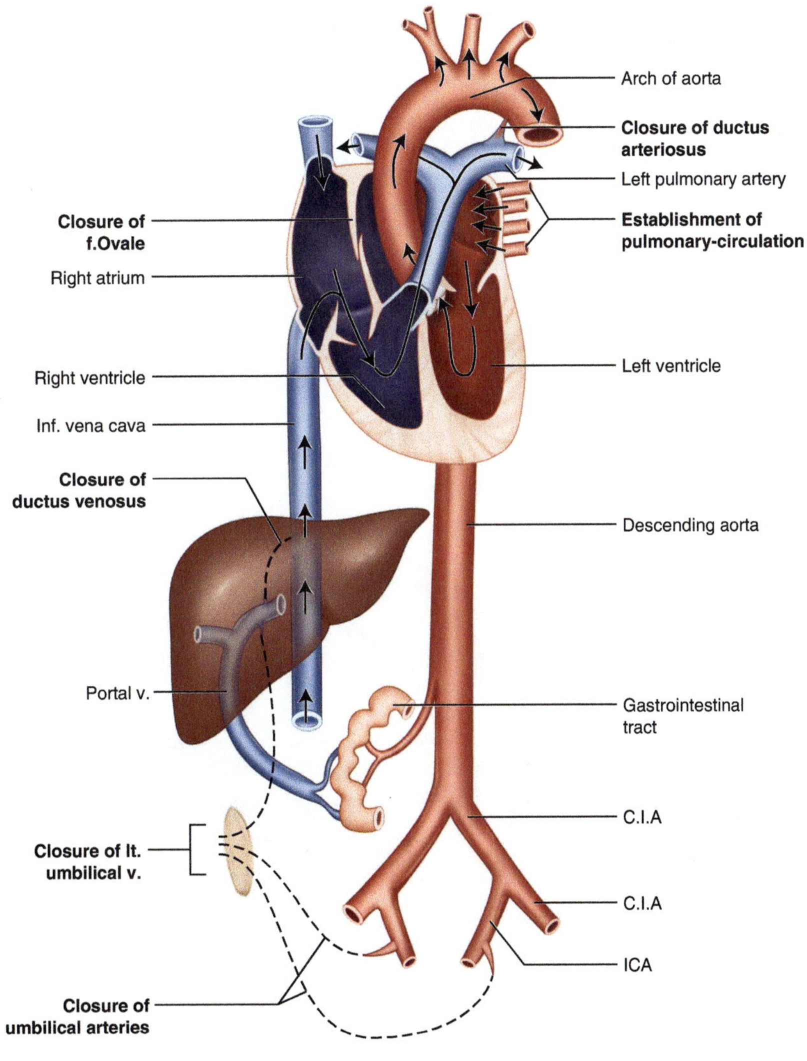

General Remarks About the Foetal Circulation

A. The tissues of the foetus receive partially oxygenated blood *due to mixing of the oxygenated and deoxygenated blood in the following sites*:
   1. **In the liver**: where mixing occurs between the blood of the left umbilical vein (oxygenated) and that of the portal vein (deoxygenated).
   2. **In the I.V.C**: the blood of the ductus venosus mixes with the venous blood returning from the lower parts of the body of the foetus.
   3. **In the Lt. left atrium**: the blood reaching it mixes with little amounts of venous blood returning from the lung buds.
   4. **In the dorsal descending aorta**: its blood mixes with the deoxygenated blood passing through the ductus arteriosus.

B. As a result of the structural changes occurring *after birth*, the heart and the circulation become divided functionally into two sides:
   1. **Rt. Right side half**: receiving and pumping deoxygenated blood only.
   2. **Lt. Left side half**: receiving and pumping oxygenated blood only.

Foetal and Transitional Circulation

There are some fundamental differences between intrauterine and extrauterine circulation. The foetus has very high pulmonary vascular resistance due to amniotic fluid in the developing lung and vasoconstricted pulmonary vasculature. This means that the right ventricle is operating in a higher pressure regime than the left ventricle. Due to the relatively small volume of pulmonary return from lungs and small atrial shunt via PFO, LV receives relatively less flow during the foetal life (approximately one third), and RV plays the role of a dominant ventricle receiving and pumping on approximately two thirds of the combined ventricular output and operating in a higher pressure regime than the left ventricle. When these high right-sided pressures persist in newborn life, it causes a problem known as persistent pulmonary hypertension of the newborn (PPHN) (Fig. 4.1).

The foetus does not need to deliver blood flow to the lung parenchyma (the oxygenation of the blood is provided by the placenta) and has relatively few metabolic demands (does not breathe, does not feed and "floats" in the amniotic fluid). There are a number of biochemical adaptations allowing foetus to deal very well with hypoxia, and due to a number of "shunts" and flow redistributions, the oxygen saturations of the blood in foetal descending aorta are between 50 and 55% and in the ascending aorta about 60–65% [1]. Some blood

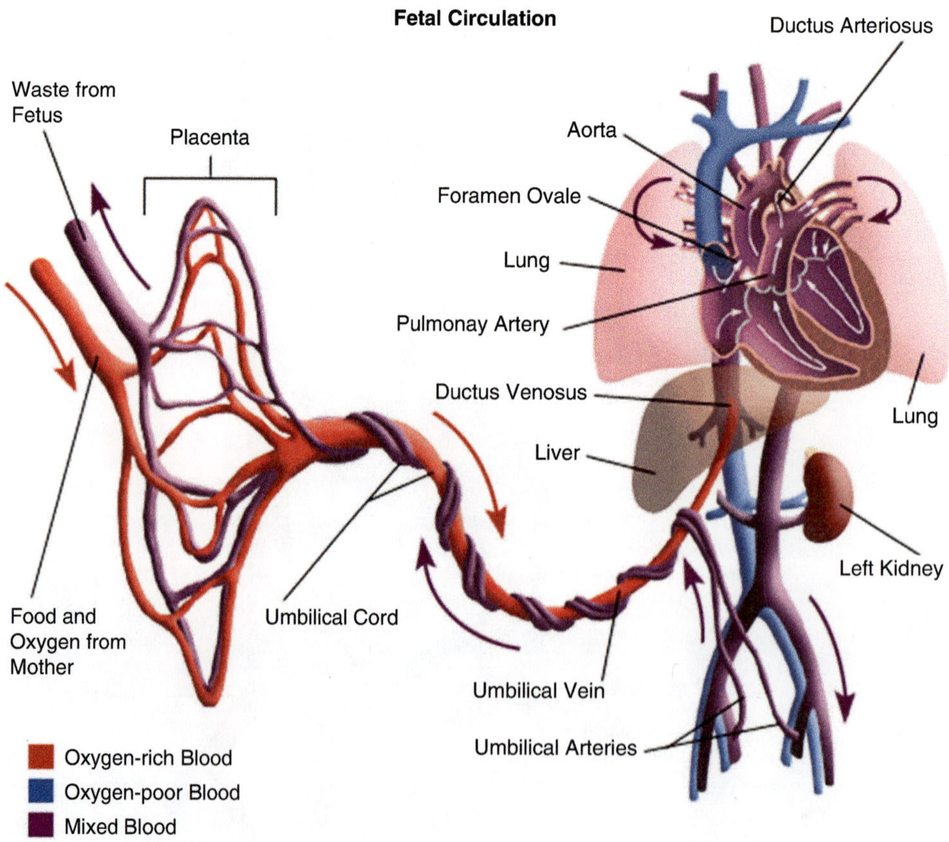

Fig. 4.1 Anatomy of foetal circulation (https://humananatomyly.com/fetal-circulation-diagram/fetal-circulation-diagram-fetal-circulation-ductus-venosus-midwifery-pinterest-nicu/)

does go through the foetal lung to provide oxygen and enable foetal lung tissue to develop.

Three "shunts" in the foetal heart, in the order encountered by the blood leaving the mother and entering foetus, are as follows:

1. The *ductus venosus* joins the umbilical vein carrying oxygenated blood from the mother directly to the inferior vena cava (IVC). It is a bypass through the liver, preventing a lot of oxygenated blood being "consumed" by the foetal liver. It closes functionally within minutes after birth and structurally within the first week of life, becoming the *ligamentum venosum* of the liver.
2. The *foramen ovale* is a shunt from the RA which receives oxygenated blood from IVC directed towards foramen ovale by the Eustachian valve, to LA, allowing oxygenated blood to enter systemic circulation via LV. Normally, the foramen ovale closes functionally within the first few weeks of life and structurally within the first few months of life; in up to 20% of subjects within the general population, it does not undergo structural closure and persists as a "flap" allowing small amount of blood to pass between atria from time to time. This structure will then be called *patent foramen ovale*, or *PFO*.
3. The *ductus arteriosus*, or *arterial duct*, is a short vessel between the pulmonary artery and the aorta, allowing most of the blood in the pulmonary artery (higher pressure in foetus) to be diverted to the descending aorta (lower pressure in foetus). The majority of this consists of superior vena cava blood, which is less well oxygenated and from the right atrium is preferentially directed to the right ventricle and from there to the main pulmonary artery. After birth, it undergoes functional closure within 48 h but sometimes takes up to a week. When it undergoes structural closure, it becomes known as the *ligamentum arteriosum*. If it persists, it becomes known as a *patent ductus arteriosus (PDA)*, which can have haemodynamic consequences [2].

Many changes occur immediately after birth when the neonate takes its first breath and begins to aerate its lungs:

- Rise in systemic vascular resistance with the interruption of the umbilical flow which occurs as a result of spasm of umbilical artery.
- Increase in $PaCO_2$ and a drop in PaO_2 and pH of the arterial blood stimulating respiratory centre (placental gas exchange no longer taking place).
- First breath and expansion of the lungs.
- Fall in pulmonary vascular resistance (can take up to 2 weeks, occasionally even longer) with corresponding increase in the pulmonary blood flow.
- Increased pulmonary venous return to the left atrium raises the left atrial pressure and closes the flap-like foramen ovale, resulting in cessation of the large right-to-left shunt, increase in the PaO_2 and closure of the arterial duct.
- Decrease in heart rate.
- Any shunt that remains open should gradually change from being a *right-to-left shunt* to *bidirectional shunt* to ultimately *left to right shunt*; circulation during this time can be said to be in a *transitional phase*.

During transitional phase, the pulmonary vessels are very sensitive to hypoxia and hypercapnia, to which they react by constriction, raising the pulmonary artery pressure and predisposing to the right-to-left shunting across the foramen ovale and arterial duct. In premature babies, closure of the arterial duct may be delayed.

References

1. Rudolph AM. Congenital diseases of the heart: clinical-physiological considerations. 2nd ed. Armonk, NY: Futura Publishing Company, Inc; 2001.
2. Casans R, Lal M, Gritsatis M. The essential revision guide to paediatric cardiology. Oxford: Radcliffe Publishing Ltd.; 2013.

Teratogenesis and Infection

5

Suzie Wills

Overview of Teratogenesis

The word teratogenic is derived from the Greek teratos, meaning monster, and genesis, to be born or produced. A teratogenic agent is the one capable of causing birth defects following exposure during pregnancy. The scientific concept of teratology is a modern one but birth defects have been a source of fascination and inspiration since the ancients from the Cyclops in Homer's Odyssey to the mermaids and chimeric beasts of myth and legend across many cultures (Fig. 5.1).

Through human history beliefs about birth defects have evolved from religious and cultural to biological via scientific exploration. In ancient times superstition prevailed with malformations seen as omens or portents. These superstitions tended to incorporate the belief that events during pregnancy could influence the developing child, leaving birthmarks and defects.

Malformed specimens were studied extensively by the early anatomists who have left us their descriptions, engravings and copperplate illustrations. This era marks the study of human malformations moving from the fanciful embellished arena of myth to that of scientific study. Modern understanding of the actual process of in utero development and how it may diverge from the norm is rooted in William Harvey's 1651 work. He postulated that the palate and lip form from several elements with interruption of this process resulting in cleft lip or palate (Fig. 5.2). This was in contrast to the prevailing belief at the time of the fully preformed embryo, a concept which took until the end of the eighteenth century to completely dispel.

With the discovery of Mendelian inheritance opinion shifted towards a purely genetic cause of birth defects. Modern medicine dismissed the old wives' tales and envisaged an isolated environment in utero where the human embryo was protected from harm by the uterus and placenta.

Fig. 5.1 Homer's Cyclops. Illustration by Mr Innes Smith

Renewed recognition that the developing embryo was indeed vulnerable to external influences, with exposure to specific agents causing specific defects, came definitively in the 1940s when Norman Gregg noted the association of maternal rubella with cataracts, heart defects and deafness (Fig. 5.3). The recognition of this viral teratogen preceded the scandal of thalidomide by nearly 20 years, yet the vulnerability of the embryo was still not fully appreciated. While it was recognised that very high doses of some drugs were toxic in pregnancy, prior to thalidomide it was not appreciated that therapeutic doses could

S. Wills, MBChB, MRCPCH (✉)
(Formerly) Royal Hospital for Children, Glasgow, UK

R. Carachi, S. H. E. Doss (eds.), *Clinical Embryology*, https://doi.org/10.1007/978-3-319-26158-4_5

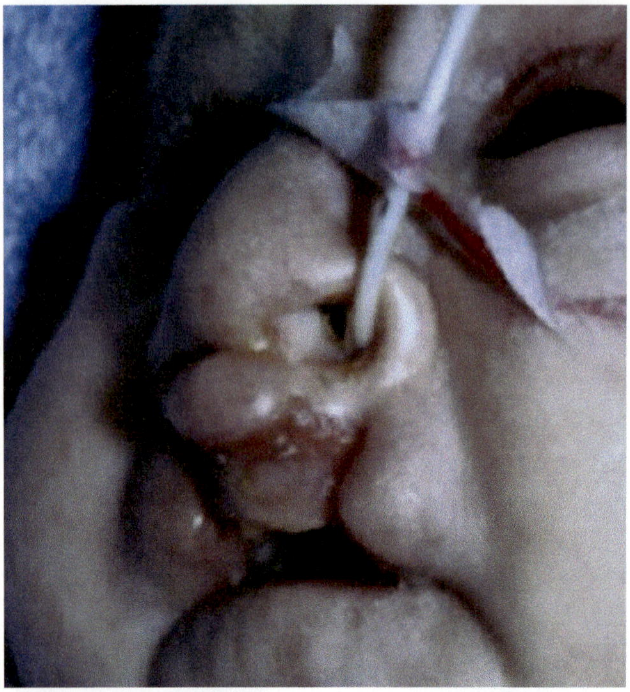

Fig. 5.2 Bilateral cleft lip

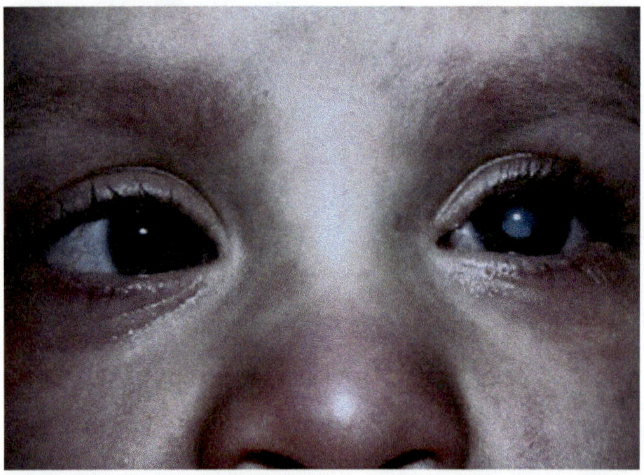

Fig. 5.3 Rubella cataract

cause the devastating malformations seen. Nowadays we are all aware of this special vulnerability and pregnant women are bombarded with advice restricting diet, activities and medications which may present risks to their developing offspring.

Agents with teratogenic qualities include drugs, environmental agents, infections, toxins and nutrients (Table 5.1).

Principles of Teratology

A teratogen acts by altering the fate of rapidly developing cell lines changing the subsequent development of an organ or body part's size, morphology or function. Mechanisms include cell death or altered migration and proliferation. This happens via the inhibition of specific biochemical processes such as cell signalling and metabolism.

Table 5.1 Examples of teratogenic agents

Drugs	• Thalidomide • Sodium valproate • Methotrexate • Lithium • Warfarin
Environmental agents	• Ionising radiation • Fever
Infections	• Rubella • Cytomegalovirus (CMV) • Parvovirus • Syphilis
Toxins	• Alcohol • Phenylalanine • Lead • Mercury
Nutrients	• Folic acid • Vitamin A

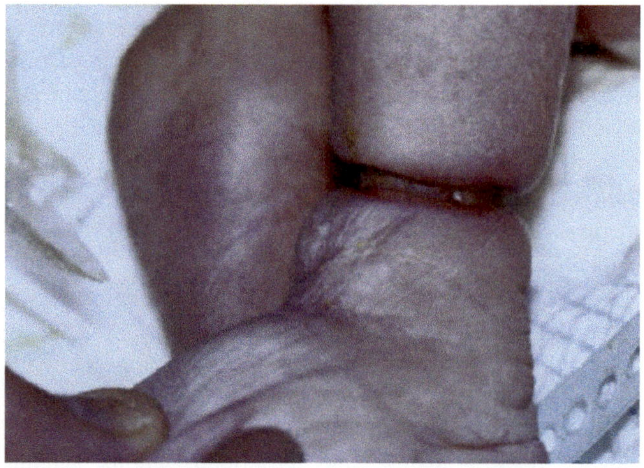

Fig. 5.4 Amniotic banding

The effect expressed falls into one of two main categories: the primary malformation of an organ or body part, e.g. neural tube defects and cleft palate, or the disruption of the formed but developing part, e.g. bone lesions in syphilis and amniotic bands (Fig. 5.4).

The scientific study of birth defects has led to the development of the main principles of teratology which state that the probability of a malformation being produced depends on:

- The dose of the agent
- The stage at which the embryo is exposed
- The genotype of the embryo and mother

These factors interact with each other to influence the outcome which may range from a minor anomaly, to major disruptions, to death in utero. Teratogenic malformations can occur along a spectrum of severity, related in part to the dose and duration of the teratogenic agent. The timing of exposure is crucial, and its effects can be understood when considering which embryological events are occurring at that time. Earlier exposure generally has more profound effects as fundamental structural processes such as neural tube or limb bud formation are taking place. Later exposure is less likely to cause major structural abnormalities but can influence ongoing processes such as neuronal proliferation and migration.

Maternal genetics influence drug metabolism, immunological competence and the biochemical milieu of the developing embryo. The genetic makeup of the conceptus likewise determines its resilience to adverse influences, for example, an embryo genetically destined to close the neuropore later than most will be more vulnerable to interruption of this process. The outcome will therefore depend on the interplay of all these influences, and most malformations are multifactorial in origin, in many cases demonstrating a threshold for disruption.

Drugs

The infamous thalidomide scandal changed attitudes to the safety of medicines during pregnancy and led to new legislation on drug testing and approval around the world. There is no means of establishing the absolute safety or otherwise of new drugs in pregnancy as randomised controlled trials in human subjects would be deeply unethical. Animal models are not robust as teratogenic effects vary significantly between species. Most drugs are therefore not licensed for use in pregnancy due to a lack of data. The risks and benefits of continuing use in pregnancy for a pre-existing condition must be weighed up on an individual basis.

Many drugs are known to be teratogenic with some causing well-defined syndromes (Table 5.2). Recreational drugs have also been reported to cause malformations. Good evidence exists that amphetamines and possibly cocaine can cause cleft palate, heart defects, intestinal atresias and structural brain abnormalities. It is not always clear whether reported effects of recreational drugs are due to the drug in question or toxic contaminants.

Table 5.2 Examples of teratogenic drugs

Drug	Effects
Phenytoin	Foetal hydantoin syndrome—facial cleft, cognitive impairment
Sodium valproate	Neural tube defect, cleft palate, atrial septal defect, hypospadias, polydactyly, craniosynostosis
Lithium	Ebstein's anomaly
Diethylstilbestrol	Genitourinary malformations
Aminopterin	Anencephaly, hydrocephalus, cleft lip and palate
Mycophenolate mofetil	Microcephaly, heart defects, cleft lip and palate, microtia
Warfarin	Nasal hypoplasia, limb hypoplasia, optic atrophy, bone abnormalities, neurological impairment
Amphetamines	Cleft palate, heart defects, intestinal atresias and structural brain abnormalities

Thalidomide

Thalidomide is a particularly potent teratogen which appears to cause major disruption to up to half of embryos exposed to just one 50 mg dose. Prior to the thalidomide experience it was believed the placenta protected the conceptus from therapeutic doses of drugs.

Thalidomide was marketed in Germany from the 1st of October 1957 as a sedative and anti-emetic and was available without prescription. Embryonic sensitivity to thalidomide occurs between 20 and 36 days after fertilisation (34–50 days after the start of the last menstrual period), so its use for morning sickness was particularly damaging.

Most memorably it produced devastating limb abnormalities but also caused congenital heart disease, eye and ear abnormalities (microphthalmia, coloboma, abnormal pinnae), intestinal atresias, renal malformations and facial palsies (Fig. 5.5). Exposure to thalidomide early on in the sensitive period (days 20–24) affects the ears and eyes, then the upper limb (days 24–31) and the lower limb (days 27–33).

Recent work suggests exposure after the time-sensitive period still has a detrimental effect and has been linked with both autism and epilepsy. The exact mechanisms of its actions are still not clear, but its anti-angiogenic properties are thought to be important.

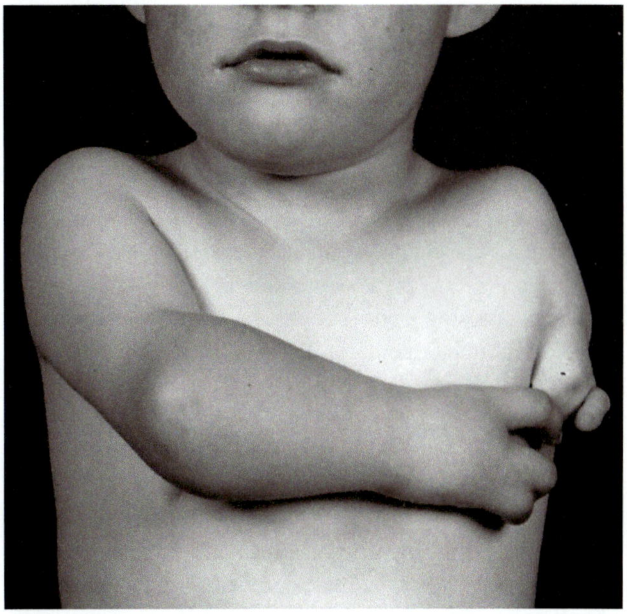

Fig. 5.5 Phocomelia and amelia

It was sold in 46 countries under various brand names and is thought to be responsible for between 10,000 and 20,000 cases worldwide. Around 50% of those affected died in infancy, mostly those with complex cardiac, renal or gastrointestinal malformations. The drug was licensed in the UK in 1958, sold under the brand name Distaval and withdrawn in 1961 by which time around 2000 UK babies were affected. Ironically its major selling point was its safety in overdose in contrast to barbiturates. The striking nature of the defects led to thalidomide's almost simultaneous identification as a teratogen by several individuals including Lenz in Germany, Macleod in Australia and the Scottish paediatrician Alexander Spiers. It was withdrawn from sale around the world by 1962.

Nowadays thalidomide is used to treat leprosy and multiple myeloma and may have a role in the treatment of other conditions. Tragically with renewed legitimate use, there has been a resurgence of thalidomide embryopathy, particularly in Brazil where sharing of prescription drugs is common.

Vignette Dr. Alexander (Sandy) Speirs OBE (1921–2008)

A Scottish paediatrician working in Stirling was the first to discover the dangers of thalidomide. In 1959 he studied the drug history of ten mothers who delivered babies with severe limb deformities and found that they took the drug Distival (thalidomide) in early pregnancy to relieve morning sickness. He reported this in 1962 in a seminal article 'Thalidomide and Congenital Abnormalities' published in the Lancet. The drug was then withdrawn in Britain and other countries soon followed. His pioneering work earned him widespread recognition and he was awarded an OBE in 1979.

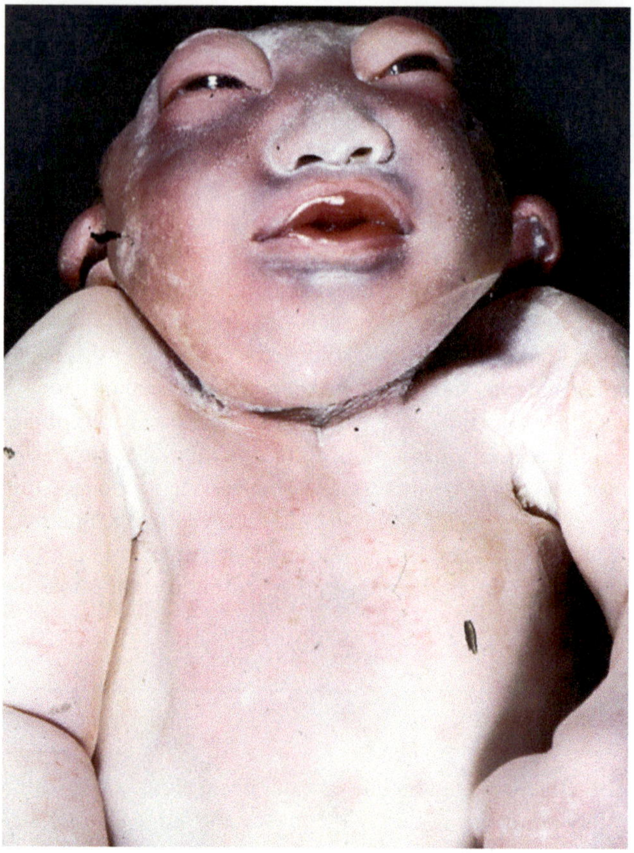

Fig. 5.6 Anencephaly

Nutrients

Adequate overall nutrition and calorie intake is important for pregnancy outcome as shown by an increase in preterm birth, very low birth weight and neonatal mortality following the Dutch famine of the Second World War. Specific nutrients which, in deficiency or excess, can cause birth defects include folic acid, iodine, zinc and vitamins A, B12 and K. Indeed various teratogens have been found to mediate their effects via alteration of the metabolism of these key molecules. An example is warfarin which induces vitamin K deficiency causing optic atrophy, bone abnormalities and neurological impairment.

Folic Acid

Folic acid availability is critically important for DNA synthesis and cell proliferation and crucial to the developing central nervous system. Folate deficiency including the use of anti-folate drugs in early pregnancy is teratogenic, classically resulting in neural tube defects but also cardiac defects and cleft lip and palate. Most cases can be prevented by adequate folic acid supplementation, but timing is crucial as the neuropore closes early in embryonic development, before the pregnancy is apparent. For this reason periconception folic acid must be taken for effective prophylaxis. In some countries supplementation of staple foods such as flour is undertaken in an attempt to ensure adequate intake. Current recommendations are for all women of childbearing age to take 400 μg per day and for those with a family history of neural tube defects to take 4 mg per day.

Neural tube defects range in severity from anencephaly to myelomeningocele, meningocele and spina bifida occulta (Figs. 5.6 and 5.7a, b). Anencephaly results from a failed closure of the rostral end of the neural tube and is characterised by a total or partial absence of the cranial vault and cerebral hemisphere. This disorder demonstrates the way in which a primary malformation—the open cephalic neuropore—disrupts further development as brain tissue exposed to amniotic fluid regresses in this unintended environment. Neural tube defects are familial in some cases with a recurrence risk of 3–4% after one affected child rising to 10% after two affected children. A genetic influence on folic acid metabolism has been postulated to account for this. Neural tube defects are also associated with trisomies 13 and 18, with various genetic syndromes and with maternal diabe-

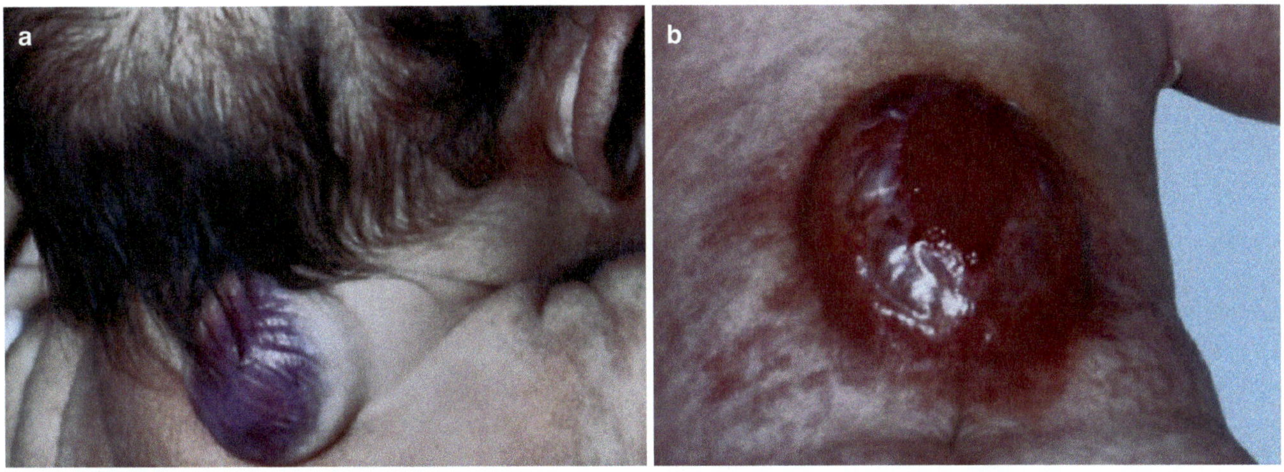

Fig. 5.7 (**a**) Meningocele. (**b**) Myelomeningocele

tes, obesity, hyperthermia and anti-folate drugs such as valproate and aminopterin.

Vitamin A

Vitamin A and its biologically active isomers play a pivotal role in the regulation of gene expression and control of cellular proliferation and differentiation. It is therefore not surprising that a deficiency or an excess of vitamin A in pregnancy can have serious adverse effects.

Excessive intake of vitamin A in pregnancy either in the diet, as supplements, or in the form of retinoid drugs can cause an embryopathy with widespread effects on many organ systems. The risk of foetal abnormality when isotretinoin is taken by a pregnant woman is 25% with a sensitive period between 4 and 10 weeks of gestation. Facial dysmorphism may occur in the form of low-set, atretic ears and external auditory meatus, microphthalmia, epicanthic folds, low nasal bridge and small jaw with cleft palate. Cardiac defects include ventricular septal defect, truncus arteriosus, double-outlet right ventricle, interrupted aortic arch and patent ductus arteriosus. There may also be hydrocephalus, Dandy-Walker malformation, corticospinal tract malformations and lissencephaly with resultant neurological problems including hypotonia and severe learning difficulties.

As well as being teratogenic in excess deficiency of vitamin A can result in malformations including microphthalmia and anophthalmia along with cardiac, lung and urogenital abnormalities.

Infections

Many agents cause congenital infection in the newborn but several are recognised specifically as teratogens when vertically transmitted in early pregnancy (Table 5.3). Infection within cells causes dysfunction and, in the growing embryo and foetus, alters subsequent tissue development and thereby structure and function.

Infections known to produce congenital defects have been described with the acronym TORCH (toxoplasma, others, rubella, cytomegalovirus [CMV], herpes) (Fig. 5.8). Rubella was the first to be identified as the cause of a specific constel-

Table 5.3 Infectious teratogenic agents

Infectious agent	Teratogenic effects
Rubella	• Cardiac (patent ductus arteriosus and peripheral pulmonary artery stenosis) • Sensorineural deafness • Eye (cataracts, glaucoma, pigmented retinopathy) • Microcephaly
Cytomegalovirus	• Microcephaly/ventriculomegaly • Periventricular calcification • Growth retardation • Progressive sensorineural deafness
Varicella zoster	• Skin scarring • Limb hypoplasia • Visceral, neurological and eye lesions
Treponema pallidum (syphilis)	• Saddle nose • Hutchinson teeth • Maxillary hypoplasia • Frontal bossing • Bone lesions • Sensorineural deafness • Neurological sequelae
Toxoplasmosis	• Intracerebral calcification • Microcephaly • Hydrocephalus • Microphthalmia, chorioretinitis • Hearing loss • Neurological sequelae
Herpes simplex	• Microcephaly • Microphthalmia • Hydrocephalus
Zika virus	• Microcephaly • Intracranial calcifications • Hypoplastic brainstem, cerebellum and corpus callosum

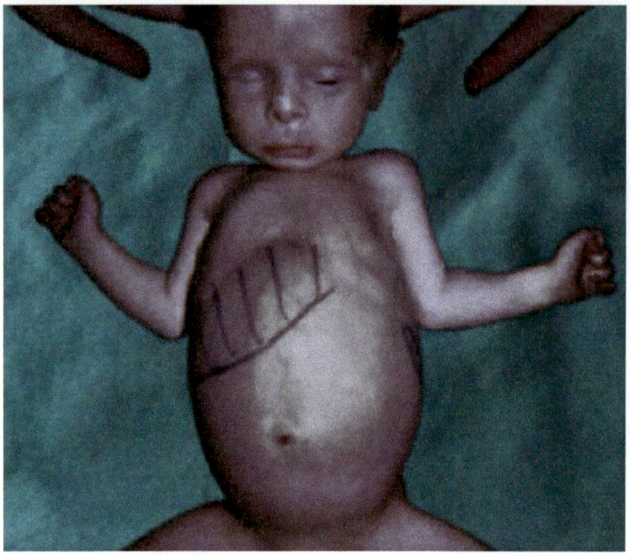

Fig. 5.8 Toxoplasmosis

lation of birth defects but since the advent of near-universal rubella immunisation CMV has become the leading cause of infection-related congenital neurological damage. The recent recognition of microcephaly associated with Zika virus infection is a growing concern.

Cytomegalovirus (CMV)

Cytomegalovirus is an endemic herpes virus which displays characteristic latency and reactivation. More than half of all people have been infected by adulthood leaving an estimated 40% of pregnant women at risk of primary infection. This is often asymptomatic or causes only a mild illness which may go unnoticed. Primary infection poses a 30% risk of vertical transmission in early pregnancy with a 20% rate of serious sequelae. Primary infection later in pregnancy has higher rates of transmission, but less severe effects. Reactivation of CMV or reinfection with a different strain in pregnancy is less likely to infect the foetus but can still cause long-term neurodevelopmental problems.

CMV infects many different cell types and organ systems with a predilection for the central nervous system. The virus has direct cytopathic effects causing focal necrosis and subsequent calcification, as is evident on brain imaging both antenatally and postnatally (Fig. 5.9). The virus also provokes a damaging inflammatory response accompanied by a vasculopathy. An actively infected symptomatic neonate can present with pneumonitis, hepatitis, colitis, a sepsis-like syndrome and a distinctive 'blueberry muffin' rash secondary to thrombocytopenia (Fig. 5.10). However, it is the teratogenic effects which account for the greatest long-term burden of this disease. These may be evident in the form of microcephaly or ventriculomegaly with encephalomalacia and growth retardation and become apparent in the longer term in the form of sensorineural deafness, cognitive impairment, cerebral palsy and visual impairment.

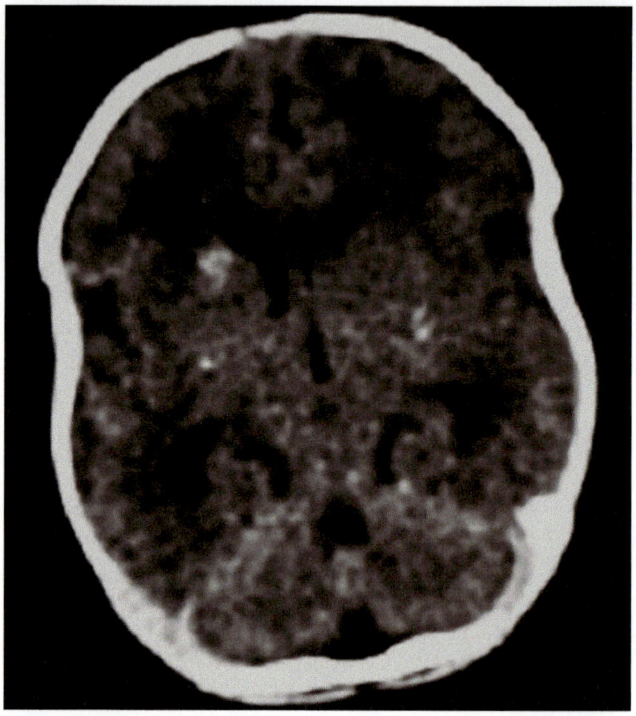

Fig. 5.9 Intracranial calcification

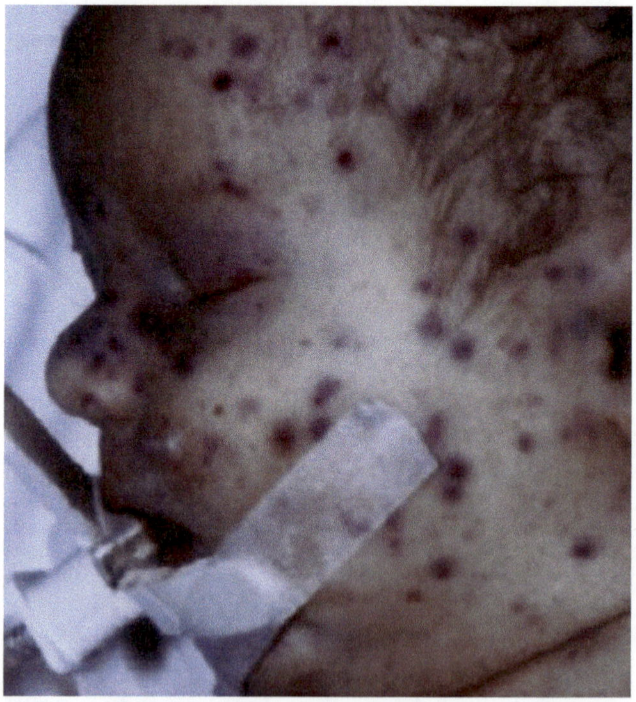

Fig. 5.10 Blueberry muffin rash

Currently there is no effective CMV vaccine and treatment options in pregnancy are limited. There is however a growing drive to identify infected infants in whom postnatal ganciclovir is likely to reduce progressive sensorineural deafness.

Rubella

The risk of congenital rubella syndrome is as high as 85% following exposure in the first trimester. This falls to around 25% between weeks 13 and 18, with little risk of major abnormalities after 18 weeks. This high level of teratogenicity aided in its identification as a teratogen in the 1940s. In addition to the classical triad of cataracts, heart defects and deafness, rubella infection may result in microcephaly, microphthalmia, glaucoma and motor and cognitive impairment.

Varicella Zoster

Varicella causes chickenpox with non-immune women susceptible to infection. Primary infection before 20 weeks carries a 1% chance of congenital varicella syndrome. This can cause cerebral atrophy with ventriculomegaly or microcephaly, microphthalmia, cataracts, limb abnormalities and skin scarring (Fig. 5.11).

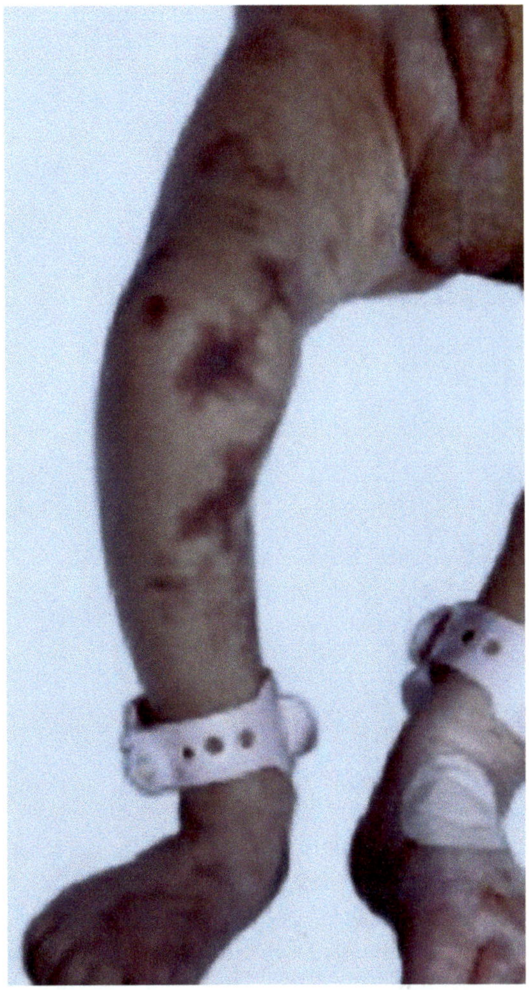

Fig. 5.11 Varicella

Zika Virus

Zika virus has been identified as the most likely cause of a sudden increase in cases of severe microcephaly observed in 2015 in South America where well over 1000 babies have been born with the condition in the Brazilian state of Pernambuco. Various reasons for this increase have been investigated but epidemiological and laboratory evidence points to Zika infection in the first trimester being responsible. A strong temporal relationship exists between an outbreak of the mosquito-borne virus in March 2015 and babies born with microcephaly in that October. IgM to the virus has been isolated in the CSF and neural tissue of affected infants and an unusual pattern of intracranial calcifications on imaging studies mark out these cases as being distinct from other causes of microcephaly. As yet only those affected with severe microcephaly are apparent; however it is reasonable to anticipate there may be milder effects on neurocognitive development among others exposed in utero which will come to light as time goes on.

The illness itself may be mild and go unnoticed or may bring symptoms of fever, rash, joint pain and conjunctivitis. Cases have been reported across South America and the Caribbean, and on the 1st of February 2016, the World Health Organization (WHO) declared Zika virus a Public Health Emergency of International Concern (PHEIC).

Toxins

Tobacco smoke contains numerous toxins such as cadmium which accumulates in the placenta. Cigarette smoking in pregnancy is associated with cleft lip and palate, intrauterine growth retardation and preterm delivery. Lead, mercury and industrial solvents are also known to be teratogenic, but the most widespread and harmful toxin is alcohol.

Alcohol

It has long been recognised that alcohol exposure during pregnancy can adversely affect the developing embryo and foetus. Since 1973 a distinctive pattern of malformations has been formally recognised and termed foetal alcohol syn-

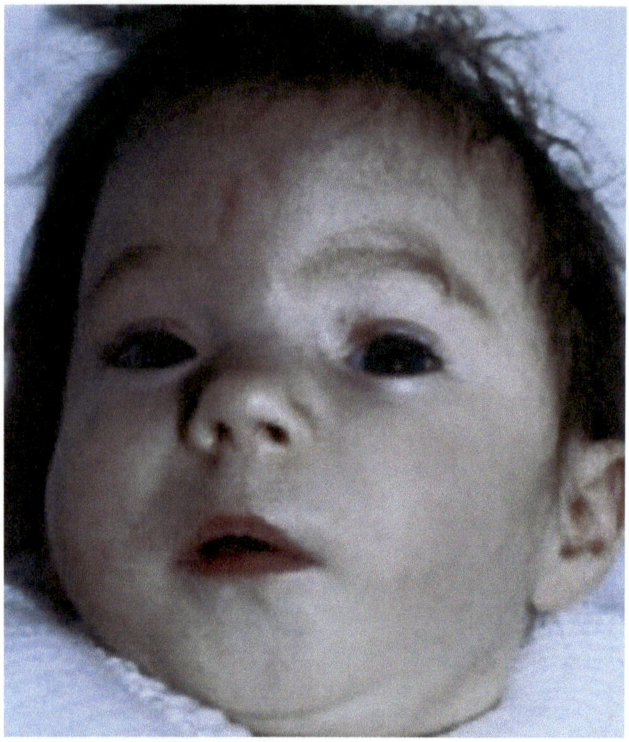

Fig. 5.12 Foetal alcohol syndrome

drome. Features include growth retardation, microcephaly, facial dysmorphism (short palpebral fissures, maxillary hypoplasia, epicanthic folds, thin upper lip) and cardiac and limb abnormalities (Fig. 5.12). More recently the term foetal alcohol spectrum disorder (FASD) has been used to highlight that many without the full blown syndrome are nonetheless affected by prenatal alcohol exposure. It is thought that FASD remains under-diagnosed and may underlie many cases of developmental delay.

As with all teratogens, the pattern and severity of effects depend on the dose, timing and duration of the alcohol exposure. No single mechanism of alcohol-induced damage has been identified, but both alcohol and its metabolites have been demonstrated to cause necrosis and apoptosis in vitro. Alcohol alters the absorption and metabolism of both folic acid and vitamin A which play such important roles in embryogenesis. Indeed the effects of vitamin A deficiency in pregnancy can result in a syndrome of birth defects much like FAS. Neural tube defects can be seen in offspring of animal models of alcohol-induced folate deficiency.

Maternal Conditions

Since at least half of pregnancies are unplanned, a pre-existing maternal condition means a woman may not only enter pregnancy on drug treatment but also with poor disease control which can be severely detrimental. Both of these factors make it crucially important for such women to plan their pregnancies.

Diabetes

Pre-existing type 1 or type 2 diabetes carries a risk of major congenital malformations two- to threefold above the general population. The spectrum of diabetic embryopathy includes anomalies of the cardiovascular system (transposition of the great arteries, ventricular septal defect, univentricular heart), central nervous system (anencephaly, encephalocele, meningomyelocele, holoprosencephaly), musculoskeletal system (sacral agenesis, caudal regression), genitourinary system (renal agenesis, multicystic dysplasia) and gastrointestinal system (anorectal malformations, hypoplastic left colon).

The exact teratogenic mechanisms of maternal diabetes are not yet understood, but the occurrence of anomalies is closely related to periconceptional and pregnancy glycaemic control as measured by glycosylated haemoglobin. Tight glycaemic control with avoidance of hypoglycaemic episodes both before and during pregnancy modifies the risk considerably.

Gestational diabetes, which occurs in around 4% of pregnancies, carries only a small increased risk of malformations, likely to represent a number of undiagnosed pre-gestational diabetics whose embryos were subject to early teratogenic hyperglycaemia. All diabetes in pregnancy risks foetal macrosomia with obstructed labour and birth injuries.

Phenylketonuria (PKU)

PKU is an inherited metabolic condition commonly screened for at birth. An individual with PKU must follow a special diet low in phenylalanine during infancy and childhood to prevent the toxic accumulation of phenylalanine which causes neurological damage. The diet is restrictive and difficult, so understandably many do not maintain it in adulthood once their own growth and development is complete. However when a woman with PKU is pregnant, high phenylalanine levels are teratogenic in the first trimester, causing microcephaly, heart defects and growth retardation. With blood phenylalanine levels greater than 1200 μmol/L, the likelihood of these defects is very high, approaching 75–90% for microcephaly and 15% for congenital heart disease. There is a dose response relationship with fewer abnormalities seen at lower levels. If a woman lowers her levels to between 120 and 360 μmol/L (2–6 mg/dl) before conception, a normal pregnancy outcome can be achieved. Some benefit can be seen when the low phenylalanine diet is restarted after conception, but blood levels take a while to fall and no benefit is seen after 13 weeks.

It is devastating, having prevented serious impairment in one generation through newborn screening, to see the next generation damaged by high phenylalanine levels in an unplanned pregnancy.

Further Reading

Beckwith JB. Congenital malformations: from superstition to understanding. Virchows Arch. 2012;461(6):609–19.

Coll O, Benoist G, Ville Y, Weisman LE, Botet F, Anceschi MM, et al. Guidelines on CMV congenital infection. J Perinat Med. 2009;37(5):433–45.

Detrait ER, George TM, Etchevers HC, Gilbert JR, Vekemans M, Speer MC. Human neural tube defects: developmental biology, epidemiology, and genetics. Neurotoxicol Teratol. 2005;27(3):515–24.

Gregg NM. Congenital cataract following German measles in the mother. Trans Ophthalmol Soc Austr. 1941;3:35–46.

Jones K, Smith D, Ulleland C, Streissguth A. Pattern of malformation in offspring of chronic alcoholic mothers. Lancet. 1973;301(7815):1267–71.

Lumley J, Watson L, Watson M, Bower C. Periconceptional supplementation with folate and/or multivitamins for preventing neural tube defects. Cochrane Database of Systematic Reviews. 2001;(3):CD001056. https://doi.org/10.1002/14651858.CD001056.

Muralidharan P, Sarmah S, Zhou FC, Marrs JA. Fetal alcohol spectrum disorder (FASD) associated neural defects: complex mechanisms and potential therapeutic targets. Brain Sci. 2013;3(2):964–91.

Newman CGH. The thalidomide syndrome: risks of exposure and spectrum of malformations. Clin Perinatol. 1986;13(3):555–73.

Smithells RW, Newman CGH. Recognition of thalidomide defects. J Med Genet. 1992;29(10):716–23.

Vargesson N. Thalidomide embryopathy: an enigmatic challenge. ISRN Dev Biol. 2013;2013:241016, 18 pages.

Vianna FS, Schuler-Faccini L, Leite JC, et al. Recognition of the phenotype of thalidomide embryopathy in countries endemic for leprosy: new cases and review of the main dysmorphological findings. Clin Dysmorphol. 2013;22:59–63.

Embryological Basis of Congenital Anomalies

6

Srinivas Annavarapu

Introduction The normal developmental processes hold the key to our understanding of the biological basis of congenital anomalies. It is remarkable to note how a single fertilized cell, the *zygote*, undergoes sequential cell divisions followed by coordinated cell proliferation, migration and differentiation, to create something as intricately complex as a human foetus [1]. Interestingly, the various stages of the human embryonic development (see Table 6.1) share homologous developmental trajectories with other vertebrates [1].

This elaborate developmental blueprint is housed within the chromosome as *genes*, which orchestrate the production of various cellular proteins (gene products) in a defined spatio-temporal manner [1, 3]. Complex interactions of these proteins drive diverse regulatory and functional pathways that are crucial in growth and development. Any aberration of these pathways, either primary failure or secondary interruption, may cause congenital anomalies [3].

Table 6.1 The developmental period in humans involves a sequential series of stages that are crucial for the normal development [2]

I	Pregenesis		All stages of development from the separation of germline stage to fertilization
II	Embryogenesis	Blastogenesis	All stages of development from fertilization to the blastocyst stage (Day 28)
III		Organogenesis	All stages of development from Day 28 to the end of 8th week postconception (Days 55–56 or 10th gestational week)
IV	Metamorphosis		Transformation from the end of 8th week postconception to the end of 24th week postconception
V	Phenogenesis		From the end of 24th week postconception to the end of 38th week postconception

Definition of Congenital Anomalies

Congenital anomalies (birth defects) are defined as structural or functional abnormalities that occur during the *intrauterine period*, irrespective of when they are identified—prenatally by ultrasound, at birth or later in life as an incidental finding [1, 3].

Aetiology of Congenital Anomalies

Congenital anomalies may arise as a result of gene defects (5%), chromosomal aberrations (7%), epigenetic events (3%) or environmental factors (10%); in 20% of the cases, the cause is multifactorial where the individual has a genetic susceptibility and is exposed to an environmental trigger such as intrauterine viral infection (e.g. rubella in congenital heart defects) or teratogen exposure (alcohol, phenytoin, thalidomide, etc.) [2, 3]. Surprisingly, still in about 60% of the cases, the underlying aetiology of congenital anomalies is unknown [2].

Genetic Defects

Genetic defects include single-gene disorders, chromosomal aberrations or epigenetic defects [2, 3].

Inheritance patterns for single-gene disorders are classified based on whether they are autosomal or X-linked and whether they have a dominant or recessive pattern of inheritance. Autosomal chromosomes have two identical copies of genes (allelic pair), one inherited from each parent. Thus, autosomal gene defects usually affect both sexes similarly. On the other hand, sex chromosomes (XX in female and XY in male), depending on whether the gene defect is on X or Y chromosome, would affect males and females differently [2, 3].

S. Annavarapu (✉)
Department of Cellular Pathology, Royal Victoria Infirmary, Newcastle-Upon-Tyne, UK

R. Carachi, S. H. E. Doss (eds.), *Clinical Embryology*, https://doi.org/10.1007/978-3-319-26158-4_6

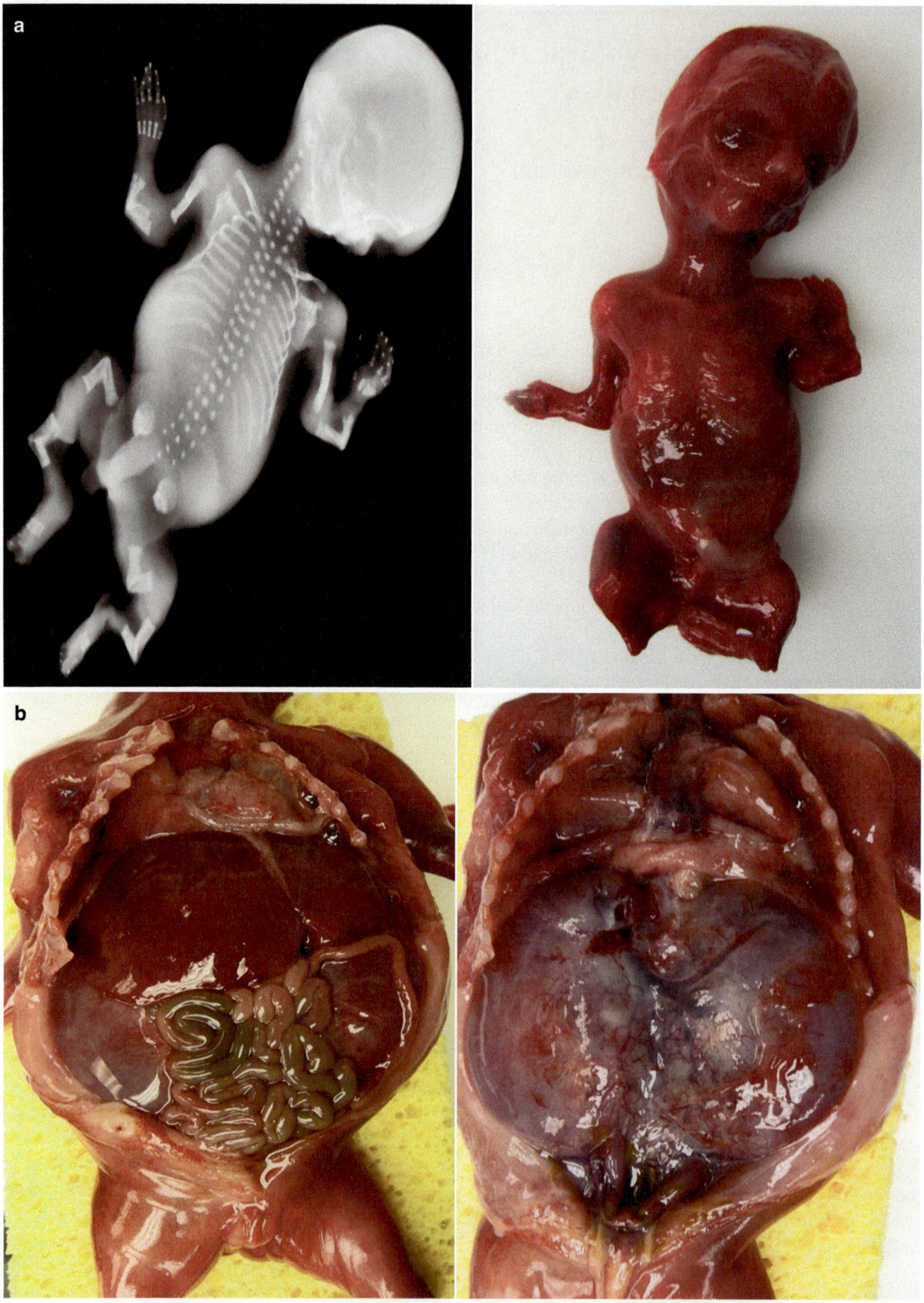

Fig. 6.1 (**a**) Osteogenesis imperfecta, type IIa, lethal perinatal disease characterized by skeletal dwarfism and multiple fractures of the ribs and long bones due to AD mutation of *COL1A1 gene.* (**b**) Bilateral enlarged polycystic kidneys presenting with massive abdominal distention, narrow chest wall and hypoplastic lungs, in keeping with AR polycystic kidney disease

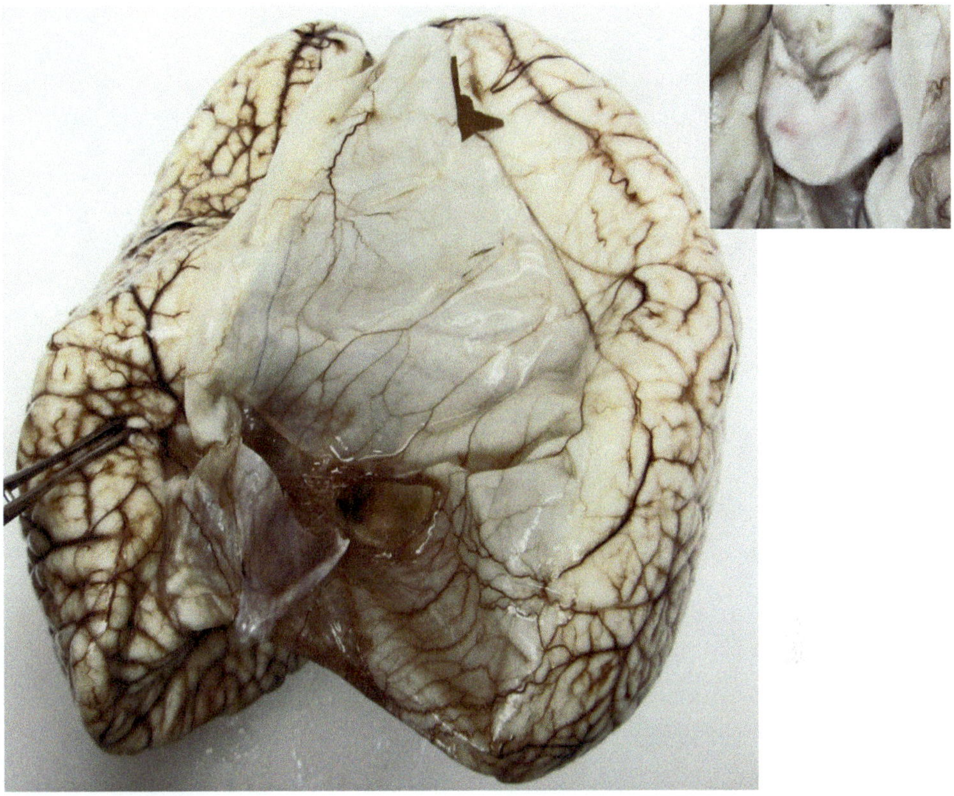

Fig. 6.2 Obstructive hydrocephalus due to congenital aqueduct (of Sylvius) stenosis (inset) caused by X-linked recessive *L1CAM gene* mutation

- Autosomal dominant inheritance: the defective gene is inherited from one/both parent who may harbour the mutated gene copy, but may or may not have the disease (disease predisposition). Only one copy of a disease allele is necessary for an individual to express the phenotype; hence there is a 50% chance of inheriting the defective gene, e.g. thanatophoric dysplasia, achondroplasia, Huntington's disease, etc. [2, 3].
- Autosomal recessive inheritance: the defective gene is inherited when both the parents harbour and transmit the defective gene, but do not have the disease (disease carriers). There is a 25% chance that a person may inherit both the copies of defective genes from parents and may express the phenotype, e.g. cystic fibrosis, AR polycystic kidney disease (ADPKD), etc. [2, 3].
- X-linked dominant inheritance: only one copy of a disease allele on the X chromosome is required for an individual to be susceptible to an X-linked dominant disease. Both males and females can be affected, although males may be more severely affected because they only carry one copy of genes found on the X chromosome, e.g. fragile X syndrome [2, 3].
- X-linked recessive inheritance: two copies of a disease allele on the X chromosome are required for an individual with two X chromosomes (a female) to be affected with an X-linked recessive disease. Since males are hemizygous for X-linked genes (they have only one X chromosome), any male with one copy of an X-linked recessive disease allele is affected [2, 3], e.g. defects include congenital aqueduct stenosis due to L1 cell adhesion molecule (L1CAM) that is crucial for the neuronal development/organization, the myelin sheath formation and the formation of synaptic communications [4]. L1CAM mutations lead to fatal obstructive hydrocephalus (Fig. 6.2). X-linked recessive inheritance implies males (XY) will suffer if they carry the mutation. Females (XX) are carriers and only exceptionally will suffer if they inherit two copies of mutations [2].

Chromosomal Aberrations

Chromosomal aberrations (e.g. gene deletion, duplication, translocation, isochromosome, etc.) may involve loss of a group of related and/or unrelated genes, resulting in wider defects [1–3].

- Aneuploidy (variations in chromosome numbers) may be due to errors in cell division from cellular non-disjunction in mitosis that may result in monosomy (e.g. Turner's syndrome/monosomy X) or trisomy (Down's syndrome/trisomy 21, Edwards syndrome/trisomy 18, Patau's syndrome/trisomy 13, etc.) (Fig. 6.3a, b) [2, 3]. Increased maternal age (>35) may cause abnormalities of mitotic spindle fibres, leading to increased incidence of cellular non-disjunction leading to an increased risk of trisomy 21, hence the rationale for the triple test (serum levels of beta-hCG, alpha-fetoprotein estriol) for Down's syndrome and neural tube defects [5].

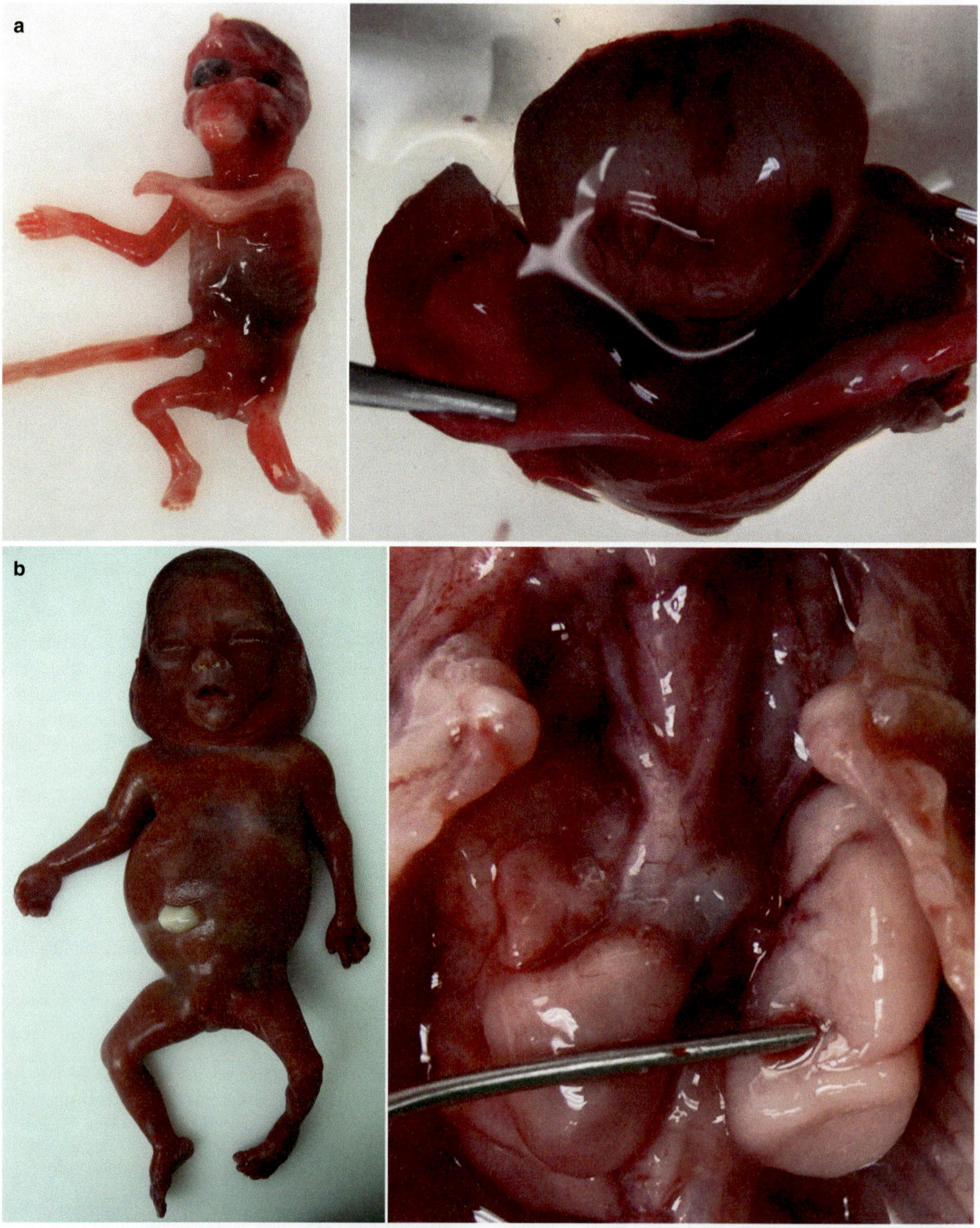

Fig. 6.3 (**a**) (Left) Patau's syndrome/trisomy 13 with typical facial dysmorphism—microcephaly, hypotelorism (close-set eyes), absent nose and proboscis on forehead. (Right) Fused cerebral hemispheres with a single ventricle (holoprosencephaly). (**b**) (Left) Turner's syndrome/monosomy X with typical massive bilateral cystic hygroma. (Right) Tubular hypoplasia of aorta with post-ductal coarctation

- Microdeletion of a short chromosomal segment may lead to a contiguous loss of an important set of genes, e.g. 22q11 segment is needed for the normal cardiac and craniofacial development [6]. The loss of this segment therefore leads to classic conotruncal cardiac defects along with the absence of thymus and parathyroid glands (DiGeorge syndrome) with or without significant craniofacial defects (Fig. 6.4) [6].

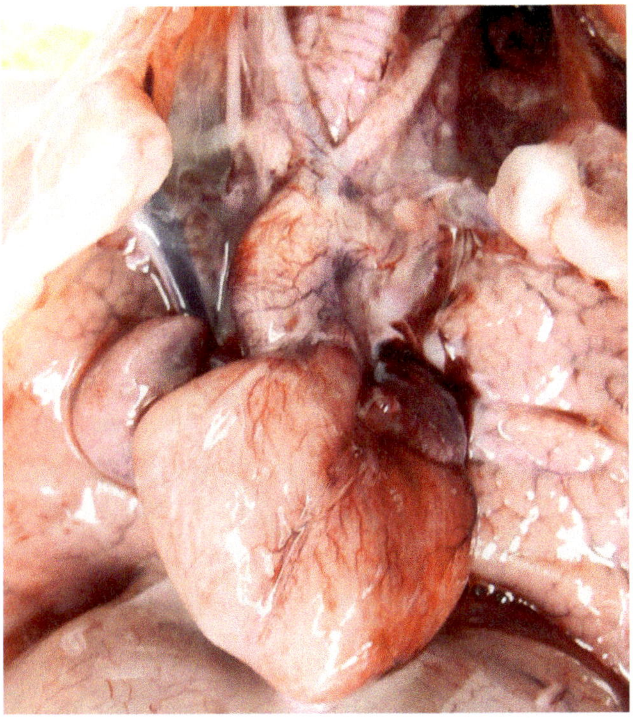

Fig. 6.4 DiGeorge syndrome (22q11 microdeletion) showing typical conotruncal cardiac defect where the great vessels fail to separate and form a common trunk (truncus arteriosus)

Epigenetic Modifications

Gene silencing due to hypermethylation of the promoter site, abnormalities in DNA folding regulated by histone deacetylation, short inhibitory RNA (siRNA) and genomic imprinting are other mechanisms by which gene function may be affected [7]. In addition, alternative gene splicing may lead to genetic polymorphisms which may be the underlying causes of non-hereditary associations and natural variations.

Environmental Factors

Some congenital defects may be due to exposure of chemical mutagens and or teratogens during embryogenesis [1–3]. Folate deficiency is known to be associated with neural tube defects (Fig. 6.5); glucose exposure before ovulation and during embryogenesis may act as a teratogen [8], possibly explaining why diabetic mothers are more predisposed to having babies with congenital defects, especially genitourinary defects (Fig. 6.6) [2, 3, 8]. Exposure to alcohol in pregnancy causes abnormal brain development, causing mental retardation, facial abnormalities and small cranium (microcephaly) seen as a part of the foetal alcohol syndrome [9].

The Concept of Developmental Field

The drivers for each of the developmental stages comprise a set of genes that are expressed and regulated in a very synchronized spatio-temporal and hierarchical manner to create a *developmental field*—the morphogenic unit of embryo (Opitz

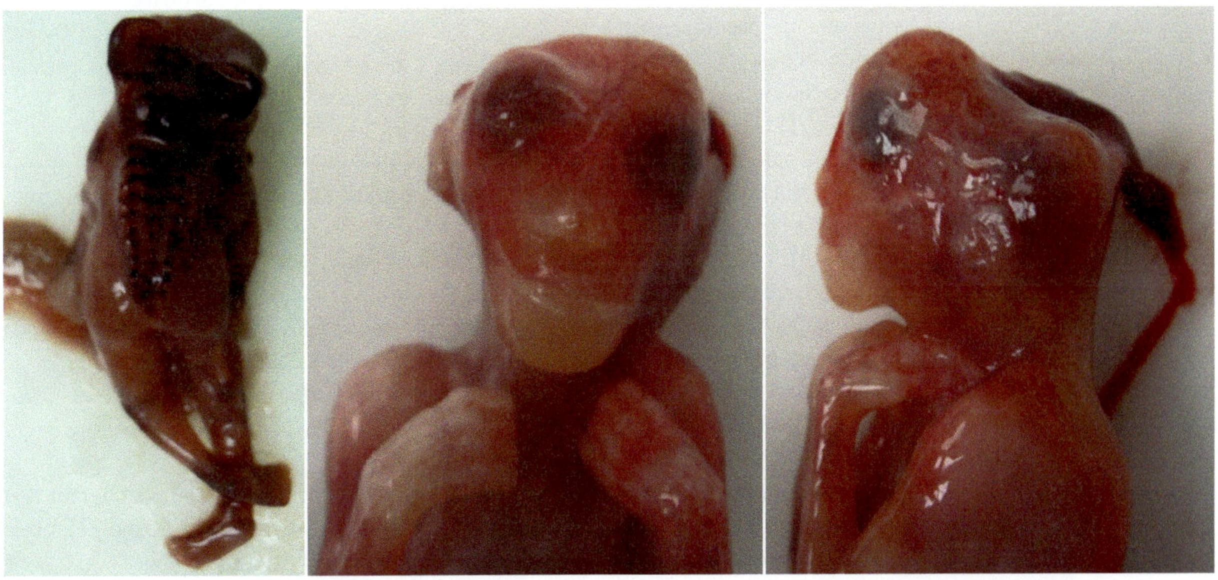

Fig. 6.5 Neural tube defect due to deficiency of folic acid during pregnancy. (Left) Lack of skull formation (anencephaly) along with failure of the vertebral column to fuse posteriorly (cranio-rachischisis); (centre, right) facial profile of anencephaly—lack of cranial vault, proptosis, low-set ears and a prominent triangular jaw

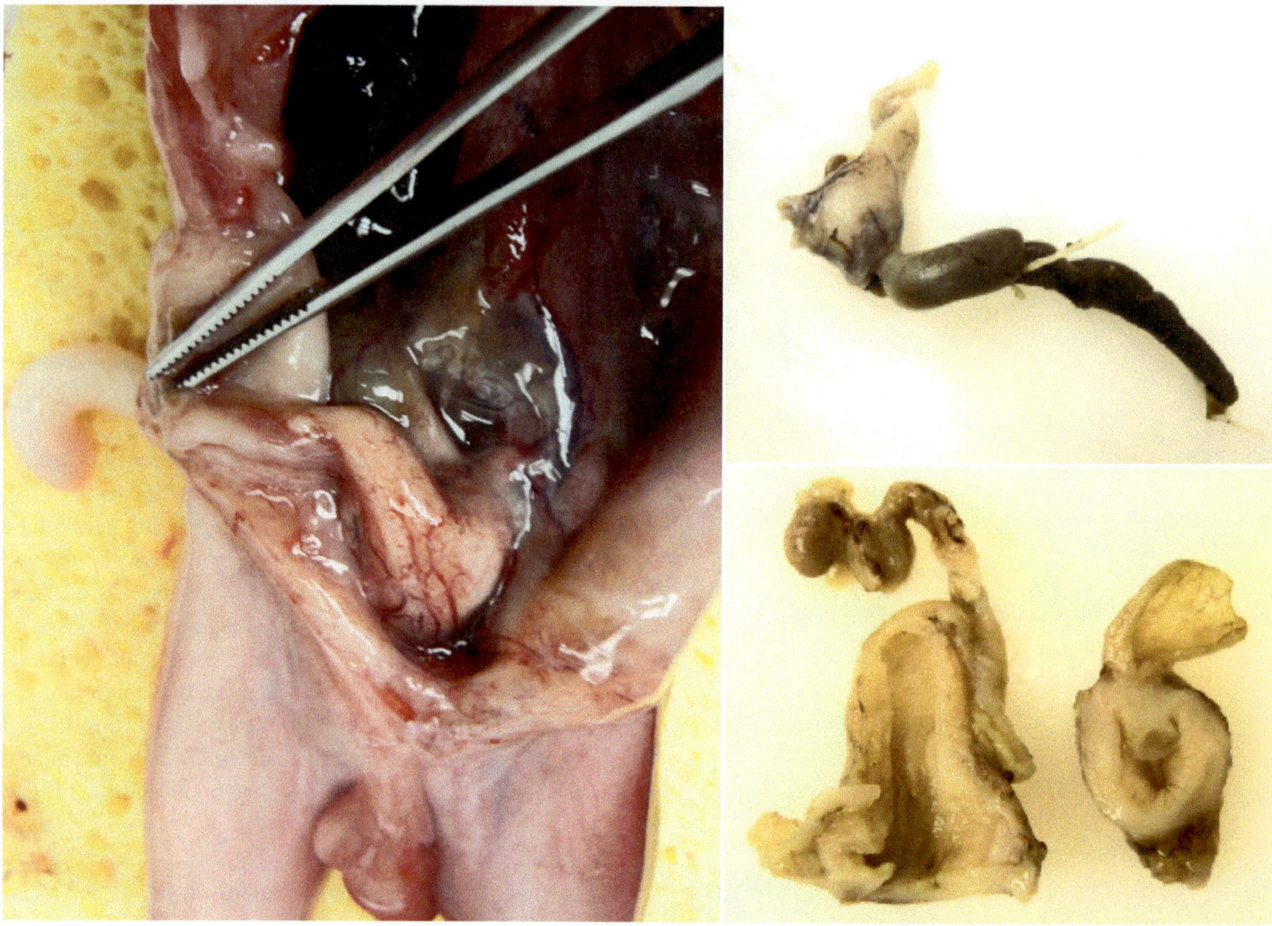

Fig. 6.6 Increased frequency of genitourinary congenital anomalies is seen in diabetic mothers due to the teratogenic effect of excess glucose. (Left) The colon opens directly into the urinary bladder (rectovesical fistula); (top right, bottom right) fusion of colon and bladder. This will also lead to imperforate anus

JM, 1985) [10]. Any disruption in the gene expression sequence would compromise not only the formation of a specific developmental field but may also have repercussions in contiguous monotopic fields (e.g. cleft lip with cleft palate, etc.) or distant polytopic fields (e.g. VACTERL association) [11].

Multiple gene sets may be involved in a particular developmental field (gene redundancy) [1, 2]. Similarly, a gene may be involved with several contiguous or non-contiguous developmental fields (gene pleiotropy) [1, 2]. Thus, a single-gene disorder may cause congenital anomalies in multiple developmental fields (e.g. loss of sonic hedgehog/*Shh* gene in laterality defects, agenesis of pancreas and congenital absence of femur). FGF cooperates with homeobox (Hox), Shh and Tbx genes to regulate apical ectodermal ridge (AER) at the limb bud to form upper and lower limbs. Abrogation of this interaction leads to terminal transverse limb reduction defects (Fig. 6.7) [12]. Congenital diaphragmatic hernia and omphalocele are other developmental field defects that may occur as an isolated lesion or may be associated with other congenital anomalies (Figs. 6.8 and 6.9) [13].

Terminology of Congenital Anomalies

Morphologically, the congenital abnormalities may be classified as primary or secondary (see Table 6.2) [2].

Primary Congenital Defects

These arise from genetic (intrinsic) defects resulting in basic alteration of the structure (<10 gw). They may further be classified as single or multiple defects. The structural abnormalities may be in the physical appearance of a structure or organ system (e.g. renal aplasia, horseshoe kidney) (Fig. 6.10) [2]. These developmental defects are called *malformations*.

Another instance of a primary congenital defect is *dysplasia*, where there is disorganization of the cells resulting in abnormal histomorphogenesis (e.g. skeletal dysplasia) (Fig. 6.11). Abnormality of FGFR3 receptor can lead to abnormal ossification, leading to lethal

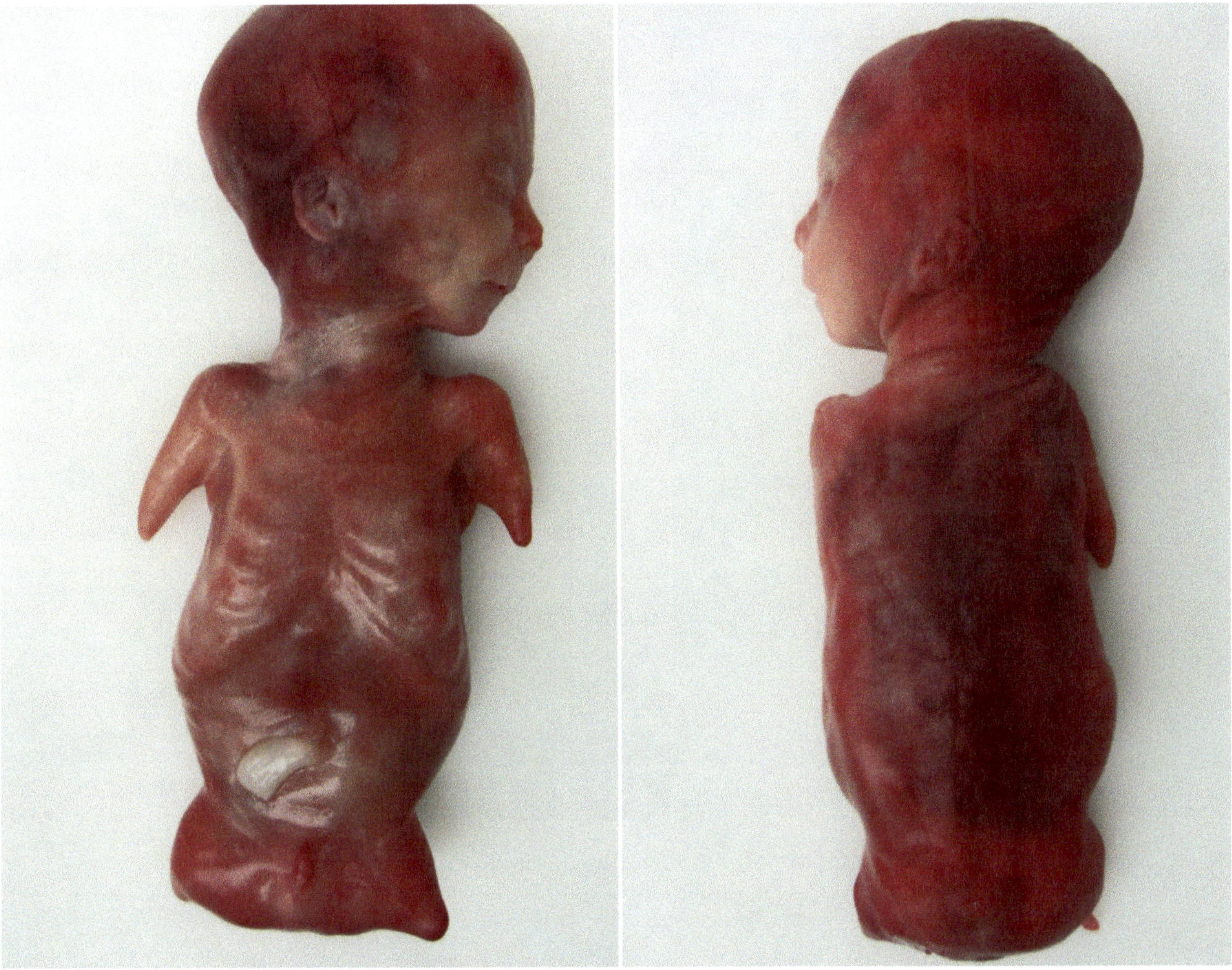

Fig. 6.7 Congenital limb deficiency defect affecting all limbs. These defects occur due to primary failure of the apical ectodermal ridge (AER) on the lateral wall of the embryo 4 weeks post fertilization due to abrogation of FGF, homeobox (Hox) and sonic hedgehog (Shh) signals

thanatophoric dysplasia/TD (type II) with distinctive clinical, radiological and pathological findings. There is skeletal dwarfism with a large head (cloverleaf shaped), shortening of long bones with straight femora, narrow chest with short ribs and increased distance between vertebral bodies which are thin (platyspondyly) (Fig. 6.11). Depending upon the mutation of the FGFR3 gene, at least six different types of skeletal dysplasia phenotypes can arise [13]: TD type I, TD type II, achondroplasia, hypochondroplasia, platyspondylic skeletal dysplasia and SADDAN (severe achondroplasia with developmental delay and acanthosis nigricans), the so-called *FGFR3 skeletal dysplasia group* [13].

Syndrome is a set of multiple primary congenital anomalies (often diverse) that are causally related to the underlying aetiology, e.g. Patau's syndrome (Fig. 6.3a) [1], Turner's syndrome, etc. (Fig. 6.3b). *Sequence* is a set of multiple anomalies that may be derived from single anomaly or mechanical factor (prior known), e.g. anhydramnios sequence (Fig. 6.12). The term *association* is related to non-random occurrence of two or more congenital anomalies that are not known to be a polytopic developmental field defect, syndrome or sequence, e.g. VACTERL association that includes vertebral, anorectal, cardiac, tracheoesophageal, renal/radial and limb abnormalities; one needs at least four of these to make a diagnosis [2, 11].

Secondary congenital defects arise from non-genetic (extrinsic) defects. These include *disruption*, where there is a morphological defect arising from the external breakdown of or an interference with an originally normal developmental process (e.g. amniotic band disruption sequence) (Fig. 6.13), and *deformation*, where a variation in structure (shape, size or position) occurs in a normal tissue in response to external mechanical forces (talipes equinovarus defect in anhydramnios) (Fig. 6.12) [2].

Complex is a non-random occurrence of two or more congenital anomalies that may be related to a primary syndrome, a sequence (e.g. amniotic band disruption sequence) or a

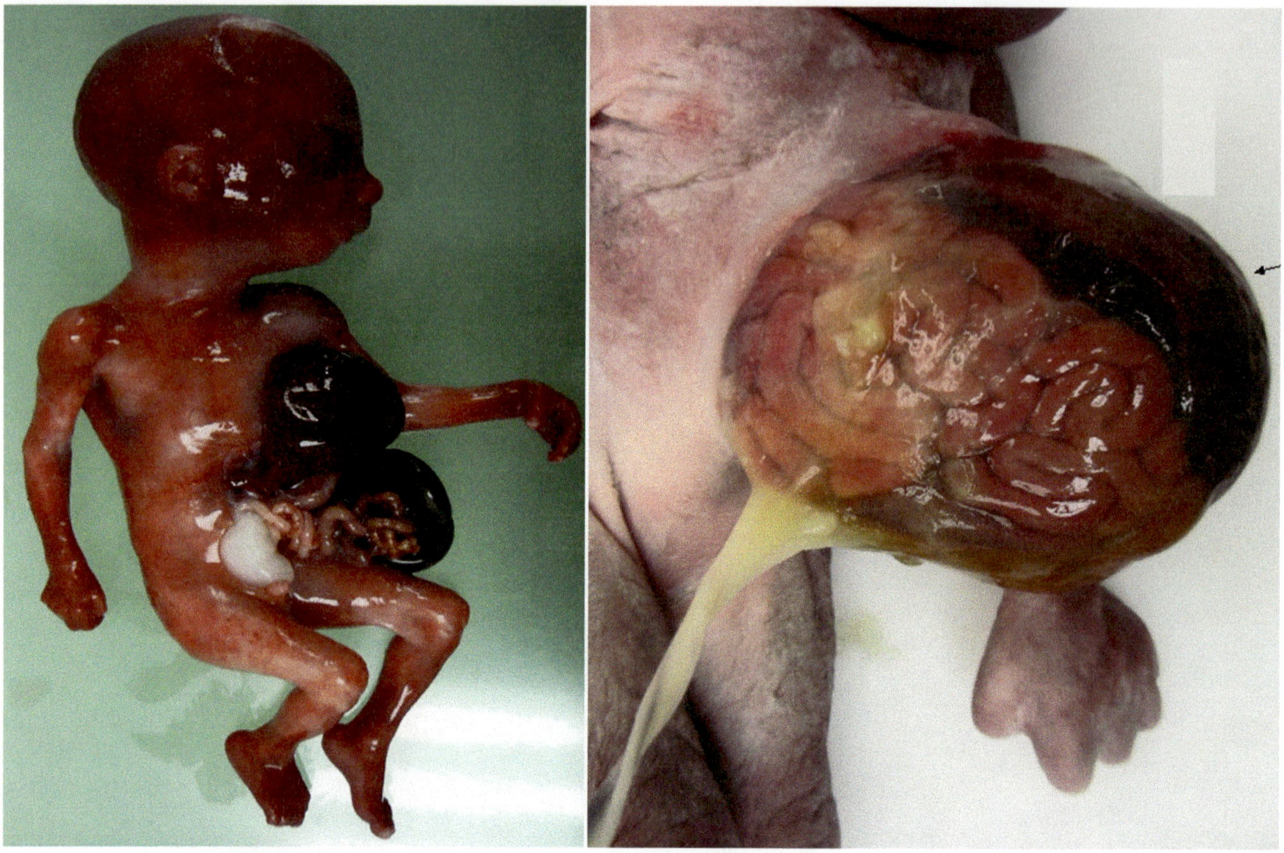

Fig. 6.8 Anterior abdominal wall defects: (left) limb-body wall defect. (Right) Omphalocele

Fig. 6.9 Congenital left diaphragmatic hernia due to posterolateral defect in the left crus of diaphragm leading to herniation of abdominal content into the left thoracic cavity, with resultant right shift of the central mediastinal structures including the heart

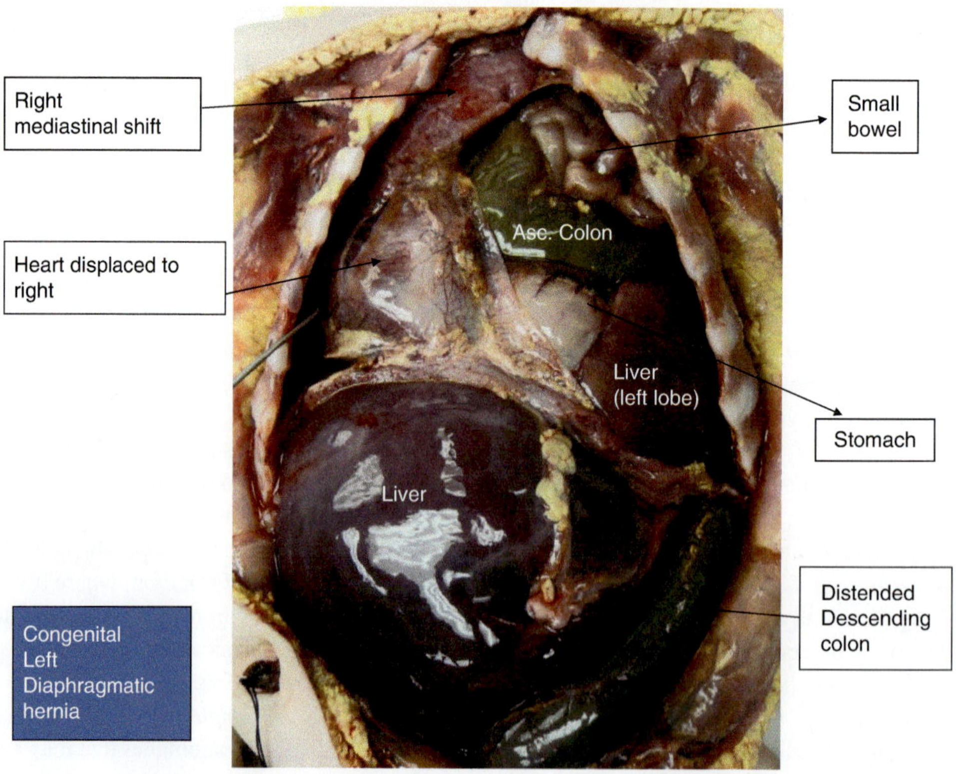

polytopic developmental field defect (e.g. OEIS complex) (Fig. 6.13) [1].

The embryological development is a dynamic continuum of a myriad of different regulatory and structural proteins that act in an orchestrated manner to give rise to the human foetus. It appears that the cellular processes crucial for development are evolutionarily conserved and are protected by genes with overlapping functions (gene redundancy) to minimize the impact of a particular gene loss on a given developmental field [10]. However, intrinsic failure of this developmental blueprint either due to genetic/chromosomal events or due to secondary interference from external physical or chemical agents can damage, distort or modify the developmental field, leading to a wide variety of congenital anomalies [2].

Table 6.2 Classification of congenital anomalies

	Single defects	Multiple defects
Primary	Malformation Dysplasia	Syndrome Sequence
Secondary	Deformation Disruption	Sequence Association Complex

Abnormalities of Monozygous Twinning

Dizygotic or fraternal/non-identical twinning occurs when two separate eggs are ovulated and fertilized. They may be of different sexes. Monozygotic or identical twinning on the other hand occurs when a single egg is fertilized, but dividing cells at the blastocyst stage split into two groups and develop into two individuals [1, 2].

The timing, symmetry and the completeness of this splitting after fertilization are critical. Cleavage of the blastocyst between days 4 and 8 results in complete separation of the monochorionic diamniotic twins, provided that it splits symmetrically and equally. If the blastocyst cleaves symmetrically and equally, but after Day 8, then the separation is not complete, and this results in conjoined twins, who can be joined at the head (craniopagus), thorax (thoracopagus) (Fig. 6.14) or pelvis (pygopagus) [1].

If the blastocyst cleavage is asymmetrical, then one twin may be normal, whilst the other twin if external may become ectoparasitic (acardiac/acephalic) (Fig. 6.15) or dramatically if internal may become endoparasitic, the so-called foetus in foetu (see Table 6.3).

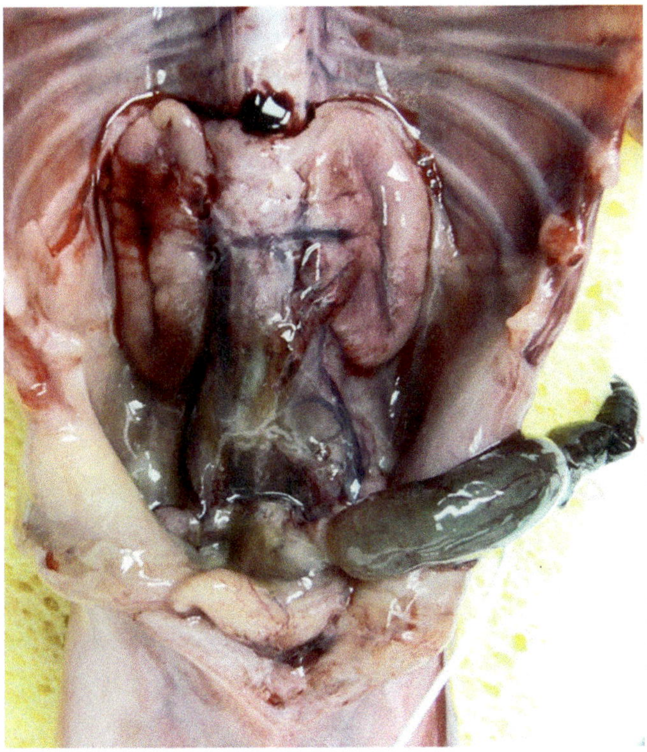

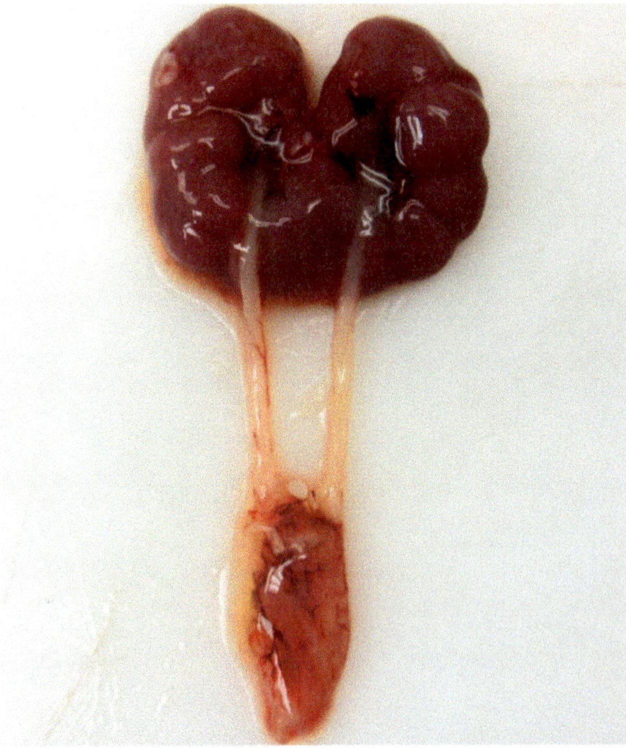

Fig. 6.10 (Left) Bilateral renal agenesis with associated rectovesical fistula; note the enlarged discoid adrenal glands that now occupy the renal fossa. (Right) Horseshoe kidney with anteriorly displaced ureters and small bladder

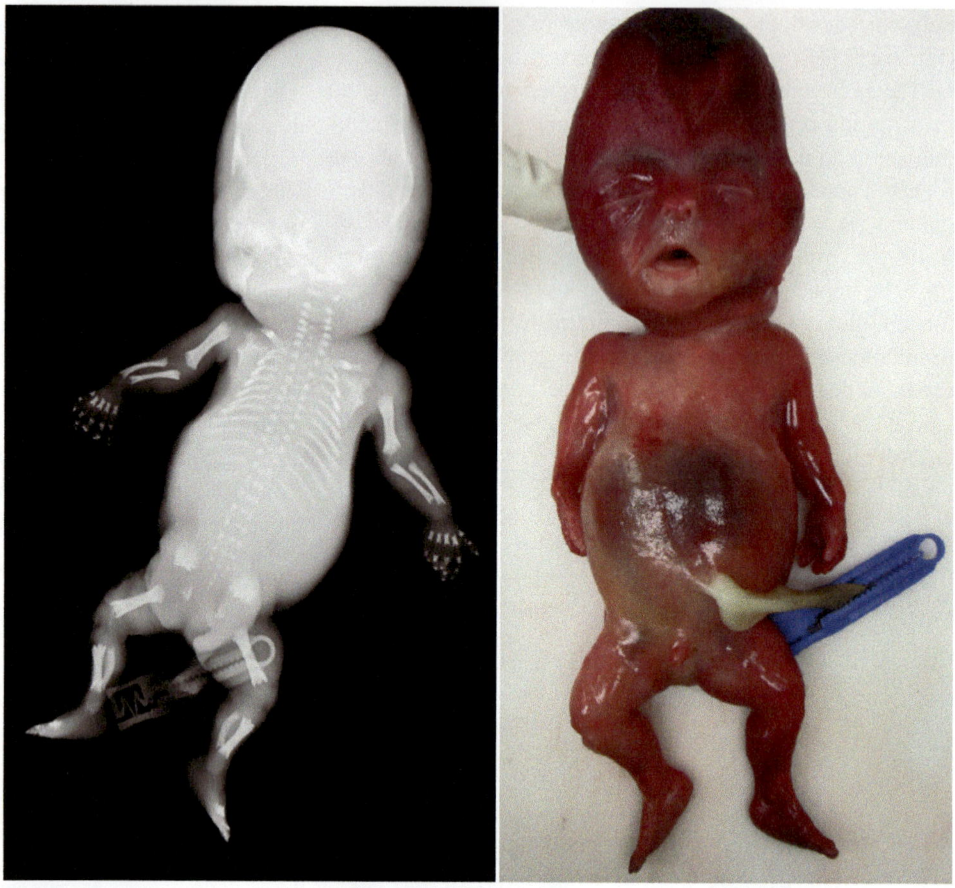

Fig. 6.11 (Left) Typical radiological features of type II thanatophoric dysplasia. Note the straight femora (in TD type I, they are typically bowed), short ribs and trilobed skull. (Right) Large head with a cloverleaf pattern; shortening of long bones

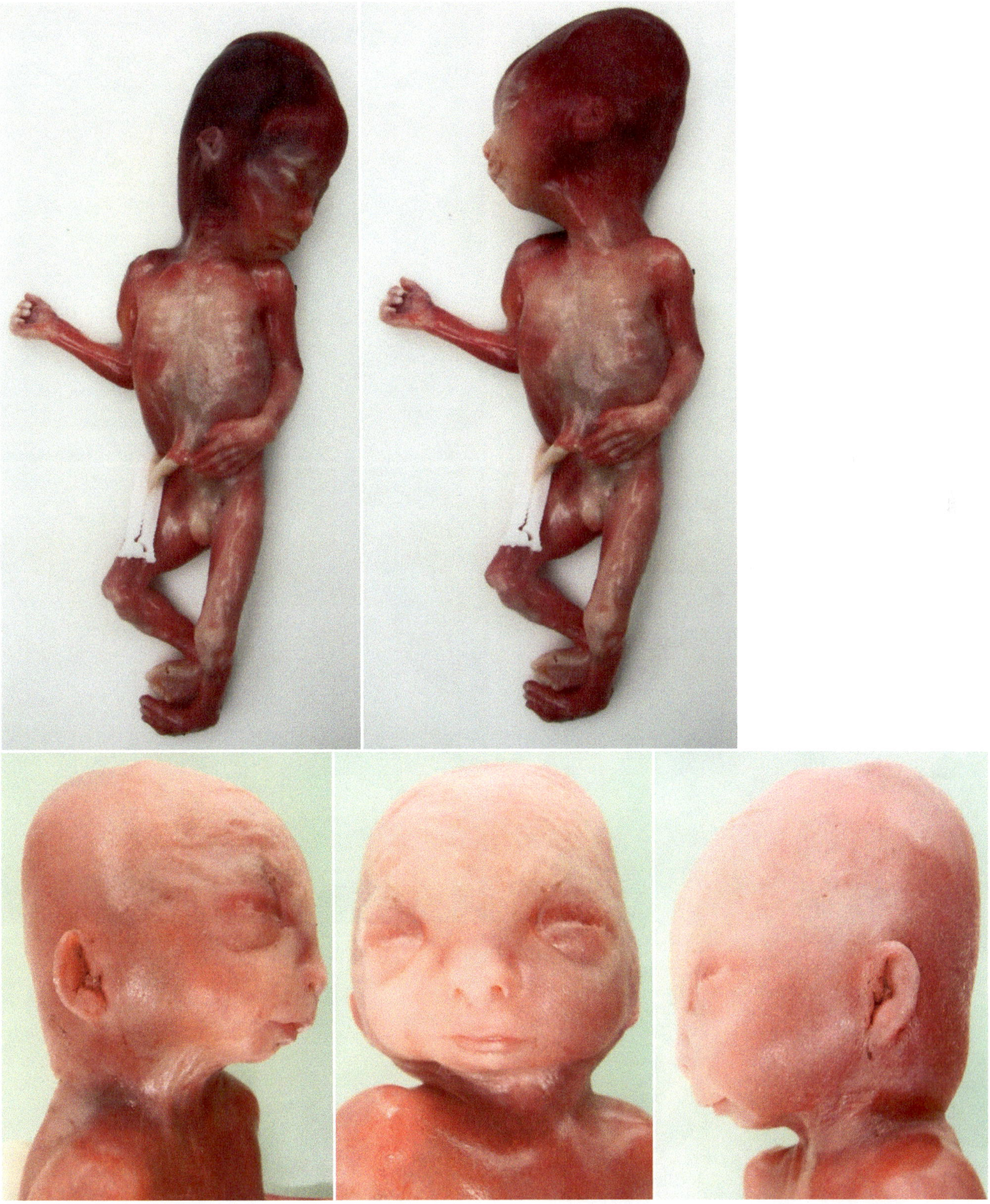

Fig. 6.12 Anhydramnios/oligohydramnios sequence occurs when there is a lack of/markedly reduced amount of amniotic fluid around the baby from congenital (renal agenesis, urethral atresia) or acquired causes (spontaneous rupture of membranes). The resultant compression of the foetus and lack of movement, thereof, leads to a predictable constellation of findings. (Top) The feet show talipes equinovarus deformity. (Bottom) The head is compressed side to side (dolichocephaly) with wide-set eyes (hypertelorism), small nose, flat nasal bridge, low-set ears and small jaw (micrognathia)—the so-called Potter facies

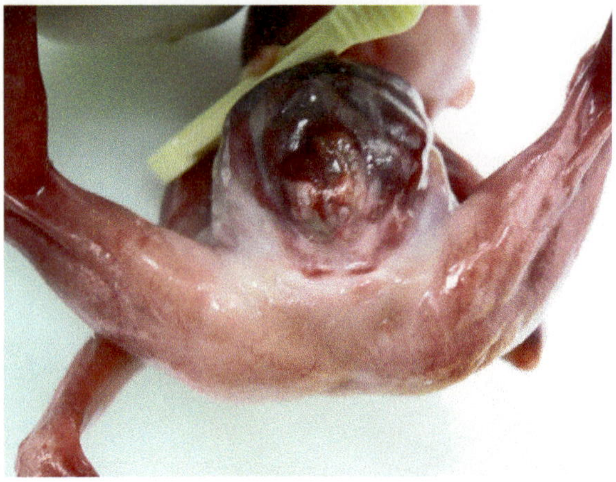

Fig. 6.13 Typical spectrum of findings in OEIS complex. OEIS includes omphalocele, exstrophy of bladder, imperforate anus and sacral hypoplasia

Table 6.3 Pathology of monozygous twinning [1]

Symmetric blastocyst cleavage	Separate	Normal MCDA twins
	Conjoined	Y, X, λ (according to the site of fusion)
Asymmetric blastocyst cleavage	External	Acardiac/acephalic twin
	Attached	Ectoparasitic twin
	Internal	Endoparasitic twin (foetus in foetu)

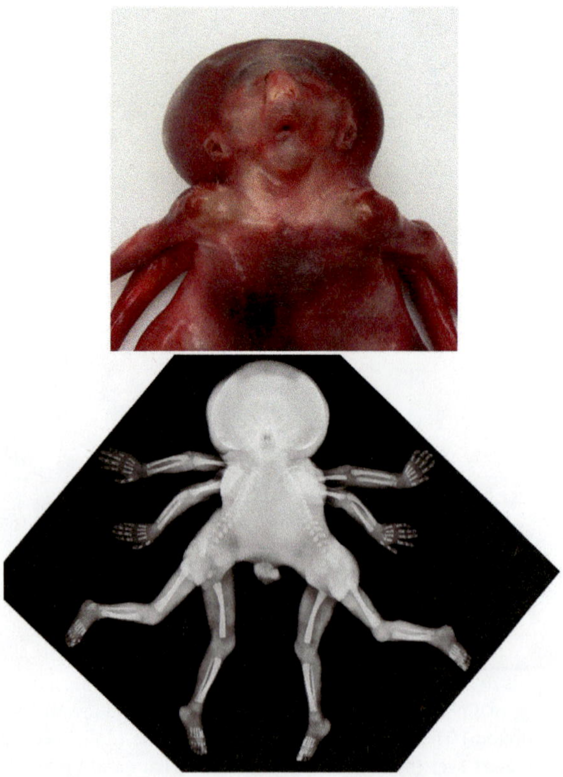

Fig. 6.14 Typical clinical and radiological features of thoracopagus, fused at the level of the head and thorax (λ-type). Both foetuses had separate thoracic structures including the heart and lungs; the crania were fused

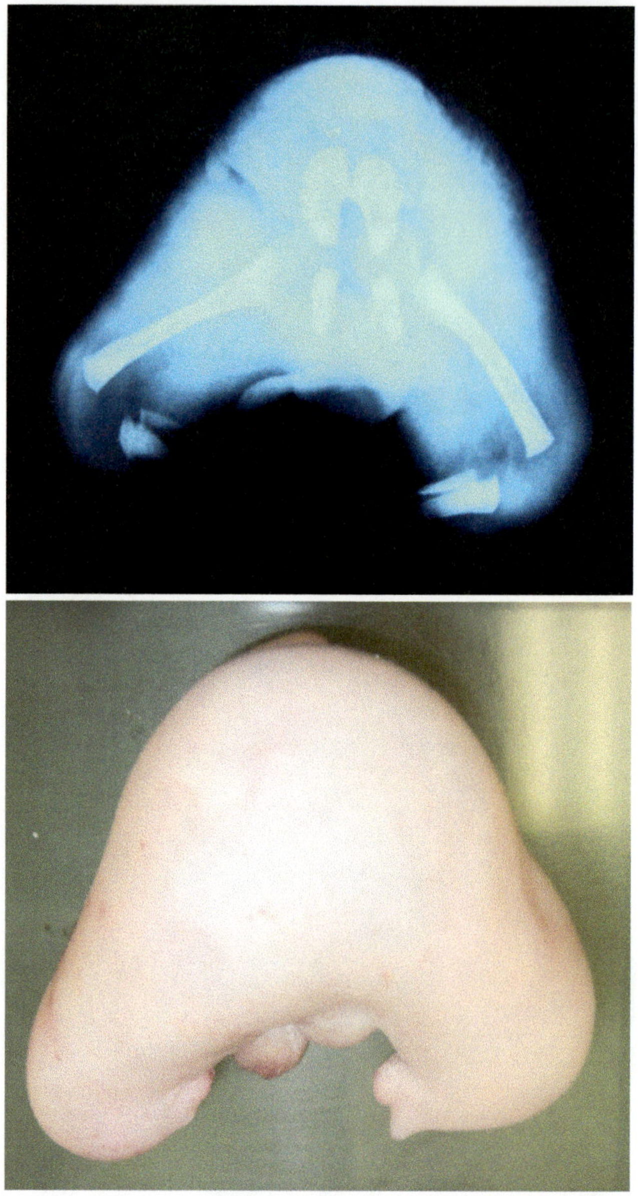

Fig. 6.15 Acardiac twin due to asymmetric blastocyst cleavage. The parasitic twin depends upon the other cardiac twin to perfuse it. Since the perfusion occurs in the revered direction, it is also referred to as TRAP (twin reversed arterial perfusion)

To conclude, if a genetic event results in complete loss of gene function, it may cause arrested embryonic development due to a breakdown of the developmental field, leading to congenital anomalies. On the other hand, if the genetic event results in an activating gene mutation, it may not only interfere with normal development but also drive the cells into uncontrolled proliferation (imparting stem cell phenotype), leading to tumour formation (see Chap. 49).

References

1. Moore KL, Persaud TVN, Torche MG. The developing human: clinically oriented embryology. 10th ed. Philadelphia: Elsevier; 2015.
2. Gilbert-Barness E, Debich-Spicer D. The human embryo and growth disorganization. In: Gilbert-Barness E, Debich-Spicer D, editors. Embryo and fetal pathology: color atlas with ultrasound correlation. Cambridge: Cambridge University press; 2004. p. 1–22.
3. Maitra A. Genetic and pediatric diseases. In: Kumar V, Abbas AK, Aster JC, editors. Robbins basic pathology 9e. Philadelphia: Elsevier; 2013. p. 243–98.
4. Ochando I, Vidal V, Gascón J, Acién M, Urbano A, Rueda J. Prenatal diagnosis of X-linked hydrocephalus in a family with a novel mutation in L1CAM gene. J Obstet Gynaecol. 2016;36(3):403–5.
5. Yoon PW, Freeman SB, Sherman SL, Taft LF, Gu Y, Pettay D, Flanders WD, Khoury MJ, Hassold TJ. Advanced maternal age and the risk of Down syndrome characterized by the meiotic stage of chromosomal error: a population-based study. Am J Hum Genet. 1996;58(3):628–33.
6. Boyarchuk O, Volyanska L, Dmytrash L. Clinical variability of chromosome 22q11.2 deletion syndrome. Cent Eur J Immunol. 2017;42(4):412–7.
7. Peters J. The role of genomic imprinting in biology and disease: an expanding view. Nat Rev Genet. 2014;15(8):517–30.
8. Allen VM, Armson BA, Committee G, Committee MFM. Teratogenicity associated with pre-existing and gestational diabetes. J Obstet Gynaecol Can. 2007;29(11):927–34.
9. Sulik KK. Fetal alcohol spectrum disorder: pathogenesis and mechanisms. Handb Clin Neurol. 2014;125:463–75.
10. Opitz JM. The developmental field concept. Am J Med Genet. 1985;21(1):1–11.
11. Solomon BD, Bear KA, Kimonis V, de Klein A, Scott DA, Shaw-Smith C, Tibboel D, Reutter H, Giampietro PF. Clinical geneticists views of VACTERL/VATER association. Am J Med Genet A. 2012;158A(12):3087–100.
12. Noramly S, Pisenti J, Abbott U, Morgan B. Gene expression in the limbless mutant: polarized gene expression in the absence of Shh and an AER. Dev Biol. 1996;179(2):339–46.
13. Geister KA, Camper SA. Advances in skeletal dysplasia genetics. Annu Rev Genomics Hum Genet. 2015;16:199–227.

7 Antenatal Screening/Prenatal Diagnosis

Nicola Brindley

Introduction

As care providers for pregnant women and their babies, the ultimate goal is always a healthy mother and a healthy baby.

However, when a foetal anomaly is diagnosed prenatally, this goal becomes achieving the best possible outcome for the neonate without compromising the mother's health.

With this philosophy in mind, this chapter discusses the current practice of antenatal screening and prenatal diagnosis to detect aneuploidy and structural anomalies in the developing foetus.

This chapter describes examples of the antenatal counselling of parents whose foetus has had an anomaly detected.

Aim

The aims of prenatal diagnosis relate to identification of a foetal abnormality in order to:

- Reassure parents by reducing the likelihood of an undiagnosed foetal anomaly
- Maximise the information available if a foetal anomaly is detected to assist in decision-making
- Allow parents of pregnancies in which a foetal anomaly is identified to prepare for the birth and perinatal events if they choose to continue
- Allow appropriate perinatal management and potential in utero therapy

N. Brindley (✉)
Department of Surgical Paediatrics, Queen Elizabeth University Hospital, Glasgow, UK
e-mail: Nicola.Brindley@ggc.scot.nhs.uk

Epidemiology of Foetal Anomalies

Congenital abnormalities are numerous, although the overall prevalence of disorders is approximately 2 per 100 pregnancies.

Congenital anomalies include:

1. Malformations—Abnormalities of tissue development.
2. Deformities—Abnormalities caused by mechanical stressors such as oligohydramnios and amniotic bands.

Chromosomal abnormalities (aneuploidy) are common and occur in 1:160 live births, with trisomy 21 (Down's syndrome), trisomy 18 (Edward's syndrome) and trisomy 13 (Patau syndrome) accounting for the majority.

Antenatal screening and diagnosis have evolved exponentially over the last decade to improve obstetric care and foetal outcome.

The programmes comprise two main parts, with screening and diagnostic components [1–3].

Firstly: screening for chromosomal abnormalities, mainly trisomy 21, at **11–13** weeks.

Secondly: screening for a variety of structural but also chromosomal and genetic foetal anomalies, by ultrasound at **16–20** weeks gestation.

While screening tests are not diagnostic, they enable the selection of pregnancies to which diagnostic tests may be done, which allows earlier detection of conditions, and thus facilitate treatment or termination.

A variety of screening tests are available, incorporating maternal age, ultrasound, and biochemical markers. The schedule of prenatal-testing and screening can vary somewhat among centres and countries, but generally conforms to a routine pattern as outlined below:

The distinction between screening and diagnosis is often unclear.

Screening tests are performed on all pregnant women in order to identify the subset of patients who are at an increased risk of an abnormality. These tests do not confer any risk to

R. Carachi, S. H. E. Doss (eds.), *Clinical Embryology*, https://doi.org/10.1007/978-3-319-26158-4_7

the pregnancy and are undertaken for disorders which carry a relatively high prevalence and for which there are accurate prenatal diagnostic confirmatory tests.

Diagnostic tests are usually carried out on a selected group of patients whose screening tests have identified them as high risk for an anomaly. These tests are usually invasive and carry a risk of miscarriage, albeit small.

Routine Testing

First Trimester Screening

Timing: The optimal timing for first trimester screening is 10–14 weeks of gestation.

Definition: First trimester screening uses ultrasound for measuring nuchal translucency (NT), fluid at the nape of the foetal neck, and two serum proteins—plasma protein A (PAAP-A) and the free beta subunit of human chorionic gonadotropin (B-hCG). Increased nuchal translucency, greater than 3 mm between 10 and 14 weeks gestation, and reduced PAAP-A and B-hCG levels indicate an increased risk for Trisomy 21, 18 and 13. Increased NT alone indicates 10–15% risk for foetal cardiac anomalies.

Interpretation: Elevated or low levels of first trimester maternal serum marker levels and increased NT indicate a need to follow up with other diagnostic tests such as ultrasound, chorionic villus sampling or amniocentesis. These diagnostic techniques are discussed later in this chapter [4].

Second Trimester Screening

Timing: The optimal timing for AFP, Triple or Quadruple marker screening is 16–18 weeks of gestation.

Definition: The screening test involves sampling the maternal serum for the following different biochemical markers:

- MSAFP—Maternal serum alpha foetoprotein
- hCG—Human chorionic gonadotropin
- uEST—Unconjugated oestriol
- Inhibin A—Placental hormone

Combining the three markers AFP, hCG and uEST with maternal age (*triple test*) allows a detection rate of some trisomies of 60–70%. Low MSAFP, low uEST and high hCG are associated with Down syndrome with approximately 60% accuracy in women under 35 years of age and with a 75% accuracy in those over 35 years and older. Low values in all three markers are associated with trisomy 18 with 70–80% accuracy. There is however a 7% false positive rate.

In addition to the trisomies, the triple test marker detects 85–90% of the open neural tube defects [5, 6].

The addition of Inhibin A to the marker screen (*quadruple test*) can further enhance the detection rate of Down's syndrome in women younger than 35 years of age. Low inhibin A levels indicate the possibility of Down's syndrome [7].

The most common chromosomal abnormality that is screened for is Trisomy 21-Down Syndrome.

There are two opportunities for Down syndrome screening

1. First trimester screening—Preferred option—higher detection rate and lower false positive rate—uses a combination of variables to provide an estimate of risk of Trisomy 21
   (a) Maternal age
   (b) Nuchal translucency
   (c) Biochemical markers: PAAP-A, B-hCG
2. Mid-Trimester screening
   (a) 15–20 weeks
   (b) Maternal age
   (c) Biochemical markers: AFP, HCG, Oestriol, Inhibin (quadruple test)

Second-trimester screening confers a detection rate for Down syndrome of 81% [14]. Detection rate for Down syndrome using the above tools ranges from 82 to 90%. Adding evaluation of the nasal bone, which is absent in 60–70% of foetuses with Down syndrome, can increase the detection rate to 95%. There is a recent tendency to delay the routine anomaly scan until 20–22 weeks as this improves the opportunity to diagnose cardiac and late developing microcephaly.

When a screening test is positive, the diagnosis can be confirmed or outruled by an invasive diagnostic procedure such as chorionic villus sampling (CVS) or amniocentesis. A brief description of the invasive tests is given below.

Amniocentesis

A thin needle is passed trans-abdominally under ultrasound guidance into the amniotic cavity.

A small amount of amniotic fluid is removed, which contains foetal fibroblasts. This test is usually performed at or after 15 weeks gestation; the procedure-related miscarriage risk is 1%. Although it is technically feasible to perform amniocentesis at an earlier gestation, it is generally avoided as it is associated with higher cell culture failure rates, neonatal talipes and respiratory difficulties [8].

Chorion Villus Sampling

A thin needle is passed trans-abdominally or trans-cervically under ultrasound guidance into the placenta (chorionic plate). Chorionic villi, which are foetoplacental in origin, are aspirated through this needle. This test is usually performed at or after 10 weeks gestation. The miscarriage rate after CVS is thought to be higher than that following amniocentesis (2–3%), this is because the natural spontaneous miscarriage rate is higher at 10 weeks. The procedure-related miscarriage rate of CVS is the same as for amniocentesis, 1%.

Although it is technically feasible to perform CVS at earlier gestations, it is generally avoided as it is associated with a higher rate of cleft lip/palate and digital amputation abnormalities [9].

Laboratory Analysis

Cytogenetic Analysis

Cells obtained from invasive prenatal diagnostic tests are cultured until adequate cells in mitosis are available to make a cytogenetic diagnosis. The more rapidly the tissue divides, the quicker the results are available. Hence the results of amniocentesis may be available within 2–3 weeks, and results of CVS available in 1–2 weeks. With CVS, the sampled chorionic villi have many cells already in mitosis so that a direct result may be available in 24–48 h. This may be adequate to detect or exclude aneuploidy, but is not of sufficient quality to permit G-banding; hence, chromosomal aberrations such as deletions or inversions may not be detected/diagnosed.

The use of fluorescence in situ hybridization (FISH) can facilitate rapid results with amniocentesis. This technique detects and localises specific DNA sequences directly in interphase and metaphase and thus cell culture is not required. It allows rapid diagnosis within 24–48 h of the three major aneuploidies for chromosomes 13, 18, 21 and XY.

DNA Analysis

Foetal DNA obtained from invasive tests can be used for DNA probe (cystic fibrosis and sickle cell disease), polymerase chain reaction (PCR) (fragile X syndrome, congenital cytomegalovirus and toxoplasmosis), or linkage analysis (fragile X syndrome).

Biochemical and Enzymatic Analysis

When DNA analysis is not possible, biochemical or enzymatic assays can be performed for specific diseases such as mucopolysaccharidoses and congenital adrenal hyperplasia.

Diagnosis of Structural Abnormalities

Neural Tube Defects

When a parent or previous sibling has had a NTD, the recurrence risk is 5%. Mid trimester maternal serum AFP levels are increased in the presence of an open NTD.

Anencephaly and encephaloceles can be detected on first trimester ultrasound if an adequate examination of the cranial vault is performed. Spina bifida detection requires a systematic detailed examination of the foetal spine at the routine 20-week anomaly scan. The 'lemon' (shape of the skull) and 'banana' (absent cerebellum) signs are highly indicative. The sensitivity of ultrasound for both open and closed NTDs is greater than 95% [10, 11].

Congenital Heart Defects

When previous sibling or a father is affected by a congenital heart defect, the risk is 2%.

When two siblings or a mother is affected by a congenital heart defect, the risk is 10%.

The risk for offspring of women with type 1 diabetes mellitus is 20%, but more than 90% of foetuses with CHD are from pregnancies without any of such risks.

Although 90% of CHDs can be detected antenatally by specialist foetal echocardiography, most units performing routine anomaly scans at 20 weeks will have a detection rate closer to 30%.

Brief Historical Perspective

Most foetal investigative techniques were developed in 1960s and 1970s and were classified into two main groups:

1. Images from ultrasound
2. Biochemical samples
   (a) Amniocentesis
   (b) Chorionic villous sample
   (c) Maternal blood/serum
   (d) Foetal blood/serum

Investigations Based on Ultrasound

Ultrasound was derived from the SONAR maritime technique to measure distances under water using sound waves.

Ultrasound was used for the first time in Obstetrics by Ian Donald in Scotland in 1950s Images were of poor quality but users could detect pregnancy failure, multiple pregnancies and placental location, and estimate liquor volume [12].

This improved technique of visualising the developing foetus enabled detection of foetal anomalies such as hydrocephalus in 1961 and anencephaly in 1964, albeit in the third trimester and resulted in foetal death [13].

Improved techniques and images during the 1970s enabled second trimester detection and diagnosis of anencephaly (1972) and spina bifida (1975) by Stuart Campbell in England [10].

Improved microprocessor and electronic technologies in the 1980–1990s resulted in an increase number of anomalies being detected. Benacerraf, a North American ultrasonographer, identified foetuses at 15–20 weeks gestation who had increased nuchal translucency and short femur length [14].

Evolution of high resolution scanners enabled Nicholaides, in London in 1992, to detect foetal trisomies during the first trimester between 11 and 14 weeks [15].

The first routine foetal scanning programs started in Sweden in the 1970s and the 20/40 anomaly scan has been standard practice since 1990.

Three-Dimensional Ultrasound

Advances in imaging technology have permitted real-time three-dimensional reconstruction of data acquired by conventional 2-D ultrasound machines. This technology permits the increased resolution required for certain foetal malformations such as cleft lip and palate, and cardiac abnormalities.

Foetal MRI

Foetal MRI is being used to detail more clearly conditions that have been detected with ultrasound such as congenital diaphragmatic hernia and foetal masses/tumours to help with prognosis and delivery planning. The development of ultrafast MRI sequences to overcome foetal movement artefact has resulted in significant improvement in image quality and usefulness [16, 17] (Table 7.1).

Table 7.1 Classification and incidence of common congenital abnormalities

Congenital abnormality	Example	Incidence per 1000 births
Structural	Congenital heart disease	4–6
	Neural tube defects	2–6
	Cleft lip and palate	1–2
Chromosomal	Trisomy 21	1.5
	Turners syndrome	0.3
	Other trisomies 18 and 13	0.3
Genetic	Cystic fibrosis	0.5
	Sickle cell disease	Depends on ethnicity

References

1. Palomaki GE, Kloza EM, Lambert-Messerlian GM, Haddow JE, Neveux LM, Ehrich M, et al. DNA sequencing of maternal plasma reliably identifies trisomy 18 and trisomy 13 as well as Down's syndrome: an international collaborative study. Genet Med. 2012;14:296–305.
2. NHS Fetal anomaly screening programme Annual Report. NHS FASP April 2010-March 2011. London: UK National Screening Programmes Directorate, 2011.
3. Ward P, Soothill P. Fetal anomaly ultrasound scanning: the development of a national screening programme for England. Obstet Gynaecol. 2011;13:211–7.
4. Khalil A, Mahmoodian N, Kulkarni A, Homfray T, Papageorghiou A, Bhide A, Thilaganathan B. Estimation of detection rates of aneuploidy in high-risk pregnancy using an approach based on nuchal translucency and non-invasive prenatal testing: a cohort study. Fetal Diagn Ther. 2015;38(4):254–61.
5. Spencer K, Spencer CE, Power M, Moakes A, Nicolaides KH. One stop clinic for assessment of risk for fetal anomalies: a report of the first year of prospective screening for chromosomal anomalies in the first trimester. BJOG. 2000;107:1271–5.
6. Baer RJ, Flessel MC, Jelliffe-Pawlowski LL, Goldman S, Hudgins L, Hull AD, et al. Detection rates for aneuploidy by first-trimester and sequential screening. Obstet Gynecol. 2015;126:753–9.
7. Gil MM, Revello R, Poon LC, Akolekar R, Nicolaides KH. Clinical implementation of routine screening for fetal trisomies in the UK NHS: cell-free DNA test contingent on results from first-trimester combined test. Ultrasound Obstet Gynecol. 2016;47:45–52.
8. Anandakumar C, Wong YC, Annapoorna V, Arulkumaran S, Chia D, Bongso A, et al. Amniocentesis and its complications. Aust NZ J Obstet Gynecol. 1992;32:97–9.
9. Odibo AO, Dicke JM, Gray DL, Oberie B, Stamilio DM, Macones GA, et al. Evaluating the rate and risk factors for fetal loss following CVS. Obstet Gynecol. 2008;112:813–9.
10. Campbell S. Early prenatal diagnosis of neural tube defects by ultrasound. Clin Obstet Gynecol. 1977;20:351–9.
11. Chaoui R, Nicolaides KH. From nuchal translucency to intracranial translucency, towards the early detection of spina bifida. Ultrasound Obstet Gynecol. 2010;35(2):133–8.
12. Donald I, Mac Vicar J, Brown TG. Investigation of abdominal masses by pulsed ultrasound. Lancet. 1958;1:1188–95.
13. Donald I. Clinical applications of ultrasonic technology in obstetrical and gynecological diagnosis. Br J Obstet Gynecol. 1962;69:1036.
14. Bencerraf BR, Barss VA, Laboda LA. A sonographic sign for the detection in the second trimester of the fetus with Down's syndrome. Am J Obstet Gynecol. 1985;151:1078–9.
15. Nicolaides KH, Azar GB, Byrne D. Nuchal translucency: ultrasound screening for chromosomal defects in the first trimester of pregnancy. Br Med J. 1992;304:867–9.
16. Hubbard AM. Ultrafast fetal MRI and prenatal diagnosis. Semin Pediatr Surg. 2003;12:143–53.
17. Levine D, Fetal MRI. J Matern Fetal Neonatal Med. 2004;15:85–94.

Antenatal Counselling and Genetics

8

Nicola Brindley

Introduction

With the development of prenatal ultrasound and of foetal medicine, the paediatric surgeon has extended his knowledge of the natural history of surgical malformations. He/she is a part of the prenatal team and parents should always be referred to him/her when a surgical malformation is suspected, even when termination of pregnancy is planned because of an expected poor prognosis.

Direct contact between the prenatal medicine specialist and the paediatric surgeon is also highly recommended to ensure continuity in the messages delivered to the parents. Postnatal counselling does not compare with prenatal counselling, and the paediatric surgeon has learned from the obstetrician to modulate his talk by including other conditions that might affect the outcome of the foetus, especially genetically determined syndromes.

When the foetal malformation is diagnosed very early, especially in the first trimester, it therefore seems important for the consultation with the paediatric surgeon to be scheduled when the complementary exams required by the anomalies diagnosed are done, in order to avoid later contradictory messages. Repeated consultations should be favoured as they allow provision of more precise information regarding changes in ultrasound and/or MRI images and so decrease parental anxiety and help them to take their decision [1].

Foetal surgery, which has reached various stages of development in different countries, requires paediatric surgeons and obstetricians to join forces to optimise procedures and evaluate their benefit/risk ratio. Since 2004, the National Rare Disease Plan in France has allowed the creation of Rare Disease Centres, which deal with congenital malformations and produce recommendations for the health care pathway of these patients by means of a multidisciplinary approach.

The use of the Internet by most couples has to be taken into account and has certainly modified prenatal counselling over the last 10 years. Parents generally use the time between the first consultation, where the malformation has been suspected, and the consultation with the foetal medicine specialist to search for information on the Internet. Their understanding of the information they retrieve may vary greatly depending on the quality of the websites explored, the family story, and their cultural and intellectual background. The role of the foetal medicine specialist is first to delineate what the parents have correctly understood and to correct potential misinterpretations [1, 2].

In this chapter, a short collection of true clinical cases is outlined as a way of illustrating how we, as paediatric surgeons counsel expectant parents whose foetus has a suspected or detected anomaly at a specialist foetal medicine unit.

Case 1

Sixteen-year-old girl, 24 weeks pregnant, attends the specialist foetal medicine department, accompanied by her mother and her boyfriend's mother for a follow-up ultrasound scan of her foetus who has been identified at her local hospital with an isolated gastroschisis.

Both grandmothers have a list of questions having searched the Internet about abdominal wall defects:

How long will the baby be in hospital?

What clothes can the baby wear in hospital?

Can she have a natural birth?

Often the questions that expectant parents ask relate very little to the clinical and surgical aspects of their babies condition, but illustrate the worries and concerns that they have when they have a baby with an anomaly and that he or she cannot go home in the early post natal period. We need to be sensitive to these psychosocial aspects and concerns and try to counsel in a general, holistic manner.

N. Brindley (✉)
Department of Surgical Paediatrics, Queen Elizabeth University Hospital, Glasgow, UK
e-mail: Nicola.Brindley@ggc.scot.nhs.uk

R. Carachi, S. H. E. Doss (eds.), *Clinical Embryology*, https://doi.org/10.1007/978-3-319-26158-4_8

Meeting the paediatric surgical team that will be looking after their baby immediately after birth is an anxious time for expectant, worried parents. Following introductions, we find it helpful to open the dialogue by asking what the couple already understand about their baby's anomaly. This can then guide us as to how detailed and elaborate a description of the condition and its management we give at the initial meeting. Often young expectant mothers and their partners are overwhelmed by the detail and description of the likely perinatal/postnatal events and do not recall much information nor wish to know the fine detail.

Using gastroschisis as our example, we describe the management in three time periods:

Labour Suite Management

Fluid resuscitation
Care of herniated bowel/viscera and their blood supply-wrapping baby in cling film
Bowel decompression using a nasogastric tube
Temperature regulation

Bowel Reduction

Primary closure versus staged silo closure
75–80% achieve a primary closure, with the remainder closed using a preformed silo, staged over 5–7 days

Recovery and Feeding

A variable period of ileus is virtually always present which dictates time to full enteral feeds. Reduction in volume of and lessening of the bilious nature of the nasogastric tube aspirate helps guide a gradual introduction of feed and time to full enteral feeds and thus discharge home varies from 28 to 72 days on average.

Recovery can be complicated by recurrent episodes of central line sepsis, necrotising enterocolitis, TPN-associated cholestasis and issues relating to associated atresias. The presence of bowel atresias in 10–15% of babies with gastroschisis will usually be predictive of a longer period of parenteral nutrition and hospitalisation compared to babies in whom atresias are not present.

However, despite episodic predictable complications during the first year, 95% of babies born with gastroschisis are eating birthday cake at their first birthday!!

Thus in answer to the anxious grandmothers' questions:

1. We cannot say prenatally or even perinatally how long the baby will be hospitalised for, but advise that the average length of stay is 47 days
2. Initially, the baby will be nursed in a nappy and a hat in a closed incubator for careful temperature regulation. It may be 2–3 weeks before typical baby wear is appropriate.
3. There is an unexplained high risk of late foetal demise after 37 weeks gestation in gastroschisis pregnancies, and hence a date for induction is usually offered at or just prior to 37 weeks. However many will deliver early spontaneously with the mean gestation 35.9 weeks [3, 4].

Case 2

A 35-year-old, primigravida, university lecturer and her 38-year-old husband have recently been informed that their baby has had a lumbo-sacral neural tube defect detected, in association with significant bilateral ventriculomegaly, in keeping with a diagnosis of spina bifida and hydrocephalus (Chiari malformation). She is 21 weeks pregnant and both she and her husband are understandably devastated with this news.

They both have a list of questions for the paediatric surgeon/neurosurgeon:

Will he walk or always be in a wheelchair?

Will he always wear nappies?

Will I be able to return to work when my maternity leave finishes in 6 months?

Will he be able to go to university?

Again, we notice that many of the questions that parents ask relate to the social aspects of life and the future plans for their unborn child. We may not be able to accurately answer many of these types of questions but we have evidence and knowledge that can help assuage their anxieties about the future for their baby.

In counselling parents whose baby has a NTD and hydrocephalus, we describe four aspects:

1. Operative closure of the open back lesion and management of hydrocephalus usually with a ventriculo-peritoneal shunt—acute neonatal period
2. Management of the neuropathic bladder and protection of the upper renal tract—long-term surveillance
3. Musculoskeletal and mobility issues—involving paediatric orthopaedic teams
4. Management of the neuropathic bowel—long-term surveillance

So to answer these parents' questions, we can say that:

Will he walk or always be in a wheelchair?

Whether he can walk independently or with aides (calipers, rollators, etc.) depends on the level of the spinal lesion and can be more accurately determined at birth by detailed clinical assessment and muscle charting. Babies with flaccid paralysis and talipes are not likely to walk unaided. The majority of those that can walk either unaided or with orthotics will usually be dependent on a wheelchair by the age of 20 years, despite independent mobility in earlier childhood. This may be due to associated musculoskeletal problems including kyphoscoliosis, hip dislocations and the grim fact that life as an ambulant spina bifida patient is often too difficult without a wheelchair as one reaches adulthood.

Will he always wear nappies?

We have many strategies to manage the neuropathic bladder and bowel to achieve social continence. This will usually involve intermittent catheterisation of some description and complex urological reconstruction usually done in late childhood.

We have similar continence strategies for bowel continence but the majority of babies with spina bifida will not attain the usual bladder and bowel continence as their peers do between 2 and 3 years of age.

Our highly skilled continence nurse specialists will become an important feature in their child's care.

Will I be able to return to work when my maternity leave finishes in 6 months?

When asked this particular question, I struggled to give a concise answer.

Yes, is an obvious answer, however in order to return to her high pressured full-time academic career, this mum and family would need to significant help and services to enable her return to work. This is always possible, but needs planning and financial input, with agreement and cooperation from nursery staff/carers/family to attend the child's daily needs such as intermittent catheterisation.

Will he be able to go to university?

Children with spina bifida/myelomeningocoele (SBM) have a rather different cognitive profile than typically developing children. However, many of the cognitive deficits may be related to hydrocephalus rather than on SBM per se. Some studies have found impairments on all three IQ measures when studying children with SB.

Thus, we cannot say that this unborn baby will not become a high achieving academic like his parents.

Therein lies the difficulty we have in counselling parents whose foetus has a complex anomaly where there are so many factors that we cannot determine antenatally or even immediately postnatally and must wait for time and development to evolve as we help them on their journey [5, 6].

References

1. Winckworth LC, Raj R, Draper L, Leith W. Antenatal counselling: documentation and recall. J Pediatr Child Health. 2013;49:422–3.
2. Skari H, Malt UF, Bjornland K, Egeland T, Bjornstad Ostensen A, Emblem R. Prenatal diagnosis of congenital malformations and parental psychology and distress – a prospective longitudunal cohort study. Prenat Diagn. 2006;26(11):1001–9.
3. Alijaladi A, Mohajerani N, Skarsgard ED. Effect of timing of enteral feeding on outcome in gastroschisis. J Pediatr Surg. 2013;48(5):971–6.
4. South AP, Stutey KM, Meinzem-Derr J. Metaanalysis of the prevalence of intrauterine fetal death in gastroschisis. Am J Obstet Gynecol. 2013;209(2):114.el–13.
5. Wills KW. Neuropsychological functioning in children with spina bifida and/or hydrocephalus. J Clin Child Psychol. 1993;22(2):247–65.
6. Danielsson AJ, Bartonek A, Saraste H. Association between orthopedic findings ambulation and health- related quality of life on children with myelomeningocoele. J Child Orthop. 2008;2(1):45–54.

The Embryological Basis of Behavioural and Psychiatric Conditions

9

Laxmi Kathuria

Introduction

In writing this chapter, I became acutely aware of a mantra forming in my head:

> The developing foetus becomes a person.

With that development comes emotional awareness, cognitive and affective processing and social interaction with others. Having an awareness of the person behind the pathology is a key skill for any human being, including professionals who can get caught up with pathological processes and how to fix them. It is also important to remember how the environment can shape a person and help enhance or destroy quality of life. Psychopathology can indeed debilitate just as much as physical pathology. Clinicians have to consider a holistic approach at all times.

In this chapter, I hope to take the readers through the development of a human being's identity and emotion and then look at how a person copes and deals with the reactions of being different.

The Development of Sense of Self

There has always been a lively debate about whether a human being is born with inherited knowledge or is born as a "tabula rasa" – a blank slate upon which nothing is written. Studies such as Rubia Vila [1] suggest that the infant is born with knowledge inherited through the evolutionary processes and certain aspects of development are innate. One example of inherited knowledge is the recognition of familiar faces, especially that of the mother's.

Language development has been an indicator of the child's self-awareness. First words usually appear in the first year of life. As language acquisition develops, the use of words like "I", "me" and "my" indicates that the child is self-aware.

The mirror/rouge test is used to test self-awareness. A red dot is typically marked on the infant's face. The infant is then placed in front of a mirror. The infant will typically touch the dot on his or her face between the age of 15 and 24 months, indicating that she or he is aware of his or her own existence.

Meltzoff and Moore [2], Rochat and Goubet [3], Rochat and Hespos [4], Rochat [5] and Legrain et al. [6] have all described a developmental path in acquiring self-awareness. The five stages have been summarised within Table 9.1. Table 9.1 describes the human journey from differentiation to self-consciousness.

Table 9.1 The five stages of the developmental path to self-awareness

Stage	Description
Stage 1 – differentiation (from birth)	Right from birth infants are able to differentiate the self from the non-self. A study using the infant rooting reflex found that infants rooted significantly less from self-stimulation, contrary to when the stimulation came from the experimenter
Stage 2 – situation (by 2 months)	In addition to differentiation, infants at this stage can also situate themselves in relation to a model. In one experiment infants were able to imitate tongue orientation from an adult model. Additionally, another sign of the differentiation is when infants bring themselves into contact with objects by reaching for them
Stage 3 – identification (by 2 years)	At this stage the more common definition of "self-awareness" comes into play, as described above where infants can identify themselves in a mirror through the "rouge test" as well as begin to use language to refer to themselves
Stage 4 – permanence	This stage occurs after infancy when children are aware that their sense of self continues to exist across both time and space
Stage 5 – self-consciousness or meta-self-awareness	This also occurs after infancy. This is the final stage when children can see themselves in third person or how they are perceived by others

L. Kathuria (✉)
Stobhill Hospital Glasgow, Glasgow University, Glasgow, UK
e-mail: laxmi.kathuria@nhs.net

R. Carachi, S. H. E. Doss (eds.), *Clinical Embryology*, https://doi.org/10.1007/978-3-319-26158-4_9

Development of a Sense of Others

The infant's first response to others is the social smile. This typically appears between 6 and 8 weeks of life. This smile is used to invite others to interact. As interaction from others is gleaned, the infant will use mimicry to show his or her awareness of the facial expressions of others. A solid and positive sense of others develops when the infant receives reinforcements from people smiling back and joining in.

Many parents have described their child going through "the terrible twos". One theory is that this is a result of the infant trying to separate from the caregiver between 12 and 20 months. Oppositionality is often the first sign of this occurring.

The preschool child will have an understanding of basic human emotions such as happiness, anger, sadness or fear. He or she is able to describe these feelings in self and also recognise that others may be feeling the same. At the age of 5 or 6, the child starts to wonder if others feel these emotions in relation to them. They start to ask the question; "will they like me?"

Piaget conducted his famous Mountain Range Experiment in 1956. The results indicate that empathy develops at approximately 7 years of life. A child is shown a three-dimensional model of three mountains. All three are different sizes with different features. A doll is introduced and is placed in a position so that it is looking at the mountains from a different perspective compared to the child. The child is asked what the doll can see and indicates his or her answer by choosing from a range of pictures. Being able to imagine what the doll can see indicates an understanding of others. Subsequent experiments repeated by Whitehurst and Sonnenschein [7] indicated that empathy develops at age 4 or 5. King expanded on this to develop a framework to further understand the development of empathy [8].

Development of Self-Esteem

Self-esteem develops once a child has a concrete sense of himself/herself and others. It is a value concept that is internalised and projected out in the way we portray ourselves to the world. In summary, it is how we value ourselves, judge our self-worth and place our own standing in society. Some attributes of a positive self-esteem include confidence and ability to handle ambiguity, problem-solve and face things with an overriding sense of optimism. Self-esteem can be situationally dependent and can be changeable across the life span.

Development of self-esteem increases rapidly in early adolescence and then levels out in adulthood. Sieving and Zirbel-Donisch [9] postulated that it is higher in those who are conscientious, extroverted and emotionally stable. A good sense of self-worth develops when the child receives positive reinforcement from others. Erol and Orth [10] stated that predictors for positive self-esteem included a high sense of mastery, low risk-taking and having good physical health. It therefore stands to reason that young people with physical pathology and visible differences may be more vulnerable to having poor self-esteem particularly if their sense of self-worth is tied up with frequent hospital appointments and pain. Low self-esteem may predispose to psychiatric illnesses such as anxiety or depressive disorders.

Development of Body Image [11]

Body image also develops from a sense of self and others. Its relationship to eating disorders has been extensively researched and is not explored in this chapter. Less has been written about development of body image in those who look different or feel different as a result of pain or illness.

Body image is an internal representation of an individual's body shape, weight, size or other features related to physical appearance. The face plays an integral part in the sense of self, identity and body image. Body image consists of a person's perception of their body in external space and cognitive schemas related to self. Body image is dependent on emotional and motivational states.

The perceptual body image is related to one's estimation of body size. Distortion of this can result in repeated checking behaviours. Those with body dysmorphia often have grossly abnormal perceptual body image. Those with disfigurements may report an overestimation of the visible difference and a complete preoccupation with this difference. This will have an impact on emotional wellbeing.

Cognitive body image is related to language and is also called "verbal body image". The development of this is dependent on interaction with others and can become abnormal directly through childhood teasing and bullying. Cognitive schemas can be stable across the lifetime, and therefore negative associations can cause body image to become critically abnormal for most of an individual's life. Common schemas in those with disfigurements are that they are less beautiful, less sexually attractive, less worthy, less likeable or less personable.

The role of the media is not directly causative as there are a number of biological and environmental influences which can make someone particularly predisposed to becoming influenced. Nevertheless, the media is a crucial regulator in child development.

Is Fear of Difference Normal?

Some fear responses are innate and some are learned. Meltzer et al. [12] stated that the concept of the tabula rasa was untrue for fear of snakes and faces. The response to these stimuli is thought to be reflexive and mediated by the amygdala and thalamus. Fear is also related to the age and stage of the child. Responses are only thought to be appropriate if they are proportionate to the intensity of the perceived threat. Beyond this, they are called phobias. Children tend to fear things like animals, monsters and the dark. Older children and adults tend to have fears that are rooted within reality – such as fear of their own mortality or of war. Girls are thought to be more fearful than boys. Socio-demographic and socio-economic factors play a role in the stability of fear for a stimulus, and there is a cultural and familial pattern to learned responses and reactions to fear. Fear of people who look different has not been extensively investigated. Interpretation bias also plays a part in the maintenance of anxiety. If a stimulus is interpreted in a negative way, then there is a greater likelihood that the cognitive processes relating to being fearful become stable across time [13].

An evolutionary hypothesis has been put forward, stating that infants prefer faces that are symmetrical and therefore more "beautiful". Should we therefore be considering a fear of visible difference as evolutionarily protective? If we are scared of those that look different, are we less likely to mate with them and pass on the abnormal genes?

Ohman [14] stated that the fear of snakes and the fear of faces (especially threatening faces) are connected. Both evoke a similar physiological response. Threatening faces have attentional priority within the brain. Behavioural avoidance and facial expressions exhibiting disgust were equivalent when subjects were presented with those who had influenza and those who had a birthmark. There is said to be implicit avoidance even when people know that a birthmark is not contagious. [15]. There is also a greater level of disgust displayed facially with increasing levels of facial disfigurement [16]. So is disgust something human beings cannot help and is it protective because we are somehow threatened by people who look different?

The answer may not be so simplistic. We know that an infant does not need a "normal" face in order to elicit a caregiving response. Neither does the carer need to look "normal" in order to care for the child. Other factors such as tone of voice, smell, proximity, temperament and positive reinforcement are equally important. Human beings have also shown to have higher-order cognitive skills which allow them to manipulate information and learn from experience. Therefore, the predatory defence system may work for avoiding snakes, but it stands to reason that not everyone who has a visible difference is dangerous. We also have the ability to learn not to look disgusted and to think of people with visible difference as being contagious. This can and should be learned through the caregiver response, the media and education. Allowing this innate fear response to flourish may be responsible for the development of prejudice, stereotyping and xenophobia. This in turn causes untold damage to the individuals experiencing this.

Considering Psychopathology

It is important to consider underlying psychiatric disorders when addressing the needs of the individual. Any concerns could be referred to liaison psychiatry services within inpatient settings. Different conditions have different psychiatric sequelae, and this has not been discussed here in detail. Professionals should aim to do a thorough assessment which may include a cognitive assessment. The timing of any assessment may yield different results. For example, an individual may show signs of anxiety and low mood related to an adjustment disorder or an acute stress reaction at the point of diagnosis and may require psychological support to help them deal with the myriad of emotions.

It is well known that some conditions have a higher risk of learning disability and neuropsychiatric disorder. Social communication disorders such as childhood autism and Asperger's syndrome should be considered. Depression and anxiety disorders are more common in those who have learning difficulties, epilepsy and social communication disorders [17].

Bradbury [18] discussed that psychosocial adjustment and long-term psychological functioning can be impaired in those with visible difference. Individuals are faced with the challenges of social reactions and their own psychological responses to looking different. There is no simple linear relation between the degree of disfigurement and the degree of experienced distress. Factors that influence an individual's ability to cope include the social meaning of the disfigurement, life history, social and family support and developmental stage.

Situations that involve meeting new people, managing curiosity and assumptions and transitioning into school or employment may cause distress. Those with avoidant or introverted personalities do less well than those who have extroverted personality traits. Van den Elzen et al. [19] suggested that those who feel internally satisfied with how they look irrespective of their visible difference have more optimal social functioning. A person's subjective appearance is the best predictor of social functioning. Avoiding stress caused by stigmatisation and uncertainty about reactions of others leads to less frequent interpersonal behaviours in adults with facial disfigurement. Masnari et al. [20] also suggested that perceived stigma is also an independent factor for psychological adjustment. If an individual knows that he or she is going to be judged, she or he is more likely to internalise these feelings and has a poorer psychological wellbeing.

Myths of Visible Difference (www.changingfaces.org.uk)

The Myth of Success

> There is no way you can be a success looking like that – a second-rate life is inevitable.

It is easy to believe this myth when you consider that most people publicly associated with success (celebrities, film stars, sporting heroes and models) are considered to be "good-looking" even if their appearance has been air-brushed.

So, it is not surprising that we might believe that people with disfigurements would have little chance of success. This belief is damaging because it can lead to lowered expectations and wasted potential. In reality, many people with disfigurements lead successful lives with interesting careers, fulfilling relationships and a family.

The Myth of Surgery

> It's amazing what surgeons can do these days.

Modern surgical and medical techniques can make some disfigurements less noticeable, but a disfigurement can rarely be removed completely. Simplistic media stories about the latest miraculous breakthrough in reconstructive surgery added to exaggerated advertisements about the brilliance of cosmetic surgery lend credence to this myth.

If you believe it, you will assume that someone who has a condition, mark or scar that affects their appearance is going to get "it" fixed soon – or expect them to do so – possibly adding to the pressure on them to seek surgery to become more "acceptable" and to disappointment when hopes are not realised.

Realistic understanding of surgical and medical treatments and their limits is essential, and it does not help if they are promoted in an exaggerated way.

The Myth of Heroism

> You are so brave – you're a real hero to live with scars like those.

This myth assumes looking different disfigurement is one of the worst things that can happen to someone, and so anyone who deals with it must be exceptionally brave.

This is patronising. The fact is there is very little choice available if you are born with or acquire a disfigurement. It can take courage to face difficult situations, but bravery does not come into it. Those who have come to terms with their scars, marks or unusual features often say that they have gained new insights and life skills.

The Myth of Horror

> I don't like the look of him, he looks frightening!

People are often scared of the unknown or difference. This fact has prompted the use of disfigurement as a device to portray evil in horror films, comic strips, religious imagery and fairy tales. There are many examples in current society that reflect this from Star Wars to the Lion King.

As a result, some people are scared of anyone who has a disfigurement and seem to feel free to resort to ridicule by hurling insults like "Phantom", "Scarface" or "Freddie Krueger" from popular films.

Visible difference has not influence whatsoever on moral character, and people play many important roles in society, a fact that is rarely reflected in the media.

The Myth of Learning Difficulties

> People think that because I have an unusual face, I also have learning difficulties.

People with disfigurements often report that other people talk down to them, speak very slowly or ignore them altogether and talk to the person next to them.

This behaviour seems to reflect an assumption that disfigurement is a visible manifestation of some form of learning difficulties, perhaps because people who have conditions like Down's syndrome, for example, have well-recognised facial features.

A small minority of people with unusual faces may have learning difficulties, but it is a fallacy to generally associate disfigurement with learning difficulties.

The Myth that "Appearances Don't Matter"

> People think that because I look unusual, I'm not interested in my appearance.

It is a myth to assume that appearances don't matter to people with disfigurements and they are not interested in looking good.

Initial judgements are often made in the first few minutes of a meeting, and being well-presented can send out positive messages as well as boosting everyone's self-confidence and self-esteem. People who look different may enjoy wearing make-up, jewellery and fashionable clothes to positively enhance their appearance. They know appearance matters – in the first few minutes at least.

Strategies for Individuals and Families (www.changingfaces.org.uk)

The charity Changing Faces has been pioneering work that supports individuals and their families in dealing with visible difference. The remaining parts of this chapter encompass their philosophy and strategies in helping parents, children, adults and families deal with their own situation.

Parents/caregivers may experience a variety of emotions such as loss, grief, joy, guilt, sadness, love, confusion, shame, anger, fear and protectiveness. These emotions are all common and may be experienced together and at different times.

Advice for caregivers includes making a list of the professionals involved, thinking about a script to say to friends/family/strangers and finding out about support groups that are available. It may be difficult to talk about things at first, but understanding the reactions of others may be protective.

It is important to help a child build on his or her self-awareness, self-esteem and body image (the development of which has been detailed above). It is important for caregivers to praise their child, build their self-esteem and reassure them that nothing is insurmountable whilst explaining to them the realities of what might happen when interacting with others. The use of social stories, visual cue cards and relaxation techniques may prepare a child for a situation by knowing what to expect and allowing them to practice responses in advance.

The acronym SCARED (www.changingfaces.org.uk) has been summarised within Table 9.2. It was developed to describe how people might think and how this might make them act.

Changing Faces has described five helpful skills to deal with a disfigurement when thinking about self and others. Table 9.3 summarises these skills and how to use them as directed towards self and others. Changing Faces also advises individuals to come up with a motto that they can repeat to themselves to make them feel more confident.

Table 9.2 SCARED and what this means (www.changingfaces.org.uk)

If someone feels…	→	They might…
*S*orry or *s*hocked	*S*	*S*tare or be speechless
*C*urious or *c*onfused	*C*	Be *c*lumsy
*A*nxious	*A*	*A*sk questions or act *a*wkward
*R*epelled (put off)	*R*	Be rude or *r*un away
*E*mbarrassed	*E*	Act *e*vasive – ignore you
*D*istressed	*D*	Be distressed or worried

Table 9.3 The five helpful skills (www.changingfaces.org.uk)

Explain-Reassure-Distract-Assert-Humour		
	To yourself	To others
Explain	Keep explaining to yourself why things are the way they are or why they have happened	Explain your condition to others to help them understand what has happened
Reassure	Reassure yourself to make sure you are OK	Reassure others by giving them more information
Distract	Distract yourself in a difficult situation by thinking about or doing something else	Distract others by talking about something else
Assert	Assert yourself to show you are in control	By asserting yourself the other person may be surprised or embarrassed
Humour	If you can use your sense of humour, do so to lighten the situation or to stand up for yourself	Using humour may make the other person laugh, respond to the situation or feel embarrassed

The Correct Use of Language (www.changingfaces.org.uk)

It's not just our assumptions about "disfigurement" that we need to look at again. The language we use can also have an undermining effect – and it often articulates our underlying attitude. Indeed disfigurement itself is not a term that many individuals and their families feel comfortable with. The charity encourages everyone to find the best term for them such as looking different, visible difference and the name of the condition. Here are a few commonly used phrases that may not be helpful.

"It's the inside that counts"

This is one of the most popular and well-meant expressions, but it is actually incredibly discounting of a person's face or appearance which after all is a major part of who they are and counts very much!

"Burn victim," "suffering from a cleft lip and palate"

Phrases about being victims or suffering from their condition are unhelpful. "Victim" and "suffering" may be words that individuals might choose to apply to themselves, but it is inappropriate to assume that they are applicable to everyone. Many people who look different are not in pain from their scars or marks or unusual features – and they prefer not to be pitied by other people. "Burn survivor" and the simple "has a cleft lip and palate" are far more accurate and less judgemental. Table 9.4 encapsulates some "do's and don'ts" regarding language related to difference. It also offers some reasoning behind why the use of language is important in positively reframing and empowering individuals and society.

Table 9.4 Some helpful tips on do's and don'ts when describing those with difference (www.changingfaces.org.uk)

Don't use	Do use	Because
Facial deformity Facial abnormality Facial defect	Facial disfigurement	Disfigurement is more positive and has a less of a sense of there being something "medically wrong" with the whole person
Disfigured people Scarred people	People who have disfigurement People living with a disfigurement/ person with a disfigurement	This respects the person first rather than labelling them by their disfigurement
Victim of… Suffering from…	Survivor He/she has…	The words in the second column are more empowering and factual
Terribly scarred Horribly disfigured Grotesque Ugly	She was scarred She was disfigured She has a visible difference	The use of adverbs liked those used in the first column describes disfigurement/scars in a negative way, whereas the statements in the second column are more factual

Case Study

X is 29 years old who lives with his mum and his twin. They both have neurofibromatosis type 1. X has extensive facial disfigurement due to tumour growths on his face. His brother does not look visibly different. X's brother has some cognitive difficulties.

The facial tumours started to manifest when X was aged 5. Since the age of 7, he has had 30 major operations and will require a lifetime of essential (not cosmetic) surgery. As a result of the tumours, he is blind in one eye and partially deaf. His life expectancy is normal and he does not have cognitive impairment.

X recalls extensive bullying at school and beyond. People have always tended to treat him as if he is less intelligent, and he has been called "spastic", "elephant man", "fat face" and "big head". In secondary school, a supposed friend told him that a teacher wished to see him. When X arrived at the location, 12 peers were waiting for him. They spat at him repeatedly. His mother recalls that his blazer was soaking wet from the assault.

In adulthood, X has become used to people asking him what is wrong, treating him in an infantile way or totally avoiding him. On a night out, somebody was convinced that he was wearing a mask and began tugging at his face. When they realised the truth, they walked away but without apology.

X's mother, friends and family and Changing Faces have been crucial in his positive development. His extroverted personality and self-confidence have also played a part. His mother has encouraged him to embrace his individual difference. Despite having an identical twin, X's and the family's attitude was not to dwell on how X may have looked but to focus on the individual he was becoming. X states "it is important to live the life you have, not mourn the one you don't". X has gone to university and has secured a job in London working with people.

X is a living testament for what is possible. He also serves as a good example to show the importance of personality characteristics, social support and positive reinforcement in helping "normalise" those who have so often been alienated.

Acknowledgements A special thanks to Dr. Helen Minnis and Professor Carachi for their valued input and also to Henrietta Spalding at Changing Faces – an organisation to be proud of.

References

1. Rubia Vila FJ. The myth of "tabula rasa". An R Acad Nac Med. 2000;117(3):543–51.
2. Meltzoff AN, Moore MK. Early imitations within a functional framework: the importance of person identity, movement and development. Infant Behav Dev. 1992;15(4):479–505.
3. Rochat P, Goubet N. Development of sitting and reaching in 5-6 month old infants. Infant Behav Dev. 1995;18:53–68.
4. Rochat P, Hespos SJ. Differential rooting response by neonates: evidence for an early sense of self. Early Dev Parent. 1997;6(2):150.1–8.
5. Rochat P. Five levels of self-awareness as the unfold early in life. Conscious Cogn. 2003;12:717–31.
6. Legrain L, Cleermans A, Destrebecqz A. Distinguishing three levels in explicit self-awareness. Conscious Cogn. 2011;20(3):578–85.
7. Whitehurst GJ, Sonneschein S. The development of communication: attribute variation leads to contrast failure. Journal of Experimental Child Psychology. 1979;25(3):490–504.
8. Alligood MR, Evans GW, Wilt DL. Kings interacting systems and empathy. In: Frey MA, Sieloff CL, editors. Advancing king's systems framework and theory of nursing. Thousand Oaks, CA: Sage Publications, Inc.; 1996. p. 66–78.
9. Sieving RE, Zirbel-Donisch ST. Development and enhancement of self esteem in children. J Paediatr Health Care. 1990;4(6):290–6.
10. Erol RY, Orth U. Self esteem development from age 14 to 30 years: a longitudinal study. J Pers Soc Psychol. 2011;101(3):607–19.
11. Cash TF. Women's body images. In: Wingood GM, Di Clemnte RJ, editors. Handbook of women's sexual and reproductive health. New York, NY: Kluwer Academic/Plenum Publishers; 2002. p. 175–94.
12. Meltzer H, Vostanis P, Dogra N, Dool L, Ford T, Goodman R. Children's specific fears. Child Care Health Dev. 2009;35(6):781–9.
13. Field ZOE. How trait anxiety, interpretation bias and memory affect acquired fear in children learning about new animals. Emotion. 2013;13(3):409–23.
14. Ohman A. Of snakes and faces: an evolutionary perspective on the psychology of fear. Scand J Psychol. 2009;50(6):543–52.
15. Ryan S, Oaten M, Stevenson R, Case T. Facial disfigurement is treated like an infectious disease. Evol Hum Behav. 2012;33(6):639–46.
16. Shanamugarajah K, Gaind S, Clarke A, Butler PE. The role of disgust emotions in the observer response to facial disfigurement. Body Image. 2012;9(4):455–61.

17. Versnel S, Plomp R, Passchier J, Duivenvoorden H, Mathijssen I. Long-term psychological functioning of adults with severe congenital facial disfigurement. Plast Reconstr Surg. 2012;129(1):110–7.
18. Bradbury E. Meeting the psychological needs of patients with facial disfigurement. Br J Oral Maxillofac Surg. 2012;50(3):193–6.
19. van den Elzen M, Versnel S, Hovius S, Passchier J, Duivenvoorden H, Mathijssen I. Adults with congenital or acquired facial disfigurement: impact of appearance on social functioning. J Craniomaxillofac Surg. 2012;40:777–82.
20. Masnari O, Schiestl C, Landolt M, et al. Stigmatization predicts psychological adjustment and quality of life in children and adolescents with a facial difference. J Pediatr Psychol. 2013;38(2):162–72.

Bibliography

Prior J, O'Dell L. "Coping quite well with a few difficult bits": living with disfigurement in early adolescence. J Health Psychol. 2009;14(6):731–40.

Websites

www.changingfaces.org.uk

10 Animal Models

Piotr Hajduk

Introduction

Paediatric surgeons continue to face the challenge of treating a variety of congenital anomalies and neonatal disorders for which the aetiology and pathophysiology are poorly understood. In addition to the fact that many of these defects are rare in overall incidence and lack a known genetic cause, there has been a general dearth of biological or animal models to specifically study the conditions that result from developmental failures. The desire to comprehend the basic developmental errors from which congenital malformations arise drives clinicians and developmental biologist to seek ways to examine these conditions in an embryological context. The establishment of animal models of congenital anomalies is a valid means to circumvent the lack of human research materials [1]. Animal models in paediatric surgery allowed us to study the aetiology and pathogenesis of complex congenital malformations, allowing the developmental basis of the birth defects to be comprehended, and have led to major advances in the surgical and therapeutic management of these conditions.

By identifying or including pathology of interest in a small mammal with multiple progeny per breeding cycle, short gestation period and low maintenance costs, a feasible vehicle for research purposes can be established. However it is now necessary to understand not only comparative mammalian, in particular murine, development but also the development of *Xenopus*, *chick*, *zebra fish* and *Drosophila* in order to appreciate recent findings. This is because animals share homologous and developmental mechanisms. Currently, mice are the developmental biologists' mammal of choice, providing greater availability of molecular tools and techniques and transferable knowledge from mutant mice, and therefore become the most important and common animal models for human disease. Rats are also widely use in research, as they are particularly suitable for physiological, pharmacological and behavioural research. They also contributed greatly to research involving surgical techniques and the study of many teratogens. Although vast majority of animals used for research purposes today are rodents, in the past, the chicken was an important surgical model to study embryological processes. It has been widely used especially in the field of epithelial/mesenchymal interactions [2]. Paediatric surgeon used chicks model to study morphological processes involved in intestinal atresia [3, 4], gastroschisis [5], omphalocele [6] and Hirschsprung's disease [7].

Animals model can be divided into spontaneous and induced based on how they were generated or how the phenotypes developed. Obviously, the animals in which disease occur naturally are ideal models to study disease pathogenesis, as there is no interference to the animal prior to the study. There are three main types of induced animal models: surgically created, teratogen induced and transgenic animals.

Surgical Models

Most surgical generated experimental models were mainly used in order to demonstrate the feasibility of possible treatment strategies including foetal intervention. One of the hypothesis concerning the pathogenesis of pulmonary hypertension in congenital diaphragmatic hernia (CDH) is that results from intrathoracic herniation of the abdominal viscera, compromising pulmonary development. Based on this idea, surgical animal models were created to study both lung pathogenesis and treatment options. The sheep model was introduced by Delorimier and Parker in 1967 [8]. The hernia is surgically created during early stages of foetal development. The abdominal bowel is positioned into the chest to mimic human CDH. However, foetal lamb is the most commonly used experimental model, but some researchers have also used rabbits for similar study [9, 10]. Series of interventions including prenatal administration of corticosteroids, in

P. Hajduk (✉)
Our Lady's Children's Hospital, Crumlin, Dublin, Ireland
e-mail: hajdukp@tcd.ie

R. Carachi, S. H. E. Doss (eds.), *Clinical Embryology*, https://doi.org/10.1007/978-3-319-26158-4_10

utero repair of the diaphragmatic defect and foetal tracheal occlusion were investigated. Based on the findings in animal model and further experiments on nonhuman primates, Harrison et al. [11] performed the first human surgical repair in utero. Following report showing that hypoplastic lungs in the surgically induced lamb model of diaphragmatic hernia could be accelerated to grow by occluding the foetal trachea [12], a decade later, Harrison et al. reported the first randomized controlled trial of open hysterectomy-guided foetal endoscopic tracheal occlusion in human [13]. Nowadays, fetoscopic endoluminal tracheal occlusion (FETO) is being subjected to a new randomized clinical trial [14], and the foetal lamb model is widely used for the refinement of fetoscopic and open foetal surgical technique for trainees.

Oesophageal atresia (OA) has been studied using surgical model as far back as the 1970s when Oh and Jacobson [15] described microsurgical technique for oesophageal reconstruction in puppies. Canine model was widely used to investigate the post-operative results following oesophageal resection or transection with subsequent re-anastomosis utilizing an autologous jejunal mucosa transplant [16], a tubular musculopleural pedicle graft [17] or a variety of myotomies with or without delayed oesophageal reconstruction [18, 19]. Further studies in canine model compared manometric findings following oesophageal transection and anastomosis versus oesophageal vagotomy alone [20]. The findings suggested that post-operative dysmotility might arise from disruption of the vagus nerve as a part of congenital abnormality or secondary to surgical intervention. Canine puppies were also used as a model to investigate effect of indwelling silicone transanastomotic tube (TAT) on the healing oesophageal anastomosis. However there was no significant differences in stenosis rate whether a TAT was used or not; study shows a shelf of stenotic tissue on the posterior wall of the oesophagus at the site where the intraluminal silk knots were placed, compared with a thin linear scar on the anterior wall where knots had been tied extraluminally [21]. Porcine (piglets) model of OA have been explored by paediatric surgeons because of animal size and anatomical similarity to human neonates. Studies involved the evaluation of surgical technique for oesophageal replacement comparing gastric tube interposition with gastric tube in continuity and gastric transposition [22].

Surgically created model of Hirschsprung's disease has used the chick embryos where aganglionosis was induced by surgical ablation of the premigratory neural crest [23]. The model was used to investigate possible treatment strategies and show successful colonization aganglionic bowel with neural crest cells transplanted from dorsal neural tube which formed enteric ganglia after incorporation into bowel wall [24, 25].

The aetiology of intestinal damage in infants with gastroschisis is uncertain and is probably multifactorial. Studies in chick embryos, foetal mice, rats, rabbits and lambs have demonstrated similar changes in the affected intestine to those seen in human neonates with gastroschisis. These studies have suggested that both exposure to amniotic fluids and possible constriction of the bowel at the level the abdominal wall defect play a role in the changes affecting the bowel [26].

The aetiology of small intestinal atresia as a consequence of an intrauterine mesenteric vascular accident was investigated by Louw and Barnard [27]. They produced the entire spectrum of neonatal intestinal obstruction, ranging from stenosis, diaphragmatic webs, atretic bands, and segmental absence of intestine and mesentery by ligating the mesenteric vasculature of foetal puppies. Observed severity of the intestinal atresia was greater, where more proximal mesenteric vasculature was compromised. Similar observations have been made by Abrams [28], who additionally concluded that extent of dilatation and muscular hypertrophy of proximal obstructed segment depends upon the time of development.

In summary, surgical animal models are useful in investigating interventional therapies but are less instructive in studying the aetiology and pathogenesis of congenital malformations in human.

Teratogenic Models

Exogenous agents (chemicals, infectious agents or physical condition) responsible for disruption of morphogenesis and subsequent function are called teratogens. The embryo is most susceptible to teratogenic agents during periods of rapid cell division and differentiation. During the first trimester, organ morphogenesis predominates, while later trimesters are devoted to organ growth and maturation. Therefore sensitivity to teratogens is held to peak during the first trimester. Based on this knowledge, the same potent teratogens are used to induce congenital anomalies in animals to study aetiology and pathogenesis of human congenital malformations (Table 10.1).

Table 10.1 Example of teratogen animal models developed to investigate congenital anomalies

Congenital anomaly	Teratogen	Animal	Reference
OA/TOF	Adriamycin	Mice, rat	[29, 30]
CDH	Nitrofen	Rat, mice	[31, 32]
Gastroschisis	Cadmium	Chick	[33]
Hirschsprung's disease	Benzalkonium chloride	Rat	[34]
Anorectal malformations	Retinoic acid	Rat, mice	[35, 36]
	Ethylene thiourea	Rat	[37]
Cloacal exstrophy	Suramin	Chick	[38]
Neural tube defects	Valproic acid	Mice	[39]
Cryptorchidism	Flutamide	Rat	[40]

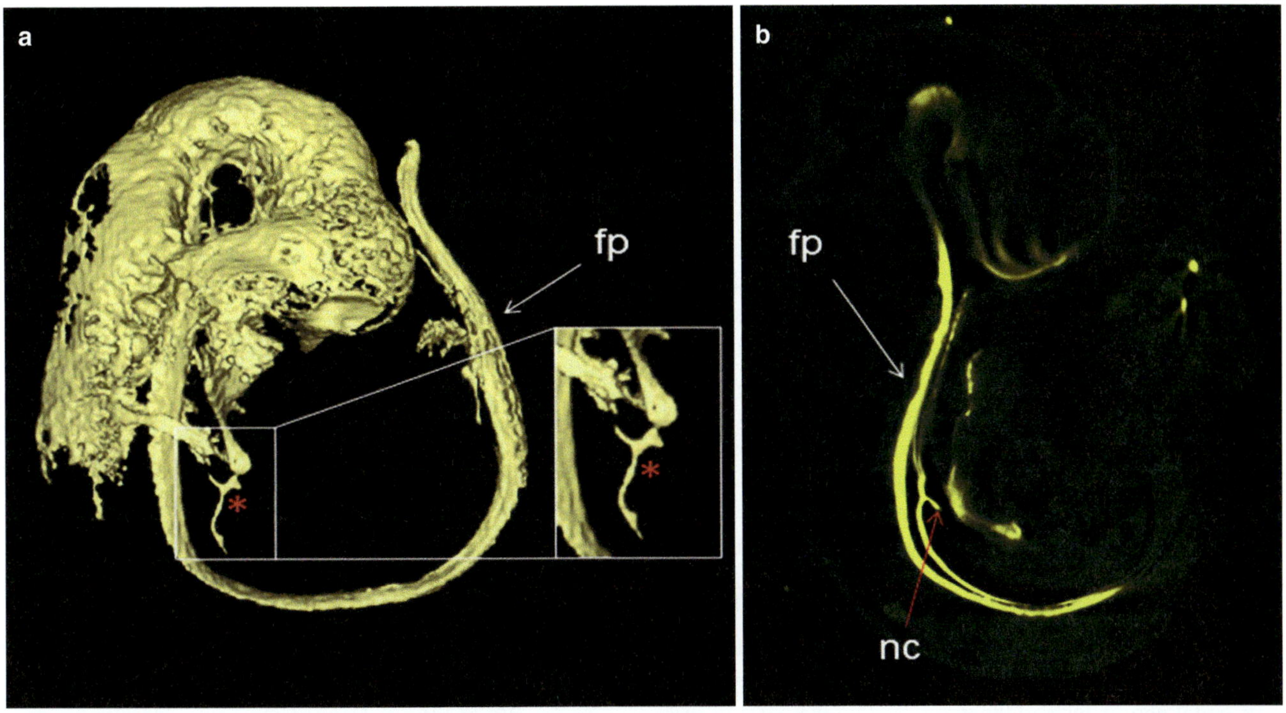

Fig. 10.1 Adriamycin mouse-treated embryos show association between notochord branching (red stars) and foregut atresia (**a**) and altered sonic hedgehog (Shh) expression in the abnormal branched notochord (**b**). *fp* floor plate, *nc* notochord [41]

The use of the teratogens causes large-scale mutation; therefore these types of models are particularly useful for identifying new genes and pathways that contribute to disease (Fig. 10.1). The challenge is to separate out which are the important signals.

Animal models that use virus infections to produce malformations are exceptional. Murine model of extrahepatic biliary atresia (EHBA) uses newborn BALB/c mice infected with rhesus rotavirus group A [42]. As a result the full spectrum of EHBA develops, as seen in the newborn with this disease. However, this model is not a model to mimic failed embryology, but it highlights the possibility that the same malformations are caused by foetal or even postnatal catastrophes.

Genetic Models

As opposed to the use of teratogens, transgenesis is a direct approach that attempts to make specific mutations in animal model genomes. Both knock-out and knock-in models are ways to target a mutation to a specific gene locus. Knock-out mice carry a gene that has been inactivated, which creates less expression and loss of function; knock-in mice are produced by inserting a transgene into an exact location where it is overexpressed. In recent years the accumulation of data arising from genetically modified mouse studies is gradually building up the picture of the key events during normal morphogenesis. Experimental embryology and molecular biology studies have revealed that there are complex spatially and temporally coordinated signalling systems, mediated by transcription factors and secreted proteins, that control the proper stereotypic morphological patterning and differentiation of the organ development. Moreover, a number of mouse mutants replicate similar phenotypes of human diseases lending some valuable information about the possible mechanisms that might be disturbed in those conditions. Deregulation of certain signalling pathways results in reproducing certain congenital anomalies. In the foregut, for example, targeted deletion of Shh gene in homozygous Shh-null mutant mice causes oesophageal atresia, tracheoesophageal fistula and tracheal/lung anomalies [43]. In the hindgut, the deletion of Shh caused the formation of cloacas, while Gli2 gene mutant mice demonstrated the anorectal malformations [44]. Ret-null mutant mice exhibit total intestinal aganglionosis and renal agenesis [45]. Mutation of the Ret gene has been demonstrated to be a major gene causing Hirschsprung's disease in human. In this way some of the key candidate gene implicated in pathogenesis of congenital anomalies can be identified and selected for further functional study as it was successfully used on lung explants to determinate the basis for lung branching morphogenesis [46] or foregut culture in liquid medium to investigate the effect of signalling manipulation on oesophageal development [47].

Future Study

In recent years the accumulation of data arising from animal studies is gradually building up a picture of key events that might lead to congenital anomalies in children; however we appear to have only scratched the surface in terms of understanding the aetiology of those anomalies in human. It is likely that the enormous advances in our understanding of the human genome will allow us to make best use in future of the information generated from the study of mouse genetics. The identification of genes associated with congenital malformations offers the prospect of improvements in genetic counselling for those disorders. For example, the sequencing of genes in the cohort of patients and looking for mutations in candidate genes can be carried out on a large scale on whole-genome amplified DNA. The goal of these studies would be to complete the 'cytogenetic map' for each congenital anomaly. Genes and/or proteins identified to play a role in aetiology of anomalies could be then used as a marker for prenatal genetic testing. DNA chips could be used for high-risk pregnancies, such as those with familial history of congenital anomalies and those selected on the basis of prenatal ultrasound examination. This could provide important information about short- and long-term prognosis, thereby aiding in parental counselling and preparation of medical care. The databases that store information on all patients with congenital anomalies are pivotal in research of aetiology of these malformations. In Europe, a number of birth defects registries have joined forces in EUROCAT, the European network of population-based registers for epidemiological surveillance of congenital anomalies (http://www.eurocat-network.eu/). The registries provide valuable information on trends in frequency of occurrence of birth defects and patterns of anomalies seen to facilitate early warning of teratogenic exposures. This international collaboration is especially valuable for rare exposure or rare anomalies, making it possible to combine data and draw more firm conclusions faster. The pooling of the data enables research questions to be answered on a large scale, making results more stable. A similar idea, expanding the data by collection of DNA from biological samples (blood, buccal swabs, urine, etc.) from the children with congenital anomalies and possibly their parents, could provide a DNA bank combined with detailed patient characteristics. This offers the big advantage for further genetic and epidemiological study of these malformations. For example, studies on gene-environment interactions may lead to more targeted prevention strategies for groups or even individuals who are at higher risk because of the genotype. With better aetiological knowledge, we could recommend preventive measures, such as the advice to use folic acid in the periconceptional period as a means to lower the risk of neural tube defects in offspring.

References

1. Rosenthal N, Brown S. The mouse ascending: perspectives for human-disease models. Nat Cell Biol. 2007;9(9):993–9. https://doi.org/10.1038/ncb437.
2. Roberts DJ, Smith DM, Goff DJ, Tabin CJ. Epithelial-mesenchymal signaling during the regionalization of the chick gut. Development. 1998;125(15):2791–801.
3. Schoenberg RA, Kluth D. Experimental small bowel obstruction in chick embryos: effects on the developing enteric nervous system. J Pediatr Surg. 2002;37(5):735–40.
4. Tibboel D, van Nie CJ, Molenaar JC. The effects of temporary general hypoxia and local ischemia on the development of the intestines: an experimental study. J Pediatr Surg. 1980;15(1):57–62.
5. Aktuĝ T, Hoşgör M, Akgür FM, Olguner M, Kargi A, Tibboel D. End-results of experimental gastroschisis created by abdominal wall versus umbilical cord defect. Pediatr Surg Int. 1997;12(8):583–6.
6. Thompson JM, Bannigan JG. Omphalocele induction in the chick embryo by administration of cadmium. J Pediatr Surg. 2012;42(10):1703–9.
7. Meijers JH, van der Sanden MP, Tibboel D, van der Kamp AW, Luider TM, Molenaar JC. Colonization characteristics of enteric neural crest cells: embryological aspects of Hirschsprung's disease. J Pediatr Surg. 1992;27(7):811–4.
8. Delorimier AA, Tierney TD, Parker HR. Hypoplastic lungs in fetal lambs with surgically produced congenital diaphragmatic hernia. Surgery. 1967;62:12–7.
9. Flemmer AW, Jani JC, Bergmann F, Muensterer OJ, Gallot D, Hajek K, et al. Lung tissue mechanics predict lung hypoplasia in a rabbit model for congenital diaphragmatic hernia. Pediatr Pulmonol. 2007;42(6):505–12.
10. Wu J, Yamamoto H, Gratacos E, Ge X, Verbeken E, Sueishi K, et al. Lung development following diaphragmatic hernia in the fetal rabbit. Hum Reprod. 2000;15(12):2483–8.
11. Harrison MR, Adzick NS, Longaker MT, Goldberg JD, Rosen MA, Filly RA, et al. Successful repair in utero of a fetal diaphragmatic hernia after removal of herniated viscera from the left thorax. N Engl J Med. 1990;322(22):1582–4.
12. Wilson JM, DiFiore JW, Peters CA. Experimental fetal tracheal ligation prevents the pulmonary hypoplasia associated with fetal nephrectomy: possible application for congenital diaphragmatic hernia. J Pediatr Surg. 1993;28(11):1433–40.
13. Harrison MR, Keller RL, Hawgood SB, Kitterman JA, Sandberg PL, Farmer DL, et al. A randomized trial of fetal endoscopic tracheal occlusion for severe fetal congenital diaphragmatic hernia. N Engl J Med. 2003;349(20):1916–24.
14. Losty PD. Congenital diaphragmatic hernia: where and what is the evidence? Semin Pediatr Surg. 2014;23(5):278–82.
15. Oh C, Jacobson JH. Ii. Microsurgical technique for esophageal reconstruction in growing puppies. Arch Surg. 1972;104(2):203–5.
16. Halsband H. Esophagus replacement by free, autologous jejunal mucosa transplantation in long-gap esophageal atresia. Prog Pediatr Surg. 1986;19:22–36.
17. Moutsouris C, Barouchas G, Karayannacos P, Dontas I, Salakos C, Skalkeas G. Tubular musculopleural pedicle grafting of esophageal long gaps in dogs. J Pediatr Surg. 1987;22:117–9.
18. Rosselló PJ, Lebrón H, Franco AA. The technique of myotomy in esophageal reconstruction: an experimental study. J Pediatr Surg. 1980;15:430–2.
19. Shoshany G, Kimura K, Jaume J, Sterman H, Birnbaum E, Stein T, Levine J. A staged approach to long gap esophageal atresia employing a spiral myotomy and delayed reconstruction of the esophagus: an experimental study. J Pediatr Surg. 1988;23:1218–21.
20. Shono T, Suita S. Motility studies of the esophagus in a case of esophageal atresia before primary anastomosis and in experimental models. Eur J Pediatr Surg. 2008;7(03):138–42.

21. Carachi R, Stokes KB, Brown TC, Kent M. Esophageal anastomosis – an experimental model to study the anastomotic lumen and the influence of a transanastomotic tube. J Pediatr Surg. 1984;19:90–3.
22. Samuel M, Burge DM, Moore IE. Gastric tube graft interposition as an oesophageal substitute. ANZ J Surg. 2001;71(1):56–61.
23. Meijers JH, Tibboel D, van der Kamp AW, van Haperen-Heuts IC, Molenaar JC. A model for aganglionosis in the chicken embryo. J Pediatr Surg. 1989;6:557–661.
24. Rothman TP, Le Douarin NM, Fontaine-Perus JC, Gershon MD. Developmental potential of neural crest-derived cells migrating from segments of developing quail bowel back-grafted into younger chick host embryos. Development. 1990;109(2):411–23.
25. Thiery JP, Duband JL, Delouvée A. Pathways and mechanisms of avian trunk neural crest cell migration and localization. Dev Biol. 1982;93(2):324–43.
26. Mortell A, Montedonico S, Puri P. Animal models in pediatric surgery. Pediatr Surg Int. 2006;22(2):111–28.
27. Louw JH, Barnard CN. Congenital intestinal atresia; observation on its origin. Lancet. 1955;19(296):1065–7.
28. Abrams J. Experimental intestinal atresia. Surgery. 1968;64(1):185–91.
29. Thompson DJ, Molello JA, Strebing RJ, Dyke IL. Teratogenicity of adriamycin and daunomycin in the rat and rabbit. Teratology. 1978;17(2):151–7.
30. Diez-Pardo JA, Baoquan Q, Navarro C, Tovar JA. A new rodent experimental model of esophageal atresia and tracheoesophageal fistula: preliminary report. J Pediatr Surg. 1996;31(4):498–502.
31. Tenbrinck R, Tibboel D, Gaillard JL, Kluth D, Bos AP, Lachmann B, Molenaar JC. Experimantal induced congenital diaphragmatic hernia in rat. J Pediatr Surg. 1990;25(4):426–9.
32. Wickman DS, Siebert JR, Benjamin DR. Nitrofen-induced congenital diaphragmatic defects in CD1 mice. Teratology. 1993;47(2):119–25.
33. Thompson JM, Bannigan JG. Omphalocele induction in the chick embryo by administration of cadmium. J Pediatr Surg. 2007;42(10):1703–9.
34. Sato A, Yamamoto M, Imamura K, Kashiki Y, Kunieda T, Saakata K. Pathophysiology of aganglionic colon and anorectum: an experimental study on aganglinosis produced by a new method in the rat. J Pediatr Surg. 1978;13(4):399–435.
35. Bitoh Y, Shimotake T, Sasaki Y, Iwai N. Development of the pelvic floor muscles of murine embryos with anorectal malformations. J Pediatr Surg. 2002;37(2):224–7.
36. Hashimoto R, Nagaya M, Ishiguro Y, Inouye M, Aoyama H, Futaki S, et al. Relationship of the fistulas to the rectum and genitourinary tract in mouse fetuses with high anorectal malformations induced by all-trans retinoic acid. Pediatr Surg Int. 2002;18(8):723–7.
37. Arana J, Villanueva A, Guarch R, Aldazabal P, Barriola M. Anorectal atresia. An experimental model in the rat. Eur J Pediatr Surg. 2002;11(03):192–5.
38. Männer J, Kluth D. A chicken model to study the embryology of cloacal exstrophy. J Pediatr Surg. 2003;38(5):678–81.
39. Ehlers K, Stürje H, Merker HJ, Nau H. Valproic acid-induced spina bifida: a mouse model. Teratology. 1992;45(2):145–54.
40. van der Schoot P. Disturbed testicular descent in the rat after prenatal exposure to the antiandrogen flutamide. J Reprod Fertil. 1992;96(2):483–96.
41. Hajduk P, Sato H, Puri P, Murphy P. Abnormal notochord branching is associated with foregut malformations in the adriamycin treated mouse model. PLoS One. 2011;6(11):e27635.
42. Petersen C, Biermanns D, Kuske M, Schäkel K, Meyer-Junghänel L, Mildenberger H. New aspects in a murine model for extrahepatic biliary atresia. J Pediatr Surg. 1997;32(8):1190–5.
43. Litingtung Y, Lei L, Westphal H, Chiang C. Sonic hedgehog is essential to foregut development. Nat Genet. 1998;20(1):58–61.
44. Motoyama J, Liu J, Mo R, Ding Q, Post M, C-c H. Essential function of Gli2 and Gli3 in the formation of lung, trachea and oesophagus. Nat Genet. 1998;20(1):54–7.
45. Schuchardt A, D'Agati V, Larsson-Blomberg L, Costantini F, Pachnis V. Defects in the kidney and enteric nervous system of mice lacking the tyrosine kinase receptor Ret. Nature. 1994;367(6461):380–3. https://doi.org/10.1038/367380a0.
46. Weaver M, Batts L, Hogan BLM. Tissue interactions pattern the mesenchyme of the embryonic mouse lung. Dev Biol. 2003;258(1):169–84.
47. Que J, Okubo T, Goldenring JR, Nam K-T, Kurotani R, Morrisey EE, et al. Multiple dose-dependent roles for Sox2 in the patterning and differentiation of anterior foregut endoderm. Development. 2007;134(13):2521–31.

Embryology Education for Nurses and Midwives

11

Gian Battista Parigi and Gloria Pelizzo

The aim of this chapter is to provide nurses and midwives assisting at the birth of an infant with congenital malformation some practical knowledge on how to handle the event.

The widespread dissemination of prenatal diagnosis (see Chap. 7)—based on foetal imaging (ultrasound, 3D ultrasound, echocardiography, MRI), biochemical screening, foetal sampling and genetic diagnosis—has allowed to diagnose promptly the vast majority of birth defects, so as to make very unlikely their fortuitous discovery at delivery.

On the other hand, in developing countries or even in Western countries in particular social settings today that are increasingly expanding (nomads, marginalised, immigrants), it still can happen that a pregnant woman does not undergo any control during pregnancy, so to have an unexpected congenital malformation presenting at birth.

This chapter will therefore describe the situation in which a prenatal diagnosis of malformation has already been formulated, the one in which the malformation represents an intra partum "surprise", and will be then detailed some practical measures useful to tackle such a demanding situation in the delivery room and immediately after.

Malformation Diagnosed Before Birth

A prenatal diagnosis of congenital anomaly allows not only to properly plan the delivery but also to duly inform and prepare the parents to the demanding task of facing such a situation, involving them also in the difficult decisions about the planning of the delivery itself.

For some specific malformations—such as myelomeningocele or congenital diaphragmatic hernia—there is also the possibility of an early correction through *foetal surgery*, a technique able to modify the natural history of the malformation preventing or reducing organ or apparatus damage and thus positively interfering with the natural evolution of the pathology. Detailed presentation of foetal surgery indications and techniques goes far beyond the scope of this chapter.

Planning of the delivery involves the decision on where, when and how the birth should best take place to secure to the newborn competent and comprehensive care and to avoid unnecessary and potentially dangerous postnatal transports. This planning must come from an interdisciplinary effort involving prenatal medicine experts, obstetricians, neonatologists, paediatric surgeons, paediatric cardiologists or neurosurgeons, according to the malformation involved. Although not immediately involved in this decisional process, nurses and midwives must know the basics underlying the decision, also in order to be able to interact knowledgeably with the parents.

Where?

In case of known foetal malformation, an *in utero transfer* must be considered, that is, moving the mother before the delivery rather than transferring the baby after birth. In this way the delivery will be carried out in a centre appropriate to deal with the expected malformation, i.e. a tertiary centre equipped with all needed, high-level facilities.

When?

As mentioned above about foetal surgery, every malformation has its own natural history: some of them can actually undergo a progressive worsening in utero, hence the question whether to prevent this phenomenon by an induced premature delivery. The scientific evidence for the optimal time of birth in foetal malformations is very low; there are hardly any prospective studies that have examined the advantages and disadvantages of induced premature birth or caesarean section.

The most studied medical benefits of an early delivery are those related to a premature birth in gastroschisis and in

G. B. Parigi (✉) · G. Pelizzo
Paediatric Surgery, University of Pavia, Pavia, Italy

IRCCS Policlinico "S.Matteo", Pavia, Italy
e-mail: gbparigi@unipv.it

R. Carachi, S. H. E. Doss (eds.), *Clinical Embryology*, https://doi.org/10.1007/978-3-319-26158-4_11

myelomeningocele: in both cases, prolonged exposition to amniotic fluid causes a progressive damage of intestinal loops or meningeal sac, hence the opportunity of an early delivery.

In other foetal malformations, no clear benefit for the child from an early delivery has been so far demonstrated: also for this reason, the presumed benefits of an early pregnancy termination must be carefully weighed against the risks of prematurity, actually justified only in rather few cases.

How?

Alternative is between vaginal delivery, either spontaneous or induced, and caesarean section. The advantage offered by a caesarean section is mostly related to the delivery room management, allowing to choose the day and hour of delivery when all needed specialists are ready in stand-by. From the medical point of view, no benefits for the baby have been scientifically proven, although some malformations positively benefit from a caesarean delivery (e.g. in omphalocele or gastroschisis, this allows not to expose the sac to the trauma of being "squeezed" during vaginal passage, although also caesarean section is not fully safe: actually, some cases of "bucket handle" complete avulsion of the gut during caesarean section in gastroschisis foetuses have been published) [1]. Also in this case, the benefits must be carefully and wisely weighed against the risks of a caesarean section, also in view of possible future pregnancies. Decision on the modality of delivery must always be interdisciplinary and take into account the individual patient and the specific local situation.

Table 11.1 summarises the answers to these questions for the more common malformations amenable to a prenatal diagnosis.

Table 11.1 Delivery planning of the most common congenital malformations detectable prenatally

	Where	When	How
Head and neck malformations			
Cleft lip and palate	Indifferent	At term	Indifferent
Choanal atresia	Indifferent if the team knows how to deal with the problem	At term	Indifferent
Severe neck lymphangioma, congenital high airway obstruction syndrome (CHAOS), masses obstructing the mouth or trachea	3L with ENT expertise available	At term	CS mandatory; EXIT procedure possibly needed
Thoracic malformations			
Chest deformities	Indifferent	At term	Indifferent
Lung malformation	3L advisable	At term	Indifferent
Congenital diaphragmatic hernia	3L mandatory	Around 38 GW	No statistically significant differences between birth modalities
Oesophageal atresia	3L advisable	At term	Indifferent
Abdominal malformations			
Gastroschisis	3L mandatory	Advisable slightly preterm	CS advisable
Omphalocele	3L mandatory	At term	CS advisable
Enteric cysts and duplications	Indifferent	At term	Indifferent
Intestinal atresia	3L advisable	At term	Indifferent
Meconium ileus	3L advisable	At term	Indifferent
Intestinal ischemia and necrosis secondary to volvulus or meconium ileus	3L mandatory	Preterm	Indifferent
Anorectal malformations	Indifferent	At term	Indifferent
Abdominal masses (neuroblastoma, nephroblastoma, choledochal/mesenteric/duplication/ovarian cysts)	Indifferent	At term	Indifferent
Urogenital apparatus/pelvis			
Sacrococcygeal teratoma	3L advisable	At term	CS advisable
Uropathies (e.g. unilateral multicystic dysplastic kidney, hydronephrosis, hydroureteronephrosis, posterior urethral valves)	According to the gravity of the malformation	Preterm (?)	Indifferent
Cloacal/bladder exstrophy	3L mandatory	At term	Indifferent
Defects usually requiring pregnancy termination			
Anencephaly, hydranencephaly, alobar holoprosencephaly, severe chromosomal anomalies, bilateral renal agenesis, infantile polycystic renal disease, severe untreatable metabolic disorders, lethal bone dysplasias			

3L third-level centre; *CS* caesarean section; *ENT* ear, nose and throat; *EXIT* ex utero intrapartum treatment; *GW* gestational week

Malformation Diagnosed Only at Birth

In the unlikely event of a malformation discovered only at birth, the first step is obviously to properly recognise the malformation. Table 11.2 gives some hints for this demanding task.

Practical Indications

As soon as the baby is born, a careful diagnostic workup must be carried out to assess the malformation/malformations in all its aspects. It must always be borne in mind that a malformation comes either from a genetic factor, usually a chromosome deletion or translocation, or from an (in utero) insult due to external causes (smoke, drugs, hypotension, etc.), particularly if this happens during the first 2 months of pregnancy, when organogenesis takes place. In the instance of a genetic factor, it is possible to observe a problem in all the organs codified on the missing part of the genome (e.g. the Beckwith-Wiedemann syndrome involving hemihypertrophy, macroglossia and Wilms tumour, among others, and coming out from a deletion of chromosome 11). In the instance of an "in utero" insult, all the organs and apparatuses under development at the very moment of the external insult can be affected: a perfect example of this situation is given by the VACTERL (Vertebrae, Anorectal, Tracheoesophageal, Ribs and Limbs) syndrome in which a problem involving the foetus at around the 40th day of pregnancy can affect all these organs.

In both cases therefore, it is mandatory, if a malformation is immediately evident at birth, to examine thoroughly the newborn to detect possible other associated malformations not so obviously evident.

This diagnostic workup includes first of all an accurate clinical exam of the newborn, including a test of patency of the oesophagus, of the choanae and of the anus (see later).

Table 11.2 Clinical presentation of the most common congenital malformations detectable at birth

Head and neck malformations	
Cleft lip and palate	Immediately evident at birth (cleft lip); to be searched for, examining the open mouth of the baby when he/she is crying (cleft palate)
Choanal atresia	Not evident at birth, it must be searched for with a nasogastric tube insertion in children with respiratory distress (see later)
Severe neck lymphangioma, congenital high airway obstruction syndrome (CHAOS), masses obstructing the mouth or trachea	Immediately evident at birth
Thoracic malformations	
Chest deformities	Immediately evident at birth
Lung malformation/congenital diaphragmatic hernia	They show their presence with respiratory distress. See Fig. 11.1 for the relevant algorithm
Oesophageal atresia	Not evident at birth, it must be searched for with a nasogastric tube insertion (see later)
Abdominal malformations	
Gastroschisis Omphalocele Ruptured omphalocele	Immediately evident at birth, but: • If there is a sac, it is an omphalocele • If there is no sac and the intestinal loops are oedematous and dilated, it is a gastroschisis • If there is no apparent sac or just some remnants and the herniated intestinal loops are apparently normal, it is an omphalocele with sac ruptured during or immediately before delivery
Enteric cysts and duplications	Cannot be clinically diagnosed at birth but only prenatally with US scan
Intestinal atresia Meconium ileus	Presenting with the classical symptoms of intestinal occlusion: • Abdominal distention • Biliary vomit • No meconium
Anorectal malformations	Immediately evident at birth, provided that this area is not neglected in the initial assessment
Abdominal masses (neuroblastoma, nephroblastoma, choledochal/mesenteric/duplication/ovarian cysts)	Sometimes evident at birth with alteration in the abdominal shape if they are really huge
Urogenital apparatus/pelvis/back	
Sacrococcygeal teratoma	Immediately evident at birth
Uropathies (e.g. unilateral multicystic dysplastic kidney, hydronephrosis)	Cannot be suspected at birth. Suspect diagnosis in case of oligohydramnios or anhydramnios; reduced or missing urinary output
Cloacal/bladder exstrophy	Immediately evident at birth
Myelomeningocele	Immediately evident at birth

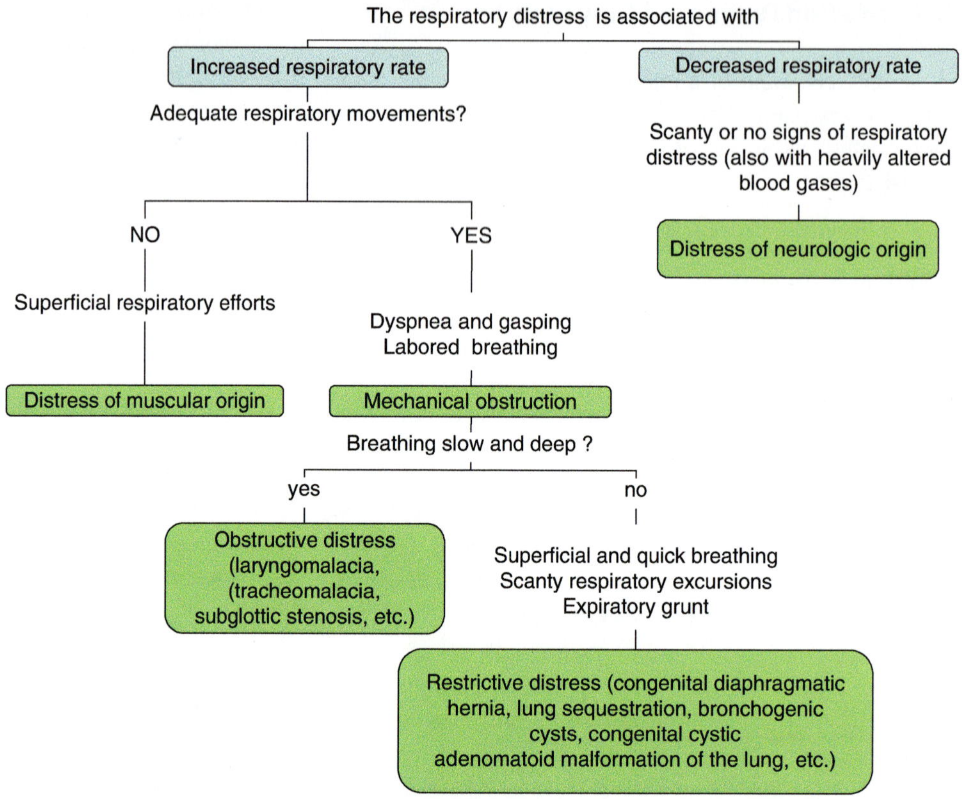

Fig. 11.1 Newborn in acute respiratory distress

This will be followed by an echocardiography and an echoencephalography to exclude cardiac or encephalic defects. An ultrasound of the thorax and abdomen will then be performed, to rule out malformations in these compartments. When suggested by the clinical situation, also genetic tests will be performed (e.g. in the duodenal atresia (often linked with Down syndrome) or in the meconium ileus (related to mucoviscidosis)).

Choanal Atresia

The larynx in newborns is located higher up in the neck than in the child or in the adult, in such a way to exclude the respiratory tree from the alimentary one during suction, thus avoiding milk inhalation.

With this anatomical arrangement, it is mandatory for the newborn to breathe only through the nose (*obligate nose breather*): in case of bilateral choanal atresia, the passage of air from the nose to the larynx is blocked, and the baby therefore feels something like being suffocated. As a reaction he starts crying and breathing through the mouth, temporarily overcoming the problem; once quiet he closes his mouth, and the vicious circle starts again. This peculiar symptoms' evolution should induce to suspect a choanal atresia, which can be easily confirmed if a catheter cannot be passed through the nose into the pharynx. To temporarily overcome the problem, it is suggested to maintain open newborn's mouth with some tricks, such as a Mayo cannula of adequate size or simply a silicon pacifier with the extremity cut away, actually acting as "home-made" Mayo cannula.

Cleft Lip and Palate

Cleft lip and cleft palate represent the most common and immediately recognisable craniofacial anomalies. They can be diagnosed prenatally on ultrasound, if the case completed by an MRI, but not infrequently (≈25% of the total), these malformations are first noted in the delivery room. The infant may have a cleft that is unilateral, bilateral, complete or incomplete and that may involve the lip only, the palate only or both. Cleft lip and cleft palate are often isolated nonsyndromic occurrences; however, when associated with other abnormal physical findings, a recognisable syndrome may be present and sought for.

The malformation doesn't need any specific procedure during or immediately after the delivery; the most demanding task for the nurse/midwife in this case is how to communicate the news to the mother. Of course nothing must be given by mouth, due to the high risk of inhalation deriving from the impaired anatomy of the area. Bottle feeding will be given only with the teat specifically designed for these babies.

Treatment of the malformation needs to follow detailed guidelines that usually start from the very day of delivery: it is therefore mandatory for those attending such children to have at hand the flowchart detailing the steps to follow without losing any time.

Severe Neck Lymphangioma

Formerly known as cystic hygroma, it is the result of an alteration in the lymphatic system development leading to the formation of multiple cysts full of lymph in the neck area. Its association with chromosomal abnormalities is frequent and may lead to foetal chylothorax and foetal hydrops. Voluminous cases leading to possible respiratory failure due to compression of the larynx or trachea immediately after birth can be easily detected prenatally, allowing to properly arrange for the delivery. In these cases all equipment for an immediate endotracheal intubation must be at hand, as well as for an emergency tracheotomy. In cases particularly severe, it should be considered an approach via the EXIT (ex utero intrapartum treatment) method, a technique used during caesarean section with the uterus in complete relaxation and the umbilical circulation maintained (placenta in situ and umbilical cord unclamped and left attached to the baby). This method allows the anaesthesiologist to intubate and secure the airway, or the surgeon to remove the obstruction or to carry out a tracheotomy, while baby's oxygenation is maintained by the mother via the placenta. On top of huge neck lymphatic malformations, the EXIT procedure has been successfully adopted to manage laryngeal obstruction in congenital high airway obstruction syndrome (CHAOS), to reverse tracheal occlusion after foetal surgery for congenital diaphragmatic hernia, to repair the trachea, to resect large cervical tumours such as giant teratomas and to place vascular cannulas for ECMO (EXIT to ECMO).

Congenital Diaphragmatic Hernia

It is somewhat paradoxical the observation that children with prenatally diagnosed congenital diaphragmatic hernia have a worse prognosis compared to infants diagnosed after birth. Actually, this observation can be easily explained considering that the malformation can be diagnosed prenatally only when herniated viscera occupy almost all the thorax, reducing the volume of the lung. The comparison between the simultaneous sonographic measurement of this volume and the one of the head (lung to head ratio, LHR), and particularly the observed vs expected ratio (O/E LHR), is considered a good prognostic predictor.

Foetuses with an O/E LHR below 35% have lower survival rates with higher needs for ECMO or use of patch to close the "hole" in the diaphragm: for these newborns the birth outside a highly specialised perinatal centre is an independent risk factor in mortality.

In the immediate postnatal care of an infant symptomatic at birth, it is mandatory to remember that the malformation is a *physiologic* and not a *surgical* emergency: every effort to stabilise the cardiorespiratory system must be weighed against the need to avoid any iatrogenic injury coming from too a forceful reanimation.

The first step is the introduction of an endotracheal tube for ventilation, which must be very gentle with low peak pressures (<25 cm H_2O), permissive hypercapnia, elective high-frequency oscillatory ventilation, nitric oxide, surfactant and extracorporeal membrane oxygenation (ECMO) available as necessary. The use of mask and Ambu bag is contraindicated because of a possible overdistention of the stomach and intestines, dislocated in the left thorax. To prevent this distention, it is also mandatory to insert a nasogastric tube immediately after birth.

Arterial and venous access can be secured through the umbilical vessels; pushing the umbilical venous catheter across the liver into the right atrium allows to monitor the central venous pressure. Postductal arterial blood gas specimens can be obtained via the catheter in the umbilical artery, preductal ones via a transcutaneous saturation probe.

Any stress to the baby should be carefully avoided, also through a pharmacological sedation: proper temperature regulation, glucose homeostasis, volume status and metabolic acid-base alterations must be strictly monitored and any alteration immediately reported.

Oesophageal Atresia

The presence of an oesophageal atresia can be suspected prenatally in foetuses with polyhydramnios or with a small or non-visualised stomach. Overall one third of cases can be diagnosed before birth, but this number raises up to 85% in atresia type 1 (15% of all cases, no tracheoesophageal fistula present, stomach non-visualised) and goes down to 25% in all other types of oesophageal atresia with fistula. Doubtful cases can be assessed by prenatal MRI, displaying a sensitivity around 70% [2].

It is therefore rather common for this malformation not to be diagnosed prenatally, but an oesophageal atresia unnoticed at birth could lead to two possible devastating consequences:

- A perinatal reanimation with mask ventilation in a child with tracheoesophageal fistula inflates the stomach, with resulting abdominal distention and forceful passage of acid gastric content into the lungs through the fistula, leading to chemical pneumonia (cyanosis, cough and worsened respiratory distress)[3].
- Feeding a child with atretic oesophagus leads to a flood of milk into the lungs and consequent pneumonia.

The mandatory need for a diagnosis of oesophageal atresia immediately after birth is therefore obvious, moreover because it can be easily excluded simply introducing a nasogastric tube into the stomach. In case of an atresia, the tube typically stops at 9–13 cm from the gums; otherwise, it reaches the stomach after some 17–20 cm. Particular attention must be paid to the possible coiling of the tube into the upper pouch of the oesophagus, thus simulating a normal progression into the stomach. To avoid this mistake, it is enough to insufflate some air into the tube while listening with a stethoscope if it bubbles into the stomach, thus verifying that the tip of the tube actually reached it. For the same reason, nothing should be given per mouth to a neonate until surely proven he/she is not suffering of this malformation.

Sometimes a tracheoesophageal fistula is present with no oesophageal atresia, passed unnoticed while inserting the nasogastric tube that normally reaches the stomach. In this case the observation of coughing and choking after every feeding of the baby should raise the suspect of a fistula, to be immediately searched for before a pneumonia is induced by the milk and gastric contents passing through the fistula.

Gastroschisis

In foetuses with gastroschisis, a "hole" in the abdominal wall is present through which the guts herniate into the amniotic sac. Continuous exposure to amniotic fluid leads to inflammation and oedema of intestinal loops: final outcome is closely related to the severity of inflammatory "peel" caused by this prolonged contact, hence the need of a close ultrasound monitoring of the level of bowel suffering, of intestinal dilation and of thickening of the intestinal wall, in order to perform a preterm delivery when signs of chronic inflammatory damage of the herniated intestinal loops become evident. A severely damaged and dilated gut hinders intestinal return to the abdomen and makes staged repair necessary in many cases, and also after abdominal closure, intestinal loops remain hypoperistaltic with consequent oral toleration difficulties, long periods of parenteral nutrition and increased hospital stay.

Although there are still some controversies on the actual benefit of such an approach, most of the babies with prenatal diagnosis of gastroschisis are delivered at an average gestational age of 36 weeks (range 35–38). There is no clear advantage offered by a delivery from caesarean section vs spontaneous or induced vaginal delivery: worth of mention is nevertheless the report of some cases of "bucket handle" avulsion of the intestinal loops during caesarean sections for gynaecological reasons in unexpected gastroschisis babies [1].

The delivery, either caesarean or vaginal, should be in a third-level centre with the surgical team ready to intervene to guarantee the baby the most timely intervention (an ad hoc study showed an overall survival rate of 76.5% vs 43.3% for infants, respectively, inborns and outborns, in a perinatal centre) [4].

Stabilisation in the delivery room is of outmost importance and includes:

1. A secure airway with endotracheal intubation and no mask ventilation (to avoid further stomach and bowel distention). Oxygen supplement only as dictated by the general conditions.
2. Insertion of a large-bore nasogastric tube to prevent further bowel distention.
3. Insertion of a rectal tube to evacuate meconium, thus decreasing the abdominal distention.
4. Prevention of thermic and fluid imbalances (exposed bowel heavily dissipate heat and water) through immediate wrapping of the exposed bowels in sterile gauzes soaked with saline solution kept at 37 °C in a sterile plastic bag.
5. A venous line to provide maintenance fluids should be secured preferably in the upper extremities, because the intra-abdominal pressure deriving from forced surgical reduction of intestines could impair the venous return from the lower limbs in the immediate postoperative period.

Particular care must be given not to overstretch the herniated loops with consequent kinking of the inferior vena cava and altered blood flow return to the heart, possible particularly when the child lies on his/her flank. To avoid this it is suggested to hang on the bag wrapping the sac to the "ceiling" of the incubator through the vent at its top.

In emergency cases when no sterile bag is immediately available, a good surrogate is represented by a drip-tubing package: once the drip is removed, the inner part of the bag is sterile and can usefully serve the purpose.

Due to the frequent association of gastroschisis with many other malformations, a quick but not hurried check-up must be performed.

Once these immediate steps are adopted, the child must be prepared for the surgical treatment, required as soon as possible not to allow the intestinal loops to further dilate, thus making the reduction even more difficult. Nurses and midwifes must know and be ready to prepare with no delay all the needed equipment and organise for the procedures to be performed.

It has also been suggested to attempt an immediate reduction of the herniated bowel directly in the delivery room without anaesthesia: if this is the surgeon's choice, everything must be at hand beforehand to keep the patient strictly monitored during the manoeuvre. It can be useful to insert a urethral catheter connected to a pressure monitor to check via the bladder the abdominal pressure induced by squeezing the bowels inside the peritoneal cavity.

Omphalocele

Prenatal and perinatal omphalocele management doesn't differ much from the one of gastroschisis (higher risk of associated malformations in omphalocele, eventually influencing the final outcome). In this malformation, easily detectable before birth at sonographic evaluation and furthermore studied with foetal MRI, the prolapsed organs are protected from the contact with amniotic fluid by a sac and by Wharton's jelly: for this reason there is no need of hurrying the delivery.

A caesarean section is suggested only in giant omphaloceles, containing more than 75% of the liver, with defined risk of rupture at delivery (transforming the omphalocele into something roughly similar to a gastroschisis).

Steps to be taken are identical to those of gastroschisis: also the omphalocele sac must be wrapped into sterile gauzes (no particular need of having them soaked in warm saline, due to the presence of the sac). Final surgical treatment can be safely planned with no hurry (but without unnecessary delays) in the operating theatre; therefore, there is no need of preparation for an immediate intervention in the delivery room.

Intestinal Occlusion

Prenatal sonography sensitivity to detect intestinal occlusion varies from 25% to 85% according to the specific underlying malformation. Usually these malformations are caused by ischemic phenomena and therefore characterised by an in utero evolution such as to hinder a correct diagnosis before the third trimester of pregnancy or also after birth. Most frequent findings are represented by a "double bubble" endoabdominal cystic image (duodenal atresia); multiple cystic images with hypoechogenic liquid content, representing dilated intestinal loops (jejunal or ileal atresia); and dilated isoechogenic intestinal loops with hyperechogenic contour (meconium ileus). Differential diagnosis has to be done with choledocal cysts, duplication cysts and hydronephrosis.

Therefore, in a rather high number of cases, the presence of intestinal occlusion can pass unnoticed until birth when there is an observation of a newborn with abdominal distention (usually developing after the baby has swallowed air, immediately present at birth only in meconium ileus); no passage of meconium and biliary vomit must immediately call for further diagnostic workup.

In the meanwhile it is highly advisable to:

1. Insert a nasogastric tube.
2. Avoid ventilation by mask not to overinflate the stomach and intestinal loops.
3. Check accurately if and when the meconium was evacuated and have it noted on the baby's medical report.
4. Give nothing by mouth until a proper diagnosis is made.
5. Secure an IV line if general conditions deteriorate.
6. Insert a rectal tube to deflate the colon and to check if meconium is present (a little bit of whitish faecal content must not be taken as meconium and doesn't allow to exclude an intestinal atresia).

Anorectal malformations represent a peculiar aspect of intestinal occlusion; they cannot be detected prenatally but in any case would not need any particular attention as far as "when", "where" and "how" birth must happen. Moreover, they do not present any particular symptom of occlusion immediately after birth: therefore, paradoxically, what at birth the malformation needs most is to be properly detected with just a simple inspection of the area. The amount of anorectal malformations not properly detected at birth is actually worrisomely high and can sometimes lead to lawsuits for malpractice.

Actually, the only type of anorectal malformation that cannot be diagnosed at inspection is the isolated rectal atresia with normal anal canal, absolutely rare, which can be suspected if a rectal tube cannot be inserted for more than few cm above the anus.

Sacrococcygeal Teratoma

At prenatal ultrasound it appears as a well-delimited mass external to the foetal pelvis, with a solid, cystic or mixed aspect, more or less developed outside the sacrum according to the different types. In one case out of five, associated malformations are present.

Sensitivity of ultrasound in identifying foetal teratoma is around 100%, and the false-positive rate is very low (around 3%).

Once a sacrococcygeal teratoma is diagnosed, pregnancy must be strictly monitored with repeated sonographic controls to check the growing pattern of the tumour, the development of polyhydramnios and a possible foetal hydrops from high-output heart failure secondary to extremely high blood flow through the tumour (fewer than 20% of cases). A further risk during pregnancy is represented by a life-threatening bleeding within the tumour.

This monitoring allows to carefully plan the delivery that sometimes must be premature by caesarean section, to cope with complications or also in case of oversized tumours.

Once born the foetus must be handled with extreme care due to the risk of haemorrhage within the tumour, with resulting hypovolemia; everything needed for an emergency transfusion must be at hand. The child must preferably lie on the side.

Ultimate prognosis in these newborns depends on the location and size of the mass (and the related technical problem to remove a mass in extreme cases as big as the foetus' body itself) and on associated perinatal complications. Associated chromosomal abnormalities and malignant invasion are of very rare occurrence.

Uropathies

Prenatal sonography can reveal a dilatation of the renal pelvis and/or the ureters either as a transient phenomenon of no pathological significance or as a direct consequence of an obstruction of the urinary tract at various levels or of a significant vesicoureteral reflux. In this case the increased pressure in the urinary tract can cause significant damage to the renal parenchyma. Unilateral or bilateral findings, dilated or normal ureters, filling state of the bladder, amniotic fluid volume and sex of the foetus can give before birth some clues on the cause of hydronephrosis and influence the birth plan.

"Where" and "when" to deliver the child is suggested by the severity of the findings detected prenatally. For highly pronounced and bilateral hydronephroses, the birth should take place in a perinatal centre, with paediatric urologist and nephrologist ready to intervene.

The observation that the longest the time in which there is an increased urine pressure in the urinary tract, the worse is the renal damage has suggested a preterm delivery, although there is no proven evidence for this. Therefore, a spontaneous vaginal delivery at term seems to be the more practical and safe option.

Also in case of severe congenital hydronephrosis, there are no immediate measures to take in the delivery room as far as kidneys are concerned. On the other hand, being the malformation often the cause of oligohydramnios or anhydramnios, a sometime severe lung hypoplasia with corresponding problems of ventilation and oxygenation must be expected and dealt with urgently after the delivery. As usual particular attention must be directed to possible associated malformations.

Cloacal/Bladder Exstrophy

Prenatal sonography report based on the absence of bladder filling, low-set umbilicus, small genitalia and lower abdominal mass is pathognomonic for bladder exstrophy; low-set umbilicus, omphalocele, diastasis of the pubic rami and split vulva are suggestive of classic cloacal exstrophy; and the presence of bladder filling and the missing characteristic elephant trunk appearance of the usually prolapsing terminal ileum suggest the diagnosis of covered cloacal exstrophy. More accurate anatomical information is provided by foetal MRI performed on axial, sagittal and coronal planes and echo gradient in the best plan for acquisition.

There is no evidence of any benefit in a premature delivery of the foetus; delivery in a tertiary centre is advisable but not mandatory, allowing the malformation enough time to plan for a postnatal transfer in a perinatal care centre.

Exstrophied bladder mucosa must be protected with sterile gauzes moistened with warm saline, covered with sterile plastic wrapping kept in place by a loose elastic bandage. Baseline renal function, electrolyte and hematologic status should be determined. Sometimes gender is not immediately obvious on examination; karyotyping must be therefore planned, and particular care must be adopted while presenting the problem and discussing with parents regarding gender assignment.

Myelomeningocele

This malformation is among the ones receiving maximum benefit from an early closure of the defect, well before its natural history will lead to irreversible damage at around the 28th–30th week of gestation. Actually, altered cerebrospinal fluid dynamics result in the hindbrain herniation (Chiari II malformation) and hydrocephalus, while trauma and longstanding exposure of the spinal cord to the amniotic fluid cause damage resulting in lifelong lower extremity neurologic deficiency, faecal and urinary incontinence, sexual dysfunction and skeletal deformities.

Due to the fact that in the presence of a myelomeningocele the function of the anal sphincter is impaired, it is often possible to detect meconium in the amniotic fluid, further damaging the spinal cord and the exposed nerve fibres (actually, in children with open myelomeningocele, meconium has been detected all the way up to the brainstem).

A very early correction of the defect such as the one offered by foetal surgery (before the 25th gestational week) could therefore offer an outcome much better than a surgical correction performed at around the 36th week of life, after induced preterm delivery (not necessarily caesarean section).

The birth should preferably be in a tertiary centre able to offer close interdisciplinary cooperation between neonatologists, neurosurgeons, paediatricians, paediatric nephrologists, paediatric urologists and orthopaedists [5].

Immediately after birth, the myelomeningocele should be protected with gauzes soaked in warm saline to prevent desiccation of the exposed neural tissue. Over the gauzes, a plastic wrap to prevent heat loss should be placed, while the neonate should be kept prone to prevent rupture of the sac and to avoid trauma to the neural placode. Once the herniated spinal cord is protected, an intravenous catheter has to be placed, to administer broad-spectrum antibiotics for the prevention of central nervous system infections.

How to Interact with Parents

Communicating to parents the presence of a malformation of their baby is always a highly upsetting and tough experience both for who gives and for who receives the unexpected news.

Problems and approach are different if the diagnosis is made prenatally or just after the delivery [6].

A prenatal diagnosis of foetal malformation is demanding and represents a decisive moment in the life of the whole family. They can imply the decision about to carry on the pregnancy, to opt for termination or to give indications about how far to go with resuscitation procedures after the birth of a severely disabled baby.

Explanation of foetal disease and malformation requires interaction with parents. Parents need as much information as they require about the malformation, its treatment, the plan for care of the infant after delivery, the perspectives of treatment and the final prognosis. A prenatal diagnosis of foetal malformation usually triggers in the parents a jumble of questions on how to prepare for the birth of their child, on advantages and disadvantages of prenatal diagnosis, on the compulsive use of the Internet to gather as many information as possible (often awfully unreliable) and on views on abortion and genetic testing, among other issues. Three major themes are common to the experience for pregnant women after prenatal diagnosis of foetal malformation: time is good, but it is also the enemy; you grieve, but you do not grieve; and my baby's not perfect, but he/she is still mine.

If parents' opinion is to carry on with the pregnancy, a somewhat paradoxical reaction is documented, with knowledge of the foetal diagnosis leading to both positive and negative consequences. Parents' decision-making has been shown not to ground on physicians' predictions of expected morbidity and mortality, but rather on religion, spirituality and hope.

In this delicate moment, it is to the medical staff welcoming, listening and processing the parental thoughts since the announcement and along pregnancy to delivery. The medical/paramedical staff attending the parents has to explain several times the foetal malformation, its natural history and postnatal outcome. Appropriate counselling is highly recommended and includes:

- A precise prenatal *prognostic evaluation*, although the prognosis of many anomalies remains difficult to define and almost 30–40% of malformations remain prenatally still undiagnosed.
- A prenatal *joint consultation* with the obstetrician and the paediatric surgeon representing the best impact on the overall acceptance of potential sequelae and complications with the malformation, especially in case of undiagnosed associated anomalies.
- A *perinatal project of care*: the parents want to know "who" will take care of their baby at birth (it is also important to make them visit the intensive care unit and the paediatric surgery department to meet all the medical and nurse staff and see the ward where the baby will be admitted).
- A *long-term project of care* since birth to adolescence is extremely important to be defined with parents; it could help to make them more confident for the future of their child, and their anxiety could also be contained.

During prenatal counselling, even the nurse and midwife, faced with the malformation, have to stand face to suffering without trying to run away from it. Communication of the diagnosis does not need the use of images of newborn disability: usually the parents do not believe the pictures shown to them. Pictures of children with the same disability, black and white pictures with disguised eyes (like accused criminals in newspapers), do not exactly represent what every parent wants for his/her child. Offering a "model of disease" captured in a picture from a medical book leads to the high risk of blaming a child still unborn an inadequate, unhuman aspect, already destroyed before birth.

No matter what parents' decision is during pregnancy, parents' suffering needs to be addressed and understood not only as a sensory experience but also as an emotional one. Foetal prenatal images help parents to discover the foetus through "invisible" details: during pregnancy they build an imaginary aspect of their child. At the same time, the high number of prenatal consultations induces a significantly high level of parental anxiety, while just two prenatal counselling sessions are considered to be enough to give less anxiety at the birth of the child.

Postnatally, the newborn in the incubator, although disabled and affected by a malformation, appears different with respect to the prenatal period; for parents, the baby in the incubator becomes their "so cute" baby, and they stare with love at their child; and they always see their baby and not only his/her disabilities, or the "black and white pictures" the doctor showed them during pregnancy, nor the image of another baby affected with the same malformation.

The words to use in foetal medicine to describe the suspicion of a foetal malformation have to be accurately checked by, in particular by the attending medical staff.

A wrong verbal communication among doctors, nurses, midwives and parents before and after the birth may lead to destroy the love, the natural love that parents feel for their newborn child. Even after the baby is born, the incorrect, one-sided negative prognosis and statements with incorrect or inconsistent diagnoses from doctors and nurses could remain branded as unforgettable in parents' memory. Even when a perinatal resuscitation is needed, some parents feel as preferable not to take any decision about it but instead to "leave things in God's hands." Even in this condition, the perception of parents' pain and its psychological, interpersonal and somatic consequences need a delicate support and a delicate management by doctors and nurses.

Parents need hope from medical staff words: the lack of empathy and loss of hope from these words could make parents passive and down. Never say to the parents "lethal malformation": survival beyond the neonatal period is recurrent in congenital malformations often described as lethal; also the term "incompatible with life" should be avoided from pre- and perinatal counselling.

The term "palliative approach" is ethically appropriate and well accepted from parents in pre- and postnatal period for foetuses having a poor prognosis; never say "If you let the child be treated you will only prolong his suffering. This would be selfish and egoistic"; "the child will die anyway"; and never give advice to leave the baby die.

Nurses and midwives are strictly engaged in parents' management before and after birth: they are involved in the care of a pregnant mother carrying a malformed foetus and must be ready to cope with this situation after delivery.

A diagnosis of congenital malformation made only at birth needs a special parental support in a global perspective of care along infancy and adolescence, with a multidisciplinary approach. A team having all specialities involved in the future care of the child has to meet the parents as soon as possible after birth in order to share their experience in the treatment of the malformation and prepare a delicate project of care for the child. Especially in the case of a baby born with permanent disability, the couple has to be supported to endure this extremely difficult period: the immediate, friendly, empathic help of a nurse or a midwife offered since the very first moments of the baby's life could be of paramount impact. Every nurse and midwife should prepare himself/herself to this demanding, engaging and appealing task.

References

1. Dinatti LA, Meagher DP Jr, Martinez-Frontanilla LA. "Bucket handle" avulsion of intestine in gastroschisis. J Pediatr Surg. 1993;28(6):840.
2. Spaggiari E, Faure G, Rousseau V, Sonigo P, et al. Performance of prenatal diagnosis in esophageal atresia. Prenat Diagn. 2015;35(9):888–93.
3. Pramanik AK, Rangaswamy N, Gates T. Neonatal respiratory distress: a practical approach to its diagnosis and management. Pediatr Clin North Am. 2015;62(2):453–69.
4. Du L, Pan WH, Cai W, Wang J, Wu YM, Shi CR. Delivery room surgery: an applicable therapeutic strategy for gastroschisis in developing countries. World J Pediatr. 2014;10(1):69–73.
5. Werlein A, Goebert P, Maier RF. Geburtsplanung bei fetalen Fehlbildungen aus neonataligischer Sicht [planning delivery in cases of congenital malformations from the neonatologist's point of view]. Z Geburtshilfe Neonatol. 2014;218(1):27–33.
6. Boss RD, Hutton N, Sulpar LJ, West AM, Donohue PK. Values parents apply to decision-making regarding delivery room resuscitation for high-risk newborns. Pediatrics. 2008;122(3):583–9.

Further Readings

Boss RD, Hutton N, Sulpar LJ, West AM, Donohue PK. Values parents apply to decision-making regarding delivery room resuscitation for high-risk newborns. Pediatrics. 2008;122(3):583–9.

Grünebaum A, McCullough LB, Sapra KJ, Brent RL, Levene MI, Arabin B, Chervenak FA. Early and total neonatal mortality in relation to birth setting in the United States, 2006-2009. Am J Obstet Gynecol. 2014;211(4):390.e1–7.

Pramanik AK, Rangaswamy N, Gates T. Neonatal respiratory distress: a practical approach to its diagnosis and management. Pediatr Clin North Am. 2015;62(2):453–69.

Tudehope D, Papadimos E, Gibbons K. Twelve-year review of neonatal deaths in the delivery room in a perinatal tertiary centre. J Paediatr Child Health. 2013;49(1):E40–5.

Werlein A, Goebert P, Maier RF. Planning delivery in cases of congenital malformations from the neonatologist's point of view. Z Geburtshilfe Neonatol. 2014;218(1):27–33.

Oral and Dental Malformations

12

Eiling Wu, Catherine Wicks, Tom W. M. Walker,
and Robert Carachi

Embryology of the Oral Cavity

The mouth cavity develops from two sources:

1. *The stomodeum which is a depression lined with "ectoderm" resulting from the enlargement of the pericardium and the forebrain*
2. *The floor of the anterior end of the foregut which is lined with "endoderm"*

The ectodermal stomodeum and the endodermal anterior end of the foregut are separated by the buccopharyngeal membrane (which is bilaminar).
At the end of the third week, the buccopharyngeal membrane ruptures, and the primitive mouth cavity is formed by the two components.

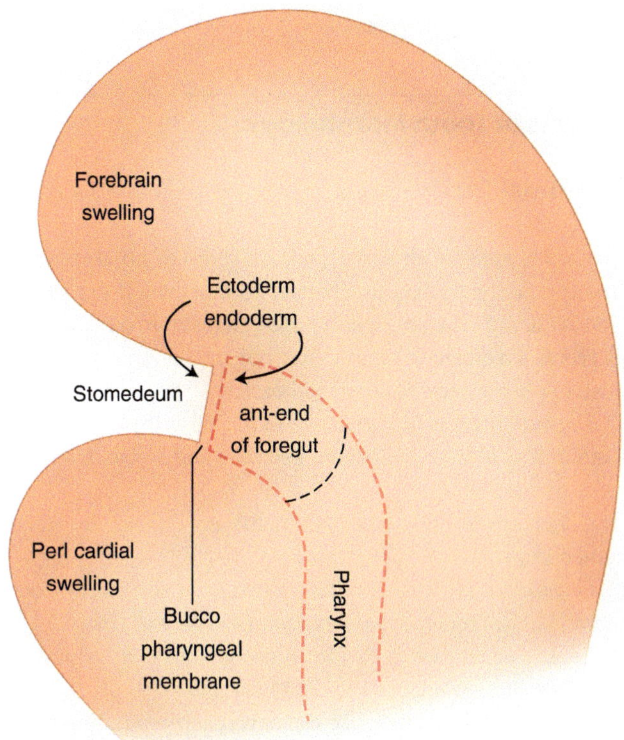

Embryology of the Oral Cavity

E. Wu
Queen Elizabeth Hospital, Birmingham, UK

C. Wicks · T. W. M. Walker (✉)
University Hospitals Bristols NHS Foundation Trust, Bristol, UK
e-mail: Tom.walker@bristol.ac.uk

R. Carachi
University of Glasgow, Glasgow, UK

R. Carachi, S. H. E. Doss (eds.), *Clinical Embryology*, https://doi.org/10.1007/978-3-319-26158-4_12

The demarcation line between the ectoderm and the endoderm is a line just lingual to the gums.

Introduction

Tooth development begins at approximately day 30 of gestation and ends around 19 years of age. The teeth do not all begin developing at the same time or rate. Therefore developmental disruption has multiple opportunities to occur. If the anomaly is genetic, it will likely affect all the teeth. However, if it is exogenous such as medication or infection, it is more likely to affect the teeth developing at that time.

The pathoembryology is not well understood, and due to complex interactions, even the same genetic mutation can result in differing phenotypes.

Dental anomalies are rarely detected antenatally although some syndromes associated with dental anomalies may be. Dental anomalies, in particular the presence of unerupted teeth, may be an incidental finding but may also indicate the presence of other syndromes such as Gardner's syndrome.

Dental anomalies themselves tend not to cause an increase in mortality, but the management can still be complex especially if associated with syndromes and may require the expertise of multiple specialties.

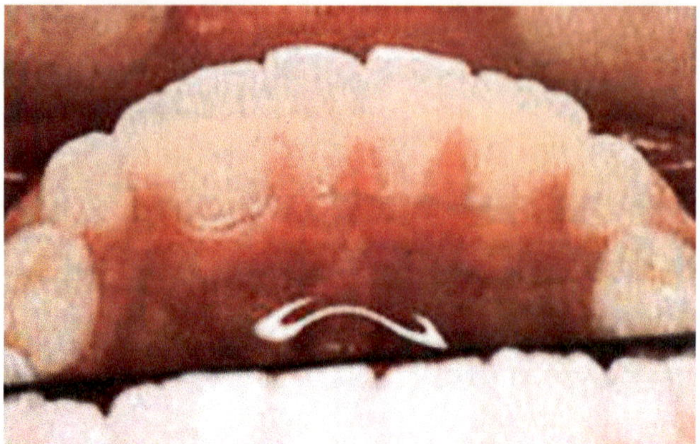

Fig. 12.1 Gemination

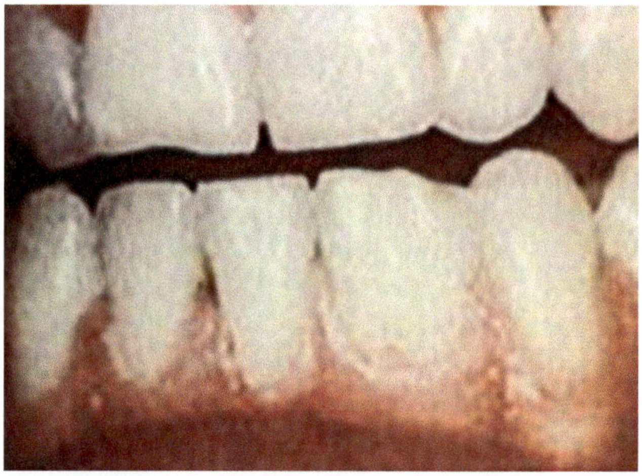

Fig. 12.2 Fusion

Defects of Tooth Morphology

Abnormal Tooth Shape

The tooth appears abnormal but, if there are no associated structural defects, may function as normal even if the occlusion is altered. The abnormal shape can result in defects such as grooves which can become areas of stagnation and predispose to plaque accumulation and caries.

Taurodontism is historically used to describe only molar teeth that have a vertically enlarged pulp and apical displacement of the furcation.

Concresence is when the teeth are joined by their cementum.

Gemination is a tooth that has divided into two; therefore one morphological root has two morphological crowns (Fig. 12.1).

Fusion occurs when two teeth merge into one; it may be complete or incomplete (Fig. 12.2).

Dens evagination describes an extra cusp protruding from a tooth, usually the premolars. The tooth should be monitored as it can lose its blood supply and therefore may need root canal treatment.

Dens invaginatus describes a tooth where invagination of the epithelium occurs resulting in coronal tissue in the pulp space. It is most commonly observed in maxillary lateral incisors (Fig. 12.3).

Talon cusp describes an extra cusp which is talon-shaped.

Odontomas occur due to aberrant development of dental lamina. Strictly speaking this also encompasses teeth with a large cingulum, extra roots and extra cusps, when the tooth is otherwise normal. *Composite odontomes* have no resemblance to a tooth. They are an irregular, haphazard development of dental tissue and occasionally form a large mass.

Enamel pearls are uncommon and minor abnormalities. Ameloblasts are displaced below the amelocemental junction, and this results in a nodule of enamel attached to the dentine. It may contain a horn of pulp which would be exposed if the pearl were to be removed.

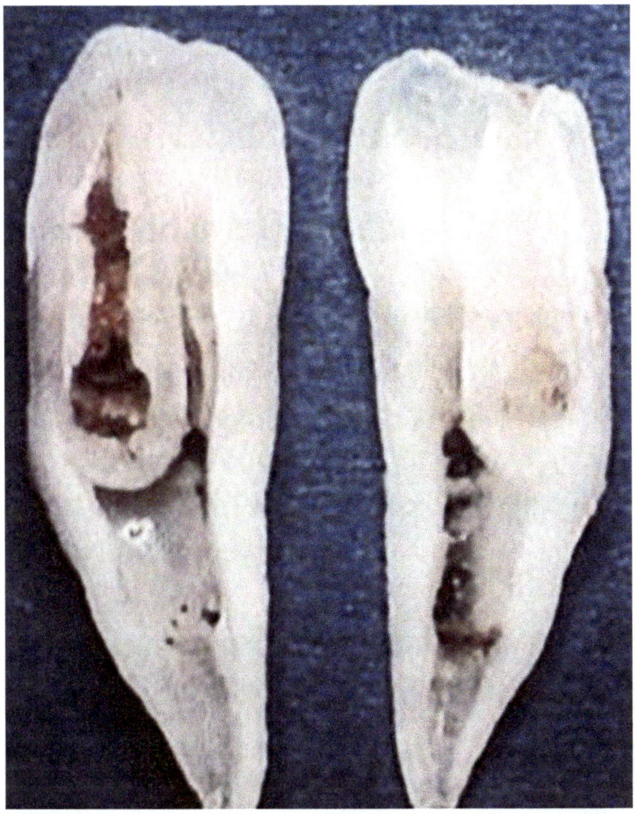

Fig. 12.3 Dens invaginatus

Abnormal Tooth Structure

The tooth may appear normal in shape, but due to faults in the structure of any of the materials that compose the complex architecture of the tooth, it may not function normally.

Enamel Defects

The outer layer of the tooth is a hard protective semitranslucent layer. It is formed in two stages. The first stage, called the secretory stage, involves proteins and an organic matrix forming a partially mineralised enamel. The second stage, called the maturation stage, completes enamel mineralisation (Figs. 12.4 and 12.5).

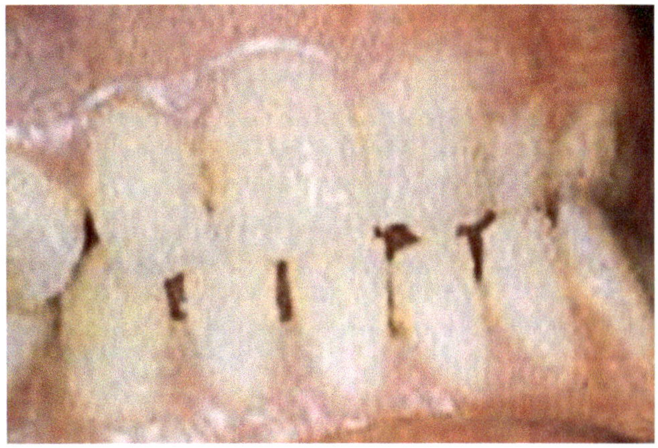

Fig. 12.4 Enamel defects

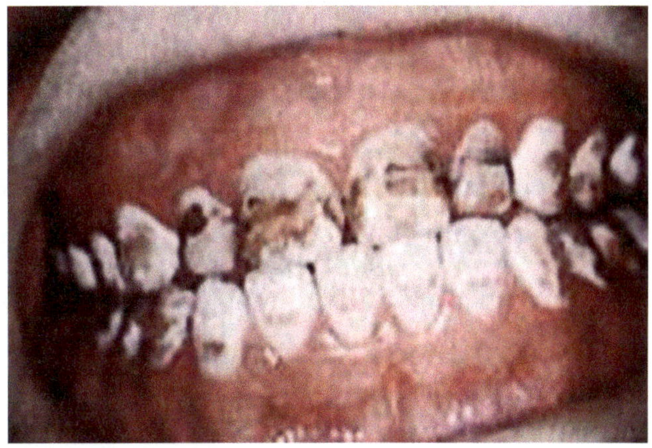

Fig. 12.5 Enamel defects

Amelogenesis Imperfecta (AI) [1–4]

Amelogenesis imperfecta affects 1:7,000–14,000 individuals. It is a disorder in development and formation of the enamel that results in defective or missing enamel and malocclusion. The malocclusion is often an open bite as the teeth are not uniform. There are many different classifications based on inheritance, phenotype or an amalgamation of both.

The abnormal enamel alters the appearance and hardness of the tooth. The tooth will lose its uniform smoothness and is prone to damage. Radiographically the enamel is less opaque and may be of similar appearance to dentine. The consequence of this defect is loss of enamel and exposure of the underlying dentine, which can result in dental sensitivity and increased caries risk.

A commonly used classification, Table 12.1 divides the condition into four main subtypes based on phenotype, and then subdivision by appearance and inheritance results in a total of 14–16 different types of AI.

The causative defect is in the genes encoding enamel matrix proteins that regulate enamel. The proteins are amelogenin, enamalin, ameloblastin and tuftelin. Although their role is not clear, the former two act as a scaffold around which enamel is laid, and the latter two regulate the formation.

AMELX (*Xp22.31-p22.1*) has at least 23 known mutations. It codes for the protein amelogenin which seems to separate and support the crystals as they grow. ***ENAM*** (*4q13.3*) has at least 14 known mutations. It codes for enamalin. ***MMP20*** (*11q22.3*) has at least seven mutations, and it codes for enamelysin, a protein which cleaves amelogenin and ameloblastin. If these proteins are not cleaved and removed, they weaken the structure. ***FAM83H*** (*8q24.3*) has 20 mutations. Although the most commonly mutated gene, its function is unknown.

The same mutation causes no consistent phenotype. The tooth may have hypoplasia, hypomaturation, hypocalcification or hypomineralisation.

Hypoplastic Amelogenesis Imperfecta

This describes defects in the amount of enamel caused by a defect in matrix formation.

It is usually an autosomal dominant or X-linked condition; rarely both inheritance patterns can be present. The resulting phenotype is often worse in males where there is almost no enamel. In females the teeth are more commonly ridged.

The enamel appears hard and translucent, or there may be random pitting, staining or sections of thin enamel. The thin sections may stain easily and are prone to caries.

It is difficult to classify due to the varied appearance but shows a clear inheritance pattern.

Hypomaturation Amelogenesis Imperfecta

This describes defects in the final growth and maturation of the crystallites within the enamel matrix. The tooth can look normal, or it may be white or brownish yellow, but the enamel is softer and vulnerable to attrition.

Hypocalcification Amelogenesis Imperfecta

This is the most common form of AI. It is a defect in the initial crystalline formation of the enamel followed by defective growth. The quality of the enamel matrix is normal, but the lack of calcium produces a chalky appearance and produces the softest and therefore weakest enamel of the AIs. The teeth are easily stained and worn away.

Hypomaturation-Hypoplasia and Taurodontism Amelogenesis Imperfecta

In this form of enamel defect, teeth are smaller than normal with enlarged pulp chambers. The teeth appear mottled with pits and grooves.

Dentine Defects [5, 6]

Dentine is the main supporting calcified structure of the tooth. It underlies the enamel layer and surrounds the pulp cavity. It is a vital tissue containing the cytoplasmic extensions of odontoblasts. There are three recognised types: primary, secondary and tertiary. Primary dentine is formed until the tooth becomes functional when it erupts. Thereafter secondary dentinogenesis develops throughout the lifetime of an individual and is laid down in layers by odontoblasts from the pulp cavity. Tertiary dentine (sometimes reparative/reactionary) is formed by odontoblasts as a defensive response to a noxious stimulus (thermal/chemical/bacterial/mechanical).

The hereditary dentine disorders, dentinogenesis imperfecta (DGI) and dentine dysplasia (DD) form a group of autosomal dominant genetic conditions. The characteristic abnormal dentine structure can affect either the primary or

Table 12.1 Witkop classification + inheritance pattern

Type	Clinical features	Subtype	Inheritance	Gene
1. a–g Hypoplastic	Variable crown size Normal or opaque white-brown	a—Diffuse pitted	AD	LAMB3
		b—Localised pitted	AD	ENAM
		c—Localised pitted	AR	ENAM
		d—Diffuse smooth	AD	
		e—Diffuse smooth	X-linked	AMELX
		f—Diffuse rough	AD	
		g—Enamel agenesis	AR	FAM20A
2. a–d Hypomaturation	Creamy white-very yellow/brown Soft, rough surface	a—Diffuse pigmented	AR X-linked	MMP20 KLK4 WDR72 C4ORF
		b—Diffuse pigmented	X-linked AD	AMELX AMELX
		c—Snow capped d—Snow capped		
3. a and b Hypocalcified (most common)	Opaque white-yellow/brown Soft, rough surface	a	AD	FAM83H
		b	AR	SLC24A4
4. a and b Hypomaturation-hypoplasia and taurodontism	White-yellow/brown Mottled	a—Tricho-dento-osseous syndrome	AD	DLX3
		b	AR	

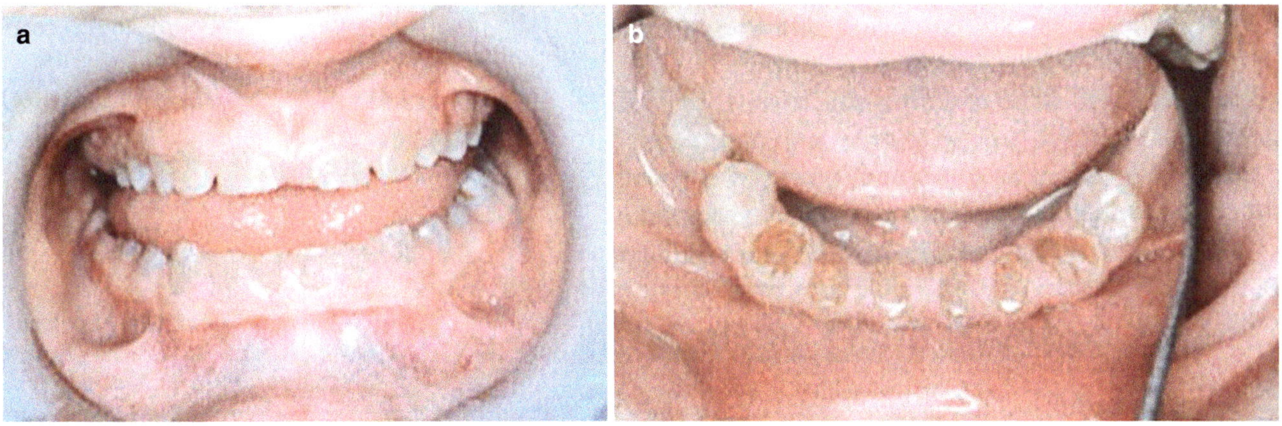

Fig. 12.6 (a, b) Translucent dentinogenesis imperfecta teeth

the primary and secondary dentitions. The result is a soft dentine with abnormally high water content which results in the normal overlying enamel being weakly attached and chipping away easily. The enamel appears normal but discoloured due to the underlying dentine.

In severe cases, the teeth can become quickly worn down.

The incidence of DGI is between 1:6000 and 1:8000. The incidence of DD (type 1) is 1:100,000. The most commonly used classification system by Shields identifies three subtypes of DGI and two subtypes of dentinal dysplasia (Table 12.2).

NB: Clinical features are not confined to classification, and a spectrum of features can be seen across the above dentine disorders.

Dentine is more susceptible to tooth decay than enamel. Therefore, all patients, regardless of disease spectrum, require regular dental assessments, and prevention of tooth decay (oral hygiene, dietary advice and fluoride) is essential. Where diagnosis occurs at an early age and treatment guidance is followed, potential psychosocial and nutritional effects can be minimised, and good aesthetics and function can be achieved.

Defects in Tooth Number

Whether there is a problem with too many teeth or too few, teeth in abnormal positions in the arch can result in consequences for oral hygiene and occlusion, and complex dental/orthodontic treatment may result. Too many teeth can often lead to plaque stagnation, and thus adjacent teeth can become carious. Also eruption patterns can be disturbed, and extraction may be required to prevent effects on the normal dentition.

Table 12.2 Dentinogenesis imperfecta subtypes

Shield's classification (1973)	Synonyms	Clinical features	Gene
Dentinogenesis imperfecta type 1	Hereditary opalescent dentine; osteogenesis imperfecta type I with DGI (OI type IA); osteogenesis imperfecta with opalescent teeth; Brandywine type DGI; syndromic DGI, non-syndromic DGI; DGI without OI, opalescent teeth without OI; capdepont teeth	Associated osteogenesis imperfecta (short stature, blue sclera); 1^0 and 2^0 dentitions, amber, translucent, short roots, pulpal obliteration, variable severity between teeth	*COL1A1/COL1A2*
Dentinogenesis imperfecta type 2		Similar to DG type 1 but do not have osteogenesis imperfecta, but penetrance complete, normal teeth never found, bulbous crowns. Rarely sensorineural hearing loss	*DSPP (dentine sialophosphoprotein)*
Dentinogenesis imperfecta type 3		Found in tri-racial population Maryland/Washington DC, similar features but variable, "shell" teeth	*DSPP*
Dentine dysplasia type 1	Rootless teeth; radicular dentine dysplasia	Clinically normal shape/form/consistency. Radiographic constricted roots, pulpal obliteration/crescent-shaped pulp remnant	*COL1A1/COL1A2*
Dentine dysplasia type 2		1^0 dentition features similar to DG-II, 2^0 unaffected	*DSPP*

Hypodontia is when one to five teeth (excluding wisdom teeth) are missing. This is the most common dental anomaly (1:18) and is more common in females than males. It is rare in the primary dentition. Commonly missing teeth are the second premolars and lateral incisors. Missing lateral incisors are more noticeable due to the pointed nature of the canines which erupt more centrally.

Oligodontia is when six or more teeth are missing. It may be sporadic or syndromic. The incidence is approximately 1:625–1250.

Anodontia is the complete absence of the secondary dentition and is very rare. It usually occurs with severe or fatal syndromes. If this occurs, the primary dentition can be retained for many years. However, in time these teeth will wear and become carious. Systemic defects that are associated with hypodontia/anodontia include anhidrotic (hereditary) ectodermal dysplasia and Down's syndrome.

Hyperdontia exists when there are additional numbers of teeth and is of unknown aetiology. The additional teeth are caused by excessive, but organised, growth of the dental lamina and are described differently depending on their morphology (supernumerary vs. supplemental). Hyperdontia may be diagnosed as an incidental finding on a radiograph or secondary to failure of eruption of teeth. Approximately 75% of hyperdontia occurs in the maxilla. The incidence varies with race and is as high as 3% in the Asian population. Most syndromes associated with hyperdontia are rare, but it is well observed in cleidocranial dysplasia.

- *Supernumerary teeth* are often conical or malformed. Those occurring in the midline are known as mesiodens; 50% do not erupt. Most commonly they occur in the incisor or molar regions. Supernumerary teeth are more common than supplemental.
- *Supplemental teeth* are usually well formed and look normal. Most often there are extra maxillary incisors, premolars and rarely wisdom teeth.

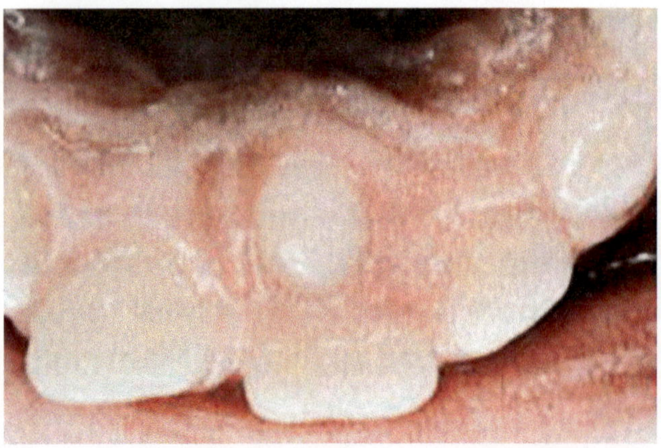

Fig. 12.7 Supernumerary teeth

Defects in Eruption of Teeth

Abnormal eruption of a tooth/teeth is not often caused by a congenital defect. It is more commonly caused by a physical obstruction such as loss of space due to overcrowding or retention of primary teeth. It can also be caused by extra teeth and scarring of the gingiva secondary to trauma or surgery. A careful history is required to elicit these latter causes, as the space may be present, but the tooth does not erupt. Unerupted teeth rarely cause significant problems, but they may cause hypercementosis, resorption of adjacent teeth and cysts.

Syndromes Associated with Dental Defects

Autosomal Dominant

Ectodermal Dysplasia

Ectodermal dysplasia is group of 170+ syndromes resulting from faulty development of the ectodermal germ cell layer during embryogenesis. They exhibit a wide range of phenotypes involving hair, nails, skin, sweat glands and variably other ectoderm derived organs such as teeth causing any combination of the afore mentioned dental anomalies. Some syndromes are associated with cleft lip and or palate. They are traditionally grouped into the anhydrotic and hydrotic syndromes, according to whether they include the absence or severe deficiency of sweat glands. The various syndromes have different inheritance patterns; autosomal dominant, recessive or X-linked trait.

Marfan's Syndrome

Marfan's syndrome affects 1.5–10:10,000 individuals. This is a connective tissue disorder with many subtypes.

Temporomandibular joint pathology is common. There is often dental crowding, elongated teeth and short roots with retrognathia.

Osteogenesis Imperfecta

Osteogenesis imperfecta is characterised as types 1–8 (1–5 autosomal dominant, 6–8 autosomal recessive) and affects 6–7:100,000 individuals. This disorder is characterised by weak bones; it is commonly due to a defect in collagen with mutations in the COL1A1, COL1A2, CRTAP and P3H1 genes. It may be associated with weak teeth due to dentinogenesis imperfecta (see above) or malocclusion secondary to jaw size or position.

Gardner's Syndrome

Gardner's syndrome affects 1:1400–12,000 individuals (APC suppressor gene (c5q21) mutation) it is a subtype of familial adenomatous polyposis. Extra-oral manifestations include lipomas and colonic polyps with a 100% rate of malignant change. Oral manifestations include multiple jaw osteomas and dental abnormalities (30%) such as abnormal morphology, hypodontia or hyperdontia and odontomas.

Cleidocranial Syndrome

Cleidocranial syndrome affects 1:1 million. Autosomal dominant gene RUNX2 affects the development of bones and teeth. It results in atypical facies (hypoplastic maxilla), under developed or absent clavicles and open fontanelles. Despite commonly beign associated with hyperdontia it is one of the few identifiable causes of delayed eruption of permanent teeth and associated dentigerous cysts.

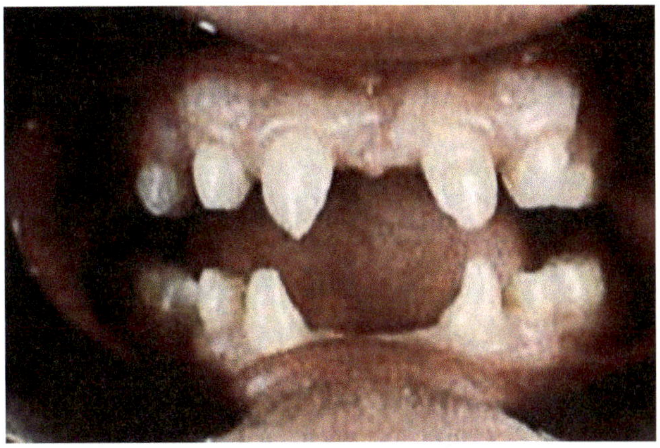

Fig. 12.8 Ectodermal dysplasia

Autosomal Recessive

Down's Syndrome

Down's Syndrome is a trisomy of chromosome 21 of which 1:1000 are sporadic mutations. Ninety percent are associated with missing third molars. There is often oligodontia, delayed eruption of permanent teeth and hypocalcification/hypoplastic dental defects. However, the latter could be related to childhood illness due to immune suppression.

Examination may show macrodont primary teeth but microdont permanent teeth with a narrow palate and large tongue. Periodontitis is thought to be secondary to immunosuppression.

X Linked

Anhydrotic (hereditary) ectodermal dysplasia.

This is usually an X-linked genetic disease. In severe cases there is complete anodontia, but usually there are a few permanent teeth. The morphology of the teeth is abnormal, usually peg-shaped or conical, giving a Dracula-like appearance (Fig. 12.8).

If there is anodontia, there will also be reduced alveolar bone. This can lead to difficulty in constructing dentures and placing implants. They also have hypotrichosis and anhydrosis.

Periodontal Tissue Malformations

The periodontium refers to the supportive tissues of the teeth including the gingiva (junctional epithelium, sulcular epithelium, free gingiva and attached gingiva) and gingival and periodontal fibres. Destruction of any of these tissues eventually leads to loss of tooth stability and eventually tooth loss.

Periodontitis

The majority of periodontal disease is caused by plaque-related inflammation; however there are some recognised systemic causes of premature periodontal destruction, the pathological mechanisms of which are diverse. In those with immunodeficiencies (Down's syndrome, leucopenia, diabetes mellitus (if severe and uncontrolled), HIV infection, Papillon-Lefevre syndrome), it is thought to be an acceleration of the usual bacterial process; however in the genetic syndromes (hypophosphatasia, Ehlers-Danlos syndrome type VIII), there are structural defects in the periodontal tissues.

Hereditary Gingival Fibromatosis

- Autosomal dominant. 1:350,000.
- Gingival fibromatosis, hypertrichosis, epilepsy and mental retardation. It may also be coarse facial features simulating acromegaly.
- Gingivae can be enlarged enough to completely bury teeth.

Oral Mucosa Malformations

White Sponge Naevus

This is a developmental anomaly, autosomal dominantly inherited and caused by mutation in keratin genes. It is usually bilateral and can involve the entire oral mucosa, with shaggy, white thickening with indiscrete borders.

Eruption Cyst

A blue discolouration of the pink/red attached oral mucosa can signify a normal eruption cyst. This is a normal cyst that exists around the erupting tooth.

Oral Malformations

Mouth

Tongue Tie (Fig. 12.10)

Tongue tie (ankyloglossia) is due to a short frenulum that tethers the tongue to the base of the mouth and prevents it from being extruded. The tongue curls on itself. Parents are concerned that it will cause problems with feeding, but this is rarely the case, and reassurance is all that is needed in most instances. Another fear is that it delays speech and causes lisps. This may occur but should not be blamed entirely on a tongue tie. If the tongue cannot be protruded beyond the teeth, the frenulum can be divided transversely under a general anaesthetic.

Short maxillary frenulum

A short frenulum of the upper lip is invisible and rarely causes a disability (Fig. 12.11). In later life it may be associated with a wide gap between the central incisor teeth. If the frenulum extends between the incisor teeth and if the deformity persists in the second dentition, the frenulum may be excised completely.

Epulis

This is an enlargement on the gingival/alveolar border of the mouth, usually of fibrous origin of no real significance, and once removed it is a benign tumour of the gum (Fig. 12.12).

Ranula

This is a soft, translucent, bluish swelling under the tongue usually confined to one side of the frenulum (Fig. 12.13). It can vary in size from a small pea to a large swelling deforming the tongue. A ranula may be subject to recurrent infections and become painful. Rarely it may be the cause of dysphagia. De-roofing of the cyst and suturing of its edges (marsupialisation) will cure this condition. Mucous glands are present in abundance in the oral

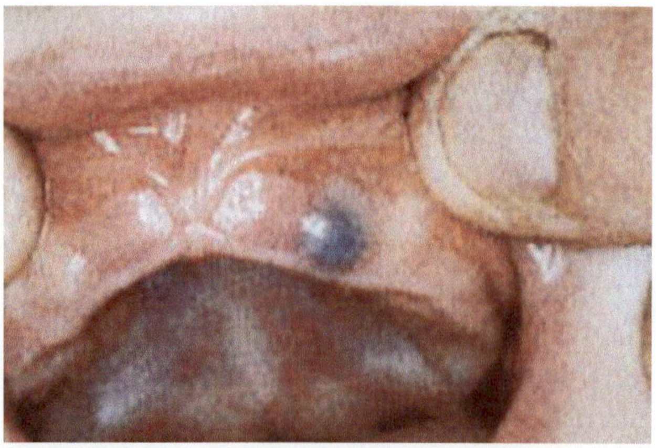

Fig. 12.9 Eruption cyst

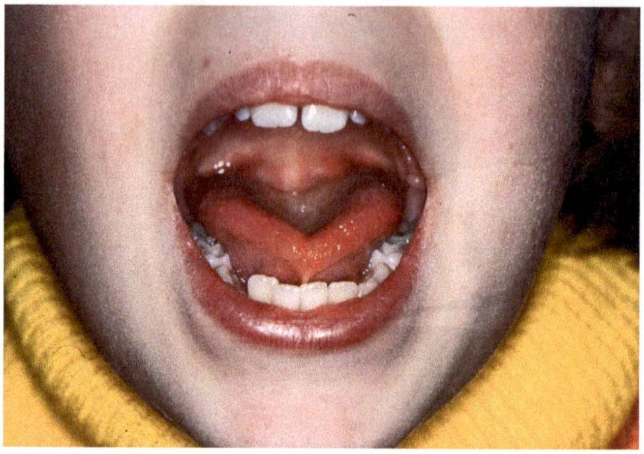

Fig. 12.10 Tongue tie

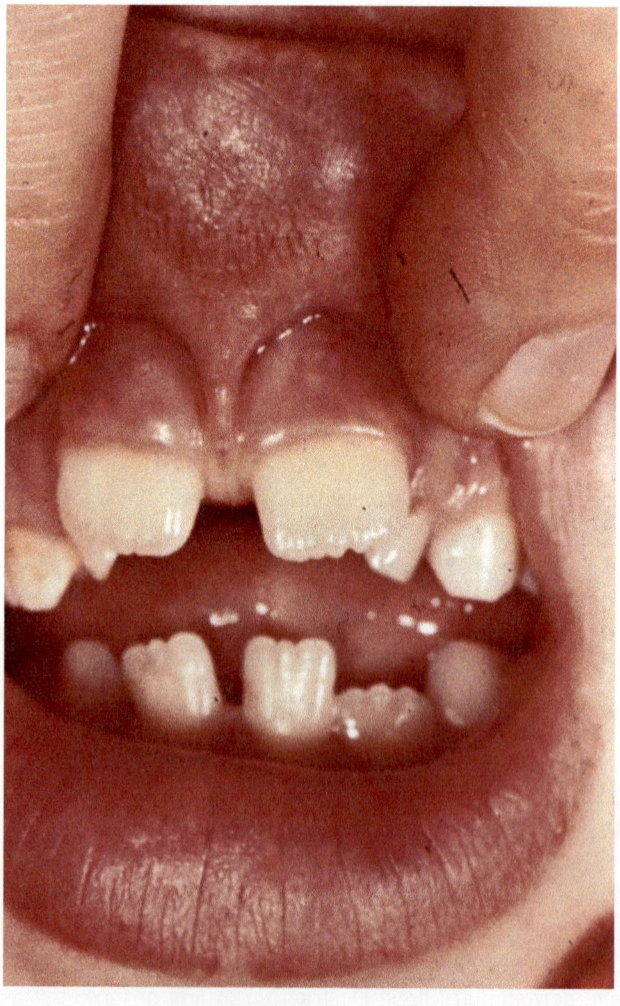

Fig. 12.11 Short maxillary frenulum

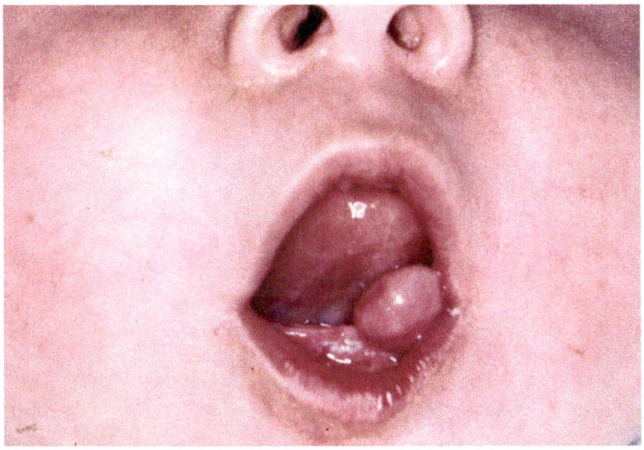

Fig. 12.12 Epulis

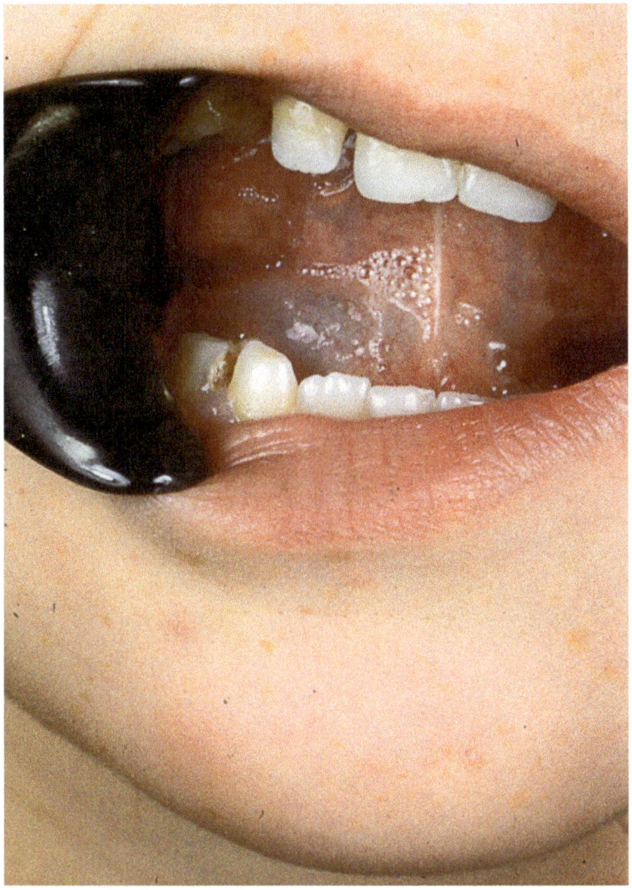

Fig. 12.13 Ranula

cavity and may become blocked in which case the mucus is retained as a mucous retention buccal cyst or a mucocele (Plate 3:4). These appear on the mucosal aspect of the lips and may disappear spontaneously. Treatment of persistent mucoceles is simple excision. Congenital pits in the lower lip rarely cause any clinical problems but can be associated with a cleft lip. If they cause an unsightly appearance, they may be excised before school age.

Macroglossia

An enlarged tongue can be a feature of a number of conditions in the neonate including Beckwith-Weidemann syndrome, Down's syndrome, hypothyroidism, and Hurler's syndrome. It can interfere with deglutition and sometimes breathing. Surgical excision can be curative, i.e. hemiglossectomy (Fig. 12.14).

Haemangioma/Lymphangioma of the Tongue

This condition can be very disabling as it can change as the haemangiomatous and lymphangiomatous elements enlarge and become infected or bleed. Surgical excision is often the only treatment possible (Fig. 12.15).

Mouth Duplication

This is a very rare condition; embryological hypotheses include a defective midline development, split notochord system and the abnormal development and duplication of the totipotential cells from the first branchial arch.

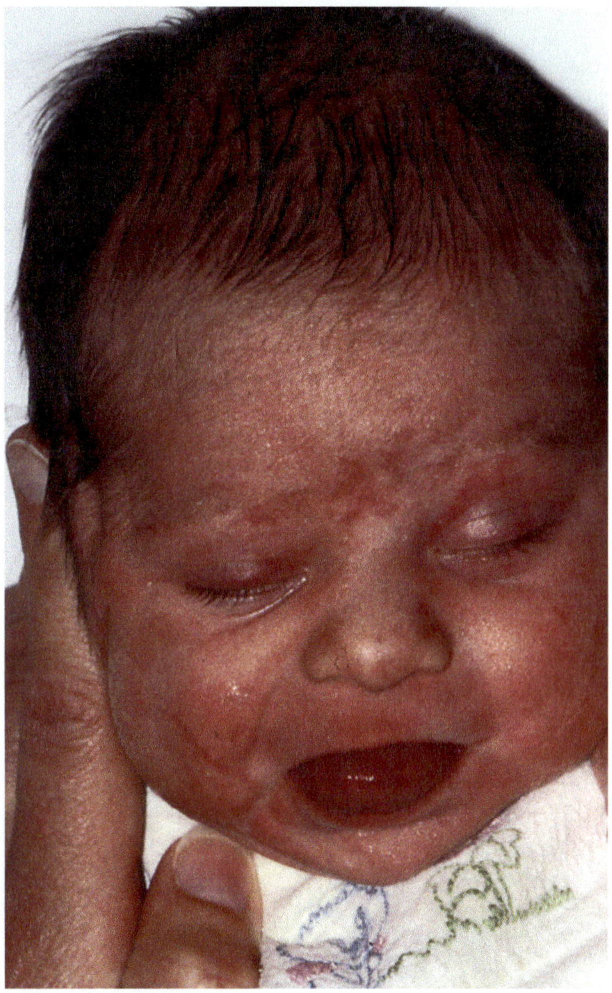

Fig. 12.14 Macroglossia

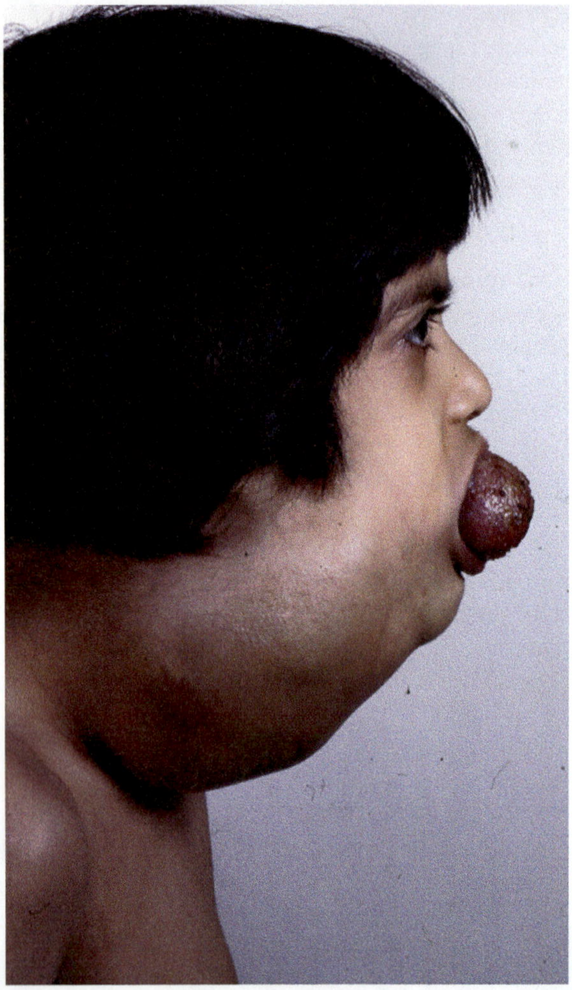

Fig. 12.15 Large mixed lymphangioma, haemangioma of tongue and oral cavity

References

1. Crawford PJM, Aldred M, Bloch-Zupan A. Amelogenesis imperfecta. Orphanet J Rare Dis. 2007;2:17. https://doi.org/10.1186/1750-1172-2-17.
2. Aldred M, Savarirayan R, Crawford P. Amelogenesis imperfecta: a classification and catalogue for the 21st century. Oral Dis. 2003;9:19–23. https://doi.org/10.1034/j.1601-0825.2003.00843.x.
3. https://rarediseases.org/rare-diseases/amelogenesis-imperfecta/. Reviewed 12/12/13.
4. http://ghr.nlm.nih.gov/condition/amelogenesis-imperfecta. Reviewed May 2015. Published: January 11, 2016.
5. Barron MJ, McDonnell ST, MacKie I, Dixon MJ. Hereditary dentine disorders: dentinogenesis imperfecta and dentine dysplasia. Orphanet J Rare Dis. 2008;3:31. https://doi.org/10.1186/1750-1172-3-31.
6. de La Dure-Molla M, Fournier BP, Berdal A. Isolated dentinogenesis imperfecta and dentin dysplasia: revision of the classification. Eur J Hum Genet. 2015;23:445–51.; advance online publication, August 13, 2014. https://doi.org/10.1038/ejhg.2014.159.

Further Reading

Cawson RA, Odell EW. Cawson's essentials of oral pathology and oral medicine. 8th ed. New York: Elsevier; 2008.
https://www.raredisease.org.uk/.

13 Development of Cleft Lip and Palate

Ambika Chadha and Alistair R. M. Cobb

Primary Palate (Fig. 13.1)

The primary palate forms from the fusion of the maxillary processes with the frontonasal prominence during weeks 6 and 7. It contains the incisive fossa.

Secondary Palate

The remaining part of the palate.

It is formed by two shelf-like processes called the palatine shelves which arise from the maxillary process.

The two palatine shelves grow medially and forwards to fuse:

1. *With each other*
2. *With the primary plate anteriorly (in a V-shaped manner)*

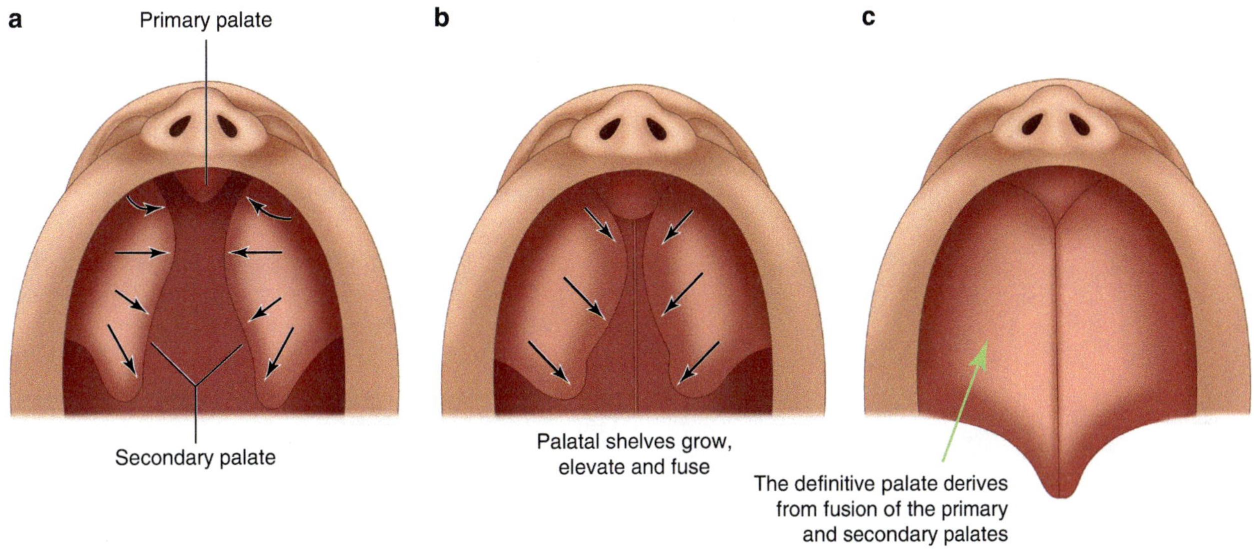

Fig. 13.1 The primary palate fuses with the secondary palate by week 9 to separate the oral and nasal cavities. (**a**) Primary palate and secondary palate, (**b**) Palatal shelves grow, elevate and fuse, (**c**) The definitive palate derives from fusion of the primary and secondary palates

A. Chadha (✉)
Department of Perinatal Imaging and Health, Kings College London, London, UK

A. R. M. Cobb
Southwest UK Cleft Service, University Hospitals Bristol NHS Foundation Trust, Bristol, UK

R. Carachi, S. H. E. Doss (eds.), *Clinical Embryology*, https://doi.org/10.1007/978-3-319-26158-4_13

Nasal Septum (Fig. 13.2)

The nasal septum derives from tissue within the embryonic nasal cavity and contacts the palate shelves as they grow, elevate and fuse.

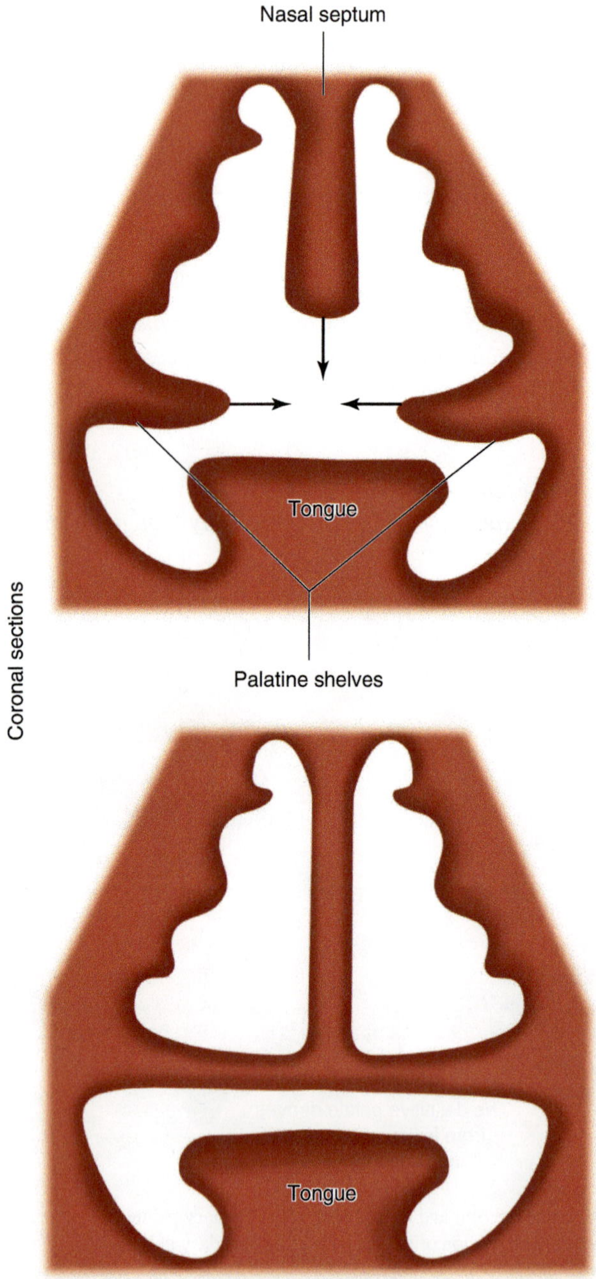

Fig. 13.2 As the tongue descends, the palatal shelves grow and elevate till they eventually fuse in their midline and with the developing nasal septal tissue

Ossification

Occurs in the anterior part of the palate, forming the bony palate, while the posterior part remains fleshy and forms the soft palate.

Definitions

Clefts of the lip and palate represent a heterogeneous group of disorders affecting the lips and oral cavity. From an embryological, aetiological and epidemiological perspective, there are two distinct entities [1]:

1. *Cleft lip* (CL) is defined as a congenital abnormality of the primary palate, i.e. anterior to the incisive foramen. It may extend to involve a palatal cleft (CL ± P).
2. *Isolated cleft palate* (iCP) is a congenital abnormality of the secondary palate.

Recent epidemiologic data challenges this historical division of orofacial clefting entities suggesting that cleft lip only may have a unique aetiology and genetic association, whereas some individuals with cleft palate only show evidence of subclinical cleft lip [2].

A cleft is part of a *syndrome* if there is more than one malformation involving more than one developmental field. A cleft is *non-syndromic* if there is only one defect or multiple anomalies that result from a single initiating event or primary malformation.

Incidence and Epidemiology

CL ± P is the most common congenital malformation of the head and neck. Overall, orofacial clefts affect approximately 1 in 700 live births but significant variation exists:

Geographical Variation

International data from registries suggests and reveals a marked geographical variation in CL ± P likely. Rates of CL ± P have been shown to be higher in Latin America and Asia (China, Japan) and lower in Israel, South Africa and Southern Europe. Variation in rates of iCP is even more marked with particularly high rates in Canada and parts of Northern Europe and lower rates in parts of Latin America and South Africa. Studies comparing rates of CL ± P among ethnic groups within the USA and UK and of immigrants to the USA from Japan and China indicate that migrant groups have rates of CL ± P more reflective of their place of origin than of their place of current abode [3]. This suggests that geographical variations in orofacial clefting are a surrogate for racial variation.

Sex Variation

Across all ethnic groups, CL ± P is more frequent in males, whereas iCP is more frequent in females. Interestingly the sex predilection varies with ethnic origin as well as with severity of cleft and the presence of additional congenital malformations. In the Caucasian population, the male/female ratio for CL ± P is approximately 2:1, and this becomes more apparent with increasing severity of cleft but less apparent when other congenital abnormalities are present [4].

Associated Defects

Orofacial clefts are not uncommonly associated with other congenital abnormalities, whether or not as part of a recognized syndrome. Overall, additional congenital abnormalities are more associated with iCP than with CL ± P. In a European study of almost 4000 individuals with iCP, 55% of cases were isolated, 18% were associated with other congenital anomalies, and 27% were part of recognized syndromes. In a report of more than 500 patients with CL ± P, 71% of cases were isolated, and 29% were seen in association with other anomalies [5]. Indeed, the presence of other congenital abnormalities at birth may lead to the diagnosis of iCP that can often elude diagnosis until a later stage.

Facial Topography of Parents and Siblings

Recently studies have been undertaken that use sophisticated software to analyse the facial features of parents and siblings of individuals born with orofacial clefts. Significant differences have been identified in the faces of normal relatives of cleft patients and the rest of the population at both a phenotypic and genotypic level [6].

Clinical Features

Types of Cleft Defect

Cleft Lip

Can be:

- *Unilateral or bilateral*
- *Incomplete*—a small gap or indentation in the lip that does not involve the nose, whether or not the alveolus is involved
- *Complete*—a gap or indentation in the lip that continues into the nasal floor, whether or not the alveolus is involved
- *Microform*—a mild form of CL that manifests as a small groove in the vermillion of the lip or as a scar-like lesion traversing the lip up to the nostril

Cleft Palate

Can be:

- *Unilateral or bilateral.*
- *Incomplete*—a cleft affecting usually the soft palate only.
- *Complete*—a cleft affecting both the soft and hard palate.
- *Submucous*—describing a palate that initially seems normal but where underlying musculature is abnormal. Classic signs are a bifid uvula, central lucency of the soft palate (zona pellucida) and a bony defect in the hard palate that can be palpated as a notch in the posterior margin of the hard palate. The condition is often missed during routine neonatal screening and consequently children often present late with speech difficulties.

Associated Problems

Although not a major cause of mortality in the developed world, CLP does carry a significant burden of morbidity for affected children. These problems arise from both the:

- Primary pathology
- Timing and type of intervention—particularly surgery

Cleft Lip (Figs. 13.3 and 13.4)

This is associated with:

1. *Aesthetic issues*—pertaining not only to the lip and its repair but also to the inevitable degree of nasal deformity that accompanies all types of CL as part of the primary

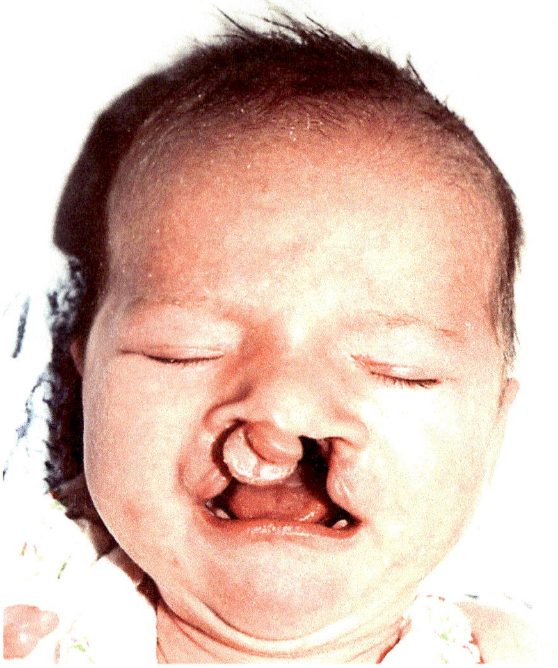

Fig 13.3 Cleft lip bilateral

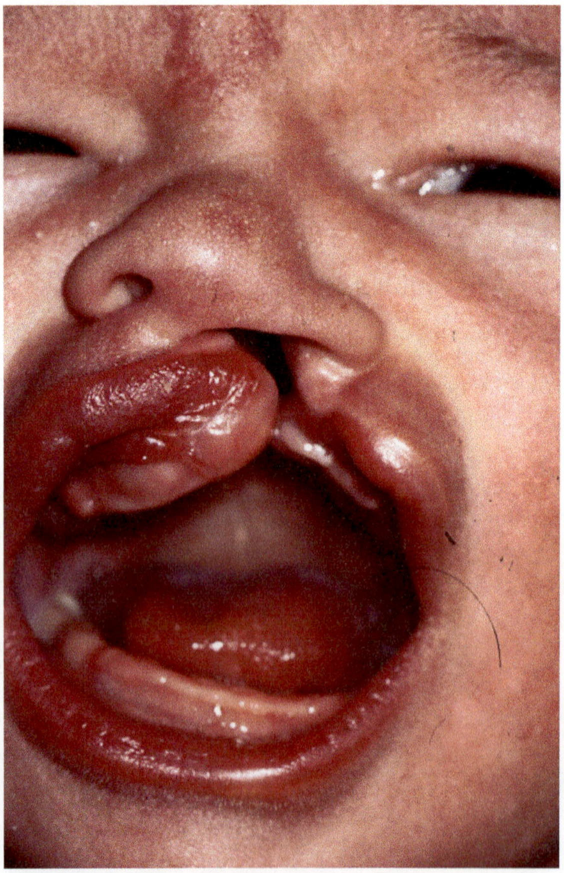

Fig. 13.4 Cleft lip intact palate

defect. Additionally, facial disharmony can be exacerbated by surgical intervention requiring joint orthodontic/orthognathic input at a later stage.
2. *Impaired production of labial sounds (pa, pi)*
3. *Dental irregularities*—these are likely to occur if there is alveolus involvement in the cleft, ranging from a notch to a complete alveolar cleft. Often there is a missing lateral/incisor tooth—occasionally an extra tooth—as part of the primary pathology. Additional dental irregularities may arise from the maxillary retrusion that progressively manifests inherent to the primary pathology and/or as a consequence of surgical intervention.
4. *Psychosocial issues* of identity and societal integration as a result of any or all of the above

Babies with cleft lip alone can feed as normal from the bottle or breast although the formation of an adequate seal is challenged.

Cleft Palate (Fig. 13.5)
This is associated with:

1. *Speech issues*—the muscles on either side of the CP are attached to the posterior edge of the hard palate, inhibiting elevation of the soft palate and the formation of an airtight seal that is necessary to prevent airflow into the nasopharynx during speech.
2. *Feeding issues*—the communication between the oral and nasal cavities in CP precludes the creation of a vacuum during a baby's suck. Breastfeeding is usually not possible, and feeding is achieved using specially adapted

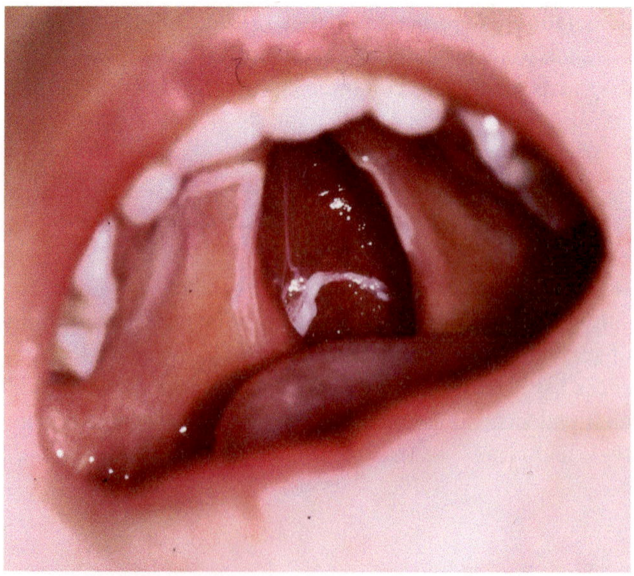

Fig. 13.5 Cleft palate

bottles and upright position to minimize nasal regurgitation.

3. *Velopharyngeal incompetence*—Even if the hard and soft palates are repaired by apposition of the two edges of the CP, the musculature of the soft palate may be abnormal resulting in a poorly functioning soft palate. A series of speech impediments can arise reducing intelligibility including:
   (a) Hypernasal speech
   (b) Backing, glottal or pharyngeal speech
   (c) Nasal turbulence
   (d) Nasal emission
4. *Middle ear effusion ('glue ear')*—impaired Eustachian tube function as a result of compromised soft palate musculature can lead to accumulation of fluid in the middle ear (glue ear), impairing the function of the tympanic membrane and predisposing to infections of the middle ear.

Developmental Pathogenesis (Patho-embryology)

The craniofacial region derives from neural crest cells that delaminate from the neural folds and migrate through the mesenchymal tissue. When considering clefts of the lip and palate, it is noteworthy that the embryonic origins of the primary and secondary palate are different.

The *primary palate* consists of the anterior nasal spine, columella, medial portion of the upper lip and premaxilla. It develops during the fourth to seventh week of embryogenesis.

The *secondary palate* consists of the hard and soft palate and develops during the seventh to twelfth week of embryogenesis.

Anatomically, division of the primary and secondary palate is marked by the incisive foramen in the roof of the mouth and bilateral sutures that extend anteriorly from this midline foramen to the junction between the maxillary lateral incisor and canine tooth.

Development of the Lip and Primary Palate

Summary Timeline of Normal Embryogenesis

- *4th week*:
  - Early oral pit (stoma) appears
  - Frontonasal prominence, paired maxillary processes and paired mandibular processes surrounding the stoma develop
- *End of 4th–5th week*:
  - The nasal placodes (ectodermal thickenings) develop dividing the lower portion of the frontonasal prominence into paired medial and lateral nasal processes.
- *End of 6th week*:
  - The medial nasal processes fuse in the midline to form the nasal tip, columella, prolabial segment and primary palate.
  - The maxillary processes fuse with the lateral aspect of each medial nasal process to form the lateral components of the upper lip.
  - Just prior to fusion, the lateral nasal processes undergo a peak of cell division that is particularly susceptible to teratogenic insults and interruptions to the fusion process.

Molecular Events

Much of our knowledge of the molecular mechanisms involved in the development of the lip and primary palate is derived from studies of the embryonic chick [7].

Events for which the molecular basis is known include:

1. *Outgrowth of facial processes*: controlled by the interaction of fibroblast growth factors (FGFs), sonic hedgehog (SHH), bone morphogenetic proteins (BMPs), the homeobox-containing genes *Barx1* and *Msx1*, the distal-less homeobox-containing genes (*Dlx*) and local retinoic acid gradients.
2. *Fusion events*: comprise an interplay of apoptosis and epithelial-mesenchymal transformations. These are controlled by an array of signalling pathways involving SHH, MSX1 and MSX2, BMPs, FGF and the TP63 genes, implicated in a number of disorders that feature orofacial clefting.

Theories of Cleft Lip Formation [8]

1. *Classic theory*: during week 6 there is failure of the medial nasal processes to fuse with the maxillary prominences resulting in a cleft of the lip and/or primary palate.
2. *Streeter's merging theory*: the epithelial layers of the medial nasal processes and maxillary processes fuse but due to a lack of mesenchymal penetration within this union to support it, the epithelial layers break down resulting in cleft formation.
3. *Dynamic fusion theory*: two possible reasons account for cleft lip formation: lack of epithelial/mesenchymal fusion and mesenchymal hypoplasia. In either case mesenchymal growth may be the driving force behind dynamic fusion.

Development of the Secondary Palate

Summary Timeline of Normal Embryogenesis

- *6th week*:
  - Outgrowths develop from the maxillary processes to form a pair of palatal shelves.
  - These are initially orientated vertically either side of the developing tongue.
- *7th–12th week*:
  - The palatal shelves elevate into a horizontal position to contact each other and fuse their epithelium in the midline.
  - This epithelial seam subsequently degenerates to allow palatal mesenchyme to migrate across the midline.
  - Palatal mesenchyme differentiates to form bony and muscular components of the hard and soft palate, respectively, so forming the 'secondary' palate.
  - This secondary palate eventually fuses with the primary palate and nasal septum, separating the oral and nasal cavities.

Molecular Events

Much of the knowledge of secondary palate development has been derived from experiments on murine models [7].

Events for which the molecular basis is known include:

1. *Vertical outgrowth of palatal shelves*: entails complex signalling cascades involving transcription factors, growth factors and the relevant receptors such as *Osr2*, *Lhx8*, *Msx1*, *Fgf10*, *Fgfr2b*, *Tgfb2* and *Tgfbr2*.
2. *Elevation of palatal shelves*: this is driven by an intrinsic shelf force directed by extracellular matrix components (mainly hyaluronic acid) and local epithelial changes.
3. *Time-sensitive adhesion competence*: this ensures that during elevation of the palatal shelves, they only acquire adhesion capability once they are raised above the tongue. Jagged 2 (JAG2) is important in this process.
4. *Palatal fusion*: molecules involved in this process include cell adhesion molecules, e.g. nectin 1, desmosonal components, growth factors, epidermal growth factor receptor (EGFR) and members of the transforming growth factor β superfamily, e.g. TGFβ3.
5. *Degeneration of the midline epithelial seam*: molecules involved play a role in apoptosis including caspase 3 and TUNEL (terminal deoxynucleotidyl transferase nick-end labelling)

Theories of Cleft Palate Formation

1. Failure of palatal shelves to fuse
2. Abnormal growth of palatal shelves
3. Cell death after fusion
4. Failure of mesenchymal consolidation and differentiation

Aetiology

CL ± P and iCP occur in the setting of multiple genetic and environmental influences [2, 9].

Environmental Risk Factors

Several studies have implicated the role of chemicals in the pathogenesis of orofacial clefting. Most of these studies involve experimental animals, yet it is well recognized that teratogens are species-specific in their effect and extension to humans is not automatic.

1. *Maternal smoking during pregnancy*
   (a) This is consistently linked with an increased risk of CL ± P and iCP with a dose-response relationship noted.
2. *Maternal alcohol consumption during pregnancy*
   (a) Although acknowledged as a cause of foetal alcohol syndrome, the role of alcohol in the aetiology of clefts remains unclear with a dose-dependent relationship likely being associated with orofacial clefting.
3. *Maternal nutrition and multivitamin supplementation*
   (a) Despite suggestions from observational studies of a role for maternal nutrition in cleft aetiology, study design to demonstrate cause and effect remains challenging.
   (b) Use of multivitamins in early pregnancy, however, has been shown in a meta-analysis to be associated with a 25% reduction in prevalence of orofacial clefts.
4. *Folate deficiency*
   (a) Folate antagonists are associated with a higher incidence of orofacial clefts in humans.
   (b) Likewise, folate deficiency in animals is associated with clefts.
   (c) Nationally adopted strategies of folate fortification have produced inconsistent trends in cleft incidence in different countries.
5. *Other nutrients*
   (a) Zinc deficiency causes iCP and other malformations in animals.
   (b) Riboflavin may play a part in orofacial clefting.
   (c) Vitamin A may also play a part in orofacial clefting given that foetal exposure to retinoid drugs (Vitamin A derivatives) can cause severe craniofacial anomalies.
6. *Maternal exposure to organic solvents*
   (a) This has been inconsistently associated with CL ± P and iCP.

7. *Maternal exposure to anticonvulsants*
   (a) Exposure to diazepam, phenytoin and phenobarbital has been associated with an increased risk of orofacial clefting.

Genetic Factors

CL ± P features in more than 200 specific genetic syndromes and iCP features in twice as many. However, the majority of patients with cleft lip and/or palate (70%) present non-syndromically and with unpredictable inheritance.

A deeper understanding of the genetic contribution to orofacial clefting has been gained through a combination of epidemiologic, candidate gene and Genome-Wide Association Studies (GWAS) as well as animal experimental models. Major advances have been made in identifying the genetic mutations underlying syndromic forms of clefting [2]. In contrast, progress has been challenged in non-syndromic clefting by the genetic heterogeneity in combination with epigenetic effects which, in turn, require larger data sets for meaningful studies. To this end, the Cleft Collective cohort studies conducted within the UK are likely to amass sufficient genetic data to power further study into the genetic and epigenetic components of CL ± P aetiology [10].

Antenatal Diagnosis

Methods of Antenatal Diagnosis [11]

Ultrasound technology has traditionally been used to detect foetal anomalies during pregnancy, in particular the second trimester scan as the foetal face is not imaged adequately in first trimester screenings. Whilst *2D trans-abdominal ultrasound* is the routine modality used in most parts of the world, detection rates for orofacial clefting vary considerably and are operator-dependent. In general, iCP is more difficult to visualize than CL ± P with the former eluding detection until the neonatal check immediately after birth.

Improvements in ultrasound technology by way of *3D and 4D ultrasound* have further improved accuracy and image capture and are popular with would-be parents. iCP detection, however, remains challenging even with these methods. Furthermore, all ultrasound methods are complicated by false positives due to shadowing.

A few cleft centres have adopted *magnetic resonance imaging* (MRI) screening for those cases where oroclefting is suspected on ultrasound or there is a strong family tendency towards clefting. Such imaging can potentially inform infant prognosis thus influencing maternal management, delivery timing and enabling discussion of foetal surgery or EXIT (*ex utero intrapartum* treatment) procedures.

Antenatal Management

Once a diagnosis of a cleft is made, the parents are referred to a cleft team for antenatal counselling. Studies confirm that the majority of parents prefer to arm themselves with knowledge about clefts prenatally. Prognosis and quality of life estimates are tempered according to whether the cleft is diagnosed within the context of other anomalies or chromosomal abnormalities. Feeding and breathing concerns as well as the typical surgical timeline are among the priorities to be discussed. Input from a geneticist and social services may also be of value.

Disease Registries and Patient Groups

These include:

1. *CRANE* https://www.crane-database.org.uk/The CRANE Database collects information on adults and children born with cleft lip and/or cleft palate throughout England, Wales and Northern Ireland. This information is stored electronically at the Royal College of Surgeons of England. An independent body, the Cleft Development Board, which represents patient representative groups, clinicians and commissioners, has the overall responsibility for running the database.
2. *EUROCRAN* http://cordis.europa.eu/result/rcn/38751_en.htmlEUROCRAN (European Collaborations on Craniofacial Anomalies) began in 2000 and is funded by the European Union (EU). Its aim is to advance several lines of research arising from the Eurocleft Cohort Study and related work in the Eurocleft clinical network.
3. *WHO International Collaborative Research on Craniofacial Anomalies Project* http://www.who.int/genomics/anomalies/cfaproject/en/This project aims to establish an international network for the purpose of encouraging research in craniofacial anomalies (CFA) and a global registry of such conditions. It focuses on four main areas:
   (a) Genetic basis of craniofacial anomalies
   (b) Hereditary and environmental interactions involved in craniofacial anomalies
   (c) Prevention of craniofacial anomalies
   (d) Optimal treatment of craniofacial anomalies

References

1. Fraser FC. Thoughts on the etiology of clefts of the palate and lip. Acta Genet Stat Med. 1955;5(4):358–69.
2. Dixon MJ, Marazita ML, Beaty TH, Murray JC. Cleft lip and palate: understanding genetic and environmental influences. Nat Rev

Genet. 2011;12(3):167–78. PubMed PMID: 21331089. Pubmed Central PMCID: 3086810

3. Croen LA, Shaw GM, Wasserman CR, Tolarova MM. Racial and ethnic variations in the prevalence of orofacial clefts in California, 1983-1992. Am J Med Genet. 1998;79(1):42–7. PubMed PMID: 9738868
4. Mossey PA. Epidemiology of oral clefts: an international perspective. In: Wyszynski DF, editor. Cleft lip and palate: from origins to treatment. Oxford: Oxford University Press; 2002. p. 127–58.
5. Calzolari E, Bianchi F, Rubini M, Ritvanen A, Neville AJ, EUROCAT Working Group. Epidemiology of cleft palate in Europe: implications for genetic research. Cleft Palate Craniofac J. 2004;41(3):244–9. PubMed PMID: 15151454
6. Boehringer S, van der Lijn F, Liu F, Gunther M, Sinigerova S, Nowak S, et al. Genetic determination of human facial morphology: links between cleft-lips and normal variation. Eur J Hum Genet. 2011;19(11):1192–7. PubMed PMID: 21694738. Pubmed Central PMCID: 3198142
7. Mossey PA, Little J, Munger RG, Dixon MJ, Shaw WC. Cleft lip and palate. Lancet. 2009;374(9703):1773–85. PubMed PMID: 19747722
8. Senders CW, Peterson EC, Hendrickx AG, Cukierski MA. Development of the upper lip. Arch Facial Plast Surg. 2003;5(1):16–25. PubMed PMID: 12533133
9. Watkins SE, Meyer RE, Strauss RP, Aylsworth AS. Classification, epidemiology, and genetics of orofacial clefts. Clin Plast Surg. 2014;41(2):149–63. PubMed PMID: 24607185
10. Cleft Collective: Cleft Collective. University of Bristol. Available online at http://www.bristol.ac.uk/dental/cleft-collective/, accessed on April 2nd 2019.
11. Goudy SL, BL. Chapter 1: genetics, prenatal diagnosis and counseling, and feeding in complete cleft care. 2014.

Facial Clefts

14

Tom W. M. Walker, Ben C. Green, Caroline Mills, and Peter Ayliffe

Congenital Anomalies of the Face and Palate

1. *Inclusion dermoid* (dermoid = skin-like) (Fig. 14.1a):
   - It is a cystic swelling which may be found along the lines of fusion of the face. The commonest site is the lat-end of the upper eyebrow (called external angular dermoid).
2. *Macrostomia* (large mouth) (Fig. 14.1b):
   - It is due to incomplete fusion between the maxillary and mandibular processes (too little closure of the stomodeum).
3. *Microstomia* (small mouth) (Fig. 14.1c):
   - Due to excessive fusion between the maxillary and mandibular processes leading to great closure of the stomodeum.
4. *Oblique facial cleft* (Fig. 14.1d):
   - It is a rare condition resulting from failure of fusion between the maxillary process and frontonasal process. There is a cleft lip which extends along the side of the nose to reach the med-angle of the eye.
5. *Lateral cleft upper lip (cleft lip)* (Fig. 14.1e, f):
   - It is due to failure of fusion between the maxillary process and the philtrum (of frontonasal process). It may occur on one side (unilateral cleft lip) or on both sides (bilateral cleft lip).
6. *Cleft palate*: Due to failure of fusion between the different segments of the palate. It may be partial or complete, unilateral or bilateral (Fig. 14.1g, h, i):
   (a) *Unilateral complete cleft palate*: A cleft runs between the two palatine shelves and then between the premaxilla and one palatine shelf.
   (b) *Bilateral complete cleft palate*: The cleft between the two palatine shelves extends anteriorly in a V-shaped manner separating the premaxilla from the two palatine shelves
   (c) *Partial cleft palate*: May affect the soft palate alone or extends to the post-part of hard palate.
   (d) *Cleft uvula*: The cleft affects the uvula alone.

Introduction

Facial clefts are a group of rare transgressional craniomaxillofacial deformities that affect both hard and soft tissues of the face. They often cause severe deformity and are a challenge to all who work with them. Their origin is felt to be at the junction of facial processes and the clefts, often centre around the oral cavity and connect with the nose, eye, maxilla, ear and cranial vault – however, are often partial or incomplete. They are complex three-dimensional dysplasias that vary in severity and extent. They often also present with a malformation of structures within the skin (hair follicles, eyebrows, eyelashes) and the dental arch and can be associated with skin tags and notching of the eyelids or nostril. If these signs present alone, bony clefts should be sought radiographically.

There are multiple complex and elaborate classification systems that are in use. The *anatomical* (clinical) classification system by Tessier [1] has become the most accepted internationally. It is important also to draw attention to the *embryological* classification system put forward by Van Der Meulen [2], as well as a more simple directional classification system by Grob. The latter (transverse or lateral, oblique, median (upper lip) and median (lower lip) is a useful starting point [3].

T. W. M. Walker (✉)
University Hospitals Bristol NHS Foundation Trust, Bristol, UK
e-mail: tom.walker@bristol.ac.uk

B. C. Green
Department of Oral and Maxillofacial Surgery, Queen Victoria Hospital, East Grinstead, West Sussex, UK

C. Mills
Great Ormond Street Hospital, London, UK

P. Ayliffe
Great Ormond Street Hospital, London, UK

King Edward VII Hospital, Marylebone, UK

R. Carachi, S. H. E. Doss (eds.), *Clinical Embryology*, https://doi.org/10.1007/978-3-319-26158-4_14

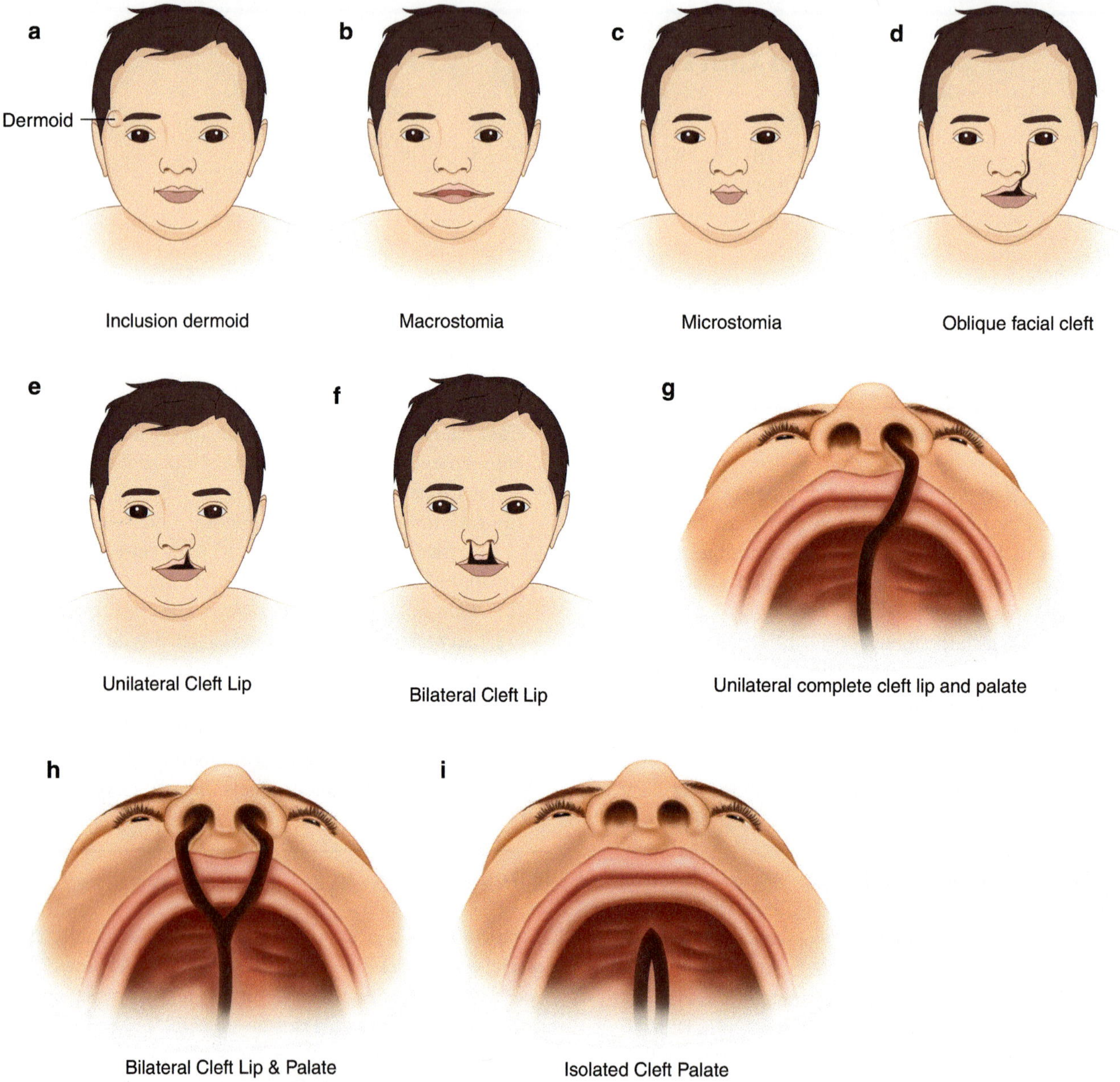

Fig. 14.1 (**a–f**) Congenital anomalies of the face and palate

Epidemiology

Individual (Tessier) facial clefts are rare. They are often grouped directionally (Grob) to report their epidemiology and often reported as a fraction of all facial cleft patients. Lateral facial clefts represent 1/100–350 clefts and 1/50,000 to 175,000 live births. However some authors suggest it may be more common 1/3000 and 1/5000 live births. They are more frequent in males. Oblique facial clefts, however, affect 1/250–1300 clefts and 2–5/100,000 live births. It is reported that 20% of oblique facial clefts are bilateral, if unilateral, they are more common on the right. There is no gender predilection. The median upper lip and median lower lip clefts are both very rare. The former account for 0.2% of clefts and the latter 4–5/million live births [1–5].

The documentation of these cases in low to middle income countries will not be rigorous. The global burden of these malformations is small. However this should not take away from the effect of these malformations on the children who suffer them and their families.

Aetiology

The pattern of facial clefts is generally consistent around the world; this provides support to the theory that polygenetic, environmental factors affecting the development of tissue in the face play a role. Environmental factors implicated include antenatal exposure to radiation, viral infections, metabolic abnormalities and tetrogenic compounds. It is felt that changes in severity relate to the timing of the causative insult and the susceptibility of the developing area [1–6].

Clinical Features

The cleft will affect the structure of the part of the face through which it runs. The extent will vary and can affect all tissues. They can be bilateral in 20–35% of cases and are often associated with standard cleft lip and palate. They will present with short stature, sparse eyebrows and eyelashes, abnormal hair follicle patterns, lower lid coloboma, abnormal nose, involvement of nasolacrimal duct and malar hypoplasia. There will be dental abnormalities in terms of structure and eruption of teeth. There may also be mental retardation, anophthalmia/microphthalmia and hemimelia.

Clefts that extend in to the frontal bone (oblique) can have encephaloceles, craniosynostosis and orbital hypertelorism. Midline (upper lip) facial clefts are part of a spectrum of holoprosencephaly, and cerebral defects may also be present in this group. Midline or oblique clefts can be associated with severe nasal deformity including aplasia, proboscis and nasoschizis. Lateral (transverse) clefts can have microtia and accessory ear tags.

Up to 40% can have limb abnormalities ranging from talipes equinovarus to intrauterine amputation, constriction rings and pseudosyndactyly.

Classification Systems

Facial clefts were first described by Morian in 1887 and divided into type I, medial to the infraorbital foramen, and type II, lateral to the infraorbital foramen. The American Association of Cleft Palate Rehabilitation (AACPR) described a classification system in 1962 that divided facial clefts in to four major groups (Table 14.1). This system does not include midline facial clefts or underlying bony defects so has become obsolete. Grob (1957) used the terms lateral (transverse), oblique and median. Boo-Chai in 1970 subdivided oro-ocular clefts into type I and type II (medial and lateral) [1–5].

Table 14.1 AACPR classification of facial clefts

Group 1	Mandibular process clefts
Group 2	Naso-ocular clefts
Group 3	Oro-ocular clefts
Group 4	Oro-auricular clefts

In 1976 Paul Tessier [1] proposed an anatomical orbitocentric descriptive classification system (Fig. 14.2). The system consists of facial clefts numbered in relation to a central "zero" line (0–14 (a line running from the between the upper central incisors up to the skull) and 0–30 (a line running from the lower central incisors running through the mandibular symphysis)). The orbit is the central point of the classification and separates the facial structures from the cranium. Clefts 0–7 are facial clefts, and 8–14 can be considered the cranial counterparts of the facial clefts. The clefts can connect the mouth and lips to the eye and also continue to the cranial structures. They can be associated with or independent of a cleft lip. Clefts of the soft tissues and bone do not always coincide – and often several different clefts coexist.

In 1983 Van der Meulen [2] proposed a classification system based on focal foetal dysplasias of the face. It fits well with the embryogenesis of the face – the union of facial processes.

Associated Anomalies

Lateral (Transverse)

Micrognathia, nasal dermoid, tetralogy of Fallot and intestinal abnormalities, accessory maxilla, limb abnormalities, hemifacial microsomia and Treacher Collins syndrome

Oblique

Hernia, genitourinary abnormalities, talipes, cutaneous tags, spinal, costal and limb abnormalities, encephalocele, hydrocephalus, mental retardation, anophthalmos

Median (Upper Lip)

Oral-facial-digital syndrome, hamartomas, double frenum, Ellis-van Creveld syndrome, non-syndromic upper limb deformity.

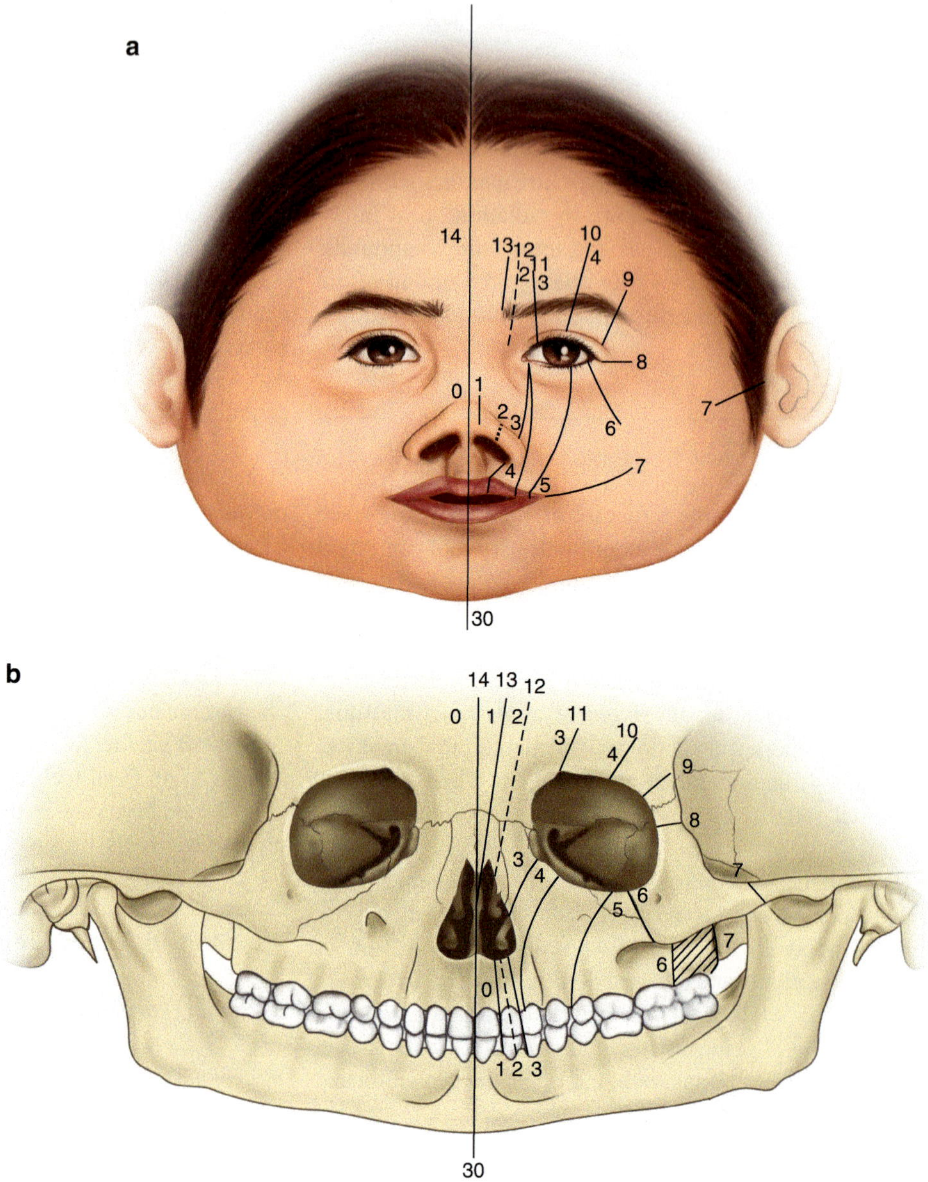

Fig. 14.2 Tessier clefting system. (**a**) Soft tissue clefts. (**b**) Bony clefts. Dotted lines represent uncertain localisation or uncertain clefting. Note that the northbound cranial line has different number than its counterpart southbound facial line. This system is descriptive and anatomic. It avoids aetiology and pathogenic speculation. Tessier P, J Maxillofac Surg 4:69, 1976) [1]

Median (Lower Lip)

Symbrachydactyly, digital contractures, cleft palate, congenital heart anomalies, ear tags, midline dermoids, epiglottis aplasia, absent tongue.

Treatment Principles

Multidisciplinary treatment will be required and is often led by the oral and craniomaxillofacial surgical team. The first priority is to ensure that the potential to harm any sight is reduced due to a cleft preventing adequate closure of the eyelids. The second priority is to ensure a safe and effective upper aerodigestive tract to allow infection-free respiration and adequate nutrition. Often both of these stages happen at the same time.

There may be a requirement to treat malformation of the spinal cord, brain and heart also – these clearly take precedence over much of the facial surgery.

The reconstructive surgical process is long and will involve many specialties, including orthodontics, speech pathology, psychology and surgery. Surgical principles centre around restoration of the dynamic muscular sphincters (oral, orbital) (3–6 months) and separation of oral, nasal and orbital spaces (1–5 years) and with time bone grafts to the dental arch to allow eruption of the teeth and to support the facial skeleton (10–12 years), orthodontics and further orthognathic/craniofacial surgery to adjust the facial profile and dental occlusion (16–20 years).

Often these children are treated in low to middle income countries by surgical missions, and only soft tissue surgery is completed. Figures 14.3, 14.4, 14.5, 14.6 and 14.7 show examples of facial clefts.

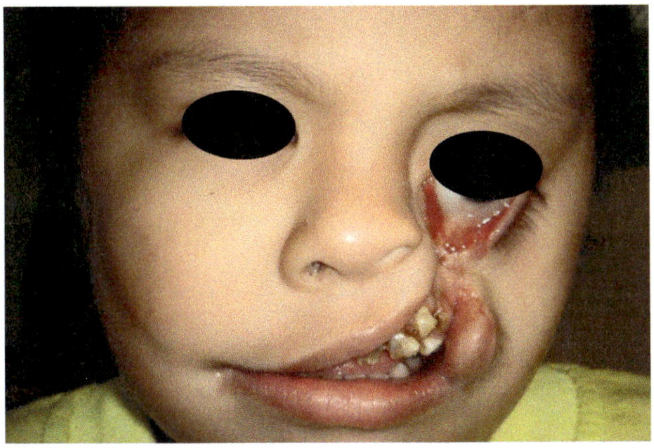

Fig. 14.3 Child with a right-sided Tessier 7 cleft with macrostomia and ridge in soft tissue of cheek, left-sided Tessier 4 cleft. Note the nostril elevation, absent eyelashes medially and hypoglobus. The cleft involves the maxilla and orbit

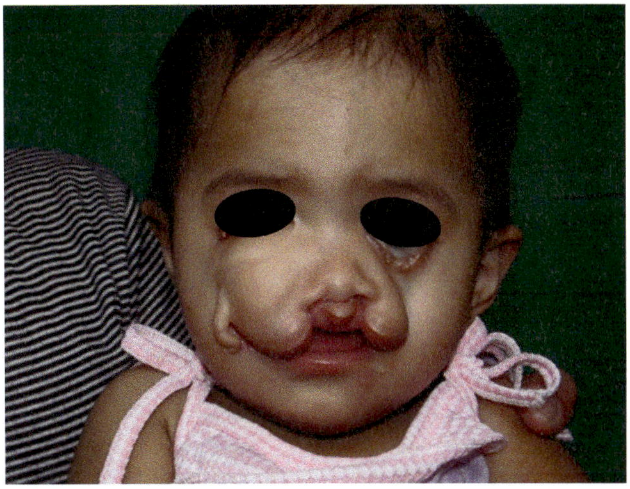

Fig. 14.5 Child with right Tessier 7 and left Tessier 5. Note the bilateral cleft lip (and palate) and right-sided skin tag

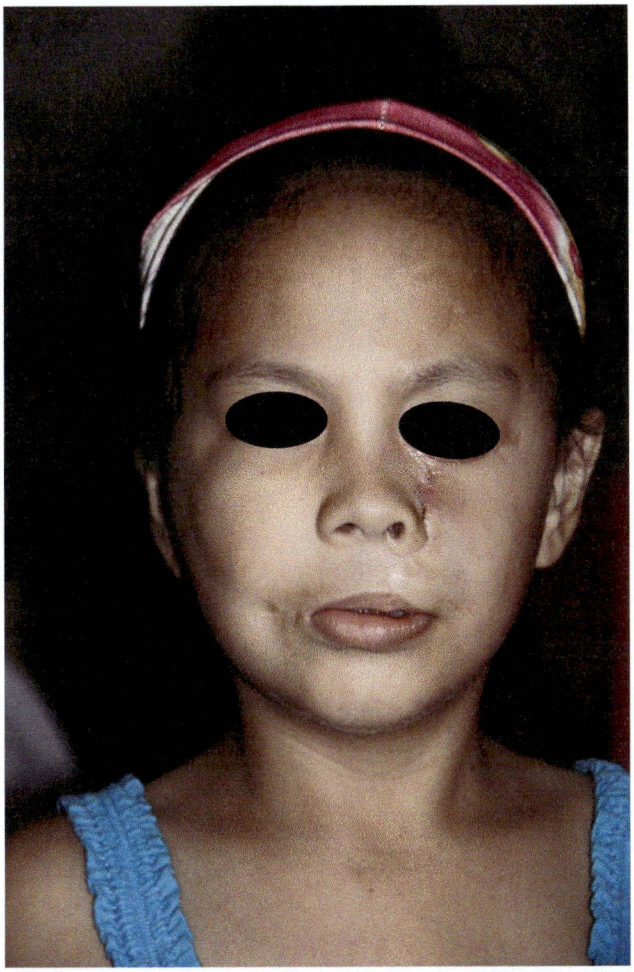

Fig. 14.4 Following the first stage of correction. Priority should be to correct vital structures such as the eye

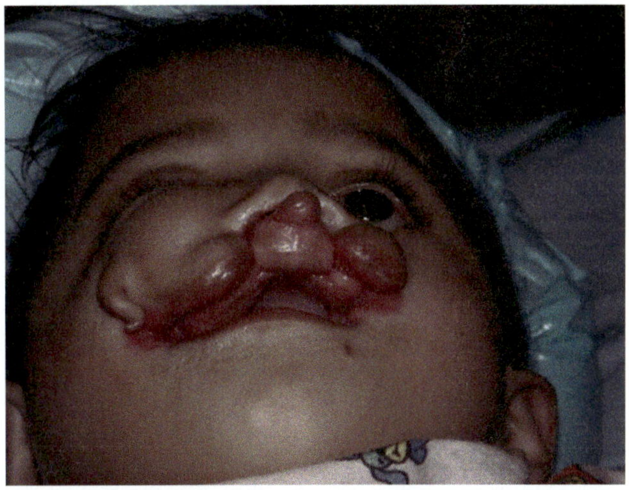

Fig. 14.6 Child with right Tessier 7 and left Tessier 5. Note incomplete bilateral cleft lip (and palate) and right-sided skin tag

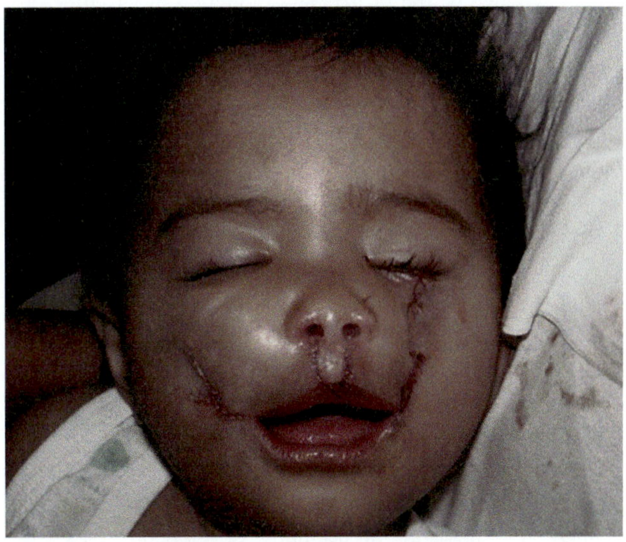

Fig. 14.7 Child in 14.5 and 14.6 after initial surgery

The Patho-embryology of Facial Clefts

There are several theories that attempt to explain the patho-embryology of facial cleft [5]. The classic theory was first proposed by Dursy and His [7, 8]. This theory argues that clefts result from the failure of fusion of the facial processes [6]. It suggests that the face forms as the finger-like ends of the maxillary processes meet and coalesce with the united paired globular processes beneath the facial pits. Once epithelial contact is established, mesenchymal penetration completes the fusion, and the lip and hard palate are formed. Clefting arises when this sequence is disturbed.

There is also the mesodermal penetration theory, first developed by Pohlmann [5] and Veau [9] and was later advocated by Stark and Saunders [10, 11]. This theory argues that free-end facial processes do not exist and that the face consists of a bilateral ectodermal membrane and the epithelial seams demarcate the major processes. The mesenchyme migrates into the double wall of ectoderm, penetrates it and smooths the seams. If mesenchymal penetration fails, the unsupported wall dehisces resulting in cleft formation. The severity of the cleft is inversely proportional to the success of mesodermal penetration, leading to varying degrees of cleft.

The third theory is the neuromeric theory of embryological development. The clinical significance of this model is that you can map out the anatomical sites of origin for all zones of ectoderm and mesoderm supplied to a given zone of the nervous system [12]. The craniofacial skeleton is derived predominantly from neural crest cells. There are several mechanisms by which neural crest cell movement is interrupted account for the type of abnormalities observed. Premigratory losses result from errors within the rhombomere [13]. There can also be defective migration, inadequate neural crest cells or the inability of neural crest cells to induce a neurovascular supply that can lead to anomalies. There can also be the insufficient induction of vascular support. Ewings and Carsten argued in 2009 that the clinical observations made by Tessier were derived from empirical observations but closely match patterns of neural crest migration as explained by the neuromeric theory [13].

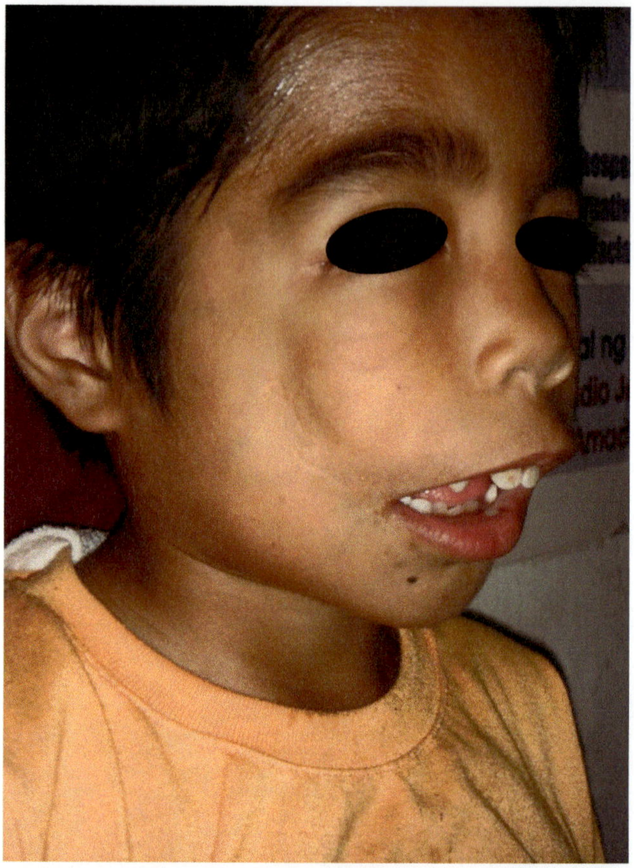

Fig. 14.8 A child with a Tessier 7 cleft lateral (transverse) unilateral macrostomia, ridge in cheek and slanting palpebral fissure can be seen

Clefts 1–9 (Fig. 14.8)

Clefts 1–9 are organised around the sensory branches of the maxillary branch of the trigeminal nerve. The key structure is the pterygopalatine ganglion. The series of infraorbital clefts (clefts 1–6) are organised around solitary mesenchymal fields.

The mesenchyme giving rise to cleft 1 is found at the junction of rhombomeres 1 and 2 and so lies caudal to rhombomere 1. Therefore, a normally developed pre-existing ethmoid plate is needed for the proper formation of the premaxillary segment. The loss of this branch or the neural crest cells giving rise to the mesenchyme may lead to loss of tissues within zone 1 [13].

The zone 1 cleft represents a defect of the central incisor portion of the premaxilla and the vomer. The premaxilla develops before the vomer, and so the premaxillary fusion to the neighbouring maxillary field fails due to the abnormal premaxillary field.

The zone 3 cleft spans three fields: the palatine bone, maxillary palate and inferior turbinate. The mesenchymal origin is from r2. The maxillary palate and palatine bones form at the same time. Therefore in the zone 3 cleft, the inferior turbinate may be absent whilst the palatine bone may be present. The footplate of the lacrimal bone rests on inferior turbinate and, if the bone is not present, will lead to disruption of lacrimal system. All skin medial to the nasolacrimal duct is derived from mesenchyme from the telencephalon. This is an example of a developmental field anomaly where there is complete absence of the developmental field.

The zone 4 defect has tissue that is derived from the mesenchyme of r2. There can be loss of the medial third and orbital rim and floor resulting in medial and inferior globe prolapse. The infraorbital musculature is disrupted as a result of failure of myoblast migration through the cleft in the developing embryo. The zone 5 cleft origin also arises from the r2 mesenchyme but caudal to that of the zone 4 cleft. Cleft 6 represents deficiency of the maxillary wall and also receives mesenchyme from r2.

The zone 7 cleft can be seen as a part of several syndromic conditions that involve multiple r2-derived fields as in Treacher Collins (TCS) and Goldenhar syndromes where the pathogenesis is the result of abnormal migration of neural crest cells (TCS) or loss of relevant blood supply (Goldenhar). The zone 8 cleft can also be seen as an isolated condition (loss of r2 mesenchyme) or as a component of syndromic craniofacial disorders. Soft tissue defects are common in Goldenhar's, whereas osseous defects are more common in TCS.

Deficiency of the greater wing of the sphenoid is a component of zone 9 cleft. It is unique in that it occurs at a watershed area between p5 (frontal-derived mesenchyme) and mesenchyme derived from r2.

Clefts 10–14 (Fig. 14.9)

These skull series of craniofacial clefts are unique as they result from pathological involvement of multiple interacting fields. The r1 sphenoethmoid and the p5 (telencephalon-derived) frontal, nasal and lacrimal fields are all potentially affected by mesenchymal irregularities in any field. These overlapping zones can lead to different anatomical outcomes due to the lamination process in the case of cleft zones 10 and 11 and stacking that occurs in zones 12–14 during the folding and development of the embryo.

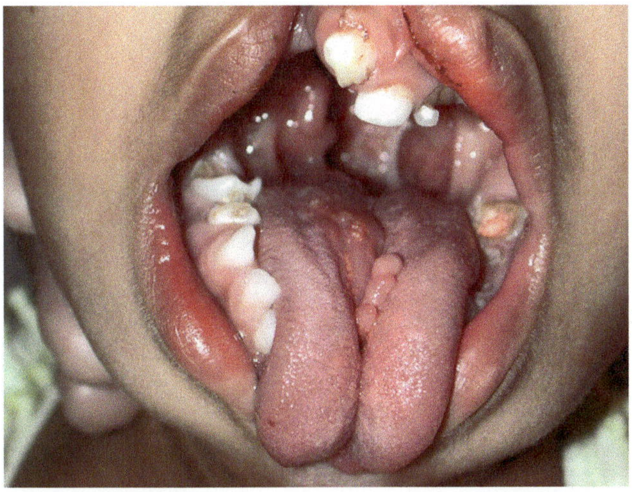

Fig. 14.9 Child with upper left cleft lip and lower lip median cleft with associated bifid tongue and notched mandible

Tessier Cleft 30 (Median Mandibular Cleft)

The embryological basis of median mandibular cleft is generally believed to lie in the failure of coaptation of the free ends of the first branchial arches in the midline. However, there is a theory which suggests that the embryonic mandibular prominences do not fuse in the midline but the furrow that is generated separating the two prominences is gradually pushed to the surface by increased proliferation of neural crest cells subadjacent to the furrow [5]. This is based on the concept that the mandible is a single bulging process formed chiefly by subepithelial neural crest cells that migrate inferiorly from the neural tube to populate the mandibular region from both sides. It has also been hypothesised that in some cases, it may be that the symphyseal cartilages in the two halves of the midline do not fuse adequately, thereby leaving a defect [14].

Once the branchial arches are formed through neural crest cell migration, vascularisation and mesodermic myoblastic ingrowth, growth centres are organised at the tips that are responsible for closing the final gap and coalescing the two sides. The incisor teeth are often missing along the medial mandibular margins, suggesting that partial or complete failure of growth centre differentiation and development may be responsible for these defects as opposed to a simple merging and contact maintenance. Presumably, inferior cervical defects would result from similar mechanisms in the second and third branchial arches. If there is a superior midline defect (cleft palate) with a median mandibular cleft, it is likely the result of a broader defect in epithelial mesenchymal merging is occurring or that the tongue defect that can accompany it affects palatal shelf closure [14–16]. The bifid tongue could result from a persistent intermandibular groove [17].

References

1. Tessier P. Anatomical classification of facial, craniofacial and Latero-facial clefts. J Oral Maxillofac Surg. 1976;4:69–92.
2. van der Meulen JCH. Oblique facial clefts: pathology, aetiology and reconstruction. Plast Reconstr Surg. 1985;76:212–24.
3. Gorlin RJ, Cohen MM, RCM H. Syndromes of the head & neck. 4th ed. Oxford: Oxford University Press; 2001.
4. David DJ, Moore MH, Cooter RD. Tessier clefts revisited with a third dimension. Cleft Palate J. 1989;26:163–85.
5. Hunt JA, Hobar PC. Common craniofacial anomalies: facial clefts and encephaloceles. Plast Reconstr Surg. 2003;112:606.
6. Kawamoto HK. The kaleidoscopic world of rare craniofacial clefts: order out of chaos (Tessier classification). Clin Plast Surg. 1976;3:529.
7. Dursy E. Zur Entwicklungsgeschichte des Kopfes des menschen und der Hoeheren wirbelthiere.T übingen. Germany: Verlag der H. Lauppschen-Buchhandlung; 1869. p. 99.
8. His W. Die Entwickelung der menschlichen und tierischer Physiognomen. Arch Anat Entwicklungsgesch. 1892;384.
9. Veau V. Hasencharten menschlicher Keimlinge auf der Stufe 21–23 mm SSL. Z Anat Entwicklungsgesch. 1938;108:459.

10. Stark RB. The pathogenesis of harelip and cleft palate. Plast Reconstr Surg. 1954;13:20.
11. Stark RB, Saunders DE. The first branchial syndrome: the oral-mandibular-auricular syndrome. Plast Reconstr Surg. 1962;29:299.
12. Carstens MH. Neural tube programming and craniofacial cleft organisation. 1. The neuromeric organization of the head and neck. Eur J Paediatr Neurol. 2004;8:181–210.
13. Ewings EL, Carstens MH. Neuroembryology and functional anatomy of craniofacial clefts. Ind J Plast Surg. 2009;42(suppl):S19–34.
14. Cosman B, Crikelair GF. Midline bronchogenic syndromes. Plast Reconstr Surg. 1969;44:41.
15. Ranta R. Incomplete median cleft of the lower lip with cleft palate, the Pierre Robin anomaly or hypodontia. Int J Oral Surg. 1984;13:555.
16. Eppley BL, van Aalst JA, Robey A, et al. The spectrum of orofacial clefting. Plast Reconstr Surg. 2005;101e:115.
17. Pohlmann EH. Die embryonale Metamorphose der Physiognomie und der Mundhoehle des Katzenkopfes. Morphol Jahrb Leipzig. 1910;41:617.

TMJ and Mandibular Congenital Malformations

15

Nabeela Ahmed and N. Shaun Matthews

Malformations of the TMJ and mandible are more frequently seen in the acquired and developmental setting. This chapter will explore the congenital malformations of the TMJ and mandible which by their very nature are often associated with a syndrome. The most commonly encountered malformations of this region will be discussed individually, to include the more commonly known and documented syndromes.

Development of the Mandible

At about week 4 in utero, the pharyngeal pouches are laid down. The first of these leads to development of the mandibular arch.

Each one of these five branchial arches has four components:

1. Central cartilage rod
2. Muscular component
3. Vascular component
4. Neural component

The TMJ is a joint which is largely still under development at birth and is the last joint to start developing at about week 7 in utero. At birth, the articular surfaces are covered with fibrous connective tissue, which is converted to fibrocartilage as the fossa evolves and allows formation of the condyle to occur. Development of the mandible itself occurs largely in utero and continues after birth. The coronoid process is formed at about week 14 when secondary cartilage is laid down and subsequently fuses with the rest of the expanding intramembranous bone of the ramus which then disappears before birth.

N. Ahmed (✉)
Department of Oral and Maxillofacial Surgery, Queens Medical Centre, Nottingham, UK

N. Shaun Matthews
Department of Oral and Maxillofacial Surgery, University of North Carolina School of Dentistry, Chapel Hill, NC, USA

Hemifacial Microsomia

This is the second most common facial anomaly after cleft lip and palate.

Hemifacial microsomia occurs in about 1 in 5600 births. On average, 1 in 10,000 babies is born with microtia or atresia. The syndrome varies in severity but always involves a degree of microtia or atresia or both.

In order to better understand HFM, some consideration needs to be applied to the most commonly used classification system for it, known as the OMENS system [1].

O = Orbital asymmetry
M = Mandibular hypoplasia
E = Ear deformity
N = Nerve involvement
S = Soft tissue deficiency

Each component is graded from 1 to 3 depending upon the degree of deformity present, as assessed by physical, clinical and radiographic evaluation.

It is possible for hemifacial microsomia to be diagnosed antenatally. Work undertaken shows the gestational age at diagnosis ranges from 14 to 35 weeks, and almost half of the cases were associated with either poly- or oligohydramnios.

Facial structures are involved in 52% of cases and include microphthalmia, ear anomalies, facial asymmetry and a facial cleft. Central nervous system defects occur in 47% and include hydrocephalus, occipital encephalocele and cerebellar hypoplasia. Congenital heart defects, primarily atrioventricular septal defects, occur in 19% [2].

R. Carachi, S. H. E. Doss (eds.), *Clinical Embryology*, https://doi.org/10.1007/978-3-319-26158-4_15

Given the frequency with which this syndrome occurs, there is still discussion about optimal timing for treatment of such defects, in order to minimise complications and relapse based on whether the patient is still growing. Whether any final result is likely to be stable or require any further soft tissue augmentation [3] remains a contentious subject.

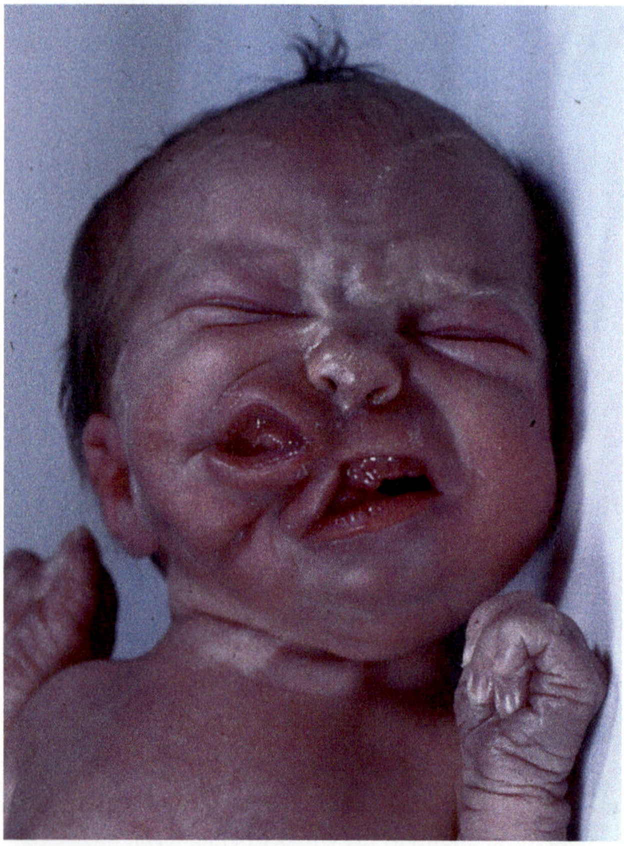

Fig. 15.1 Mouth duplication

As with cleft defects, these patients should be managed in departments with a multidisciplinary input. There is a wide spectrum to the phenotypical appearance of this syndrome and its progression, and treatment planning to manage such cases needs to be individualised [4].

Pierre Robin Syndrome

This heterogeneous birth defect has a prevalence of approximately 1 per 8500 live births. The male-to-female ratio is 1:1, except in the X-linked form.

It was first described in 1891 as a triad of features [5] and is characterised by:

- Micrognathia
- Posterior displacement or retraction of the tongue often secondary to the micrognathia and known as glossoptosis
- Upper airway obstruction
- A cleft palate (which is present in the majority of patients and is commonly U-shaped)

These patients are often diagnosed in utero and may require urgent surgery after birth by the provision of a surgical tracheostomy in order to prevent airway obstruction.

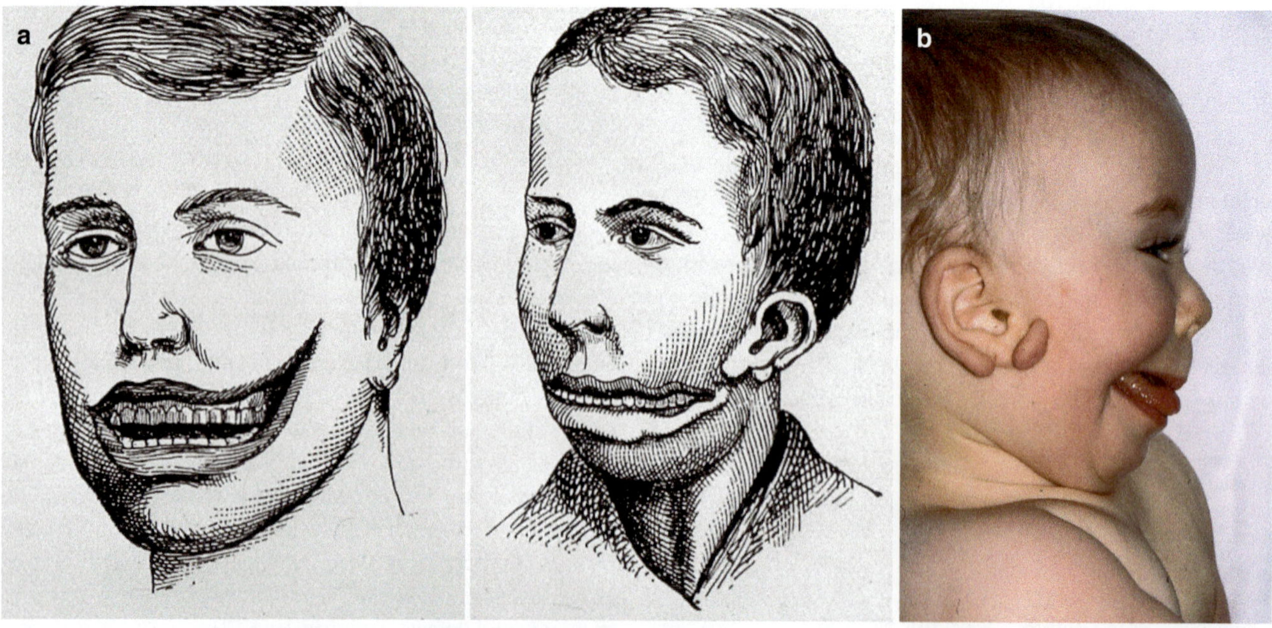

Fig. 15.2 (**a**) Macrostoma (nineteenth century). (**b**) Macrostoma

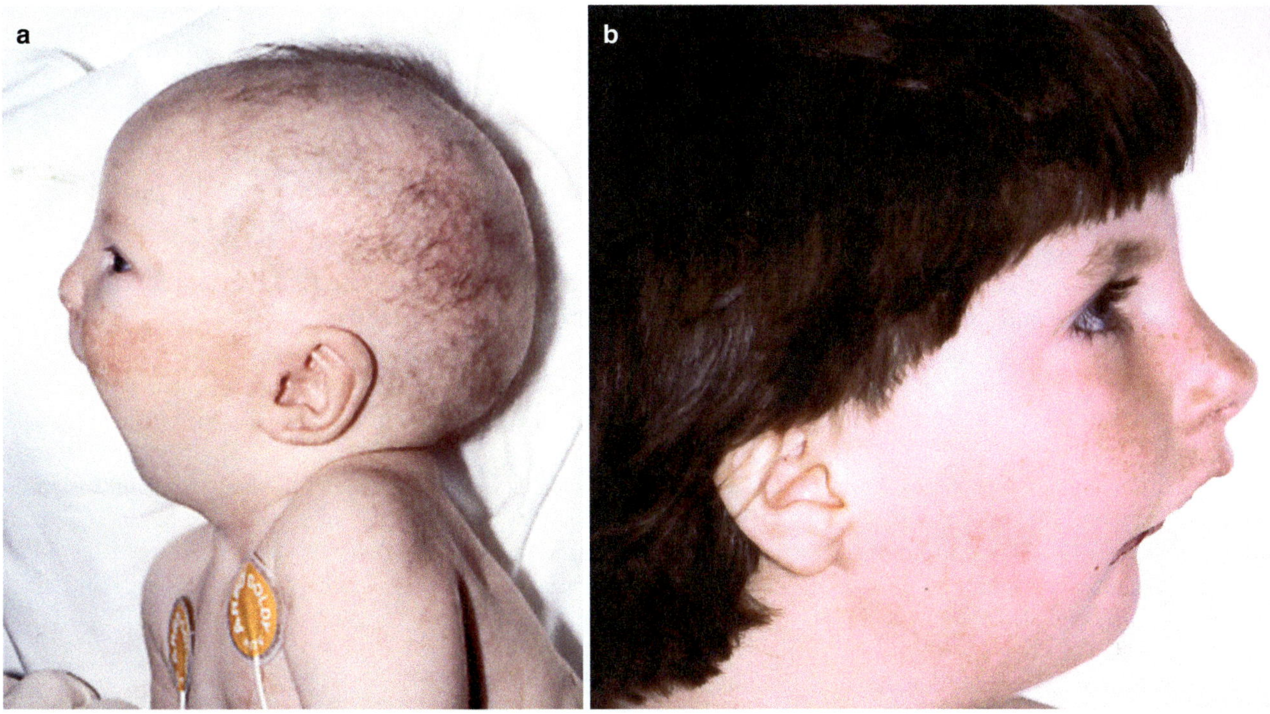

Fig. 15.3 (**a**) Micrognathia. (**b**) Pierre Robin syndrome

Treacher Collins Syndrome

(Also known as mandibulofacial dysostosis or Franceschetti-Zwahlen-Klein syndrome)

This was first reported in 1846 by Thompson [6]. It earnt the name it is more often to referred by in 1900 having been fully described by an ophthalmologist of the same name [7]. Treacher Collins syndrome (TCS) results from mutations in the TCOF1 or POLR1D gene. It is considered an autosomal dominant condition and affects an estimated 1 in 50,000 people.

Characteristics include:

- Down-slanting eyes.
- Notched lower eyelids.
- Malar, zygomatic and orbital hypoplasia or aplasia.
- Mandibular hypoplasia.
- Underdeveloped, malformed and/or prominent ears.
- Despite the facial appearance, most children with Treacher Collins have normal development and intelligence; however, it is important that they undergo early hearing tests.

Surgical management of these cases is complex and requires management in a specialist centre with a multidisciplinary approach [8]. Early operative intervention is guided by the need to protect and establish a safe airway and protect the eyes. Subsequent input is then required to correct airway, facial and aural deficiencies.

Goldenhar Syndrome

This was first reported in 1952 by Dr. Maurice Goldenhar [9].

Goldenhar syndrome (also known as oculo-auriculo-vertebral syndrome) is a rare congenital defect characterised by incomplete development of the ear, nose, soft palate, lip and mandible.

It is associated with anomalous development of the first and second branchial arches and is sometimes considered to be a form of hemifacial microsomia with ocular dysfunction [10].

Hurlers Syndrome

Hurler syndrome has an overall frequency of 1 per 100,000. The mucopolysaccharidoses as a whole have a frequency of 1 in every 25,000 births, and this one is of particular interest as it has a characteristic facial appearance given the bilateral condylar hypoplasia which is evident from birth [11]. There are also associated clinical features which appear later in life which may be of interest to the reader but beyond the remit of this chapter to explore [12].

Hallermann-Strieff Syndrome (HSS)

Hallermann-Streiff syndrome was first reported in 1948 [13].

It is characterised by a typical skull shape (brachycephaly with frontal bossing), hypotrichosis, microphthalmia, cataracts, beaked nose, micrognathia, skin atrophy, dental anomalies and proportionate short stature. Mental retardation is present in a minority of cases [14].

HSS is a rare genetic condition which involves multiple congenital abnormalities chiefly affecting the head and the face. Around 150 cases have been reported in the literature worldwide and given its rarity, the condition requires management in a multidisciplinary team environment [15].

Nager Syndrome

This is an extremely rare inherited syndrome. It has an autosomal dominant trait caused by mutations in the *SF3B4* gene on chromosome 1q12-q21, and to date there are just over 100 cases documented worldwide.

Characteristics include:

1. Hypoplasia of the zygomatic complex and mandible
   (a) Downward slanting palpebral fissures
2. Lack or absence of the lower eyelashes
3. Lack of development of the internal and external ear
4. Possible cleft palate
5. Underdevelopment or absence of the thumb
6. Shortened forearms and poor movement in the elbow

Auriculocondylar Syndrome

A hallmark of this condition is the so-called 'question mark ear'.

Auriculocondylar syndrome (ARCND) is an autosomal dominant disorder of the first and second pharyngeal arches and is characterised by malformed ears (the so-called question mark ears), prominent cheeks, microstomia, abnormal temporomandibular joint and mandibular condyle hypoplasia [16].

DiGeorge Syndrome/Velocardiofacial Syndrome

This was first described in 1968 by the paediatric endocrinologist Angelo DiGeorge [17]. It is an autosomal dominant condition.

This syndrome is caused by the deletion of a small piece of chromosome 22 and, as such, is often referred to as the 22q11.2 deletion syndrome.

Approximately 1 in 4000 children are born with Di George syndrome every year and characteristics include:

- A long face with a prominent upper jaw
- Flattening of the cheeks
- An underdeveloped lower jaw
- A bluish colour below the eyes
- A prominent nose with narrow nasal passages
- A long thin upper lip and a down-slanting mouth
- Cleft palate or submucous cleft palate

Classification Systems

The most commonly used classification system for mandibular anomalies (mainly in the context of hemifacial microsomia) is the Pruzansky classification [18], which more recently has been modified by Kaban [19]. Variations in both of these works have been proposed over the years, but these remain the most utilised.

Pruzansky reported a grading system of progressive mandibular deficiency:

Grade I	Minimal hypoplasia of the mandible
Grade II	Functioning but deformed temporomandibular joint with anteriorly and medially displaced condyle
Grade III	Absence of the ramus and glenoid fossa

Kaban proposed a modification of Pruzansky's system which is now the most commonly used classification system and is called the Pruzansky-Kaban classification system. The advantage of this system is that it allows treatment planning to be based largely on the type of anomaly seen.

Type I	All mandibular and temporomandibular joint components are present and normal in shape but hypoplastic to a variable degree
Type IIa	The mandibular ramus, condyle and temporomandibular joint are present but hypoplastic and abnormal in shape
Type IIb	The mandibular ramus is hypoplastic and markedly abnormal in form and location, being medial and anterior, and with no articulation with the temporal bone
Type III	The mandibular ramus, condyle and temporomandibular joint are absent, and the lateral pterygoid muscle and temporalis, if present, are not attached to the mandibular remnant

Using this classification system, the first two types (I and IIa) allow for treatment using conventional orthognathic or distraction techniques alone. The later types (IIb and III) require reconstruction of the missing component, alongside possible orthognathic surgery, as well as soft tissue augmentation and reconstruction. Simplifying the patients for treatment planning purposes, by converting a Pruzansky III to a IIa with the use of costochondral graft which can then be used for multi-vector distraction, has been described [20].

Other classification systems have been proposed and given consideration to other congenital anomalies [21].

References

1. Vento AR, LaBrie RA, Mulliken JB. The OMENS classification of hemifacial microsomia. Cleft Palate Craniofac J. 1991;28(1):68–77.
2. Castori M, Brancati F, Rinaldi R, Adami L, Mingarelli R, Grammatico P, et al. Antenatal presentation of the oculo-auriculo-vertebral spectrum (OAVS). Am J Med Genet A. 2006;140(14):1573–9.
3. Sugar AW. The management of syndromic mandibular asymmetry, especially hemifacial microsomia. Br J Oral Maxillofac Surg. 2009;47(6):503–4.
4. Ongkosuwito EM, van Neck JW, Wattel E, van Adrichem LN, Kuijpers-Jagtman AM. Craniofacial morphology in unilateral hemifacial microsomia. Br J Oral Maxillofac Surg. 2013;51(8):902–7.
5. et Ménard L. Affections congénitales. I. Tête et Cou. 1891.
6. Thomson A. Notice of several cases of malformation of the external ear, together with experiments on the state of hearing in such persons. Monthly J Med Sci. 1846;7:420.
7. Collins ET. Case with symmetrical congenital notches in the outer part of each lower lid and defective development of the malar bones. Trans Ophthalmol Soc U K. 1900;20:190.
8. Cobb AR, Green B, Gill D, Ayliffe P, Lloyd TW, Bulstrode N, et al. The surgical management of Treacher Collins syndrome. Br J Oral Maxillofac Surg. 2014;52(7):581–9.
9. Goldenhar M. The association between the ear malformations, and particularly the syndrome of epibulbar dermoids appendices and congenital auricle fistulas, and their relationships with the mandible-facial dysostosis. J Genet Hum. 1952;1:234–82.
10. Feingold M, Baum J. Goldenhar's syndrome. Am J Dis Child. 1978;132(2):136–8.
11. MacLeod SP, Macintyre DR. Bilateral hypoplasia of mandibular condyles in Hurler's syndrome. Oral Surg Oral Med Oral Pathol. 1993;75(5):659–60.
12. Gardner DG. The oral manifestations of Hurler's syndrome. Oral Surg Oral Med Oral Pathol. 1971;32(1):46–57.
13. Hallermann W. Vogelgesicht und Cataracta congenita. Klin Monbl Augenheilkd. 1948;113:315–8.
14. Gorlin RJ, Cohen MM, Hennekam RC. Syndromes of the head and neck. New York: Oxford University Press; 1990.
15. Sushil G, Pai N, Gagandeep V. Hallermann-Streiff syndrome in a neonate:-a case report. Br Biomed Bull. 2014;2(4):644–9.
16. Masotti C, Oliveira KG, Poerner F, Splendore A, Souza J, da Silva Freitas R, et al. Auriculo-condylar syndrome: mapping of a first locus and evidence for genetic heterogeneity. Eur J Hum Genet. 2008;16(2):145–52.
17. DiGeorge AM. Congenital absence of the thymus and its immunologic consequences: concurrence with congenital hypoparathyroidism. Birth Defects. 1968;4(1):116–21.
18. Pruzansky S. Not all dwarfed mandibles are alike. Birth Defects. 1969;5:120–9.
19. Kaban LB, Padwa BL, Mulliken JB. Surgical correction of mandibular hypoplasia in hemifacial microsomia: the case for treatment in early childhood. J Oral Maxillofac Surg. 1998;56(5):628–38.
20. Chadha A, Shorafa M, Evans B, Ayliffe P. Distraction of bone grafts to treat severe mandibular deficiency in hemifacial microsomia—a case series. Br J Oral Maxillofac Surg. 2011;49:S8.
21. Kaneyama K, Segami N, Hatta T. Congenital deformities and developmental abnormalities of the mandibular condyle in the temporomandibular joint. Congenit Anom. 2008;48(3):118–25.

16 Craniofacial Syndromes

Elizabeth Anne Gruber and Michael Stephen Dover

Development of the Face

Formation of Five Processes Around the Stomodeum

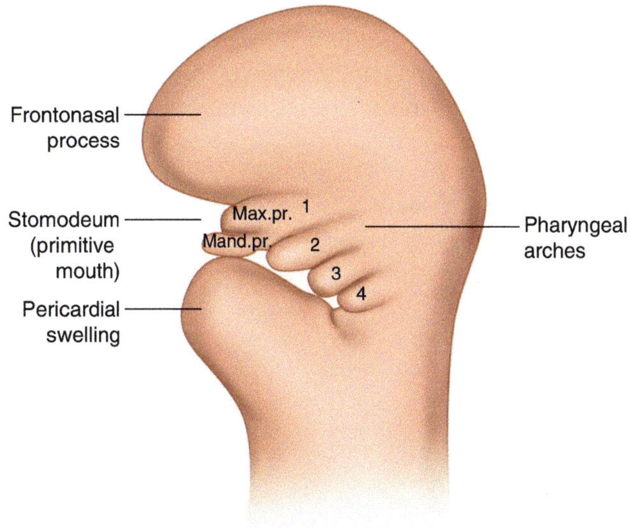

1. The upper part of the head fold projects downwards and forwards to form the **frontonasal process**.
2. The **pericardial swelling** projects upwards.
3. A depression called the **stomodeum** (primitive mouth) is formed between the previous two swellings.
4. **Pharyngeal arches** appear on either side of the pharyngeal gut.
5. The first pharyngeal arch develops two processes

   ↗ Mandibular process ventrally.

   ↘ Maxillary process dorsally.

6. The stomodeum becomes surrounded by five processes:
   (a) Frontonasal process—cranially
   (b) Two maxillary processes—on each side
   (c) Two mandibular processes—caudally

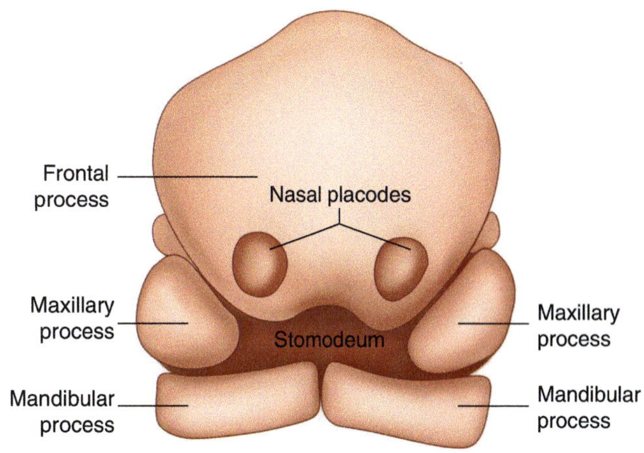

E. A. Gruber (✉)
Oral and Maxillofacial Surgery Department, Queen Elizabeth Hospital, Birmingham, UK

M. S. Dover
Oral and Maxillofacial Surgery Department, Queen Elizabeth Hospital, Birmingham, UK

Birmingham Children's Hospital, Birmingham, UK

R. Carachi, S. H. E. Doss (eds.), *Clinical Embryology*, https://doi.org/10.1007/978-3-319-26158-4_16

Differentiation and Fusion of the Five Processes

The Frontonasal Process

1. *Two nasal placodes* (buds) develop on either side of the frontonasal process.
2. Each nasal placode becomes invaginated to form a *nasal pit*.
3. The edges of each nasal pit form *medial-* and *lateral-nasal folds*.
4. The nasal pits become deeper forming the *nasal cavities* which will later open into the pharynx posteriorly.
5. Each *lateral-nasal fold* will form the ala of the nose.
6. The *two medial-nasal folds* unite together:
   (a) *On the surface*: to form the middle part of the nose and the philtrum of the upper lip
   (b) *At a deeper level*: to form the premaxilla which includes
   - The anterior-part of the upper jaw (carrying the incisor teeth)
   - The primary palate (the anterior-triangular part of the palate carrying the incisive fossa)

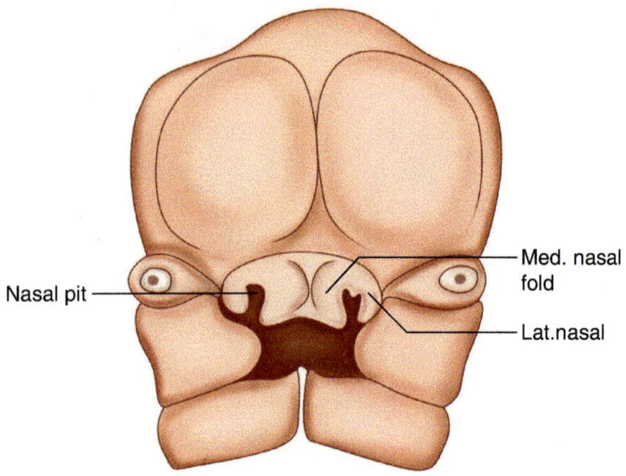

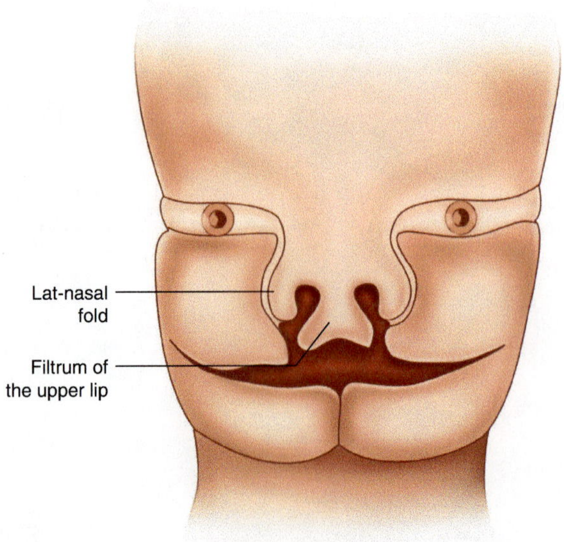

The Two Maxillary Processes

1. Each maxillary process grows medially and approaches the medial- and lateral-nasal folds but remains separated from them by the nasolacrimal groove which later will form the nasolacrimal duct.
2. The two maxillary processes fuse with the medial nasal folds of the frontonasal process to form the upper lip (except the philtrum which is formed by fusion of the two medial-nasal folds).
3. Each maxillary process unites:
   (a) Anteriorly: with the lateral-nasal fold along the side of the nose
   (b) Posteriorly: with the mandibular process to form the cheek

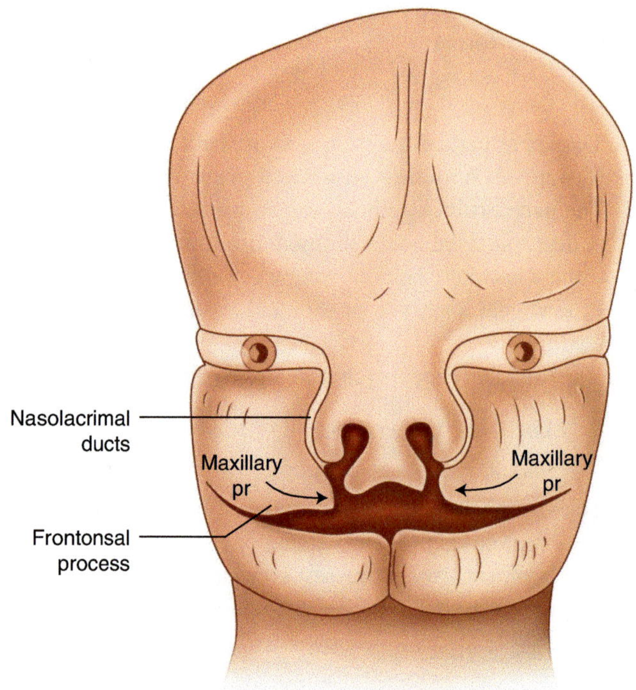

The Two Mandibular Processes

1. They fuse above with the maxillary process forming the cheeks.
2. They fuse with each other medially to form the lower lip and chin.

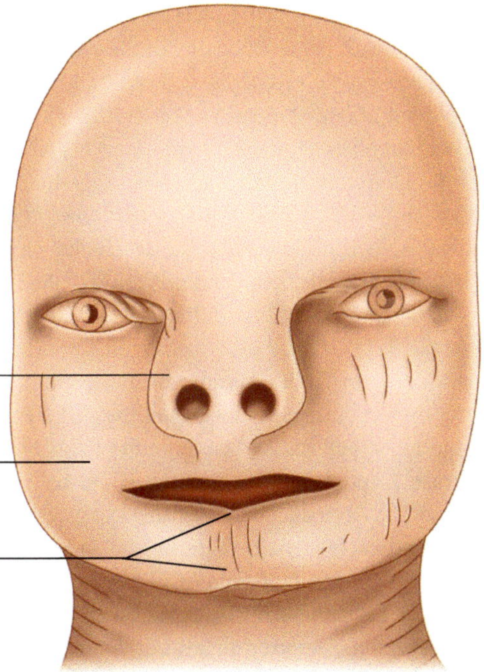

N.B:

1. The degree of fusion between the maxillary and mandibular processes determines the width of the mouth.
2. The *nasolacrimal duct* develops at the line of fusion between the maxillary and frontonasal processes (at the nasolacrimal groove) as a solid cord which becomes canalized. The lids and conjunctival sac appear at the upper end of the nasolacrimal duct.

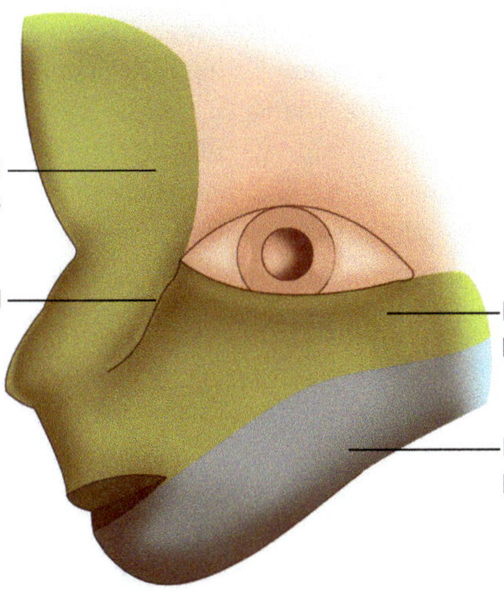

Introduction

A craniofacial condition involves both the cranium, with its contents, and the face. Craniosynostosis is the premature fusion of cranial sutures. The overall prevalence of craniosynostosis has been estimated at between 1 in 2100 and 1 in 2500 live births. It is most often an isolated finding, affecting the sagittal or coronal sutures, but can also occur as part of a syndrome, with additional findings such as limb abnormalities and developmental delay. It is termed 'simple' synostosis if a single suture is involved. All multiple suture synostoses ('complex' cases) are syndromic. More than a hundred syndromes associated with craniosynostosis have been described. The commoner ones are described in this chapter.

Treacher Collins is a nonsynostotic syndrome that affects facial development, and individuals show a spectrum of characteristic features. These facial anomalies result from malformations of structures arising from the first and second pharyngeal arch, groove and pouch.

Encephaloceles occur due to herniation of the intracranial contents outside the cranial cavity and may be present as part of a syndrome.

Genetics

Clinical assessment of the patient is confirmed with genetic testing (Table 16.1).

The *FGFR2 gene* encodes a transmembrane receptor tyrosine kinase. Heterozygous mutations of FGFR2 cause three craniosynostosis syndromes, those of Crouzon, Apert and Pfeiffer. All exhibit a characteristic 'crouzonoid' facial appearance (exorbitism, flattened malar region and beaked nose).

Mutations in FGFR2 and FGFR3 tend to encode highly localised, recurrent missense substitutions encoding proteins with gain-of-function properties. The cellular consequences of mutation are complex, including enhancement of proliferation, differentiation and apoptosis of osteoblasts bordering the cranial suture mesenchyme. This is probably the main factor leading to craniosynostosis [1].

Nearly all Apert syndrome mutations arise de novo and have been shown to originate exclusively from the father (increasing with paternal age) [2]. FGFR2 mutations in Pfeiffer syndrome overlap those in Crouzon syndrome, but the majority of severe cases are caused by a small subset of substitutions. Crouzon syndrome is usually the mildest of the FGFR2-associated disorders.

Table 16.1 Genetics of craniofacial syndromes

Syndrome	Mode of inheritance	Gene mutation	Chromosome
Crouzon	Autosomal dominant, variable expression	FGFR2	10q26
	Crouzon with acanthosis nigricans	FGFR3	4p16.3
Apert	Autosomal dominant, most sporadic	FGFR2	10q26
Pfeiffer	Autosomal dominant, variable expression	FGFR2 (95%) FGFR1 (5%)	10q26 8p11.23-p11.22
Muenke	Autosomal dominant, variable expression	FGFR3 (Pro250Arg)	4p16.3
Saethre-Chotzen	Autosomal dominant, variable expression	TWIST-1	7p21
Carpenter	Autosomal recessive	RAB23 MEGF8	6p11 19q12
Cranio-frontonasal dysplasia	X-linked	EFNB1	Xq12
Treacher Collins	Autosomal dominant (TCOF1, POLR1D) variable expression Autosomal recessive (POLR1C)	TCOF1 POLR1C POLR1D	5q32 6p21.1 13q12.2

FGFR3 encodes a protein that has a domain structure closely resembling FGFR2. Two heterozygous mutations cause specific craniosynostosis syndromes, Muenke syndrome and Crouzon syndrome with acanthosis nigricans [3]. The mutation that causes Muenke syndrome substitutes the amino acid arginine for the amino acid proline at position 250 in the FGFR3 protein. This results in ligand-dependent gain of function; however the reasons for the association with craniosynostosis are not fully understood currently. Crouzon syndrome with acanthosis nigricans is characterised by the Ala391Glu substitution.

Heterozygous mutations in the *TWIST1* gene are linked to Saethre-Chotzen syndrome. This encodes a transcription factor belonging to the basic helix-loop-helix family. Non-penetrance for craniosynostosis does occur. TWIST1 has a key role in maintaining the boundary between neural crest and cephalic mesoderm at the site of the developing coronal suture [4].

The protein encoded by the ephrin-B1 gene (*EFNB1*) is a cell surface transmembrane ligand for Eph-related receptor tyrosine kinases. These are crucial for cell migration, repulsion and adhesion during neuronal, vascular and epithelial development. They are thought to be important in constraining the orientation of longitudinally projecting axons. Paradoxical to other X-linked conditions, with craniofrontonasal dysplasia, females are more severely affected than males. This is due to the process of X-inactivation in females, where at random either the maternal or paternal X-chromosome is inactivated in a cell. Due to this process, the body's tissues contain either cells with normal EFNB1 or the mutated EFNB1. This mosaic pattern of cells interferes with the functionality of the cell-cell interactions, as a result causing the severe physical malformations in females [5].

About 200 mutations in the *TCOF1* (Treacle) gene have been identified in people with Treacher Collins syndrome. Most of these mutations insert or delete a small number of base pairs in the TCOF1 gene. As a result, the production of ribosomal RNA (rRNA) is reduced, which likely triggers the apoptosis of neural crest cells and prefusion of the neural folds [6].

At least 6 mutations in the *POLR1C* gene and 20 mutations in the *POLR1D* gene have been identified in individuals with Treacher Collins syndrome. These mutations appear to alter the structure and function of the POLR1C and POLR1D proteins, which reduce the amount of functional RNA polymerase I and RNA polymerase III in cells. Consequently, less rRNA is produced, with the same results as described previously. This could underlie the specific problems with facial development found in Treacher Collins syndrome; however, it is unclear why these effects are limited to facial development. Approximately 40% of cases of Treacher Collins syndrome are familial. Exposure to isotretinoin in utero in a murine model can reproduce craniofacial abnormalities similar to Treacher Collins [7].

Epidemiology (Table 16.2)

Clinical Features

Crouzon Syndrome (Fig. 16.1)

Clinical features have a wide phenotypic variability. Premature fusion of sutures may occur at the level of the cranial vault, with bilateral coronal suture fusion being the most common. The result is a brachycephalic head shape (short and wide skull). The anterior cranial base and facial sutures are variably affected, resulting in symmetrical hypoplasia of the orbits, zygomas and maxilla, leading to exorbitism. Mandibular growth is normal, but relative mandibular prognathism is present due to maxillary hypoplasia. There may be subtle skeletal abnormalities affecting the limbs, but they are rarely obvious or symptomatic. Intelligence is normal.

Table 16.2 Incidence of craniofacial syndromes

Syndrome	Incidence
Crouzon	1:25,000
Apert	1:100,000–1:160,000
Pfeiffer	1:100,000
Muenke	1:10,000
Saethre-Chotzen	1:25,000–1:50,000
Carpenter	Rare
Craniofrontonasal dysplasia	1:100,000–1:120,000
Treacher Collins	1:50,000

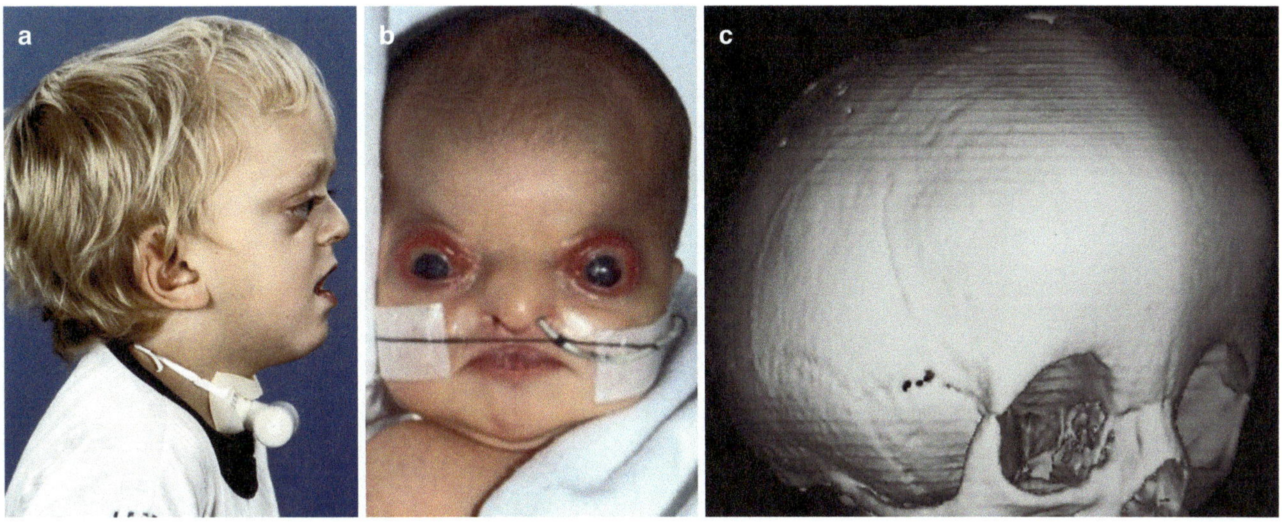

Fig. 16.1 (**a**) Picture of face from side, showing characteristic facies with exorbitism and midface hypoplasia. (**b**) Neonatal diagnosis of Crouzon Syndrome. (**c**) 3D CT head showing pancraniosynostosis

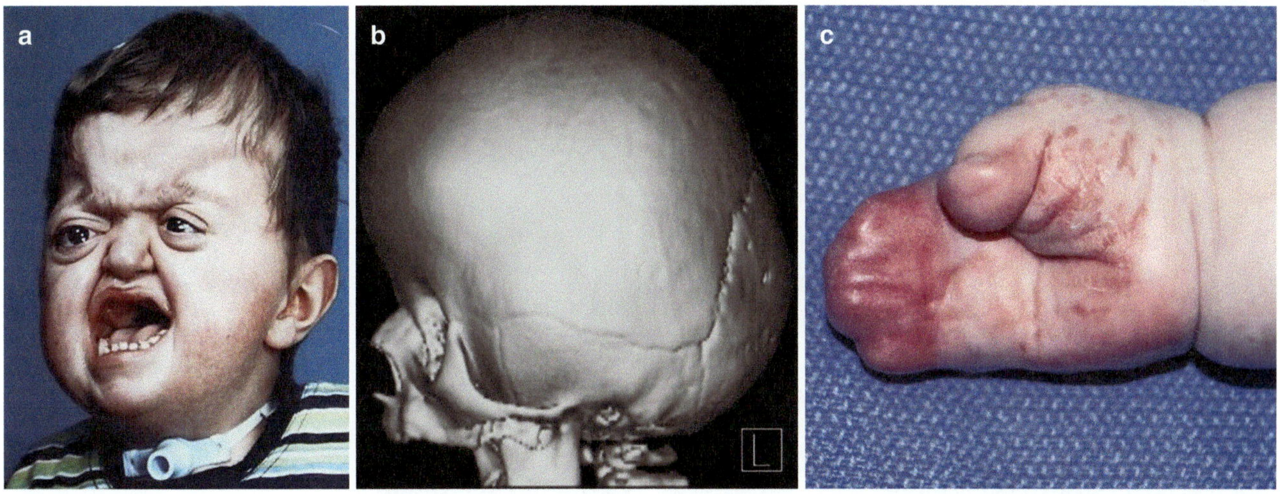

Fig. 16.2 (**a**) Picture of face from front and side showing typical facies. (**b**) 3D CT head showing bicoronal synostosis. (**c**) Picture of hands/feet showing complex syndactyly

If associated with acanthosis nigricans, this usually develops during childhood, so specific genetic testing should be requested (for FGFR3 mutation) and if positive should prompt careful neurosurgical assessment, as hydrocephalus is frequently associated.

Apert Syndrome (Fig. 16.2)
This is characterised by bicoronal craniosynostosis, resulting in a tall skull flattened anteriorly and occipitally (turribrachycephaly). Individuals demonstrate variable hypertelorism with downward slanting palpebral fissures and exorbitism due to shallow orbits and a shortened skull base. The maxilla is also hypoplastic leading to midface retrusion and a beaked nose.

The most obvious extracranial feature is a variable degree of hand and foot abnormalities with bilateral symmetrical complex syndactyly. Cleft palate is a common association (up to 44%), otherwise the palate is narrow and high-arched. Intellectual impairment is more common than in Crouzon syndrome.

Pfeiffer Syndrome (Fig. 16.3)
This may be difficult to differentiate from Crouzon syndrome when assessing the craniofacial features as craniosynostosis and exorbitism are present in both. Pfeiffer syndrome is usually characterised by broad, medially deviated thumbs and/or big toes, sometimes with cutaneous syndactyly. The FGFR1 mutation causes a mild form of Pfeiffer syndrome.

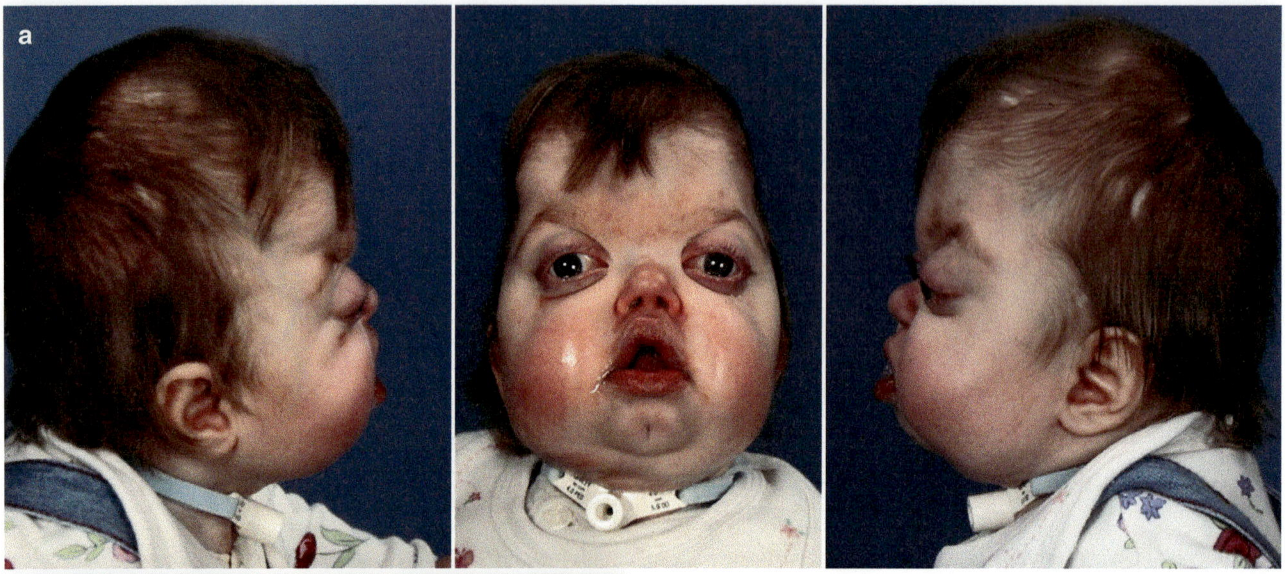

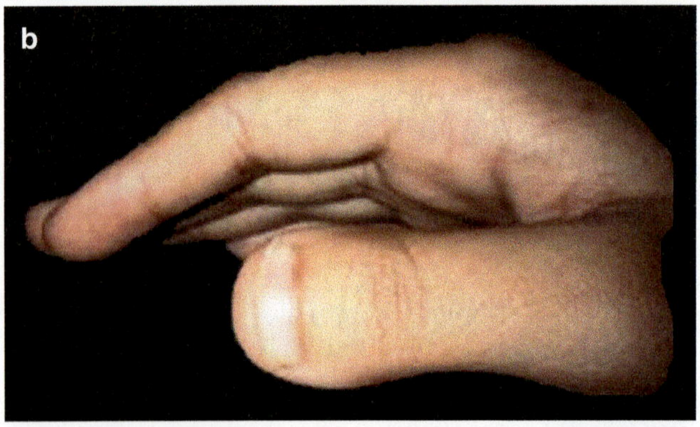

Fig. 16.3 (**a**) Picture of face from front and sides. (**b**) Picture of hand showing broad and short thumb

Muenke (Pro 250Arg) Syndrome

Features are unicoronal or bicoronal synostosis and minor digital abnormalities (especially brachydactyly and sometimes carpal and tarsal fusions). Sensorineural hearing loss may be present and developmental delay is variable. The phenotype is highly variable, and up to 20% of individuals do not have any characteristic features of the disorder.

Saethre-Chotzen Syndrome (Fig. 16.4)

This may not present until adulthood as features can be mild. Features are unicoronal or bicoronal synostosis. Most do not have midface retrusion. Particular characteristics are ptosis of the upper eyelids, tear duct abnormalities, low frontal hairline and a parrot-beaked nasal shape. A prominent superior crus of the ear may be present. Abnormal extremities (broad, laterally deviated first digits, cutaneous syndactyly and brachydactyly) are variable. Individuals have normal intelligence.

Carpenter Syndrome (Acrocephalopolysyndactyly)

This is characterised by craniosynostosis. Any combination may be seen. The main facial features are flat nasal bridge, downward slanting palpebral fissures, low set and abnormally shaped ears, underdeveloped jaws and visual problems. Cutaneous syndactyly (most common between middle and ring fingers), brachydactyly and polydactyly (frequently next to big or second toe or fifth finger) may be present. Intellectual disability is common, but not in all individuals.

Other features in some individuals include umbilical hernia, hearing loss, kyphoscoliosis, genu valgum, genital abnormalities (cryptorchidism), heart defects and situs invertus/dextrocardia.

Complex craniosynostosis may lead to a cloverleaf (trilobular)-shaped skull (Kleeblattschädel deformity) associated with significant mortality (Fig. 16.5). This is a physical finding and not a syndrome and is aetiologically and pathogenetically heterogeneous.

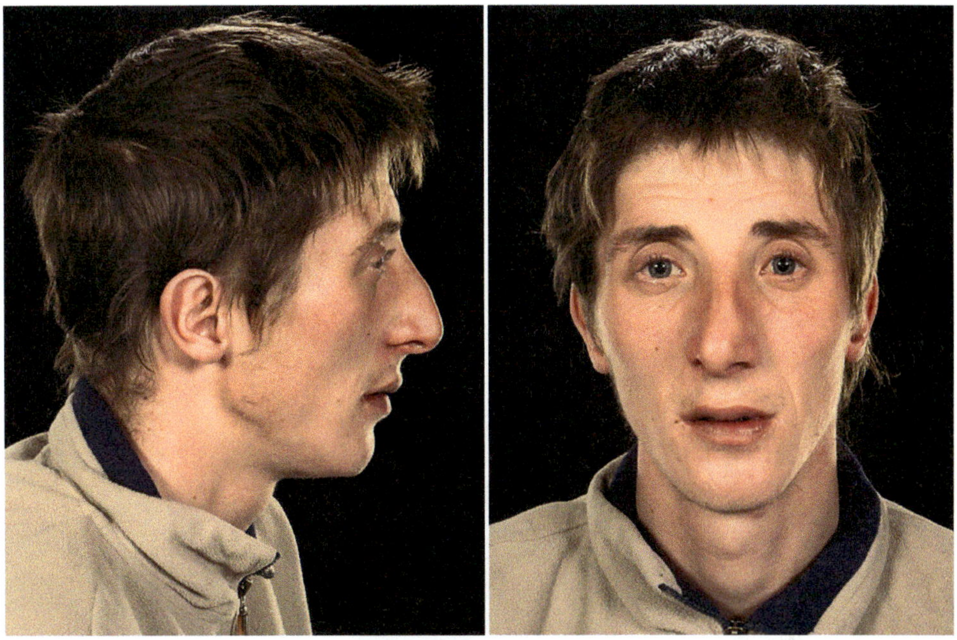

Fig. 16.4 Picture of face showing typical appearance with ptosis

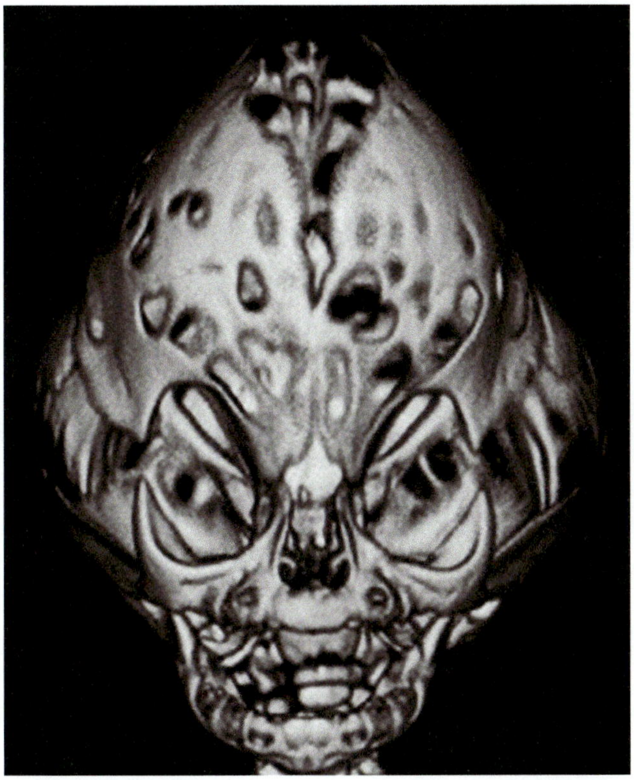

Fig. 16.5 Picture of 3D CT skull showing cloverleaf skull (Kleeblattschädel deformity)

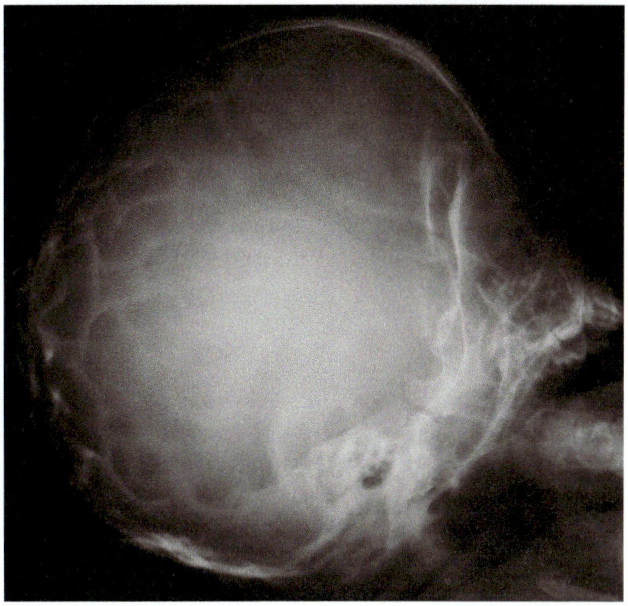

Fig. 16.6 Picture of plain radiograph with copper beaten appearance

Elevated intracranial pressure is one of the most important functional problems related to craniosynostosis. It affects approximately one-third of patients with craniofacial dysostosis syndromes. The late effects of this may be seen on plain radiographs as a copper beaten appearance (Fig. 16.6) or on CT scan as bony erosions of the skull.

Craniofrontonasal Dysplasia (Fig. 16.7)

Phenotypic expression varies greatly amongst affected individuals, with females being more severely affected. Characteristics are unicoronal or bicoronal synostosis, hypertelorism, downward slanting palpebral fissures and clefting of the nasal tip. Longitudinally grooved fingernails and toenails are usual.

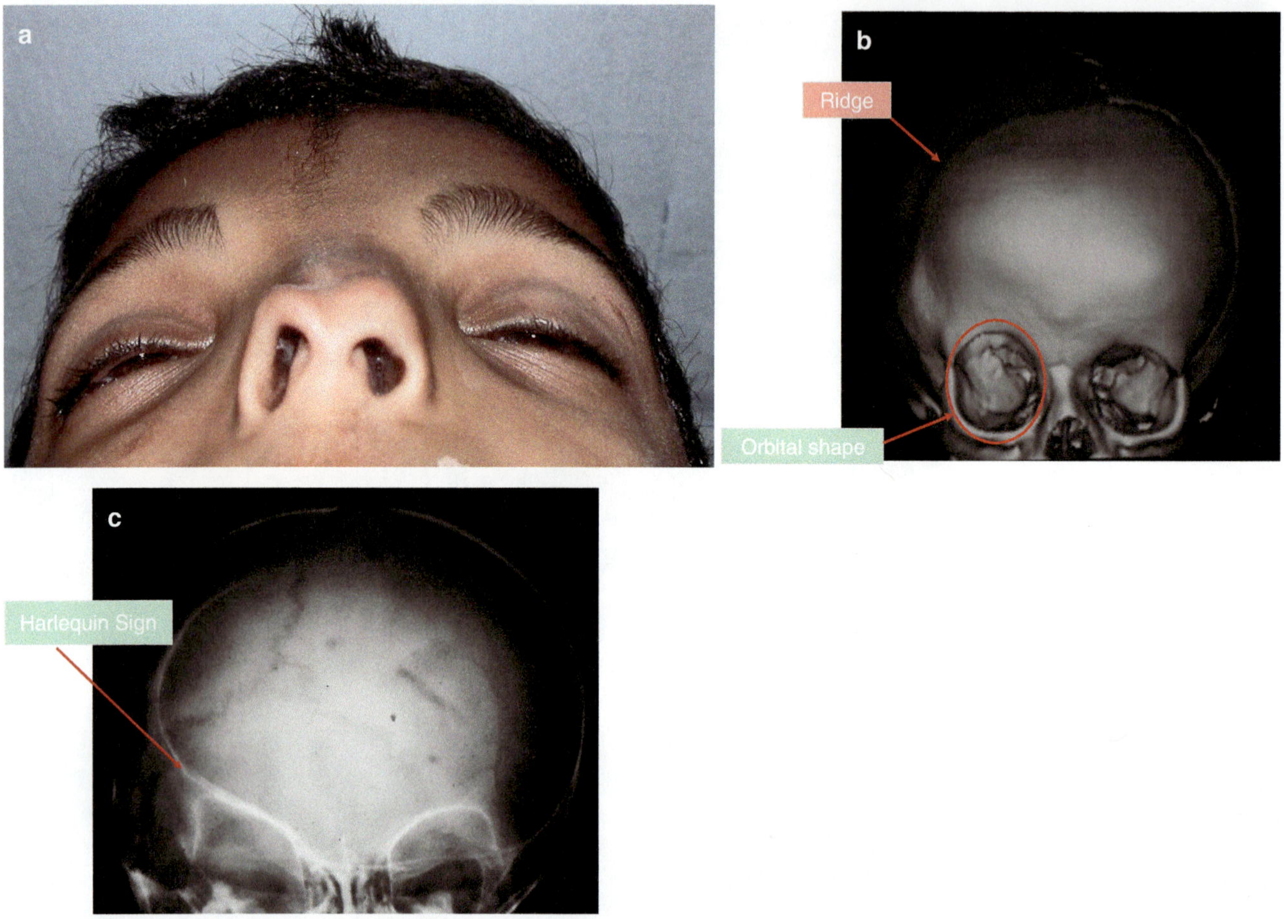

Fig. 16.7 (**a**) Picture of face showing brachycephaly, hypertelorism and bifid nasal tip. (**b**, **c**) 3D CT skull showing unicoronal synostosis, harlequin sign, orbital hypertelorism and facial asymmetry

Role of Antenatal Detection in Craniosynostosis

Where a molecular abnormality has already been identified in a family, the issues surrounding prenatal or preimplantation diagnosis are similar to those for other genetic disorders. It is more difficult to diagnose craniosynostosis prenatally by ultrasound or other imaging techniques. The cranial sutures form late (around 16 weeks gestation) compared with most other embryonic structures. Therefore at the time of most routine ultrasound scans (around 20 weeks gestation), growth distortion of the skull has not yet occurred, except in the most severe cases. While other signs of a syndrome, such as syndactyly, may be seen, the diagnosis is usually made in the early postnatal period.

Treacher Collins Syndrome (Mandibulofacial Dysostosis) (Fig. 16.8)

The development of the bones and soft tissues of the face is affected and leads to functional and cosmetic problems. The signs and symptoms vary greatly, ranging from almost unnoticeable to severe, but tend to be symmetrical. Most affected individuals have underdeveloped facial bones, particularly the zygomas and mandibular hypoplasia with retrogenia. Shortened posterior vertical height and anterior open bite are seen. The lack of malar development results in compensatory growth accounting for the hyperprojection and clockwise rotation of the maxilla. The maxilla has a high-arched palate or cleft in around 30% of cases. In severe cases, micrognathia may cause airway obstruction requiring tracheostomy. Skull abnormalities include brachycephaly with bitemporal narrowing.

Abnormal pinnae (size, shape and position), atresia of the external auditory canal and ossicle abnormalities of the middle ear can lead to hearing loss in around half of individuals with Treacher Collins syndrome. Downward slanting of the palpebral fissures is a characteristic. True colobomas (a gap in the structure of the eye) are present in around 25% of cases and pseudocolobomas in 50%. These tend to occur in the lateral third of the lower eyelid, along with a lack of eyelashes. Refractive errors and amblyopia are also common. People with Treacher Collins syndrome usually have normal intelligence, but developmental delay has been attributed to hearing loss.

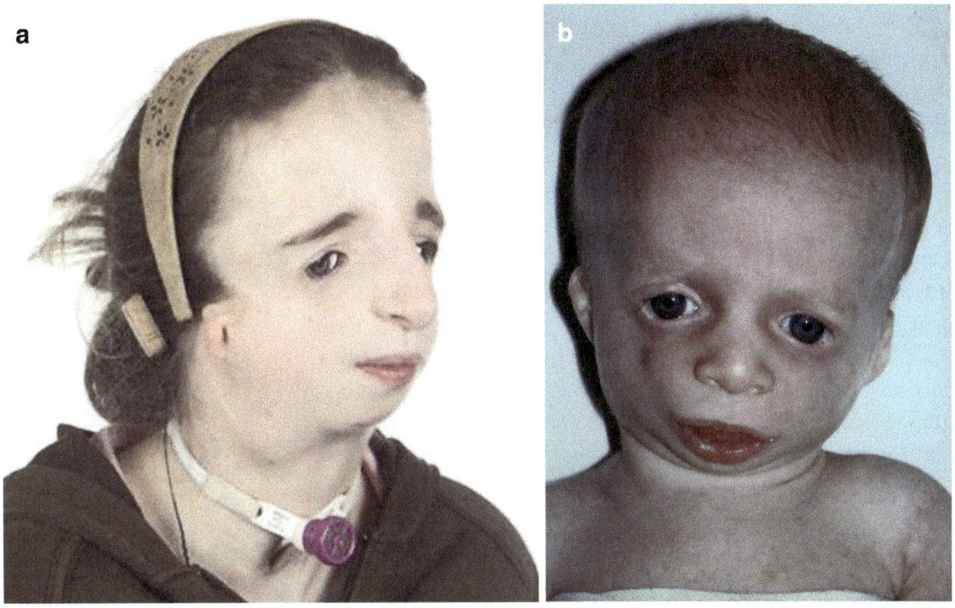

Fig. 16.8 (**a**, **b**) Picture of face from front and side showing underdeveloped zygomas, mandible and downward slanting palpebral fissures

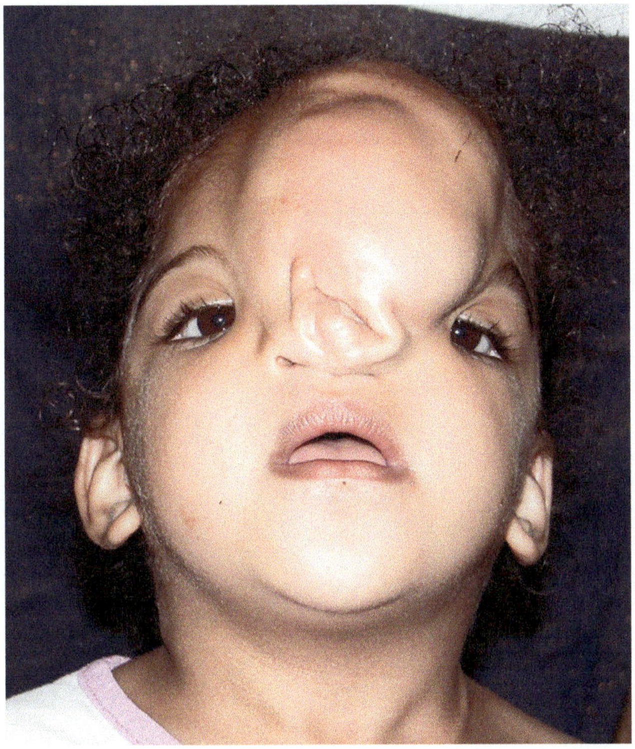

Fig. 16.9 Clinical picture of encephalocele anterior (face)

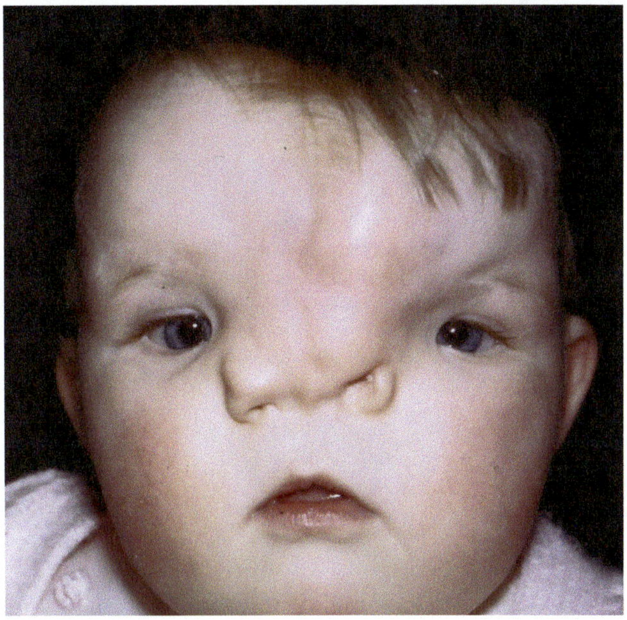

Fig. 16.10 Frontal meningocele

It is possible to identify patients with Treacher Collins syndrome on prenatal ultrasound scan, but often only late in the pregnancy.

Encephaloceles (Figs. 16.9 and 16.10)

Encephaloceles are characterised by herniation of the meninges alone (meningocele), or with brain (encephalomeningocele), through a bony skull defect. These are either the result of neural tube defects that are caused by failure of the embryonic neural tube to close properly between days 24 and 28 of gestation or may be part of a Tessier craniofacial clefting disorder. The size of the defect and the sac is extremely variable. Encephaloceles most commonly occur at the base of the skull and can also occur between the forehead and nose or at the midline of the upper part of the skull (nasofrontal, nasoethmoidal or nasoorbital). Encephaloceles are often accompanied by other craniofacial abnormalities and brain malformations. The prognosis, symptoms and severity all depend upon the type of brain tissue involved, the location of the encephalocele and the accompanying brain malformations.

It is likely that the aetiology is multifactorial: genetic, maternal nutrition (folic acid) and exposure to toxins in early fetal life. The main symptoms and signs are microcephaly, ataxia, developmental delay, hydrocephalus, visual problems and seizures.

Up to 40% of encephaloceles are associated with chromosomal abnormalities. The Dandy-Walker malformation is enlargement of the fourth ventricle with a cyst that extends to the posterior fossa. A Chiari malformation occurs when the cerebellar tonsil herniates into the spinal canal, blocking the flow of cerebrospinal fluid. Meckel-Gruber syndrome comprises an occipital encephalocele associated with microcephaly, cleft lip and palate, polydactyly, polycystic kidneys and ambiguous genitals.

Encephalocele may be diagnosed on first trimester ultrasound scan, which demonstrates a cranial defect with varying degrees of herniation. A prenatal MRI may be performed to more accurately delineate the defect and any associated anomalies. Amniocentesis can be performed for chromosomal anomalies. The deformity is generally apparent at birth; however, smaller encephaloceles of the nasal or forehead region may remain undetected for many years.

References

1. Iseki S, Wilkie AOM, Morriss-Kay GM. *Fgfr1* and *Fgfr2* have distinct differentiation and proliferation-related roles in the developing mouse skull vault. Development. 1999;126:5611–20.
2. Goriely A, MacVean GAT, van Pelt AMM, et al. Gain-of-function amino acid substitutions drive positive selection of *FGFR2* mutations in human spermatogonia. Proc Natl Acad Sci U S A. 2005;102:6051–6.
3. Johnson D, Wilkie AOM. Craniosynostosis (Practical Genetics). Eur J Hum Genet. 2011;19:369–76.
4. Merrill AE, Bochukova EG, Brugger SM, et al. Cell mixing at a neural crest-mesoderm boundary and deficient ephrin-Eph signalling in the pathogenesis of craniosynostosis. Hum Mol Genet. 2006;15:1319–28.
5. Wieacker P, Wieland I. Clinical and genetic aspects of craniofrontonasal syndrome: towards resolving a genetic paradox. Mol Genet Metab. 2005;86:110–6.
6. Dixon J, Ellis I, Bottani A, Temple K, Dixon MJ. Identification of mutations in TCOF1: use of molecular analysis in the pre- and postnatal diagnosis of Treacher Collins syndrome. Am J Med Genet. 2004;127A:244–8.
7. Dauwerse JG, Dixon J, Sealnd S, et al. Mutations in gene encoding subunits of RNA polymerases I and III cause Treacher Collins syndrome. Nat Genet. 2011;43:20–2.

Further Reading

Elden LM, Zur KB, editors. Congenital malformations of the head and neck. New York: Springer; 2014.

Forrest CR, Hopper RA. Craniofacial syndromes and surgery. Plast Reconstr Surg. 2013;131(1):86e–109e.

Hennekam RCM, Allanson JE, Krantz ID. Gorlin's syndromes of the head and neck. 5th ed. Oxford: Oxford University Press; 2010.

Patient Groups

Headlines (patients' support group) http://www.headlines.org.uk.

Changing Faces (support group for patients with disfigurement) http://www.changingfaces.org.uk.

17 External, Middle, and Inner Ear

Frank G. Garritano and Vito C. Quatela

Development of the Ear

External Ear

1. ***The auricle*:** develops from six mesenchymal swellings → around the post-part of the first pharyngeal cleft.
2. ***Ext-auditory meatus*:** develops from the first pharyngeal cleft.
3. ***Ear drum*:** consists of three layers:
   (a) Outer ectodermal layer: develops from the bottom of the first pH-cleft.
   (b) Inner endodermal layer: from the tubotympanic recess.
   (c) Intermediate mesodermal layer: derived from the mesoderm in between the first cleft and first pouch.

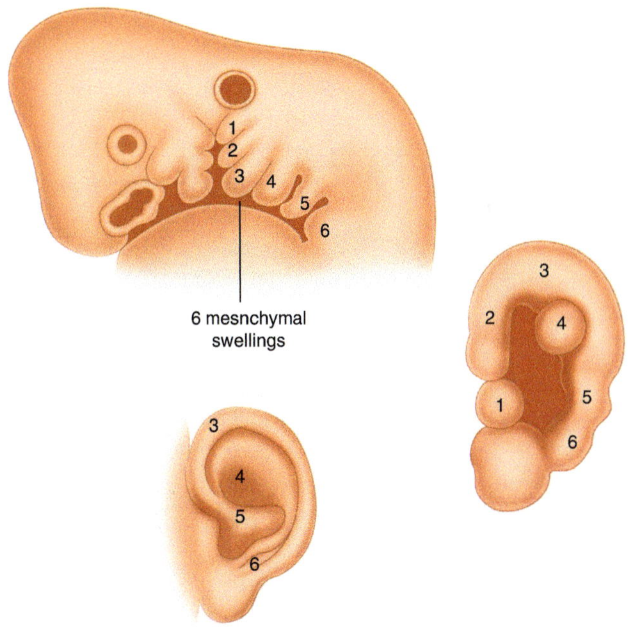

Middle Ear

1. *Middle-ear cavity*: develops from the dorsal part of the tubotympanic recess which is derived from the first pH-pouch.
2. *Ossicles of the middle ear*:
   (a) Malleus and incus: derived from Meckel's cartilage of the first arch.
   (b) Stapes: derived from Richert's cartilage of the second ph-arch.
3. *Muscles of the middle ear*:
   (a) Tensor tympani: derived from the first arch and supplied by mandibular n.
   (b) Stapedius m.: derived from the second arch and supplied by facial n.

F. G. Garritano (✉)
Facial Plastic and Reconstructive Surgery, The Permanente Medical Group, Oakland, CA, USA

V. C. Quatela
Quatela Center for Plastic Surgery, Rochester, NY, USA

University of Rochester, Rochester, NY, USA

R. Carachi, S. H. E. Doss (eds.), *Clinical Embryology*, https://doi.org/10.1007/978-3-319-26158-4_17

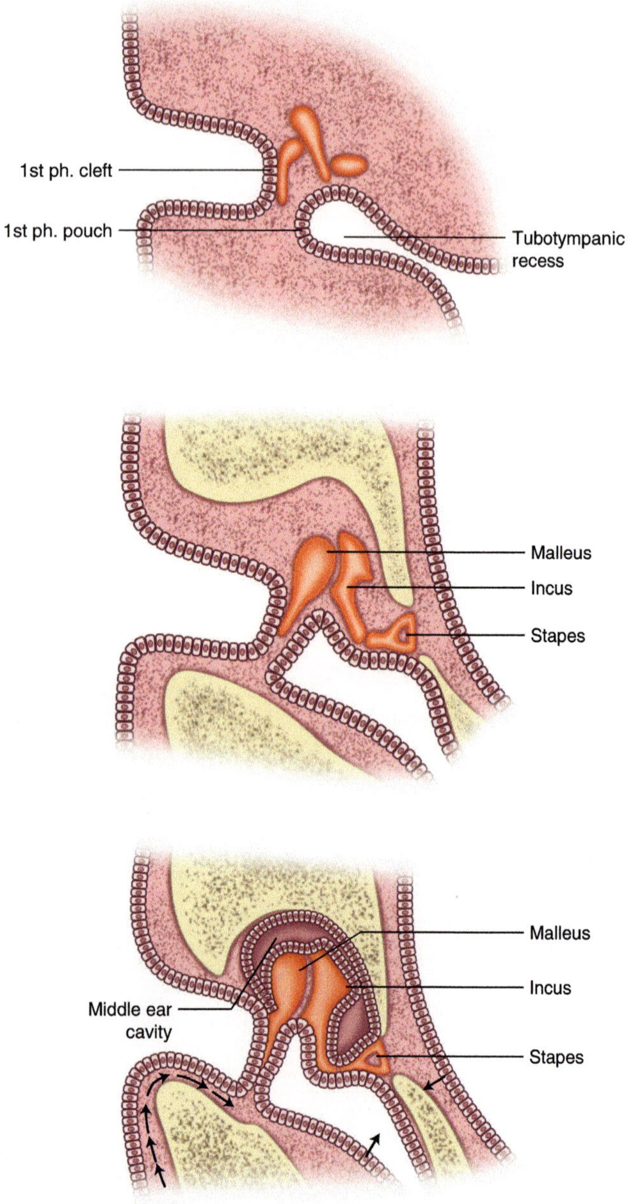

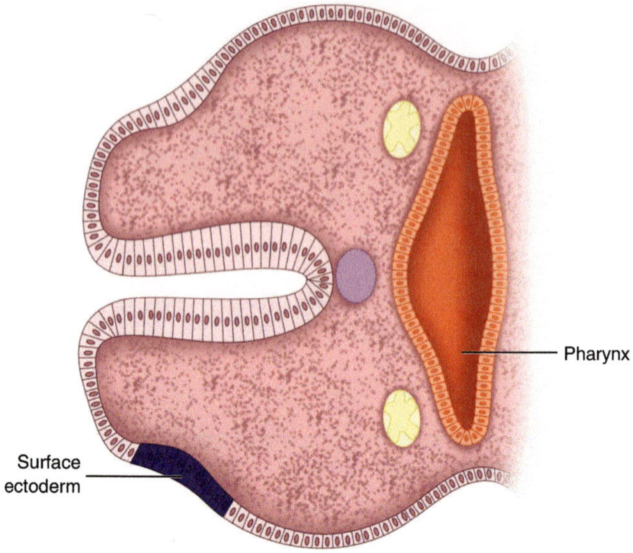

Internal Ear

1. ***The membranous part:*** develops from ectoderm as follows:
   (a) *Otic placode* develops a thickened part of surface ectoderm opposite the hindbrain. It becomes invaginated forming *otic pit*.
   (b) The otic pit separates from the surface forming *otic vesicle*.
   (c) The otic vesicle enlarges in a dorsoventral direction and divides into:
   - Ventral cochlear part which is coiled to form cochlea and saccule.
   - Dorsal utricular part which forms the utricle and the three semicircular canals.
2. ***The bony part:*** develops from the mesoderm which surrounds the membranous part.

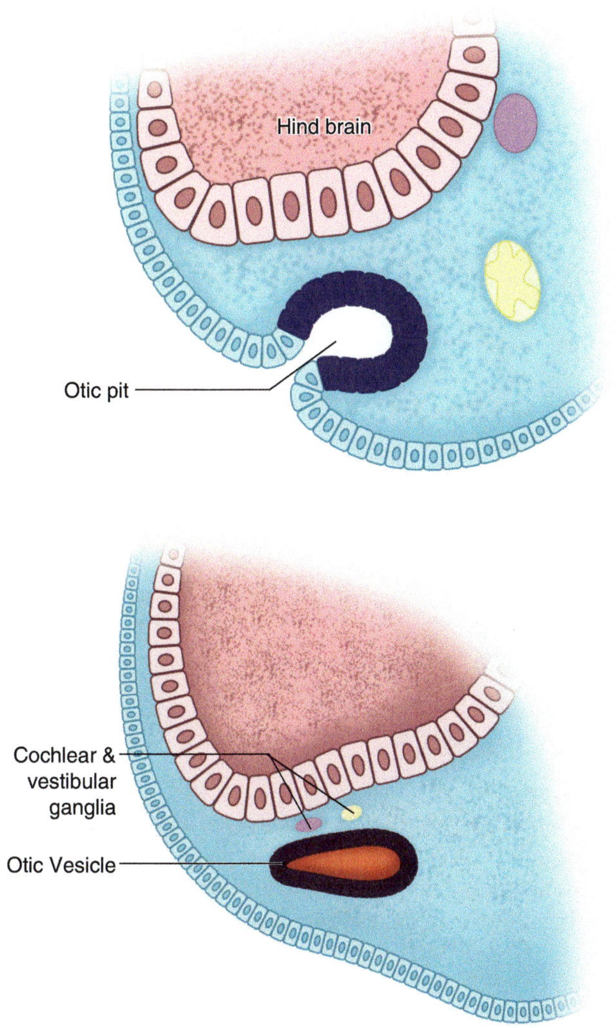

The embryogenesis of the external, middle, and inner ear is a highly complex process. The external and middle ear are largely formed as part of the branchial apparatus, whereas the inner ear develops separately from the otic placode. Deviations from the normal pattern of embryogenesis at varying stages of development can give rise to a wide variety of congenital malformations. Because of how closely these structures are linked, a malformation affecting the external ear can also frequently affect the middle and internal ear and vice versa. A clear knowledge of the processes that lead to these congenital malformations helps inform our understanding of these pathologic conditions.

External Ear Malformations

Microtia

The external ear is embryologically derived from six distinct swellings that form on the outer surface of the embryo. These swellings, known as the hillocks of His, are derived from the first and second branchial arches and first appear at around weeks 5–6 of embryogenesis. The six hillocks together give rise to the normal anatomic features of the external ear, and interruptions in this process can result in an absent, malformed, or small external ear—a condition known as microtia.

Microtia is estimated to affect approximately 1–2 people per 10,000 births [1] . There is a significant racial variability, with Caucasians demonstrating lower incidence than Asians or Hispanics [1]. In unilateral microtia, the right ear is affected more often than the left. Other risk factors include male gender, increasing birth order, and exposure to teratogens in utero (e.g., tretinoin, mycophenolate mofetil) [1, 2]. Microtia can occur in isolation or with any number of other associated congenital malformations, including facial clefts, cardiac anomalies, limb defects, and renal anomalies [1].

Microtia can occur in both familial and sporadic forms. In one study of 145 cases, approximately two-thirds of cases occurred sporadically, and one-third occurred in patients with a positive family history and were suspected to be genetic in origin [3]. Microtia is commonly seen in patients with particular genetic syndromes, such as Treacher Collins syndrome, Goldenhar syndrome (oculo-auriculo-vertebral syndrome), and hemifacial microsomia.

Microtia is classified into four grades depending on the severity of the defect. In grade I the external ear is slightly smaller than the unaffected side, but the normal external anatomical structures are all present (helix, antihelix, tragus, etc.). In grade II the external ear is only partially formed, with absence of several of the normal anatomic features. In grade III, the most common form of microtia, the ear is severely malformed with the presence of only a small, peanut-like vestige of the auricle. Grade IV entails complete anotia or complete absence of the external ear (Fig. 17.1).

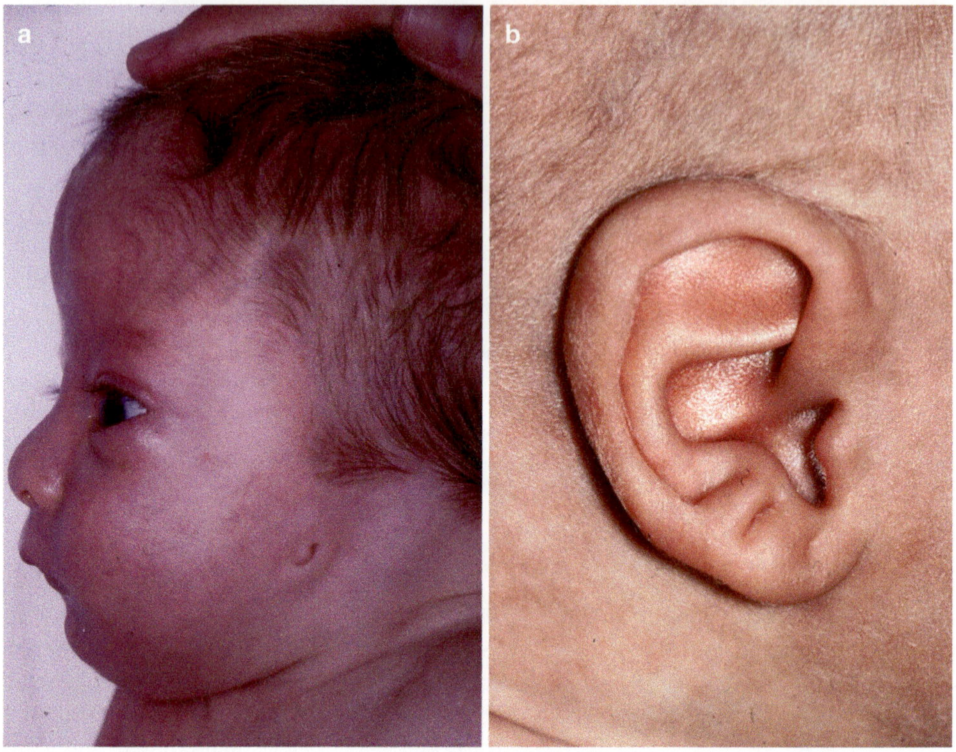

Fig. 17.1 (**a**) Agenesis of the ear. (**b**) Beckwith-Wiedemann syndrome transverse creases in ear

Patients with microtia should be carefully examined to rule out the presence of other congenital malformations. As microtia is often associated with other anomalies of the external, middle, and inner ear, all patients should undergo audiologic testing to assess hearing status, and early audiologic amplification should be provided as necessary to allow for optimal language development. Treatment of microtia is complex, frequently requires multiple surgical procedures, and is often delayed until the child is 6–10 years of age.

External Auditory Canal Atresia

The ear canal is formed from an invagination of surface epithelium of the first branchial cleft beginning in the fifth week of gestation. This meets with a layer of endoderm invaginating from the first branchial pouch (which forms the middle ear) trapping a layer of mesoderm in between the two and forming the three layers of the tympanic membrane. The ear canal is plugged with a solid core of squamous epithelial cells during this time, and recanalization of the ear canal does not complete until 28 weeks gestation. Failure of development or of subsequent recanalization can result in external auditory canal stenosis or atresia. This frequently occurs in association with microtia in addition to other middle ear anomalies. This can cause varying degrees of conductive hearing loss, particularly if present bilaterally. Imaging studies such as CT scans can help identify the degree of stenosis or atresia, the presence of membranous or bony components, and any other associated middle ear anomalies. Treatment options include hearing amplification and surgical reconstruction of the external auditory canal.

Minor External Ear Anomalies

The lop ear deformity causes abnormally prominent ears due to underdevelopment of the antihelical fold and scapha as a result of malformation of the auricular cartilage during embryogenesis. It causes a minor cosmetic deformity and can be corrected with surgical otoplasty.

Accessory auricular appendages can be composed of the skin, subcutaneous tissue, or cartilage and can be present anywhere around the auricle. Goldenhar syndrome (oculo-auriculo-vertebral syndrome), which is caused by anomalous development of the first and second brachial arch, can manifest with accessory auricular appendages in addition to incomplete development of the ear, nose, soft palate, lip, and mandible on one side of the body. Accessory auricular appendages can cause a minor cosmetic deformity and can be surgically removed (Fig. 17.2).

Preauricular pits and sinuses are characterized by the presence of a dimple or fistula tract and can be present anywhere around the external ear. They are fairly common, can be present bilaterally in up to 50% of cases, and can develop sporadically or be inherited [4]. The inherited form typically

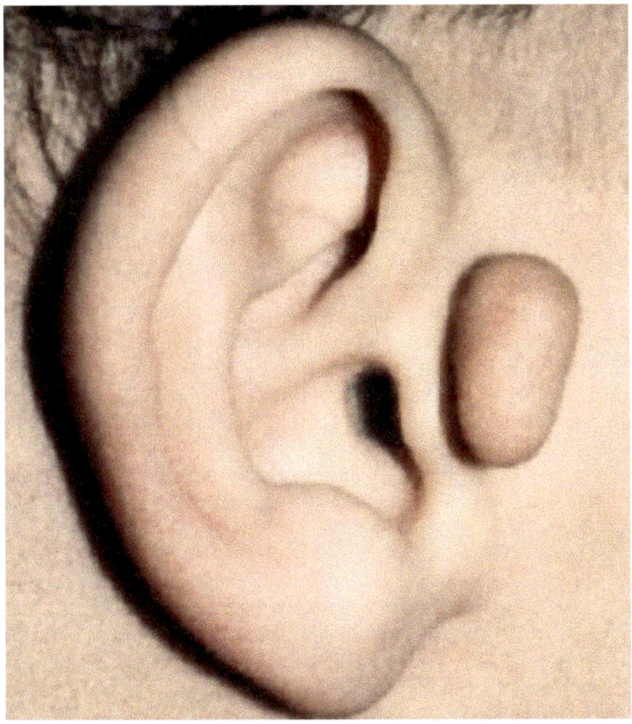

Fig. 17.2 Accessory auricle

occurs in an autosomal dominant fashion with variable penetrance [4]. In 3–10% of cases, there is an association with another congenital malformation, primarily with branchio-oto-renal syndrome [4]. They can become recurrently infected, and this is the most common indication for surgical excision, although recurrences are common.

Middle Ear Malformations

The development of the middle ear is complex and closely related to the development of the external ear. The first and second branchial arches (Meckel's and Reichert's cartilage) give rise to the majority of the ossicular chain during weeks 16–30 of development, including the incus, malleus, stapes, tensor tympani muscle and tendon, and stapedius muscle and tendon. The cartilaginous ossicular chain undergoes endochondral ossification and at the time of birth is almost completely ossified and nearly adult size.

While the exact mechanisms are poorly understood, disruption of the branchial apparatus during week 16–30 of development can result in one of several middle ear malformations, such as malformation or fixation of the ossicular chain, discontinuity of the ossicular chain, and aplasia or dysplasia of the oval or round windows. These malformations can occur in isolation, or in association with other malformations of the external and middle ear, and they have been broadly classified into four major classes based on their suitability for surgical correction [5]. Middle ear malformations occur frequently in association with other congenital syndromes such as the CHARGE (coloboma, heart defects, choanal atresia, retardation of growth/development, genital anomalies, ear anomalies) syndrome, VACTERL (vertebral anomalies, anal atresia, cardiac defects, tracheoesophageal fistula, renal anomalies, limb defects) association, Treacher Collins syndrome, branchio-oto-renal syndrome, and Beckwith-Wiedemann syndrome.

Middle ear malformations uniformly present with conductive hearing deficits, ranging from 10 to 60 dB, and can cause significant morbidity, particularly if present bilaterally. In one study the stapes was the most commonly malformed auricle, and the anomaly was bilateral in half of patients [6]. In patients with a conductive hearing loss and suspected middle ear malformation, imaging such as computed tomography (CT) scanning is frequently performed to attempt to identify the underlying pathologic process. Treatment for these malformations can range from simple hearing amplification to surgical correction of the underlying hearing loss.

Inner Ear Malformations

The inner ear develops separately from the middle and external ear. Whereas the latter are formed as part of the branchial apparatus, the inner ear is formed from a set of otic placodes in the ectoderm layer of the germinating embryo. This process begins around the twenty-second day of embryonic development. The otic placodes invaginate into the underlying mesoderm layer eventually forming the otic vesicle. The otic vesicle eventually is immersed in and surrounded by the mesodermal layer, creating an island of epithelial tissue in the mesenchyme that eventually gives rise to the inner ear structures, including the cochlea by the eighth week of gestation, the vestibular organs by the eleventh week, and the semicircular canals between the nineteenth and twenty-second week of gestation

Malformations of the inner ear can be broadly placed into two categories: malformations involving only the membranous labyrinth and malformations involving both the bony and membranous labyrinth. Malformations of the membranous labyrinth, such as complete membranous labyrinth dysplasia, are generally not identifiable on radiologic imaging and require histopathologic examination for diagnosis [7]. Malformations of the bony and membranous labyrinth, by contrast, frequently lend themselves to identification with radiologic imaging, such as CT and MRI.

Complete labyrinthine aplasia, or Michel aplasia, is the most severe form of inner ear deformity and is characterized by complete absence of the structures of the inner ear resulting in a total hearing loss in the affected ear. It is caused by arrested development of the otic placode during the third week of gestation [8]. It is extremely rare and accounts for only 1% of all inner ear malformations [9]. By contrast, the Mondini deformity is the most common type of cochlear malformation. It accounts for more than 50% of all cochlear deformities, and it occurs due to developmental arrest in the seventh week of gestation [8]. The Mondini deformity is characterized by cochlear growth that is interrupted at less than the normal 2.75 turns and is associated with total hearing loss in the affected ear and other vestibular anomalies.

There are varying degrees and subtypes of labyrinthine aplasia, including cochlear aplasia, cochlear hypoplasia, common cavity deformity, incomplete partition anomalies, enlarged vestibular aqueduct, and malformations of the vestibule and semicircular canal. These are the result of interruptions at various stages of embryogenesis and can result in varying degrees of hearing and vestibular dysfunction. CT and MRI are critical to differentiate among the various types of inner ear malformation and can inform prognosis and management.

Conclusion

The development of the external, middle, and inner ear is a complex process, and interruptions of this process at varying stages of development can give rise to a wide variety of congenital malformations.

References

1. Harris J, Källén B, Robert ES. The epidemiology of anotia and microtia. Med Genet. 1996;33(10):809.
2. Ang GS, Simpson SA, Reddy AR. Mycophenolate mofetil embryopathy may be dose and timing dependent. Am J Med Genet A. 2008;146A(15):1963.
3. Llano-Rivas I, González-del Angel A, del Castillo V, Reyes R, Carnevale A. Microtia: a clinical and genetic study at the National Institute of Pediatrics in Mexico City. Arch Med Res. 1999;30(2):120.
4. Scheinfeld NS, Silverberg NB, Weinberg JM, Nozad V. The preauricular sinus: a review of its clinical presentation, treatment, and associations. Pediatr Dermatol. 2004;21(3):191.
5. Teunissen EB, Cremers WR. Classification of congenital middle ear anomalies. Report of 144 ears. Ann Otol Rhinol Laryngol. 1993;102(8):606–12.
6. Swartz JD, Faerber EN. Congenital malformations of the external and middle ear: high-resolution CT findings of surgical import. Am J Roentgenol. 1985;144(3):501–6.
7. Joshi VM, Navlekar SK, Kishore GR. CT and MR imaging of the inner ear and brain in children with congenital sensorineural hearing loss. Radiographics. 2012;32(3):683–98.
8. Jackler RK. Congenital malformations of the inner ear. In: Cummings CW, Flint PW, Harker LA, et al., editors. Cummings otolaryngology: head and neck surgery. 4th ed. Elsevier Mosby: Philadelphia, PA; 2005. p. 4413–4.
9. Ozgen B, Oguz KK, Atas A, Sennaroglu L. Complete labyrinthine aplasia: clinical and radiologic findings with review of the literature. AJNR Am J Neuroradiol. 2009;30(4):774–80.

Nose and Paranasal Sinuses

18

Catherine Lau and Steve Goudy

Development of the Nose and Paranasal Sinuses

Formation of Five Processes Around the Stomodeum

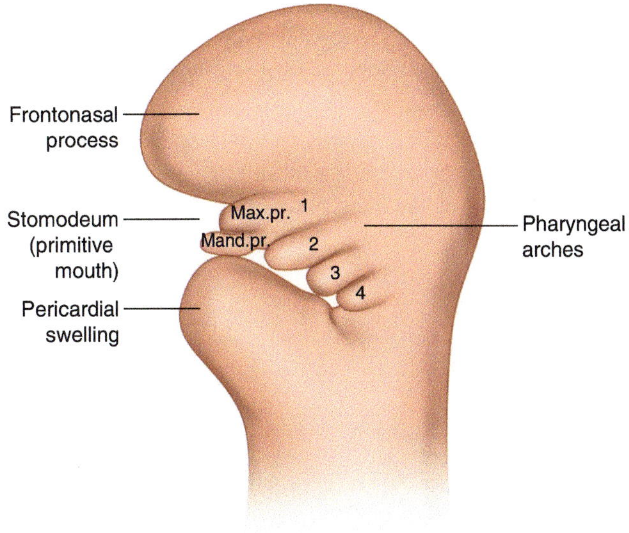

1. The upper part of the head fold projects downwards and forwards to form the **frontonasal process.**
2. The **pericardial swelling** projects upwards.
3. A depression called the **stomodeum** (primitive mouth) is formed between the previous two swellings.
4. **Pharyngeal arches** appear on either side of the pharyngeal gut.
5. The first pharyngeal arch develops two processes

   ↗ Mandibular process ventrally.

   ↘ Maxillary process dorsally.

6. The stomodeum becomes surrounded by five processes:
   (a) Frontonasal process—cranially
   (b) Two maxillary processes—on each side
   (c) Two mandibular processes—caudally

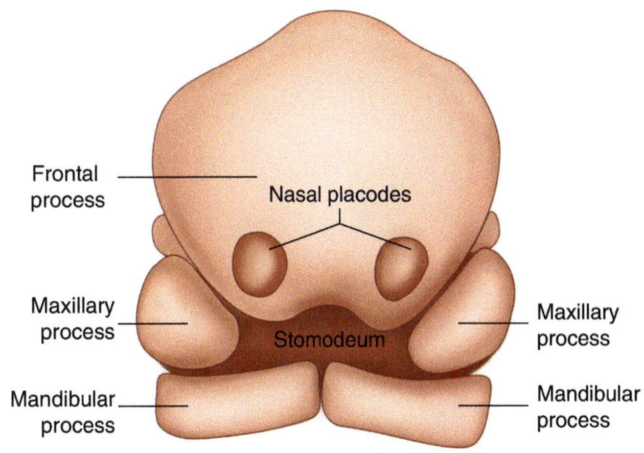

C. Lau
Oral & Maxillofacial Surgery, Royal London Hospital, London, UK

S. Goudy (✉)
Paediatric Otolaryngology, Emory School of Medicine, Atlanta, GA, USA
e-mail: steven.goudy@emory.ed

R. Carachi, S. H. E. Doss (eds.), *Clinical Embryology*, https://doi.org/10.1007/978-3-319-26158-4_18

Differentiation and Fusion of the Five Processes

The Frontonasal Process

1. *Two nasal placodes* (buds) develop on either side of the frontonasal process.
2. Each nasal placode becomes invaginated to form a *nasal pit.*
3. The edges of each nasal pit form *med-* and *lat-nasal folds*.
4. The nasal pits become deeper forming the *nasal cavities* which will later open into the pharynx posteriorly.
5. Each *lat-nasal fold* will form the ala of the nose.
6. The *two med-nasal folds* unite together:
   (a) *On the surface*: to form the middle part of the nose and the philtrum of the upper lip
   (b) *At a deeper level*: to form the premaxilla which includes
   - The ant-part of the upper jaw (carrying the incisor teeth)
   - The primary palate (the ant-triangular part of the palate carrying the incisive fossa)

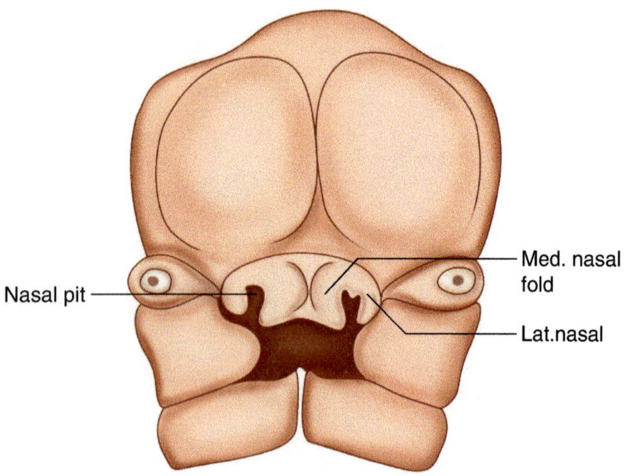

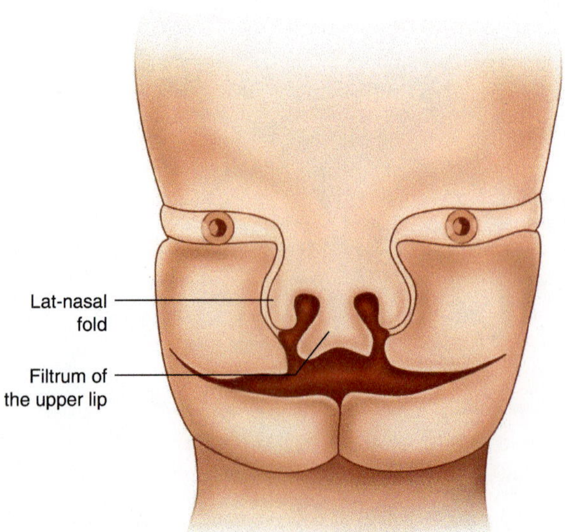

The Two Maxillary Processes

1. Each maxillary process grows medially and approaches the med- and lat-nasal folds but remains separated from them by nasolacrimal groove which later will form the nasolacrimal duct.
2. The two maxillary processes fuse with the medial nasal folds of the frontonasal process to form the upper lip (except the philtrum which is formed by fusion of the two med-nasal folds).
3. Each maxillary process unites:
   (a) Anteriorly: with the lat-nasal fold along the side of the nose
   (b) Posteriorly: with the mandibular process to form the cheek

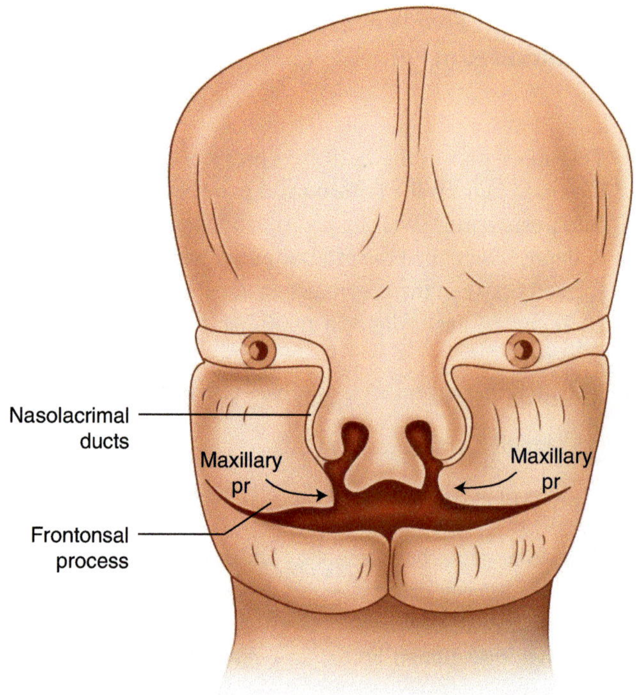

The Two Mandibular Processes

1. They fuse above with the maxillary process forming the cheeks.
2. They fuse with each other medially to form the lower lip and chin.

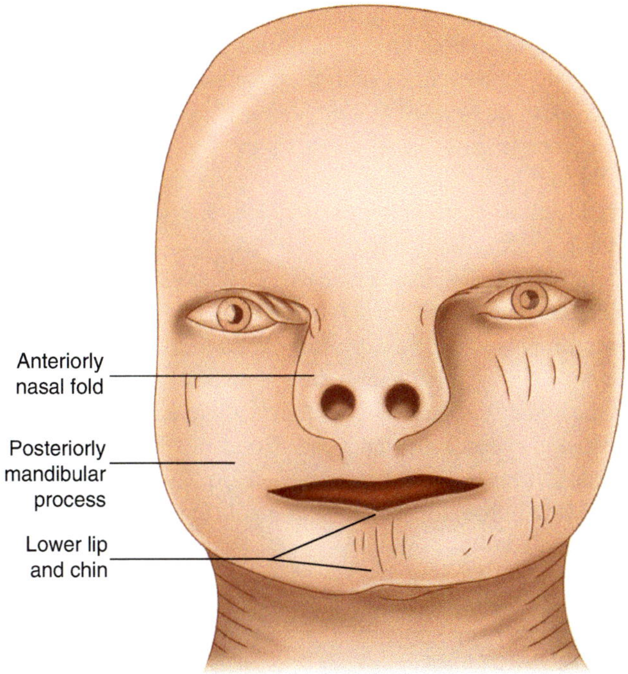

N.B:

1. The degree of fusion between the maxillary and mandibular processes determines the width of the mouth.

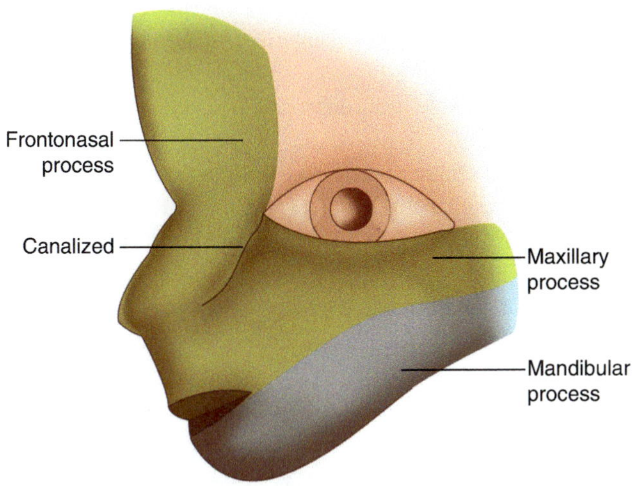

Congenital malformations of nose and paranasal sinuses are rare manifestations of developmental disorder of the foetal aerodigestive tract. The developments of nose and paranasal sinuses are interlinked. A teratogenic influence may occur anytime up to birth, resulting in its susceptibility to be involved in many other facial anomalies. Nasal anomalies may arise as primary embryologic defects or as secondary to defects in other facial units, for example, in cases of cleft lip and palate. Many rare nasal anomalies are associated with distinct genetic syndromes, and some follow known exposures of the foetus to teratogenic drugs or environmental conditions at critical times.

Overall, multiple migrational pathways are possible for mesodermal reinforcement of tissues derived from the neural crest, but a preordained pattern must occur if an anomaly is to be prevented.

Congenital nasal deformities are classified using the Losee classification system (Losee) [1].

Type I—Hypoplasia and atrophy
Type II—Hyperplasia and duplications
Type III—Clefts
Type IV—Neoplasms and vascular anomalies

Nasal hypoplasia is often present with many craniofacial syndromes. These include Apert, Fraser, Binder, craniofacial microsomia, and Goldenhar syndromes.

Nasal Dermoids

Nasal dermoids are frontonasal inclusion cysts or sinus tract caused by embryological errors localised to the anterior neuropore. These contain variable numbers of epidermal appendages, including hair follicles, sebaceous glands, and eccrine glands. Dural attachments may exist, creating a tract between the nasal skin and the dura that passes through the prenasal space through the fonticulus frontalis. Dermoids are the most common congenital nasal anomaly and midline masses. These dermoids occur sporadically with a slight male predominance.

Nasal dermoids can be associated with the following congenital anomalies (Balasubramanian) [2]:

1. Aural atresia
2. Pinna deformity
3. Mental retardation
4. Hydrocephalus
5. Branchial arch anomalies
6. Cleft lip and palate
7. Hypertelorism
8. Hemifacial microsomia

Nasal dermoids account for 3.7–12% of dermoids in the head and neck region. They are generally situated in the midline, most commonly on the nasal dorsum manifests as a nasal pit with a hair extruding from it or mass with a widened nasal bridge (Fig. 18.1). Mass lesions are firm, lobulated and noncompressible; these are usually associated with a sinus opening with intermittent discharge of caseous material. These can be located anywhere from columella base to the glabella. Dermoids can be single or multiple and typically

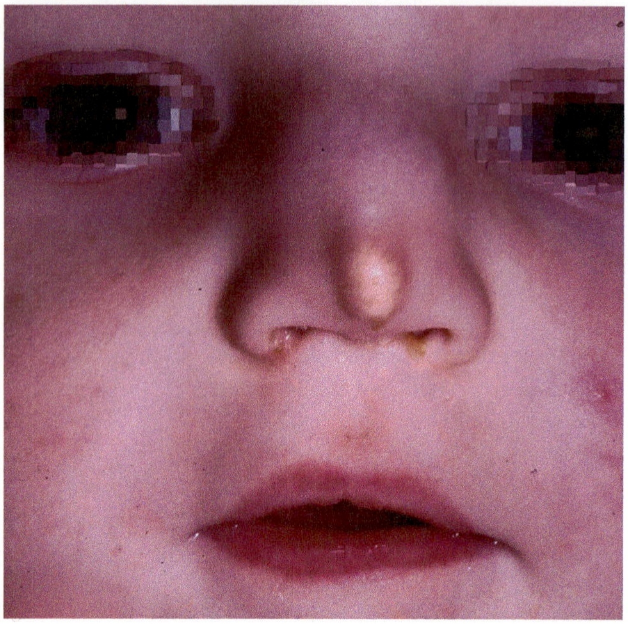

Fig. 18.1 Nasal anomaly

evident within the first month of life with 73% diagnosed by the age of one. Dermoids may extend intracranially and should be differentiated from encephaloceles using CT and MRI. Nasal dermoids do not enlarge when the child is crying, and it does not transilluminate.

Gliomas

Gliomas are unencapsulated collections of glial cells situated in an extradural site. These account for approximately 5% of all congenital nasal swellings. They are said to be more common in males and usually present in childhood as intranasal (30%), extranasal (60%) or combined masses (10%). Five to 20% of gliomas retain fibrous communications with the central nervous system or subarachnoid space but no cerebrospinal fluid connection. These intermediate lesions may be indicative of a common development aetiology for glioma and encephalocele.

Possible theories of gliomas embryogenesis have been proposed.

1. Sequestration of glial tissue of the olfactory bulb entrapped during cribriform plate fusion
2. Ectopic neural tissue cells
3. Pinched encephalocele
4. Inappropriate closure of the anterior neuropore (fonticulus frontalis), with failure of mesoderm to enter the region, resulting in inadequate bone formation

Unlike dermoids, they do not necessarily occur in midline or attach to sinuses or skin. Gliomas are smooth, noncompressible rubbery mass with grey, yellow or purple surface that does not increase in size on Valsalva testing or jugular vein compression (Furstenberg's sign) and does not transilluminate. Most are notice at birth or early childhood. No sex predilection exists and heredity does not appear to be a factor.

Intranasal gliomas are often associated with middle turbinate and may mimic nasal polyp. This type of gliomas presents as polypoidal pale masses protruding from the nasal cavity. They arise commonly from the lateral nasal wall close to the middle turbinate [3]. Combined gliomas have both intranasal and extranasal components. Combined intra-/extranasal gliomas have a typical dumbbell shape with a connecting band (Medscape) [4]. Assessment should include MRI prior to surgery.

Encephaloceles

An encephalocele is an extracranial herniation of the meninges and brain through a defect in the skull. They may contain meninges (meningocele) or brain matter and meninges (encephalomeningocele), or they may communicate with a ventricle (encephalomeningocystocele). Encephaloceles have an aetiology similar to that of gliomas. No familial pattern or sex predilection has been demonstrated with these lesions. Some congenital encephaloceles are secondary to a generalised dysplastic process or mesodermal migrational failure, for example, neurofibromatosis and Ehlers-Danlos syndrome.

Some geographic variations are seen with incidence of 1 in 6000 live births in southeast Asia and Russia and 1 in 35,000 in western Europe, the USA, Australia, Japan, China, and India.

Classification of meningoencephaloceles (Chap. 15: congenital anomalies of the nose Michael Cinnamond) [5]

Occipital	**Basal**
Cranial vault	Transethmoidal
Interfrontal	Sphenoethmoidal
Anterior fontanelle	Transsphenoidal
Interparietal	Fronto-sphenoidal
Posterior fontanelle	Sphenoorbital
Temporal	**Cranioschisis**
Frontoethmoid/nasal	Cranial upper facial cleft
Nasofrontal	Basal lower facial cleft
Nasoethmoidal	Occipito-cervical cleft
Nasoorbital	Acrania and anencephaly

Classification of nasal encephaloceles

Sincipital 60%	Nasofrontal (40%)	• Exits cranium between the nasal and frontal bones • Present as glabellar masses • Cause telecanthus and inferior displacement of nasal bones • Bony defect lies between the orbitals and in between the nasal and frontal bones	Sincipital encephaloceles typically present as soft compressible masses over the glabella.
	Nasoethmoidal (40%)	• Exits between the nasal bones and nasal cartilages • Presents as dorsal nasal masses causing superior displacement of nasal bones and inferior displacement of alar cartilages • Masses transverse via the foramen caecum deep to the nasal bones and becomes superficial at the cephalic end of upper lateral cartilage	
	Nasoorbital (20%)	• Exits through a defect in the maxillary frontal process • Usually manifest as orbital masses causing proptosis and visual disturbances • Masses pass through foramen caecum deep to the frontal and nasal bones and enter the orbit via a defect in the medial orbital wall	
Basal 40%	Transethmoidal	• Exits through the cribriform plate into the superior meatus, extending medial to the middle turbinate • Presents with nasal obstruction, hypertelorism, broadening of nasal vault and unilateral nasal mass	Basal encephaloceles may remain undetected for years
	Sphenoethmoidal	• Exits through the cribriform plate, between the posterior ethmoid cells and sphenoid, to present in the nasopharynx • Presents with nasal obstruction, hypertelorism, broadening of nasal vault and unilateral nasal mass	
	Sphenoorbital	• Enters the orbit via the superior orbital fissure and may produce exophthalmos • Presents as a nasopharyngeal mass causing bilateral nasal obstruction • Usually associated with a cleft palate	
	Transsphenoidal	• Herniates in the nasopharynx via defects posterior to the cribriform plate • Causes unilateral exophthalmos, diplopia and other visual defects	

Both sincipital and basal forms expand with the Valsalva manoeuvre, have a positive Furstenberg sign and transilluminate. Their reducibility and positive cough reflex plays a vital role in the diagnosis and thereby distinguishing them from gliomas. Patients may have a history of rhinorrhoea or recurrent meningitis and may have a broad nose or hypertelorism (dystopia canthorum). MRI is recommended to determine the presence, location and extend of an encephalocele.

Nasal Clefts

Failure of frontal processes to develop appropriately or to merge with other facial processes results in various malformations. Numerous facial cleft classification systems exist. These include DeMyer, Sedano and Tessier classifications.

Nasal clefting can be medial or lateral. Rare in isolation, it often is associated with other congenital anomalies, or it constitutes manifestation of a syndrome, such as frontonasal dysplasia, Goldenhar and Gorlin syndrome. Nasal clefts can vary from simple groove to complete separation of either side of the nose, or they can present as a large furrow involving the ipsilateral alum and medial canthus. The incidence of other associated regional defects increases with the severity of deformity.

The frontonasal syndrome combines some aspects of medial and lateral nasal clefts. Usually lateral alar notching and deformity of the nasal cartilage and nasal bones occur on the affected side. Hypertelorism and frontal bone defects of the skull also often exist. The incidence of anterior encephalocele, haemangioma of the septum and cerebral tumour is high in these patients.

Proboscis Lateralis and Supernumerary Nostrils

Proboscis lateralis (congenital tubular nose) is an extremely rare anomaly in which there is a unilateral failure of the external nose to develop and is replaced by a tubular structure originating from the medial canthus. In this deformity, the medial and lateral processes and the globular processes are absent. The maxillary process on the affected side fuses with the opposite nasal and globular process, creating nasal closure with absence of the nasal cavity, choanae, ethmoid, nasal bones and nasolacrimal duct. Proboscis lateralis may be associated with other congenital anomalies particularly those of the central nervous system.

Supernumerary, or accessory, nostrils are a very rare type of congenital nasal anomaly. They can be unilateral (most cases) or bilateral and may be associated with malformations such as facial. The accessory nostril may communicate with the ipsilateral nasal cavity.

Arhinia

Arhinia is the congenital absence of the external nose, nasal cavities and olfactory apparatus. This can be bilateral or unilateral. Both conditions are extremely rare condition. The pathogenesis of arhinia is poorly understood. This may occur as an isolated defect or with other anomalies. In all reported cases, the nasal chambers are incomplete or absent, and the palate is highly arched and hypoplastic. Arhinia is clearly evident at birth; the facial appearance is varied although usually distinctive with hypertelorism or hypotelorism and normal labial development. This extremely rare entity is often associated with anomalies of the ocular and central nervous systems. It has been associated with inversion and trisomy of chromosome 9. Prenatal diagnosis of total arhinia using obstetric ultrasound is possible by demonstration of flattened midface without a distinguished nose and a prominent upper lip (Thornburg) [6].

Polyrrhinia

This extremely rare anomaly is characterised by two completely formed noses. The duplication of media nasal processes during embryogenesis is believed to cause polyrrhinia.

Nasopharyngeal Teratoma and Epignathus

Teratomas are very rare lesions that contain tissues originating from the three embryological germ layers. An epignathus is an extremely rare form of oropharyngeal teratoma that arises from the oral cavity, most commonly from the palate and is associated with a high mortality secondary to airway obstruction requiring endotracheal intubation or tracheostomy in the neonatal period. An ex utero intrapartum treatment (EXIT) procedure may be considered to secure the foetal airway while uteroplacental circulation is still maintained [7].

Possible theories of pathogenesis include pathogenic development of germ cells (does not applied to head and neck teratoma), escaped totipotent cells from embryological organisation and conjoined twin theory.

Choanal Atresia

Bilateral or unilateral choanal atresia can occur as an isolated congenital anomaly, but 60% are associated with other defects such as Treacher Collins syndrome, branchial arch anomalies and cardiac or gut abnormalities (Kaplan) [8]. Choanal atresia has been linked with a limited number of specific defects—the CHARGE association [9] :

C colobomatous blindness
H heart disease
A atresia of the choanae
R retarded growth or development, including the central nervous system
G genital hypoplasia in males
E ear deformities, including deafness

Choanal atresia is not a common anomaly but it is not a rarity. It occurs in 1 of every 5000–7000 live births and is twice as common in females as in males. Bilateral choanal atresia is associated with airway distress in the newborn, as they are obligate nasal breathers, and requires immediate intubation and surgical intervention. It is unilateral more often than bilateral, and it occurs twice as often on the right side as the left side. An incomplete atresia is termed a *choanal stenosis.*

Choanal atresia may be bony (90%) or membranous (10%) and is situated just in front of the posterior end of the nasal septum. In substantial atresia, a considerable proportion of the posterior nasal cavity can be obliterated by dense bone (Fig. 18.2). An incomplete atresia is termed a choanal stenosis. Facial skeleton asymmetry is common especially in unilateral cases; most patients will have a high arched palate.

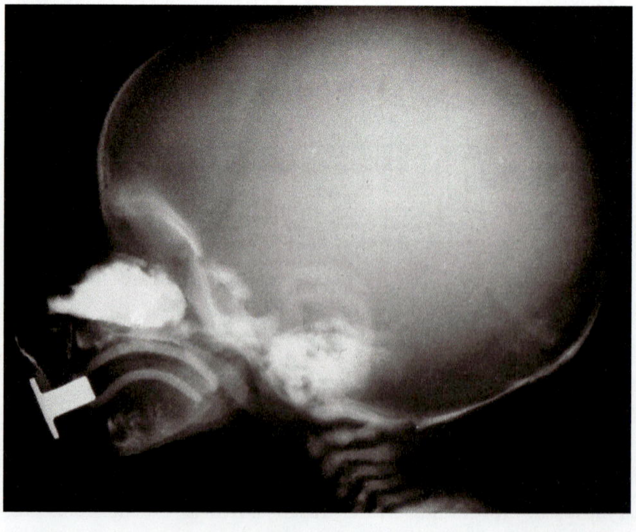

Fig. 18.2 X-ray of choanal atresia showing dye in the obstructed choanal region

Unilateral choanal atresia may not be evident until late childhood or even in adulthood unless specifically sought at birth. Feeding difficulties may occur, especially during breast-feeding when the functioning nostril is occluded by contact with the breast. More commonly it presents as a unilateral nasal discharge. Bilateral atresia almost always presents as a respiratory emergency and is therefore apparent at birth.

Congenital causes of nasal swelling in childhood (Chap. 15: congenital anomalies of the nose, Michael J Cinnamond) [5]

Cystic	Solid
Meningoencephalocele	Glioma
Meningocele	Haemangioma
Dermoid cyst	Lymphangioma
Epidermoid cyst	Neurofibroma
	Neuroblastoma
	Rhabdomyosarcoma
	Chordoma
	Craniopharyngioma

References

1. Losee JE, Kirschner RE, Whitaker LA, et al. Congenital nasal anomalies: a classification scheme. Plast Reconstr Surg. 2004;113(2):676–89.
2. Balasubramanian T. Congenital malformations of nose and paranasal sinuses. http://otolaryngology.wdfiles.com/local%2D%2Dfiles/rhinology/60225605-Congenital-malformations-of-nose-and-sinuses.pdf. Accessed 8 Sep 2015.
3. Okumura M, Francisco RP, Lucato LT, Zerbini MC, Zugaib M. Prenatal detection and postnatal management of an intranasal glioma. J Pediatr Surg. 2012;47(10):1951–4.
4. Tewfik TL, et al. Congenital malformations of the nose. http://emedicine.medscape.com/article/837236-overview#a1. Accessed 8 Sep 2015).
5. Cinnamond MJ, Adams DA. Chapter 15: congenital anomalies of the nose. In: Paediatric otolaryngology: Butterworth-Heinemann Ltd.; 1996.
6. Thornburg LL, Christensen N, Laroia N, Pressman EK. Prenatal diagnosis of total arhinia associated with normal chromosomal analysis: a case report. J Reprod Med. 2009;54(9):579–82.
7. Chiu HH, Hsu WC, Shih JC, et al. The EXIT (Ex Utero Intrapartum Treatment) procedure. J Formos Med Assoc. 2008;107(9):745–8.
8. Kaplan LC. Choanal atresia and its associated anomalies. Further support for the CHARGE Association. Int J Pediatr Otorhinolaryngol. 1985;8(3):237–42.
9. Pagon RA, et al. Coloboma, congenital heart disease, and choanal atresia with multiple anomalies: CHARGE association. J Pediatr. 1981;99(2):223–7.

Congenital Abnormalities of the Thyroid and Parathyroid

19

John D. Collin and Ceri Hughes

The thyroid primordium appears at the fourth week as a median endodermal proliferation in the floor of the pharynx between the tuberculum impar and the hypobranchial eminence. This is at the site of the foramen caecum in the adult tongue.

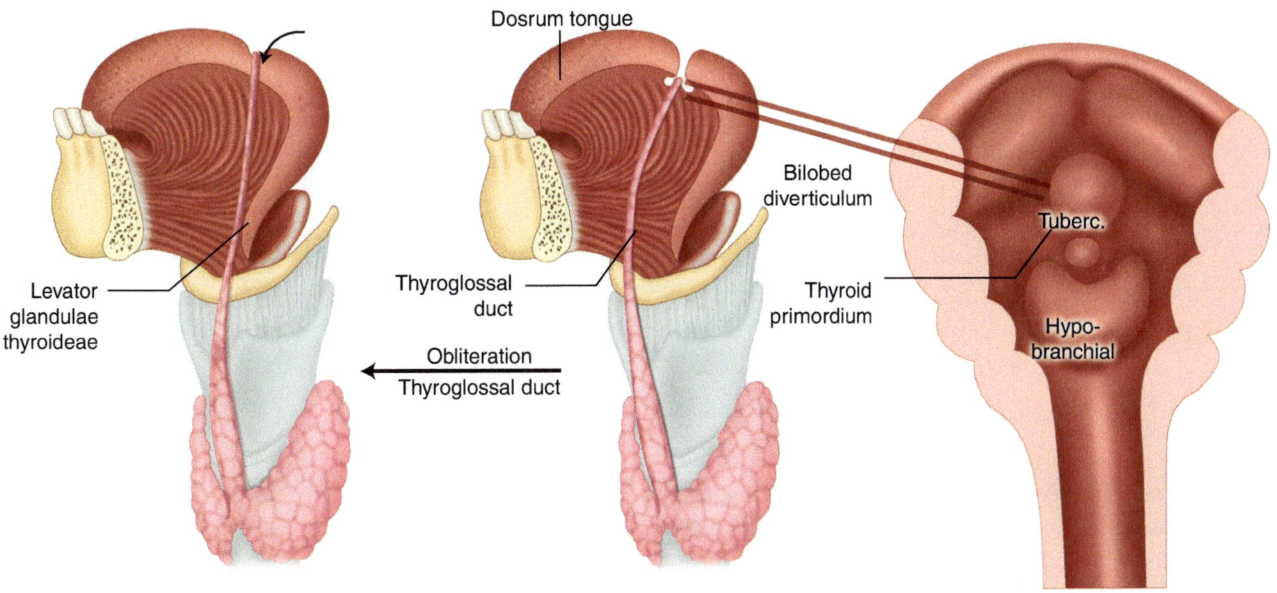

J. D. Collin (✉) · C. Hughes
Department of Oral and Maxillofacial Surgery, University Hospitals Bristol NHS Foundation Trust, Bristol, UK
e-mail: jcollin@doctors.org.uk

R. Carachi, S. H. E. Doss (eds.), *Clinical Embryology*, https://doi.org/10.1007/978-3-319-26158-4_19

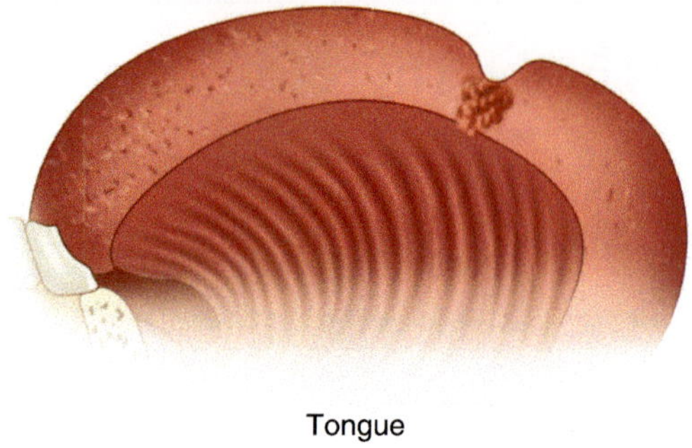

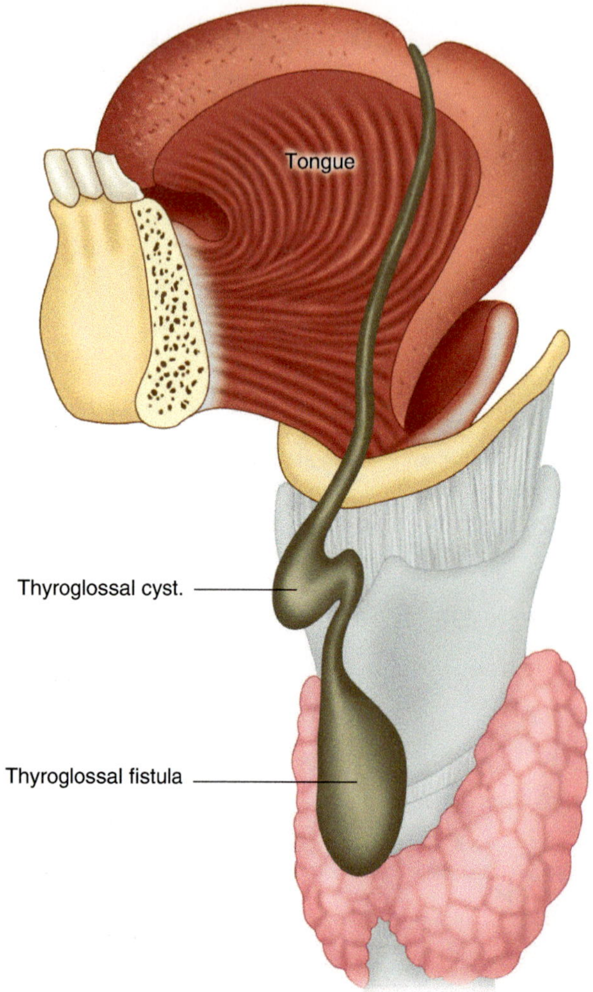

The thickening is evaginated to form a bilobed diverticulum which descends in front of the hyoid bone and laryngeal cartilages and remains connected to the dorsum of the tongue by the thyroglossal duct. The thyroid gland finally reaches its position in front of the thyroid cartilage and upper part of the trachea, and then the thyroglossal duct disappears or becomes fibrosed to form the lavatory glandulae thyroideae.

The parathyroid gland arises as a derivative from the third and fourth pharyngeal pouches.

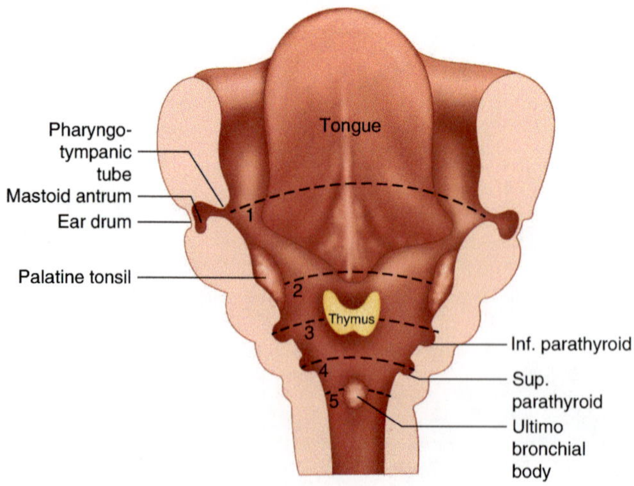

The dorsal part of the pouch forms the inferior parathyroid gland, and this will migrate caudally to eventually lie below the superior parathyroid gland. The dorsal part of the fourth pouch gives rise to the superior parathyroid gland.

Thyroglossal Duct Cysts

The thyroglossal duct normally invaginates and atrophies by the tenth week post-fertilisation, following descent of the thyroid gland. Failure of this process results in a persistent thyroglossal duct and potentially an open connection between the foramen caecum of the tongue and the thyroid gland. This can result in fluid accumulation and infection. Remnants of the duct following incomplete atrophy can also undergo cystic change even where there is no oral communication. These thyroglossal duct cysts are most common cause of a congenital neck mass (Fig. 19.1).

A few cases are familial, mainly showing an autosomal dominant inheritance with incomplete penetrance and a female predilection. There is an association with ectopic thyroid glands and, in 11–33% of cases, thyroid cancer (usually papillary). Malignant conversion of the cyst itself is seen in around 1% of cases.

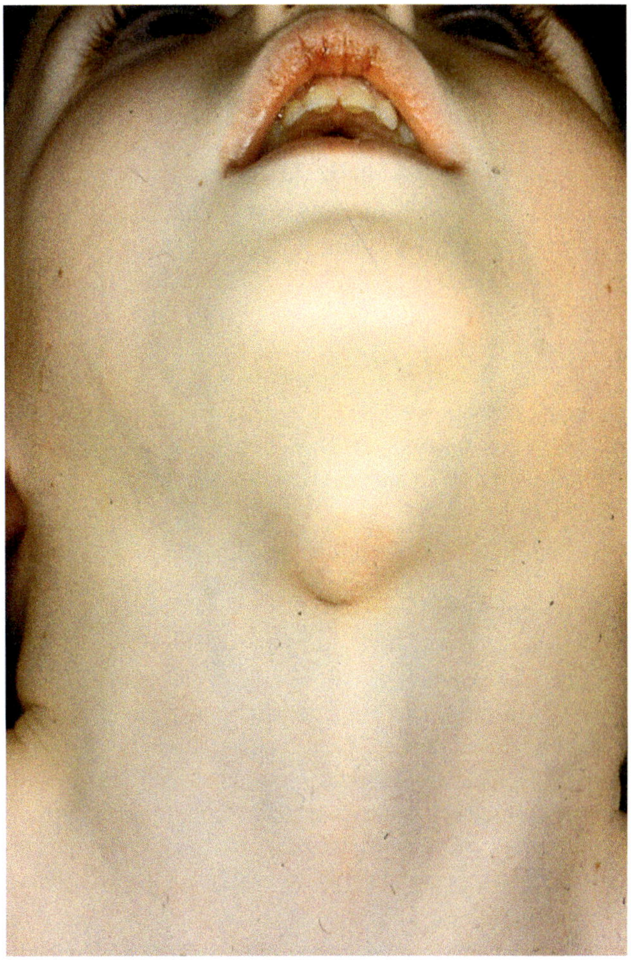

Fig. 19.1 Thyroglossal cyst

Epidemiology

Thyroglossal duct cysts account for around 70% of all congenitally acquired neck masses. Adult cadaveric studies suggest a prevalence of approximately 7%, although most are asymptomatic. Many are detectable during childhood, with a mean age at presentation of 6 years and around 50% of patients presenting before age 20.

Clinical Features

Thyroglossal duct cysts are normally asymptomatic and often only become apparent when observers notice a swelling that moves when the patient swallows. Typically they are 1–4 cm in size, with around 70% located in the midline, and almost all within 2 cm of the midline. Around 80% are located at or below the hyoid bone. Inferiorly arising cysts are more likely to be lateralised, in which case they tend to be adjacent to the thyroid cartilage.

Secondary infection gives rise to increased swelling, pain, erythema and eventually symptoms of sepsis. Cutaneous and/or pharyngeal fistulae may develop. The rare lingually located cysts may be associated with dysphagia and airway embarrassment.

Classification

Thyroglossal duct cysts can be classified according to their anatomical location. Approximately 65% are located between the hyoid bone and thyroid gland. The low incidence of lingual thyroglossal duct cysts may be because atrophy of the thyroglossal duct commences from the oral side.

- Lingual: 1–2%
- Suprahyoid: 20–25% (less common in adults ~5%)
- At level of hyoid bone: 15–50%
- Infrahyoid: 25–65%

Diagnosis

Ultrasound or sectional imaging will often strongly support the clinical diagnosis. Ultimately definitive confirmation is given by histopathological examination of the excised cyst. Contrast CT or MRI studies will show a well-delineated cystic lesion with capsular enhancement and intimate relation or attachment to the hyoid bone.

Prognosis

The majority of cysts are subclinical. Patients with symptomatic cysts have an excellent prognosis after excision, even if malignant change has occurred as metastasis outside of the cyst is rare.

Management

Sistrunk's procedure to remove the cyst and the portion of hyoid bone to which it is attached is usually appropriate. In cases of malignant transformation, hemi- or total thyroidectomy with cervical lymph node dissection in addition may be required. Failure of surgical treatment can be due to the presence of multiple, irregular microscopic branches of the duct in the suprahyoid region that open onto the pharyngeal and/or lingual mucosa.

Ectopic Thyroid

An ectopic thyroid gland is located in a location other than the normal position anterior to the thyroid cartilages. This occurs as a result of aberrant migration from the foramen caecum at the sulcus terminalis of the tongue, to its normal location. Migration can be prematurely halted at any point along the thyroglossal duct or can stray completely from the usual path to an alternative site in the neck or other sites in the body.

Pathology

A lingual thyroid results from failure of the normal caudal migration of the thyroid from foramen caecum down to its normal location anterior to the larynx and upper trachea. Thyroid tissue may be found anywhere along the course of the thyroglossal duct; however, complete arrest with thyroid tissue located at the base of tongue is most common and represents 90% of all cases of ectopic thyroid. Microscopic deposits of thyroid tissue along its route of descent have been identified in up to 10% of the population, representing small amounts of tissue being 'left behind' during normal development. The thyroid tissue is histologically and functionally normal. Carcinoma of a lingual thyroid has been reported, but the incidence is presumably no higher than occurs in orthotopic thyroid glands.

Epidemiology

The overall prevalence is around 1 per 100,000–300,000 people. The male to female ratio is 1:3.

Lingual Thyroid

The most common location of an ectopic thyroid is near its embryological origin at the foramen caecum, resulting in a lingual thyroid. This is the site for 90% of cases of ectopic thyroid glands. This specific type of ectopic thyroid results from lack of normal caudal migration of the thyroid gland.

Clinical Presentation

Most patients are asymptomatic, and the diagnosis is incidental as a result of imaging the tongue or attempting to image an absent thyroid. A lingual thyroid may, however, result in dysphagia, bleeding from mucosal ulceration or airway compromise, particularly in infants. Direct or endoscopic examination may reveal an erythematous nodular red mass that can range in size from a few millimetres to around 4 cm. Thyroid function tests are usually normal but can demonstrate hypothyroidism of varying severity in up to one-third of cases.

Other Sites

In addition to the foramen caecum, ectopic thyroid has been described in numerous other sites, between the base of the tongue and its final pretracheal position. Rarely, thyroid tissue has been identified in sub-diaphragmatic locations and cranial locations.

Neck

Rarely, when the cells of the lateral thyroid anlage do not join those of the median, a lateral ectopic or aberrant thyroid gland can form. These are usually found in the submandibular region, presenting as a lateral, palpable, mobile, painless mass in the carotid or submandibular triangles. This variant has a female preponderance and is more commonly found on the right side of the neck. In most cases, patients are euthyroid with orthotopic thyroid tissue also present. Some cases do present with the ectopic gland as the only functional thyroid tissue, which has obvious implications if considering excision.

Ectopic thyroid tissue can also be found along the course of the usual embryological migratory path. Histological examination of thyroglossal duct cysts identifies ectopic thyroid within the cyst wall in up to 5% of cases. The tissue is often functional, although patients may still be hypothyroid. Hashimoto's thyroiditis and papillary carcinoma of these variants have also been described.

Intra-tracheal thyroid tissue is rare and may be the result of either division of the developing thyroid by the trachea and its cartilaginous rings or ingrowth of thyroid tissue into the tracheal lumen. The latter may be facilitated by a developmental defect of the mesenchymal tissue between the thyroid and the trachea, allowing the primitive thyroid to adhere to the trachea. Intra-tracheal ectopic thyroid can present at any age, but predominantly between the ages of 30 and 50 years, mainly in females. Presenting symptoms include cough, dysphagia, dyspnoea, haemoptysis and stridor. Alternatively patients may be asymptomatic. A normally functioning orthotopic thyroid usually coexists, and, hence, patients are euthyroid.

Intrathoracic

Intrathoracic ectopic thyroid tissue is also rare and accounts for around 1% of all mediastinal tumors. The usual location is in the anterior mediastinum, but two cases have been found posteriorly. Intrathoracic ectopic thyroid within the lungs or heart is extremely rare, but has been reported to cause dry cough, dyspnoea and haemoptysis. Other patients have presented with dysphagia or superior vena cava syndrome.

Intrathoracic thyroid may also be revealed incidentally on chest radiographs, during cardiothoracic surgery or at autopsy. Orthotopic tissue usually coexists, and affected patients are euthyroid.

Intracardiac thyroid mainly involves the right ventricle, and the tumour is usually identified by cardiac ultrasonography to investigate the resultant dyspnoea. Larger tumours have been reported to cause right ventricular outflow tract obstruction. Para-cardiac thyroid tissue attached to the ascending aorta has also been reported, presenting with chest pain and palpitations, due to pericarditis and right atrial compression.

Peri- or intracardiac thyroid tissue probably arises due to the close embryological relationship between the thyroid primordium and developing myocardium. In the early embryo, the cardiac mesoderm is closely apposed to the ventral pharyngeal endoderm. As the heart descends, the thyroid is drawn caudally, leading to various anomalies of its final position. This embryological intimacy may explain why cardiac malformations are the birth defects most frequently associated with thyroid dysgenesis.

Sub-diaphragmatic

Ectopic thyroid tissue occurs extremely rarely at sites including the ovaries, adrenals, gallbladder, pancreas, duodenum and mesentery of the small intestine. Ovarian thyroid tissue, also known as 'struma ovarii', develops into a teratoma containing large amounts of thyroid tissue that is histologically and functionally identical to normal thyroid. Struma ovarii represent 1% of all ovarian tumors and 2–4% of ovarian teratoma, with the thyroid proportion usually accounting for the majority of the tumour mass. The mean age at diagnosis is 45 years, and often they are detected incidentally on ultrasonography. Symptoms can include lower abdominal pain and/or palpable mass and menorrhagia. Malignant transformation occurs in around 15% of cases and thyrotoxicosis in 5–15% of patients.

A few cases of intra-adrenal thyroid tissue have been reported in the literature, in women of middle age. The patients were euthyroid, and ectopic gland was diagnosed on histological examination after adrenalectomy. Enlarged adrenals, sometimes with a cystic compartment, were detected either incidentally or after investigation for secondary hypertension.

Similarly, ectopic thyroid within or adjacent to the gallbladder, pancreas and duodenum has been described as an incidental finding during surgery or histology for other unrelated disease. Occasionally symptoms due to mass effect of intra-abdominal thyroid occur or hyperthyroidism despite thyroidectomy. Finally, unique cases of ectopic thyroid include uterus, pituitary fossa and sphenoid sinus.

The possible mechanism for these rare variants of ectopic thyroid is either highly aberrant migration or heterotopic differentiation of uncommitted endodermal cells. Orthotopic thyroid gland usually coexists, and the possibility of a metastatic spread from an occult orthotopic thyroid carcinoma should always be excluded.

Investigation

Ultrasound

Ultrasound is useful in determining the presence or absence of thyroid tissue in the normal location and confirming ectopic tissue at sites within the neck in conjunction with FNAC.

Computerised Tomography

CT demonstrates hyper-dense soft tissue mass of the same attenuation as normal thyroid tissue due to accumulation of iodine. There is prominent homogeneous enhancement with contrast administration.

Magnetic Resonance Imaging

MRI will usually identify a well-defined mass with no invasive features.

Signal characteristics include:

T1—iso- to hyperintense to muscle
T2—can vary from hypo- to iso- to hyperintense to muscle
T1 C+ (Gd)—homogeneous contrast enhancement

Nuclear Medicine

Thyroid scintigraphy can be used to confirm the diagnosis and detect thyroid tissue at additional sites within the neck.

Treatment and Prognosis

Often no intervention is required. In cases where surgical excision is proposed, it is essential to establish if there is any normal thyroid tissue elsewhere so that hypothyroidism is not unexpected and can be managed appropriately. In cases of ectopic lingual thyroid in particular, there is often no orthotopic thyroid.

Congenital Hypothyroidism

Congenital hypothyroidism is the result of a thyroid developmental abnormality in approximately 85% of cases. The thyroid can be ectopic, hypoplastic or completely absent (athyreosis). The remainder of cases are due to errors in thyroid metabolism or thyroid axis signalling in the presence of an anatomically normal gland A congenital goitre may be present (Fig. 19.2).

Epidemiology

The overall incidence of congenital hypothyroidism is 1 in 3–4000 live births with around 2% of cases thought to be familial. Most studies report a female-to-male ratio of around 2:1 mainly due to a higher rate of thyroid ectopy in females. The condition is more prevalent in Hispanics (1 in 1886 births), and the sex ratio is increased, with a 3:1 female-to-male ratio. In contrast the rate in black infants is around a third of that seen in white populations, and the gender difference is lower.

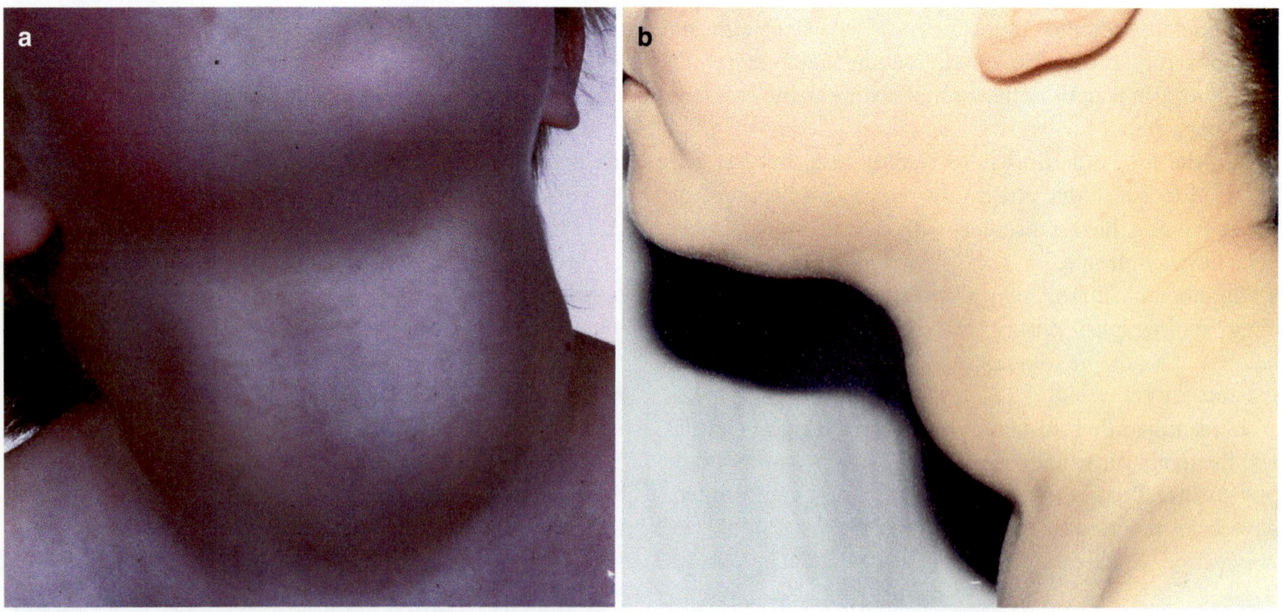

Fig. 19.2 (**a**) Enlarged thyroid, goitre. (**b**) Goitre

Twins are approximately 12 times as likely to have congenital hypothyroidism as singletons. In the majority of cases, only one twin is affected, but an in utero exposure could potentially cause congenital hypothyroidism in both.

Aetiology

The main causes of congenital hypothyroidism vary with geographical location, but approximate ranges are as follows:

Ectopic thyroid—25–50%
Thyroid agenesis—20–50%
Dyshormonogenesis—4–15%
Hypothalamic-pituitary dysfunction—10–15%

Thyroid hemiagenesis is a rare anomaly, more commonly seen on the left side (ratio 4:1) and in females (ratio 3:1).

Prognosis

Congenital hypothyroidism does not affect the all-cause standardised mortality ratio in adequately treated patients. Untreated patients can suffer from profound retardation of mental and physical development. Delay in treatment can lead to neurological problems including spasticity and gait abnormalities, dysarthria and autistic behaviour.

Congenital Hypoparathyroidism

There are several congenital diseases that result in congenital hypoparathyroidism:

- Familial isolated hypoparathyroidism
- X-linked recessive hypoparathyroidism
- Familial hypocalcaemia with hypercalciuria

Hypoparathyroidism is also associated with several syndromes including:

- Hypoparathyroidism with deafness and kidney abnormalities
- Kenny-Caffey syndrome (hypoparathyroidism with impaired growth and intellectual disability)
- 22q11 deletion syndromes
  - DiGeorge syndrome
  - Velocardiofacial syndrome
  - Conotruncal anomaly face syndrome
  - Opitz GBBB syndrome
- Kearns-Sayre syndrome
- Autoimmune polyendocrine syndrome type 1
- Storage disorders
  - Haemachromatosis (iron deposited in the parathyroids)
  - Wilson's disease (copper deposited in the parathyroids)

Pathophysiology

If the parathyroid glands do not fully develop, their ability to produce parathyroid hormone may be impaired, as in X-linked hypoparathyroidism. Alternatively, a defect in the production of parathyroid hormone, as in familial isolated hypoparathyroidism, may occur. Thirdly, congenital hypoparathyroidism can be caused by an abnormally strong signal from the calcium sensor receptors as seen in familial hypocalcaemia with hypercalciuria.

Familial hypocalcaemia with hypercalciuria is caused by a mutation in gene CASR (calcium-sensing receptor) on chromosome 3 (3q13.3-q21) and follows an autosomal dominant inheritance pattern.

Familial isolated hypoparathyroidism is usually caused by mutation of the PTH gene (11p15.3-p15.1) and has shown both autosomal dominant and recessive inheritance patterns. Mutations in the GCMb gene (glial cells missing) on chromosome 6 (6p24.2) have also been identified with an autosomal recessive inheritance pattern.

X-linked recessive hypoparathyroidism (XLHPT) is a very rare cause of congenital hypoparathyroidism reported in two multigenerational families from Missouri. The condition is caused by agenesis of the parathyroid glands due to mutation that has been mapped to chromosome Xq26–q27, in a 1.5 Mb interval flanked by markers F9 and DXS984. Affected males have true neonatal hypoparathyroidism with undetectable parathyroid hormone levels that leads to severe hypocalcaemia. They are also sterile and epileptic. Carrier females are normocalcaemic and asymptomatic. Neonatal onset and parathyroid agenesis found at autopsy in one patient suggest that the gene involved in XLHPT plays a role in parathyroid gland development.

Further Reading

Brandi ML. Genetics of hypoparathyroidism and pseudohypoparathyroidism. J Endocrinol Invest. 2011;34:27–34.
Sameer KS, Mohanty S, Correa MM, Das K. Lingual thyroglossal duct cysts—a review. Int J Pediatr Otorhinolaryngol. 2012;76:165–8.
Wassner AJ, Brown RS. Congenital hypothyroidism: recent advances. Curr Opin Endocrinol Diabetes Obes. 2015;22:407–2.
Zander DA, Smoker WR. Imaging of ectopic thyroid tissue and thyroglossal duct cysts. Radiographics. 2014;34:37–50.

Embryology of the Salivary Glands

20

Catherine Lau and Mark McGurk

Development of the Parotid Gland

The parotid gland appears behind the angle of the mouth, which is lined with buccal epithelium. This groove then closes to form the parotid duct which extends backwards. The blind distal end branches repeatedly and canalizes to form the acini of the gland. The duct recedes from the angle of the mouth to open in the vestibule of the mouth opposite the upper second molar tooth.

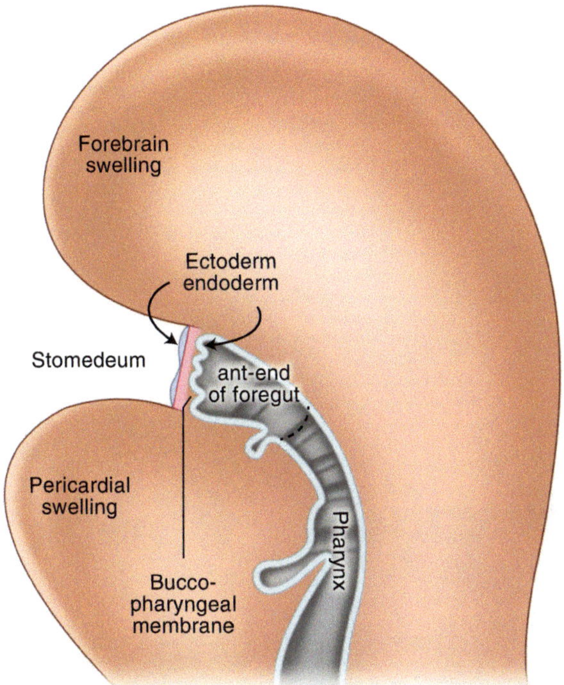

Fig. 20.1 CT scan of parotid swelling

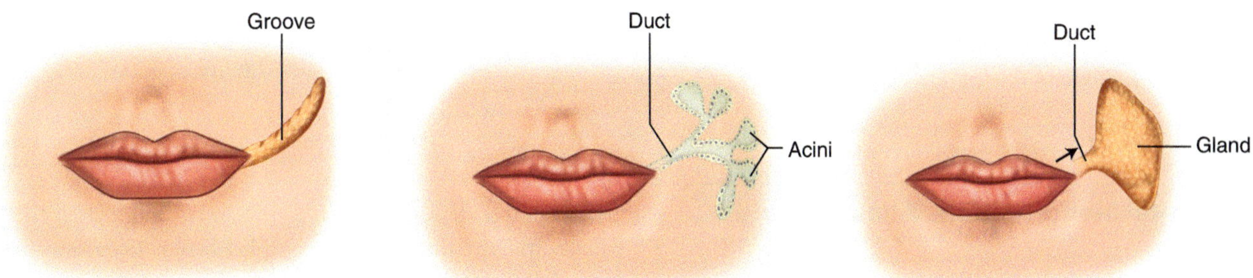

C. Lau (✉)
Royal London Hospital, London, UK

M. McGurk
University College Hospital, London, UK

R. Carachi, S. H. E. Doss (eds.), *Clinical Embryology*, https://doi.org/10.1007/978-3-319-26158-4_20

Development of the Submandibular Gland

Development begins with the appearance of a groove between the tongue and the gums, the alveololingual groove. This groove then closes to form the submandibular duct. The distal end of the duct then grows backwards along the floor of the mouth, branches repeatedly and canalizes to form the acini of the gland. The anterior end of the duct proceeds forwards to open adjacent to the frenulum of the tongue.

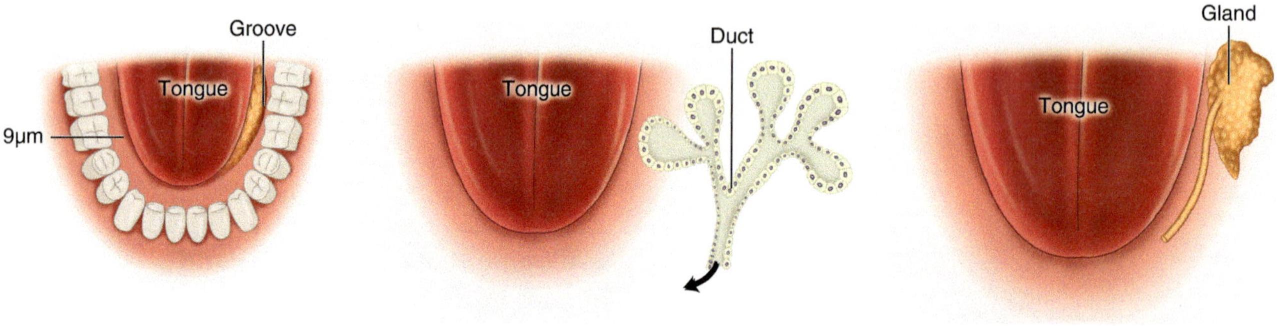

Development of the Sublingual Gland

The sublingual gland arises as a small bud developing from the alveololingual groove. These buds attain a common sheath and appear as one gland having several openings into the floor of the mouth.

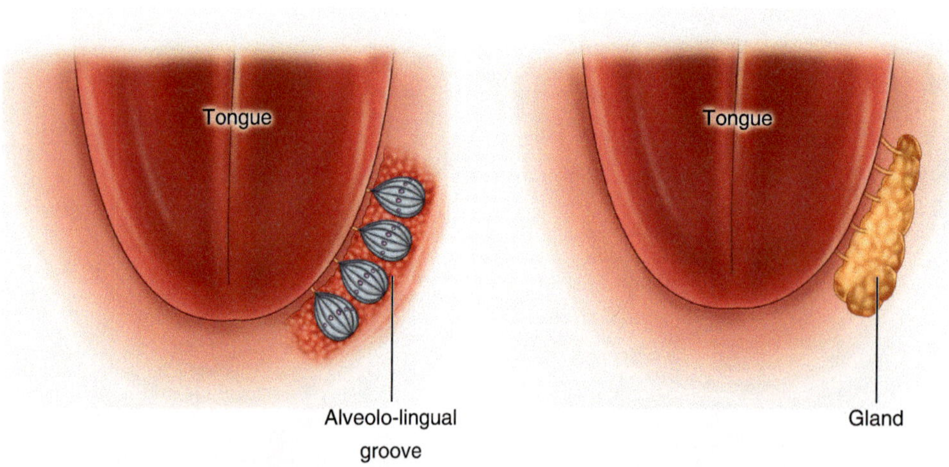

Congenital Malformations Atlas: Salivary Glands

The main developmental anomalies of the salivary glands are aplasia of glands, ductal anomalies, accessory or ectopic glands and a variety of cysts.

Aplasia of Salivary Glands

Aplasia of one or more glands has been reported but is exceptionally rare. Even more rarely, it is complete and then results in total failure of secretion of saliva and related complications. Aplasia of the salivary glands may occur in isolation or be associated with other malformations of the first branchial arch such as hemifacial microsomia, Treacher Collins syndrome and other congenital malformations of the face [1]. Aplasia and hypofunction of the salivary glands can exist with other ectodermal malformations of the lacrimal ducts, skin appendages and teeth. One of the better-known conditions is the lacrimo-auriculo-digital (LADD) syndrome. The diagnosis of hypoplasia or aplasia of the salivary gland is often delayed due to the lack of patients' perception of normal oral environment. These patients can present with rampant caries associated with dry mouth, difficulty in swallowing and excessive oral fluid intake. Early diagnosis using a combination of clinical examination, ultrasound, MRI or scintigraphy would enable early prevention against such complications.

Ductal Anomalies

Duct atresia is less uncommon in comparison to gland aplasia. This may affect the submandibular duct, and cyst may develop as a consequence.

Accessory ducts are the most common developmental anomaly associated with salivary glands. Accessory ducts of the parotid gland are usually located superior and anterior to the normal Stenson's duct orifice.

Diverticuli are small pouches of the ductal system of one of the major salivary glands. These are usually incidental findings.

Congenital fistula of the parotid gland can present as a sinus tract opening in the crease behind the pinna or in the front of tragus.

Accessory Salivary Glands

Accessory parotid tissues are extremely common and have been found to be present in one in five people [2]. These usually lie along the line of and are closely related to the parotid duct. Accessory gland has the same histological structure as the parotid gland and can suffer from the same tumours and other diseases on a less frequent basis.

Ectopic and Aberrant Salivary Tissue

Ectopic salivary tissue can form within the developmental areas of the first and second branchial arches, in the lateral part of the neck, pharynx or middle ear [3]. In some rare cases, ectopic salivary tissue can even be found situated in gingiva or the brain. Stafne bone cyst, owing its name to its radiological appearance, is an uncommon condition in which a lobe of the submandibular gland invaginates into the bone of the lingual aspect of the angle of the mandible. Ectopic or aberrant salivary tissue can be a site for development of retention cyst or neoplasm.

Polycystic Disease of the Parotid Glands

This rare anomaly described by Seifert et al. [4] may be unilateral or bilateral. Clinically, salivary gland enlargement typically presents in childhood and may be associated with symptoms of mealtime syndrome. Sialography shows a snowstorm appearance resembling punctate sialectasis. Operation and diagnosis may however be delayed until adult life. Polycystic disease of the parotids is believed to be completely benign, but removal of the glands may be necessary for cosmetic reasons or for confirmation of the diagnosis.

Congenital and Acquired Cysts of Salivary Glands Ranulae

Superficial mucous retention cysts of the floor of the mouth may be congenital or acquired. These are rarely found in the newborn but are more frequently diagnosed in adulthood. Ranula presents as a superficial translucent swelling in the floor of mouth. Large ranula can lead to breathing and swallowing difficulties; in some cases, an ex utero intrapartum treatment (EXIT) procedure may be required [5]. These cysts can rupture spontaneously or secondary to unnoticed trauma to release thick viscid mucus. A plunging ranula can present as a painless swelling in the submandibular or submental triangle associated with a swelling in the floor of mouth. A congenital defect in the mylohyoid muscle has been implicated in the pathogenesis of plunging ranulas [6]. Investigation using computerised tomography can differentiate these from similar swellings such as thyroglossal cysts, cystic hygromas or dermoid cysts.

Congenital Tumours of Salivary Glands

Congenital tumours such as sialoblastomas and rhabdomyosarcomas have been reported in the literature [7, 8]. These are extremely uncommon. Prognosis is relatively good with a combined surgical and chemotherapy treatment regime [9].

References

1. Iro H, Zenk J. Salivary gland diseases in children. GMS Curr Topics Otorhinolaryngol Head Neck Surg. 2014;13. ISSN:1865-1011. http://www.egms.de/static/pdf/journals/cto/2014-13/cto000109.pdf.
2. Polayes IM, Rankow RM. Cysts, masses, and tumors of the accessory parotid gland. Plast Reconstr Surg. 1979;64:17–23.
3. Cawson RA, Gleeson MJ, Eveson JW. The pathology and surgery of the salivary glands. Chapter 3: Developmental anomalies, cysts and infiltrations. Oxford: Isis Medical Media Ltd.; 1997. p. P22–32.
4. Seifert G, Thomsen S, Donath K. Bilateral dysgenetic polycystic parotid glands. Morphological analysis and differential diagnosis of a rare disease of the salivary glands. Virchows Arch A Pathol Anat Histol. 1981;390(3):273–88.
5. Chan DF, Lee CH, Fung TY, Chan DL, Abdullah V, Ng PC. Ex utero intrapartum treatment (EXIT) for congenital giant ranula. Acta Paediatr. 2006;95(10):1303–5.

6. Morton RP, Ahmad Z, Jain P. Plunging ranula: congenital or acquired? Otolaryngol Head Neck Surg. 2010;142(1):104–7. https://doi.org/10.1016/j.otohns.2009.10.014.
7. Salomao DR, Sigman JD, Greenebaum E, Cohen MB. Rhabdomyosarcoma presenting as a parotid gland mass in pediatric patients: fine-needle aspiration biopsy findings. Cancer. 1998;84(4):245–51.
8. Vawter GF, Tefft M. Congenital tumors of the parotid gland. Arch Pathol. 1966;82(3):242–5.
9. Prigent M, Teissier N, Peuchmaur M, El Maleh-Berges M, Philippe-Chomette P, Cardin P, Orbach D. Sialoblastoma of salivary glands in children: chemotherapy should be discussed as an alternative to mutilating surgery. Int J Pediatr Otorhinolaryngol. 2010;74(8):942–5.

Embryology of the Branchial Arches

21

Mark Wilson and Margaret Coyle

The pharyngeal or branchial arches are six curved cylindrical mesodermal thickenings on each side of the primitive pharynx. Each arch forms a swelling on the outer surface of the embryo and a swelling on the wall of the primitive pharynx internally. They are produced by the proliferation of the mesoderm of the lateral wall of the pharynx forming six arched thickenings. Each arch consists of an outer ectodermal covering, an inner endodermal lining and a mesodermal core between the two. The arches are separated from each other externally by five grooves called the pharyngeal clefts and are separated from each other internally by four grooves, the pharyngeal pouches. Each ectodermal cleft is separated from the corresponding endodermal pouch by a thin layer of mesoderm.

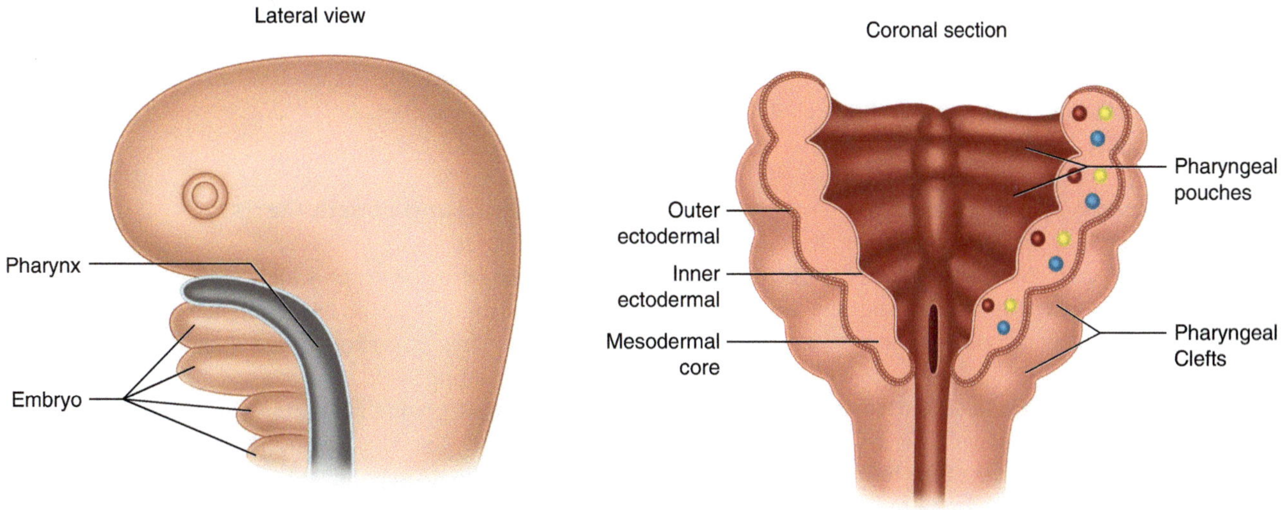

The arches begin to appear on the outer surface of the embryo at around 4–5 weeks. The cranial arches precede the caudal arches and are more prominent. After the fifth week of development, they become transformed to the bone, cartilage, ligaments, muscles and vessels of the head and neck.

The first arch—the mandibular arch—is the longest and most prominent. It divides externally into two processes, a short maxillary process and a long mandibular process. The second arch—the hyoid arch—is less prominent than the first. The ventral ends of the first and second arches reach the middle line of the floor of the pharynx.

M. Wilson, MCh, FRCSI(OMFS), MFDS RCSEd
Oral and Maxillofacial Surgeon,
University Hospital Limerick, Limerick, Ireland

M. Coyle, MFDS RCSI, MB, BCh, BAO, FRCS (✉)
Oral and Maxillofacial Surgeon,
Gloucester Royal Hospital, Gloucester, UK

R. Carachi, S. H. E. Doss (eds.), *Clinical Embryology*, https://doi.org/10.1007/978-3-319-26158-4_21

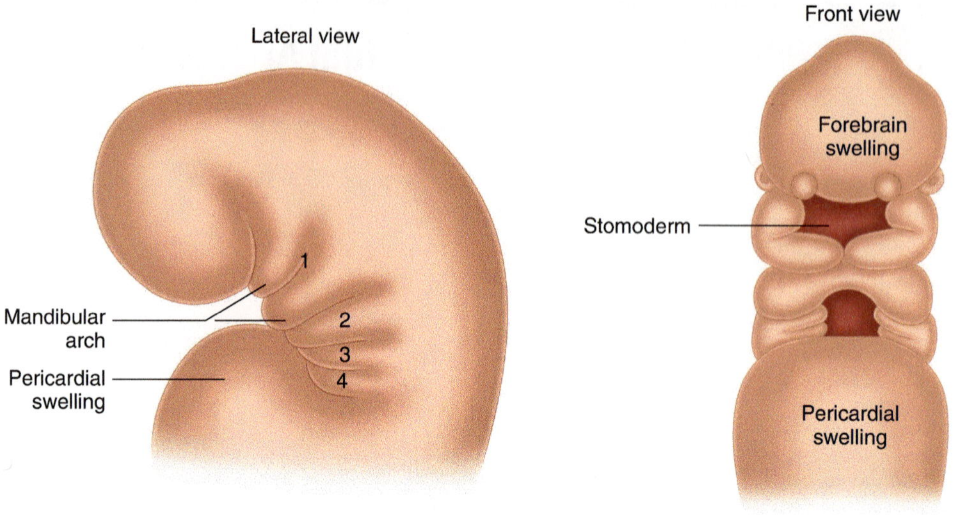

The third and fourth arches are not prominent; they lie in a depression on the surface, and their ventral ends do not reach the middle line of the floor of the pharynx but meet in a medial swelling, the hypobranchial eminence.

The fifth arch is rudimentary and disappears early, and the sixth arch is the most caudal and not prominent on the surface.

The mesodermal core of the arches differentiates into the several components.

1. Skeletal element—the arch cartilage
   - A cartilaginous bar develops from the mesodermal core of the arch and is surrounded by a perichondrial sheath. This bar will differentiate into certain cartilages, ligaments and bones of the head and neck.
2. Muscular element
   - A striated muscle mass develops from the mesoderm around the arch cartilage. This will differentiate into special visceral muscles of the head and neck.
3. Vascular element
   - An aortic arch artery develops in each arch and becomes connected to the aortic sac ventrally and to the dorsal aorta dorsally.
4. Nervous element
   - Represented by a mixed (motor and sensory) cranial nerve which arise from the lateral aspect of the hindbrain and grows into the arch, dividing into:
     (a) Motor branch—supplies the striated muscle mass of its own arch
     (b) Post-trematic sensory branch—runs with the motor nerve and supplies its own arch
     (c) Pre-trematic sensory branch—ascends to supply the preceding arch

Derivatives of the Ectodermal Clefts

First Cleft

The dorsal part deepens to form the external auditory meatus and the outer layer of the eardrum. The auricle of the ear develops by the fusion of six thickenings situated around the entrance of the first cleft.

Second, Third and Fourth Clefts

The mesoderm of the second arch grows in a downward direction pushing over the overlying ectoderm at the root of the neck, thus covering the second, third and fourth clefts and forming a cavity known as the cervical sinus which becomes gradually obliterated by apposition and fusion of its walls. The second, third and fourth clefts essentially become buried and do not give rise to any structure.

Derivatives of the Pouches

First Pouch

The ventral part is obliterated by the developing tongue, while the dorsal part forms the tubotympanic recess, which gives the pharyngotympanic tube, the middle ear cavity and mastoid antrum and the inner layer of the eardrum.

Second Pouch

The ventral part is obliterated by the developing tongue, and the dorsal part gives rise to the primordium of the palatine tonsil.

Third Pouch

The ventral part forms the thymus gland, which will migrate caudally to enter the thorax; the dorsal part forms the inferior parathyroid gland, which migrates caudally to lie below the superior parathyroid gland.

Fourth Pouch

The ventral part gives a small contribution to the sides of the thyroid gland and the thymus gland. The dorsal part forms the superior parathyroid gland.

Fifth Pouch

This gives rise to a small cellular mass called the ultimobranchial body. It is thought that this becomes incorporated into the thyroid gland forming its parafollicular cells.

Pathoembryology Branchial Arch Anomalies

When the second arch overgrows the second, third and fourth cervical grooves, the grooves lose contact with the surface of the neck and temporarily form an ectoderm lined cavity called the cervical sinus. The cervical sinus normally disappears, and failure of obliteration results in a branchial cyst. If the cyst opens on to the surface of the neck, a branchial fistula develops. Incomplete overgrowth of the second arch may also result in cysts or fistulae.

First Branchial Arch Anomalies

First branchial arch anomalies represent less than 10% [1] of all branchial anomalies and usually involve the external auditory meatus. The usual site of presentation is in the lateral neck below the external auditory canals and above the hyoid bones. Lesions associated with the external auditory canal may present as auricular swellings, fistulas, or with otorrhea. They can sometimes be misdiagnosed clinically as preauricular pits or sinuses (Figs. 21.1 and 21.2).

Work [2] classified first branchial cleft anomalies into two types based on anatomical and histological features:

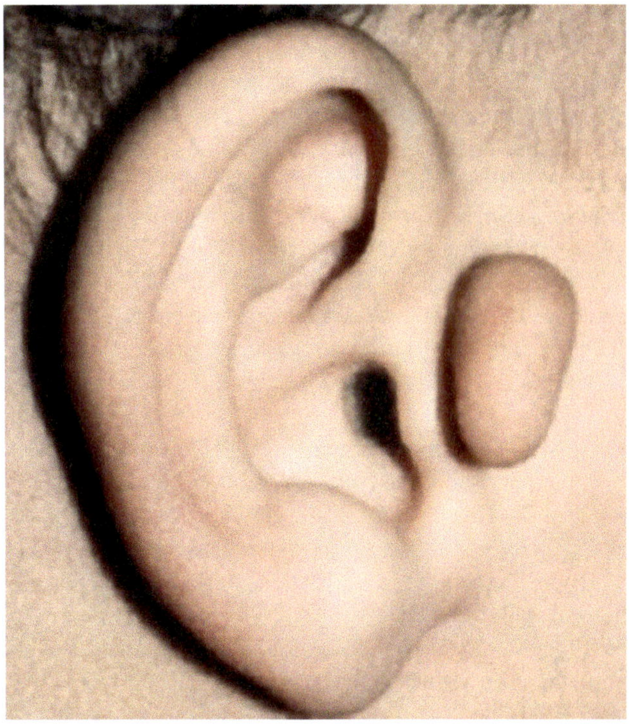

Fig. 21.1 Accessory auricle

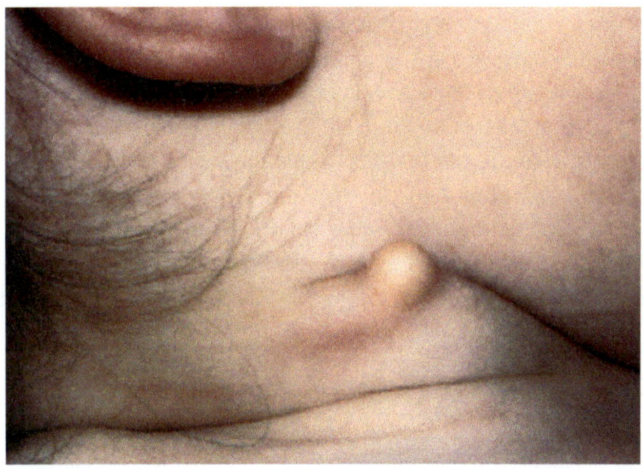

Fig. 21.2 Branchial cyst

- Work Type I lesions present as a cystic mass and are purely ectodermal in origin. They generally lie superficial to the facial nerve in close proximity to the pinna.
- Work Type II lesions can present as a cyst, sinus or fistula. They are of ectodermal and mesodermal origin, and they may contain skin adnexal structures and cartilage. They communicate with the external auditory canal or tympanic membrane and often lie medial to the facial nerve.

Second Branchial Arch Anomalies

Second arch anomalies are the most common and represent 90–95% of branchial anomalies. They most commonly develop during the second decade of life.

Cystic lesions are more common than fistulae. Fistulae tend to present as recurrent neck infections often following an upper respiratory tract infection. They usually present below the level of the digastric muscle. Cysts present as smooth soft masses in the lateral neck and are classified initially by Bailey [3] into four categories based on anatomic location:

- Type I—Cyst lies anterior to SCM at junction middle and lower thirds, deep to platysma and deep cervical fascia.
- Type II—Cyst lies in contact with the great vessels.
- Type III—Cyst passes medially between internal and external carotid vessels, extending toward lateral pharyngeal wall.
- Type IV—Cyst is located next to pharyngeal wall, medial to the great vessels at the level of the tonsillar fossa.

Third Branchial Arch Anomalies

Third branchial arch anomalies are extremely rare [4]. Fistulae open along the anterior border of SCM. The tract courses posterior to the common carotid superior to the hypoglossal nerve and inferior to the glossopharyngeal nerve before piercing the thyrohyoid membrane to enter the pyriform sinus. They may present as neck abscesses, acute thyroiditis or hypoglossal nerve palsies in adults. In neonates they may cause upper airway compromise (Figs. 21.3, 21.4 and 21.5).

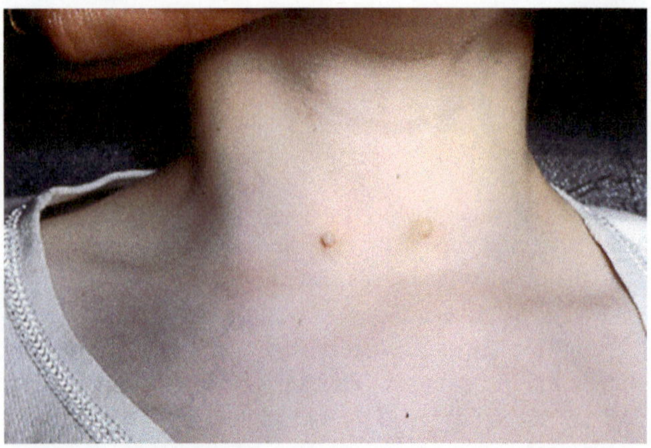

Fig. 21.3 Branchial fistulae

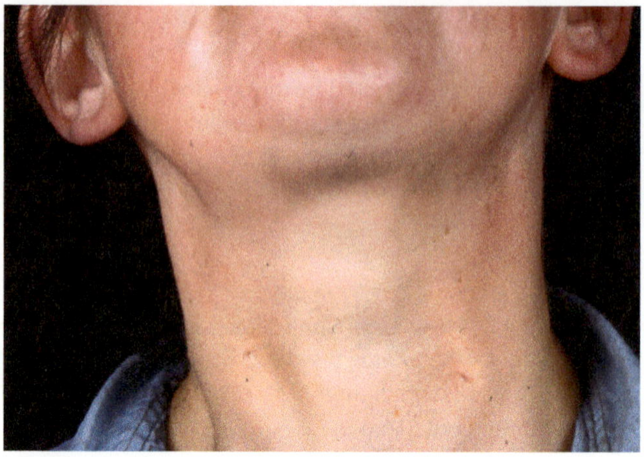

Fig. 21.4 Branchial sinuses

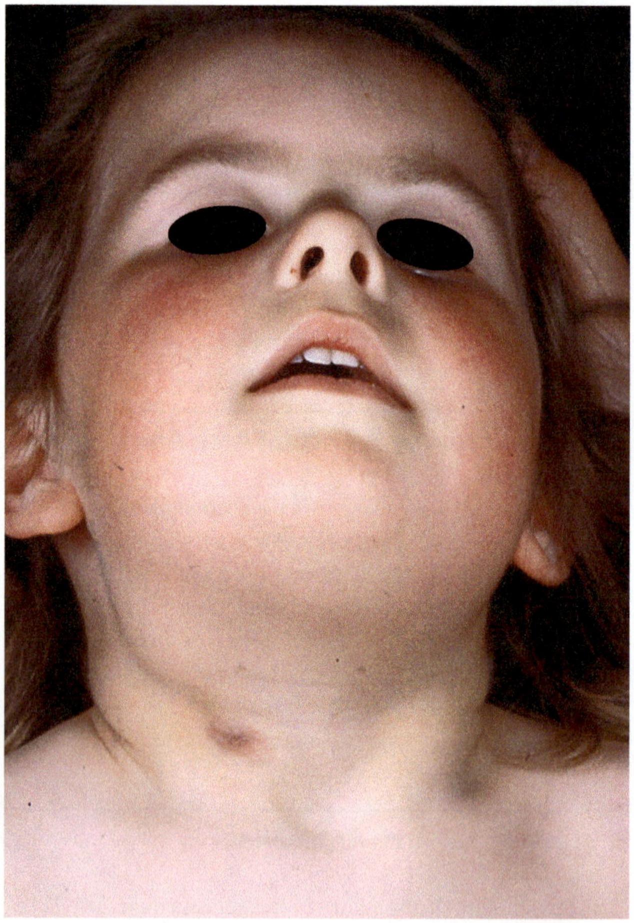

Fig. 21.5 Branchial sinus infection

Fourth Branchial Arch Anomalies

Fourth branchial arch anomalies are also extremely rare [4]. Unlike second arch anomalies, they tend to present in childhood. They tend to originate at the apex of the pyriform sinus and travel anteriorly and inferiorly to the cricothyroid muscle and thyroid cartilage. They may manifest as lateral cervical cysts with an internal fistula in the pyriform sinus. They may also present as abscesses in the region of the thyroid gland or the cervical oesophagus.

Thyroglossal Duct Cyst

Pathoembryology: A thyroglossal duct cyst develops as a result of persistence of any segment of the thyroglossal duct along its course from the foramen cecum of the tongue to the pyramidal lobe of the thyroid gland. The cyst lining usually consists of respiratory (columnar) epithelium but may also include squamous epithelium. Thyroid tissue can be found in over 60% of cases and may be normal, hyperplastic and nodular, or neoplastic.

Epidemiology: Thyroglossal duct cysts represent the most common congenital anomaly of the neck, accounting approximately 4% of all neck masses. They are most commonly present in the first decade of life but may also be seen in adults.

Clinical: They usually present as an asymptomatic midline neck mass. Occasionally they can become inflamed or infected and present as a painful neck swelling. They typically move upward on swallowing. The majority of cases occur in the midline of the neck above the thyroid isthmus but below the level of the hyoid bone. They are nearly always connected to the hyoid bone. Uncommonly, they may occur lateral to the midline but do not occur in the lateral portion of the neck, i.e. lateral to the internal jugular vein. Rarely, they can cause external compression of the airway in neonates causing respiratory compromise.

Haemangiomas and Vascular Anomalies

Pathoembryology: Haemangiomas result from the inappropriate development of vascular endothelium and channels and associated nervous components.

Epidemiology: Haemangiomas represent the most common of all congenital anomalies and can occur in up to 10% of children with a female-to-male ratio of 2:1.

Clinical: Haemangiomas present at birth, grow rapidly during the first year of life and begin to slowly involute at 18–24 months of age. Haemangiomas are seen as red or bluish masses. Typically, these masses are compressible and increase in size with straining or crying. With large haemangiomas, a bruit may be heard over the lesion. Spontaneous involution occurs in up to 90% of cases, with the majority involuting by age 5 years.

The International Society for the Study of Vascular Anomalies (ISSVA) classification for vascular anomalies is the most current classification system [5] which broadly classifies vascular lesions into:

- Vascular tumours—benign, locally aggressive and malignant
- Vascular malformations—simple, combined, involving a major vessel or associated with another congenital anomaly

Cervical Thymic Cysts

Pathoembryology: The thymus develops from the third pharyngeal pouch and descends into the chest. Cervical thymic cysts are cervical thymic tissue sequestered from the main thymic gland during its embryologic descent.

Epidemiology: Cervical thymic cysts are uncommon neck lesions. Occur slightly more often in men. The vast majority occurs during the first decade of life.

Clinical: They can be found anywhere between the angle of the mandible and the sternum, including the lateral and midline neck. Most are located on the left side of the neck [6]. Most patients present with a slow-growing, painless neck mass that transiently increases in size during a Valsalva manoeuver. Rarely, they can present with dyspnoea, dysphagia, hoarseness or pain. The surface lesion can represent an isolated cystic mass in the neck, a lesion that may extend into the mediastinum or a lesion that is in continuity with an intrathoracic thymus gland.

Congenital Muscular Torticollis

Pathoembryology: Congenital muscular torticollis (CMT) or sternomastoid tumour of infancy is characterised by a fibrous tissue mass within the sternomastoid muscle. The aetiology remains unclear, but peripartum injury, ischaemia of the muscle and intrauterine positioning have been implicated.

Epidemiology: CMT is the most common neck mass of the early perinatal period, usually seen within the first 2 months of life (Fig. 21.6).

Clinical: They typically present as a firm, painless mass with tapering ends, approximately 1–3 cm in length. They are most commonly located in the inferior to the middle third of the sternocleidomastoid muscle. Typically, these lesions spontaneously resolve in more than 80% of cases (Fig. 21.7).

Congenital Cervical Teratomas

Pathoembryology: Cervical teratomas are extremely rare germ cell tumours that occur in the neck. They are derived from multipotent primitive germ cells and can differentiate into a variety of tissues. They usually consist of tissues foreign to the site from which they arise.

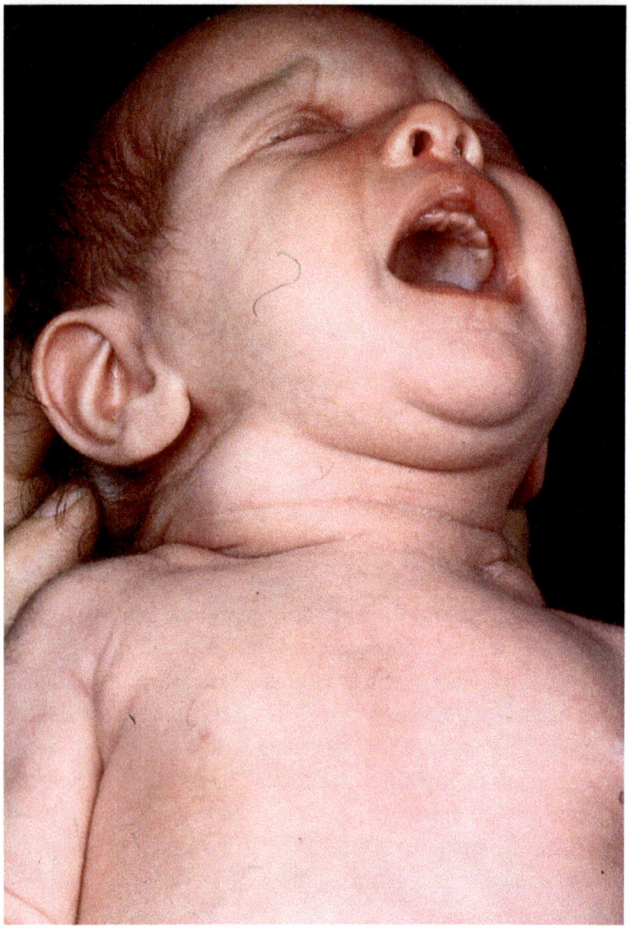

Fig. 21.6 Sternomastoid tumour

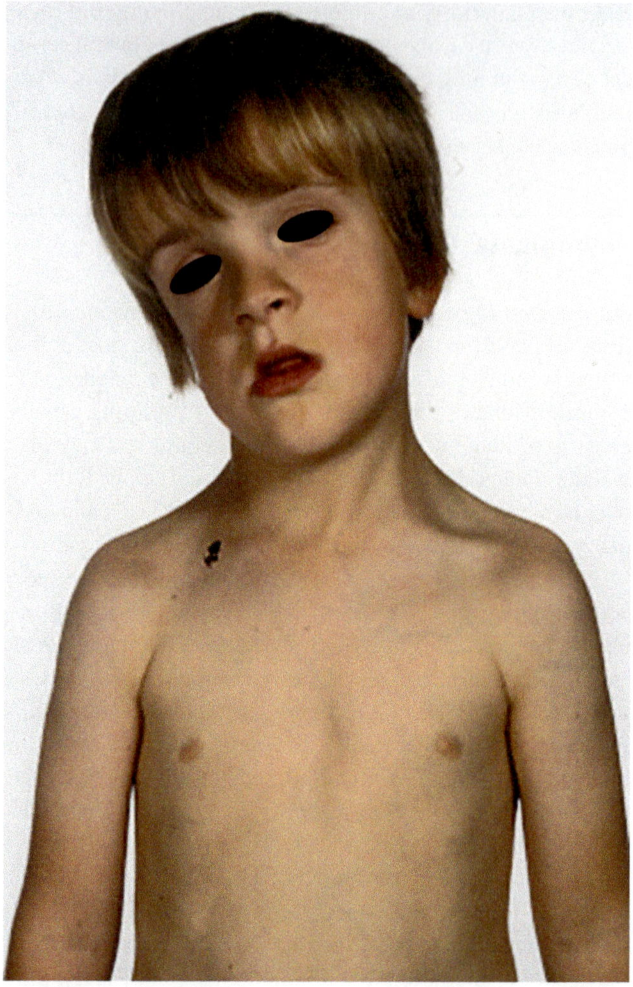

Fig. 21.7 Torticollis

Epidemiology: Cervical teratomas account for approximately 3% of all congenital teratomas and occur in approximately 1 in 20,000 to 1 in 40,000 live births.

Clinical: Most tumours are diagnosed at birth, but occasionally an in utero diagnosis can be made. Tumours detected antenatally are usually large and may be associated with polyhydramnios, hydrops and premature delivery, resulting in mortality in excess of 50%. Teratomas usually present as a large single mass, although multiple lesions may occur. Airway obstruction is the most serious complication with tracheal compression or occlusion that can be seen in 80–100% of cases.

In some patients, endotracheal intubation or tracheostomy is required. An ex utero intrapartum treatment (EXIT) procedure is a technique designed to allow partial foetal delivery via caesarean section with establishment of a safe foetal airway and can be carried out in certain centres if the lesion is identified as causing significant airway compression in utero.

Dermoid Cyst

Pathoembryology: Dermoid cysts are benign developmental cystic anomalies originating from ectoderm and mesoderm but not endoderm. They are thought to be due to epithelium entrapment in tissue during embryogenesis or by traumatic implantation.

Epidemiology: No gender predilection. Most common in the first decade of life but can present later. Approximately 34% occur in the head and neck region.

Clinical: Dermoid cysts are predominantly subcutaneous lesions but can occur at mucosal sites such as the orbit, oral cavity and nasal cavity. Typically, dermoid cysts are seen in the midline of the neck, usually in the submental region. They are attached to and move with the overlying skin and are slow growing and not associated with pain unless they become infected.

Bronchogenic Cyst

Pathoembryology: Bronchogenic cysts are rare, congenital anomalies that are commonly located in the mediastinum or within the lung parenchyma [7]. They are rarely seen in the neck and are thought to originate from buds or diverticula that separate from the foregut during the formation of the tracheobronchial tree.

Epidemiology: Bronchogenic cysts occur mainly in males with a ratio of approximately 4:1.

Clinical: Bronchogenic cysts of the neck are mostly midline but may also occur laterally. Lateral neck cysts are usually located in the lower cervical regions in close proximity to the thorax. However, cysts of the upper cervical regions are generally more midline. Sometimes a fistulous opening that drains mucoid material may be seen. The cysts can become infected and present as neck abscesses. Respiratory distress may be the initial symptoms if there is a compression on the tracheobronchial tree.

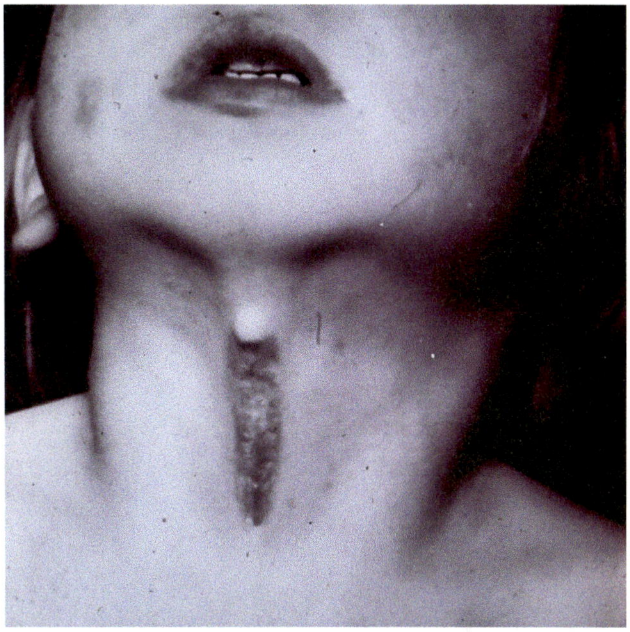

Fig. 21.8 Median cervical cleft

Cervical Lymphangioma/Cystic Hygroma

Pathoembryology: Lymphangiomas are thought to arise due to the failure of lymphatics to connect to the venous system. This can result in abnormal budding of lymphatic tissue and sequestered lymphatic rests that retain their embryonic growth potential. Lymphangioma cystica (cystic hygroma) is composed of large lymphatic cysts that invade into adjacent soft tissue planes and are well defined, circumscribed or lobulated.

Epidemiology: Cystic hygromas may be present as a birth defect or may present at any time during life. Overall they are uncommon lesions, accounting for fewer than 5% of all congenital neck masses. Cystic hygromas are associated with a number of genetic conditions including Turners syndrome, Noonan syndrome, trisomies 13, 18 and 21.

Clinical: Cystic hygromas diagnosed prenatally may progress to hydrops and eventually foetal death. Smaller prenatally diagnosed cystic hygromas may resolve leading to a webbed neck. Cervical lymphangiomas in children often present after a sudden increase in size secondary to infection or bleeding within the lesion itself. Spontaneous decompression or shrinkage is uncommon. They can occur anywhere in the neck but are classically found in the posterior triangle. As a result of acute enlargement or chronic growth, the hygromas may cause respiratory and swallowing difficulties.

Midline Cervical Clefts

Pathoembryology: Midline cervical cleft (MCC) is not considered a true cleft because no skin gap exists. The aetiology is unclear and is believed to occur as a consequence of abnormal fusion of the second and third arches or nonfusion at the level of the ectoderm [8]. They can be associated with a median cleft of the mandible, tongue or lower lip (Fig. 21.8).

Epidemiology: MCC is a rare congenital anomaly, and fewer than 100 cases have been reported in the literature.

Clinical: They typically present at birth with a cleft extending from the inferior aspect of the chin to the level of the suprasternal notch. Often the area has serous drainage, and the superior aspect of the cleft takes on the appearance of a pseudonipple.

References

1. Schroeder JW Jr, Mohyuddin N, Maddalozzo J. Branchial anomalies in the pediatric population. Otolaryngol Head Neck Surg. 2007;137(2):289–95.
2. Work WP. Newer concepts of first branchial cleft defects. Laryngoscope. 1972;82(9):1581–93.
3. Bailey H. Branchial cysts and other essays on surgical subjects in the faciocervical region. London: Lewis; 1929.
4. Rea PA, Hartley BE, Bailey CM. Third and fourth branchial pouch anomalies. J Laryngol Otol. 2004;118(1):19–24.
5. International Society for the Study of Vascular Anomalies (ISSVA). Classification of Vascular Anomalies. 2014. Accessed Sep 2015.
6. Terzakis G, Louverdis D, Vlachou S, Anastasopoulos G, Dokianakis G, Tsikou-Papafragou A. Ectopic thymic cyst in the neck. J Laryngol Otol. 2000;114(4):318–20.
7. Ustundag E, Iseri M, Keskin G, Yayla B, Muezzinoglu B. Cervical bronchogenic cysts in head and neck region. J Laryngol Otol. 2005;119(6):419–23.
8. Gargan TJ, McKinnon M, Mulliken JB. Midline cervical cleft. Plast Reconstr Surg. 1985;76(2):225–9.

Suggested Reading

Swanson VC, Gries H, Koh J. Chapter 181: Anatomy and developmental embryology of the neck. In: Haughey BH, Niparko J, Thomas Robbins K, Lesperance MM, Flint PW, Regan TJ, Lund VJ, editors. Cummings otolaryngology—head & neck surgery. 5th ed. Philadelphia, PA: Mosby Elsevier; 2010. ISBN: 978-0-323-08563-2.

Wetmore RF, Potsic WP. Chapter 198: Differential diagnosis of neck masses. In: Haughey BH, Niparko J, Thomas Robbins K, Lesperance MM, Flint PW, Regan TJ, Lund VJ, editors. Cummings otolaryngology—head & neck surgery. 5th ed. Philadelphia, PA: Mosby Elsevier; 2010. ISBN: 978-0-323-08563-2.

Congenital Malformations of the Larynx and Trachea

22

Christopher Barringer, Ramanathan Kasivisvanathan, Mark Catolico, and Steve Goudy

The trachea begins its development in week 3 of development. The epithelium develops from the endoderm of the foregut. The laryngo-tracheal groove, which is the primordium of the respiratory system, appears in the endodermal floor of the pharynx behind the hypobranchial eminence. The edges of the groove unite, which divides the upper part of the foregut into two parts: the dorsal part forms the oesophagus, and the ventral part forms the laryngo-tracheal tube. The union of the edges of the groove begins from below upwards. The tube grows caudally, its upper end forms the larynx which remains connected to the foregut by the laryngeal orifice, and the next part forms the trachea while its lower part divides into two lateral out pouches known as the lung buds.

The larynx is a derivative of the pharyngeal arches, which are mesodermal in origin.

Arch	Skeletal derivative	Muscular derivative	Nerve
First	Meckel's cartilage: differentiates into the incus and malleus, sphenomandibular ligament, anterior ligament of the malleus and the anterior part of the body of the mandible	The muscles of mastication: temporalis, masseter, lateral pterygoid and medial pterygoid Other muscles include mylohyoid, anterior belly of digastric, tensor palati and tensor tympani	Mandibular nerve and chorda tympani
Second	Reichert's cartilage: differentiates into the stapes, styloid process, stylohyoid ligament and the lesser horn and upper part of the body of the hyoid bone	Muscles of the scalp, face, the platysma muscle of the neck, stylohyoid muscle and the posterior belly of the digastric muscle and the stapedius muscle	Facial nerve and tympanic branch of glossopharyngeal nerve
Third	Greater horn and lower part of the body of the hyoid bone	Stylopharyngeus muscle	Glossopharyngeal nerve and internal laryngeal nerve
Fourth	Thyroid cartilage of the larynx	Cricothyroid muscle	External laryngeal nerve
Fifth	Degenerates		
Sixth	All cartilage of the larynx except the thyroid cartilage	All laryngeal muscles (except cricothyroid), constrictor muscles of the pharynx and the muscles of the palate (except tensor palati)	Recurrent laryngeal nerve

C. Barringer
Whipps Cross University Hospital, Barts Health NHS Trust, London, UK

R. Kasivisvanathan
The Royal Marsden NHS Foundation Trust, London, UK

M. Catolico
St Mary's Hospital, Imperial College Healthcare NHS Trust, London, UK

S. Goudy (✉)
Paediatric Otolaryngology, Emory School of Medicine, Atlanta, GA, USA
e-mail: steven.goudy@emory.ed

R. Carachi, S. H. E. Doss (eds.), *Clinical Embryology*, https://doi.org/10.1007/978-3-319-26158-4_22

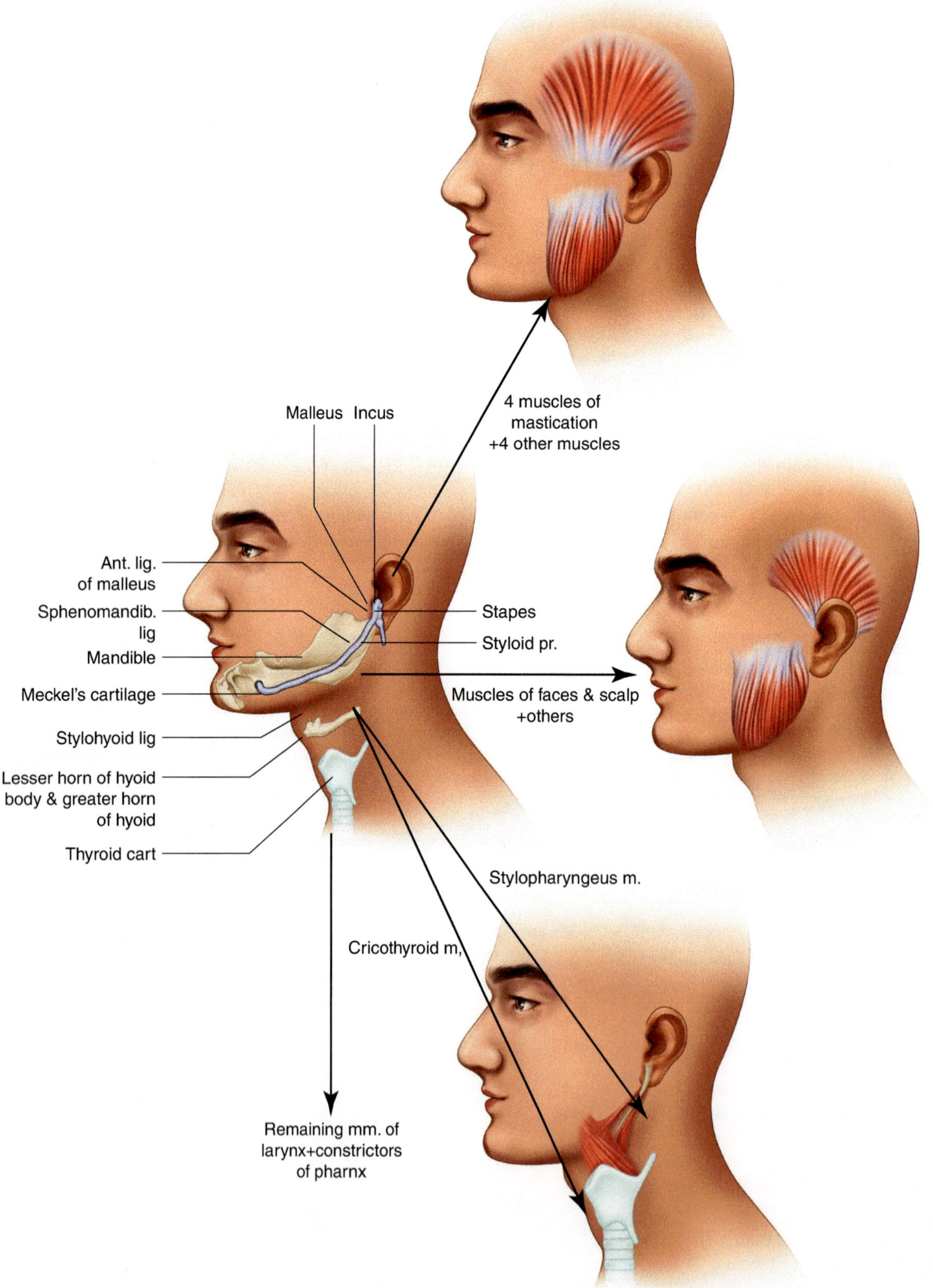
4 muscles of mastication +4 other muscles
Malleus
Incus
Ant. lig. of malleus
Sphenomandib. lig
Mandible
Meckel's cartilage
Stylohyoid lig
Lesser horn of hyoid body & greater horn of hyoid
Thyroid cart
Stapes
Styloid pr.
Muscles of faces & scalp +others
Stylopharyngeus m.
Cricothyroid m,
Remaining mm. of larynx+constrictors of pharnx

Key Conditions

Laryngomalacia, vocal cord paralysis, congenital subglottic stenosis, laryngeal webs and atresia, congenital vascular abnormalities, laryngeal cysts, laryngeal/tracheal clefts tracheoesophageal fistulae, tracheal agenesis and atresia and complete and deformed tracheal rings.

Laryngomalacia

Definition

This is a condition where the tone of the larynx is weakened and/or the tissue is redundant, leading to dynamic prolapse of the supraglottic structures during inspiration.

Incidence

Exact incidence is unknown. It is however the most frequently reported congenital laryngeal anomaly and the most common cause of congenital stridor.

Aetiology and Patho-embrology

The exact cause is unknown although there are multiple theories:

- Immaturity of cartilaginous structures.
- Maldevelopment; overgrowth of the third brachial arch resulting in a long epiglottis with lateral tethering of the aryepiglottic folds. This omega-shaped epiglottis then has a predisposition to fall posteriorly during inspiration.
- Immature neuromuscular control, resulting in arytenoids prolapse.
- Foetal warfarin syndrome caused by maternal ingestion of warfarin has been associated with laryngomalacia.

Presentation, Clinical Features and Outcome

It usually presents with inspiratory stridor within the first 2 weeks of life. Symptoms peak at 6–9 months and usually resolve spontaneously prior to 12 months. The condition is rarely fatal, and less than 10% require surgical intervention.

Associations

It is frequently seen in otherwise healthy infants. Gastro-oesophageal reflux disease is present in 80% of children with laryngomalacia. It is also associated with other congenital airway malformations, the commonest of which are subglottic stenosis, tracheomalacia, vocal cord paralysis and a number of congenital syndromes (Table 22.1).

Table 22.1 Syndromes associated with laryngomalacia

Syndromes associated with laryngomalacia	
Down	Cri du Chat
Rubinstein-Taybi	Ehlers-Danlos
Smiths	Optiz
Goldenhar	Costello
Neurofibromatosis	Alagille
Hunters	VATER
CHARGE	Arthrogryposis

Vocal Cord Paralysis

Definition

This is the inability of one or both vocal cords to move.

Incidence

Exact incidence is unknown and generally considered the second most common congenital anomaly of the larynx.

Aetiology and Patho-embrology

It can be acquired or congenital. The congenital form is normally due to abnormal nerve input to the laryngeal muscles or brain stem compression in the case of bilateral vocal cord paralysis.

This can be caused by:

- Lesions in the central nervous system—Arnold-Chiari malformation, cerebral palsy, hydrocephalus, myelomeningocele, spina bifida
- Immaturity of the central nervous system
- Lesions in the mediastinum—usually cause unilateral
- Idiopathic (most common)

Presentation, Clinical Features and Outcome

There is a spectrum of severity, from the mild unilateral form that usually goes undetected to its severest form involving bilateral vocal cord paralysis, stridor and significant airway obstruction. Congenital bilateral cord paralysis usually requires a tracheostomy for management; over half of these cases gain normal vocal cord function by 12 months.

Associations

Bilateral vocal cord paralysis can exhibit autosomal dominant inheritance; it is sometimes associated with Charcot-Marie-Tooth disease and also can be found with other cranial nerve lesions.

Congenital Subglottic Stenosis

Definition

This is a narrowing of <4 mm in a full-term infant and <3 mm in a premature infant of the cricoid region. Myers and Cotton [1] have a further classification dependent on the severity of the stenosis (Table 22.2).

Incidence

Exact incidence is unknown. It is considered the third most common congenital laryngeal anomaly and consists of malformation and narrowing of the cricoid ring, the only complete cartilaginous ring of the airway. Males are twice more likely to develop this than females.

Aetiology and Patho-embrology

The malformation occurs in the third month of gestation when the laryngo-tracheal tube fails to recanalise completely after normal epithelial fusion. There are two types:

- Membranous type: This type involves the true vocal cords and consists of submucosal hypertrophy, excess fibrous connective tissue and excess mucous glands. This is the commonest form.
- The cartilaginous type: This is usually shelf-like at the level of the cricoid and consists of abnormally shaped cartilage.

Presentation, Clinical Features and Outcome

Symptoms depend on the extent of stenosis and at their worse can result in acute airway obstruction at birth requiring tracheal intubation. Most mild to moderate cases resolve spontaneously with growth and subsequent widening of the trachea. Decision to surgically intervene is based on the symptoms and severity of the stenosis. Patient's may require a tracheostomy at birth or may undergo airway reconstruction to expand the airway, with most cases being decanulated in childhood, by the age of 4.

Table 22.2 Myers and Cotton classification

Myers and Cotton classification
I—less than 50% obstruction
II—51–70% obstruction
III—71–99% obstruction
IV—no detectable lumen

Laryngeal Webs and Atresia

Definition

This is where a membrane-like structure, or fusion, extends across structures within the larynx between the vocal cords. Laryngeal webs that involve both the vocal cords and subglottis are within the spectrum of laryngeal atresia.

Incidence

The exact incidence is unknown, but it has been reported to account for up to 5% of all congenital laryngeal abnormalities.

Aetiology and Patho-embrology

Laryngeal webs occur in the third month of gestation and are due to incomplete recanalisation of the laryngo-tracheal tube. Failure of the recanalisation leads to varying degrees of obstruction in the form of webs or atresia. They can be classified by anatomical location [2] or by thickness [3].

Presentation, Clinical Features and Outcome

Laryngeal atresia can be detected antenatally by ultrasound scan: as congenital high airway obstruction syndrome (CHAOS). This consists of a flattened diaphragm, fluid-filled dilated airway distal to obstruction, enlarged hyperechoic lungs and complete obstruction of the airway.

There is a range in severity of symptoms from mild dysphonia to complete airway obstruction that directly correlates to the severity of the web. The most severe form, laryngeal atresia, is often fatal without immediate management with a tracheostomy. In these the presence of a concurrent tracheoesophageal fistula (TOF), which is big enough to permit a passage of air, can sustain the patient's ventilation until an established airway is secured. Staged airway reconstruction of the glottis and subglottic area is required for recreation of the vocal cords and expansion of the subglottis, before decannulation can be accomplished.

Associations

Thirty percent are associated with other respiratory tract anomalies, mostly subglottic stenosis. Glottic and supraglottic atresia are associated with oesophageal atresia, TOF, limb defects and urinary tract abnormalities.

Anterior glottic webs are linked to chromosome 22q11.2 deletion syndromes such as Shprintzen syndrome, velocardiofacial syndrome and DiGeorge syndrome. Therefore a cardiac workup should be performed.

Congenital Vascular Anomalies

Definition

There are two main types, haemangiomas and vascular malformations. Haemangiomas are benign vascular tumours, while vascular malformations are anomalies in any vascular structure, i.e. venous, arterial and lymph mixed.

Incidence

Exact incidence of congenital vascular anomalies is unknown. It is estimated though that haemangiomas account for 1–1.5% of congenital laryngeal anomalies.

Aetiology and Patho-embrology

Haemangiomas arise from endothelial hyperplasia and mast cell proliferation. Haemangioma formation has been linked to low birth weight and prematurity. Females are affected twice as many times as males. Vascular malformations arise from errors in vascular morphogenesis.

Presentation, Clinical Features and Outcome

It is reported only 10–15% of anomalies are evident at birth and symptoms are dependent on the location and size of the anomaly. Neonates with haemangiomas are generally asymptomatic at term. Following this, there is a rapid phase of proliferation of the lesion between 1 and 3 months with involution by the age of 1 year. Infants with subglottic haemangiomas typically have a normal cry but can have noisy breathing and a bark-like cough—often misdiagnosed as recurrent croup.

Associations

Almost half of congenital haemangiomas are associated with cutaneous skin haemangiomas, and subglottic haemangiomas can be part of PHACES syndrome (Table 22.3).

Table 22.3 PHACES syndrome

PHACES syndrome
P—posterior fossa brain malformations
HA—haemangiomas
C—cardiac anomalies and coarctation of the aorta
E—eye anomalies
S—with or without sternal cleft

Laryngeal Cysts

Definition

Fluid- or air-filled lesions in the larynx.

Incidence

Exact incidence is unknown but is extremely uncommon.

Aetiology and Patho-embrology

There are two types: saccular and ductal. Saccular cysts are caused by blockage of the laryngeal saccule orifice in the ventricle causing retention of mucus. Ductal cysts are caused by blockage of submucosal glands.

Presentation, Clinical Features and Outcome

Congenital laryngeal cysts can present with airway symptoms and difficulties in feeding in neonates. These cysts may be cause life-threatening stridor and airway obstruction. Management of the cyst usually requires surgical intervention.

Laryngeal/Tracheal Cleft

Definitions

These are clefts that consist of a posterior and sagittal communication between the larynx and pharynx, and in the cases of tracheal cleft extend down to the trachea.

Incidence

They are very rare, with an estimated incidence between 1 in 10,000 and 20,000 live births.

Aetiology and Patho-embrology

They are thought to be caused by incomplete development of the tracheoesophageal septum and incomplete fusion of the

Table 22.4 Benjamin-Inglis classification

Benjamin-Inglis classification
0—submucosal cleft
I—limited to the interarytenoid region, above level of true vocal folds
II—cricoid lamina is partially involved, with extension below level of true vocal folds
III—total cricoid cleft, extends completely through cricoid cartilage with or without further extension into cervical trachea
IV—extends into posterior wall of thoracic trachea and may extend as far as carina

cricoid cartilage. This probably occurs during week 6 of gestation when the larynx develops from endoderm and mesenchymal from the fourth to sixth brachial arches. There are multiple theories on how these anomalies occur. They include abnormalities in intra-embryonic pressure, epithelial occlusion, vascular occlusion and differential cell growth. There is a genetic component to the disease with familial cases with autosomal dominance inheritance reported. Males are slightly more affected than females.

Presentation, Clinical Features and Outcome

There is a wide spectrum of severity. Ranging from type 0 the mild form with almost no symptoms to type IV causing swallowing disorders, aspiration and airway obstruction, with a very poor prognosis (Table 22.4). This severe form has a high mortality rate and can be an immediate survival threat in the first few hours of life due to the inability to ventilate.

Associations

Although occasionally seen as an isolated anomaly, it is frequently part of a syndrome—these include Pallister-Hall, Opitz-Frias, VACTERL, CHARGE and Di George, plus a whole host of associated malformations. Laryngeal clefts are also present in 6% of patients with tracheoesophageal fistulae.

Tracheoesophageal Fistulae (TOF)

Definition

This consists of a communication between the oesophagus and trachea.

Incidence

The incidence is estimated at 1 in 2500–4000 live births.

Table 22.5 Ladd and Gross classification

Ladd and Gross classification
A—oesophageal atresia without fistula
B—oesophageal atresia with proximal fistula
C—oesophageal atresia with distal fistula (most common)
D—oesophageal atresia with double fistula
E—tracheoesophageal fistula without atresia

Aetiology and Patho-embrology

It occurs due to an incomplete division of the budding ventral foregut. The aetiology is multifactorial. The non-isolated or syndromic forms are often caused by changes in single genes or chromosomes. Nineteen percent of cases have polyhydramnios—due to the inability to swallow fluid and may help antenatal detection.

The different anatomical variations of the components of the defect are described well in the Ladd and Gross classification (Table 22.5).

Presentation, Clinical Features and Outcome

Prenatal ultrasound can be used to make an early diagnosis of TOF, it is associated with polyhydramnios and an absent gastric bubble.

The diagnosis of congenital TOF is generally made in the early neonatal period where complications from feeding or significant respiratory events occur, including recurrent aspiration or aspiration pneumonia. Surgical management and repair is the primary closure of the TOF and is required in the neonatal period. Mortality and significant morbidity is low in high-volume centres, although many neonates will go on to develop respiratory, feeding and growth issues later in infancy and childhood.

Associations

Fifty percent are associated with malformations and/or syndromes (Table 22.6).

Table 22.6 Syndromes and malformations associated with tracheoesophageal fistula

Syndromes and malformations associated with tracheoesophageal fistula	
Trisomy 21/18/13	Patent ductus arterious
Pallister-Hall	Atrial septal defect
CHARGE	Ventricle septal defect
Opitz G	Right-sided aortic arch
Arthrogryposis	Imperforate anus
Goldenhar	Malrotation
VACTERL	Intestinal atresia
Feingold	Fanconi anaemia

Tracheal Web

Definition

This consists of a layer of tissue (which can vary in thickness) across the trachea. Unlike laryngeal webs and stenosis, there is no underlying pathology of the cartilage.

Incidence

Estimated incidence of 1 in 10,000 live births.

Aetiology and Patho-embrology

Tracheal webs are formed in a similar way to tracheal stenosis and atresia, probably resulting from unequal partitioning of the foregut.

Presentation, Clinical Features and Outcome

Symptoms usually depend on the patency of the tracheal lumen. Often neonates are asymptomatic. Later during infancy, childhood or even adulthood, a respiratory tract infection can cause symptoms such as wheeze or stridor. These usually require intervention.

Tracheomalacia

Definition

Congenital tracheomalacia is where the integrity of the tracheal wall is compromised due to intrinsic cartilaginous deficiency or extrinsic compression resulting in dynamic prolapse of the tracheal wall and constriction of the tracheal lumen.

Incidence

Exact incidence is unclear but congenital tracheomalacia is very rare. Many cases of mild to moderate tracheomalacia probably go unreported.

Aetiology and Patho-embrology

Congenital or primary tracheomalacia is caused by cartilaginous weakness. Secondary tracheomalacia is more common and caused by extrinsic compression on the tracheal wall. This compression can be from other congenital tracheal abnormalities such as vascular abnormalities or a tracheoesophageal fistula.

Presentation, Clinical Features and Outcome

In the neonate tracheomalacia should be considered with any episodes of unexplained cyanosis especially if these episodes occur during crying, feeding or when the neonate is distressed. In infants and children, tracheomalacia could present with stridor or recurrent cough. In most instances congenital tracheomalacia improves by the age of 2 as the size of the trachea increases. For severe intrinsic tracheomalacia, the use of a tracheostomy tube and ventilator to stent the airway obstruction may be sufficient, or external 3-D printed scaffolds may also relieve airway obstruction. For extrinsic compression of the trachea, aortopexy with removal of the thymus may decompress the airway obstruction.

Associations

Secondary tracheomalacia is associated with tracheoesophageal fistulae repair and cardiac abnormalities with vascular abberants causing external compression.

Tracheal Agenesis and Atresia

Definition

This is where the trachea is completely absent (agenesis) or partially intact but underformed (atresia).

Incidence

This condition is exceedingly rare.

Aetiology and Patho-embrology

Tracheal agenesis and atresia are almost uniformly fatal and are fortunately quite rare. In both atresia and agenesis, communication between the larynx proximally and the alveoli of the lungs distally is absent. Due to the lack of a normal continuous airway, affected newborns survive only if an alternate pathway for ventilation exists (e.g. a patent bronchoesophageal fistula).

Presentation, Clinical Features and Outcome

This is essentially a non-correctable malformation incompatible with life in most instances. Although operative techniques are available to correct the underlying abnormality, surgical attempts have yielded poor outcomes.

Associations

Associated with a pulmonary artery (PA) sling—70% of those with a PA sling have tracheal stenosis. Also associated with other cardiac anomalies: most common are atrial ventral septal defect and trisomy 21, Tetralogy of Fallot and pulmonary hypoplasia.

Complete or Deformed Tracheal Rings

Definition

This is where cartilaginous tracheal rings have no muscular posterior wall (pars membranacea).

Incidence

Incidence is estimated to be 1 in 64,500 live births.

Aetiology and Patho-embrology

The pars membranacea normally lacks tracheal rings, and its absence leads to a fixed, narrowed dimensioned trachea. It contributes to approximately 1% of all laryngo-tracheal stenosis.

Presentation, Clinical Features and Outcome

These patients are unable to cough due to the inability to compress their trachea and have coarse breath sounds. They are typically very ill and often have other associated anomalies (i.e. cardiac, pulmonary agenesis, craniofacial). The majority of these patients require surgical intervention to repair their trachea, using a cardiac patch or slide tracheoplasty technique.

References

1. Myer CM, O'Connor DM, Cotton RT. Proposed grading system for subglottic stenosis based on endotracheal tube sizes. Ann Otol Rhinol Laryngol. 1994;103(4):319–23.
2. Benjamin B. Congenital laryngeal webs. Ann Otol Rhinol Laryngol. 1983;92(4):317–26.
3. Cohen SR. Congenital glottic webs in children. Ann Otol Rhinol Laryngol. 1985;94(6 suppl):2–16.

Further Reading

Berrocal T, Madrid C, Novo S, Gutiérrez J, Arjonilla A, Gómez-León N. Congenital anomalies of the tracheobronchial tree, lung, and mediastinum: embryology, radiology, and pathology. Radiographics. 2004;24(1):e17. https://doi.org/10.1148/rg.e17.

Conway H, Wagner KJ. Congenital anomalies of the head and neck. Plast Reconstr Surg. 1965;36:71–9.

Heimann K, Bartz C, Naami A, Peschgens T, Merz U, Hörnchen H. Three new cases of congenital agenesis of the trachea. Eur J Pediatr. 2007;166(1):79–82. https://doi.org/10.1007/s00431-006-0210-4.

Ho AS, Koltai PJ. Pediatric tracheal stenosis. Otolaryngol Clin North Am. 2008;41(5):999–1021. https://doi.org/10.1016/j.otc.2008.04.006.

Hysinger EB, Panitch HB. Paediatric tracheomalacia. Paediatr Respir Rev. 2015;17:9–15. https://doi.org/10.1016/j.prrv.2015.03.002.

Krause U, Rödel RMW, Paul T. Isolated congenital tracheal stenosis in a preterm newborn. Eur J Pediatr. 2011;170(9):1217–21. https://doi.org/10.1007/s00431-011-1490-x.

Leboulanger N, Garabédian E-N. Laryngo-tracheo-oesophageal clefts. Orphanet J Rare Dis. 2011;6(1):81. https://doi.org/10.1186/1750-1172-6-81.

Stevenson RE, Hall JG. Human malformations and related anomalies. New York: Oxford University Press; 2005.

Wiatrak BJ. Congenital anomalies of the larynx and trachea. Otolaryngol Clin North Am. 2000;33:91–110.

Vascular Anomalies of the Head and Neck

23

Ross Elledge, Kevin McMillan, Andrew Monaghan, and Rhodri Williams

Aetiology

It is important at the outset to distinguish the two main groups of vascular anomalies: *vascular malformations* and *vascular tumours* (e.g. haemangiomas). Vascular malformations are localized errors of angiogenic development, which should be distinguished from haemangiomas. They may involve any portion of the vascular system—arteries, veins, capillaries and lymphatics. Vascular malformations are therefore structural anomalies present at birth, which may or may not be clinically evident. Haemangiomas by contrast are tumorous growths, with increased endothelial cell proliferation as a distinguishing hallmark [1].

Pathoembryology

Vascular Tumours

Various theories have emerged to explain the origins of *infantile haemangiomas*, including theories of retained immature tissue similar to placental tissue, endothelial progenitor cells and mesenchymal stem cells. Infantile haemangioma endothelial cells express several markers hinting at placental origins, including glucose transporter-1 (GLUT-1), Lewis Y antigen and merosin [2]. This has given rise to the theory that they may originate from placental endothelial cells reaching foetal tissues through a permissive left-to-right shunt in utero [3]. This placental theory has further garnered support from the observation of an increased incidence of infantile haemangiomas following chorionic villous sampling, placenta praevia and pre-eclampsia [2].

A separate theory is that a single somatic mutation leading to a clonal duplication of endothelial progenitor cells results in haemangiomas. This is supported by the finding of cell surface proteins CD133+ and CD34+ and the observation that stem cells from human haemangiomas can develop histopathologically similar tumours in nude mice [2].

The subsequent proliferative phase of haemangioma development is spurred on by an environment rich in cytokines such as vascular endothelial growth factor (VEGF), basic fibroblast growth factor (bFGF), insulin-like growth factor (IGF) and matrix metalloproteinase-9 (MMP-9) [2]. Involution occurs as mast cells appear following endothelial cell differentiation to produce interferon-β and other mediators (e.g. tissue inhibitor of metalloproteinase) that arrest cellular proliferation. *Congenital haemangiomas* by contrast do not stain for GLUT-1, and it would therefore seem reasonable to assume that they do not originate from placental tissues.

Vascular Malformations

Primitive vascular channels develop in the third week of gestation from mesodermally derived haemangioblasts aggregating to form blood islands, the inner cells of these being haematopoietic stem cells and the outer cells being endothelial cell precursors, or angioblasts. Angiogenesis occurs as the primary vascular plexus is reorganized into a functional vascular system in response to growth factors such as VEGF and platelet-derived growth factor-β (PDGF-β) [2].

R. Elledge (✉)
Department of Oral and Maxillofacial Surgery, Queen Elizabeth Hospital, Birmingham, UK
e-mail: rosselledge@doctors.org.uk

K. McMillan
Department of Oral and Maxillofacial Surgery, Birmingham Children's Hospital, Birmingham, UK

A. Monaghan · R. Williams
Department of Oral and Maxillofacial Surgery, Queen Elizabeth Hospital, Birmingham, UK

Department of Oral and Maxillofacial Surgery, Birmingham Children's Hospital, Birmingham, UK

R. Carachi, S. H. E. Doss (eds.), *Clinical Embryology*, https://doi.org/10.1007/978-3-319-26158-4_23

The developing vasculature passes through three stages—undifferentiated, retiform and maturation. VEGF induces the penetration of avascular epidermis by developing capillaries, and if the differentiation is abnormal, then persistence of the normal embryonic vascular system arises, resulting in vascular malformations, due to failure of the primitive venous plexus to regress and mature properly. The behaviour of the malformation is contingent upon the point during embryogenesis at which normal development was interrupted.

Defective development in the early stage of embryogenesis arrests embryonic vessels in the form of reticular networks. Thus, after birth, independent clusters of primitive venous tissue remain with no direct involvement of main vessels. The structures retain mesenchymal cell properties with the ability to proliferate in response to trauma, infection and hormonal changes. If the defective development occurs later in embryogenesis, however, such capacity is lost, as the vessels have now matured and involve the main vessels, rather than existing as isolated clusters [4]. Further to this, initial theories of growth were of hypertrophy rather than hyperplasia in response to increased hydrostatic pressure; however, recent work has demonstrated increased amounts of MMP-9 in intramuscular malformations, suggesting a role for vasculogenesis in growth of these lesions [5].

More recently, the presence of neural cells has been demonstrated in slow-flow and fast-flow vascular malformations, leading to some authorities postulating that neural elements may play a role in development of vascular malformations, although the specific role is as yet unknown [2].

Genetics

Vascular Tumours

Most vascular anomalies occur sporadically, but several genomic errors have been linked to particular subtypes of vascular malformations. Haemangiomas in particular may be familial in 12% of cases, and a linkage to locus 5q31-33 has been found. There is increasing evidence that aberrant signalling pathways at the molecular level may be responsible for alterations in proliferation and apoptosis of vascular endothelial cells [6].

Vascular Malformations

Capillary malformations may be sporadic or exhibit autosomal dominant pattern inheritance. Familial inheritance of capillary malformations has been demonstrated with a locus CMC1 on chromosome 5q being identified [7]. The candidate gene RAS p21 protein activator-1 (RASA-1) has been identified at this locus and mutations in this gene detected in families with capillary malformations [8]. Ras is a signal transducer for VEGF-mediated angiogenesis [9].

Inherited lymphoedema may be traced to a mutation in VEGF-3 on chromosome 5q35.3, with the postulated reason for lymphoedema being insufficient signalling via the VEGFR-3 receptor [7].

Venous malformations (VMs) may be sporadic, paradominant or autosomal dominant, with the latter due to mutations in TIE2/TEK on chromosome 9p21. This has been highlighted as the cause of familial mucocutaneous VMs [10]. *Glomovenous malformations* (venous malformations with glomus cells) have furthermore been linked to a mutation in the glomulin gene on chromosome 1p [11].

Epidemiology

Vascular Tumours

Haemangiomas are the most common benign soft tissue tumour of infancy, being present in 10–12% and being more common in Caucasians, females, premature infants, twins and born to mothers of higher maternal age [1, 12]. Many of these appear postnatally in the first weeks of life (infantile haemangiomas), although they may be present at birth (congenital haemangiomas). Sixty percent occur in the head and neck area [1]. The incidence is higher in preterm babies (up to 25–30%) and low birth weight [13].

Vascular Malformations

Vascular malformations by contrast are much less common, occurring in 1.5% of the population, with an incidence of 1:5,000–1:10,000 live births. Predominantly venous malformations are the commonest subtype. Capillary malformations or "port-wine stains" (also known as "angel kiss" or "stork bites") occur in around 0.3% of childbirths with an equal sex distribution [3].

Clinical Features

Vascular Tumours

Infantile haemangiomas are recognized within days to weeks of delivery and most commonly occur as single lesions [3]. Most initially proliferate for the first year of life and then generally tend to regress or involute with the common dictum being that 70% regress by 7 years of age [2]. During this course, besides their cosmetic unacceptability, they may also cause problems such as ulceration, bleeding, infection,

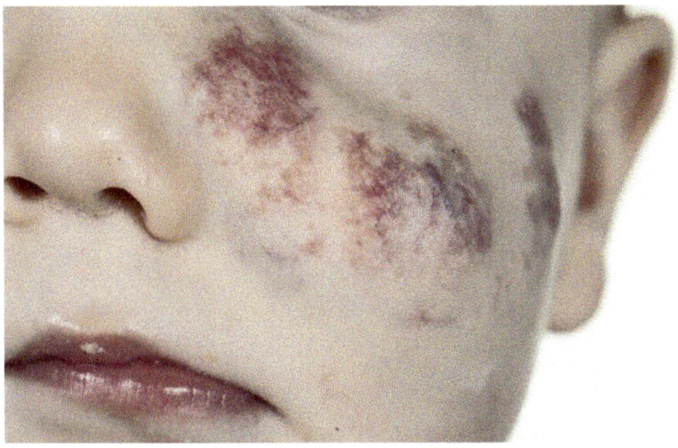

Fig. 23.1 A non-involuting congenital haemangioma (NICH) of the left cheek with characteristic violaceous hue and coarse telangiectasias

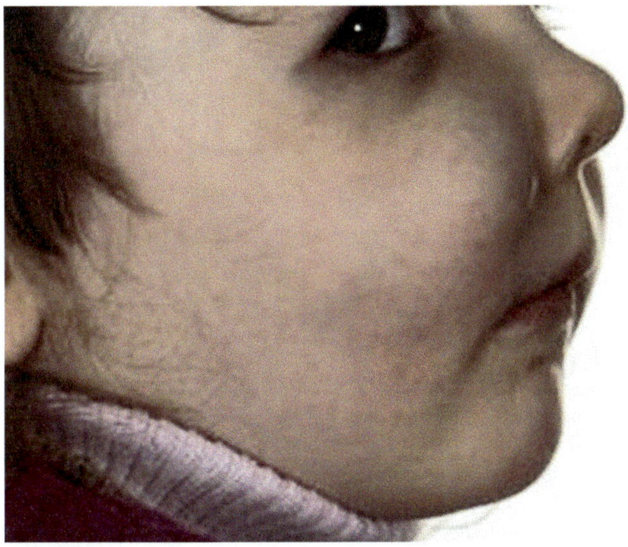

Fig. 23.2 An infantile haemangioma which is regressing or involuting as this child grows

obstruction of vision (with resultant amblyopia) and very occasionally high output heart failure (Figs. 23.1 and 23.2).

Congenital haemangiomas which are *rapidly involuting congenital haemangiomas* (*RICHs*) often begin involuting before birth and fully regress by 6–14 months of age. *Non-involuting congenital haemangiomas (NICHs)* persist as their name suggests. Both varieties present as pink or violaceous lesions with a central depression and surrounding pale rim, which may be accompanied by coarse telangiectasias. They are fast-flow lesions on Doppler evaluation [3].

Vascular Malformations

As distinct from congenital and infantile haemangiomas that may or may not involute and regress spontaneously, no proliferation or involution has been demonstrated in vascular malformations. Typically, the growth of the malformations is commensurate with the growth of the child. Patients with vascular malformations typically present with the cosmetic deformity of a birthmark of the skin or mucosa. Whilst vascular malformations are by definition present at birth, patients may not present until later in childhood or adolescence.

Fast-flow malformations such as *arteriovenous malformations (AVMs)* may have an audible bruit and palpable thrill or be pulsatile. These malformations in particular may remain stable for years, becoming more evident at times of hormonal changes such as puberty or pregnancy or following trauma or infection [14].

By contrast, slow-flow malformations may change more gradually. Slow-flow malformations are usually bluish or purple and easily compressible. Venous malformations may be painful due to phlebothrombosis and may contain phleboliths that are easily palpable [3]. Depending on their location, they may cause exophthalmia, dental malocclusions and obstructive sleep apnoea (OSA) as well as airway embarrassment.

Capillary malformations present as pink macular discolourations of the skin and may slowly change colour to a deeper purple as the patient passes into adulthood. Venous malformations` may enlarge gradually in response to trauma or infection with increasing venous ectasia. They may be made more evident on clinical examination by performance of a Valsalva manoeuvre or compression of the jugular vein.

Classification Systems

Mulliken and Glowacki first described a classification system of vascular tumours and malformations in 1982. This was subsequently adapted by the International Society for the Study of Vascular Anomalies (ISSVA) in 1996 and has undergone multiple revisions, the most recent in 2014 (Tables 23.1 and 23.2) [12]. In this system, a clear distinction is made between vascular tumours and vascular malformations. Furthermore, vascular malformations are further subdivided based on their flow characteristics (slow flow and fast flow) and the component vessels that make up the malformation (e.g. venous malformations, arteriovenous malformations, etc.).

Considerable confusion has surrounded the classification of these lesions, and the main departure made by Mulliken and Glowacki from previous systems was to distinguish lesions with endothelial cell proliferation (e.g. haemangiomas) from structural anomalies (vascular malformations). Indeed vascular malformations are not true neoplasms but rather localized defects of vascular morphogenesis.

Table 23.1 International Society for the Study of Vascular Anomalies (ISSVA) classification of vascular tumours © 2014 International Society for the Study of Vascular Anomalies

Vascular tumours
Benign vascular tumours
Infantile haemangioma
– Rapidly involuting (RICH)
– Non-involuting (NICH)
– Partially involuting (PICH)
Congenital haemangioma
Tufted angioma
Spindle cell haemangioma
Epithelioid haemangioma
Pyogenic granuloma
Others
Locally aggressive or borderline
Kaposiform haemangioendothelioma
Retiform haemangioendothelioma
Papillary intralymphatic angioendothelioma (PILA)
Composite haemangioendothelioma
Kaposi sarcoma
Others
Malignant vascular tumours
Angiosarcoma
Epithelioid haemangioendothelioma
Others

Available at issva.org/classification

Haemangiomas may be further regarded as being either segmental or multifocal and superficial (involving superficial dermis only) or deep (involving deep dermis and subcutis) [13].

Role of Antenatal Detection

The role of antenatal detection for vascular malformations is very much in its infancy and largely consists of sporadic case reports and case series within the literature.

Marler et al. [15] have demonstrated an accuracy of 59% for sonography of all vascular anomalies (including haemangiomas) detected antenatally in their series of 29 patients. They highlighted in particular the importance of distinguishing the posterior nuchal translucency (PNT) in the first and early second trimester from lymphatic malformations of the head and neck region in the second and third trimesters. Accurate prenatal diagnosis of large cervicofacial *lymphatic malformations (LMs)* in particular allows for planning of delivery and postnatal care, with possible ex utero intrapartum (EXIT) procedures giving consideration to airway control with intubation or tracheostomy prior to clamping the umbilical cord (Fig. 23.3).

There have been sporadic cases of antenatal interventions. Examples include the use of maternal corticosteroids in a case of foetal cardiac decompensation secondary to a

Table 23.2 International Society for the Study of Vascular Anomalies (ISSVA) classification of vascular malformations © (2014 International Society for the Study of Vascular Anomalies)

Vascular malformations
Slow (low) flow
Capillary malformations (CM)
– Cutaneous and/or mucosal CM
– Telangiectasia
– Cutis marmorata telangiectatica congenita (CMTC)
– Nevus simplex/Salmon patch
– Others
Venous malformations (VM)
– Common VM
– Familial VM cutaneo-mucosal (VMCM)
– Blue rubber bleb naevus (Bean) syndrome VM
– Glomuvenous malformation (GVM)
– Cerebral cavernous malformation (CCM)
– Others
Lymphatic malformation (LM)
– Common (cystic) LM
– Generalized lymphatic anomaly (GLA)
– LM in Gorham-Stout disease
– Channel type LM
– Primary lymphoedema
– Others
Fast (high) flow
Arterial malformations (AM)
Arteriovenous fistula (AVF)
Arteriovenous malformation (AVM)
Combined vascular malformations
Capillary-venous malformation (CVM)
Capillary-lymphatic malformation (CLM)
Lymphatic venous malformation (LVM)
Capillary-lymphatic-venous malformation (CLVM)
Capillary-arteriovenous malformation (CAVM)
Capillary-lymphatic-arteriovenous malformation (CLAVM)
Capillary-venous-arteriovenous malformation (CVAVM)
Capillary-lymphatic-venous-arteriovenous malformation (CLVAVM)

Available at issva.org/classification

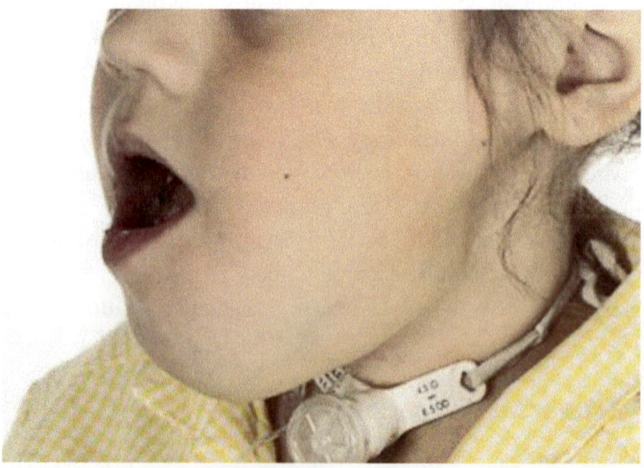

Fig. 23.3 A large microcystic lymphatic malformation (LM) causing airway compromise necessitating a permanent tracheostomy

vascular malformation [15], as well as a reported case of a foetal LM managed with intralesional OK-432 [16].

Associations

Vascular Tumours

Haemangiomas have been shown to be associated with *PHACE(S)*, a neurocutaneous disorder presenting with features responsible for the acronym, namely, *p*osterior cranial fossa malformations, cervicofacial and/or laryngeal *h*aemangiomas, *a*rterial anomalies of the head and neck, *c*oarctation of the aorta and cardiac defects and *e*ye anomalies, with or without *s*ternal defects. Cases of suspected PHACE(S) should have magnetic resonance angiography (MRA) of the brain and intracranial vasculature. Suspicion should be aroused by infantile haemangiomas greater than 5 cm in diameter in the head and neck region [17].

Tufted angioma and *kaposiform haemongioendotheliomas* in particular may be associated with sequestration of platelets and resultant life-threatening thrombocytopaenia, a condition termed *Kasabach-Merritt phenomenon* [13]. These children are at risk for intracranial, pleural, pulmonary, peritoneal and gastrointestinal haemorrhage [3]. These are often refractory to platelet transfusions due to trapping in the lesion.

Larger haemangiomas and multiple cutaneous haemangiomas (more than five) should prompt a search for visceral haemangiomas (most commonly the liver, gastrointestinal tract, lungs and/or brain) [18]. Abdominal ultrasound scans and/or magnetic resonance imaging (MRI) are the investigations of choice.

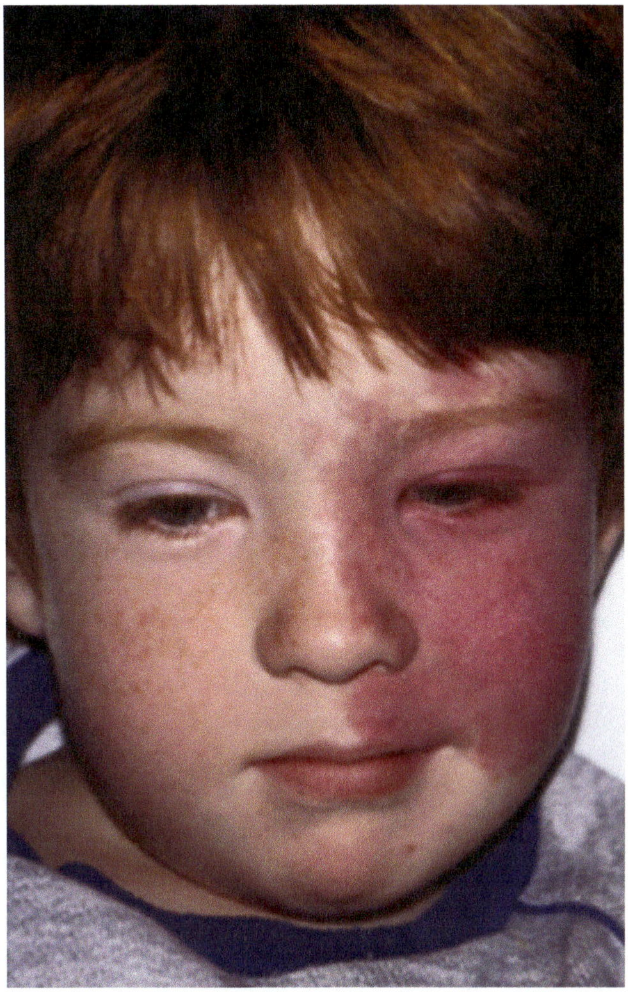

Fig. 23.4 Sturge-Weber syndrome

Vascular Malformations

Capillary malformations in the head and neck area occurring in the distribution of the trigeminal nerve have an associated risk of intracranial vascular anomalies of the leptomeninges, known as *Sturge-Weber syndrome* [19]. Leptomeningeal involvement may manifest as seizures, contralateral hemiplegia and/or motor and cognitive delays [2]. Eighty percent of patients will have their first convulsion before the second year of life. Choroidal vascular malformations may also be a feature in 70% with glaucoma in 30%. An interesting feature on computed tomography (CT) is "tram tracks" due to calcification along the cerebral sulcus owing to vascular malformations in the pia mater [17] (Fig. 23.4).

A further subset of patients with CMs have *capillary malformation-arteriovenous malformation (CV-AVM) syndrome* or *Parkes Weber syndrome*, featuring multiple CMs with or without AVMs and arteriovenous fistulas [10]. As previously mentioned this stems from a RASA-1 gene mutation with autosomal dominant inheritance and variable penetrance. Treating physicians and surgeons should be aware of this association as patients may have intracranial and spinal anomalies coexisting with cutaneous CMs (e.g. spinal dysraphism, encephalocele or ectopic meninges) [3]. It is currently unclear where screening for intracranial and spinal AVMs is warranted for all newborns with CMs, as evidence for this is restricted to case series, although the largest of these has shown an 8% incidence of intracranial AVMs in cases of CM-AVM with confirmed RASA-1 mutations [20].

Vascular malformations have also been described in *Cowden/PTEN hamartoma tumour syndrome and CLOVES syndrome* (congenital lipomatous overgrowth, vascular malformations, epidermal naevi and skeletal abnormalities) [3]. Vascular metameric syndromes are an interesting subset of conditions that revolve around the embryologic concept that an anomaly in one body segment simultaneously causes failure in nerves, skin and blood vessels within that anatomical segment. The classic example is *Wyburn-Mason syndrome*,

Table 23.3 Syndromes associated with vascular tumours and malformations © (RSNA, 2013 reproduced with permission from Nozaki et al. [17])

Tumours
Infantile haemangioma, PHACE syndrome
Malformations
Low flow
Sturge-Weber syndrome, Klippel-Trenaunay syndrome, Proteus syndrome, Cutis marmorata telangiectasia congenita, Adams-Oliver syndrome, blue rubber bleb naevus syndrome (Bean syndrome), Maffucci syndrome, Gorham-Stout syndrome
High flow
Bonnet-Dechaume-Blanc syndrome (Wyburn-Mason syndrome), Parkes Weber syndrome, Rendu-Osler-Weber syndrome (hereditary haemorrhagic telangiectasia), Cobb syndrome, Cowden syndrome, Ehler-Danlos syndrome (type 4)

where AVMs of the brain or retina are associated with facial vascular malformations in the same body segment [16]. Further rare syndromes which include low-flow vascular malformations are *Maffucci syndrome and Proteus syndrome* [17].

Multiple cutaneous venous malformations may be seen in the so-called *blue rubber bleb naevus syndrome (Bean syndrome)*, which may be combined with multiple venous malformations of the gastrointestinal tract causing gastrointestinal haemorrhage. Further complications of the condition may include anaemia, haemothorax, hypercalcaemia and widespread vascular malformations of internal organs.

A comprehensive overview of syndromes associated with vascular tumours and malformations is given in Table 23.3.

References

1. Tucci FM, De Vincentiis GC, Sitzia E, et al. Head and neck vascular anomalies in children. Int J Paediatr Otorhinolaryngol. 2009;73(Suppl):S71–6.
2. Buckmiller LM, Richter GT, Suen JY. Diagnosis and management of haemangiomas and vascular malformations of the head and neck. Oral Dis. 2010;**16**:405–18.
3. Kulungowski AM, Fishman SJ. Vascular anomalies. In: Coran AG, Adzick NS, Krummel TM, editors. Paediatric surgery. 7th ed. Philadelphia, PA: Elsevier Saunders; 2012.
4. Lee B. Venous embryology: the key to understanding anomalous venous conditions. Phlebolymphology. 2012;19(4):170–81.
5. Wang Y, Qi F, et al. Endothelial cell culture of intramuscular venous malformation and its invasive behavior related to matrix metalloproteinase-9. Plast Reconstr Surg. 2009;123:1419–30.
6. Vikkula M, Boon LM, Mullikan JB. Molecular genetics of vascular malformations. Matrix Biol. 2001;20:327–35.
7. Eerola I, Boon LM, Mulliken JB, et al. Capillary malformation-arteriovenous malformation, a new clinical and genetic disorder caused by RASA1 mutations. Am J Hum Genet. 2003;73:1240–9.
8. Yadav P, De Castro DK, Waner M, Meyer L, Fay A. Vascular anomalies of the head and neck: a review of genetics. Semin Ophthalmol. 2013;28(5-6):257–66.
9. Frigerio A, Stevenson DA, Grimmer JF. The genetics of vascular anomalies. Curr Opn Otolaryngol Head Neck Surg. 2012;20(6):527–32.
10. Orme CM, Boyden LM, Choate KA, et al. Capillary malformation—arteriovenous malformation syndrome: review of the literature, proposed diagnostic criteria and recommendations for management. Paediatr Dermatol. 2013;30(4):409–15.
11. Boon LM, Brouillard P, Irrthum A, et al. A gene for inherited cutaneous venous anomalies ("glomangiomas") localizes to chromosome 1p21-22. Am J Hum Genet. 1999;65:125–33.
12. George A, Mani V, Noufal A. Update on the classification of haemangioma. J Oral Maxillofac Pathol. 2014;18(Supp 1):S117–20.
13. Elden LM, Kirschner RE. Congenital malformations of the oral cavity and oropharynx. In: Elden LM, Zur KB, editors. Congenital malformations of the head and neck. New York: Springer; 2014.
14. Fishman SJ, Mulliken JB. Haemangiomas and vascular malformations of infancy and childhood. Paediatr Surg. 1993;40:1177–200.
15. Marler JJ, Fishman SJ, Upton J, et al. Prenatal diagnosis of vascular anomalies. J Paediatr Surg. 2002;37(3):318–26.
16. Watari H, Yamada H, Fujimo T, et al. A case of intrauterine medical treatment for cystic hygroma. Eur J Obstet Gynaecol Reprod Biol. 1996;70:201–3.
17. Nozaki T, Nosaka S, Miyazaki O, et al. Syndrome associated with vascular tumours and malformations: a pictorial review. Radiographics. 2013;33:175–95.
18. Metry DW, Hawrot A, Altman C, et al. Association of solitary, segmental haemangiomas of the skin with visceral haemangiomatosis. Arch Dermatol. 2004;140(5):591–6.
19. Graham JM, Scalding GK, Bull PD. Haemangiomas and vascular malformations. In: Pediatric ENT: Springer; 2008.
20. Revencu N, Boon LM, Mulliken JB, et al. Parkes Weber syndrome, vein of Galen aneurysmal malformation and other fast-flow vascular anomalies are caused by RASA1 mutations. Hum Mutat. 2008;29:959–65.

Eyes

24

Aaron Jamison and Gerard McGowan

Embryology of the Eye

Development of the eye starts around day 22 (embryo is 2 mm in length). While the neural folds are fusing rostrally and caudally to form the neural tube, two shallow grooves, or 'optic pits', form from neural ectoderm in the neural plate. The neural folds in this area fuse soon after, forming the primordial diencephalon. Continued evagination of the optic pits results in the formation of two symmetrical hollow hemispherical outgrowths on the lateral sides of the primordial forebrain—the primary optic vesicles. Over the next few days, the optic vesicles enlarge and become surrounded by mesenchymal cells, likely derived from the cephalic neural crest (Fig. 24.1).

At the apex of the optic vesicle, a circular area of thickened neural ectoderm (the retinal disc) lies opposed to a layer of thickened surface ectoderm (the lens placode), around day 27. At the same time, the area of the optic vesicle proximal to the forebrain constricts to form the optic stalk—through which the cavity of the optic vesicle remains continuous with the future third ventricle of the brain (Fig. 24.2).

The optic vesicle then undergoes invagination to create the optic cup—formed by a double layer of cells. The outer layer will eventually form the pigmented layer of the retina, while the inner layer will form the iris, ciliary body and all non-pigmented layers of the retina. The lens placode invaginates simultaneously to the optic vesicle, possibly aided by fine cellular bridges between the retinal disc and lens placode. This process completes around day 29. Growth of the optic cup is not perfectly symmetrical, and a groove forms at the distal and ventral aspect to form the choroidal (or optic) fissure (Fig. 24.3).

The lens vesicle separates from the overlying surface ectoderm around day 36 and becomes surrounded by a basal lamina which will form the future lens capsule. The epithelial cells enclose the lens cavity. Cells from the posterior layer later elongate to fill this space (primary lens fibres). The choroidal fissure holds the hyaloid artery, which extends anteriorly to the lens vesicle. At around this stage, the edges of this fissure oppose and fuse to form a conduit allowing the hyaloid vessels (the future central retinal artery and vein) to run through the optic stalk (Fig. 24.4).

Once the lens vesicle has separated from the overlying surface ectoderm, the latter regenerates to form the future corneal epithelium. Around day 39, mesenchymal cells pass between the optic cup and the surface ectoderm. The first of these cell migrations lies close to the optic cup and develops tight junctional complexes to form the corneal endothelium. Subsequent mesenchymal cell migrations form the remaining layers of the cornea (except the epithelium), the iris stroma and irido-corneal angle (anterior chamber angle) (Fig. 24.5).

At the end of week 8, the two layers of the retina—the thin outer 'retinal pigment epithelium' and the thicker inner 'neural retina'—are separated by a thin intraretinal (or subretinal) space. The neural retina matures to form inner and outer neuroblastic layers. Ganglion cell axons extend from the inner retina towards the optic stalk and through it (forming the optic nerve) to the brain. Elsewhere, mesenchyme surrounding the optic cup condenses and will form the future choroid—homologous in embryonic development with the pia mater. An outer layer of mesenchymal cells—homologous with the dura mater—condenses to form the sclera (Fig. 24.6).

This summary describes to the end of the embryonic period, by which time most of the important steps of eye development have occurred. Forrester et al. (2015) and Skuta et al. (2012) contain very useful descriptions of the further development of the eye following this period.

A. Jamison · G. McGowan (✉)
Tennent Institute of Ophthalmology, Gartnavel General Hospital, Glasgow, UK

R. Carachi, S. H. E. Doss (eds.), *Clinical Embryology*, https://doi.org/10.1007/978-3-319-26158-4_24

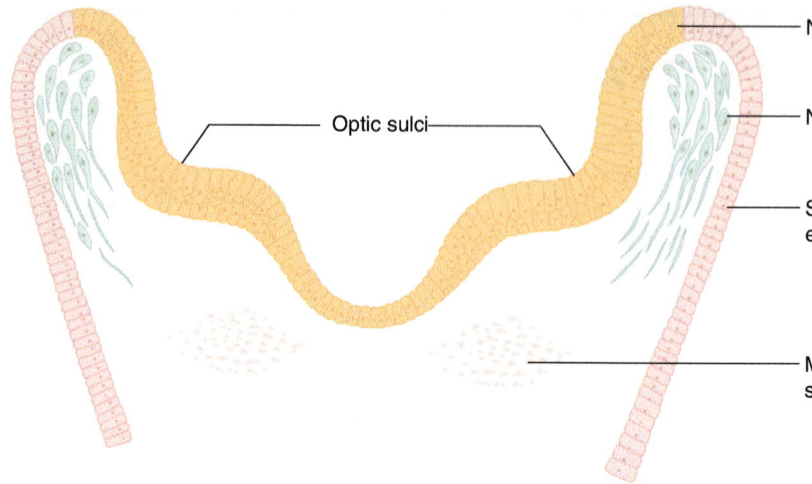

Fig. 24.1 Embryonic development: day 22

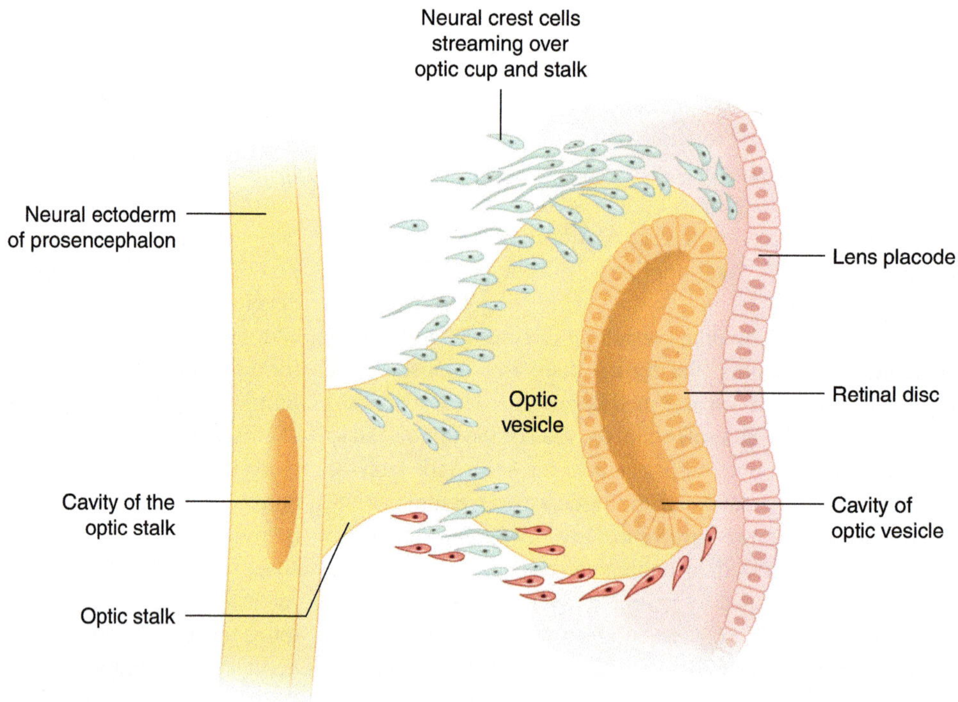

Fig. 24.2 Embryonic development: day 27

Congenital Malformations of the Eye

Coloboma

Derived from the Greek word *koloboma*, for mutilation, a coloboma is a congenital malformation caused by failure of choroidal fissure closure during the fifth to seventh weeks of gestation, which can affect the iris, ciliary body, choroid, retina and/or optic nerve. Typically, colobomata are bilateral, sporadic, and, due to the inferonasal position of the choroidal fissure, located in the inferonasal quadrant of the eye. Atypical colobomata can be found elsewhere within the eye, and their aetiology is less clear (Fig. 24.7a, b) [1–3].

The clinical appearance and significance of colobomata will depend on the ocular structures involved. Nystagmus may be present if both optic nerves or both maculae (the central portion of the retina) are involved.

- **Iris**: Usually affecting the inferonasal stroma, smooth muscle and pigment epithelium of the iris, iris colobomata produce a defect often described as a 'keyhole pupil'

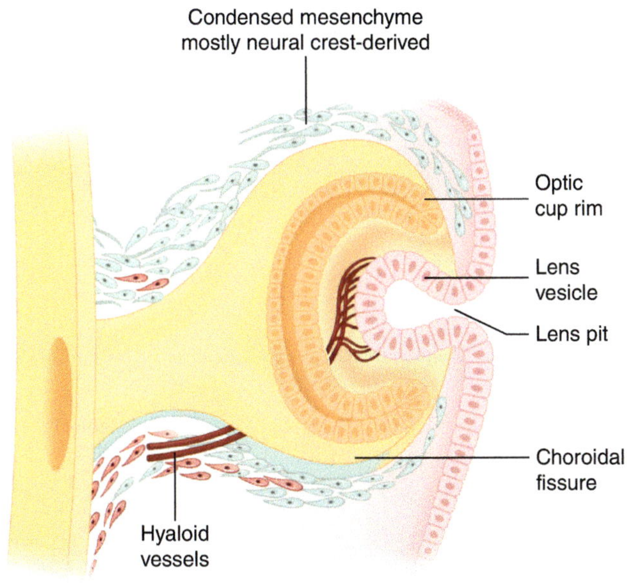

Fig. 24.3 Embryonic development: day 29

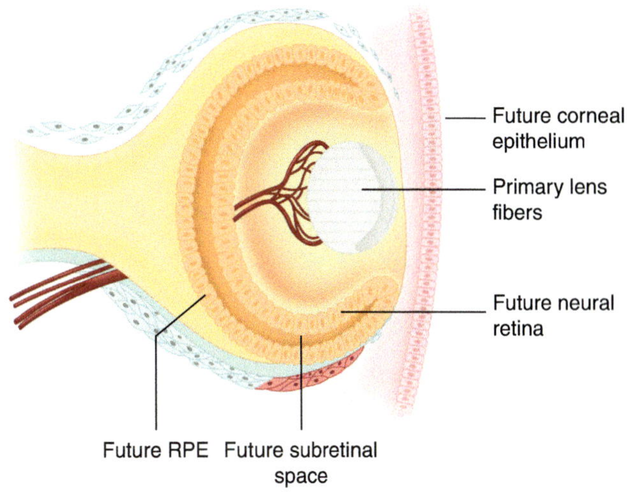

Fig. 24.4 Embryonic development: day 37

Fig. 24.5 Embryonic development: day 44

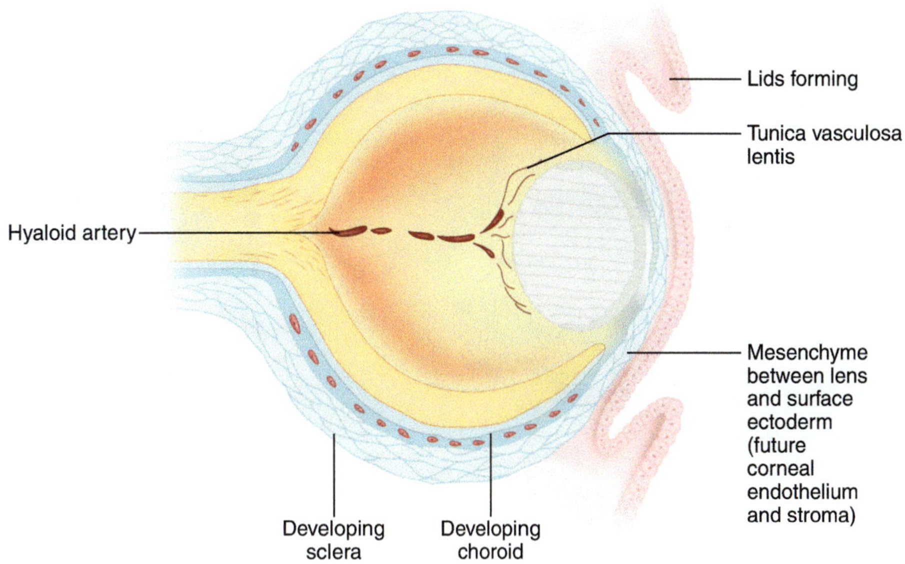

or 'inverted teardrop'. This abnormality does not usually lead to any significant complications.

- **Ciliary body**: These colobomata appear as defects filled with mesodermal tissue and blood vessels, associated with absence of the ciliary processes which can lead to an adjacent indentation of the lens.
- **Lens**: Indentation of the lens due to absent ciliary processes/zonular fibres, see *Ciliary Body*.
- **Choroid/retina**: These colobomata are primarily characterised by absence of the RPE (retinal pigment epithelium), with associated partial or complete absence of the underlying choroid. There is usually pigment clumping at the edge of the coloboma. The neural retina is both atrophic and gliotic and, due to the proliferation of neuroblastic tissue, may display rosettes. The sclera is usually intact, although if it is thinned or absent, the neural retina can herniate through to form a cyst or staphyloma. The hypoplastic retina overlying a coloboma may contain a retinal tear which can lead to the development of a (rhegmatogenous) retinal detachment (Fig. 24.8).
- **Optic nerve**: Colobomata of the optic disc are usually seen as a white bowl-shaped excavation lying slightly inferiorly within an enlarged optic disc. This defect may

Fig. 24.6 Embryonic development: week 8

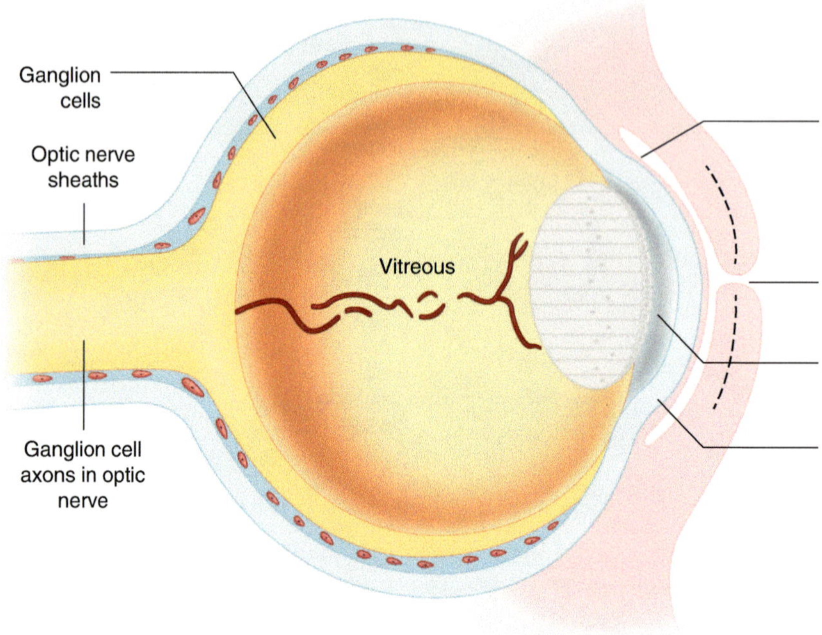

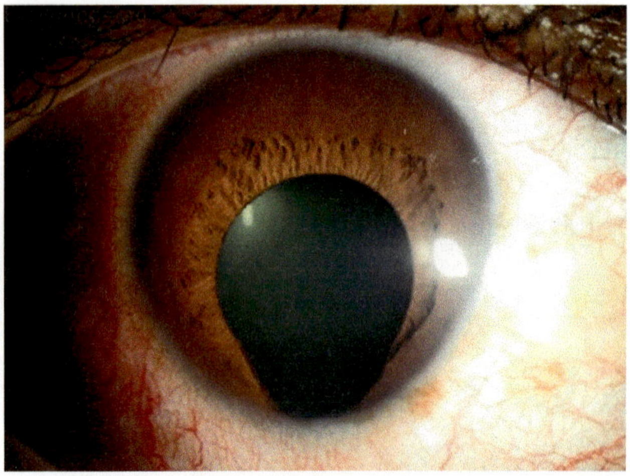

Fig. 24.7 Iris coloboma

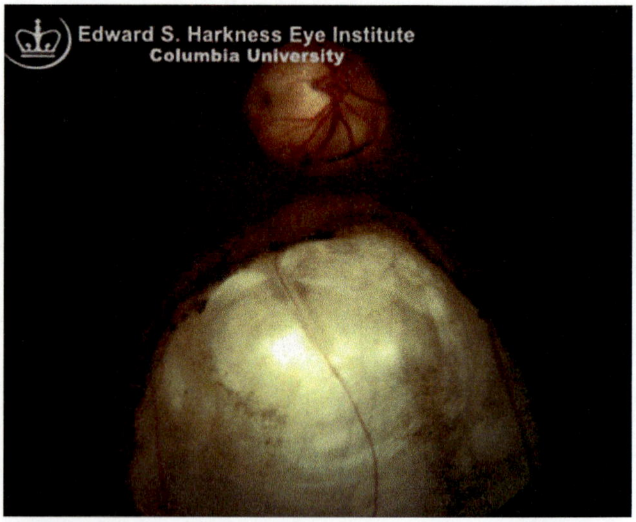

Fig. 24.8 Retinochoroidal coloboma

extend inferiorly to involve the adjacent choroid and retina. Vision may remain intact but is often limited. Eyes with optic disc colobomata are at risk of developing serous retinal detachments of the macula, in contrast to the rhegmatogenous detachments seen in chorioretinal colobomata.

- *Optic nerve pit: A form of coloboma that manifests as a small inferotemporal depression within the optic disc, which is also commonly associated with serous macular detachment and other macular changes.*
- *Morning glory disc anomaly: A coloboma of the optic nerve which is more common in females and in contrast to a typical optic disc coloboma is rarely familial and usually unilateral. In contrast to optic disc colobomata, the optic disc lies within the excavation, and the defect is symmetrical in that the optic disc lies centrally within the excavation, resembling the appearance of the morning glory flower. The morning glory disc anomaly is associated with trans-sphenoidal basal encephalocoele, for which surgery is contraindicated.*

Colobomata can be associated with a number of chromosomal abnormalities, including trisomy 13 (Patau's syndrome) and trisomy 18 (Edward's syndrome). They have also been associated with exposure to teratogens such as thalidomide, cocaine and alcohol. Ocular colobomata may be associated with the following syndromes:

- CHARGE syndrome: an inferonasal and posterior **C**oloboma of the eye is associated with **H**eart defects, **A**tresia of the nasal choanae, **R**etardation of growth and/or development, **G**enital and/or urinary abnormalities and **E**ar abnormalities and deafness
- Aicardi syndrome: in association with partial or complete absence of the corpus callosum
- Goldenhar syndrome
- Goltz's syndrome: aka focal dermal hypoplasia
- Median cleft face syndrome: aka frontonasal dysplasia
- Warburg's syndrome
- Rubinstein-Taybi syndrome
- Linear nevus sebaceous syndrome
- Meckel syndrome
- Klinefelter's syndrome
- Turner's syndrome

Patients with a coloboma should undergo a complete eye and medical examination, which should include a family history and may include examination of any available relatives. Children with ocular coloboma should be monitored for the development of amblyopia (a 'lazy eye'), anisometropia (significantly different refractive power of each eye) or retinal detachment.

Disorders of the Eye as a Whole

Anophthalmia

Anophthalmia, or anophthalmos, refers to the complete absence of ocular structures within the orbit. Histological examination of the orbit is required to differentiate anophthalmia from extreme microphthalmia, and so the term 'clinical anophthalmia' may be used to describe the condition where no eye can be found clinically. True anophthalmia is very rare (birth prevalence of 1.8–3.0 per 100,000 births) and can involve one or both eyes [4, 5].

Anophthalmia may occur due to one of three abnormal embryological processes.

- *Primary anophthalmia*: A failure of the optic pit to deepen and form an optic outgrowth (primary optic vesicle) from the forebrain, which would occur after the formation of a rudimentary forebrain (around the 2 mm stage of embryonic development). In primary anophthalmia only the ectodermal elements are missing, but the orbit, eyelids, lacrimal apparatus, conjunctiva, extraocular muscles and their nerves all develop without stimulus from the optic vesicle (Fig. 24.9).
- *Secondary anophthalmia*: There is complete suppression or abnormality of development of the entire forebrain. Absence of a primary optic vesicle is only one of many simultaneous developmental abnormalities, which in most cases are not compatible with life.
- *Degenerative or consecutive anophthalmia*: A primary optic vesicle forms initially but subsequently degenerates or disappears. Overlap between anophthalmia and severe microphthalmia occurs here, as ocular structures may still be present and detectable within the orbit or may be absent.

Anophthalmia is usually inherited sporadically, although dominant, recessive and X-linked inheritance patterns are recognised. Mutations of the SOX2, PAX6 and RX genes have all been associated with anophthalmia, and it can occur as part of several clinical syndromes including:

- 'Anophthalmos plus' syndrome: anophthalmia or microphthalmia with cleft lip or palate and sacral neural tube defects

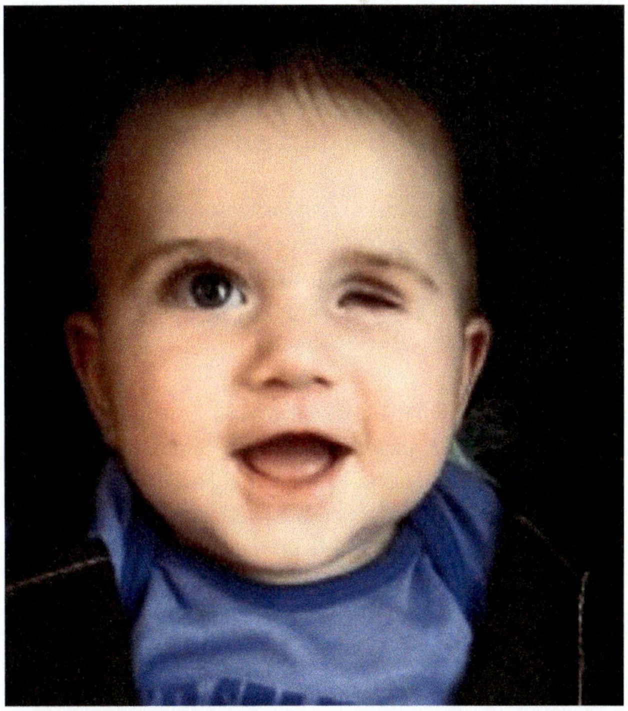

Fig. 24.9 Primary anophthalmia

- Goldenhar syndrome
- Hallermann-Streiff syndrome
- Waardenburg syndrome: usually characterised by varying degrees of sensorineural hearing loss

Isolated anophthalmia causes unilateral or bilateral blindness, due to the absence of the eye(s), but few other symptoms. Other clinical features may include a small orbital rim, a narrowed palpebral fissure (opening of the eyelids) and reduced eyelid movements. The clinical diagnosis of anophthalmia will usually rely on the use of B-scan ultrasonography, magnetic resonance imaging (MRI) and/or computed tomography (CT) imaging of the head and orbits.

Management options are primarily cosmetic in nature and may include orbital conformers (inserted into the orbital cavity to stimulate orbital bony growth and increased in size periodically), ocular prosthetics (which can overlie the orbital conformer and provide a more satisfactory cosmetic appearance) and various orbital fillers (to counteract the reduced orbital volume, offering a more symmetrical appearance in unilateral cases). The absence of one or both eye(s), with associated blindness and/or facial disfigurement, may raise a number of psychosocial issues in children, for whom referral to a psychological counselling service may be beneficial

Microphthalmia

Microphthalmia, or microphthalmos, is a condition in which the size of the eye is reduced, either unilaterally or bilaterally. Microphthalmia is present when the axial length of the eye is less than 21 mm in an adult (or less than 19 mm in a 1-year-old child, less than 15 mm at birth), which represents two or more standard deviations below that of the normal eye. It occurs in 15 per 100,000 births, with no gender or racial predilection [4, 5].

Microphthalmia results from formation of a primary optic vesicle without subsequent normal development. Three subtypes of microphthalmia exist, and the effect on the patient's vision will depend on the severity of the condition and whether it is unilateral or bilateral:

- *Simple microphthalmia*: The eye has a reduced axial length but is otherwise anatomically intact. Simple microphthalmia is usually associated with high hypermetropia (aka long-sightedness)—an optical consequence of the shortened optical pathway within the eye—and this may be the first detected clue that microphthalmia is present. Vision can be variable and is largely dictated by the degree of hypermetropia, and the level of amblyopia ('lazy eye') may develop if this refractive error is not promptly corrected with glasses.
- *Complex microphthalmia*: The eye has a reduced axial length which is associated with developmental abnormalities of the anterior segment of the eye, the posterior segment, or both. These most commonly include colobomata (see *Coloboma section*) but may also include persistent hyperplastic primary vitreous (PHPV), aniridia or congenital cataract, among others.
- *Severe microphthalmia*: The axial length of the eye is less than 10 mm at birth or less than 12 mm in a 1-year-old child. The eye may not be detectable on clinical examination, although remnants of ocular tissue will be detectable with orbital imaging techniques such as B-scan ultrasonography, CT and MRI. Again, some overlap between severe microphthalmia and degenerative anophthalmia occurs here (Fig. 24.10).

Two other clinical patterns of microphthalmia exist, classified separately to the above:

- *Microphthalmia with cyst*: Failure of the choroidal fissure to close can result in a coloboma through which a cyst may herniate. This usually presents as a bulge behind the lower lid which can be mistaken for a discrete orbital mass such as a neoplasm (Fig. 24.11).
- *Microphthalmia with cryptophthalmos*: Microphthalmia is associated with a layer of the skin overlying the eye with varying degrees of attachment to the underlying cornea. Coverage can be partial or complete and is usually bilateral (Fig. 24.12a, b).

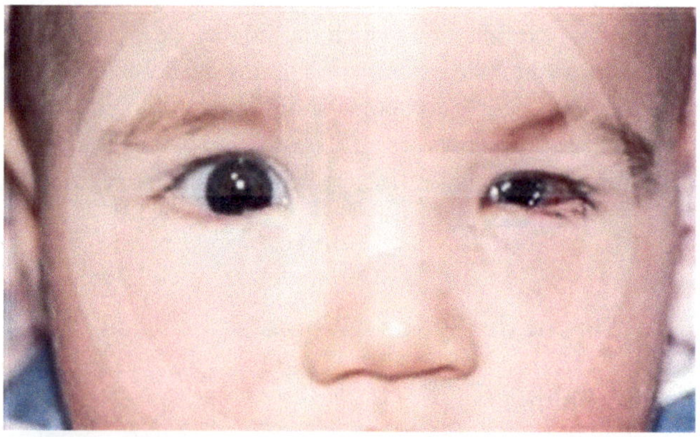

Fig. 24.10 Severe microphthalmia

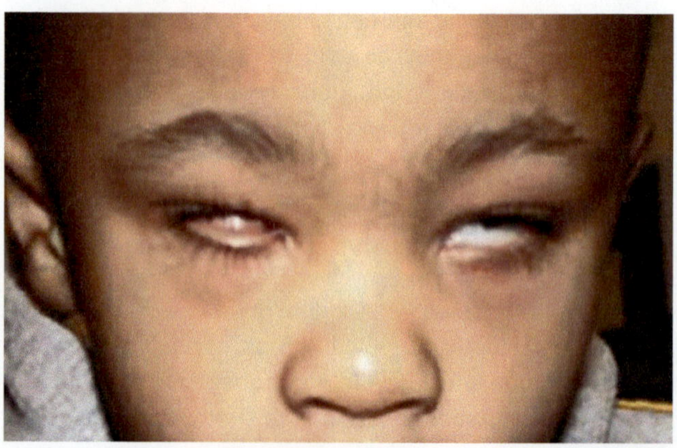

Fig. 24.11 Microphthalmia with cyst

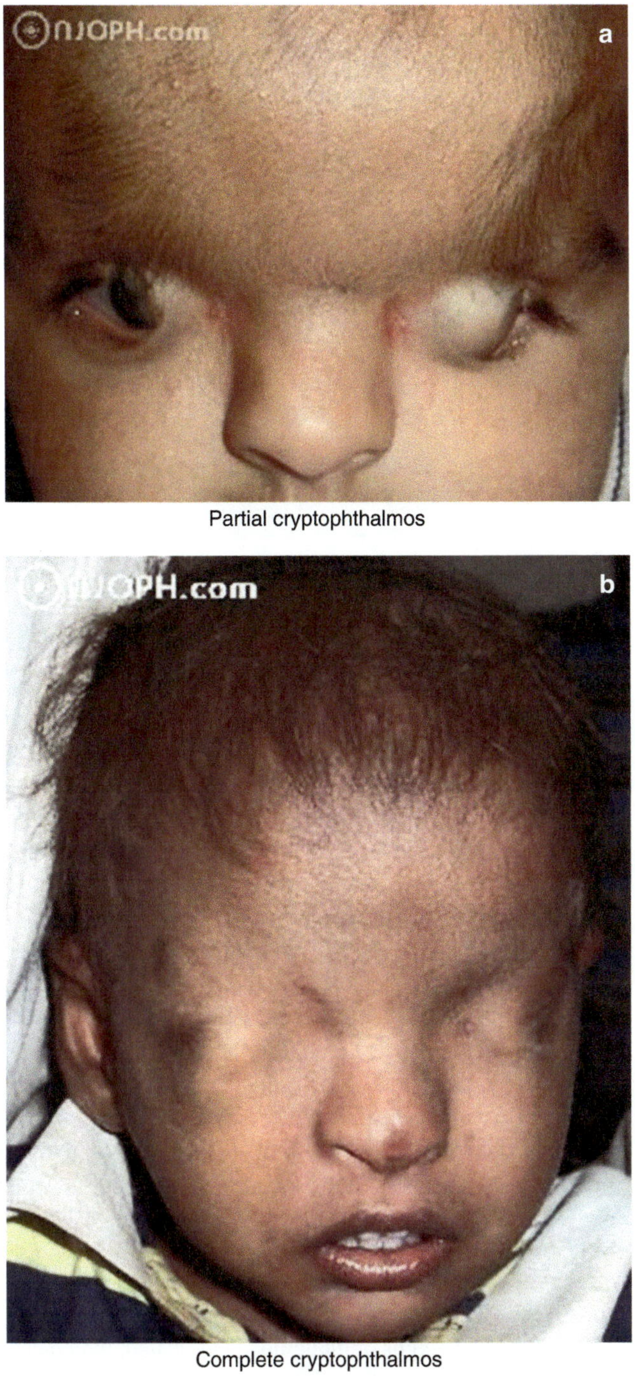

Fig. 24.12 Microphthalmia with cryptophthalmos

Microphthalmia may occur secondary to environmental factors (prenatal exposure to thalidomide, alcohol, retinoic acid or rubella), inherited disease (autosomal dominant, recessive or X-linked inheritance patterns) or chromosomal aberrations (affecting the SIX3, HESX1, SHH, CHX10 or RX genes). Microphthalmia forms part of several clinical syndromes due to single-gene disorders:

- CHARGE syndrome: **C**olobomatous microphthalmia of the eye is associated with **H**eart defects, **A**tresia of the nasal choanae, **R**etardation of growth and/or development, **G**enital and/or urinary abnormalities and **E**ar abnormalities and deafness
- 'Anophthalmos plus' syndrome: anophthalmia or microphthalmia with cleft lip or palate and sacral neural tube defects
- Aicardi syndrome: in association with partial or complete absence of the corpus callosum
- Lenz microphthalmia: an X-linked recessive syndrome in which microphthalmia is associated with mental retardation, ear malformation and skeletal abnormalities
- Goltz's syndrome: aka focal dermal hypoplasia
- Warburg's syndrome: microphthalmia associated with brain development abnormalities resulting in muscle weakness, hypotonia, mental retardation and seizures
- Norrie disease: a primarily ocular syndrome which may involve progressive hearing loss
- Meckel-Gruber syndrome
- Incontinentia pigmenti

Like anophthalmia, the clinical features of microphthalmia may include a small orbital rim, a narrowed palpebral fissure (opening of the eyelids) and reduced eyelid movements. The globe itself is reduced in size, may or may not be visibly deformed and may or may not be associated with a visibly small cornea (microcornea). The level of visual loss will depend on the severity of the condition and its associated ocular abnormalities. B-scan ultrasonography plays a role in determining the severity of microphthalmia based on the axial length of the eye, while CT and MRI imaging can help assess the internal structures of the globe, the presence of the extraocular muscles and optic nerve and the involvement of the brain.

Management of severe microphthalmia is predominantly cosmetic and similar to the management of anophthalmia, described previously. Patients with simple microphthalmia, or complex microphthalmia with vision, will usually benefit from correction of their hypermetropic refractive error (long-sightedness) and strategies to prevent amblyopia (a 'lazy eye'), such as patching the healthy eye to stimulate visual development in the affected eye.

Nanophthalmia

Nanophthalmia, or nanophthalmos, is a subtype of simple microphthalmia in which the axial length of the eye is reduced, although the eye is otherwise anatomically intact. The underlying pathogenesis is identical to that of simple microphthalmia, and although autosomal dominant and recessive patterns of inheritance have been described, most cases are sporadic, resulting in bilateral disease. Common

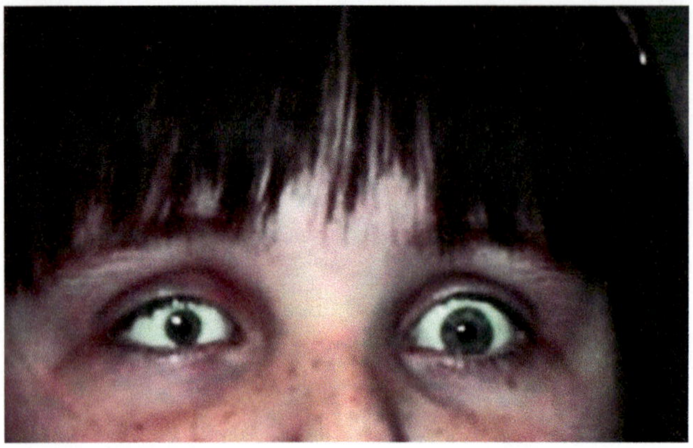

Fig. 24.13 Nanophthalmia

clinic features include high hypermetropia (long-sightedness), a thick but weak scleral wall and an increased risk of angle-closure glaucoma (acute or chronic) and post-operative or spontaneous uveal effusion (Fig. 24.13) [4, 5].

The management of associated angle-closure glaucoma is usually hampered by poor response to medical treatments, although peripheral iridotomy performed with YAG laser may be helpful. If glaucoma is detected and treated promptly, then the visual prognosis for patients with nanophthalmia is good.

Aniridia

Aniridia is a rare bilateral condition characterised by an apparent absence of the iris, although the name is a misnomer as histological examination reveals a rudimentary iris adjacent to the ciliary body (i.e. iris hypoplasia). Iris hypoplasia may be associated with foveal (the central part of the macula, responsible for central vision) and optic nerve hypoplasia, in which case visual acuity will be markedly reduced and congenital nystagmus will often be present. The presence of nystagmus and the apparent absence of the iris are often the first clues that a child has an ocular problem, although reduced visual acuity with or without photophobia (due to the inability to control the amount of light entering the eye) becomes apparent later in life. Other associated clinical features of aniridia may include limbal stem cell deficiency, cataract and glaucoma (Fig. 24.14) [6].

Aniridia occurs due to sporadic, autosomal dominant (complete penetrance with variable expression) or, rarely, autosomal recessive patterns of inherited PAX6 gene mutations. PAX6 (11p13) is a transcription factor gene crucial in oculogenesis, mutations of which have been implicated in several other ocular developmental abnormalities such as Peter's anomaly, Axenfeld-Rieger syndrome and anophthalmia.

Sporadic mutations of the PAX6 gene may also involve the neighbouring WT1 (Wilms' tumour gene) locus on 11p13, giving rise to the noted association of aniridia with Wilms' tumour. Mutations of further neighbouring genes can result in development of the AGR triad: **A**niridia, **G**enitourinary abnormalities and mental **R**etardation. WAGR syndrome (or WAGR complex) is the diagnosis given when these findings are also associated with the presence of **W**ilms' tumour.

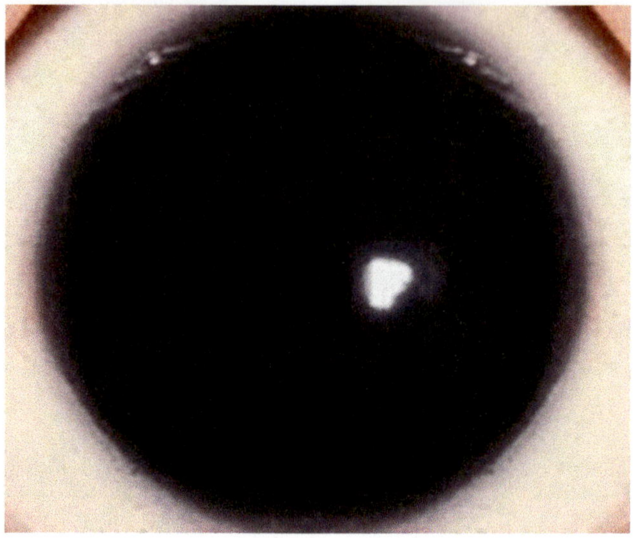

Fig. 24.14 Aniridia

Children with aniridia usually require examination under anaesthesia in order to perform a full clinical examination. Gonioscopy (a manual imaging technique used to view the anatomical angle formed by the iris and cornea) offers the best visualisation of the iris root in order to measure the extent of the condition and can detect anatomical formations that are likely to increase the risk of secondary glaucoma. Intraocular pressure measurements are useful in the detection and treatment of early glaucoma. Finally, slit-lamp examination is useful to detect corneal pannus (the growth of fine blood vessels into the clear cornea, due to limbal stem cell deficiency), which starts as a greyish superficial change of the peripheral cornea but can spread centrally resulting in reduced vision.

The treatment of aniridia is usually conservative in nature. Correction of any refractive error, with filtered lenses, can improve the visual acuity while reducing glare. Close monitoring for the early signs of glaucoma is indicated, which can be managed by a range of surgical procedures. Goniotomy (a procedure in which an opening is made in the trabecular meshwork of the anterior chamber angle) may be performed prophylactically, to allow aqueous humour to drain from the eye and reduce the risk of developing glaucoma. Cataract surgery may be indicated, although the decision to operate will be affected by the presence of other features such as foveal or optic nerve hypoplasia, which will also limit vision.

Children without a family history of aniridia should undergo cytogenetic and molecular genetic screening in order to detect WAGR syndrome which, if present, would indicate 3-monthly renal ultrasound screening until 8 years old.

Congenital Cataract

Cataract is the opacification of the lens, which in most cases is associated with increasing age. Congenital cataracts are present at the time of birth and are thought to be present in 1 in 250 newborns—accounting for nearly 10% of visual loss in children worldwide. Congenital cataracts may be unilateral or bilateral, congenital or acquired, and can range in severity, causing reduced visual acuity and even blindness due to the lens malformation itself or the subsequent amblyopia ('lazy eye') (Fig. 24.15) [7–11].

Sixty percent of bilateral cases and 80% of unilateral cases are idiopathic, while around 30% are familial in nature. Congenital cataract is associated with the following genetic, often metabolic, conditions:

- Alport's syndrome
- Fabry's disease
- Galactosemia
- Myotonic dystrophy
- Trisomy 13
- Trisomy 18
- Trisomy 21

They can also be associated with the following conditions:

- Systemic illnesses
  - Diabetes mellitus
  - Juvenile idiopathic arthritis (JIA)
  - Systemic lupus erythematosus (SLE)
  - Various malignancies
- Ocular abnormalities
  - Aniridia
  - Anterior segment dysgenesis
  - Persistent hyperplastic primary vitreous (PHPV)

Acquired congenital cataracts may be the result of maternal infection during pregnancy—the so-called TORCH infections: **T**oxoplasmosis, **O**ther infections (syphilis, varicella), **R**ubella, **C**ytomegalovirus or **H**erpes simplex virus.

In the newborn, congenital cataracts may be detected by a reduced red reflex and by visible lens opacity ranging from mild changes to leukocoria (a whitened pupil reflex). Other clinical features may include photophobia or nystagmus. As the child grows, a reduced visual acuity will become apparent, often through failure to achieve the normal visual developmental milestones (Fig. 24.16).

Examination of children with congenital cataracts involves dilated ophthalmic slit-lamp examination including fundoscopy and visual acuity testing where possible. B-scan ultrasonography may be required to assess the posterior segment of the eye where the cataract is too dense to allow fundoscopy. Again, these investigations may be more easily

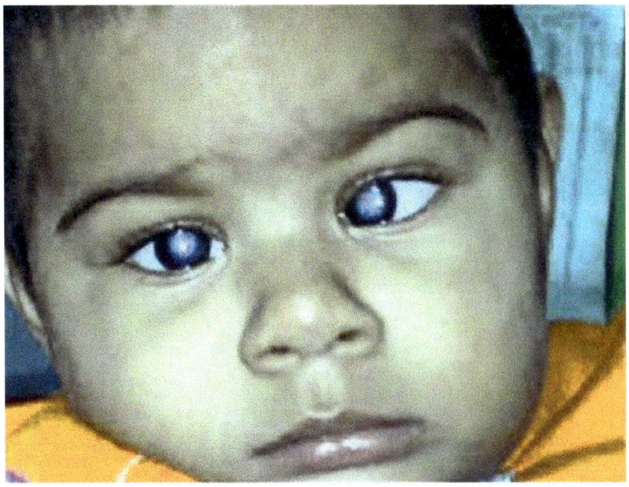

Fig. 24.15 Congenital cataract

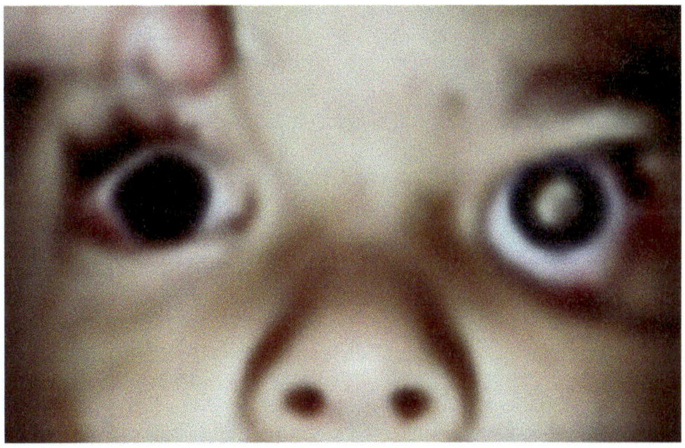

Fig. 24.16 Leukocoria - the pupil appears white, rather than the usual black

performed as an examination under anaesthesia. These children should also undergo full physical examination and, where appropriate, be referred for genetic, infectious and/or metabolic investigations.

Children with a visual acuity of 6/15 or better (Snellen acuity) may benefit from conservative management, and in these patients pupillary dilatation may helpful.

The decision to operate to surgically correct congenital cataracts is usually based on whether the lenticular opacity is sufficient to cause nystagmus, strabismus ('squint') or difficulty in obtaining a full refraction, although the appearance (size and/or density) of the opacification may also be taken into account. Cataracts may lead to amblyopia while the visual system and vision are developing (until around the age of seven), and so, in those children in which it is indicated, surgery should be performed as soon as possible after diagnosis. However, cataract surgery within the first month of life has been associated with an increased risk of glaucoma, and therefore it is common to wait 4 weeks before performing cataract surgery for newborns. The post-operative refractive error should be corrected immediately after cataract surgery (with glasses or contact lenses), and steps should be taken to prevent or reduce amblyopia (such as occlusion therapy with patches).

The long-term visual prognosis of congenital cataracts will depend on the laterality and severity of cataract, the promptness of diagnosis and treatment and the effectiveness of (and adherence to) any amblyopia prevention strategies.

References

1. Chang L, Blain D, Bertuzzi D, et al. Uveal coloboma: clinical and basic science update. Curr Opin Ophthalmol. 2006;17:447–70.
2. Onwochei BC, Simon JW, Bateman JB, et al. Ocular colobomata. Surv Ophthalmol. 2000;45:175–94.
3. Taylor D. Developmental abnormalities of the optic nerve and chiasm. Eye. 2007;21:1271–84.
4. Bardakiian T, Weiss A, Schneider AD. Anophthalmia /microphthalmia overview. In: Pagon RA, Bird TC, Dolan CR, et al., editors. GeneReviews. Seattle: University of Washington; 2007. p. 1993–2004.
5. Verma AS, Fitzpatrick DR. Anophthalmia and microphthalmia. Orphanet J Rare Dis. 2007;2:47.
6. Adeoti CO, Afolabi AA, Ashaye AO, et al. Bilateral sporadic aniridia: review of management. Clin Ophthalmol. 2010;4:1085–9.
7. Amaya L, Taylor D, Russell-Eggitt I, et al. The morphology and natural history of childhood cataracts. Surv Ophthalmol. 2003;48(2):125–44.
8. Birch EE, Stager DR. The critical period for surgical treatment of dense congenital unilateral cataract. Invest Ophthalmol Vis Sci. 1996;37(8):1532–8.
9. Vishwanath M, Cheong-Leen R, Taylor D, et al. Is early surgery for congenital cataract a risk factor for glaucoma? Br J Ophthalmol. 2004;88(7):905–10.
10. Medsinge A, Nischal KK. Pediatric cataract: challenges and future directions. Clin Ophthalmol. 2015;9:77–90.
11. Lim Z, Rubab S, Chan YH, et al. Management and outcomes of cataract in children: the Toronto experience. J AAPOS. 2012;16(3):249–54.

Recommended Reading

Ferris J. Basic sciences in ophthalmology. 2nd ed. London: BMJ Books; 1999.

Forrester JV, Dick AD, McMenamin PG, et al. The eye: basic sciences in practice. 3rd ed. China: Saunders Elsevier; 2008.

Hoyt CS, Taylor D. Pediatric ophthalmology & strabismus. 4th ed. Saunders Elsevier: China; 2013.

Mann I. The development of the human eye. 2nd ed. London: British Medical Association; 1948.

Mann I. Developmental abnormalities of the eye. 2nd ed. London: British Medical Association; 1957.

Nelson LB. Color Atlas & synopsis of clinical ophthalmology: pediatric ophthalmology (Wills Eye Institute). China: Lippincott Williams & Wilkins; 2012.

Skuta GL, Cantor LB, Cioffi GA. American Academy of Ophthalmology: basic and clinical science course—section 2: fundamentals and principles of ophthalmology. Singapore: American Academy of Ophthalmology; 2012.

Skuta GL, Cantor LB, Weiss JS. American Academy of Ophthalmology: basic and clinical science course—section 6: pediatric ophthalmology and strabismus. Canada: American Academy of Ophthalmology; 2012.

Yanoff M, Sassani JW. Ocular pathology. 6th ed. China: Mosby Elsevier; 2009.

Skin and Soft Tissues

25

Mairi Steven

Development of the Skin

* The skin consists of two components (epidermis and dermis) which arise from two sources:

1. ***The epidermis*:** arises from the surface ectoderm and receives melanoblasts (melanin-secreting cells) which migrate from the neural crest to the epidermis
2. ***The dermis and subcutaneous tissue*:** are mesodermal and arise from two sources:
   (a) The mesoderm of the dermatomes of the somites
   (b) The somatopleuric mesoderm (part of lat-plate mesoderm)
3. ***The hair*:**
   (a) The hairs begin to develop in the third month as solid ectodermal buds arising from the epidermis.
   (b) The hair buds sink into the underlying mesenchyme and its lower end becomes swollen to form the *hair bulb*.
   (c) The hair bulb becomes invaginated by a mass of mesoderm forming the *hair papilla*.
   (d) Muscle fibres become attached to the hair follicle forming *arrector pili muscle*.
   (e) The central cells of the hair bulb elongate and become keratinized forming the *hair shaft*.
   (f) The first hairs which appear on the foetus are fine and called lanugo hairs.
   (g) At birth, the lanugo hairs degenerate and are replaced later by coarser hairs.
4. ***Skin glands*** (sebaceous and sweat glands):
   (a) *Sweat glands*: appear in the fifth month as solid downgrowths from the surface ectoderm into the underlying mesenchyme. Later on, they canalize and their terminal ends become convoluted.
   (b) *The sebaceous glands*: arise as side branches from the buds of the hair follicles and then their central cells degenerate forming fatty secretion that passes out into the hair follicles.

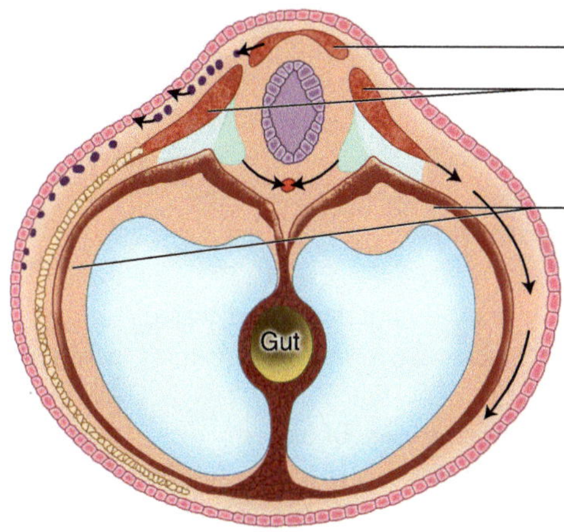

M. Steven (✉)
Department of Surgical Paediatrics, Royal Hospital for Children, Glasgow, UK
e-mail: mairisteven@doctors.org.uk

R. Carachi, S. H. E. Doss (eds.), *Clinical Embryology*, https://doi.org/10.1007/978-3-319-26158-4_25

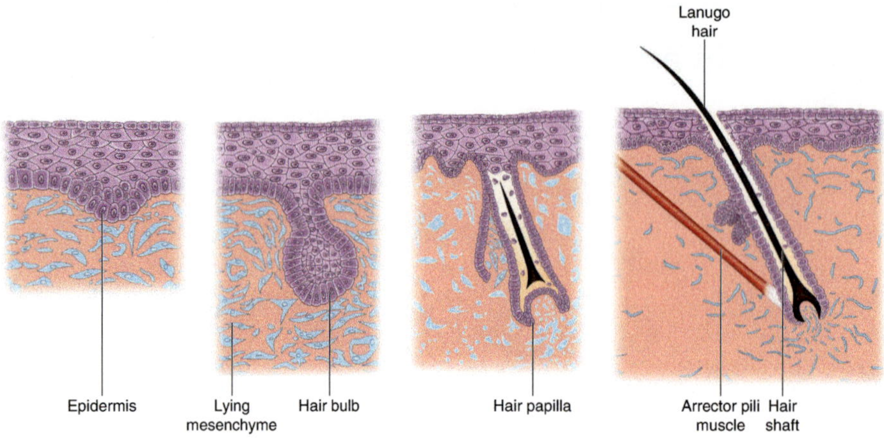

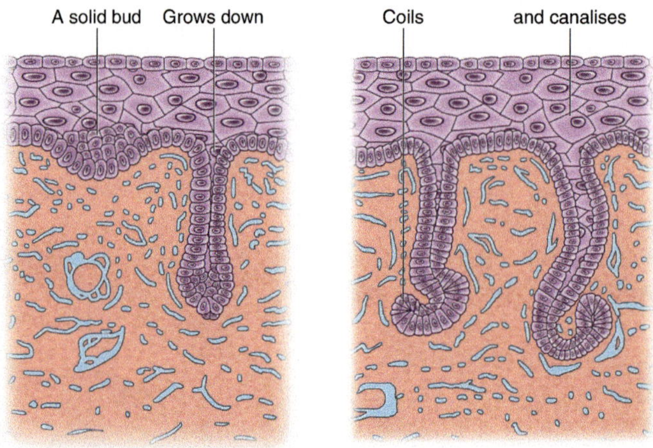

Congenital Anomalies

1. ***Ichthyosis*:** dry scaly skin (hereditary disease)
2. ***Hypertrichosis*:** excessive formation of hairs (generalized or localized)
3. ***Atrichia*:** congenital absence of hair
4. ***Sequestration dermoid*:** this is a congenital cyst caused by burying of ectodermal cells deep to the surface at the lines of closure of embryonic clefts (see p. 70)

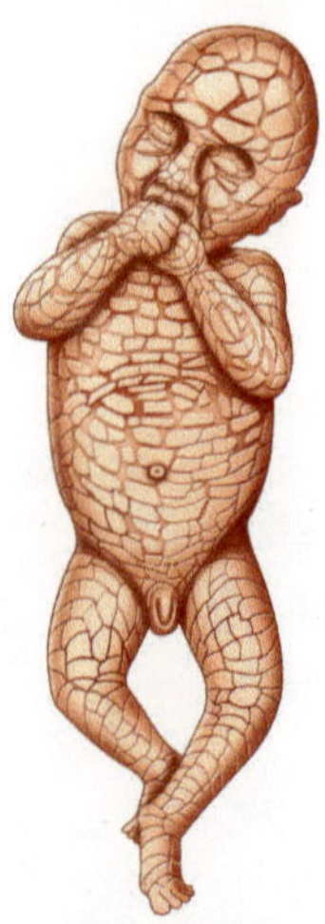
Hereditary disease

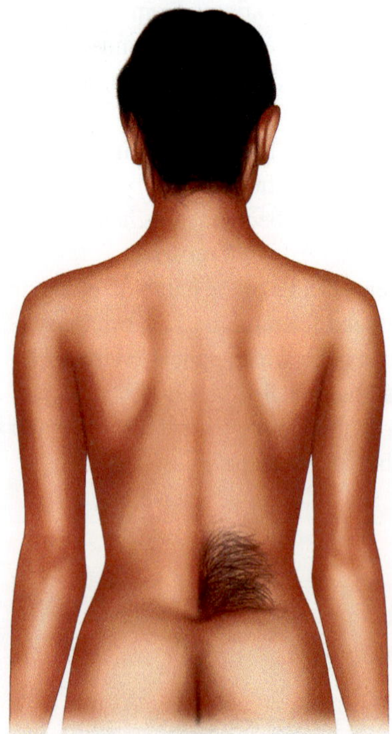
Generalized or localized

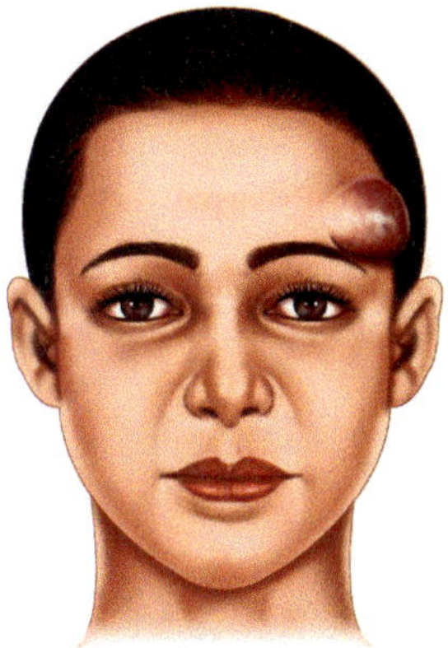

Sequestration dermoid

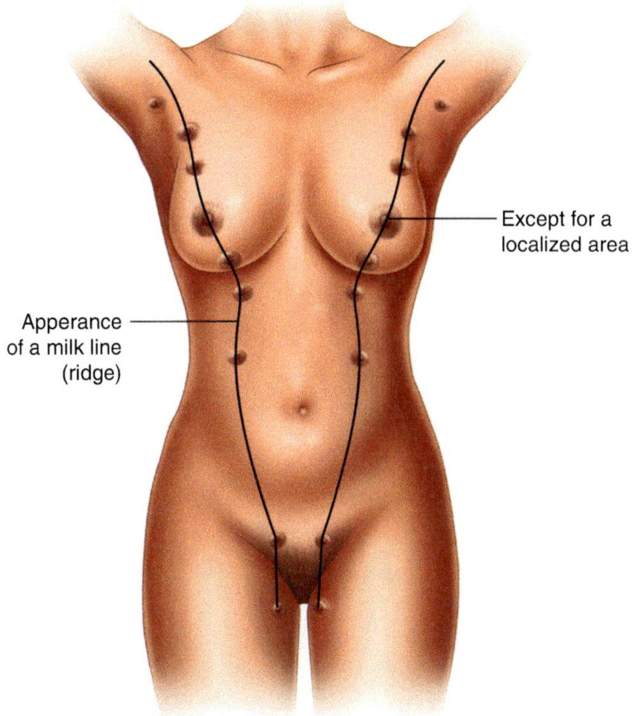

Development of the Mammary Gland

1. ***Appearance of a milk line (ridge)*:**
   (a) In the young embryo, a linear thickening of the surface ectoderm (called the *milk line*) appears on the ventral body wall extending from the axilla down to the medial part of the inguinal region.
   (b) The line disappears shortly after its formation *except for a localized area* in the pectoral region which *develops into mammary gland as follows*:
2. ***Formation of the glandular tissue*:**
   (a) 15–20 solid *ectodermal buds* arise from the remaining part of the milk line.
   (b) The buds form *solid cords* which grow into the underlying mesenchyme. (Each cord represents a future milk duct and a lobe of the mammary gland.)
   (c) The cords continue to *grow and branch* throughout the foetal life.
   (d) Shortly before birth, the cords *become colonized* to form the lactiferous sinuses, lactiferous ducts and secretory alveoli.
   (e) The lactiferous ducts open into a *depressed area* on the surface.
   (f) Shortly after birth, this depressed area becomes elevated forming the *nipple*.
   (g) *At puberty the following occur*:
   - *In the male*: the mammary gland remains rudimentary (glandular tissue consisting of ducts only without alveoli).
   - *In the female*: under the effect of the ovarian hormones, the mammary gland enlarges in size due to more branching in the lactiferous ducts and deposition of more fat between the lobules of the gland.

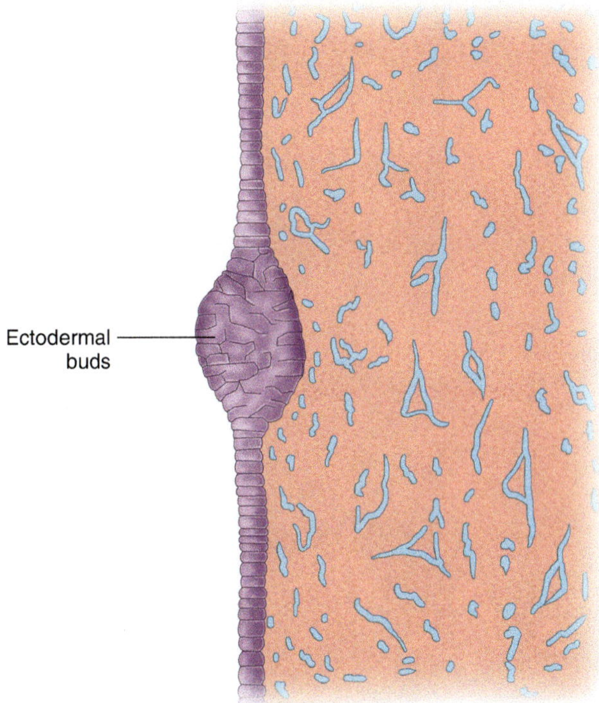

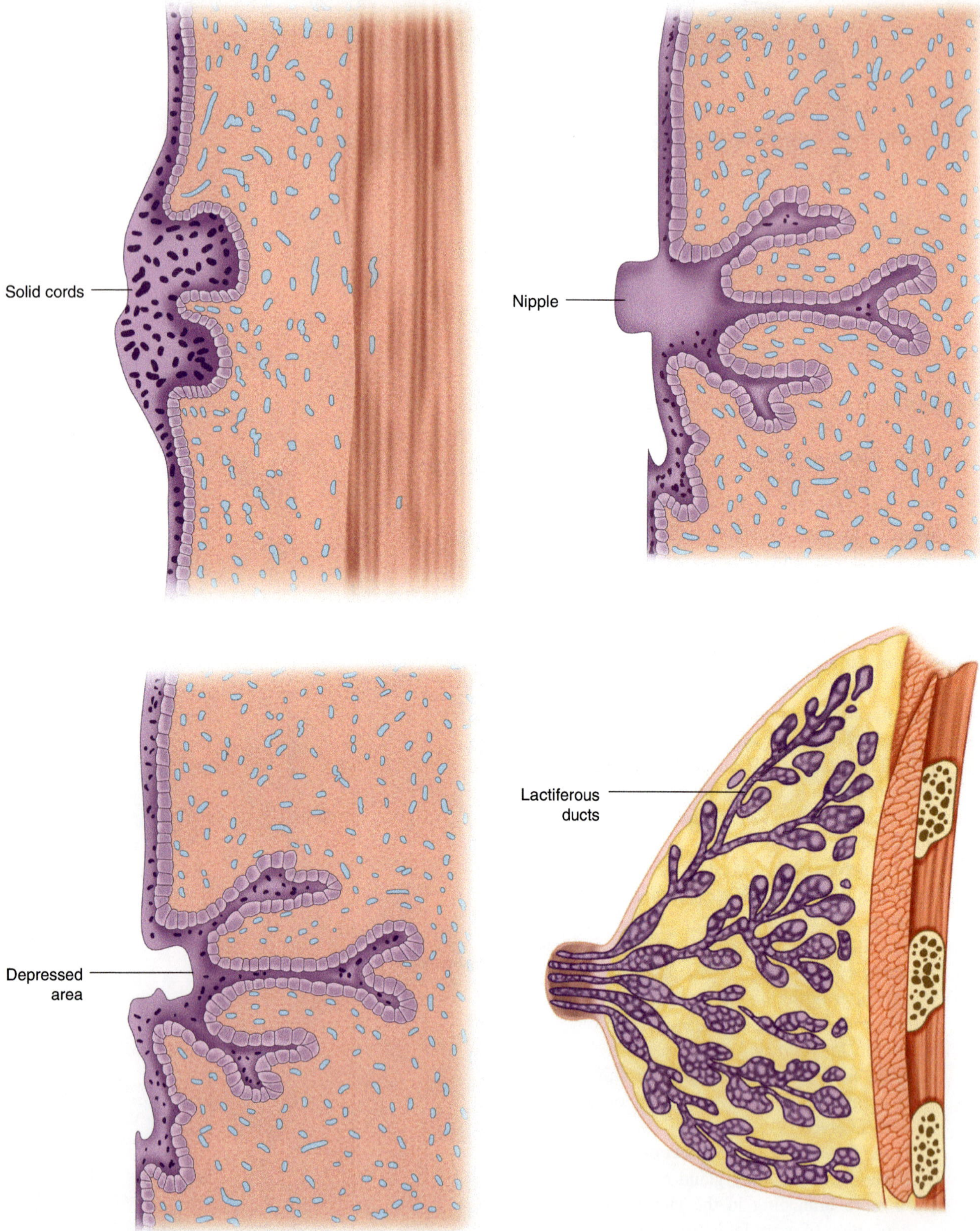
Solid cords
Nipple
Depressed area
Lactiferous ducts

Congenital Anomalies

1. *Amastia*: absence of one or both mammary glands.
2. *Micromastia*: abnormally small breast (retention of prepubertal state)
3. *Macromastia*: abnormally large breast
4. *Gynaecomastia*: a male developing a female type of breast
5. *Polymastia* : the presence of accessory brest
6. *Polythelia* : the presence of accessory nipple

   } anywhere along the mammary line
7. *Inverted nipple*: due to failure of elevation of the mammary pit

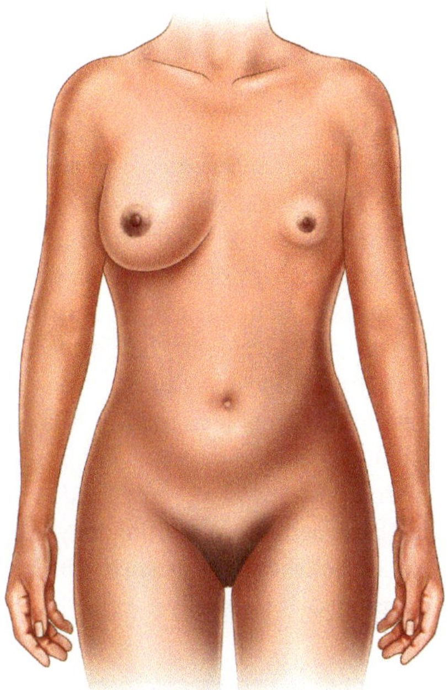

Micromastia (retention of prepubertal state)

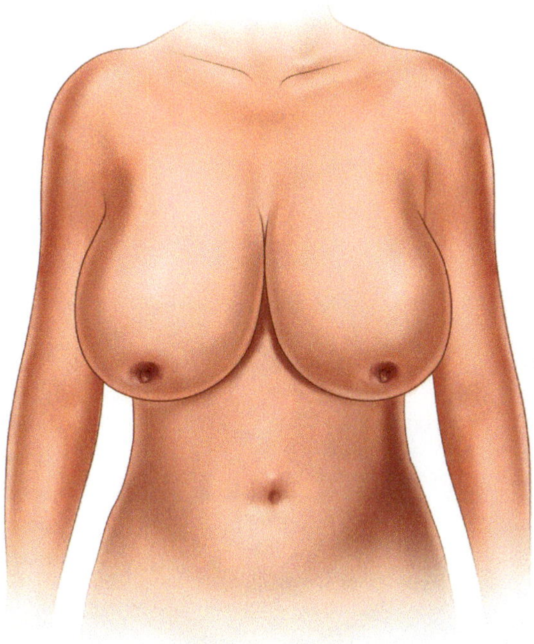

Macromastia abnormally large breast

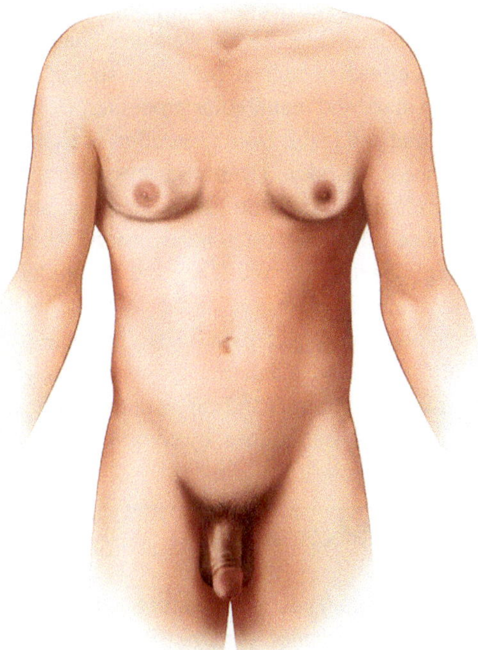

Gynaecomastia a male devloping a female type of breast

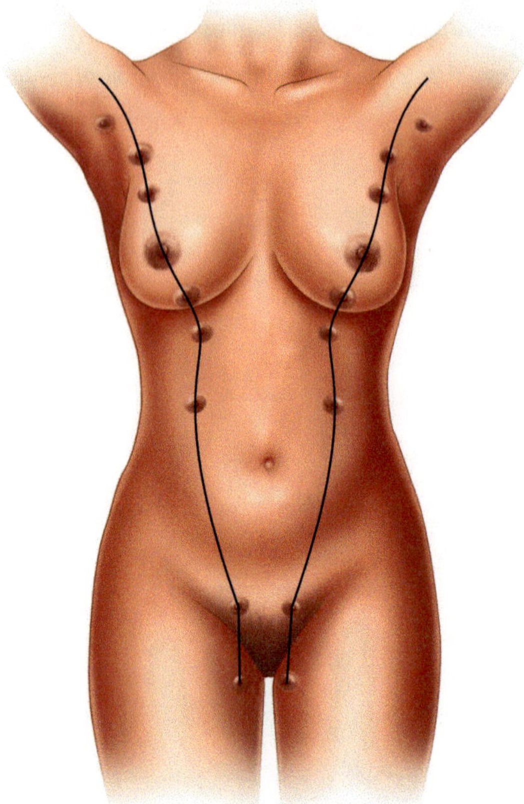

Inverted nipple

Introduction

Congenital malformations of the skin and soft tissues include a huge range of conditions which can occur in any part of the body and range from the extremely common in otherwise fit and well children to more rare conditions often associated with syndromes and chromosomal abnormalities. This chapter will endeavour to cover the main topics in a succinct manner with salient clinical features highlighted.

Cutis Aplasia

This literally means "missing skin" and can occur congenitally in any part of the body, most commonly the scalp when it is termed scalp aplasia. Cutis aplasia or aplasia cutis congenita (ACC) is thought to occur in 1 in 10,000 live births and was first described in 1767 by Cordon. Figure 25.1a shows an example of cutis aplasis affecting the limb only. Figure 25.1b shows an infant with multiple areas of the body affected. Its cause is not well understood, although certain genes have been implicated such as mutations in ribosomal GTPase BMS1 in autosomal dominant ACC, and in certain forms, it is also associated with syndromes such as Patau's syndrome (trisomy 13) and Edwards syndrome (trisomy 18). Membranous ACC usually affects the scalp and is sporadic, whereas non-membranous is often inherited [1]. A link has also been proposed between maternal drug exposure such as methimazole [2], valproic acid and misoprostol in addition to maternal infection such as herpes and varicella. Several other theories of aetiology include an association with neural tube defects, premature rupture of membranes leading to an amniotic band-like syndrome and in scalp aplasia a relation to increased tension on scalp skin at 10–15 weeks' gestation when brain development and growth is most pronounced [3]. Cutis aplasia can not only occur anywhere in the body, but its symptoms also depend on size ranging from less than a centimetre to more than ten. Lesions can also vary in depth from epidermis only to full thickness, and this can influence risk of infection, ulceration and effect on other structures such as hair follicles. There are no specific investigations needed as the diagnosis is a clinical one; however, antenatal AFP levels may be elevated, and often genetic testing is performed postnatally if there is any suggestion of genetic or chromosomal abnormality. Management is usually supportive in the form of dressing and managing any signs of infection promptly. Surgery is only needed in rare cases when lesions are large and where grafting maybe considered. In the case of small lesions not associated with any genetic abnormality, the prognosis is usually excellent.

Dermoid Cysts

Dermoid cysts or external angular dermoid are extremely common and are the most common benign periorbital tumour in children (Fig. 25.2). We know that three embryological structures fuse to form the soft tissues of the face, the frontal, maxillary and mandibular processes. Dermoid cysts are congenital and result from separation of tissue during the development of skin. Dermoid cysts form when sheets of developing skin don't fuse end on and tissue becomes trapped. They are often noted around the eye or at the lateral angle of the eyebrow. They are benign and rarely have any intracranial extension. If the mass is midline or its extent cannot be palpated, then further imaging is indicated such as ultrasound or magnetic resonance imaging. Rupture can result from trauma and can lead to an inflammatory reaction. Treatment is, therefore, by complete surgical excision [4, 5].

Pilomatrixoma

Pilomatrixoma is an uncommon, benign tumour arising in hair follicles hence its name "Pilo". It is also known as a "calcifying epithelioma of Malherbe". Pilomatrixoma is most often diagnosed in young children. It can occur anywhere on the body and is characterised by calcification within the

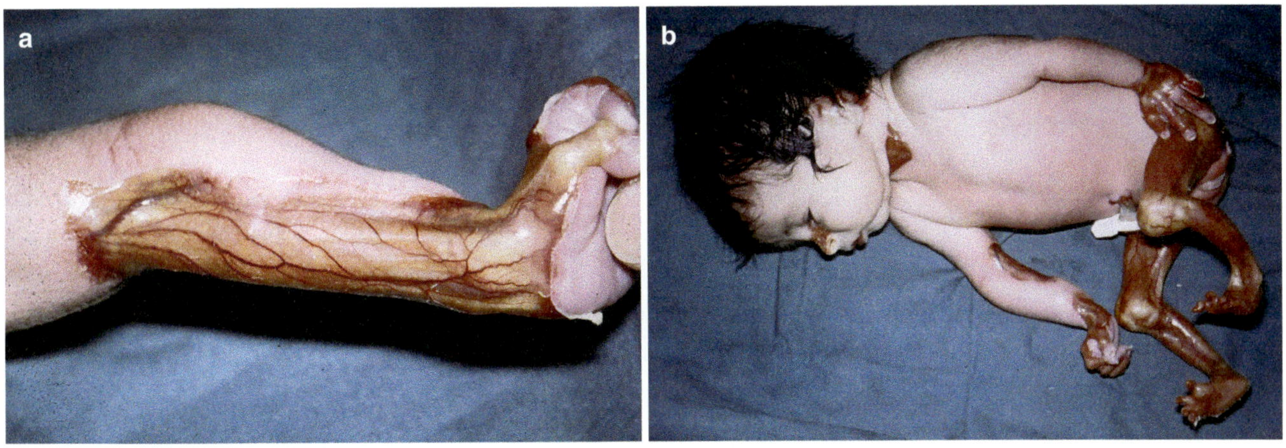

Fig. 25.1 (**a**) Cutis Aplasia of the limb only (**b**) Cutis Aplasia affecting different areas of the body

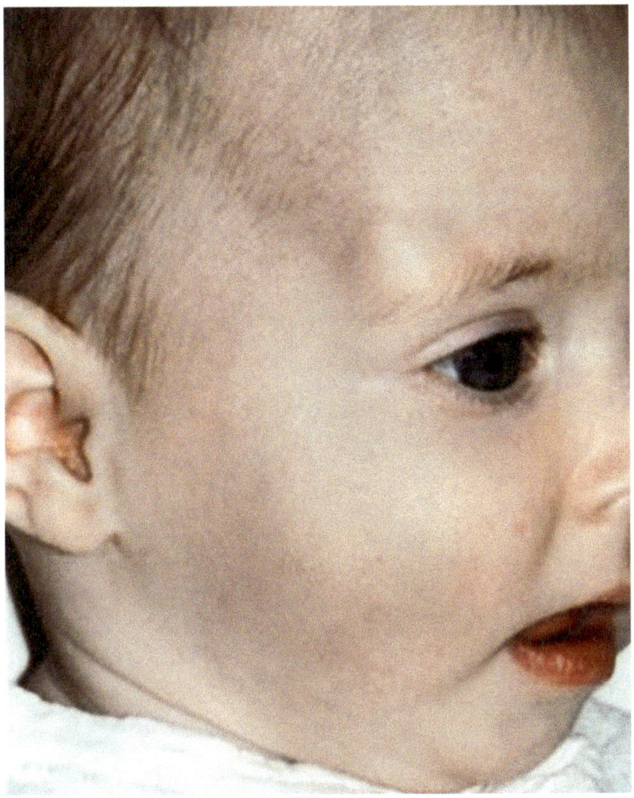

Fig. 25.2 Angular dermoid cyst

lesion, which makes it feel hard and bony, and can give it an angulated shape otherwise known as the "tent". The cause of pilomatrixoma is unknown; however, recently some genetic changes have been found in the affected hair follicles such as overactivity of the proto-oncogene BCL [6, 7].

Microscopically pilomatrixomas comprise anucleate cells otherwise known as "ghost cells", normal healthy squamous cells and multinucleated giant cells. Treatment is by simple excision.

Naevus

A naevus (Latin for birthmark) or often referred to as a "mole" is a common skin lesion caused by aggregations of melanocytes. A simple naevus (Fig. 25.2a) is not congenital but is related to sun exposure, and their number is often hereditary. A simple naevus can occur at any site on the body. They need no treatment unless there is a change in their appearance, they bleed or become raised or they lose their regular contour. This is often referred to as the ABCDE criteria: "A" for asymmetry, "B" for border irregularity, "C" for change in colour, "D" for diameter greater than 6 mm and "E" for evolution [8]. Any of these signs may indicate features of malignant melanoma result in a referral for excision. Children, however, account for only 2% of all cases of malignant melanoma with an incidence of 0.8 per million up to the age of 10 years old [9].

A child may be born with a congenital melanocytic naevus (CMN) which is felt to be a result of abnormal melanoblast development between 5 and 24 weeks' gestation. Although congenital some may be not be fully pigmented or visible until the child is older, and generally CMN are classified as small, medium and large (<1.5 cm, 1.5–19.9 cm and >20 cm). Figure 25.3b shows a large CMN affecting most of the back of the child and extending down to the natal cleft. The larger their size, the more risk there is of them developing into melanoma [10]. As a group, patients with CMN have a lifetime risk of 2.9% of developing malignant melanoma [11]. The congenital nevomelanocytic nevus (CNN), commonly known as the congenital hairy naevus, often has a rough or pebbled appearance and is noted at birth. Like the CMN it is graded by size and also has an increased risk of melanoma. If there is any concern with regard to a pigmented cutaneous lesion in a baby or child, then a paediatric dermatology opinion should be sought (Fig. 25.3b).

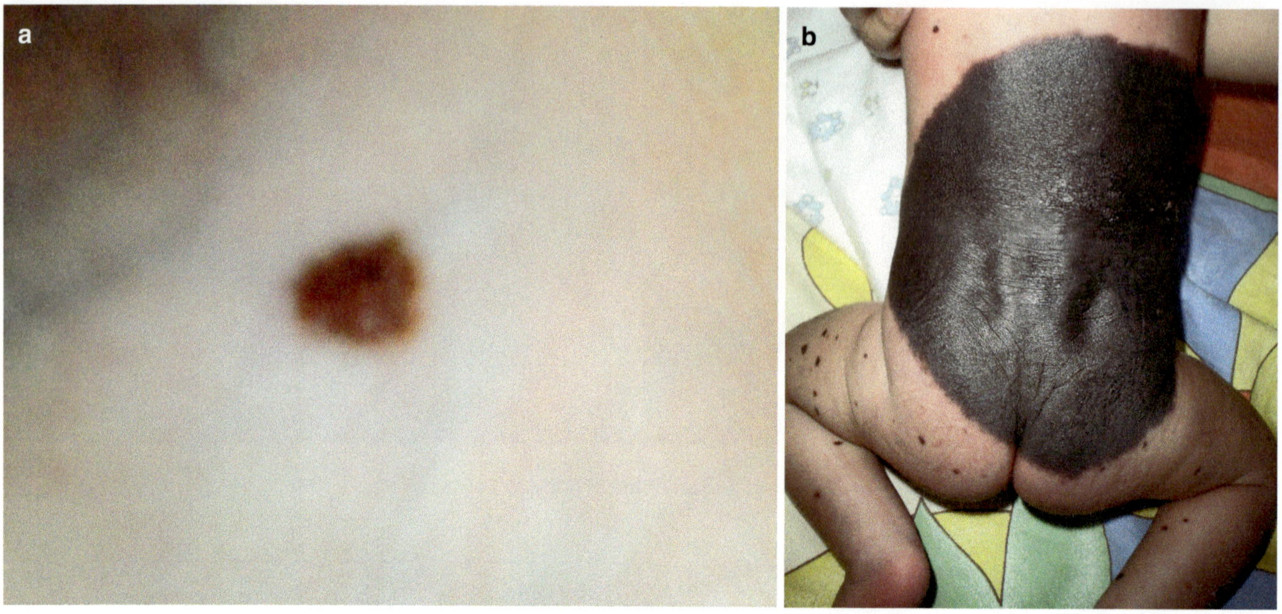

Fig. 25.3 (**a**) Naevus (**b**) a Large Congenital Melanocytic Naevus

A blue naevus or Mongolian blue spot is quite a different entity. It is a skin lesion also known as congenital dermal melanocytosis and dermal melanocytosis. It is benign and flat and presents at birth with an irregular border and often cited in the lumbosacral area or over the buttocks (Fig. 25.4). The most common colour is blue, although th ey can be blue-grey, blue-black or even brown. It results from the entrapment of melanocytes in the lower half to two-thirds of the dermis during their migration from the neural crest to the epidermis during embryonic development. It is more prevalent in certain races and in mixed race children. It normally disappears 3 to 5 years after birth and almost always by puberty. If not correctly diagnosed, it may sometimes be mistaken for a bruise and may inappropriately raise child protection concerns of nonaccidental injury.

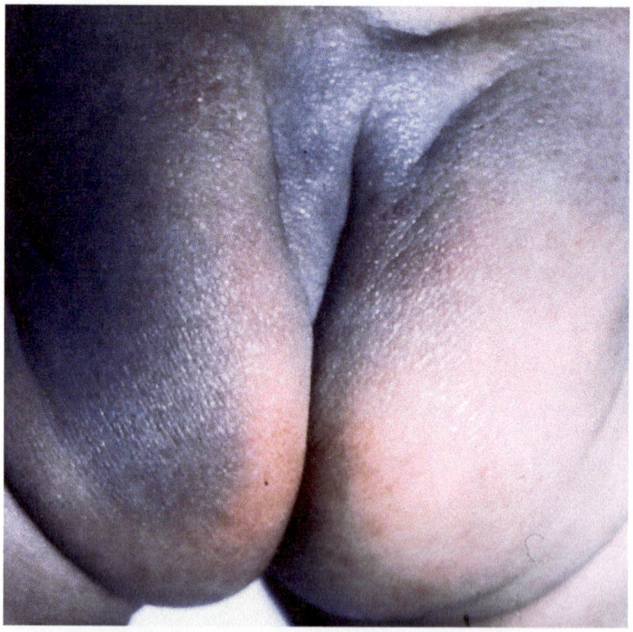

Fig. 25.4 Mongolian blue Spot

Haemangioma

Haemangiomas are very common benign tumours seen in childhood presenting at or just after birth. Figures 25.5a and b show the more common simple haemangiomas. There incidence is thought to be up to 12% in Caucasian children by 1 year of age but less common in those of Asian and African descent. The female to male ratio is about 5:1 [12].

The exact aetiology of haemangiomas is not known; however, it is hoped that through developments in molecular biology looking into expression of factors such as GLUT-1, growth factors and integrins, medical treatments may be developed to prevent the lesions [13]. A recent theory of the aetiology of haemangiomas is that they arise from ectopic placental tissue [14]. There have been many different classifications described with regard to haemangiomas and vascular malformations. Until recently no one system was regarded as the gold standard. This led to difficulty and confusion when trying to interpret data, as well as correctly diagnose lesions and treat them. The Mulliken and Glowacki nomenclature, first introduced in 1982, is now widely accepted [15]. This describes haemangiomas as lesions that characteristically display rapid growth over 3–10 months then have a phase of involution over 5–7 years. They consist of proliferated endothelial cells, in contrast to vascular malformations, which consist of dysplastic vessels with no cellular proliferation. Vascular malformations are biologically inactive; they present at birth, and their growth parallels that of the patient. Haemangiomas can occur anywhere on the body, and symptoms reflect their site. They are commonly found on the head and neck. They are often referred to as a strawberry naevus. The majority are asymptomatic and can be managed expectantly; however, if symptoms do arise or the lesion interferes with the child's development or function such as near their visual field or airway or involution is unlikely due to large size, the investigation and treatment should be sought. Figures 25.5c and d show the same child at different ages with a large facial haemangioma affecting the trigeminal nerve districution on the right side of the face. This is known as Sturge-Weber syndrome. Symptoms very much depend on site, e.g. if affecting the limbs cannot only cause bleeding but also pain and swelling and even leg length discrepancy. Rarely, haemangiomas can be associated with syndromes such as KMS which can present with bleeding and consumptive coagulopathy.

The use of different investigative modalities has changed over the years as technology and imaging has improved, and MRI is now considered the modality of choice. It is very effective at illustrating flow within the lesion and involvement of surrounding structures [16–18]. The management of haemangiomas is still debatable as some studies have contradicted the traditional view that these lesions regress spontaneously. Finn et al. claimed that many haemangiomas do not involute [19]. They found that only 50% regressed by the age of 6 years, and of that group, 38% retained a marked cosmetic deformity. In terms of medical treatment, historically steroids were employed first-line treatment; however, these

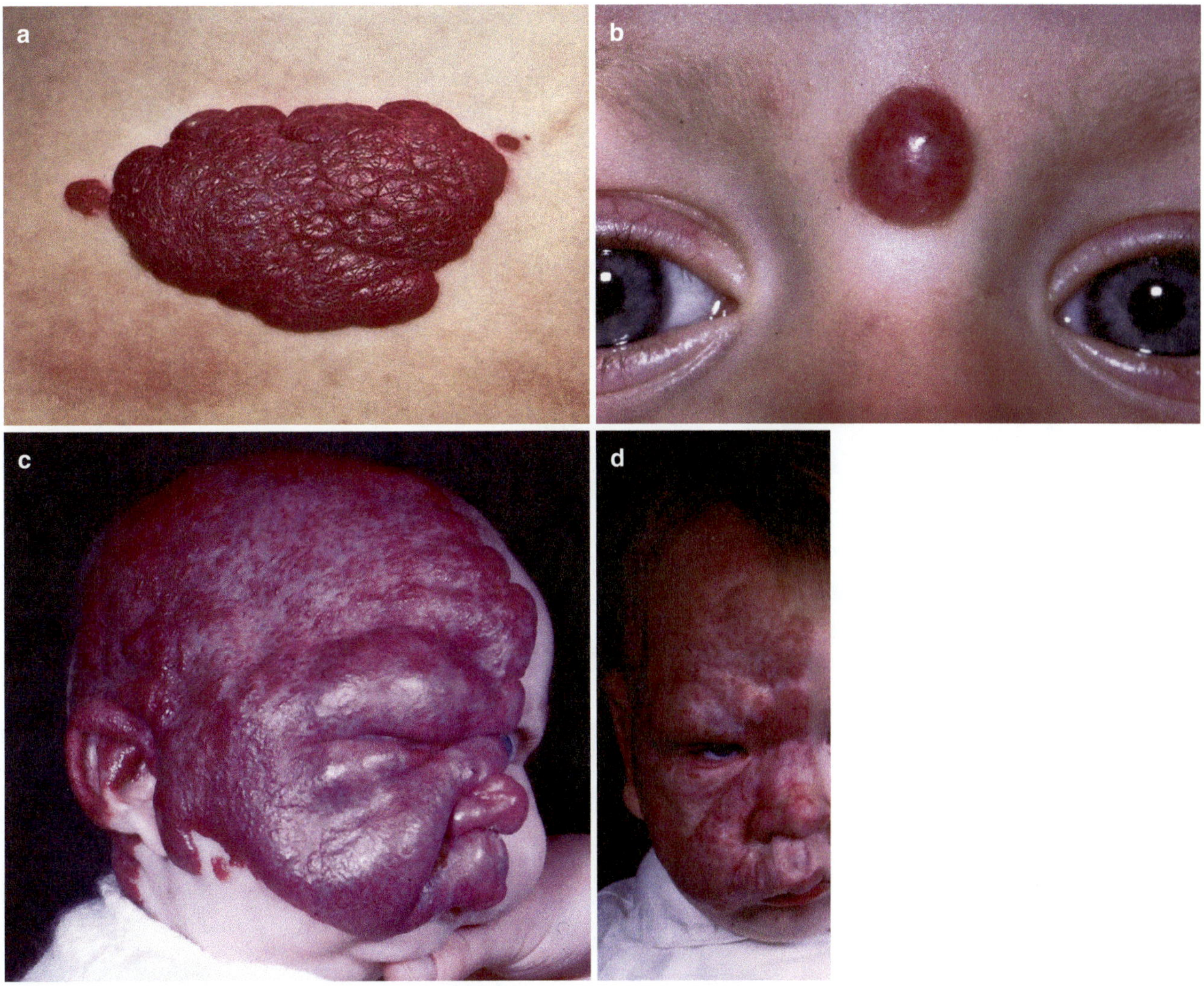

Fig. 25.5 (**a**, **b**) Simple Haemangiomas and (**c**) and (**d**) A child with Sturge-Weber syndrome

had systemic side effects to consider. Interferon was also used; however, this should be reserved for those that are life-threatening and when other measures have failed as there is the irreversible complication of spastic diplegia [20]. Sugarman et al. showed that a topical version of platelet-derived growth factor, teicoplanin, was effective in treating ulcerated haemangiomas [21]. Most recently propranolol has been shown to be effective, and a very recent multicentre, randomised, double-blinded trial has been published that showed that propranolol was effective at a dose of 3 mg per kilogram per day for 6 months with 88% of patients showing improvement by week 5 versus 5% of patients who received a placebo [22]. Surgery is only considered for lesions that do not respond or are not amenable to medical treatment.

Klippel–Trénaunay Syndrome

Klippel–Trénaunay syndrome (KTS) is a rare genetic syndrome in which blood vessels and lymphatics develop abnormally. It affects 1 in 20,000–40,000 children and is named after the two French doctors who described the condition in 1900. It comprises the triad of port wine stain, venous and/or lymphatic malformations and limb hypertrophy. It is sometimes referred to as angioosteohypertrophy syndrome and hemangiectatic hypertrophy. KTS can either affect blood vessels, lymph vessels or both. The condition most commonly presents with a mixture of the two. Impedance to blood flow causes these children pain, swelling, inflammations, and in some cases, even ulceration and infection. Those with venous involvement experience increased pain and complications. Those with large AVMs are at risk of thrombus formation, pulmonary embolism and high-output heart failure. KTS has a large spectrum of severity and as such management is individualised to the patient and focuses on controlling symptoms. Sclerotherapy and more recently ultrasound-guided foam sclerotherapy is an option as is compression therapies [23, 24] (Fig. 25.6).

In terms of surgical treatment options, debulking has been the most widely used but is only utilised as a last resort. The Mayo Clinic has reported the largest experience in managing KTS with major surgery. In 39 years at Mayo Clinic, the surgery team evaluated 252 consecutive cases of KTS, of which only 145 (57.5%) could be treated by primary surgery [25]. The immediate success rate for treating varicose veins was only 40%; excision of vascular malformation was possible in 60%, debulking operations in 65% and correction of bone deformity and limb length correction in 90%. All the procedures demonstrated high recurrence rate in the follow-up. The Mayo Clinic studies demonstrate that primary surgical management of KTS has limitations, and non-surgical approaches need to be developed in order to offer a better quality of life for these patients. Major surgery including amputation and debulking surgery does not seem to offer any benefit on a long-term basis.

Lymphangiomatosis

Lymphangiomatosis is a very broad term encompassing all abnormalities of the lymphatic system, i.e. tumours, cysts or other aberrations. Any lymphatic malformations are thought to be the result of abnormal lymphatic development occurring prior to the twentieth week of gestation.

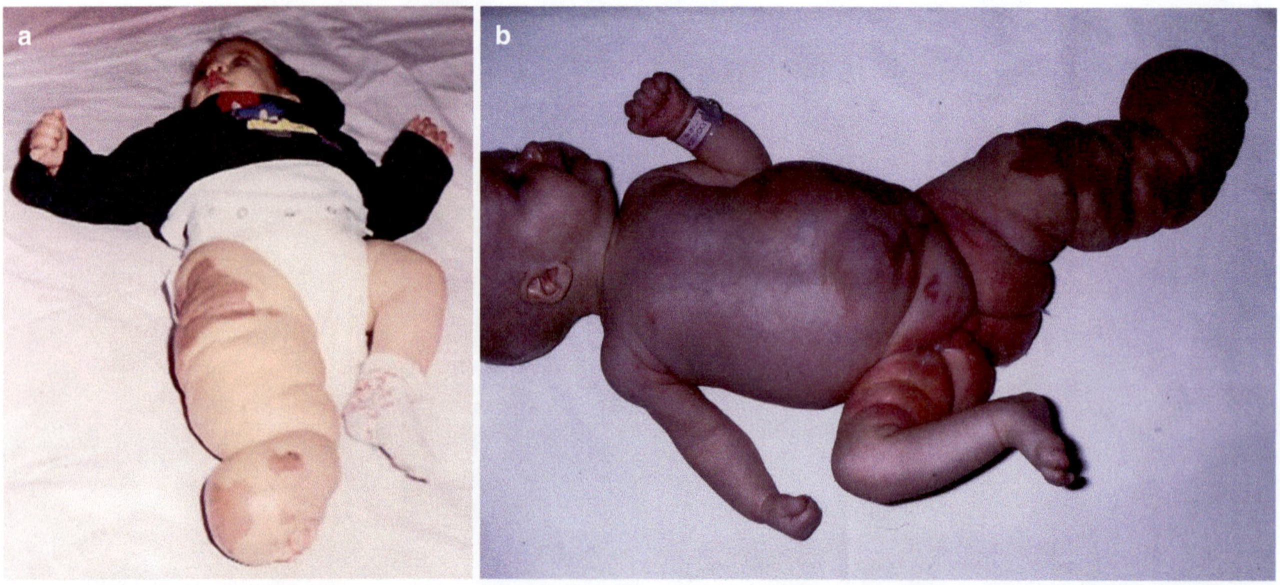

Fig. 25.6 (**a**) Klipple Trenaunay Syndrome, (**b**) Klipple Trenaunay Syndrome

Lymphangiomatosis is a condition characterised by cystic swellings of the lymphatic system which has not developed properly [26, 27]. In around three-quarters of cases, more than one part of the body is affected. It usually presents in childhood with symptoms related to pressure effect on the adjacent structures such as the lung or bone. The most common site is the skeletal and respiratory systems, and it is linked to Gorham's disease [28]. In its most severe form, it affects the lungs of very young children [29, 30]. It cannot only lead to chylothorax but also chylopericardium or in the abdomen chylous ascites. Investigations such as chest X-ray, CT scan, MRI, ultrasound, lymphangiography, bone scan and bronchoscopy all can be useful; however, biopsy remains the gold standard. As in KTS treatment is directed at managing symptoms rather than cure, and a multidisciplinary approach is generally necessary for optimal care. Surgery is again generally reserved as a last resort.

A cystic hygroma is quite a different entity and is a form of lymphangioma caused by benign aberrant development of lymphatics (Figs. 25.7a and b). Traditionally, the De Serres classification attempts to categorise lymphangiomas according to laterality and whether they are supra- or infrahyoid. They may also be classified according to their cystic element as either macrocystic, microcystic or mixed. Ultrasound can help to classify the appearance of the cystic elements (Fig. 25.7c). Patients may present at birth with airway obstruction which may be anticipated antenatally. Clinical features include a soft, doughy lump which may transilluminate. About 10–15% resolve or involute spontaneously probably after an infective episode.

It can be associated with Turners or Noonan's syndrome. Imaging such as ultrasound, CT or MRI can identify the extent of the lesion. Treatment is by surgery which may need to be at delivery (EXIT procedure). Sclerotherapy can be useful with agents such as OK 432 (Picibanil); however, this is now difficult to source and is not suitable for penicillin-allergic patients. Bleomycin is another option; however, pulmonary fibrosis is a potential side effect. Doxycycline and ethanol are alternatives, and more recently sildenafil has promising results and maybe the drug of choice in the future.

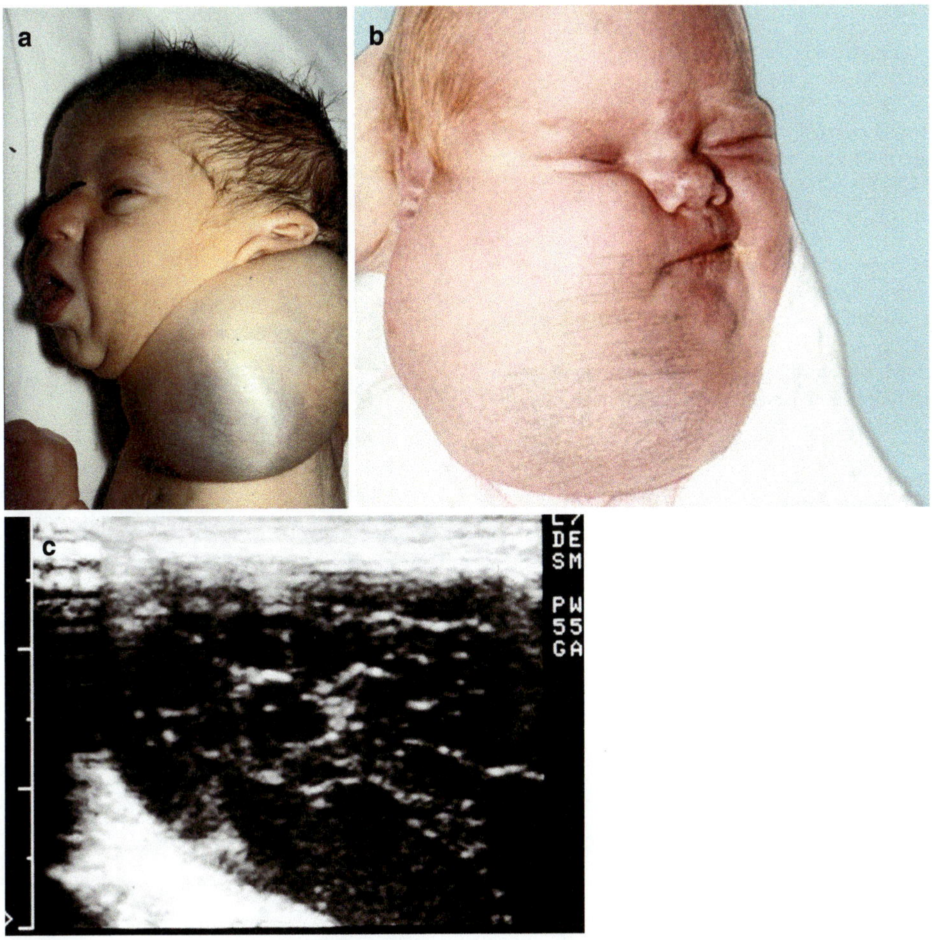

Fig. 25.7 (**a**) A neonate with a cystic hygroma affecting the neck only (**b**) A child with a classic cystic hygroma extending from the neck in the face (**c**) The ultrasound appearance of a cystic hygroma

References

1. Baselga E, Torrelo A, Drolet BA, Zambrano A, Alomar A, Esterly NB. Familial non membranous aplasia cutis of the scalp. Pediiatr Dernatol. 2005;22(3):213–7.
2. Karg E, Bereg E, Gaspar L, Katona M, Turi S. Aplasia cutis congenita after methimazole exposure in utero. Pediatr Dermatol. 2004;21(4):491–4.
3. NA V, Heymans O, Deleuze JP, Fabre G, Vranckx JJ, Van den hof B. Abdominal aplasia cutis congenita: case report and review of the literature. J Pediatr Surg. 2004;39(2):237–9.
4. Cozzi DA, Mele E, d'Ambrosio G, et al. The eyelid crease approach to angular dermoid cysts in pediatric general surgery. J Pediatr Surg. 2008;43(8):1502–6.
5. Sreetharan V, Kangesu L, Sommerlad BC. Atypical congenital dermoids of the face: a 25-year experience. J Plast Reconstr Aesthet Surg. 2007;60(9):1025–9.
6. Levy J, Ilsar M, Deckel Y, Maly A, Anteby I, Pe'er J. Eyelid pilomatrixoma: a description of 16 cases and a review of the literature. Surv Ophthalmol. 2008;53(5):526–35. https://doi.org/10.1016/j.survophthal.2008.06.007.
7. Hassanein AM, Glanz SM, Kessler HP, Eskin TA, Liu C. Beta-Catenin is expressed aberrantly in tumors expressing shadow cells. Pilomatricoma, craniopharyngioma, and calcifying odontogenic cyst. Am J Clin Pathol. 2003;120(5):732–6.
8. Abbasi NR, Shaw HM, Rigel DS, et al. Early diagnosis of cutaneous melanoma: revisiting the ABCD criteria. JAMA. 2004;292:2771–6.
9. Tannous ZS, Mihm MC Jr, Sober AJ, Duncan LM. Congenital melanocytic nevi: clinical and histopathologic features, risk of melanoma, and clinical management. J Am Acad Dermatol. 2005;52:197–203.
10. Bittencourt FV, Marghoob AA, Kopf AW, Koenig KL, Bart RS. Large congenital melanocytic nevi and the risk for development of malignant melanoma and neurocutaneous melanocytosis. Pediatrics. 2000;106:736–41.
11. Pappo AS. Melanoma in children and adolescents. Eur J Cancer. 2003;39:2651–61.
12. Macfie CC, et al. Diagnosis of vascular skin lesions in children: an audit and review. Ped Dermatol. 2008;25(1):7–12.
13. Metry D. Update on hemangiomas of infancy. Curr Opin Pediatr. 2004;16(4):373–7.
14. North PE, Waner M, Mizeracki A, et al. A unique microvascular phenotype shared by juvenile hemangiomas and human placenta. Arch Dermatol. 2001;137:559–70.
15. Mulliken JB, Glowacki J. Hemangiomas and vascular malformations in infants and children: a classification based on endothelial characteristics. Plast Reconstr Surg. 1982;69:412–22.
16. Cohen JM, Weinreb JC, Redman HC. Arteriovenous malformations of the extremities. MR imaging. Radiology. 1986;158:475–9.
17. Breugem CC, Maas M, van der Horst CM. Magnetic resonance imaging findings of vascular malformations of the lower extremity. Plast Reconstr Surg. 2001;108:878–84.
18. Breugem CC, Maas M, Reekers JA, van der Horst CM. Use of magnetic resonance imaging for the evaluation of vascular malformations of the lower extremity. Plast Reconstr Surg. 2001;108:870–7.
19. Finn MC, Glowacki J, Mulliken JB. Congenital vascular lesions: clinical application of a new classification. J Pediatr Surg. 1983;18:894–900.
20. Steven M, Kumaran N, Carachi R, Desai A, Bennet G. Haemangiomas and vascular malformations of the limb in children. Pediatr Surg Int. 2007;23(6):565–9.
21. Sugarman JL, Mauro TM, Frieden IJ. Treatment of an ulcerated hemangioma with recombinant platelet-derived growth factor. Arch Dermatol. 2002;138:314–6.
22. Léauté-Labrèze C, Hoeger P, Mazereeuw-Hautier J, et al. A randomized, controlled trial of oral propranolol in infantile hemangioma. N Engl J Med. 2015;372:735–46.
23. Cabrera J, Cabrera J Jr, Garcia-Olmedo MA, Redondo P. Treatment of venous malformations with sclerosant in microfoam form. Arch Dermatol. 2003;139(11):1409–16. https://doi.org/10.1001/archderm.139.11.1409.PMID14623700.
24. McDonagh B, Sorenson S, Cohen A, Eaton T, Huntley DE, La Baer S, Campbell K, Guptan RC. Management of venous malformations in Klippel–Trenaunay syndrome with ultrasound-guided foam sclerotherapy. Phlebology. 2005;20(2):63–81. https://doi.org/10.1258/0268355054069188.
25. Jacob AG, Driscoll DJ, Shaughnessy WJ, Stanson AW, Clay RP, Gloviczki P. Klippel-Trenaunay syndrome: spectrum and management. Mayo Clin Proc. 1998;73(1):28–36. https://doi.org/10.4065/73.1.28. PMID 9443675
26. Faul JL, Berry GJ, Colby TV, Ruoss SJ, Walter MB, Rosen GD, Raffin TA. Thoracic lymphangiomas, lymphangiectasis, lymphangiomatosis, and lymphatic dysplasia syndrome. Am J Respir Crit Care Med. March 2000;161(3):1037–46.
27. Tazelaar HD, Kerr D, Yousem SA, Saldana MJ, Langston C, Colby TV. Diffuse pulmonary lymphangiomatosis. Hum Pathol. 1993;24(12):1313–22.
28. Duffy B, Manon R, Patel R, Welsh JS, et al. A case of Gorham's disease with chylothorax treated curatively with radiation therapy. Clin Med Res. 2005;3:83.
29. Huaringa AJ, Chittari LM, Herzog CE, Fanning CV, Haro M, Eftekhari F. Pleuro-pulmonary lymphangiomatosis: malignant behavior of a benign disease. Internet J Pulm Med. 2005;5(2)
30. Aviv RI, McHugh K, Hunt J. Angiomatosis of bone and soft tissue: a spectrum of disease from diffuse lymphangiomatosis to vanishing bone disease in young patients. Clin Radiol. 2001;56(3):184–90.

26 The Brain and Central Nervous System

Roddy O'Kane and Thomas Begg

Development of the Central Nervous System

The CNS develops from the **neural tube which arises from the **ectoderm** as follows*:

- It appears in the third week as a thickening of the ectoderm known as the ***neural plate*** extending from the prochordal plate in front to the caudal end of the embryonic disc behind.
- The neural plate will form the ***neural groove*** which has two elevated edges called the neural folds.
- The neural folds fuse together transforming the neural groove into ***neural tube*** lying beneath the ectoderm in the median plane.
- The fusion of the folds is absent at the ant- and post- ends of the tube leaving two openings on the ectoderm called the ant- and the post-neuropores.
- Later on, the ant-neuropore closes on the 23rd day while the post-neuropore closes on the 25th day.

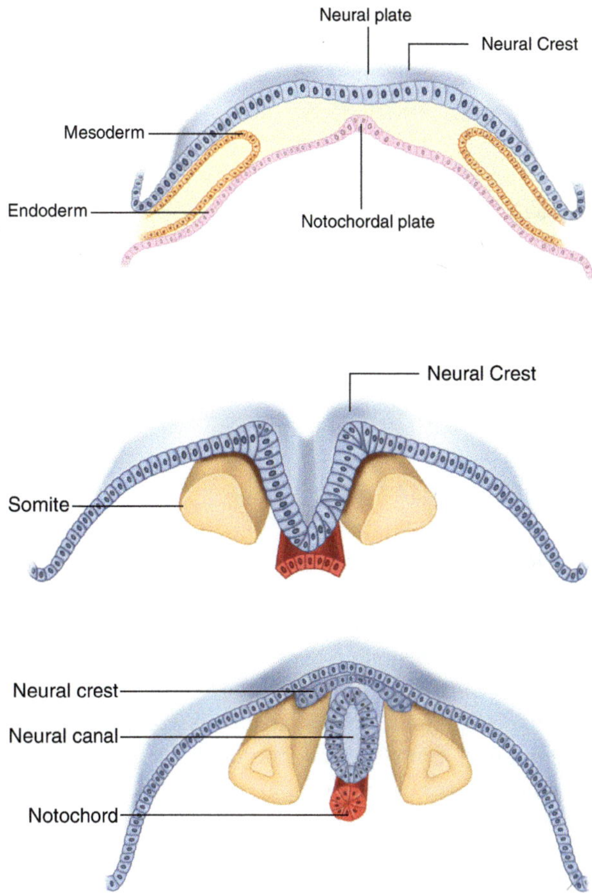

R. O'Kane (✉)
Queen Elizabeth University Hospital, Glasgow, UK
e-mail: roddy.okane@ggc.scot.nhs.uk

T. Begg
Department of Neurosurgery, Heinrich Heine University of Dusseldorf, Dusseldorf, Germany

R. Carachi, S. H. E. Doss (eds.), *Clinical Embryology*, https://doi.org/10.1007/978-3-319-26158-4_26

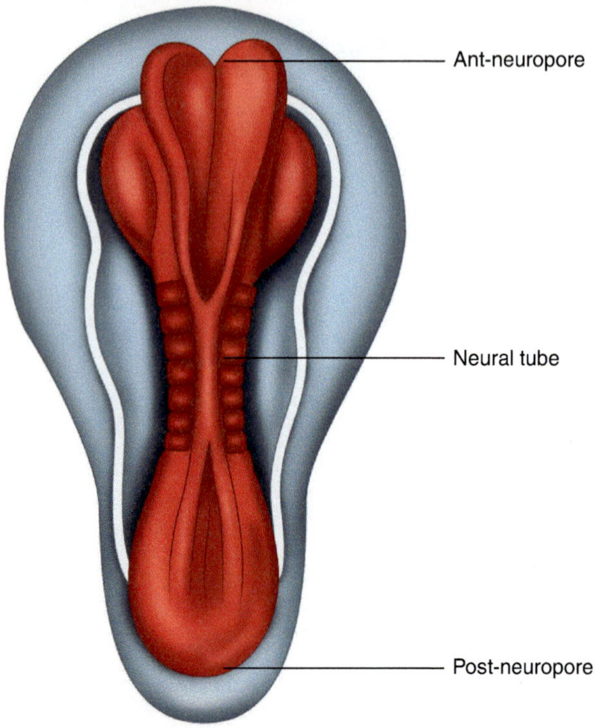

The closed neural tube develops into the spinal cord and the brain.*

The Neural Crest

Formation*:

- It arises as a strip of ectodermal cells situated along the lat-edge of the neural groove.
- As the two edges of the neural groove fuse together forming the neural tube, the two neural crests separate as two longitudinal cords that migrate ventrally to lie one on each side of the neural tube.

  Derivatives*:** The neural crest becomes segmented into masses which give the following derivatives:

1. Sensory ganglia of the cranial nerves (5, 7, 9 and 10).
2. Autonomic ganglia (both sympathetic and parasympathetic).
3. Dorsal root ganglia of all spinal nerves.
4. Neurilemmal (Schwann) cells of peripheral nerves.
5. The medulla of the suprarenal gland (chromaffin cells).
6. Melanoblasts of the skin which produce melanin pigment.
7. Arachnoid and pia mater which are ectodermal (but not the dura mater which is mesodermal).

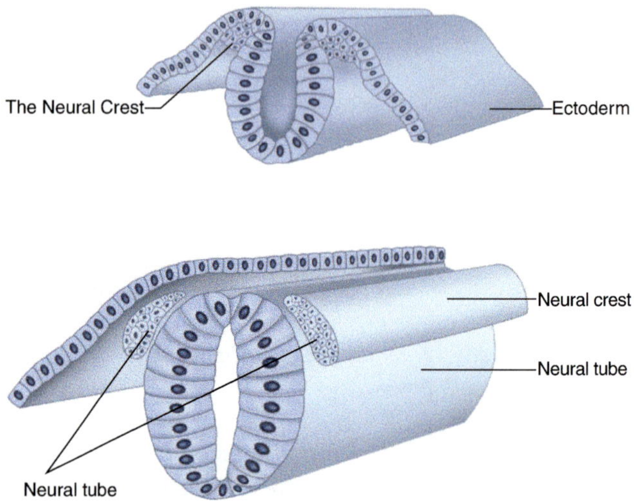

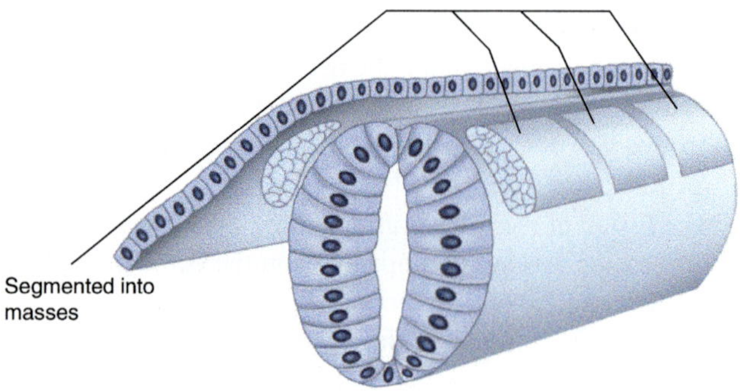

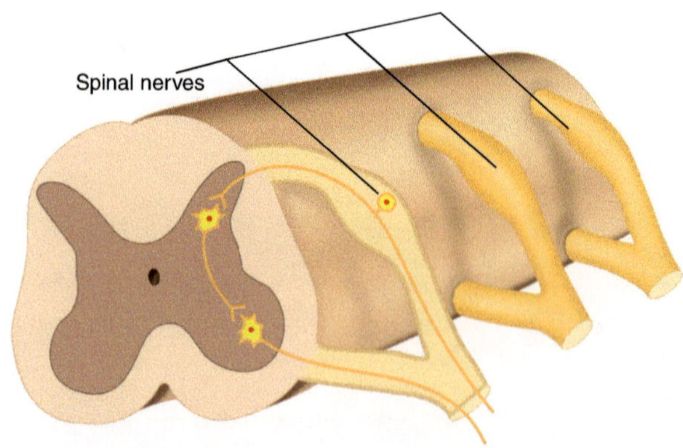

Development of the Spinal Cord

The spinal cord develops from the **caudal part** of the **neural** tube as follows:

1. At first, the neural tube is formed of one layer of simple columnar epithelium surrounding an oval central canal.
2. Later on, this layer proliferates and the neural tube becomes formed of **two thick lat-walls** connected by a thin roof plate and a thin floor plate.
3. The **lat-walls** differentiate into three layers:
   (a) ***Inner ependymal layer*:** the cells of which give rise to
      - The *ependymal cells* lining the central canal of the spinal cord.
      - The primitive nerve cells (neuroblasts) which migrate to the mantle layer.

   (b) ***Middle mantle layer*:** formed of nerve cells (neuroblasts) and neuroglial cells (spongioblasts) which form the **grey mater**
   (c) ***Outer marginal layer*:** formed of nerve fibres (ascending and descending tracts) which constitute the **white mater**
4. **A groove called *sulcus limitans*** appears on the inner surface of the lat-wall on either side dividing it into
   (a) An ***alar plate*** posteriorly which contains sensory cells and forms the post-horn of the spinal cord.
   (b) A ***basal plate*** anteriorly contains motor cells and forms the ant-horn of the spinal cord.
5. Enlargements are formed in the cervical and lumbar regions of the spinal cord and the central canal becomes marrow.
6. ***Growth of the spinal cord*:**
   (a) Till the third month the spinal cord fills the vertebral canal completely.
   (b) The vertebral column then grows at a faster rate than the spinal cord producing the following changes:
      - The caudal end of the cord is overstretched forming the filum terminale.
      - The lower end of the cord shifts upwards to lie at the level of L3 at the time of birth.

(N.B) In the adult, the spinal cord ends at the disc between L1 and L2.

7. *Myelination of the nerve fibres* in the spinal cord begins in the fourth month of intrauterine life and is completed by the end of the first year of postnatal life.
8. *Development of the spinal meninges*:
   (a) The dura mater: develops from the mesoderm of the sclerotomes which form the vertebral column.
   (b) The arachnoid and pia mater: develop from the neural crest (**ectodermal in origin**).

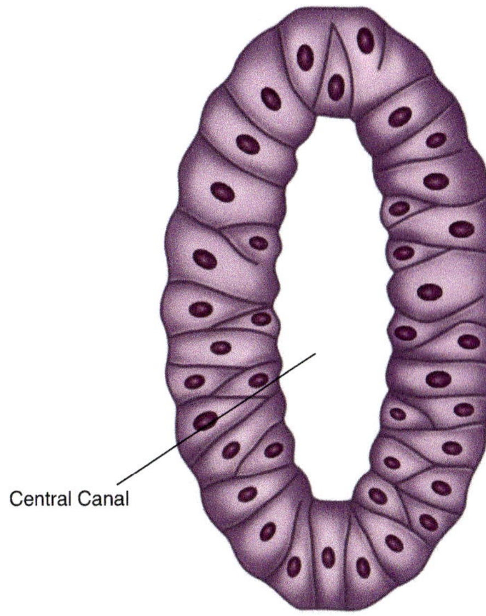

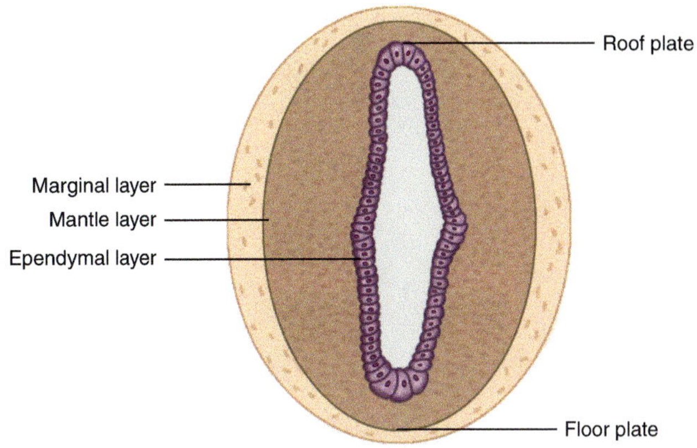

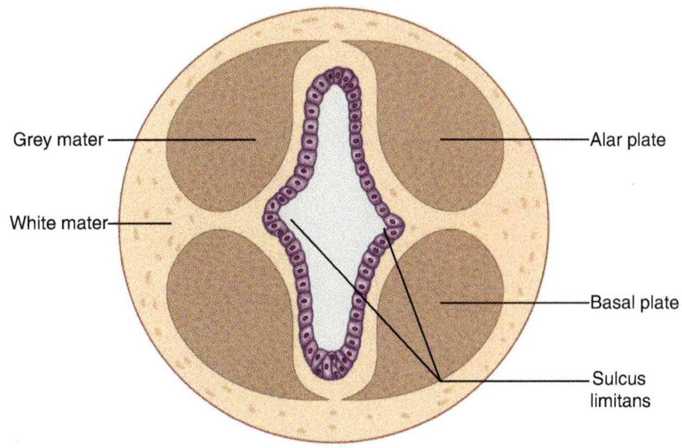

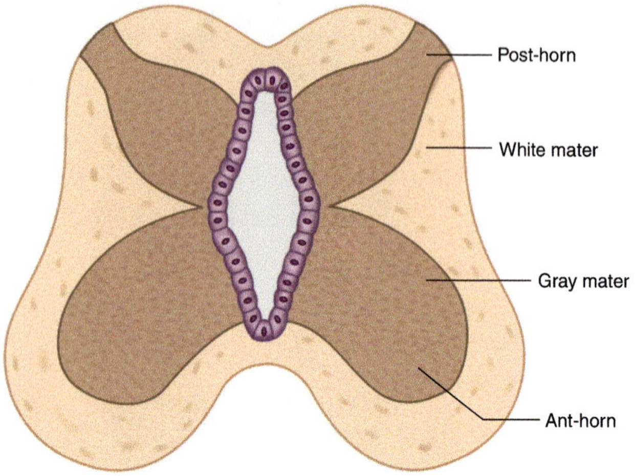

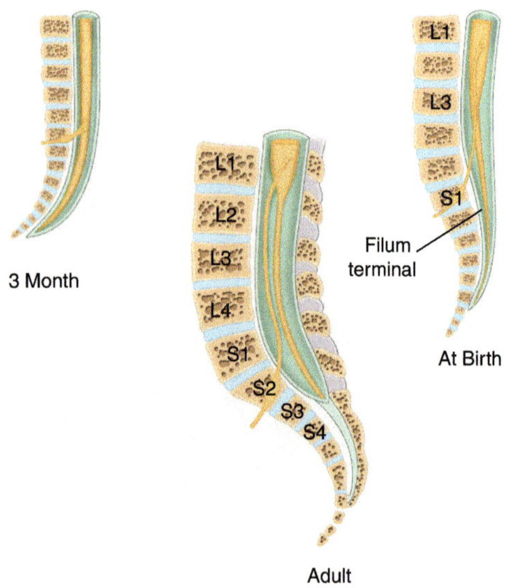

Congenital Anomalies of the Spinal Cord

1. ***Spina bifida occulta*:** due to failure of fusion of the dorsal parts of one of the vertebrae around the spinal cord (which is normal). This condition occurs commonly in the lumbosacral region and the affected site is covered by hairy skin.
2. ***Meningocele*:** due to failure of fusion of the dorsal parts of two or three vertebrae. In this anomaly, the meninges bulge through the defect and the condition is accompanied by some neurological symptoms.
3. ***Meningomyelocele*:** like the previous condition but here the spinal cord bulges through the defect. This anomaly is accompanied by severe neurological manifestations.
4. ***Myelocele*:** due to failure of closure of the neural tube and the affected part of the spinal cord remains exposed to the surface through the defect in the vertebral canal. It is the most serious anomaly.

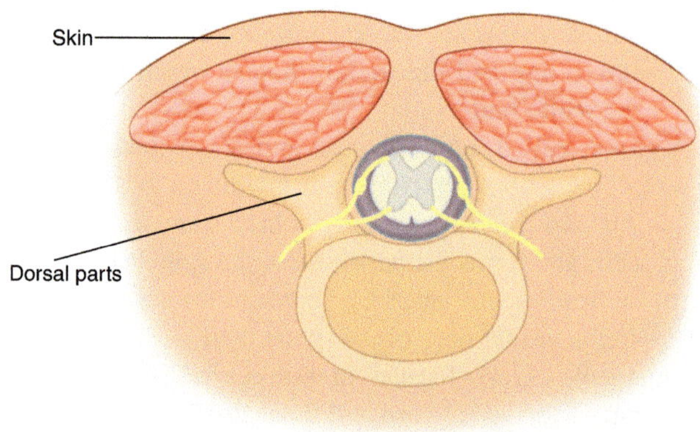

Spina bifida occulta

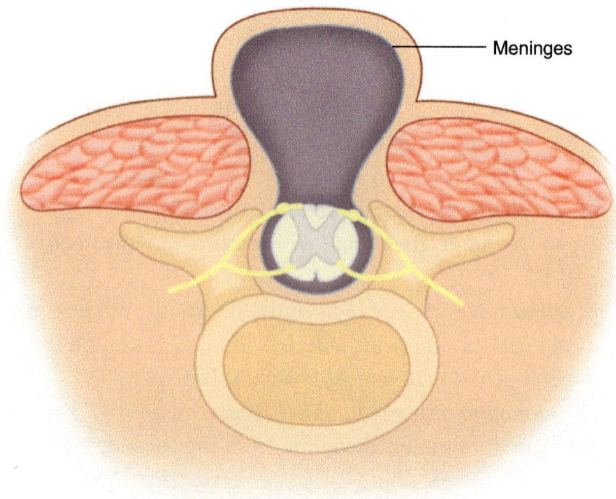

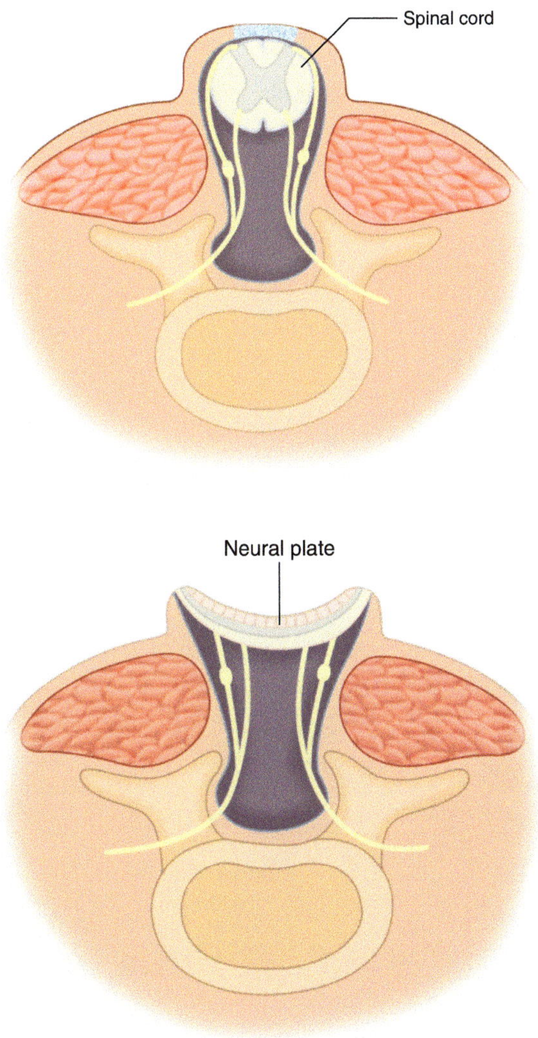

Development of the Brain

* The brain develops from the cranial end of the neural tube as follows:

1. The cranial end of the neural tube expands to form the brain swelling.
2. Two constrictions appear in the brain swelling, dividing it into three parts called brain vesicles:
   (a) Forebrain or prosencephalon.
   (b) Midbrain or mesencephalon.
   (c) Hindbrain or rhombencephalon.
3. The three brain vesicles differentiate as follows:
   (a) **The forebrain:** gives two optic vesicles (the future eyes) and then divides into a median part called the diencephalon
      - Two lat-diverticula called telencephalic vesicles (the future cerebral hemispheres)
   (b) **The midbrain:** remains undivided.
   (c) **The hindbrain:** gives rise to the following derivatives:
      - Metencephalon which forms the pons and cerebellum
      - Myelencephalon which forms the medulla oblongata

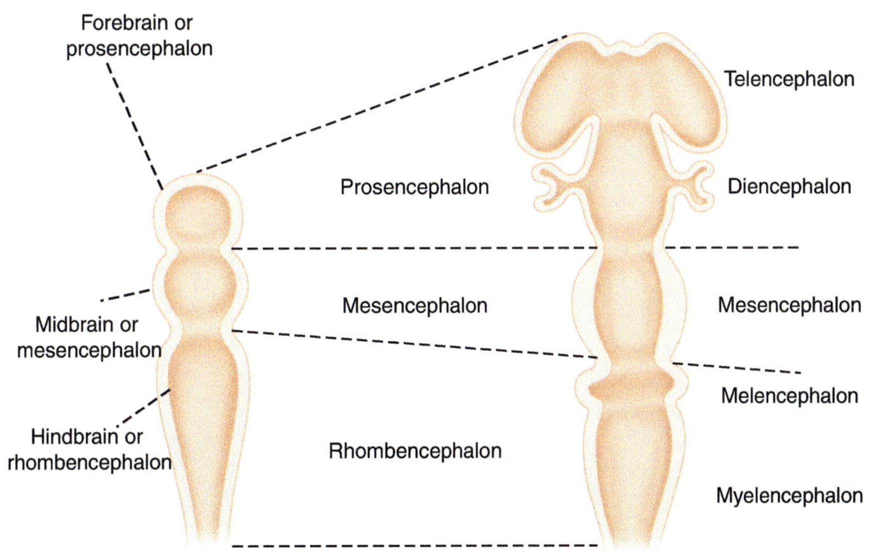

Development of the Brainstem

*As in the development of the spinal cord, the lateral walls of the brainstem are connected by a thin roof plate and a thin floor plate and will have a ventral basal lamina (containing motor nuclei) and a dorsal alar lamina (containing sensory nuclei). Each wall shows a sulcus limitans internally, separating the alar lamina from the basal lamina.

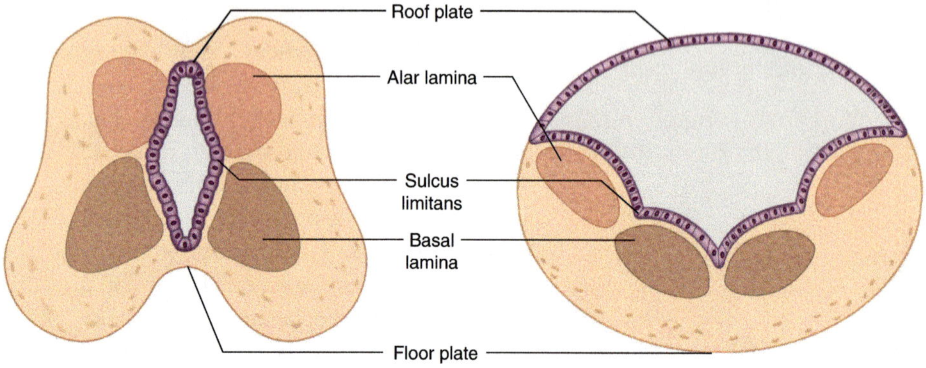

***Differentiation of the basal and alar laminae into columns of nuclei*:**

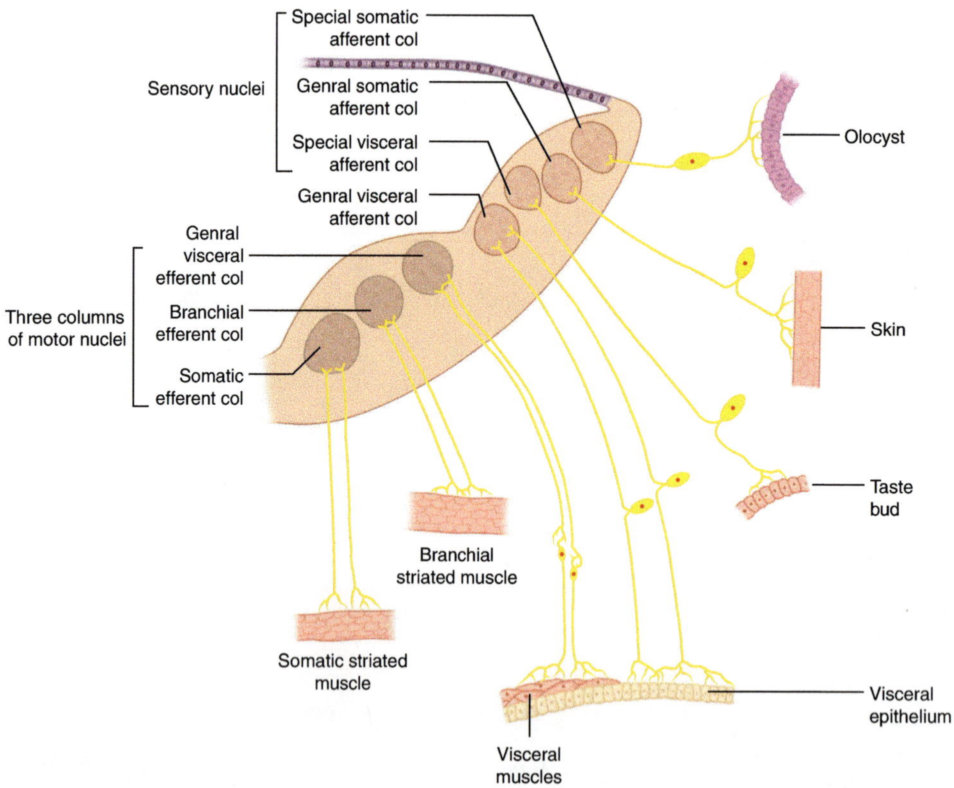

- The *alar lamina* differentiates into four columns of sensory nuclei.
- The *basal lamina* differentiates into three columns of motor nuclei.

1. ***Columns of the basal lamina***:
   (a) ***Somatic efferent column*** (most medial):
   - It *lies* close to the middle line, in line with the ant-horn cells of the spinal cord.
   - Its *efferent fibres* supply somatic striated muscles (i.e. derived from the somites).
   - It *differentiates into* the following nuclei
     - –12th n nucleus in medulla.
     - –6th n nucleus in pons.
     - 3rd and 4th nn nuclei in the midbrain.

   (b) ***Special visceral (branchial) efferent column***:
   - It lies just lateral to the somatic efferent column.
   - Its efferent fibres supply the muscles derived from the branchial (pharyngeal) arches.
   - It differentiates into the following nuclei: 5th & 7th nerve motor nuclei: in the pons.
   - 9th, 10th and 11th nerve motor n »: in the medulla (which join together forming nucleus ambiguus)

   (c) ***General visceral efferent column***:
   - *Lies* just lat- to the special visceral efferent column, in line with autonomic lat-horn grey matter of the spinal cord.
   - Its *efferent fibres* are parasympathetic supplying visceral smooth muscles and glands.
   - It *differentiates into* the following nuclei:
     - Dorsal nucleus of vagus: in the medulla
     - Inf-salivary nucleus (of glossopharyngeal n.): in the medulla
     - Sup-salivary nucleus (of facial n.): in the pons
     - Edinger–Westphal nucleus (of oculomotor n.) in the midbrain
2. ***The columns of the alar lamina***:
   The alar lamina forms four sensory (afferent) columns as follows:

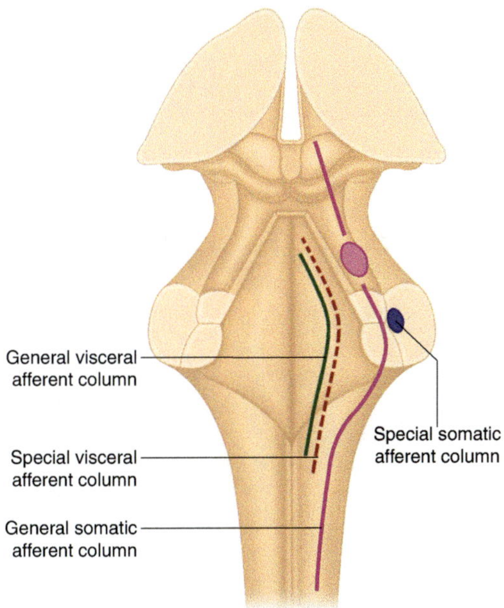

   (a) General visceral afferent column:
   - It is the most med-of the sensory columns.
   - It receives afferent sensory fibres from the viscera.
   - It is represented by the sensory component of the dorsal vagal nucleus.

   (b) Special visceral afferent column:
   - It lies just lat-to the general visceral afferent column.
   - It receives taste sensation from the tongue and epiglottis.
   - It is represented by the nucleus of tractus solitarius.

   (c) General somatic afferent column:
   - Lies lat-to the special visceral afferent column.
   - It receives afferent sensory fibres from the skin of the face and scalp + proprioceptive sensation from the muscles of mastication.
   - This column differentiates into the three sensory nuclei of trigeminal n → ↗ spinal. → main sensory. ↘ mesencephalic.

   (d) Special somatic afferent column:
   - It is the most lat-column of the alar lamina.
   - It receives afferent auditory and vestibular fibres of the vestibulocochlear n.
   - It is represented by the vestibular and cochlear nuclei.

****Migration of nuclei (neurobiotaxis):***

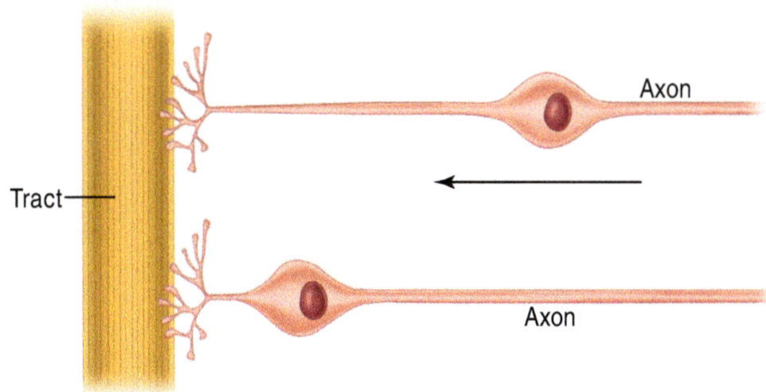

1. Some of the motor nuclei of the cranial nerves migrate from their original position towards the chief fibre tracts from which it receives maximal impulses. This is called neurobiotaxis and is manifested by the migration of the facial motor nucleus and the vagus motor nucleus (nucleus ambiguus) resulting in bending of the axons emerging from these nuclei.
2. Some of the nuclei of the dorsal alar lamina (mostly extrapyramidal) migrate ventrally. Examples of these nuclei are the following:
   (a) The olivary nucleus in the medulla
   (b) The pontine nuclei in the pons
   (c) The red nucleus and substantia nigra in the midbrain

Development of the Lumen of the Brainstem

1. ***Formation of the fourth ventricle*:**
   (a) In the rhombencephalon, the alar laminae of both sides move away from each other (as in the way of opening a book) leading to widening of the lumen, thus forming the fourth ventricle.
   (b) The thin roof plate becomes stretched and rhomboidal in shape, extending laterally to form the lat-recess of the fourth ventricle on each side.
   (c) The vascular mesenchyme lying in contact with the outer surface of the roof plate forms the pia mater and the two layers form the tela choroidea which sends vascular projections into the cavity of the fourth ventricle to form the ***choroid plexus*.**
   (d) Local resorption of the roof plate leads to the formation of a median foramen (of **Magendie**) and two lateral foramina (of **Luschka**)
2. ***Formation of the cerebral aqueduct:***
   The lumen of the mesencephalon (midbrain) becomes much reduced (due to thickening in the walls) and is transformed into a central canal called the cerebral aqueduct (of sylvius).

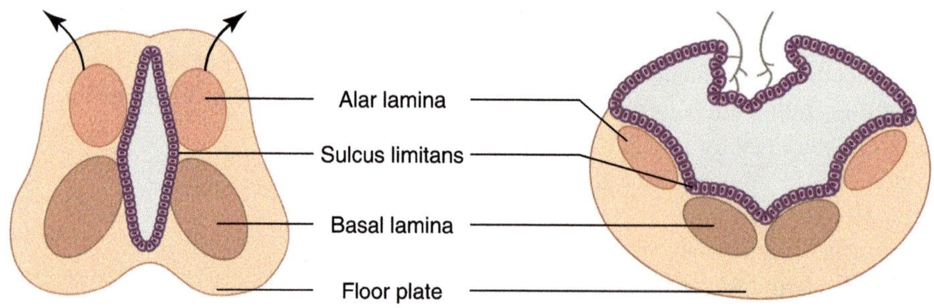

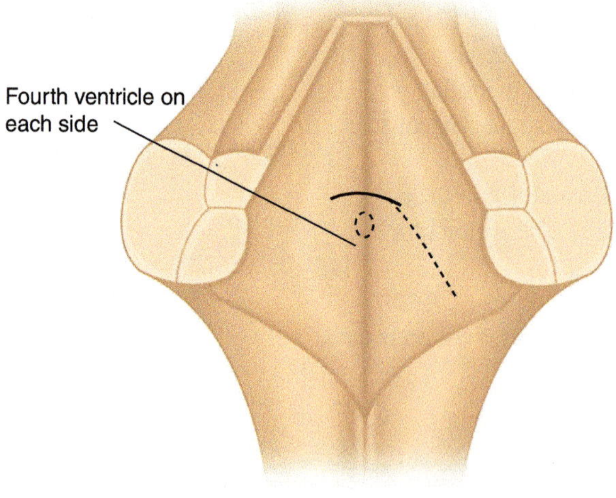

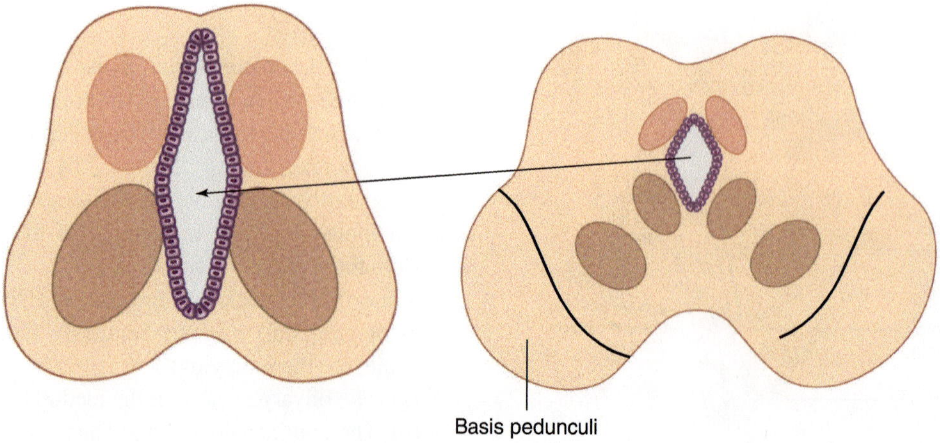

Development of the Cerebellum

*The cerebellum is formed of the dorsal part of the alar laminae of the metencephalon as follows:

1. The alar laminae of both sides bend medially to form the *rhombic lips*, each of which grows to form *medial and lateral bulges*.
2. The *med-bulges* of both sides meet each other over the roof plate of the fourth ventricle and unite forming the *vermis*.
3. The lat-bulges grow to form the cerebellar hemispheres.
4. The *cerebellar cortex* is formed by migration of neuroblasts from the mantle layer to enter the marginal layer.
5. The *dentate nucleus* develops as a collection of neuroblasts which remains deeply situated in the mantle layer.
6. The *cerebellar peduncles* develop later as the axons of the neurones of the cerebellar nuclei grow out of the cerebellum to reach the brainstem.

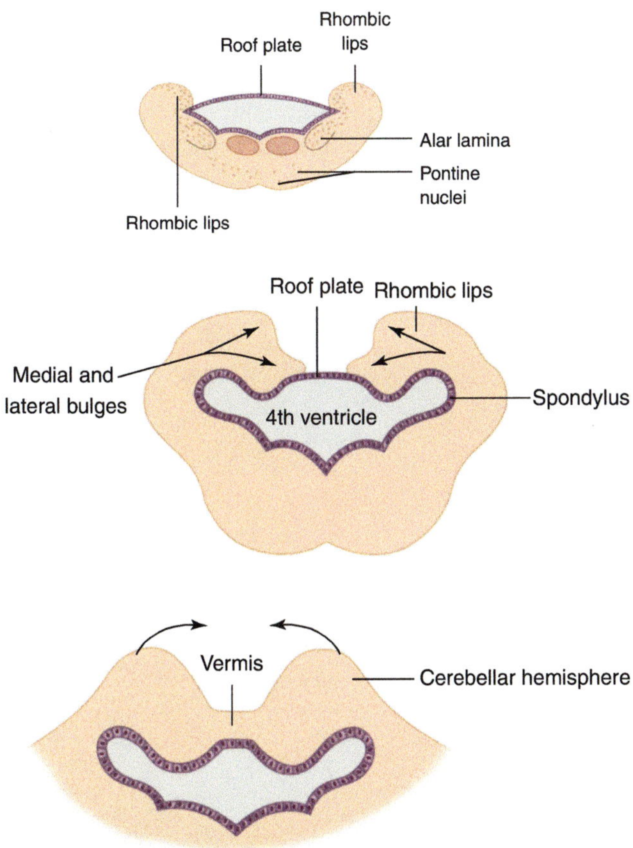

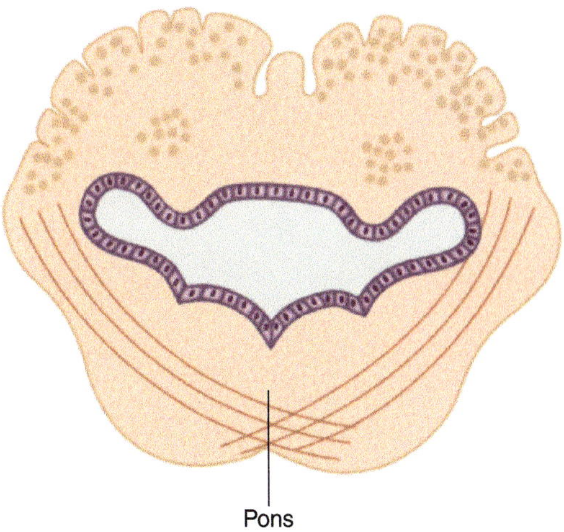

Development of the Forebrain (Prosencephalon)

*It is the most cranial of the three brain vesicles.

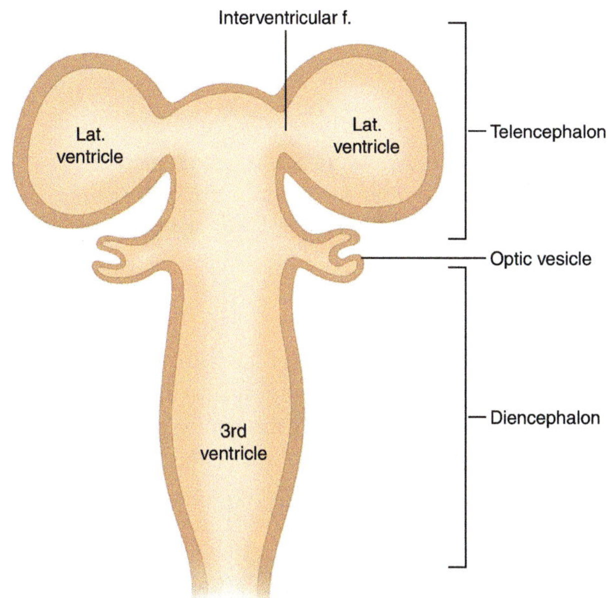

*A lat-diverticulum appears on each side of the forebrain (called the optic vesicle) subdividing the forebrain into:

(a) *Telencephalon*: the cranial part of the forebrain, including the two lat-diverticula. It gives two lat-out pocketings forming the cerebral hemispheres while optic vesicle and its stalk will form the retina and optic n.
(b) *Diencephalon*: the part immediately caudal to the optic vesicles. It develops into the thalamus, hypothalamus, epithalamus and pineal body.

Development of the Cerebral Hemispheres

*The two cerebral hemispheres arise as two evaginations from the lat-wall of the forebrain (prosencephalon).

**The cavity* of each evagination expands forming lat-ventricle.

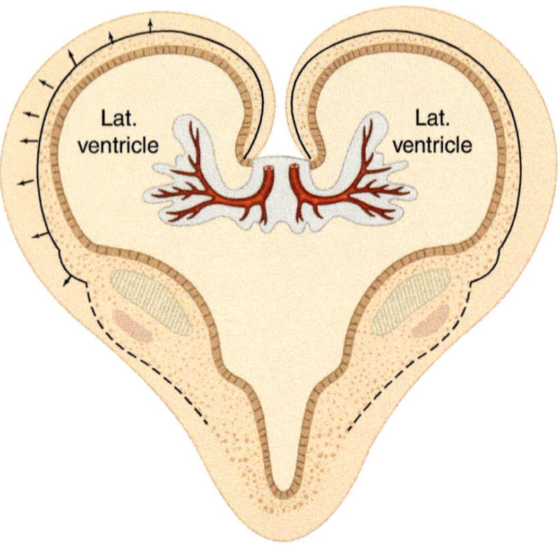

**The wall* of each hemisphere is formed of the following layers:

(a) *Outer marginal layer* formed of white matter
(b) *Middle mantle layer* formed of nerve cells (neuroblasts)
(c) *Inner ependymal layer* lining the cavity of the lat-ventricle

**As development proceeds the following changes occur*:

1. The neuroblasts of the wall of the hemisphere (except its base) migrate from the mantle layer to the marginal layer forming the *grey matter* of the cortex and the axons of these neuroblasts invade the rest of the wall forming the *white matter of the cerebral hemisphere*.

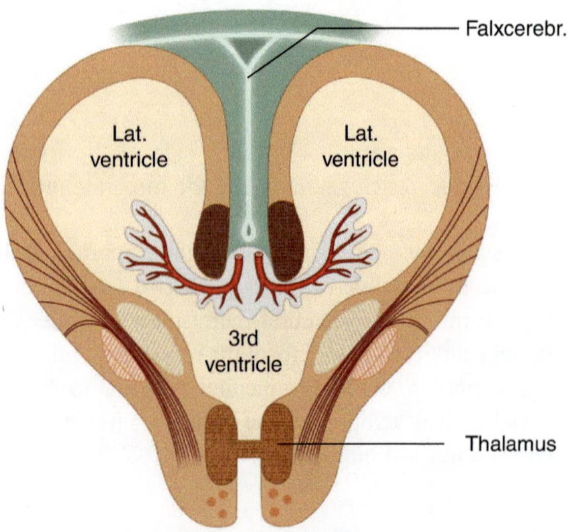

2. The neuroblasts of the mantle layer at the base of the hemisphere do not migrate to the marginal layer but remain deeply situated forming masses of grey matter (corpus striatum).
3. The part of the med-wall of the hemisphere just above the roof of the third ventricle is invaginated by choroid plexus forming the *choroid fissure*.

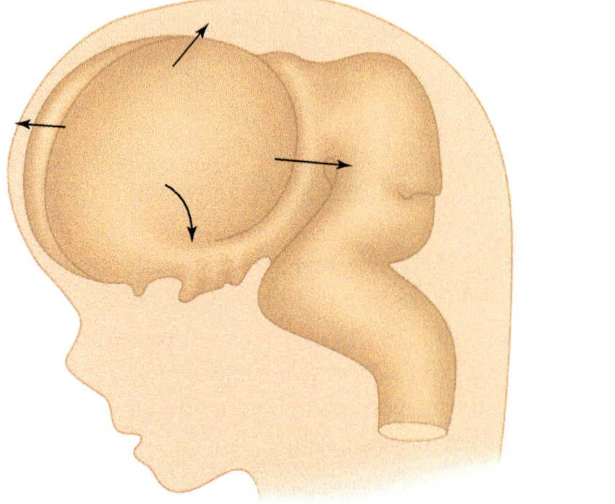

4. *Expansion of the cerebral cortex*: the cortex expands in all directions resulting in the following changes:
   (a) The med-surfaces of the two hemisphere come very close to each other.

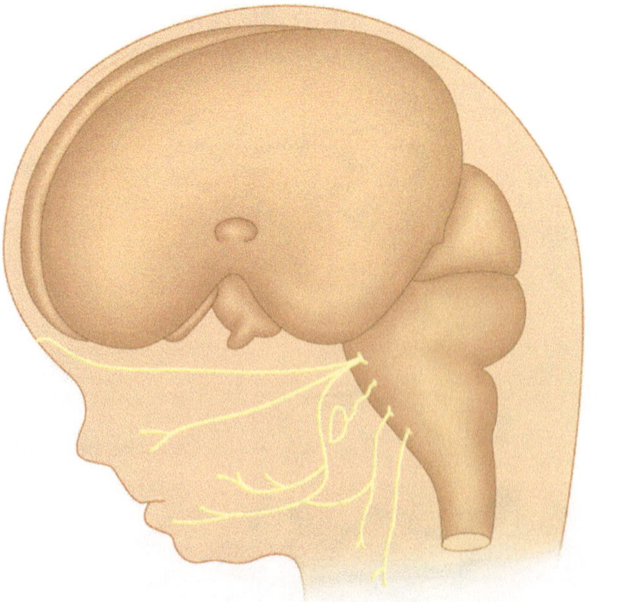

   (b) The corpus striatum at the base of the hemisphere is pushed towards the thalamus but kept separated from it by the fibers of the internal capsule.

(c) The insula which appears on the lat-surface as an area of less active growth becomes buried deep to the surface by the surrounding actively growing parts.

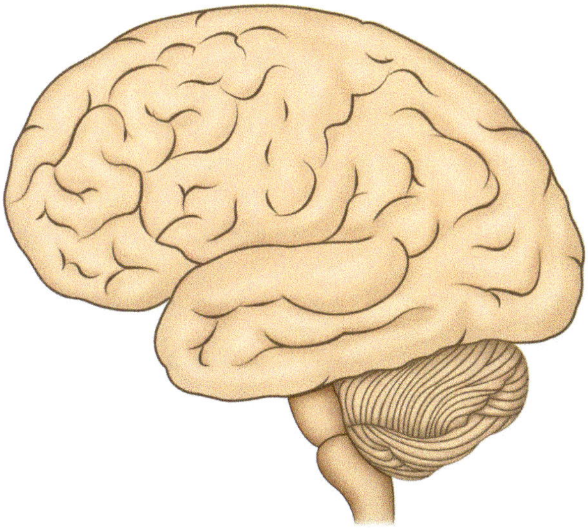

(d) The cerebral hemispheres overlap the brainstem and cerebellum.

Neural Tube Defects

The complex neural tube system which develops from infolding of the dorsum of the foetus is the site of a number of different types of maldevelopment. The global term "neural tube defects—NTD" is currently used to encompass the infants who are born with spina bifida and meningocele or myelomeningocele defects and at the cranial end of the neural tube, the meningoceles, encephaloceles, iniencephaly and anencephaly (a, b, c and d). Hydrocephalus is often associated with NTD and will be considered subsequently as it also occurs due to other causes. There is an interesting geographical variation in the incidence of the different types of NTD with the spinal anomaly of meningocele and myelomeningocele constituting 80% of NTD live-born infants in Europe, North America, Australia and some communities in South Africa. Defects of the cranial end, particularly encephaloceles, are more common in Southeast Asia than in Europe and account for 40% of the NTD found there.

Scotland, particularly the West of Scotland, has been an area of high incidence of NTD. For every live-born child with NTD, there has also been a stillbirth or early neonatal death of an infant with anencephaly. Previously, each of these accounted for approximately 3 per thousand live births, whereas over the last 15 years, the incidence of NTD in Scotland has decreased to 1 per 1000 live births. A couple who has a foetus or child

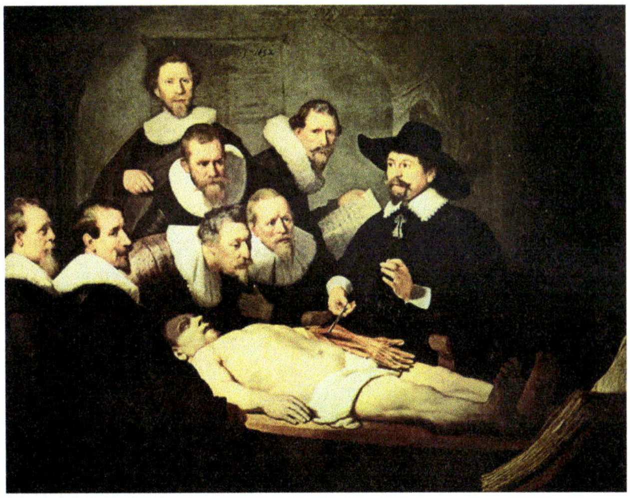

Fig. 26.1 Dr Tulp coined the term spina bifida

affected by a neural tube defect has approximately a tenfold increased likelihood of subsequent siblings being affected, and if the couple has two siblings affected, there is again a further increase by a factor of three for subsequent children. In 1992 an expert advisory group of the UK department of health recommended that to prevent recurrence of a neural tube defect, women at risk should take a daily folic acid supplement until the twelfth week of pregnancy. To prevent a first occurrence of NTD, women who are planning a pregnancy should eat more folate-rich foods and to take 4 mg folic acid daily from when they begin trying to conceive until the twelfth week of pregnancy. A Medical Research Council study published in 1991 suggested that folic acid offered a protective effect of 72 per cent in the risk of a second affected child.

Detection of elevated maternal serum alpha-fetoprotein levels and refinement in intrauterine ultrasound diagnosis have resulted in intrauterine diagnosis and pregnancy termination in many cases. A genetic predisposition to the development of neural tube defects may explain the higher incidence in those of Celtic (Irish, Welsh and West of Scotland population) than in those of Anglo-Saxon or Norse origin in the British Isles, but folate deficiency is not uncommon in these populations (Fig. 26.1).

Spina Bifida Occulta

In spina bifida occulta, there is failure of fusion of the spinous processes posteriorly. In most individuals this is a minor deviation from normal. A few children have an associated haemangiomatous or hairy patch over the site of the spina bifida occulta, and in these infants there may be some neurological deficit such as a foot-drop or a sudden alteration in continence of urine or problem deep to the lesion. Tethering of the cord is the most common problem, and producing neurological signs merits exploration and freeing of the cord. Other variations of the spine include an anomaly where a bony spur protrudes from the posterior aspect of the vertebral bodies and results in splitting of the cord (diastematomyelia). In a few patients, this may merit exploration and removal of the bony spur. Visualisation of these has been greatly enhanced with the development of magnetic resonance imaging which can give very much better definition of the intra-spinal anatomy than was previously possible.

Spina Bifida/Cystica

The spina bifida defect may result in a *meningocele* in which there is simply a protrusion of the spinal cord coverings to form a sac which contains CSF. This defect accounts for less than 10% of infants born with spina bifida. There are no neurological sequelae from the meningocele itself, and treatment is excision of the sac and closure of the dura and overlying structures. Fascial flaps can be brought across the spina bifida defect, and no attempt is made to alter the bony structure. The problem which may occur in these infants is the development of progressive hydrocephalus. Some infants already have hydrocephalus at birth, and others develop it after excision of the meningocele.

The more severe forms of spina bifida involve neural tissue and range from the *meningomyelocele* to the most severe end of the spectrum *myelocele* in which the cord lies exposed with no covering (Fig. 9.3). The most common site of these lesions is the thoracolumbar area in which case there is severe impairment of neural function in the lower half of the body. This affects both the motor and the sensory aspects to a level distal to the lesion. The next most common site is the sacral area, and affected infants may have much less severe affliction of the lower limbs, but defective neural supply to the bowel and bladder may give rise to serious problems of incontinence.

Effective management of the infant born with spina bifida depends on an initial detailed clinical examination and assessment of level of neurological impairment. Motor and sensory deficits are charted usually by physiotherapists who continue to help parents in managing the lower limb defects. The lack of muscle pull on the bones may leave them weaker than normal, and care in handling the infant is necessary to prevent fracture of their fragile bones. In consequence of the intrauterine paralysis which occurs in many of these children, there may be secondary structural problems and deformities such as loss of the normal lumbar lordosis and kyphosis. These problems progress with time as they are secondary to imbalance of muscular activity. Similar imbalanced muscle activity may result in dislocation of the hips, deformities of the knees or severe feet deformities (talipes). On the initial examination, it is also important to assess upper limb function which is normal in the majority of infants, but in a small number, due to further cord problems, upper limb problems become apparent as time goes by. A very common accompaniment of spina bifida is hydrocephalus, and this will in many infants be progressive and require active treatment.

Treatment of the infant with spina bifida is to maximise the potential of the child rather than being in any way curative. The neurological problem has been in existence for 6 or 7 months before birth, and the exposed neural tissue is irrevocably damaged. However, in some infants early closure of the back defect may be indicated to prevent infection and fibrosis affecting parts of the spinal cord which were not previously damaged. Also closure of the back has an important cosmetic aspect in that it quickly allows the back to be covered by normal skin and is therefore much easier for the parents to manage. The one drawback of early closure of myelomeningocele is that progression of hydrocephalus requiring active treatment is almost invariable, and leaving the larger defects (in which paraplegia is complete) to epithelialise spontaneously can result in the production and reabsorption of CSF becoming balanced so that active intervention is unnecessary. This occurs in one-third of the patients. Decisions have to be made according to the circumstances of each individual child and family.

Embryological classification of central nervous system disorders is a challenging and rapidly evolving field. To date there is no unifying, all-inclusive classification system that covers this area in its entirety.

It is estimated that approximately 2–3% of all newborns have major structural abnormalities; a significant number of these will involve the central nervous system{Dolk:2010jn}.

A sound knowledge of normal embryogenesis of the CNS leads to a better understanding of the malformations and as such improved management of them.

This chapter in no way gives a comprehensive coverage of the breadth and depth of the field. Instead we will focus on the more common malformations occurring at various anatomical sites that arise due to abnormalities at different stages of the embryological clock in CNS development.

Central Nervous System Dysraphism

This is a broader term encompassing abnormalities of the central nervous system affecting approximately 1 in every 100 pregnancies. The prevalence has reduced over the last

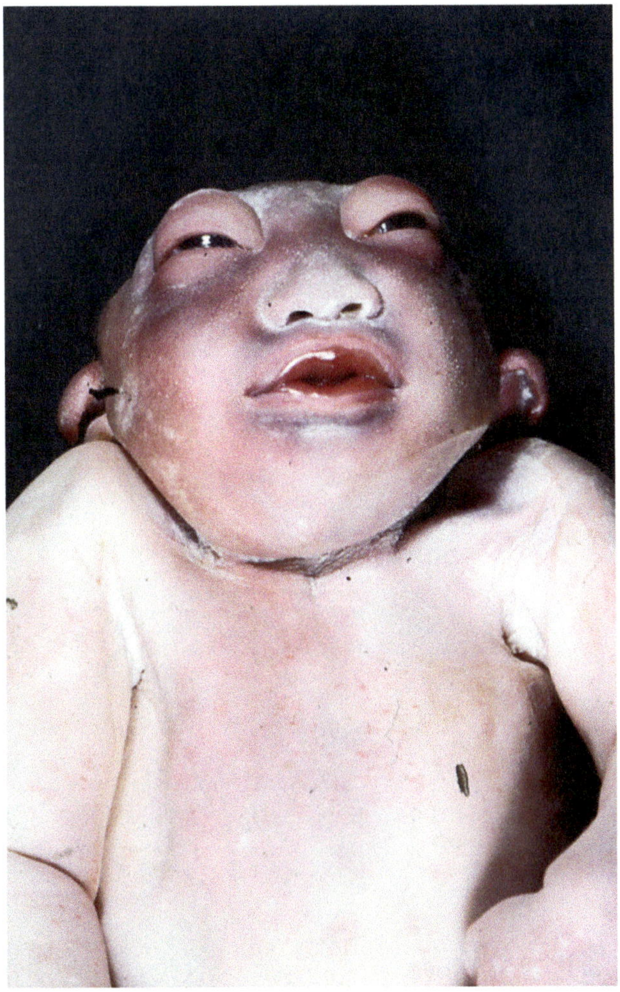

Fig. 26.2 Anencephaly

few decades. Considerable variability in prevalence exists with geographical location (Fig. 26.2).

These abnormalities of the head and spine include:

- Anencephaly
- Encephalocele
- Craniorachischisis
- Neurenteric cyst
- Myelomeningocele
- Lipomyelomeningocele
- Diastematomyelia
- Myelocystocele

They can be further divided into open or closed abnormalities depending on whether there is intact skin covering of the lesion.

Spinal Dysraphism

A useful clinico-embryological classificationof spinal dysraphism has been described by Thompson{Thompson:2014ga} (Fig. 26.3).

Myelomeningocele

This is the most common spinal neural tube defect. This is felt to arise from embryological failure in closure of the neural tube.

Failure of closure of the caudal neuropore (day 26) is the generally accepted embryological abnormality, although errors in gastrulation may also contribute.

The result is exposure of the neural placode. This is a "rolled open" spinal cord. The central groove represents the central canal of the spinal cord. The spinal nerve roots exit ventrally, the lateral root representing sensory and the more medial roots representing motor roots.

Failure of the mesoderm and ectoderm to form results in the neural placode presenting to the external surface. The dura fuses laterally with the fascia.

The majority of these defects occur in the caudal thoracolumbar spine.

The Chiari II malformation is an almost exclusive associated finding with myelomeningocele. A unified theory supporting the development of this abnormality in association with myelomeningocele proposes that failure to close the neural tube results in loss of CSF from the neural tube. This results in loss of distension and turgidity within the neural tube resulting in loss of pressure that normally facilitates development of the posterior cranial fossa. The result is loss in size of the posterior fossa and crowding of the neural structures, which phenotypically present as the constellation of Chiari II malformation.

Hydrocephalus requiring treatment develops in 80–90% of patients with myelomeningocele. Impairment of normal CSF homeostasis is thought to be multifactorial with maldevelopment of the CSF pathway in part as a result from the posterior fossa Chiari II abnormality.

Aetiology is multifactorial with both environmental and genetic factors contributing. Folate insufficiency has been implicated, and folate supplementation has been advocated periconception. It is estimated that supplementation may reduce MMC in the order of 70%. Maternal use of antiepileptic medication (valproic acid/carbamazepine), type 1 diabetes mellitus and obesity are associated with increased risk.

Various investigations can assist in the antenatal detection of myelomeningocele.

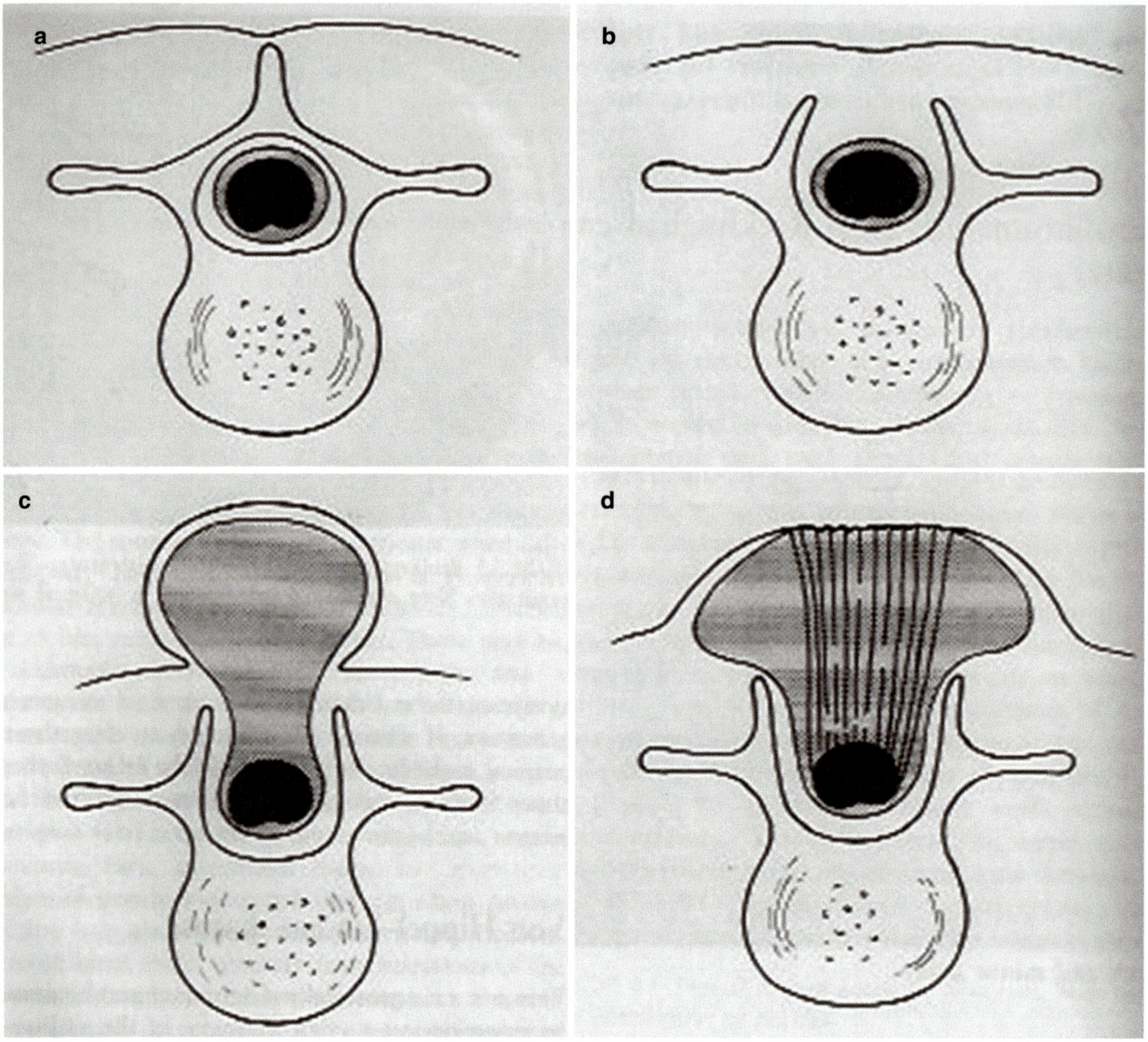

Fig. 26.3 Types of spina bifida (**a**) normal, (**b**) spina bifida, (**c**) meningocele and (d) myelomeningocele

Maternal serum alpha-fetoprotein can be used as a screening tool to detect open neural tube defects. It is calculated appropriate to gestational age. The optimal timing is between 16 and 18 weeks. The accuracy is estimated to be around 60–70%. It is important to be aware that elevated maternal serum alpha-fetoprotein is found in over 20 other foetal abnormalities. Elevation of alpha-fetoprotein can prompt further diagnostic testing in the form of ultrasonography and foetal MRI.

High-resolution US is almost 100% sensitive in detecting neural tube defects. This can also characterise the Chiari II malformation and the degree of ventriculomegaly.

Prenatal MRI is increasingly utilised to examine neurological abnormalities including MMC. It offers better resolution than US.

Today treatment for myelomeningocele is undertaken to optimise quality of life. Survival at 2 years exceeds 95% of patients. The management of these patients is multidisciplinary and includes neurosurgical, orthopedic and urological specialties.

Closure of the myelomeningocele is performed within 72 h after birth. Symptomatic hydrocephalus, if present, usually presents within the first 3 months of life. Ventriculo-peritoneal shunting is the preferred treatment modality. Endoscopic third ventriculostomy has an estimated success rate between 20% and 30%.

The recent MOMS study reported on the intrauterine repair of myelomeningocele before 26 weeks gestation. Improved outcomes in motor deficit, hydrocephalus and the Chiari II malformation were reported, although foetal and maternal risks from the procedure were not insignificant. With further development of this technique, it seems that this may play an increasing role in management in the future.

Neurenteric Cyst

Various terminologies for these abnormalities (a.k.a. enterogenous cyst, neuroepithelial cyst, endodermal cyst, foregut cyst, etc.) have been used over the years adding to the confusion of their aetiology and management.

Neurenteric cysts represent an abnormality that is felt to arise during the gastrulation phase of embryogenesis. It is postulated that either a persistent or accessory neurenteric canal gives rise to communication between ectoderm and endoderm.

They are a rare entity and occur predominantly in the spine, ventral to the cord. They are commonly associated with vertebral anomalies consistent with disruption of the notochord by the neurenteric fistula.

Symptoms usually arise from the compressive effect of the cyst on adjacent structures, notably the spinal cord. Treatment is surgical excision which can prove challenging, and recurrence is well-described.

Cranial Dysraphism

Encephalocele

The term encephalocele describes any herniation of meninges and brain tissue through a skull defect. If no neural tissue is found within the lesion, it is called a meningocele.

Congenital encephaloceles are a group of neural tube defects, which are covered by skin and predominantly found in the midline. They occur from occipital region through to basifrontal region. The most common location is the occipital region of the skull, with the second most common location being through the skull base at the anterior cranial fossa.

The incidence of encephaloceles is 1 per 10,000 births in Western countries. A higher incidence of 1:5,000 is reported for Asia and Africa. A female preponderance is observed.

Encephaloceles are thought to develop after the closure of the neural tube on the 27th to 28th day of gestation, as they are closed lesions covered by skin. It has been postulated that they arise from an error in mesodermal development. A cleft in the mesoderm that develops after primary neurulation allows contact with cutaneous and neuroectoderm. This creates the skull defect.

Risk factors for sporadic encephaloceles are folate deficiency, young maternal age, exposure to tobacco smoke and alcohol, positive family history, low educational level and ethnic origin. Several associated genetic syndromes have an association with encephaloceles including Meckel Gruber syndrome and Walker-Warburg syndrome (Figs. 26.4, 26.5, 26.6, 26.7 and 26.8).

Foetal ultrasound and MRI can detect encephaloceles prenatally, although small lesions as well as nasofrontal encephaloceles can remain undetected.

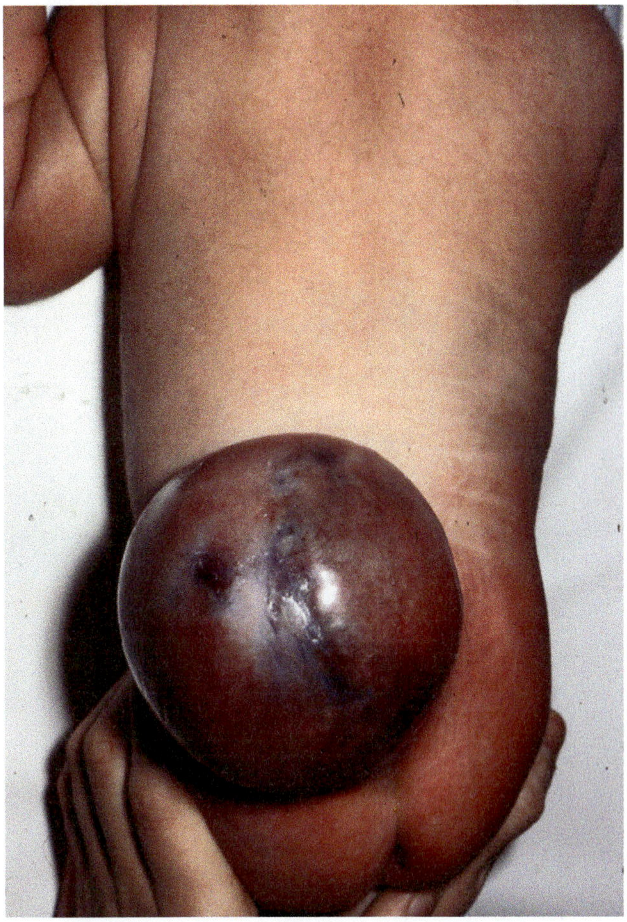

Fig. 26.4 Myelomeningocele cystic

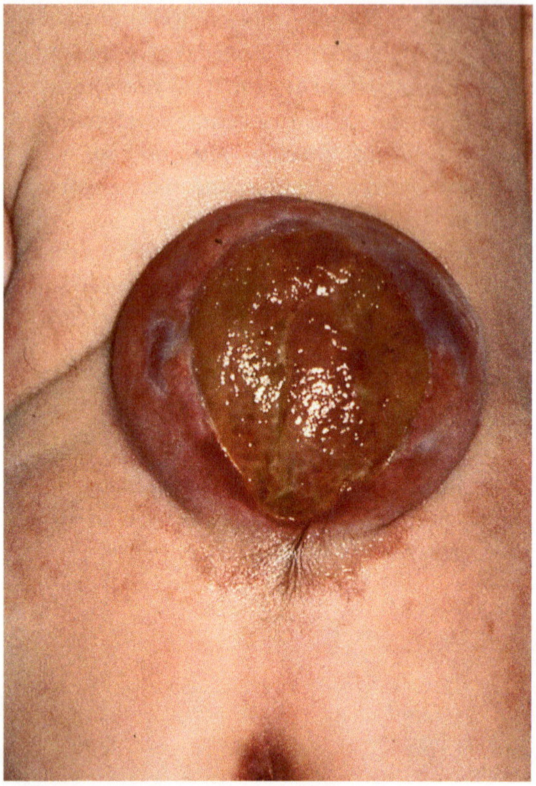

Fig. 26.5 Myelomeningocele meconium stained

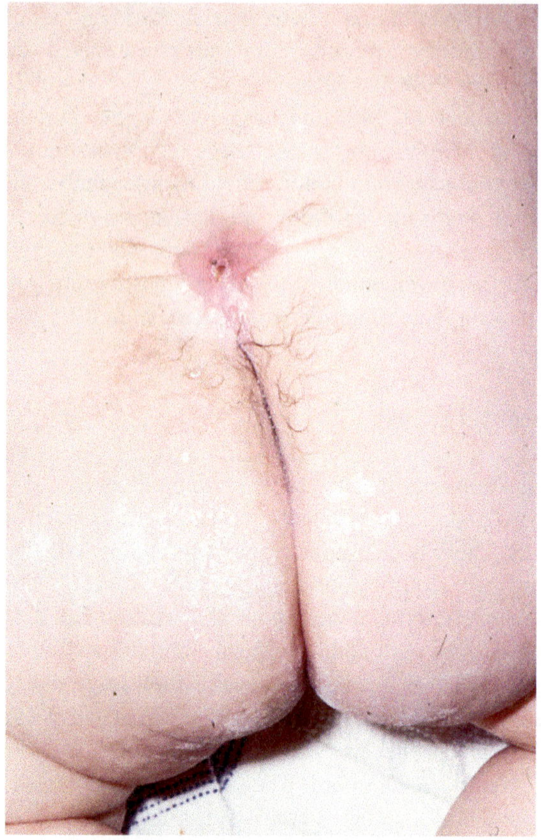

Fig. 26.7 Spina bifida occulta

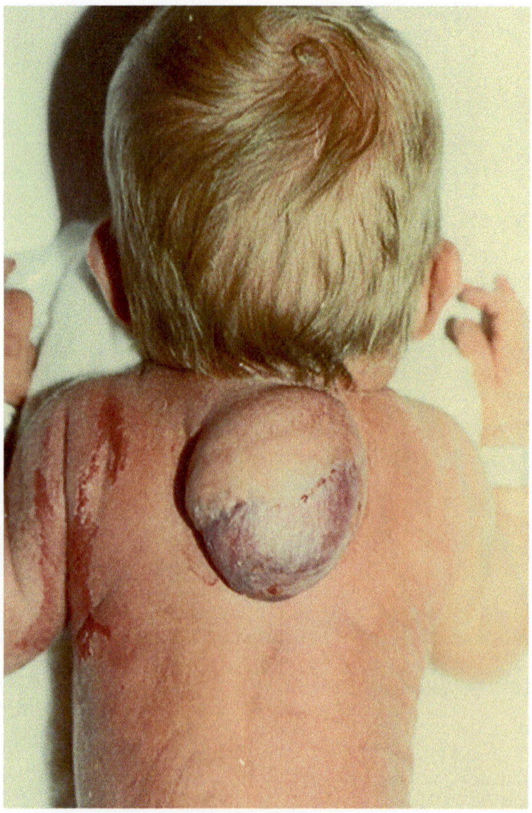

Fig. 26.6 Myelomeningocele-thoracic

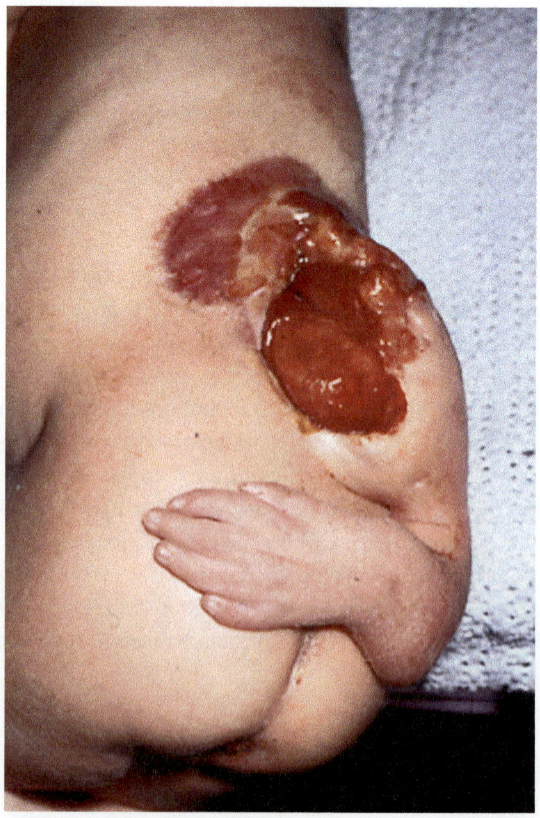

Fig. 26.8 Spina bifida extra limb foetus in fetu

Size and anatomical location guide management of these lesions. Small, atretic encephaloceles with little to no neural tissues may be managed conservatively. Larger symptomatic lesions will require surgery. The principle of surgical intervention is removing the sac, preserving where possible functional neural tissue and achieving closure that will not permit CSF leak. This may involve surgical teams from various disciplines including neurosurgeons, maxillofacial, plastic and ENT surgeons.

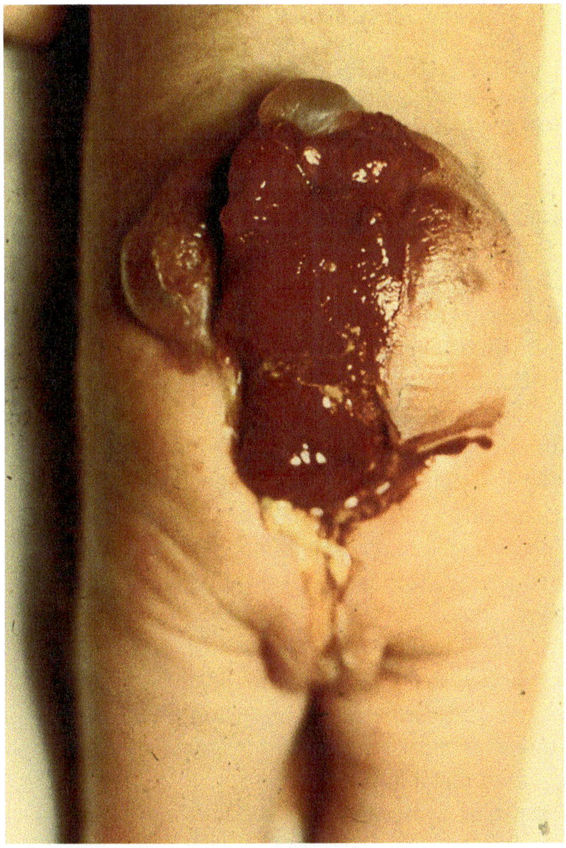

Fig. 26.9 Split notochord syndrome

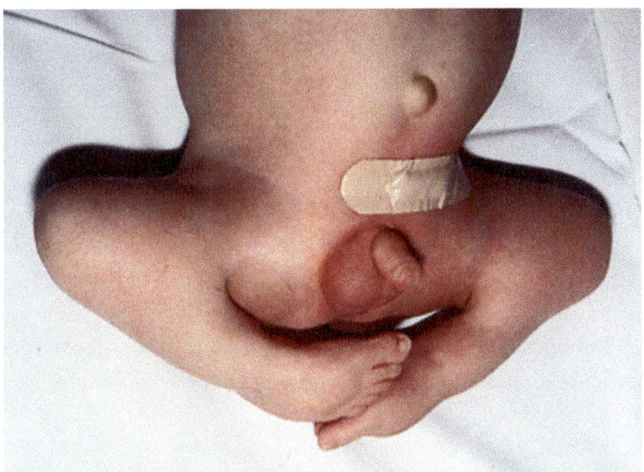

Fig. 26.10 Caudal regression syndrome

Outcome varies with the nature of each individual lesion, associated CNS and extra-CNS abnormalities. In modern practice the perioperative mortality is low. The main predictors of outcome are associated with hydrocephalus and seizure disorders. These are more commonly associated with the posterior occipital lesions than with the anterior frontal lesions. As such it is traditionally taught that anterior encephaloceles have a better outcome than posterior ones. Fifty percent of patients have a normal long-term neurological development, while 25% have mild to moderate and severe developmental delay, respectively (Figs. 26.9, 26.10, 26.11, 26.12 and 26.13).

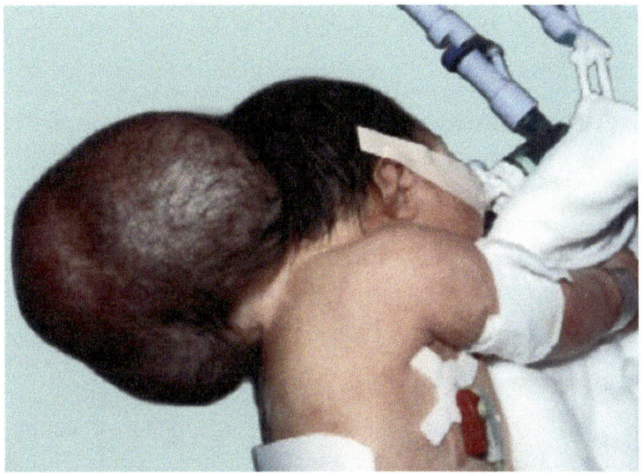

Fig. 26.11 Encephalocele

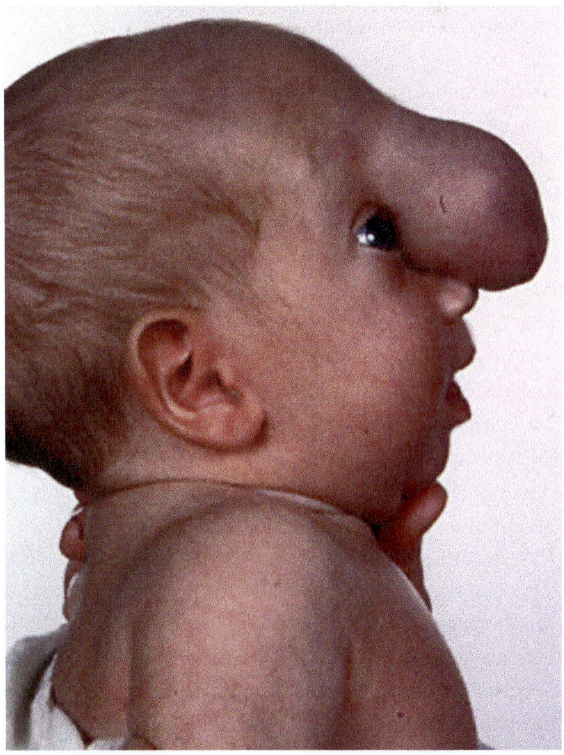

Fig. 26.12 Encephalocele anterior

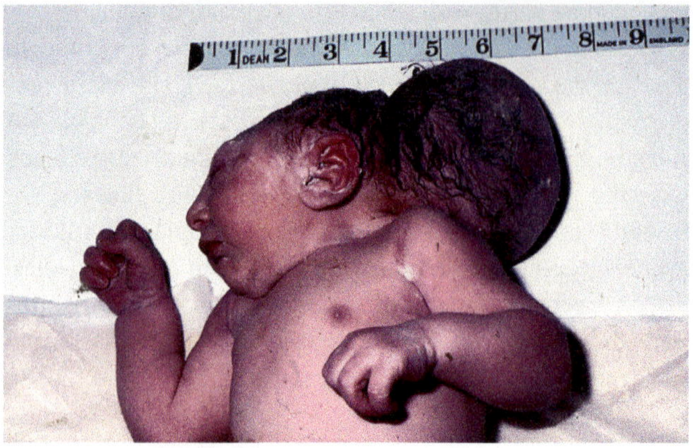

Fig. 26.13 Encephalocele with anencephaly

Table 26.1 Aetiology of non-syndromic and syndromic forms of congenital hydrocephalus

Non-syndromic	Syndromic
– *Hydrocephalus associated with neural tube defects* (e.g. spina bifida) – *Isolated hydrocephalus* (e.g. congenital aqueductal stenosis, X-linked hydrocephalus) – *Hydrocephalus associated with CNS malformations* (e.g. Dandy-Walker malformation, vein of Galen malformation, congenital intracranial arachnoid cysts, callosal anomalies, encephaloceles) – *Congenital communicating hydrocephalus* (e.g. infection, haemorrhage)	– *Cytogenetic abnormalities* (e.g. trisomy 9, 13, 18) – *Craniosynostosis* (e.g. Crouzon syndrome, Apert syndrome) – *VACTERL association*

Hydrocephalus

Hydrocephalus is a condition manifested by disturbance in CSF homeostasis. There is no universally agreed definition of the condition, and various classifications for hydrocephalus exist. For a basic introduction, we propose to classify it according to underlying cause and simplified pathophysiology.

Hydrocephalus can be considered as congenital or acquired. This chapter is focused on congenital hydrocephalus, which is by definition present at birth.

Overall, the main causes of this condition are neural tube defects (30%), aqueductal stenosis (20%) and Dandy-Walker malformation (10%). Table 26.1 summarises the most important aetiologies of congenital hydrocephalus.

The pathophysiology of hydrocephalus is complex, and our understanding of CSF hydrodynamics is incomplete. From a clinical perspective, a division into communicating and noncommunicating (obstructive) hydrocephalus can be useful in particular when considering treatment modalities.

In communicating hydrocephalus the ventricles are uniformly enlarged, and no obvious obstruction of the CSF pathways is seen on cranial imaging. In contrast, noncommunicating hydrocephalus is caused by an obstruction of the CSF pathways. In children with congenital hydrocephalus, this obstruction is most frequently located at the level of the cerebral aqueduct, leading to an enlargement of the third and lateral ventricles only.

The incidence of congenital hydrocephalus is approximately 1 per 1,000 live and stillbirths.

Risk factors for congenital hydrocephalus are maternal diabetes and hypertension as well as maternal alcohol abuse.

Mutations of the L1 gene, which is located on the X chromosome and encodes for the L1 cell adhesion molecule (L1 CAM), have been identified in a subset of male patients with congenital hydrocephalus due to aqueductal stenosis. This form of X-linked hydrocephalus accounts for a minority of male cases of congenital hydrocephalus. Clinical features include developmental delay, spasticity and adducted thumbs. An associated malformation is agenesis of the corpus callosum.

Congenital hydrocephalus can be detected on foetal ultrasound. Since ventriculomegaly usually develops after 20 weeks of gestation, the diagnosis might be missed on early ultrasound scan. When ventriculomegaly is diagnosed on ultrasound, foetal MRI may be useful in helping to identify the underlying cause of hydrocephalus and detect associated anomalies.

Other anomalies are found in up to 80% of children with congenital hydrocephalus. 20–40% of these anomalies are extracranial.

Antenatal treatment for congenital hydrocephalus has been reported in the 1980s but has fallen out of favour. Poor outcomes and the need for postnatal treatment for these patients who underwent in utero treatment have led to abandonment of foetal intervention.

Congenital hydrocephalus associated with open myelomeningocele, intrauterine repair of the spinal defect, which is only available at highly specialised centres, can indirectly treat hydrocephalus and reduce the postnatal shunt rate by up to 50%.

The vast majority of patients are treated after birth with either a ventriculo-peritoneal shunt or a neuro-endoscopic procedure (e.g. endoscopic third ventriculostomy).

Hydrocephalus has evolved from a fatal condition to a treatable, chronic condition. Patients with treated hydrocephalus report outcomes ranging from severe physical, cognitive and psychosocial disturbance to those with near normal lives.

The burden of hydrocephalus on individual patients is determined by factors related to the underlying condition itself and by factors related to treatment.

Functional concerns encountered in hydrocephalus patients include:

- Mobility (e.g. cerebral palsy)
- Cognition (e.g. developmental delay)

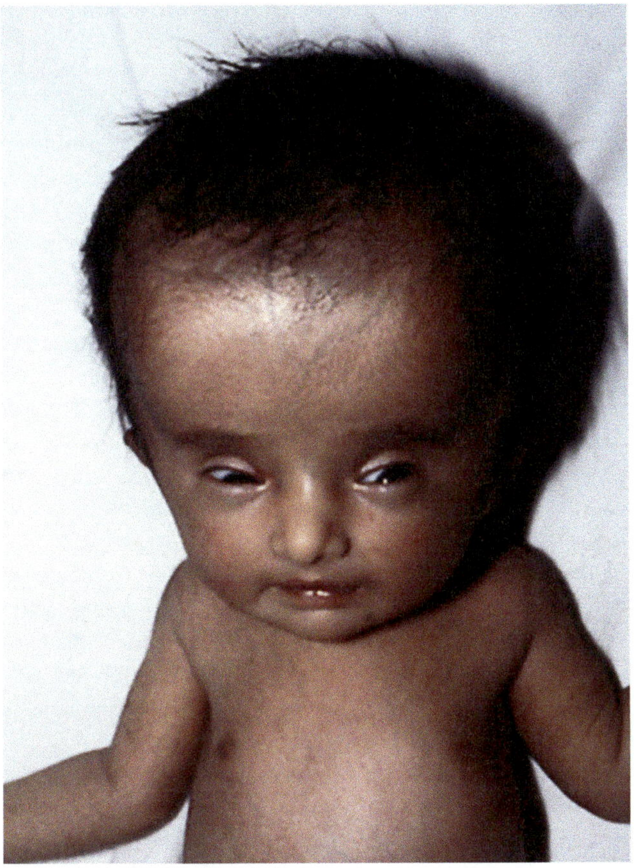

Fig. 26.14 Hydrocephalus neonatal

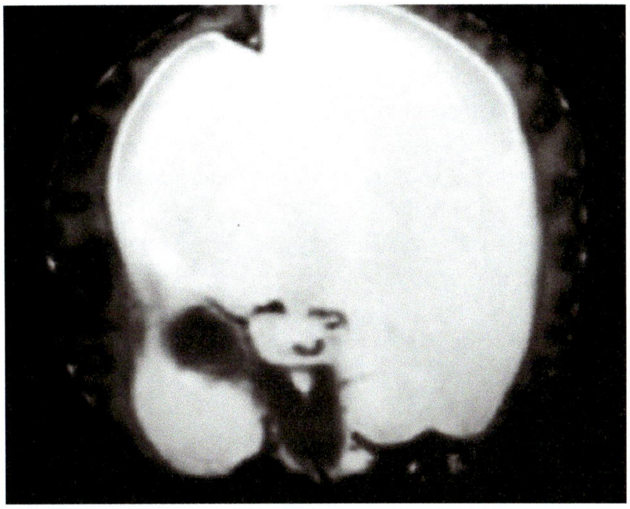

Fig. 26.15 Hydrocephalus CT scan

- Sensory (e.g. impaired vision or hearing)
- Epilepsy
- Depression
- Pain (e.g. chronic headache)

Due to the broad spectrum of aetiologies, there is a significant variation in the outcome data reported in the literature. Across all entities, approximately 45% have some degree of motor handicap, and 35% are developmentally delayed. Fifty percent of patients achieve normal schooling. Forty-five percent are able to compete in the labour market, whereas 25% work in a sheltered job. Regarding the burden of treatment, approximately four shunt revisions per patient are necessary at 20 years (Figs. 26.14, 26.15 and 26.16).

Holoprosencephaly

Holoprosencephaly covers a continuous spectrum of phenotypical abnormalities of the forebrain resulting from improper cleavage of the prosencephalon.

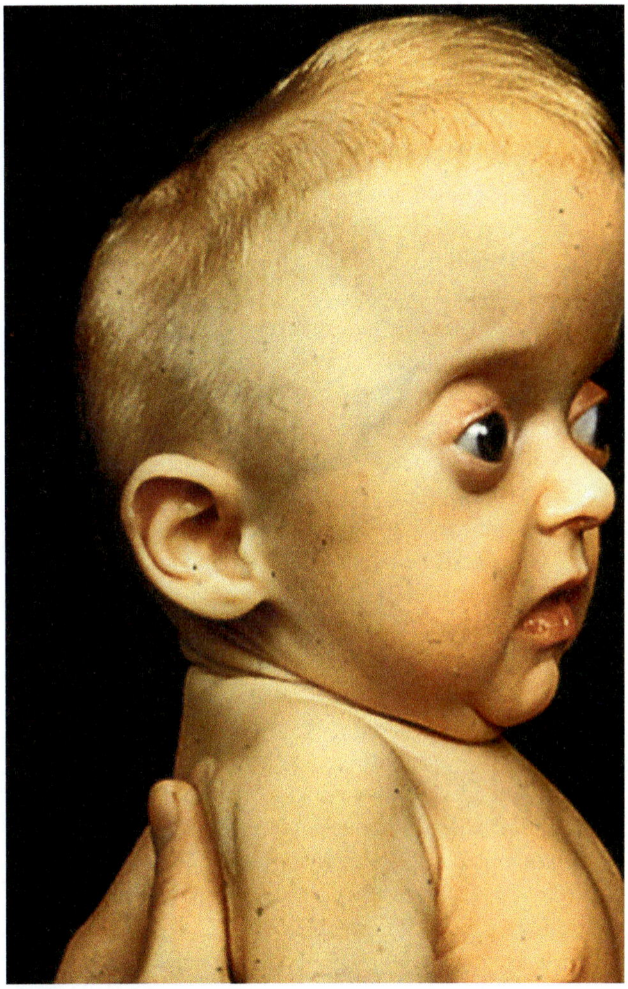

Fig. 26.16 Oxycephaly

It is commonly associated with midline facial anomalies which range from severe (cyclopia) in the alobar form to mild (cleft palate, single central incisors, iris colobomas) or absent in the lobar/mildest forms.

In general the severity of the facial abnormality corresponds with the severity of the cerebral abnormality; "the face predicts the brain".

It is generally classified by the degree of hemispheric separation into

(a) Alobar—the most severe form
- This is represented by
  - Monoventricle
  - No telencephalic cleavage
  - Dorsal interhemispheric cyst (=out pouching of third ventricle)
  - Diencephalic involvement
    - Fusion of palladium
    - Fusion of Thalami
    - No third ventricle
    - Hypotelorism or cyclopia (neural retina is outpouching of diencephalon)

(b) Semialobar
- This is represented by
  - Abnormal anterior (rostral) hemispheric cleavage
  - Incomplete formation of falx
  - Partial fusion of thalami
  - Small third ventricle
  - Hypertelorism (less severe than alobar)

(c) Lobar—the mildest form
- This is represented by
  - Incomplete formation of the anterior falx cerebri
  - Unseparated frontal horns
  - Absent septum pellucidum

(d) Septo-optic dysplasia (de Mosier syndrome)
- Often considered the mildest form
- Characterised by
  - Complete/incomplete absence of septum pellucidum
  - Optic nerve hypoplasia
  - Abnormal pituitary-hypothalamic function

Failure of cleavage of the ventral forebrain is thought to occur at day 33 gestation. Although the molecular cause of holoprosencephaly is thought to manifest during the preneurulation phase, this ultimately represents a post neurulation (day 24–28) disorder. Cleavage of the telencephalon is controlled by the prechordal plate.

There is wide variation on reported epidemiology of this spectrum of disorders. The more severe end of the spectrum is usually lethal in the embryonic/foetal stage making detection of the abnormality time/age-dependent.

Prevalence rates are reported at 1/10000 when live and stillbirths are considered but higher when termination of pregnancies is also considered. This may underestimate the prevalence as milder forms may only present later in life.

Both environmental and genetic factors have been implicated. Associations with maternal alcohol consumption and maternal diabetes have been described.

Cytogenetic confirmed abnormalities have been reported in one-third of perinatal studies. Holoprosencephaly has been associated with many chromosomal defects (e.g. Trisomy 13/Patau's syndrome). Several genes including the sonic hedgehog gene (SHH gene product expressed at prechordal plate) and its misexpression have been implicated in the development of holoprosencephaly.

First trimester diagnosis of the most severe form, alobar holoprosencephaly, is based on the visualisation of a single ventricle. Failure to identify the "butterfly sign" of the choroid plexus at 11–13 weeks has also been described as a warning sign of holoprosencephaly. The detection rate is estimated at 78% in the first trimester.

Survival beyond infancy is uncommon especially with the more severe forms. Many will have major developmental disability. Hydrocephalus requiring treatment may occur in some. Seizures are commonly associated.

Patients with septo-optic dysplasia also manifests with endocrinopathies.

Agenesis of Corpus Callosum

The corpus callosum is the largest interhemispheric commissure and is divided into genu (anterior), body and splenium (posterior). It is a bundle of topographically organised white matter fibres that mediate interhemispheric integration. Partial or complete agenesis of the corpus callosum can appear as an isolated finding or in conjunction with other congenital abnormalities. The spectrum of symptoms is broad, ranging from asymptomatic to severely impaired patients.

Development of the corpus callosum begins at 11–12 weeks and traditionally has been thought to proceed in a rostro-caudal direction with the exception that the final portion to form is at the rostrum. However current evidence suggests that the corpus callosum forms at two loci in the commissural plate and progresses bidirectionally. By 20 weeks gestation, the final shape of the corpus callosum is complete.

Many developmental processes are involved in the formation including neurogenesis, midline patterning, neuronal migration, neuronal specification, axon guidance and post-axon guidance development.

Commissural axons may form but do not cross the midline and remain ectopic. These appear as bundles of Probst.

There are many differing processes that can be disturbed that will result in partial to complete agenesis of the corpus callosum.

Agenesis of the corpus callosum is usually not an isolated abnormality. It is usually seen with other CNS abnormalities, e.g. Dandy-Walker malformation and holoprosencephaly.

Agenesis of the corpus callosum is part of the distinct pattern of CNS malformations in foetal alcohol syndrome.

Malformations of the corpus callosum are associated with aneuploidy syndromes (e.g. trisomy 13 and 18) as well as non-aneuploidy syndromes (e.g. Aicardi syndrome, Apert syndrome and Zellweger syndrome). Of note, Aicardi syndrome is characterised by agenesis of the corpus callosum, chorioretinal lacunae and infantile spasms. Moreover, abnormalities of the corpus callosum can also occur in patients with inborn errors of metabolism, such as mitochondrial defects or mucopolysaccharidoses.

As corpus callosum malformations can be asymptomatic, their prevalence can only be estimated. While the estimated prevalence in the general population is 30–70 per 10,000 births, it is 3–7 times higher in children with developmental disabilities.

Agenesis of the corpus callosum can be detected antenatally with US. Foetal MRI allows detailed assessment of all components of the corpus callosum, and associated abnormalities may also be detected.

Agenesis of the corpus callosum can lead to a broad spectrum of clinical presentations, ranging from asymptomatic patients or only subtle impairment detected on neuropsychological testing to development delay, seizures and cerebral palsy.

There is no antenatal or postnatal treatment for agenesis of the corpus callosum. However, associated malformations may require treatment.

Isolated agenesis of the corpus callosum has a more favourable prognosis.

Approximately 90% of patients diagnosed with agenesis of the corpus callosum develop symptoms, usually within the first 2 years of life. While the more subtle neuropsychological symptoms in mild cases do not significantly affect outcome, associated conditions, especially seizures and hydrocephalus, are the main predictors of outcome.

Microcephaly

Microcephaly is a descriptive term defined as a head circumference more than two standard deviations below the mean for age and gender. The most basic classification of this heterogeneous condition is (1) primary microcephaly, which is caused by a failure of the brain to grow appropriately and is thus present prenatally, and (2) secondary microcephaly, which is caused by a deceleration of growth of an initially normal brain by a variety of insults and is thus not present at birth. The most important aetiologies for both types of microcephaly are listed in Table 26.1.

Primary microcephaly is diagnosed in 0.5% of children at birth. The incidence of secondary microcephaly, which usually becomes evident in the first 2 years of life, is estimated to be higher.

Primary microcephaly occurs by 32 weeks of gestation. As the majority of neurons in the human brain are formed by 21 weeks of foetal development, primary microcephaly is generally thought to be caused by reduced production of neurons.

Epidemiological studies identified foetal exposure to maternal alcohol and tobacco use as well as intrauterine infections (TORCHES) as main risk factors for primary microcephaly. Maternal alcohol intake increases the risk for microcephaly 2.6-fold.

A specific genetic syndrome can be diagnosed in 20% of microcephalic children. Primary microcephaly occurs in patients with Trisomy 13, 18 and 21. Postnatal onset of microcephaly is observed in Rett syndrome and Ataxia telangiectasia.

Primary microcephaly can be detected on foetal ultrasound. Foetal MRI can confirm the diagnosis and demonstrate underlying or associated malformations, such as disorders of cortical development.

Neuroimaging, preferentially MRI, is recommended in microcephalic children to identify structural causes and associated anomalies. The most common findings on MRI scans of the head are migrational disorders, callosal malformations, posterior fossa abnormalities and disorders of myelination.

Mental retardation is observed in 50% of microcephalic children. Epilepsy is found in 40%, cerebral palsy in 20% and ophthalmological disorders in 20%. The severity of developmental impairment correlates with the degree of microcephaly, with 80% of children with severe microcephaly (i.e. head circumference more than three standard deviations below the mean for age and gender) being affected (Table 26.2).

Table 26.2 Aetiology of primary and secondary microcephaly

	Primary = congenital	Secondary = postnatal onset
Genetic	*Isolated microcephaly* (autosomal recessive, autosomal dominant, X-linked)	*Inborn errors of metabolism* (e.g. glycosylation/mitochondrial/ peroxisomal disorders, amino acidopathies, organic acidurias)
	Syndromic microcephaly (e.g. Trisomy 13, 18, 21)	*Syndromic microcephaly* (e.g. Rett syndrome, Ataxia telangiectasia)
Acquired	*Disruptive foetal injuries* (e.g. TORCHES infection, HIV, stroke)	*Disruptive injuries* (e.g. trauma, stroke, meningitis/encephalitis)
	Teratogens (e.g. maternal alcohol and tobacco use, maternal diabetes)	
	Deprivation (e.g. maternal folate deficiency, maternal malnutrition, placental insufficiency)	*Deprivation* (e.g. hypothyroidism, malnutrition)

Abnormalities of Pituitary Development

The formation of the pituitary gland is brought about by the joining of ectodermal stomodeum with neuroectoderm from the developing diencephalon. This complex process allows for various aberrations to develop.

Craniopharyngioma

Craniopharyngiomas are histologically benign epithelial tumours.

They are thought to arise from epithelial cell rests remnants along the embryological migration path of the anterior pituitary lobe pouch and the hypophyseal pharyngeal duct/ craniopharyngeal duct. Evidence suggests that embryonic rests of cells give rise to craniopharyngiomas.

Two histological subtypes are recognised: adamantinomatous craniopharyngiomas have a bimodal age distribution with the first peak between 5 and 15 years and the second in the 45 and 60 years of age; the papillary type is almost exclusively seen in 40–55 years of age group.

Metaplasia of the cell rests forms have been suggested as the origin of the papillary adult form of this tumour.

Annual incidence is reported between 0.5 and 2.5 cases per million population per year.

While the development of these tumours may arise in utero, the clinical manifestation is not until many years post-partum. This may reflect the benign slow growing nature and the potential requirement for metaplastic processes to occur in the cell rests.

Several case reports have described antenatal detection, but these are more likely to represent a more aggressive, early presenting phenotype of this tumour.

Their intracranial location abutting eloquent cerebral and vascular structures makes complete surgical resection difficult to achieve without inflicting damage to structures in this region resulting in long-term damage. The benign, slow growing pathology makes them less susceptible to chemotherapeutic and radiation modalities. As such their behaviour is considered to be more malignant than benign.

Rathkes Cleft Cysts

This is a non-neoplastic cyst derived from Rathke's cleft found in the intermediate lobe of the pituitary.

Rathke's pouch develops at approximately day 24 as an outpouching in the doral stomodeum. This travels to meet with the down growth of neuroepithelium that will become the posterior pituitary. The residual lumen of the pouch involutes. Enlargement of this lumen is felt to be the cause of Rathke's cleft cyst.

They are often asymptomatic and have been reported to be found in 30% of normal pituitaries at autopsy. Clinically they can cause local mass effect with visual disturbance from the optic chiasm, endocrinopathies from pituitary/hypothalamic compression.

Antenatal detection is not commonly reported.

They tend to present after the first decade of life. Treatment for symptomatic cysts is surgical excision/debulking.

Vein of Galen Aneurysmal Malformation

VGAM is an arteriovenous fistula with arterial supply from the choroidal arteries and drainage into the median vein of the prosencephalon—the precursor of the vein of Galen (Fig. 26.17).

The embryology of VGAM formation follows along the development of vascularization of the choroid plexus.

The anterior cerebral and anterior and posterior choroidal arteries develop and supply the choroid plexus. Numerous quadrageminal arteries supply numerous meningeal capillaries to the meninx primitiva. It is the choroidal and quadrageminal arteries that go on to supply the VGAM.

Growth of the choroid plexus leads to development of a midline vein draining the bilateral choroid plexus—the median vein of the prosencephalon. This vein is maintained up to approximately week 10.

With further growth of neural strictures, paired internal cerebral veins develop; these annex the venous drainage which leads to involution of the median vein of the prosencephalon where the most caudal remnant joins the internal cerebral vein and becomes the true vein of Galen.

VGAM formation occurs in the period between 6 and 11 weeks. It arises when the median vein of prosencephalon fails to involute and instead dilates.

VGAM can be detected on antenatal U/S and MRI.

Clinically they manifest in neonates with high output cardiac failure. In older children they tend to present with venous hypertension.

Treatment consists of endovascular occlusion. VGAMs and their treatment present a high risk of morbidity and mortality, although with the evolution of neuroendovascular treatments, outcomes are improving.

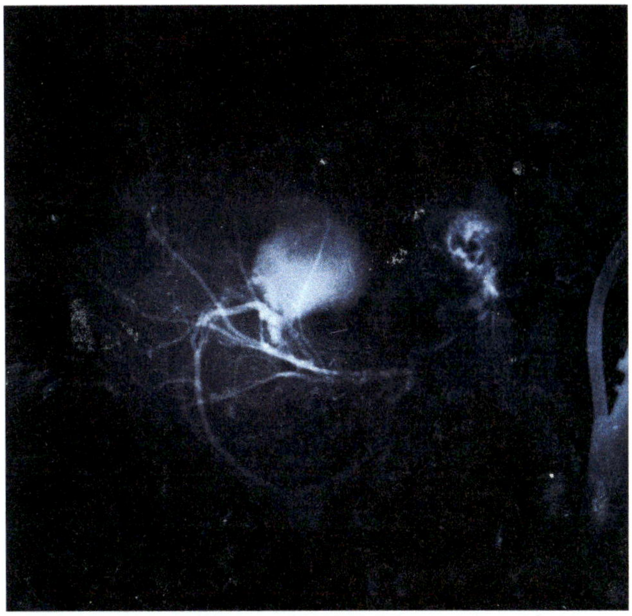

Fig. 26.17 Angiogram of vein of Galen aneurysm

Chest Wall Abnormalities

27

James Andrews

Introduction

Chest wall deformities have been well recognised since at least the sixteenth century [1] and encompass a wide range of heterogeneous conditions. Although primarily cosmetic in nature, there are other factors which may have an impact on the patient and society in broader terms.

Epidemiology

The epidemiology of chest wall deformities is difficult to ascertain, as the array of different entities make them difficult to classify. The commonest anomalies are those of pectus excavatum (Fig. 27.1) and pectus carinatum (Fig. 27.2), which account for the vast majority (over 90%) of these conditions [2].

When reviewing over 5000 patients, Acastello et al. [3] showed that 0.15% of patients had a sternal cleft (Fig. 27.3) and only 3.4% had Poland syndrome (Fig. 27.4). Other varients include Cleidocranial Dysostosis (Fig. 27.5) and bifid ribs (Fig. 27.6).

A variety of series have given wildly different rates for pectus excavatum and carinatum. Many North American studies suggest that most patients have an excavatum deformity [1]. But this is not replicated in other studies [3, 4]. It is possible that the higher proportion of excavatum from historical literature represents a bias due to treatment options, which will be rebalanced as more patients look to have their carinatum treated with bracing.

Incidence of pectus excavatum varies from 1 in a 1000 to 8 in a 1000, and there is a male preponderance of between 2:1 and 9:1 [2]. These numbers reflect the difficulty in gathering population data on such a variable entity. It is also accepted that pectus excavatum is up to five times more common in Caucasians and appears to be rarely seen in Africa [1].

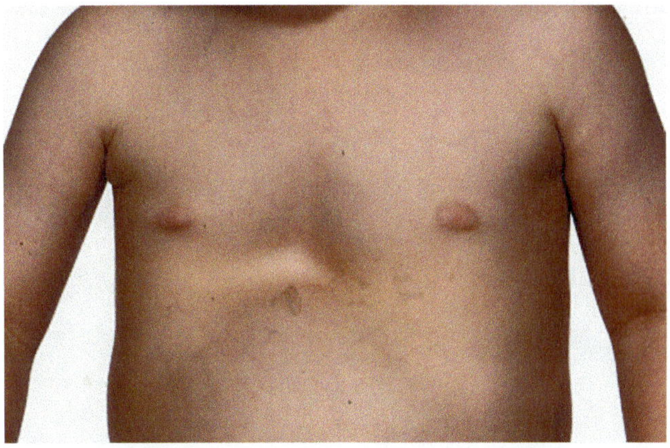

Fig. 27.1 Pectus excavatum

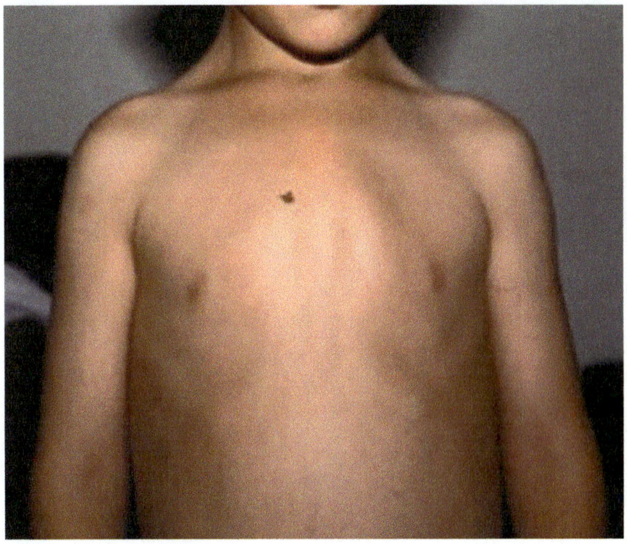

Fig. 27.2 Pectus carinatum

J. Andrews (✉)
Department of Surgical Paediatrics, Queen Elizabeth University Hospital, Glasgow, UK
e-mail: jamesandrews@nhs.net

R. Carachi, S. H. E. Doss (eds.), *Clinical Embryology*, https://doi.org/10.1007/978-3-319-26158-4_27

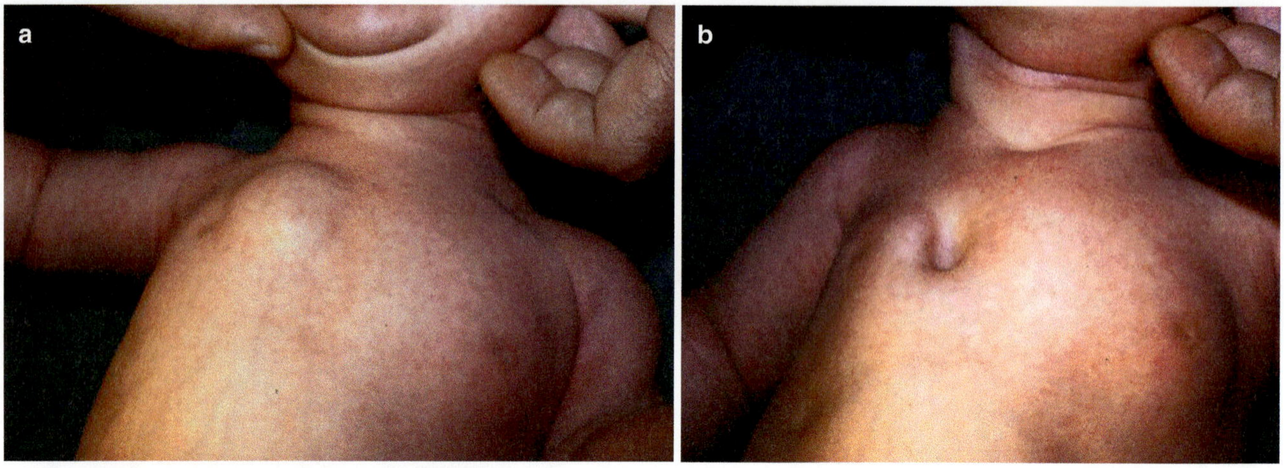

Fig. 27.3 (**a**, **b**) Cleft sternum

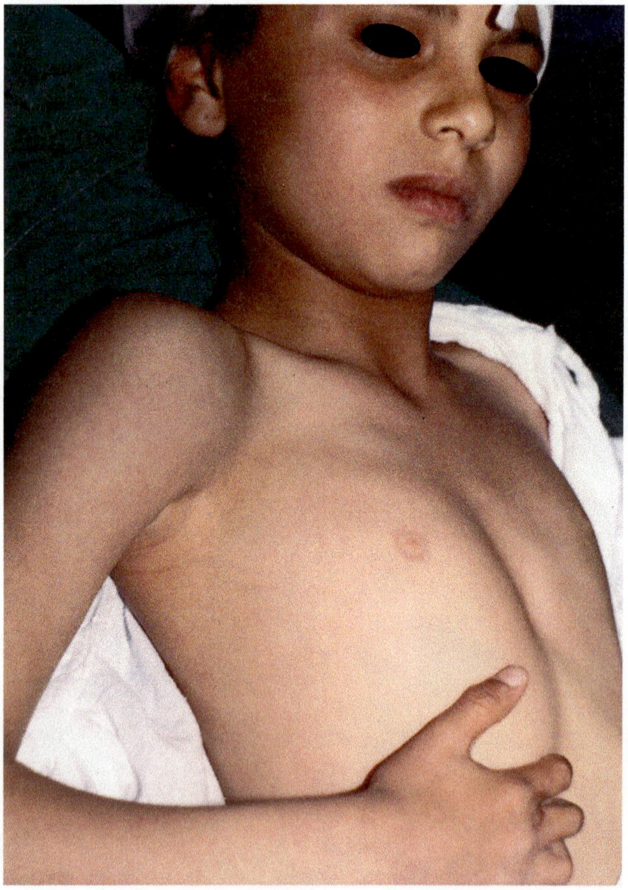

Fig. 27.4 Poland syndrome

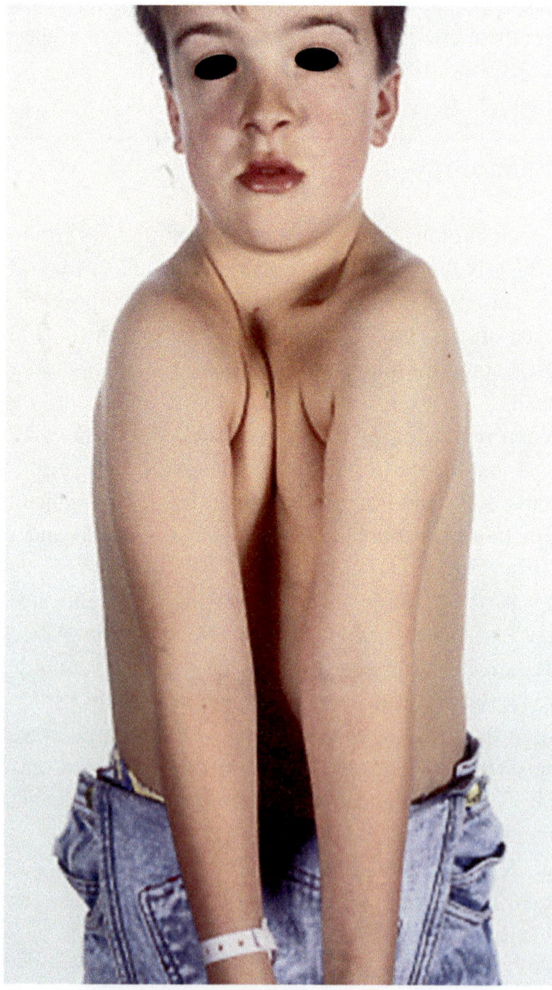

Fig. 27.5 Cleidocranial dysostosis

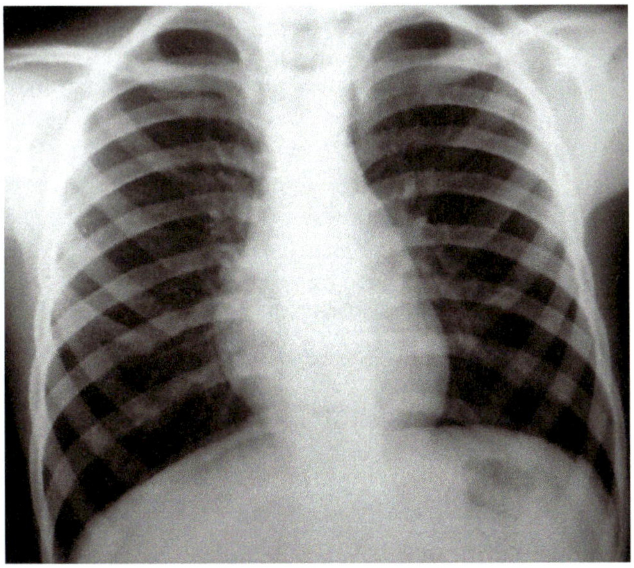

Fig. 27.6 CXR bifid right rib

Embryology

The thoracic wall forms from the development of the ribs and sternum. This commences around the fifth week of gestation. The ventromedial sclerotome segments into the primitive ribs, which then grow ventrally to reach the sternal plate. The sternum forms from two lateral mesenchymal bands. These fuse in a cranio-caudal direction, and complete fusion with the sternal plate and ribs is achieved by the end of the 6th week. Ossification of the ribs then commences from the rib angle. Ossification of the sternum commences around the 6th month of gestation and is not completed until 12 years of age [2]. There are between five and nine distinct ossification centres which correspond to the shape of the sternum [5], and it is hypothesised that the mechanical stressors placed on these centres may contribute to the development of both pectus carinatum and excavatum [1]. The cranio-caudal direction of fusion of the sternal bands may explain why an apparent majority of pectus excavatum appear to affect the lower third of the sternum and is certainly implicated in the development of sternal clefts [6].

What Causes It?

The causative factors behind chest wall deformities have not been fully elaborated. By definition, those present from birth (some forms of excavatum and carinatum, Poland syndrome and sternal clefts) must have a congenital origin. Ventral body development has been shown to be controlled by *Hoxb* gene expression in mice, and this may be implicated in some severe forms of sternal cleft, Cantrell's pentalogy or ectopia cordis [6]. Deletion of the *Gpr126* gene in mice has been shown to cause development of pectus excavatum and scoliosis [7]. *GAL3ST4* mutation has been shown to be present in a familial pectus excavatum spanning four generations [8]. There is further evidence to suggest that there are inherited forms of chest wall deformity—one large population study showed 44% of those with pectus excavatum demonstrated a family history [9].

Chest wall deformities are also associated with some syndromes. The most documented is that of Marfan's syndrome, seen in almost 3% of cases of pectus excavatum. A further 18% are described as having 'Marfanoid' features but no genetic diagnosis [9]. Three per cent are associated with Ehlers-Danlos syndrome [10]. It can be hypothesised that these patients with connective tissue disorders have abnormal ossification and growth of the sternum in childhood, leading to the variety of complex deformities that are often seen in them. Recent work has identified an association of pectus excavatum with maturity onset diabetes of the young—specifically in those with mutations or deletions to the *HNF1B* gene on chromosome 17q12 [11].

Most cases of pectus excavatum and carinatum appear to develop after birth. It can be surmised that the slow ossification of the sternum and ribs during childhood leads to an altered shape if mechanical stressors are applied to different areas—and this is reinforced by the increased incidence in patients with scoliosis, diaphragmatic hernia and chronic respiratory conditions [1, 12].

As of yet, no external teratogenic factor has been implicated in the formation of chest wall deformities.

Clinical Features

Pectus excavatum presents as a 'funnel'-shaped chest. The sternum is sunken, often shaped like a 'cup' or 'bowl'. It is often symmetrical and can be associated with outward flaring of the ribs. It most frequently appears to affect the lower half of the sternum but can extend to the manubrium [1, 2, 13]. It is often associated with discomfort in the growing child, and this probably represents costochondritis of the growing cartilage. Although many patients will present with shortness of breath or exercise intolerance [10], many will have normal pulmonary function tests. However, there is recent evidence to suggest that there may be some compromise identifiable on full cardiopulmonary exercise testing [14], and it is certainly evident that correction of pectus excavatum improves cardiopulmonary function [15, 16].

Pectus carinatum presents as a 'barrel'- or 'pigeon'-shaped chest wall. It may be symmetrical but often has an asymmetrical preference for one side of the sternum. It can appear as a very pointed sternal notch and may also be associated with rib flaring. In younger children it is usually very easy to correct with pressure as the cartilaginous ribs are very compliant. As the rib ossification progresses, this becomes more difficult [10].

Poland syndrome is characterised by absence of pectoralis major, serratus anterior and the external oblique muscles. These are associated with a wide range of deformities of the ribs, sternum and breast in females along with ipsilateral upper limb anomalies [6].

Rare Conditions

Sternal clefts vary in appearance from a simple defect in fusion of the sternum with skin coverage to complete separation of the sternal bands and extrusion of the heart outwith the thorax—so-called ectopia cordis (Fig. 27.7). The thoracoabdominal variant of this is a feature of Cantrell's pentalogy [3, 6]. Thoracopagus conjoint twins (Fig 27.8) will be discussed in Chap. 50.

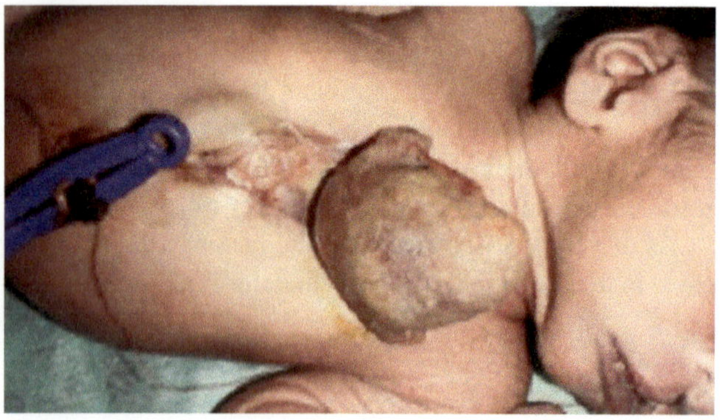

Fig. 27.7 Ectopia cordis

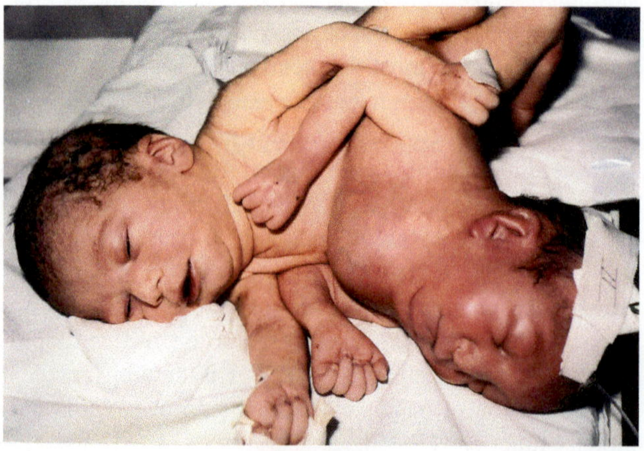

Fig. 27.8 Thoracopagus conjoint twins

Antenatal Considerations

Even the chest wall anomalies which present congenitally can be difficult to identify on antenatal ultrasound. Whereas ectopia cordis and Cantrell's pentalogy will be obvious, the subtle changes in Poland syndrome or in congenital pectus excavatum are not readily visible. In a foetus with ectopia cordis, consideration must be given to the compatibility with survival, as very few cases of the thoracic variant have been documented as surviving. In infants with obvious chest wall deformities, the possibility of a syndrome should be considered, and investigations should be performed if any clinical features, such as those evident in Marfan's syndrome, are recognised [9].

Outcomes

The majority of chest wall deformities have an excellent prognosis. Many patients with pectus excavatum or carinatum will show minimal or no progression during their years of growth. Some will demonstrate progression and present for consideration of treatment. For many, this will be simply based on cosmesis, but a number may complain of shortness of breath or pain [18].

Many centres have moved towards a multimodal model of treatment in a multidisciplinary clinic, given the small numbers of patients and myriad options for treatment [19].

A number of options exist for treating these conditions. Many mild variants of pectus excavatum will respond to simple physiotherapy. Some centres have begun using the so-called vacuum bell developed in Germany to treat the condition with some encouraging early success [20]. More severe examples can be treated surgically, and most will receive the minimally invasive procedure pioneered by Nuss. This involves placement of a stainless steel or titanium bar behind the sternum to lift the defect forward, thus correcting the excavatum. Most centres consider using two bars and leaving them in situ for up to 3 years [17].

In patients with pectus carinatum, early treatment can obviate the need for any surgery. While the chest wall remains compliant, custom-made braces can be designed, fitted and worn to compress the defect back into a more normal position. This method has proven to be very successful. In older patients with more rigid chest walls, surgery may be considered. Most surgery still seems to comprise of a modified Ravitch technique where the costochondral cartilages are divided and some form of metal work is inserted to hold the sternum in a more normal position [21]. However, several minimally invasive options have now been developed including the Abramson technique [22].

For patients with Poland syndrome, a multidisciplinary approach should be taken, including plastic surgery and paediatric surgery. Excellent cosmetic outcomes can be achieved.

In general, sternal clefts should be closed primarily, either with native tissue or a synthetic material [3]. The management of ectopia cordis and pentalogy of Cantrell is challenging and beyond the scope of this chapter.

References

1. Kelly R. Pectus excavatum: historical background, clinical picture, preoperative evaluation and criteria for operation. Semin Pediatr Surg. 2008;17:181–93.
2. Brochhausen C, Turial S, Muller FKP, et al. Pectus excavatum: history, hypotheses and treatment options. Interact Cardiovasc Thorac Surg. 2012;14:801–6.
3. Acastello E, Majluf R, Garrido P, et al. Sternal cleft: a surgical opportunity. J Pediatr Surg. 2003;38(2):178.
4. Martinez-Ferro M, Fraire C, Bernard S. Dynamic compression system for the correction of pectus carinatum. Semin Pediatr Surg. 2008;17:194–200.
5. Ashley GT. The relationship between the pattern of ossification and the definitive shape of the mesosternum in man. J Anat. 1956;90:1.
6. Engum SA. Embryology, sternal clefts, ectopia cordis, Cantrell's Pentalogy. Semin Pediatr Surg. 2008;17:154–60.
7. Karner CM, Long F, Solnica-Krezel L, et al. *Gpr126/Adgrg6* deletion in cartilage models idiopathic scoliosis and pectus excavatum in mice. Hum Mol Genet. 2015;24(15):4365–73.
8. Wu S, Sun X, Zhu W, et al. Evidence for GAL3ST4 mutation as the potential cause for Pectus Excavatum. Cell Res. 2012;22:1712–5.
9. Kelly RJ, Goretsky MJ, Obermeyer R, et al. 21 Years of experience with minimally invasive repair of pectus excavatum by the Nuss procedure in 1215 patients. Ann Surg. 2010;252:6.
10. Goretsky MJ, Kelly RE, Croitoru D, et al. Chest wall anomalies: pectus excavatum and pectus carinatum. Adolesc Med Clin. 2004;15:455–71.
11. Dubois LaForgue D, Bellane-Chantelot C, Subra J-F, et al. Pectus excavatum is part of the clinical spectrum of HNF1B MODY. Diabetes Care. 2014;37:e72.
12. Hong JY, Suh SW, Park HJ, et al. Correlations of adolescent idiopathic scoliosis and pectus excavatum. J Pediatr Orthop. 2011;31(8):870–4.
13. Dean C, Etienne D, Hindoms D, et al. Pectus excavatum (funnel chest): a historical and current prospective. Surg Radiol Anat. 2012;34:573–9.
14. Lesbo M, Tang M, Nielsen HH, et al. Compromised cardiac function in exercising teenagers with pectus excavatum. Interact Cardiovasc Thorac Surg. 2011;13:377–80.
15. Malek MH, Berger DE, Housh TJ, et al. Cardiovascular function following surgical repair of pectus excavatum: a metaanalysis. Chest. 2006;130:506–16.
16. Chen Z, Amos EB, Luo H, et al. Comparative pulmonary functional recovery after Nuss and Ravitch procedures for pectus excavatum repair: a meta-analysis. J Cardiothorac Surg. 2012;7:101.
17. Kelly RE, Mellins RB, Shamberger RC, et al. Multicentre study of pectus excavatum, final report: complications, static/exercise pulmonary function, and anatomic outcomes. J Am Coll Surg. 2013;217(6):1080–9.
18. Steinmann C, Krille S, Mueller A, et al. Pectus excavatum and pectus carinatum patients suffer from lower quality of life and impaired body image: a control group comparison of psychological characteristics prior to surgical correction. Eur J Cardiothorac Surg. 2011;40:1138–45.
19. Pierce A, Williams C. Development of the centre for pectus excavatum and carinatum. J Pediatr Surg Nurs. 2016;5:14–21.
20. Lopez M, Patoir A, Costes F, et al. Preliminary study of efficacy of cup suction in the correction of typical pectus excavatum. J Pediatr Surg. 2016;51:183–7.
21. Ravitch MM. The operative treatment of pectus excavatum. Ann Surg. 1949;129(4):429–44.
22. Abramson H, D'Agostina J, Wuscovi S. A 5-year experience with a minimally invasive technique for pectus carinatum repair. J Pediatr Surg. 2009;44:118–24.

Cardiac Abnormalities

28

Maria Ilina and Stuart Lilley

Heart Embryology

Development of the heart begins in the third week. It develops from the splanchnic mesoderm in the cardiogenic area in front of the buccopharyngeal membrane.

The primordium of the heart is represented by a group of mesodermal cells, the angiogenic cells, which reside in the anterior part of the embryonic disc in front of the buccopharyngeal membrane. Once the ventral head fold has formed, the heart primordium becomes ventral in position.

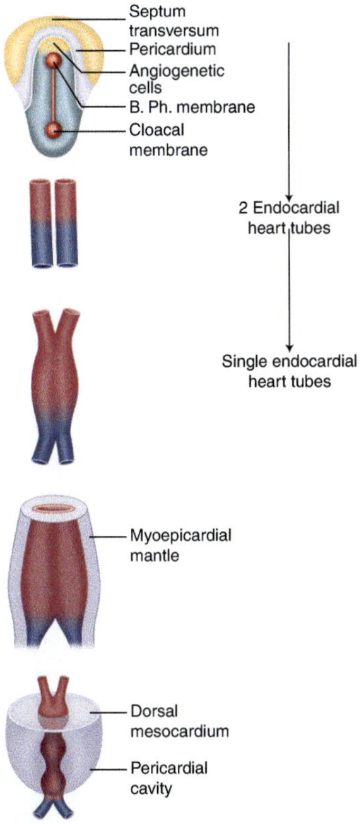

The angiogenic cells give rise to two endocardial heart tubes which lie close to each other in the midline of the embryo. These tubes will unite to firm a single heart tube, forming the endocardium of the heart. This becomes surrounded by splanchnic mesoderm, which develops into the myoepicardial mantle. This differentiates into epicardium and myocardium of the heart. Surrounding the heart tube is the pericardial cavity. This becomes suspended to the roof of the cavity by the dorsal mesocardium; this in turn disappears leading to the formation of a passage dorsal to the heart tube—the transverse sinus of the pericardium.

Differentiation of the Heart Tube

Three constrictions appear in the primitive heart tube, dividing it into four primitive chambers

(a) Bulbus cordis
(b) The primitive (common) ventricle
(c) The primitive (common) atrium
(d) The sinus venosus

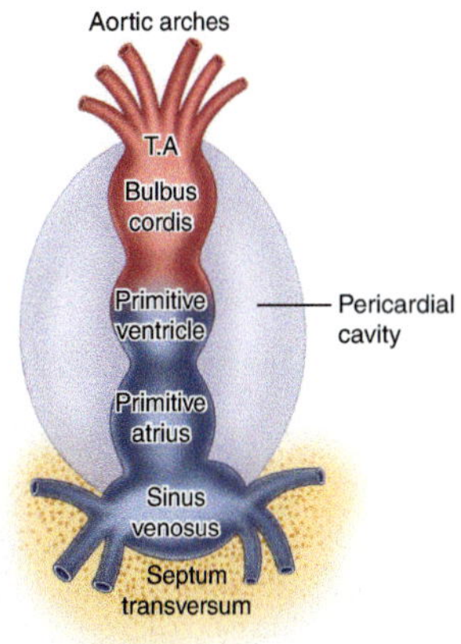

M. Ilina (✉) · S. Lilley
Royal Hospital for Children, Glasgow, UK
e-mail: Maria.Ilina@ggc.scot.nhs.uk

R. Carachi, S. H. E. Doss (eds.), *Clinical Embryology*, https://doi.org/10.1007/978-3-319-26158-4_28

The bulbus cordis is the most cranial part of the heart tube. Its upper end, the truncus arteriosus gives rise to the aortic arches. The primitive ventricle lies below the bulbus cordis, and the primitive atrium below the primitive ventricle. The most caudal part of the heart tube, the sinus venosus is formed of a median part and two horns (left and right), which receive the veins. The sinus venosus is first embedded in the septum transversum, but later separates from it.

Once the heart tube has been formed, it grows at a faster rate than the pericardium, so it becomes folded to form a u-shaped tube. The ventricle now lies below both the bulbus cordis and the primitive atrium. Rapid growth of the cardiac loop results in an s-shaped formation with:

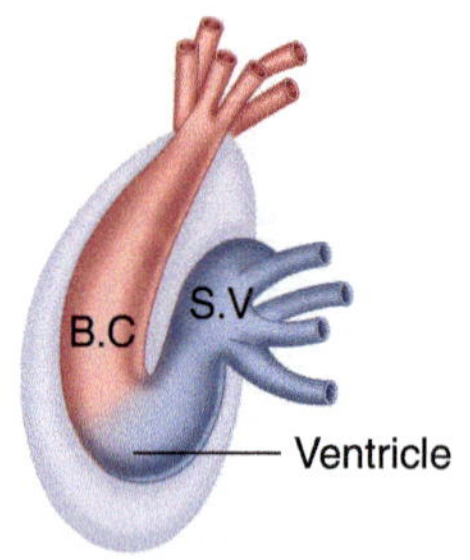

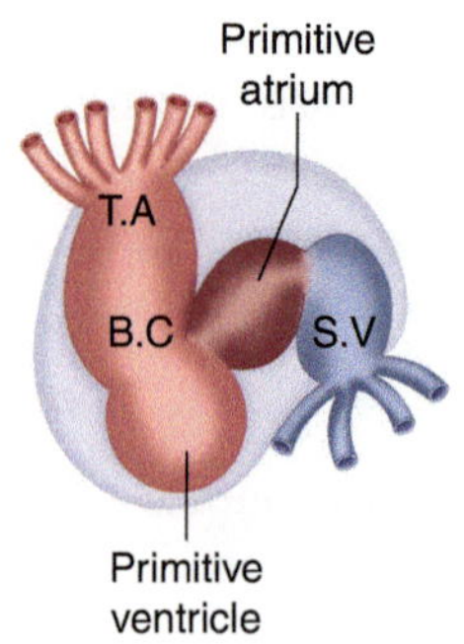

Side view

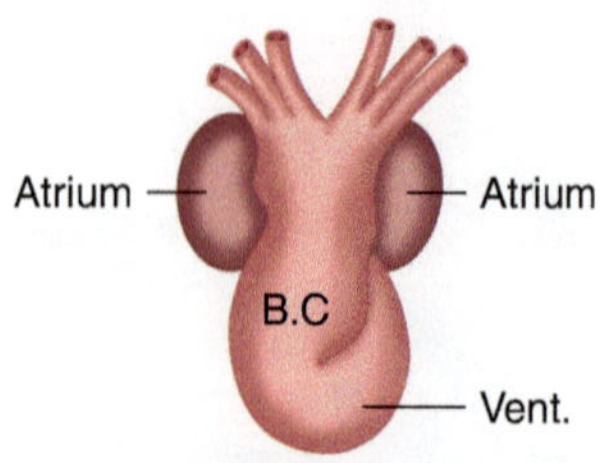

Front view

- The bulbus cordis in front and to the right hand side
- The primitive ventricle lying in front and to the left hand side of the primitive atrium
- The sinus venosus lying behind the primitive atrium

Internal Structures of the Heart

The sinus venosus is the most caudal part of the heart tube and receives all the veins of the body of the foetus. Each horn of the sinus venosus receives the following veins:

1. Vitelline vein—from the yolk sac
2. Umbilical vein—from the placenta
3. Common cardinal vein—formed by the union of the anterior and posterior cardinal veins from the body of the embryo itself

The sinus venosus opens into the primitive atrium by a sino-atrial orifice guarded by right and left valves.

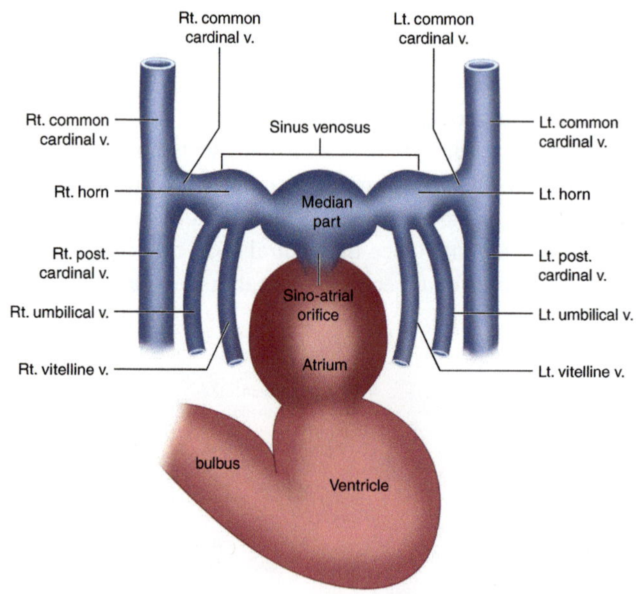

The left sinus venosus diminishes in size because the left vitelline and umbilical veins lose their connections with the sinus venosus. Transverse anastomoses develop, which shift blood from the left-sided veins to the right-sided veins. As a result of this, the left sinus horn becomes smaller and forms the coronary sinus. The left common cardinal vein forms the oblique vein of the left atrium, which ends in the coronary sinus.

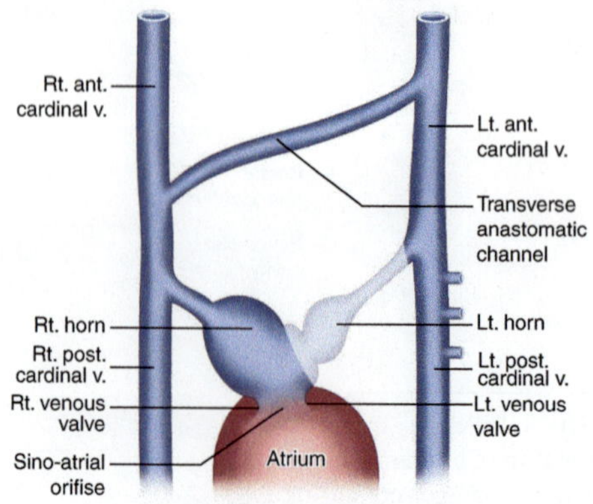

The right horn and the median part of the sinus venosus become absorbed into the right side of the primitive common atrium and will later form the post-smooth part of the right atrium. The right umbilical vein disappears completely and the right vitelline vein forms the uppermost part of the inferior vena cava. The left venous valve of the sinoatrial orifice fuses with the interatrial septum, while the right venous valve has two fates, the upper part forms the crista terminalis, while the lower part forms the valves of the inferior vena cava and the coronary sinus. The septum spurium, which is formed by the fusion of the right and left venous valves at the cranial end of the sinoatrial orifice, forms the lower portion of the crista terminalis.

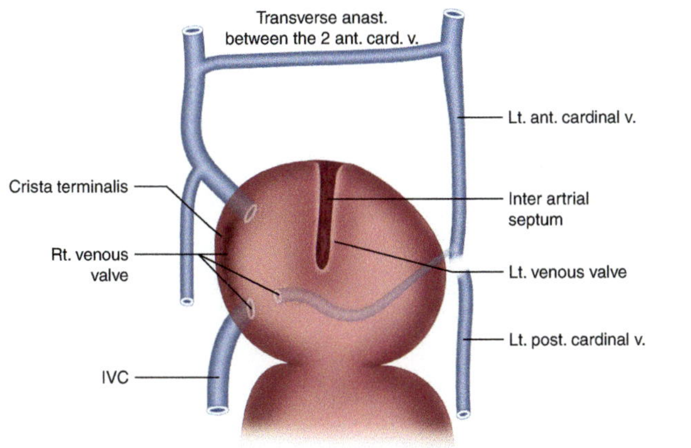

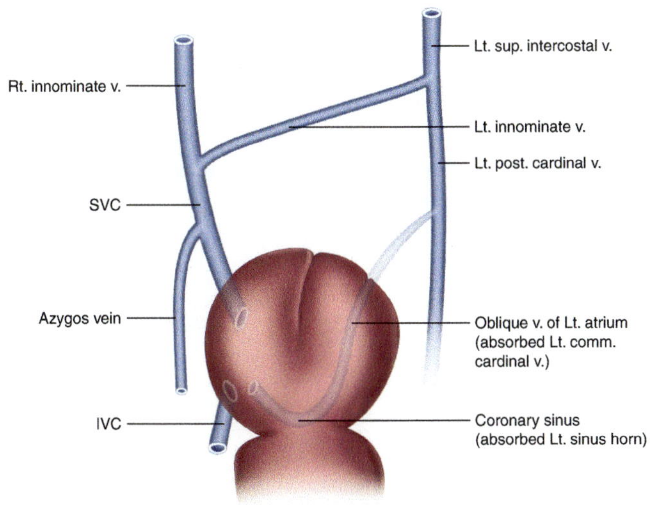

The transverse anastomosis between the two cardinal veins will form the left innominate vein. The upper part of the right anterior cardinal vein forms the right innominate vein. The lower part of the right anterior cardinal vein and the right common cardinal vein will give rise to the superior vena cava. The left common cardinal vein forms the oblique vein of the left atrium.

Development of the Atria

The atrioventricular canal becomes divided into tricuspid and mitral canals by the development of the septum intermedium. Two sub-endocardial proliferations called the ventral and dorsal endocardial cushions appear in the ventral and dorsal walls of the atrioventricular canal. The two cushions approach each other and fuse together forming the septum intermedium.

The interatrial septum develops from three sources:

(a) The septum primum (first to appear)
(b) The septum secundum
(c) The septum intermedium of the AV canal

The septum primum develops first inside the common atrium. It is sickle shaped having a lower crescentic border

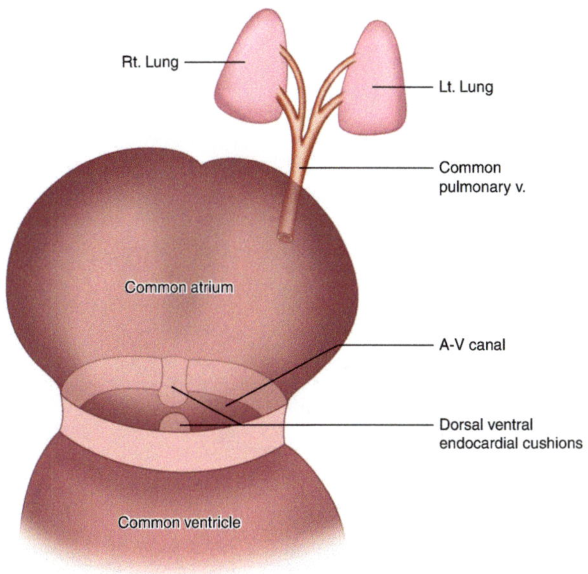

ending in two horns (ventral and dorsal). It arises from the dorsally placed roof of the common atrium on the left side of the sinoatrial orifice and descends in the direction of the septum intermedium until its ventral and dorsal horns unite with the ventral and dorsal endocardial cushions of the septum intermedium. The septum primum divides the cavity of the primitive atrium completely into a right and left half because the lower border is still separated from the septum intermedium by a gap called the ostium primum. As development proceeds, the following changes take place.

(a) The ostium primum becomes obliterated as a result of caudal growth of the septum primum and the proliferation of the endocardial cushions
(b) A second foramen called the ostium secundum appears as a result of breaking down of the cephalic part of the septum primum.

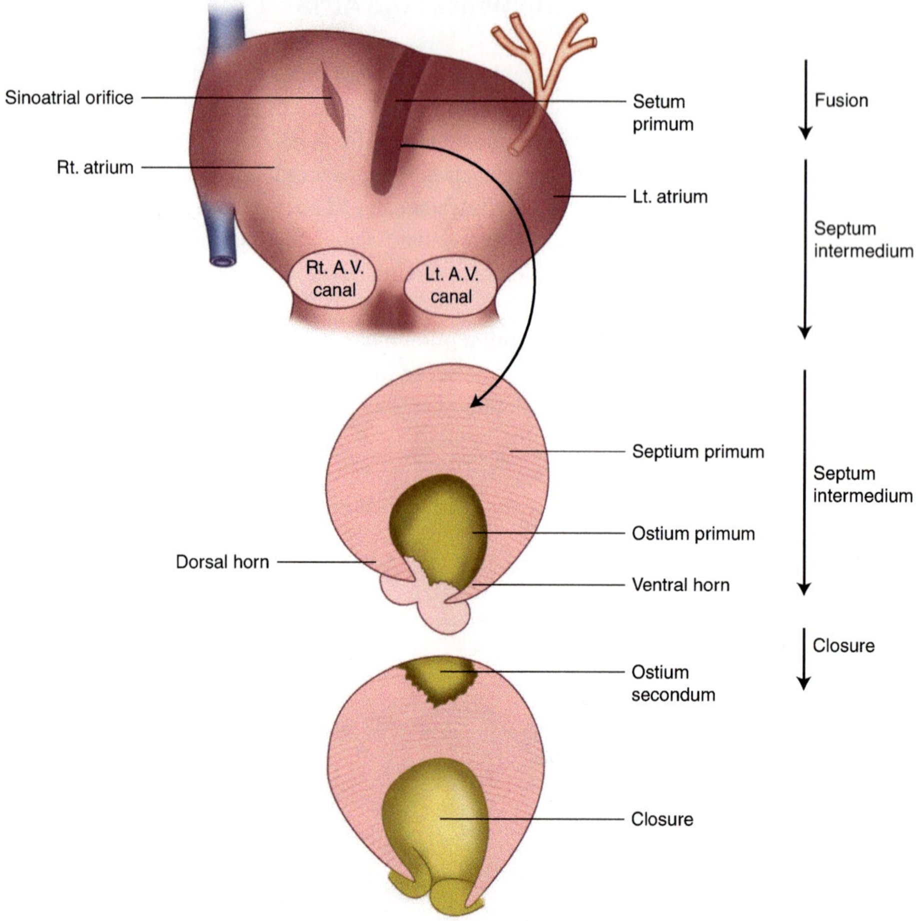

The septum secundum is similar in shape to the primum, sickle shaped with a lower crescentic edge. It arises from the roof of the common atrium on the right side of the septum primum and descends caudally in the direction of the septum intermedium until its two horns fuse with the endocardial cushions of the septum intermedium. It overlaps the ostium secundum but its caudal crescentic edge is still separated from the cephalic free edge of the septum primum by a gap called the foramen ovale. This gradually gets narrower by the caudal growth of the septum secundum.

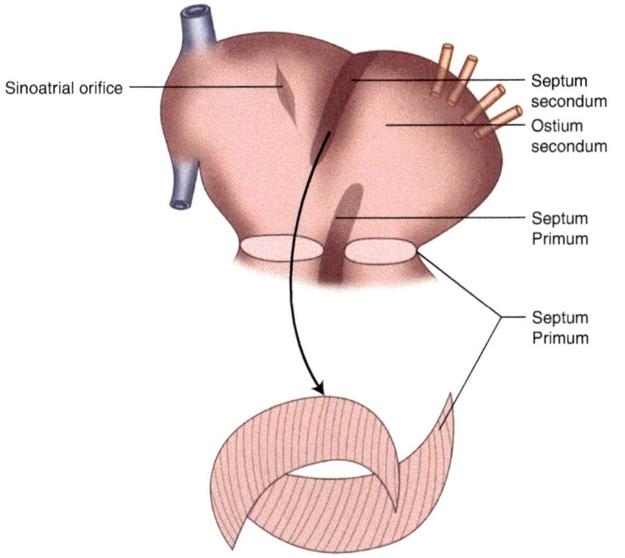

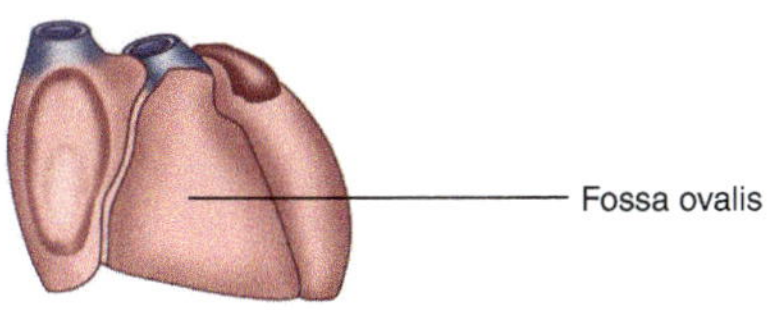

The foramen ovale finally closes at birth due to firm apposition and fusion between the septum primum and secundum to form the interatrial septum.

In the adult heart, there are embryological remnants of the foetal septa. The fossa ovalis represents part of the septum primum enclosed by the two horns of the septum secundum and the annulus ovalis represents the free caudal edge of the septum secundum.

As development progresses, each of the right and left atrioventricular canals becomes absorbed into the corresponding atrium.

Septum Anomalies

1. Premature closure of the foramen ovale
    - Occurs in intrauterine life, this serious anomaly leads to massive hypertrophy of the right atrium and right ventricle, with underdevelopment of the left atrium and ventricle

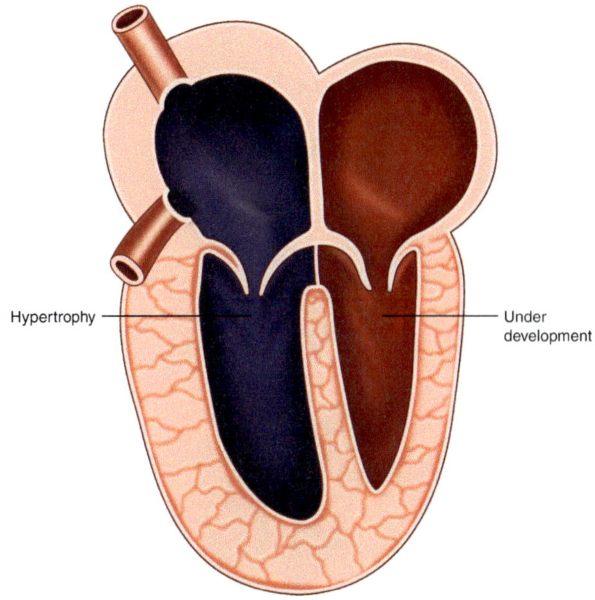

2. Patent foramen ovale
    - Normally the foramen ovale closes completely at birth, but in 20–25% of people, a probe can be passed through; the defect is often so small that blood cannot pass between the two atria, but in larger defects, surgery may be required.

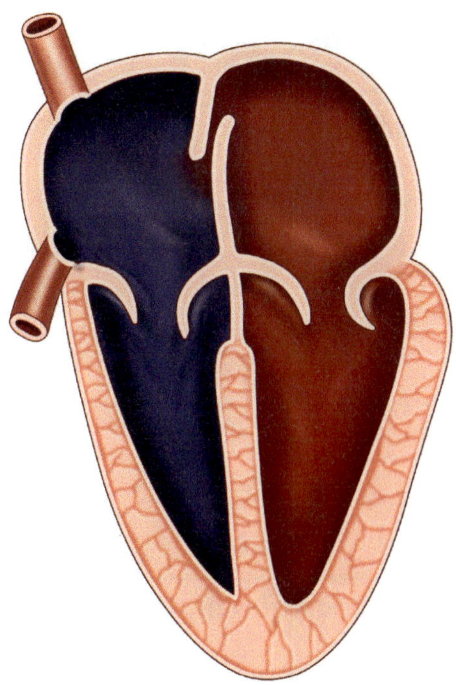

3. Ostium secundum defect
- This is a serious anomaly where there is a large opening between the two atria. It arises due to either excessive resorption of the septum primum or improper formation of the septum secundum. The anomaly leads to severe cyanosis.

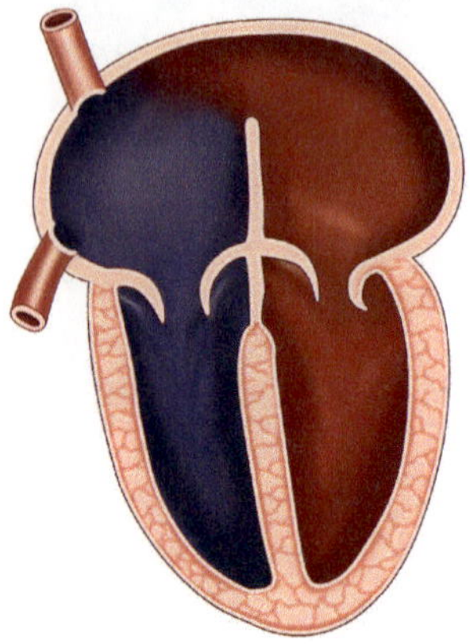

4. Complete failure of formation of the interatrial septum
   - Both septum primum and secundum fail to develop in this serious anomaly, resulting in the heart consisting of three chambers, a common atrium and a left and right ventricle, known as a triocular heart.

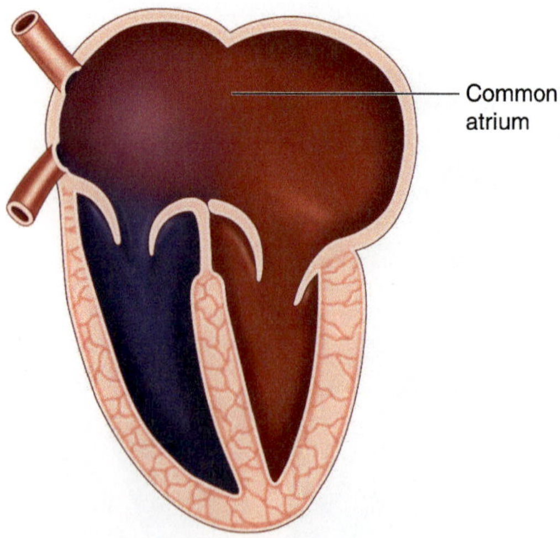

Anomalies of the Atrioventricular Canal

1. Persistent atrioventricular canal
   - Failure of fusion of the anterior and posterior endocardial cushions leads to failure in the formation of the septum intermedium which normally divides the atrioventricular canal into two halves. This anomaly is often accompanied by an atrial septal defect and an interventricular defect as the endocardial cushions share in the formation in both septa.

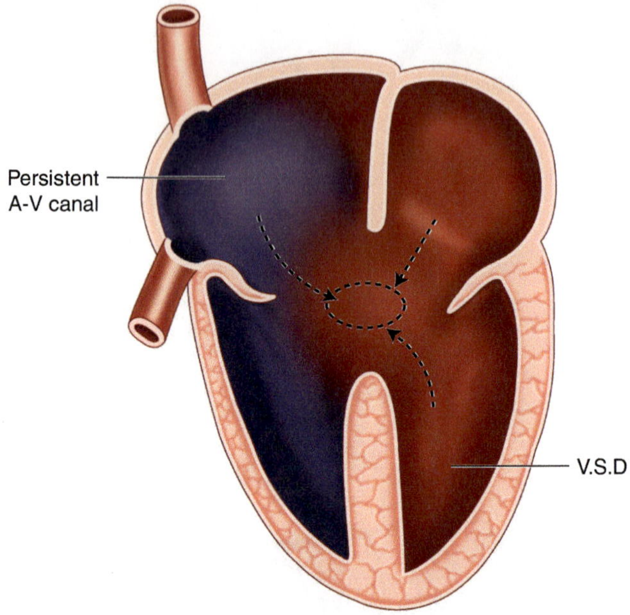

2. Ostium primum defect
   - This anomaly arises due to partial fusion of the anterior and posterior endocardial cushions leading to defective formation of the septum intermedium. The septum primum fails to join the septum intermedium leading to failure of closure of the ostium primum.

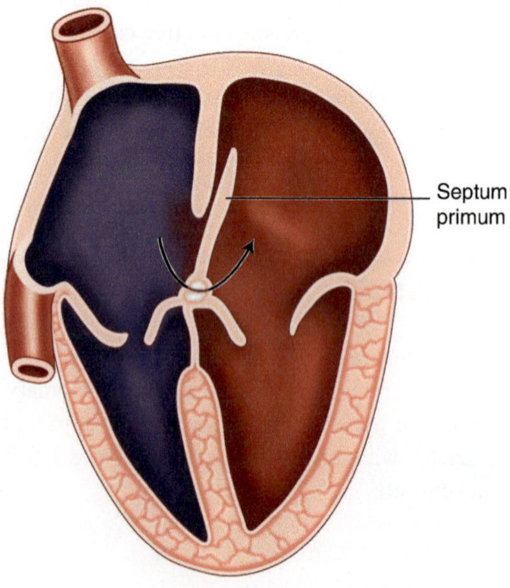

3. Tricuspid atresia
   - Fusion of the cusps of the tricuspid valve leads to narrowing. This is often accompanied by a patent foramen ovale, hypertrophy of the left ventricle, patent interventricular foramen and a small right ventricle.

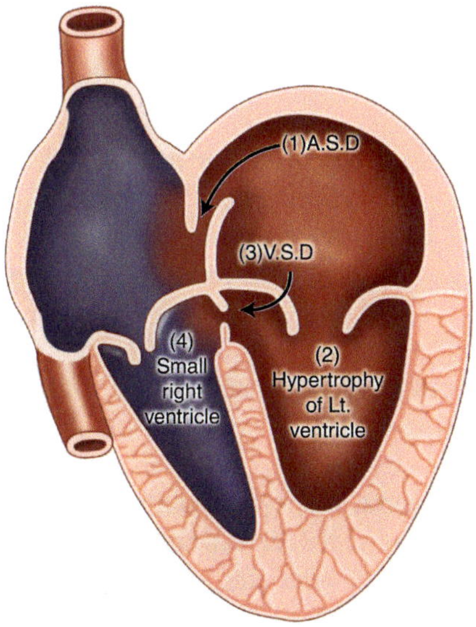

Development of the Ventricles

The bulbus cordis lies to the right side of the primitive common ventricle with a deep sulcus separating the two structures. The sulcus gradually becomes obliterated and the bulbus cordis moves to the left to lie in front of the common ventricle

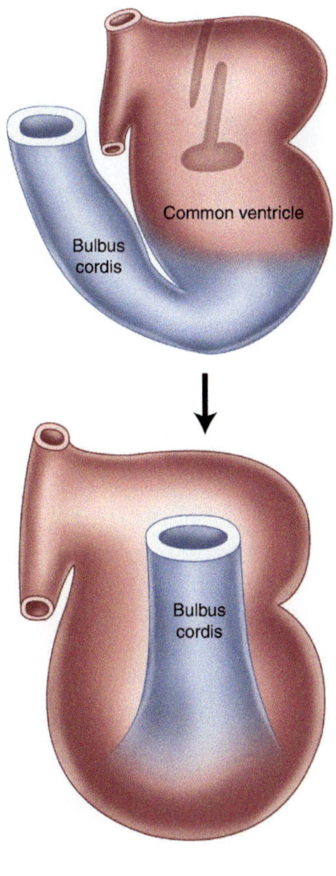

The proximal part of the bulbus cordis becomes absorbed inside the primitive ventricle to form the common bulboventricular chamber. This part will form the trabecular part of the right ventricle. The middle of the bulbus cordis gives rise to the outflow tracts of the right and left ventricles, and the distal part becomes divided by a spiral septum to form the pulmonary trunk and ascending aorta.

The muscular part of the interventricular septum develops from the floor of the bulboventricular chamber as a crescent-shaped ridge with a free upper border with anterior and posterior horns. This ridge grows cranially until the anterior horn fuses with the ventral surface of the root of the bulbus cordis, while the posterior horn fuses with the septum intermedium. This incompletely divides the cavity of the common ventricle which can now communicate with each other.

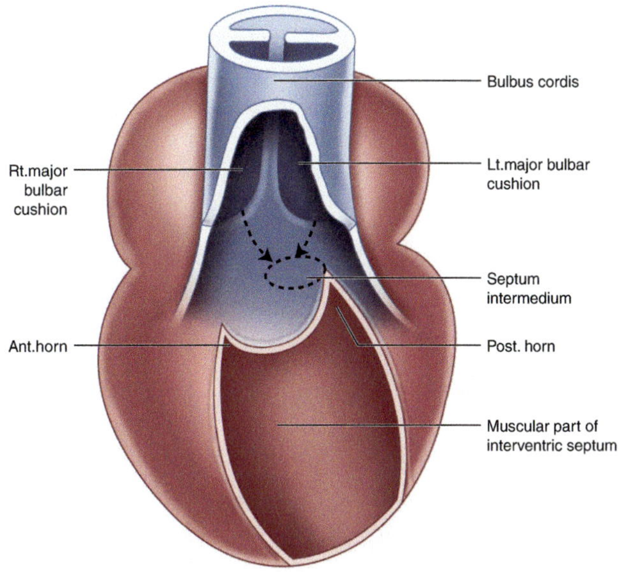

Two more ridges, the left and right major bulbar cushions develop inside the distal part of the bulbus cordis and descend in the direction of the ventricles following a spiral course. They will later fuse to form the aortico-pulmonary septum. The distal part of the bulbus cordis becomes divided by this aortico-pulmonary septum into the ascending aorta and the pulmonary trunk. These are twisted round each other due to the spiral course of the septum. The membranous part of the interventricular septum develops late to fill the gap which still remains between the two ventricles. This is formed by cell proliferation of all the cushions which surround the gap, that is, the right and left major bulbar cushions and the anterior and posterior atrioventricular cushions.

Development of the Aortic and Pulmonary Valves

The right and left major bulbar cushions unite in the region of the valves to form the distal bulbar septum.

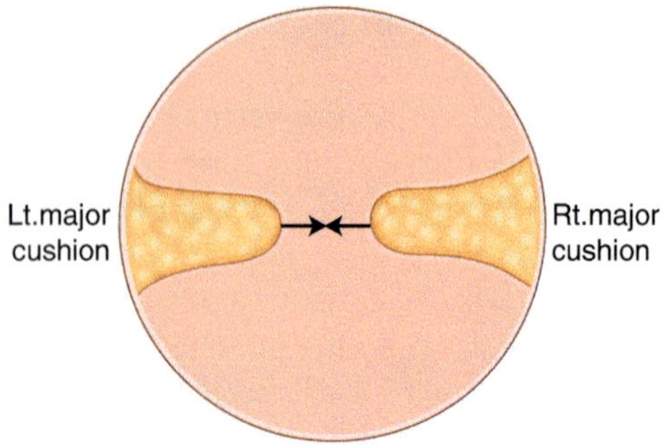

Two accessory ridges, the anterior and posterior minor bulbar cushions arise in the anterior and posterior walls of the common orifice of the bulbus cordis at a plane perpendicular to the bulbar septum. The anterior minor ridge forms the anterior cusp of the pulmonary valve. The right and left major bulbar cushions give rise to the posterolateral pulmonary cusps and anterolateral aortic cusps. The posterior minor bulbar cushion forms the posterior aortic cusp.

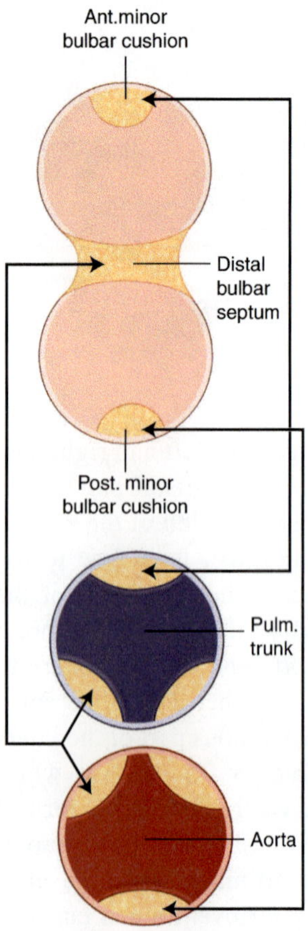

As a result of the rotation of the heart to the left around its longitudinal axis, the position of the cusps changes to acquire their adult position.

Anomalies of the Truncus Arteriosus

1. Tetralogy of Fallot

This is the commonest of all heart anomalies; it arises due to displacement of the aorticopulmonary septum anteriorly, leading to unequal division of the conus, the narrow origin of the pulmonary trunk. The condition consists of four anomalies

(a) Pulmonary stenosis
(b) Hypertrophy of the right ventricle (caused by pulmonary stenosis)
(c) Ventricular septal defect
(d) Overriding aorta, where the mouth of the aorta overlies the ventricular septal defect and receives blood from both right and left ventricles.

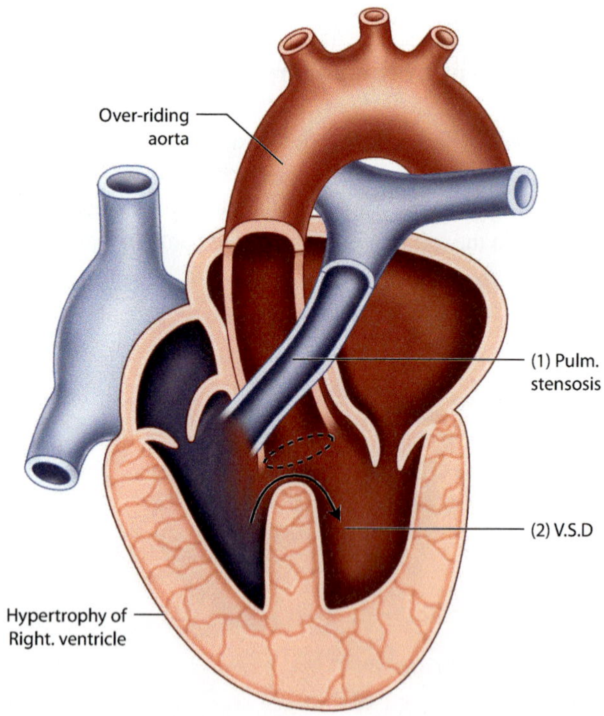

Persistent Truncus Arteriosus

This is due to failure of fusion of the two spiral bulbar cushions to form the aortico-pulmonary septum. This anomaly is usually accompanied by a ventricular septal defect in the membranous part and an overriding aorta.

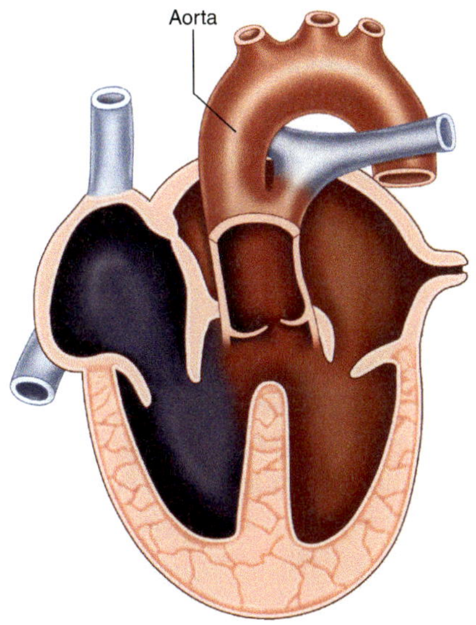

Transposition of the Aorta and Pulmonary Valves

The aortico-pulmonary septum runs in a straight course instead of following a spiral course in this anomaly. This leads to division of the truncus arteriosus into an aorta which is connected below with the right ventricle and a pulmonary trunk which is connected below with the left ventricle. The ductus arteriosus remains patent and is able to carry oxygenated blood to the aortic arch.

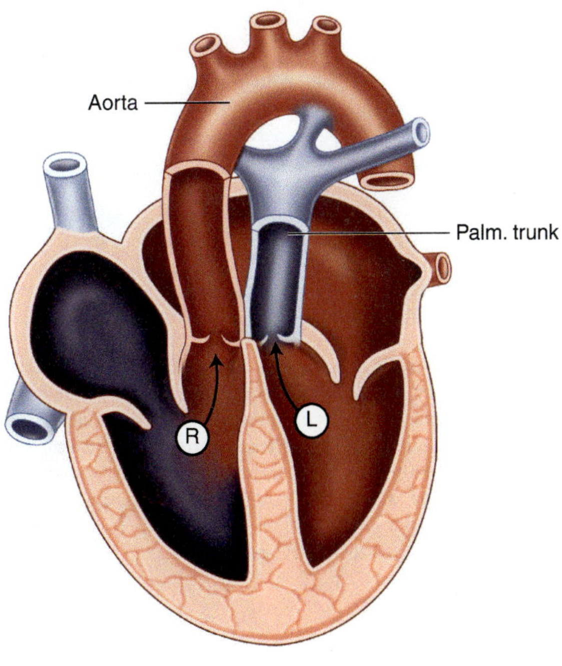

Anomalies of the Semilunar Valves

1. Pulmonary stenosis

This arises due to fusion of the cusps of the pulmonary valve. The ductus arteriosus remains patent and forms the only route to circulation to the lungs.

Aortic Stenosis

This is due to fusion of the cusps of the aortic valve. The ductus arteriosus remains patent and delivers blood to the aorta.

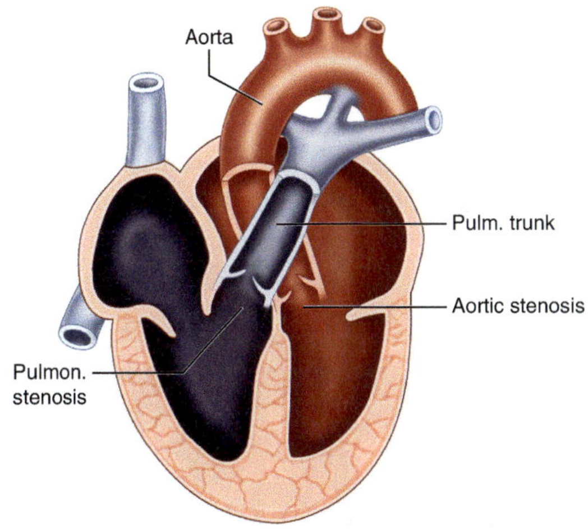

Anomalies of the Position of the Heart

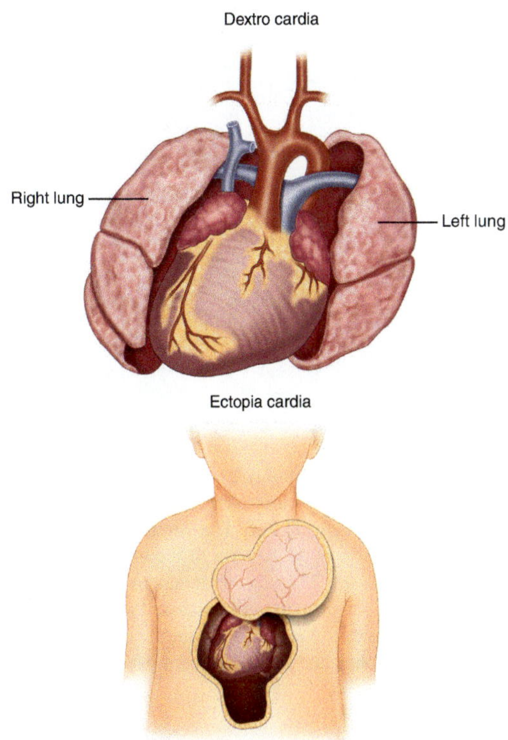

Dextrocardia is the condition where the heart and the great vessels lie as a mirror image to their normal position. It may occur alone or be associated with reversal of all the abdominal organs (situs inversus totalis). Ectopia cordis is a rare condition in which the heart is exposed to the surface of the thorax through a defect in the sternum due to failure of the embryo to close in the midline.

Heart Development

Development of a vascular system is an early necessity in the embryonic life. Experimentation with animal models such as mouse and chicken embryo has improved understanding of the cardiovascular system development and mechanisms underlying human disease. Two extraembryonic sources exist (vessels of the yolk sac and vessels of the chorion) which form the vitelline and umbilical systems, and the third, completely intraembryonic cardinal system, produced close to the developing central nervous system, eventually give rise to the heart and vasculature.

- Formation of the heart tube takes place from the horseshoe-shaped group of cells called cardiogenic crescent in the mesoderm between the yolk sack and the confluence of the left and right coelomic cavities from which all of the body cavities are later derived—pericardial, pleural and peritoneal. This occurs on the 20th day of life. The heartbeat probably begins at the straight tube stage or at the early loop stage.
- Cardiac loop formation (normally to the right) begins by 21 days of age breaking the symmetry of embryo, and forms four cardiac chambers.
- The main features of cardiovascular development from 22 to 28 days of age are (Fig. 28.1):
  - Formation of the cardiac loop
  - Beginning of the development of left and right ventricles
  - Cardiovascular septation
  - Evolution of the aortic arches
  - Commencement of the circulation
- The main features of cardiovascular development between days 29 and 35 of embryonic life are:

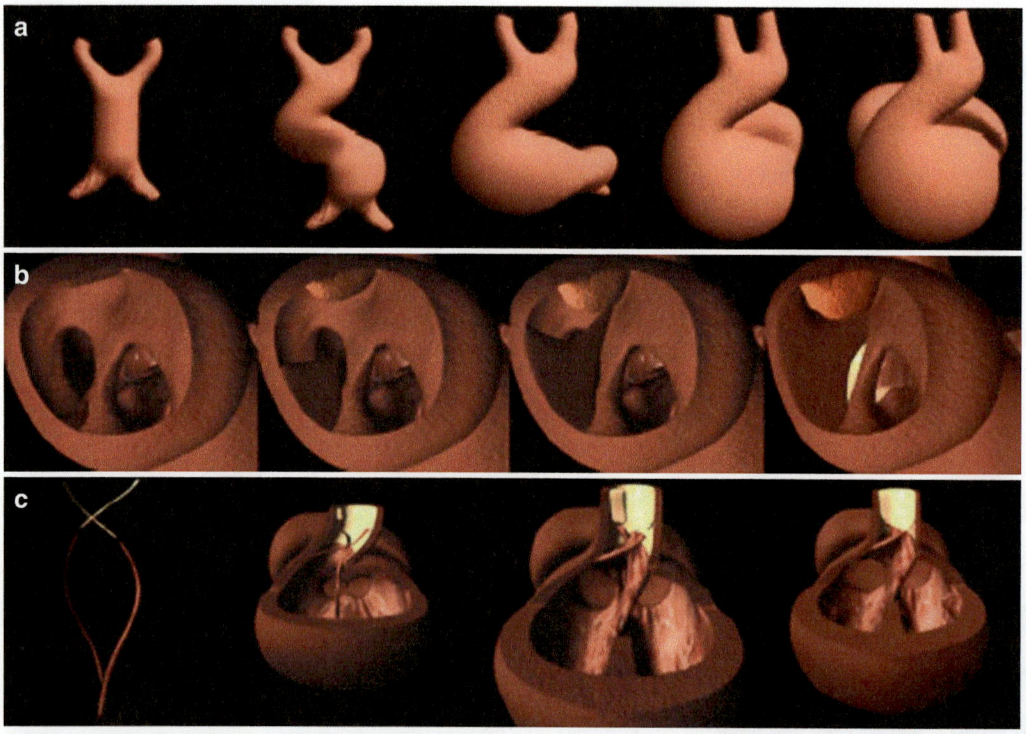

Fig. 28.1 Embryogenesis from day 21 to 28. (**a**) The cardiac loop is formed. The heart tube is folded into an S-shaped dextro-ventral convexity. (**b**) The atria are partitioned. The septum primum (in brown) grows from the inferior part of the atria to the top, leaving a foramen called the ostium primum. The septum secundum (in orange) comes from the top. The ostium primum will be closed at the end of the fifth week by an expansion of tissue coming from the endocardial cushions (in yellow). (**c**) The conus and the truncus are partitioned. The dextrodorsal and sinistroventral conus (or infundibular) ridges, which are isolated in the first picture, partition the conus (outlet part of the heart) by a helical outgrowth into two cavities: the subpulmonary and the subaortic coni (or infundibulums). The truncus is partitioned from the bottom upward from aortopulmonary swellings, leading to the formation of the aorta and pulmonary arteries (adapted from: Jean-Marc Schleich, MD, Jean-Louis Dillenseger, PhD, Virtual Imaging for Teaching Cardiac Embryology, Circulation 2001: 104: e134)

- The LV, RV and ventricular septum continue to grow and develop
- There is approximation of the aorta to the interventricular foramen, mitral valve and left ventricle
- Separation of the ascending aorta and main pulmonary artery occurs
- The right ventricle enlarges and muscular ventricular septum moves from right to left leading to the opening of the tricuspid valve into the right ventricle
- The ostium primum is closed by tissue from the endocardial cushions and separates the atria
- The ventricular apex swings leftward
- From day 30 to 36 the pulmonary valve moves from the posterior to the anterior and to the left of the aortic valve (its normal position)
- Continuous development of the aortic arches and contribution of the neural crest cells to the development of the infundibulum, great arteries and their branches (Fig. 28.2). The knowledge of the embryonic development of the aortic arches is key to understanding of the sidedness of the aortic arch and vascular rings

• The main features of the cardiovascular development between the 36th and 49th days of life (weeks 6–7 of the embryonic development) are
  - Closure of the infundibular septum
  - Closure of the membranous part of the ventricular septum (can be delayed until after birth, known as spontaneous ventricular septal defect closure)

The primary heart field was initially thought to contribute to the entire heart. In fact it gives rise to the caudal segment of the cardiac tube, eventually forming the left ventricle and part of the atria. The discovery of the second cardiac field has outlined the embryonic development of the heart as a process where undifferentiated cells, or cardioblasts, are recruited from the second lineage in response to local cues such as growth factors, differentiate into cardiac cells myocytes, endothelial cells and smooth muscle of the blood vessels. In some genetically determined conditions such as, for example, DiGeorge syndrome (Table 28.1), the defective gene operates primarily in the second lineage of cells affecting its differentiation and migration, and leads to malformations involving outlet part of the embryonic heart and arterial trunk.

Table 28.1 Embryological components of each atrium

Components of the right atrium	Components of the left atrium
Right half of the common atrium, forms the anterior part and the auricle	Left half of the common atrium, forms the auricle of the left atrium
The absorbed right atrioventricular canal	The absorbed right atrioventricular canal
The absorbed sinus venosus	The absorbed pulmonary veins

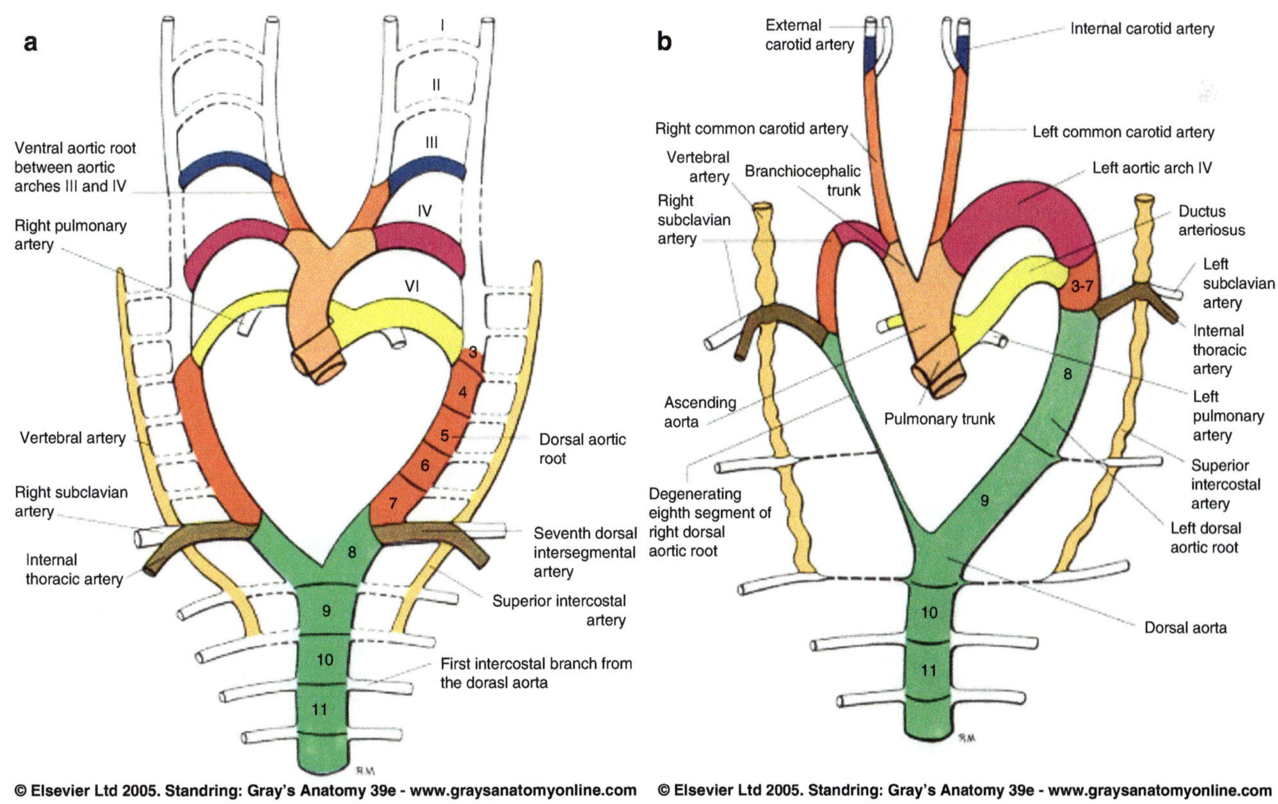

Fig. 28.2 (**a**, **b**) Development of aortic arches

Myocardium has a spongy structure consisting of a mass of trabeculations lined by endocardium. The intertrabecular spaces play a role in the formation of the myocardial vascular bed. They are encased within the myocardium by consolidation of the trabeculations. An intermittent stage exists in which a subendocardial vascular plexus within the myocardium is continuous with the cardiac lumen as well as with the ascending aorta. Endothelial sprouts form a ring next to the developing arterial trunk and invade the aorta from the outside, later developing a lumen and producing coronary arteries orifices. Differentiation of the vessel walls into arterial, capillary and venous starts from the aorta and proceeds in an apical direction.

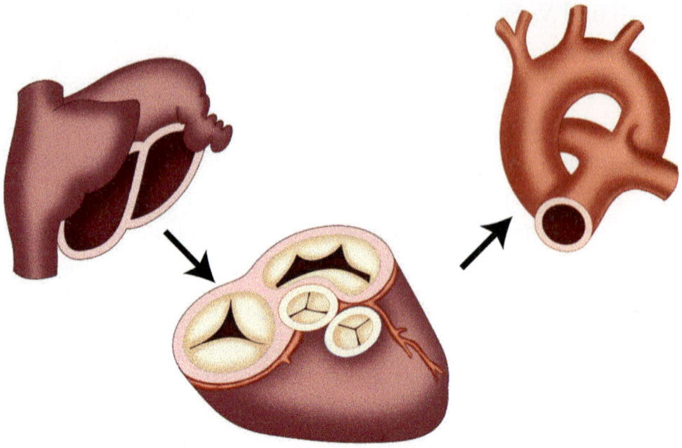

Fig. 28.3 Segmental analysis: heart is presented as three segments: atria, ventricles and great arteries connected at atrioventricular and ventriculoarterial junctions (From: Frescura C, Thiene G. The new concept of univentricular heart. Frontiers in Pediatrics, July 2014, Volume 2, Article 62)

Segmental Approach to Diagnosis

The heart has three "building blocks", namely the atriums, the ventricular mass and the arterial trunks (Fig. 28.3). The system of sequential segmental analysis used for categorisation of the heart malformations has significantly improved diagnosis and descriptive classification of complex congenital heart disease. First, atrial chambers arrangement ("atrial situs") is established:

- Solitus (LA on the left, RA on the right)
- Inversus (LA on the right, LA on the left)
- Atrial isomerism, or isomerism of the atrial appendages (right or left)

Thereafter it describes anatomical nature of the junctions between the atrial and ventricular myocardial mass, or "type of atrioventricular connection":

- Concordant (LA to LV, RA to RV)
- Discordant (LA to RV, RA to LV)
- Absent (left, right)
- Double inlet (left, right)

In addition, morphology of the valve or valves that guard the atrioventricular junctions is described (mitral, tricuspid, left AV valve, right AV valve, common AV valve).

After this, ventriculo-arterial junctions are described (separate attention is paid to the morphology of outflow tracts and great vessels relationships):

- Concordant (LV to Ao, RV to PA)
- Discordant (LV to PA, RV to Ao)
- Double outlet (left ventricle, right ventricle)
- Common arterial trunk
- Single outlet heart (pulmonary atresia with absent main PA and true central PA)

Then a "catalogue" of all associated cardiac malformations is made (e.g., ASD, VSD, AV valvar lesions, subarterial and arterial valves lesions, side of the aortic arch, presence of coarctation or patent arterial duct), and non-cardiac malformations when appropriate (bronchial morphology, abdominal organs arrangements such as asplenia, polysplenia, gut malrotation—associated with disorders of laterality, that is, isomerism, etc.).

There is a set of anatomical features present in most congenitally malformed cardiac chambers, no matter how unusual the chamber position is, that allows correct anatomical identification of each chamber or great vessel in the vast majority of cases. Common patterns of malformations exist; however, almost any combination is possible, some exceedingly rare. The term "common atrium" usually describes almost complete failure of the development of interatrial septum, whereas true absence of a ventricle (true solitary ventricle) is exceedingly rare but also has been described.

Part I—Cardiac Malformations

Ventricular Septal Defects

Introduction and Background

The term *ventricular septal defect* describes an opening in any part of the ventricular septum. Isolated ventricular septal defect is the most common congenital heart lesion (Table 28.2). It may also form part of a more complex malformation, such as tetralogy of Fallot or pulmonary atresia, and is usually a part of complex lesions resulting in univentricular circulation.

Embryology

Ventricular septal defects result from failure of the closure of the interventricular septum beyond the first 7 weeks of

Table 28.2 Epidemiology of congenital cardiac defects

Lesion	Prevalence per 1000 live births	Incidence % of all CHD
All cardiac defects	19	100%
VSD	2–3	31.4%–32.1%
ASD	0.67	5%–11.4%
AVSD	0.19	2.9%–7.4%
TAPVD	0.06–0.07	1%–1.5%
Tetralogy of Fallot (and double outlet right ventricle)	0.28	3.5%–6.8% (2.0%)
Pulmonary atresia with VSD	0.07	2.4%–3.4%
TGA	0.2–0.3	4%–7%
Coarctation of the aorta (and interrupted aortic arch)	0.08	4.6%–5.7% (0.7%)
Aortic stenosis	0.5	2.9%–6%
Common arterial trunk	0.03–0.056	<1–1.2%
Pulmonary stenosis	0.8	5%–9.0%
PDA	0.2–0.5 ("silent" up to 2)	5%–10%
Tricuspid atresia	0.06–0.08	0.7%–2.4%
HLHS	0.16–0.36 (before foetal echocardiography)	0.6%
Ebstein malformation	0.072	0.3%–0.5%
Congenitally corrected transposition	0.03	0.05%–1.1%
Single ventricle	–	<1%
Isomerism	0.1 (reports vary)	1%–2.5%

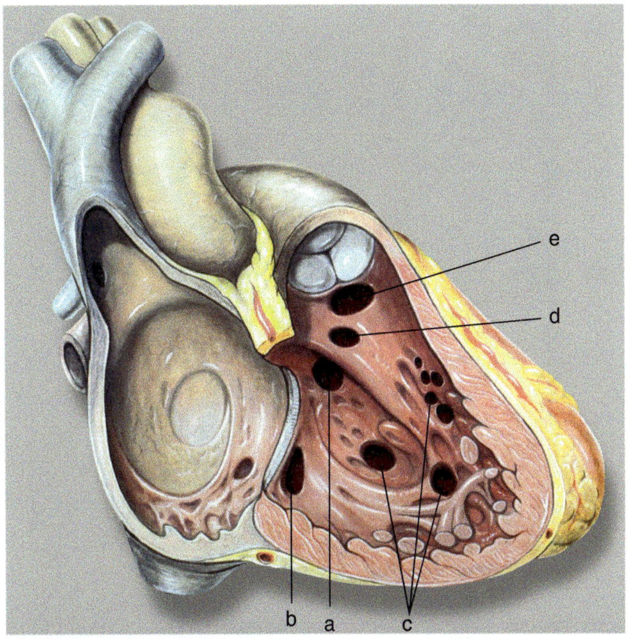

Fig. 28.4 Anatomic locations of various types of VSD, viewed with the RV anterior free wall removed. (**a**) Perimembranous VSD (**b**) inlet muscular VSD; (**c**) trabecular VSD; (**d**) infundibular or outlet muscular VSD; (**e**) doubly committed subarterial VSD

intrauterine life. The particular mechanisms vary depending on the part of the interventricular septum that the defect is located in. Perimembranous defects are likely to have a complex mode of formation that accounts for their diversity in position and orientation. The mechanisms, by the VSD type, include (a) failure of closure of the embryonic interventricular communication by the membranous septum, and deficiency of the adjacent muscular septum; (b) failure of fusion of muscular septum trabeculations; (c) failure of fusion of the muscular septum with the subpulmonary infundibular part of the septum, producing a muscular defect opening between the right and left ventricular outlets; and (d) the doubly committed subarterial defect occurs as a result of failure of formation of muscular subpulmonary infundibulum and is developmentally related to common arterial trunk.

Epidemiologic, Genetic, Morphologic and Clinical Correlations

Many genetic and chromosomal abnormalities as well as environmental factors and maternal disease such as diabetes have been shown to increase the incidence of ventricular septal defect. Subpulmonary and doubly committed subarterial defects are more common amongst Asian than Western populations. Children of parents who have a congenital cardiac lesion run a 10-fold increase of risk of being born with the defect themselves. This is true for any congenital cardiac lesion.

The ventricular septum consists of a small membranous portion and a larger muscular septum (Fig. 28.4). The muscular septum has three components: the inlet, infundibular, and trabecular (or muscular) septums. The perimembranous VSD (Figs. 28.4 and 28.5) is more common than the trabecular (70% vs. 5%–20%) (Figs. 28.4 and 28.6), infundibular (5%–7%) or inlet defects (5%–8%) (Fig. 28.4). Frequent associations are patent arterial duct (PDA) and coarctation of the aorta (CoAo). Larger VSDs are easily identifiable during antenatal screening, smaller defects are usually diagnosed after birth.

In the hearts with small VSDs and small to moderate left to right shunts, volume overload is placed on LA and LV but not on the RV. Pulmonary blood flow is increased to a varying degree depending on the size of the defect and the pulmonary vascular resistance. Children with small VSDs are asymptomatic, with normal growth and development. Amongst patients with large VSDs, delayed growth and development, recurrent respiratory infections and decreased exercise tolerance are common. If a large VSD is left untreated, pulmonary hypertension due to a large left to right shunt leads to irreversible changes in the wall of pulmonary arterioles and pulmonary vascular disease. At this stage, surgical correction is no longer recommended. Right ventricular pressure may exceed the left and right to left shunt across the VSD may develop leading to the development of cyanosis, and this life-limiting condition is called Eisenmenger's syndrome.

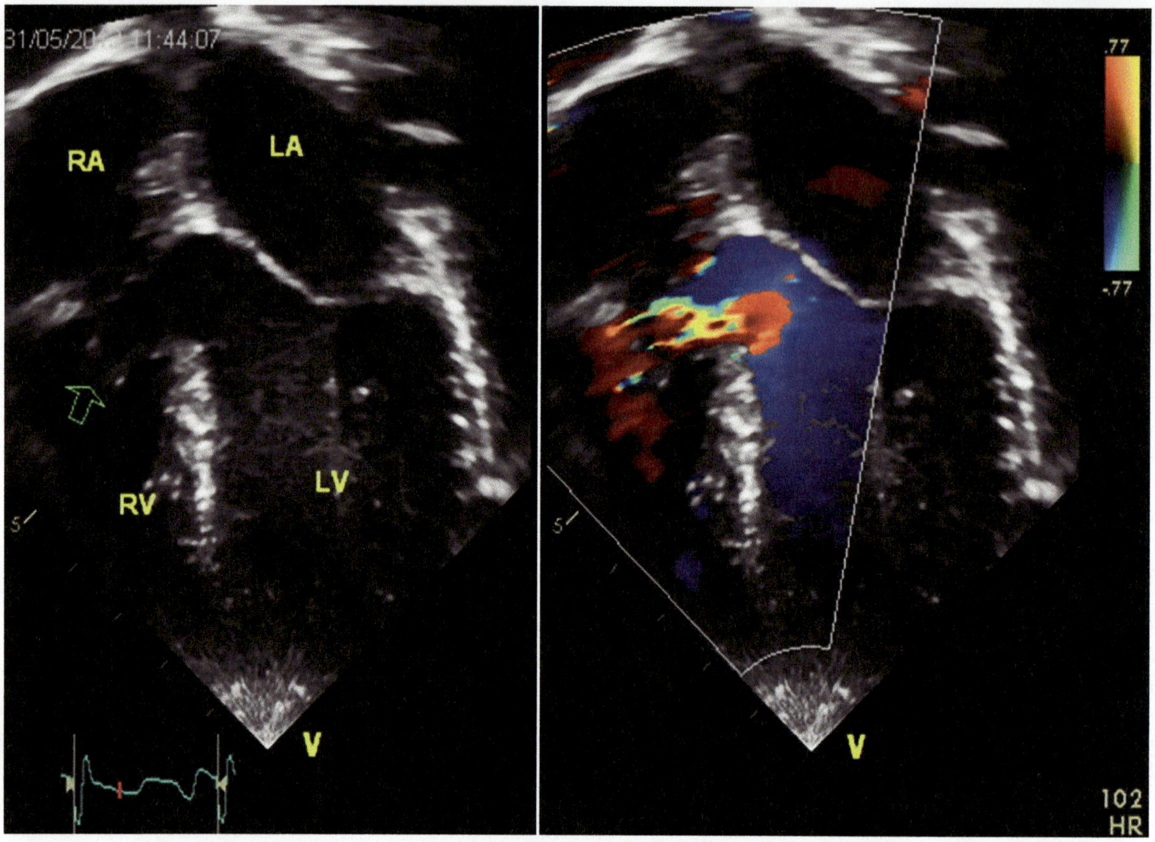

Fig. 28.5 Echo image: perimembranous VSD with aneurysmal membranous septum (apical four chambers view)

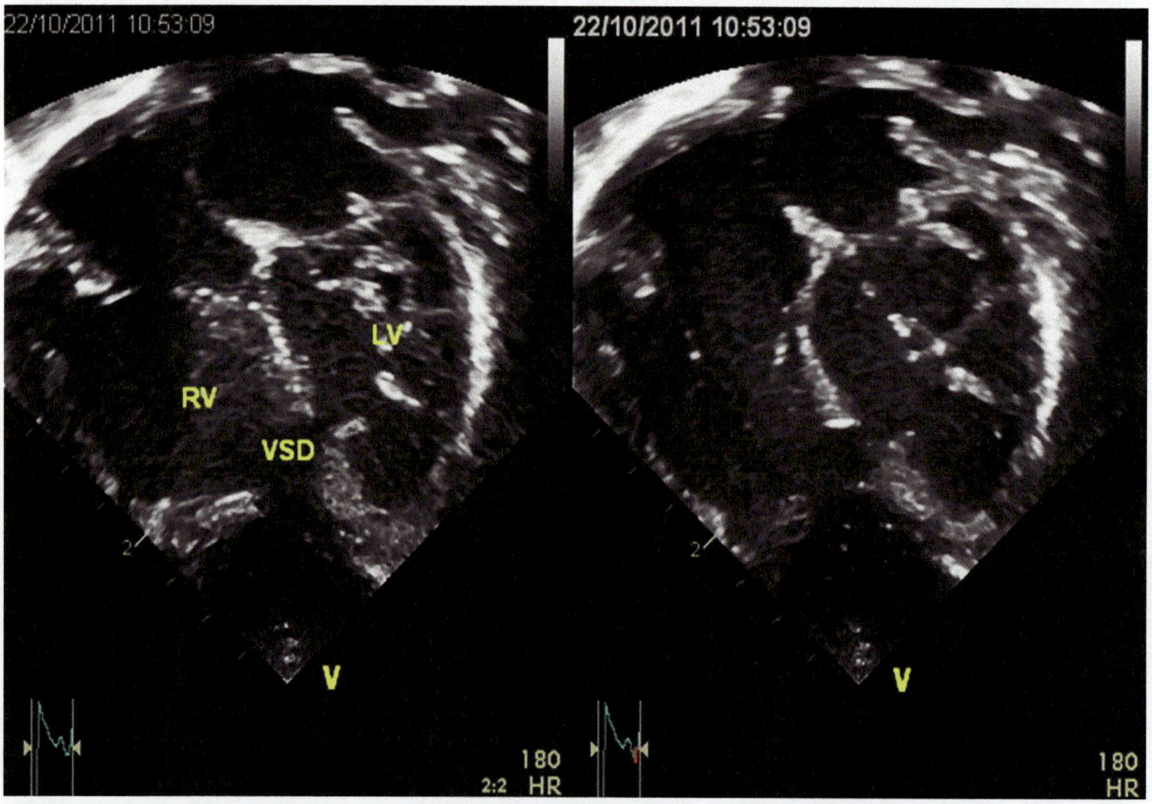

Fig. 28.6 Echo images: muscular VSD located in mid-trabecular septum

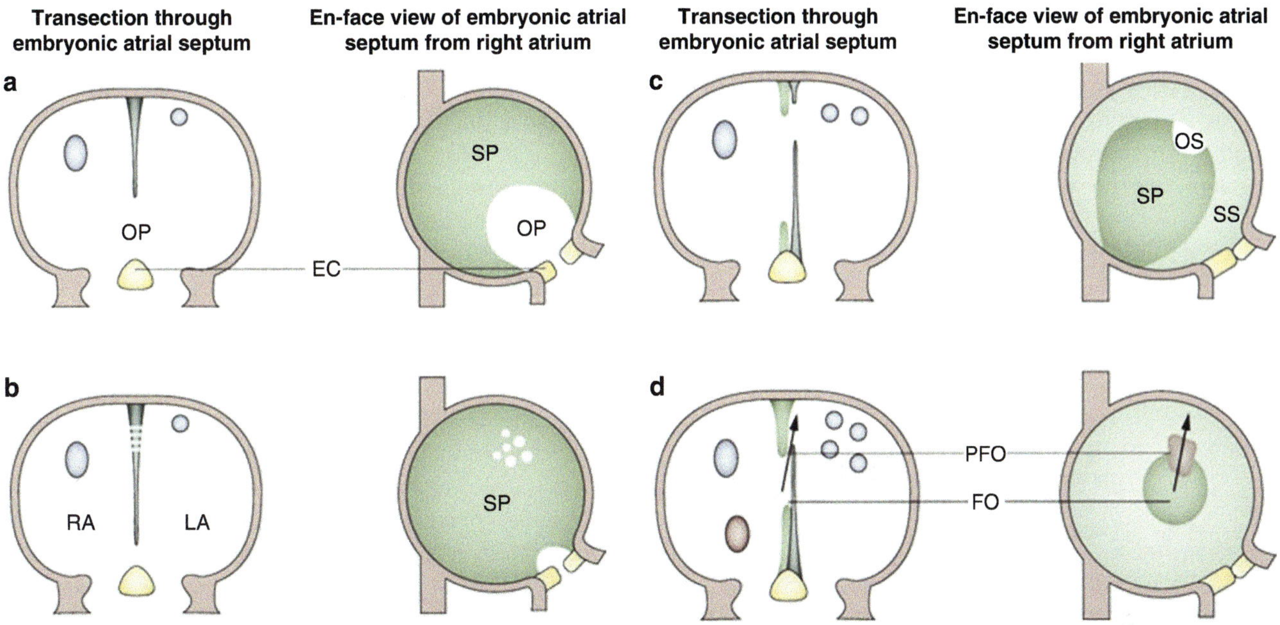

Fig. 28.7 Development of the atrial septum in utero. *OP* ostium primum, *SP* septum primum, *OS* ostium secundum, *SS* septum secundum, *RA* right atrium, *LA* left atrium, *PFO* patent foramen ovale, *FO* fossa ovale. From: Calvert PA, Rana BS, Kydd AC, Shapiro LM. Patent foramen ovale: anatomy, outcomes, and closure. Nat Rev Cardiol. 2011;8(3):148–60

Atrial Septal Defects

Introduction and Background

The term "atrial septal defect" describes any communication between the two atria. This is the third commonest congenital heart defect and an example of acyanotic heart lesion. It can present at any age, in children the most frequent presentation is an incidental heart murmur discovery, in adults as a result of investigations following a stroke.

Embryology

Cardiac chambers develop by ballooning from the lumen of the primary heart tube (Fig. 28.1a, b).

In the embryo, cell addition at the venous pole produces initially the atrioventricular canal and then the atrial chambers precursor, to which drain the systemic veins. Future atrial chambers balloon symmetrically and appear to either side of the outflow tract (Fig. 28.1a). There is another population of cells in the atrial component of the heart, called mediastinal myocardium, which makes up the part of the tube that retains its connection with the developing mediastinum. These cells will eventually form the dorsal wall of the left atrium, or pulmonary venous component. The primary atrial septum, or septum primum, grows down from the roof of the atriums. Within the atrioventricular canal, endocardial cushions grow towards each other and divide the canal into two channels. The primary septum grows towards the cushions, carrying on its leading edge a "mesenchymal cap". By the time the mesenchymal cap approach the cushions, the part of the septum, at the atrial roof, breaks down, creating the secondary interatrial foramen (Fig. 28.7b, c).

The primary foramen, or ostium primum, is the space between the mesenchymal cap and the atrioventricular endocardial cushions (Fig. 28.7a). Fusion of the mesenchymal cap with the endocardial cushions obliterates the primary atrial foramen. The base of the newly formed atrial septum is then further reinforced by growth of mesenchymal tissue. This tissue together with the mesenchymal cap on the primary septum anchors the septum firmly against the central fibrous body of the heart, formed from the fused atrioventricular cushions. This part of the interatrial septum is called secondary septum, or septum secundum (Fig. 28.7c). The septation is complete by the 9th week of embryonic development. Later on, as the pulmonary veins are incorporated into the atrial roof, the upper margin of the oval foramen becomes a fold of the flap valve. The flap itself is formed by the primary atrial septum. Once the pulmonary veins are incorporated into the atrial roof, the upper margin of the oval foramen turns into the interatrial fold which provides the buttress for the flap valve. This process is completed well after the finish of definitive cardiac septation (Fig. 28.7d).

Epidemiologic, Genetic, Morphologic and Clinical Correlations

Female preponderance of ASD is reported (male to female ratio of 1:2). Some estimate the incidence of patent foramen ovale (PFO) as high as 20% of the general population. There are no known intrauterine events which predispose to defects of atrial septation, and most cases occur sporadically. ASD is seen with a higher frequency in a number of genetically determined conditions and syndromes (Table 28.3). There are familial cases associated with mutations of the TBX5 gene and forming part of the Holt–Oram syndrome when the atrial septal defect is associated with certain skeletal abnormalities of the forearm and hand, and occasional familial cases without these associated abnormalities. Atrial septal defects also nearly always form part of complex cardiac lesions.

Table 28.3 Chromosomal and genetic disorders commonly associated with congenital heart lesions

Syndrome	Genetic marker	Heart defect
Alagille syndrome	JAG-1 NOTCH 2 20p12 microdeletion	Peripheral branch PA stenosis
CHARGE association	CHD7 gene mutations Sporadic inheritance	PDA, ASD/VSD/AVSD Tetralogy of Fallot DORV Aortic arch anomalies
Cornelia de Lange syndrome	NIPBL, SMC1A, SMC3 gene mutations	VSD Aortic stenosis
Cri du Chat syndrome	Deletion 5p	VSD, PDA, ASD
DiGeorge syndrome	22q11 microdeletion	Defects of the outflow tracts
Down's syndrome	Trisomy 21	VSD, AVSD, PDA, Tetralogy of Fallot
Edward's syndrome	Trisomy 18	VSD, PDA Coarctation of the aorta Bicuspid aortic valve
Foetal Alcohol syndrome	Ethanol teratogenicity	VSD, PDA, ASD, Tetralogy of Fallot
Goldenhar syndrome	Sporadic inheritance	VSD, Tetralogy of Fallot
Holt–Oram syndrome	TBX5 mutations	ASD, muscular VSD, TAPVC, HLHS
Kartagener syndrome (primary ciliary dyskinesia)	Genetically heterogenous	Dextrocardia
Noonan syndrome	PTPN11 mutations	Pulmonary stenosis ASD Hypertrophic cardiomyopathy
Patau syndrome	Trisomy 13	ASD, VSD, coarctation of the aorta, bicuspid aortic valve, abnormalities of cardiac position
Pierre Robin syndrome	SOX9	VSD, PDA, ASD, coarctation of the aorta, Tetralogy of Fallot
Rubinstein–Taybi syndrome	16p13.3 deletion Sporadic inheritance	PDA, VSD, ASD
Treacher-Collins syndrome	TCOF1, POLR1C, POLR1D mutations	VSD
Turner's syndrome	45XO	Aortic stenosis, coarctation of the aorta, PAPVC
VACTERL association	Sporadic inheritance	Tetralogy of Fallot VSD Tricuspid atresia
Williams syndrome	Chromosome 7 microdeletions (elastin gene)	Supravalvar aortic stenosis Branch PA stenosis

Only the defects within the oval fossa are a result of the true deficiencies of the septal components. Types of the atrial septal defect are represented in (Fig. 28.8). The *ostium secundum* defect results from deficiency of the primary atrial septum, which forms the floor of the oval fossa (Fig. 28.9). The *ostium primum* defect represents an atrioventricular septal defect in the setting of a common atrioventricular junction (Fig. 28.10). The *sinus venosus* defects are found in the mouths of the caval veins, more often the superior caval vein because of the anomalous attachment of one of the right pulmonary veins (Fig. 28.11). The *coronary sinus-type* defect is the fenestration of the wall separating the coronary sinus and the left atrium, and is almost always associated with the drainage of persistent left superior caval vein to the coronary sinus or the roof of the left atrium.

Regardless of morphology, the pathophysiology of an atrial septal defect is related to the magnitude of the shunt, influenced by the size of the defect or, when the defect imposes no restriction to flow across it, related to the relative resistance to filling, or compliance, of the right and left ventricles. A large left-to-right shunt at the atrial level leads to enlargement of both the right atrium and the right ventricle. It is generally well tolerated and heart failure rarely occurs before the age of 40 years, although occasionally a large isolated ASD may cause symptoms of heart failure in infancy. Natural history of large atrial septal defects includes a risk of developing pulmonary hypertension which is found in more than a half of patients older than 40 years, paradoxic embolism leading to a stroke, atrial arrhythmias resulting from chronic stretching of the atriums, and an association with migraine.

When signs of a significant left-to-right shunt are present, therapeutic intervention in the pre-school age (3–5 years) is indicated. Afterwards, life expectancy is normal, and with the increasing use of transcatheter devices the need for surgery is reduced. However, for certain types of defects such as sinus venosus, or coronary sinus-type atrial septal defects, there is no alternative to surgery.

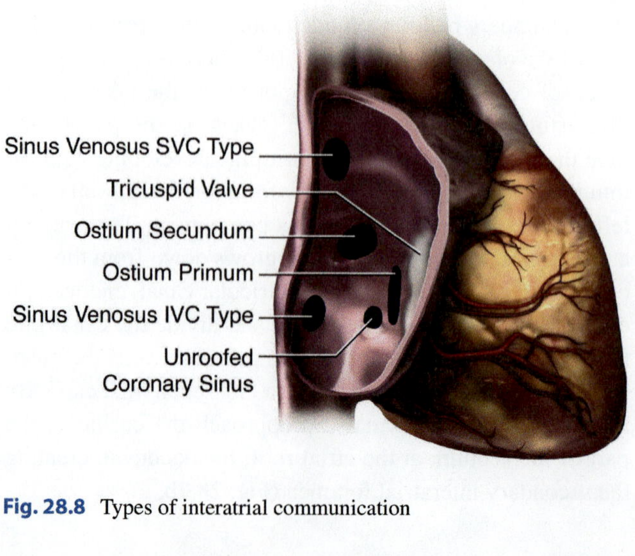

Fig. 28.8 Types of interatrial communication

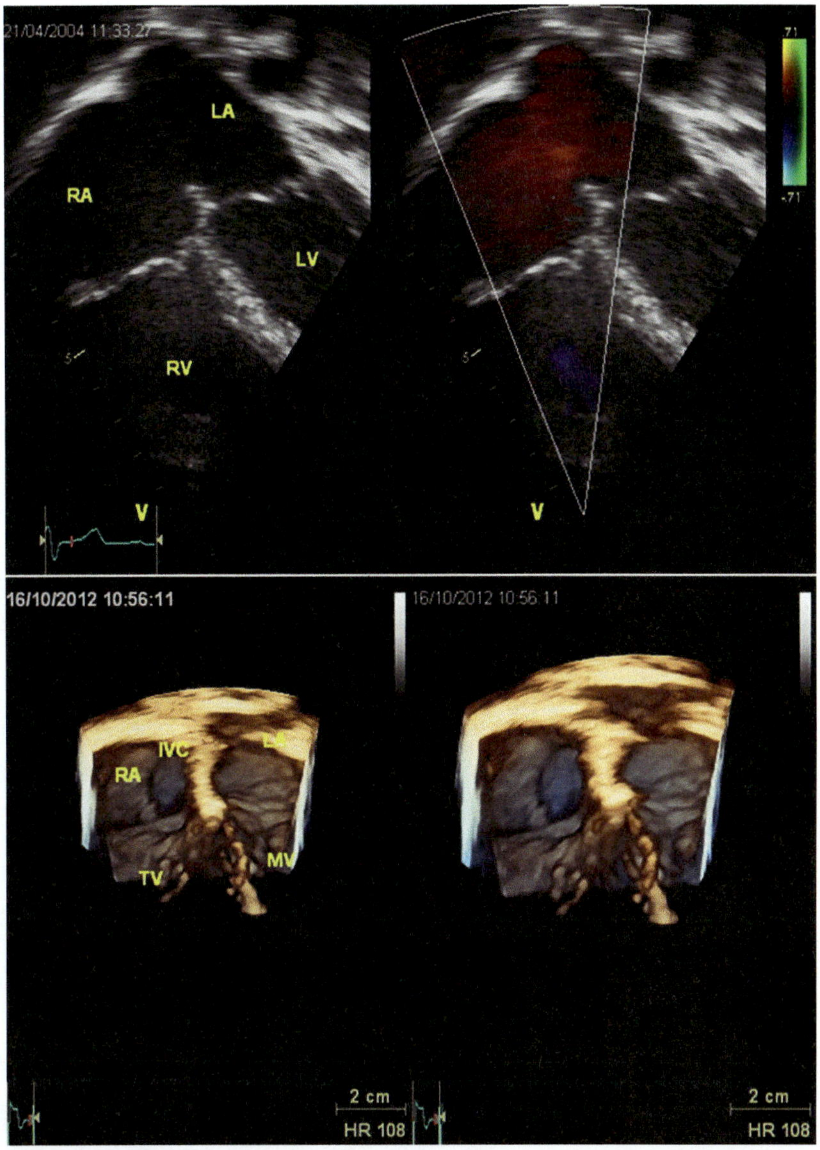

Fig. 28.9 Echocardiographic images: ostium secundum atrial septal defect 2D (top left), colour-coded Doppler (top right) and 3D (lower panel). *LA* left atrium, *RA* right atrium, *IVC* inferior caval vein, *TV* tricuspid valve, *MV* mitral valve

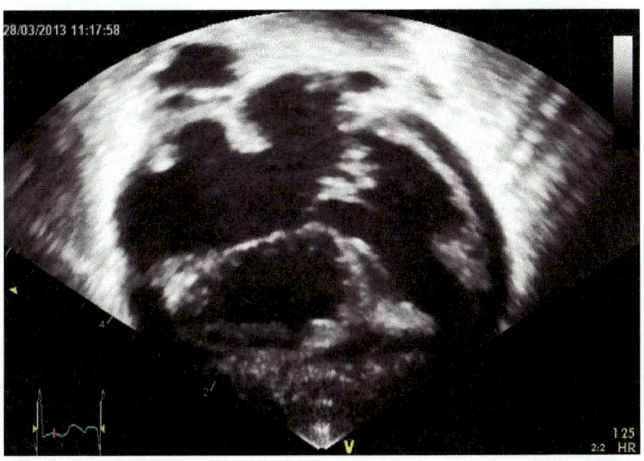

Fig. 28.10 Echocardiographic image: ostium primum atrial septal defect, 2D

Atrioventricular Septal (Endocardial Cushions) Defect

Introduction and Background

This is a group of lesions unified by the anatomical hallmark of a common atrioventricular junction coexisting with deficient atrioventricular septation and lack of a normal central fibrous body of the heart.

Embryology

In early stages, the junction between the developing atrial component and the inlet of the ventricular portion is a canal, which is later septated by fusion of the superior and inferior atrioventricular endocardial cushions. During the normal development, they merge at about 6–7 weeks of development and divide the atrioventricular canal into right-sided and left-sided channels as well as forming the central fibrous body. The research into

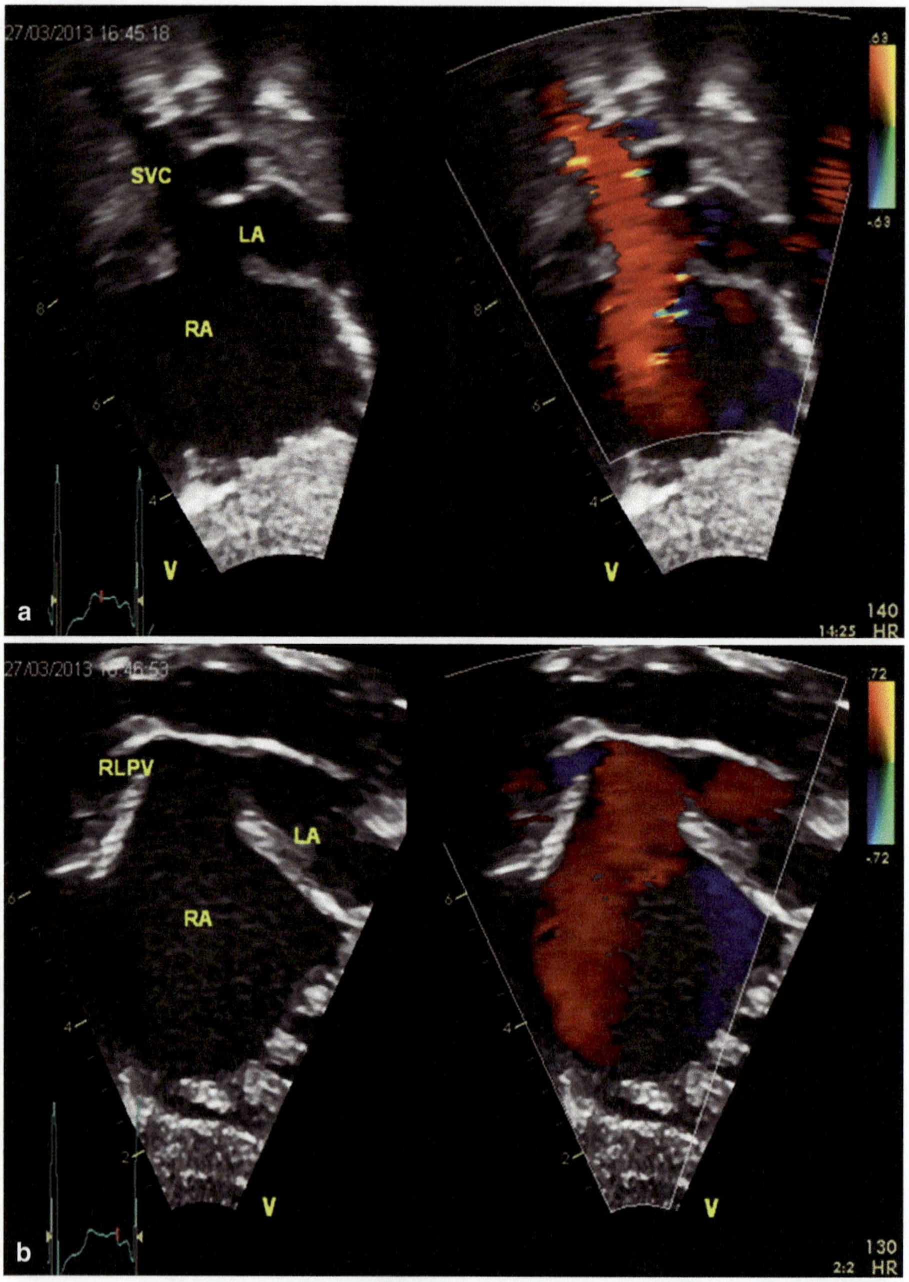

Fig. 28.11 (**a, b**) Echocardiographic images: sinus venosus atrial septal defect, subcostal views, 2D and colour-coded Doppler. *SVC* superior caval vein, *RA* right atrium, *LA* left atrium, *RLPV* right lower pulmonary vein

the developmental disturbance precluding endocardial cushion fusion is ongoing. Some aspects of AVSD embryology are also described in the ASD section of this chapter.

Epidemiologic, Genetic, Morphologic and Clinical Correlations

This defect is commonly associated with Down's syndrome, although its overall proportion in all CHD is only 2–3%. Complete and partial forms are often described: complete where both atrial and ventricular septal defect are associated with various degrees of left and right AV valves insufficiency due to a gap (some call it "cleft") in the leaflets, and (Figs. 28.12 and 28.13), partial where the common AV valve tissue is fused with the crest of the ventricular septum forming separate AV valve orifices and leaving interatrial communication (or "ostium primum"-type ASD) (Figs. 28.10 and 28.14) and variable degrees of left AV valve regurgitation. These hearts are very abnormal. There is deficiency of the membranous atrioventricular septum, the arrangement of the papillary muscles within the left ventricle and the orientation of leaflets of the valve guarding the oval-shaped common junction is nothing like in the normal heart. The left ventricular outflow tract is longer than the inlet portion and prone to development of fibromuscular obstruction. The entire AV nodal area is displaced posteriorly and inferiorly, which is reflected in universal left superior QRS axis deviation on surface ECG.

Complex forms of AVSD, such as AVSD associated with Tetralogy of Fallot, unbalanced AVSD where hypoplasia of one of the ventricles is present, and association with atrial isomerism, described later in this chapter, are rare. In the recent years, over a half of all AVSD patients are diagnosed antenatally.

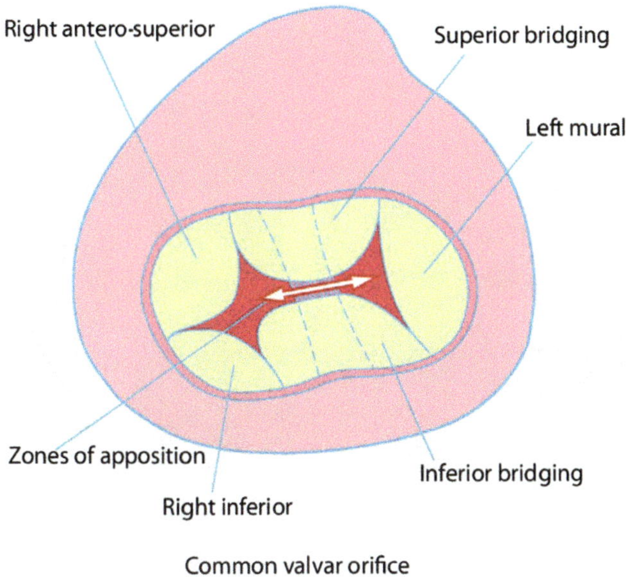

Fig. 28.12 Basic arrangement of the common atrioventricular valve leaflets

Left to right shunting or common mixing on both atrial and ventricular levels due to often very large and unrestrictive septal defects, combined with regurgitant AV valves, produces volume and pressure overloading of both ventricles. A raised pulmonary arterial pressure is a feature of patients having atrioventricular septal defects with large ventricular components, elevated pulmonary blood flow and increased right ventricular pressures are instrumental in the development of obstructive pulmonary vascular disease which may develop earlier and progresses more rapidly than in patients with isolated ventricular septal defect. Its progression is further accelerated in the presence of Down's syndrome, possibly due to unusual reactivity of pulmonary vascular bed in this cohort of patients. In the current era of routine screening of patients at high risk (Down's syndrome) and increasing likelihood of antenatal diagnosis, development of Eisenmenger's syndrome is rare. Patients may present in early infancy with signs and symptoms of congestive heart failure or incidental finding of a heart murmur; in rare instances they may not present until adolescence or early adulthood.

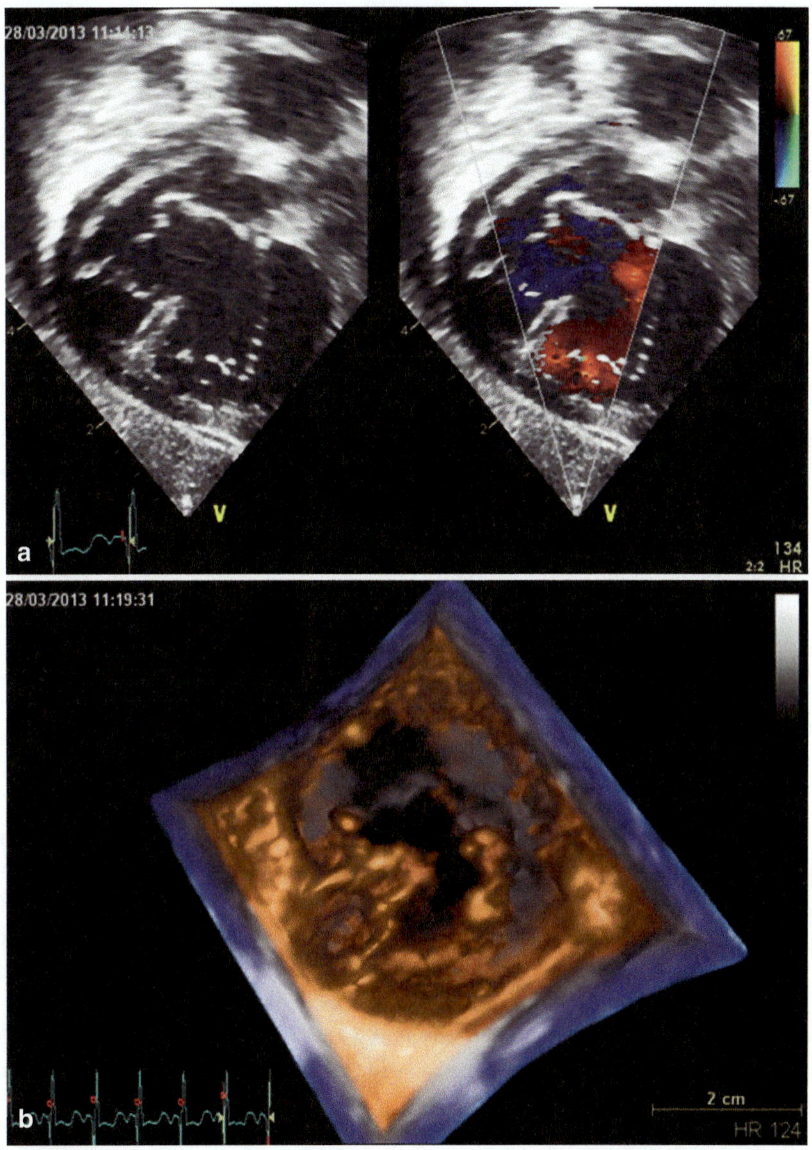

Fig. 28.13 (**a**, **b**) Echocardiographic images: atrioventricular septal defect (AVSD) with common atrioventricular (AV) valve, subcostal en-face view of the common AV valve (2D-top left, colour-coded Doppler-top right, and 3D-lower panel)

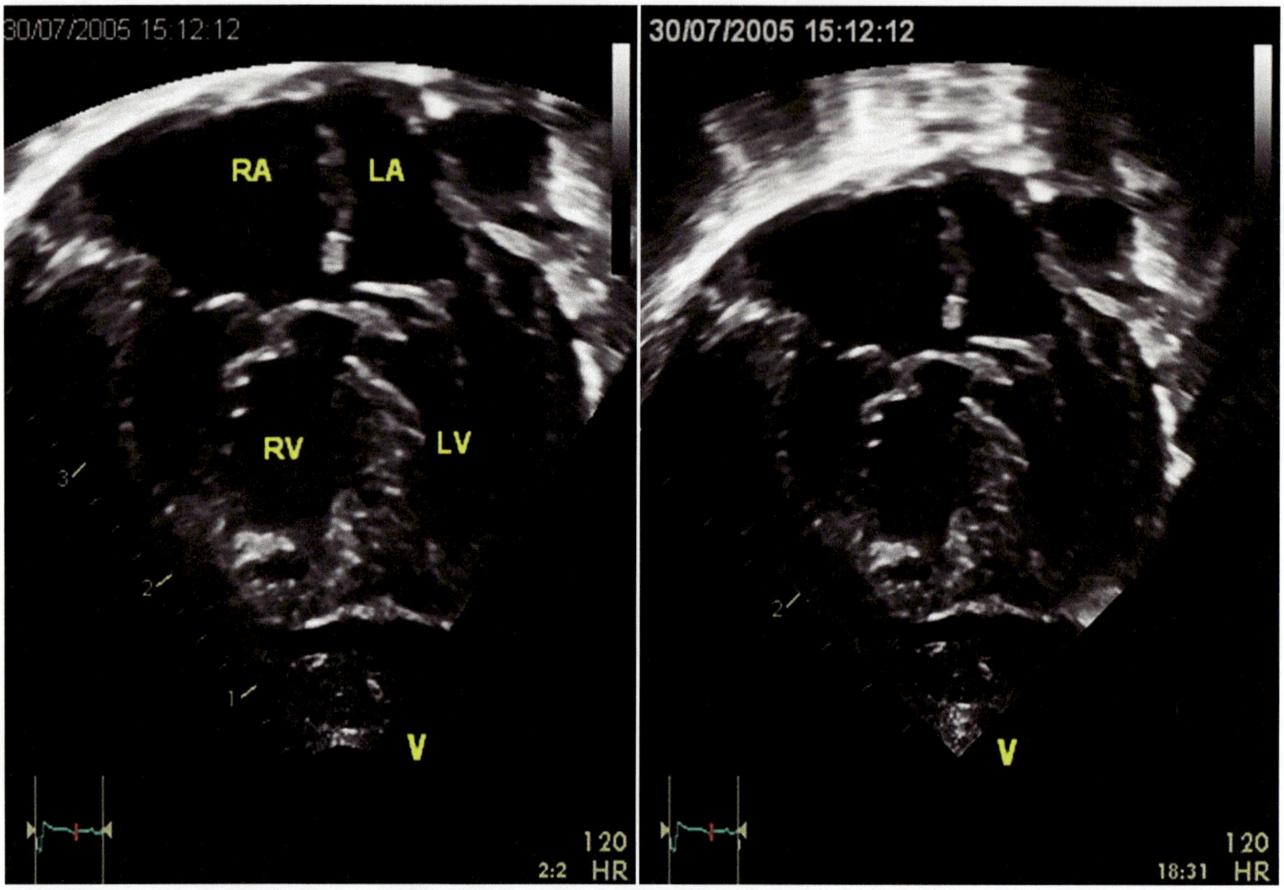

Fig. 28.14 Echocardiographic image: complete AVSD, apical four chambers view (*RA* right atrium, *LA* left atrium, *RV* right ventricle, *LV* left ventricle)

Total Anomalous Pulmonary Venous Drainage

Introduction and Background

In this condition, pulmonary veins drain into the right atrium (RA) or its venous tributaries, pulmonary and systemic venous blood mix in the RA and an ASD is necessary for survival. In rare circumstances, pulmonary veins may be connected normally to the left atrium but can drain to a systemic site due to persistence of embryological levoatrial cardinal vein taking pulmonary venous return to the SVC/RA junction. This rare form is associated with mitral stenosis and ASD. There can be partial forms of anomalous pulmonary drainage, when a solitary pulmonary vein or veins from one lung are connected anomalously to a site other than the left atrium. An example is *scimitar syndrome* when one or all of the right pulmonary veins drain into the inferior caval vein, and there's an abnormal feeding arterial vessel arising from the descending aorta and supplying some segments of the right lung.

Embryology

The developing atrium is initially connected to the mediastinum through the dorsal mesocardium, and, as the lung develops, the plexus of intrapulmonary veins joins with a pulmonary venous channel in the developing mediastinum. This leads to formation of primary pulmonary vein and it joins the atrial component of the developing heart tube. After normal fusion with the developing intrapulmonary venous plexuses, the pulmonary veins form the venous components of the left atrium. This occurs relatively late in embryonic development and when this process fails, the primitive connections of the pulmonary veins to the systemic venous system persist and enlarge, eventually resulting in anomalous pulmonary venous connection. An anastomosis with the anterior cardinal venous system results in supracardiac anomalous connection, with the systemic venous sinus producing cardiac connection, while infradiaphragmatic connection is the consequence of anastomosis with the omphalomesenteric system.

Epidemiologic, Genetic, Morphologic and Clinical Correlations

It is an uncommon lesion accounting for just over 1% of all congenital lesions and is commonly associated with isomerism of atrial appendages discussed later in this chapter. Autopsy studies demonstrated changes in the pulmonary venous walls which take on arterial characteristics, as well as an increase in the muscularity of the pulmonary arteries with age.

Types of TAPVD, in order of decreasing frequency (Figs. 28.15 and 28.16):

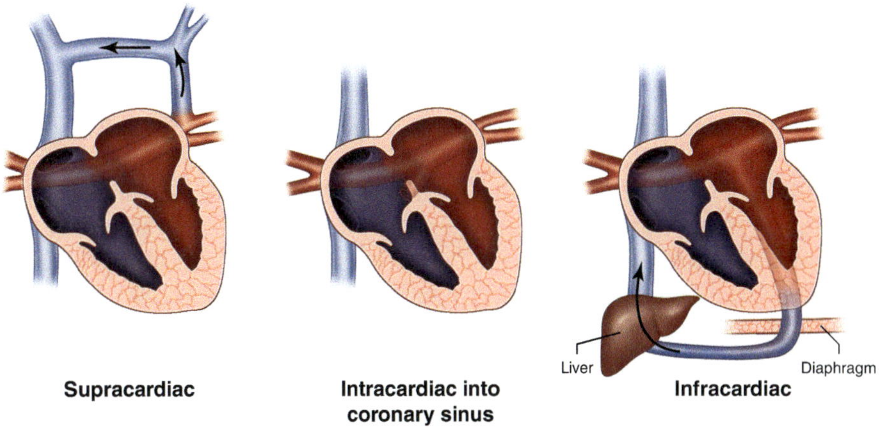

Fig. 28.15 Anatomic types of total anomalous pulmonary venous drainage (TAPVD)

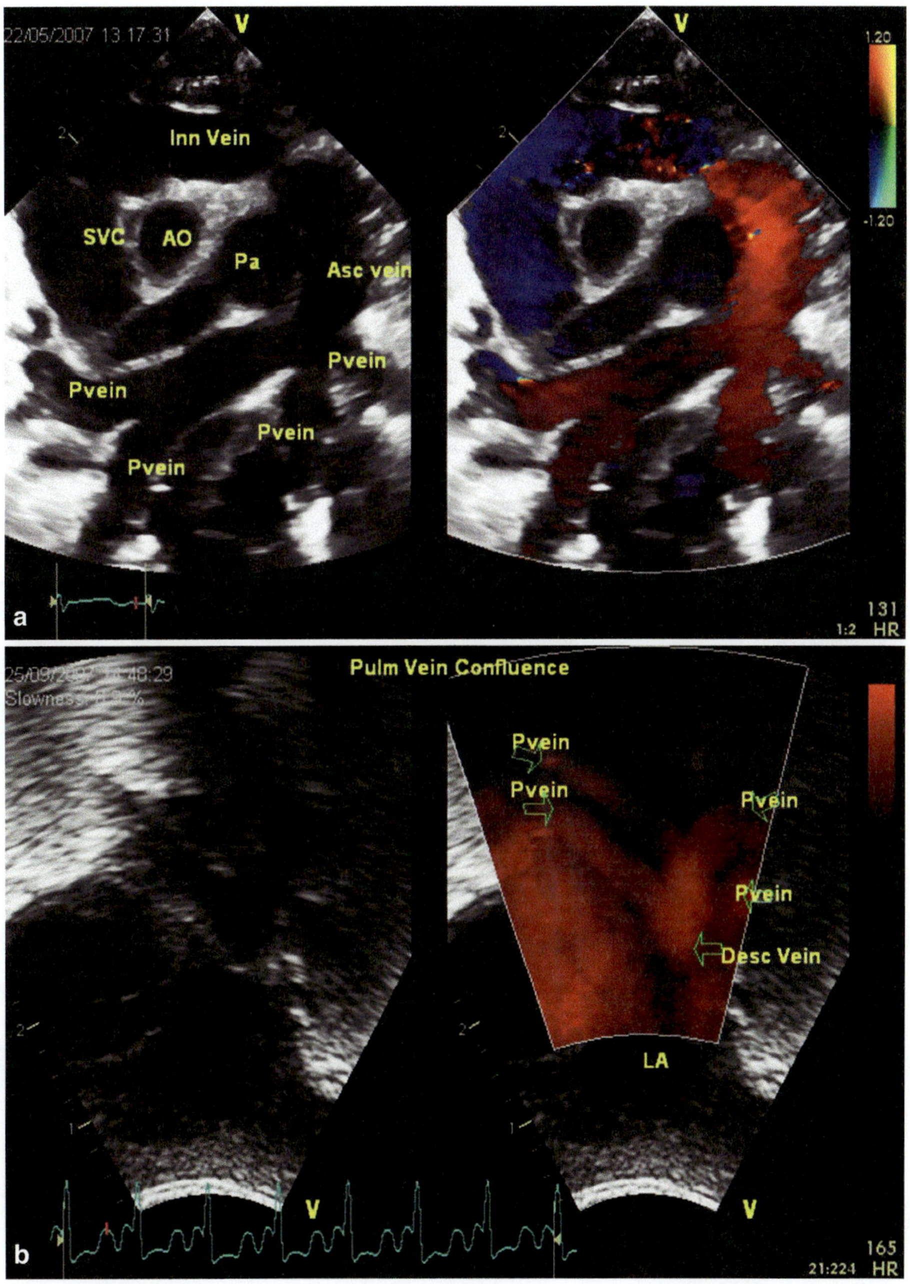

Fig. 28.16 (**a**, **b**) Echocardiographic image: supracardiac total anomalous pulmonary venous drainage, TAPVD. (**a**) Top left-2D, top right-colour-coded Doppler, (**b**) 2D and corresponding power Doppler image

- Supracardiac or supradiaphragmatic (commonest) where the PV collector, or confluence, drains into SVC via the ascending vertical vein and innominate vein. In this form of TAPVC obstruction commonly occurs by compression of the vertical vein between bronchus and pulmonary artery
- Intracardiac: the PV collector drains into RA via coronary sinus (Fig. 28.17), or, very rarely, pulmonary veins are connected to RA individually
- Infracardiac or infradiaphragmatic: the pulmonary venous collector drains via a descending vertical vein into the portal vein, venous duct, hepatic vein or IVC. Obstruction in this form commonly occurs due to postnatal closure of the venous duct. A significant preponderance of males (4:1) is reported in this type.
- Mixed

In TAPVC where no obstruction to the pulmonary venous return is present, obligatory left-to-right shunt always occurs, and as long as right ventricular compliance is good and diastolic pressure is low, blood from the right atrium flows preferentially into the right ventricle rather than across the PFO to the left atrium; thus pulmonary blood flow is significantly increased and the patient will demonstrate failure to thrive, signs of respiratory distress and may or may not display mild cyanosis. A systemic output can only be maintained if there is a right-to-left shunt, which is almost always at the atrial level.

The tendency for foetal patterns of flow is maintained by the valvar mechanism of the oval foramen; thus in infradiaphragmatic forms most pulmonary venous return coming back to RA from the IVC is directed towards the left atrium and systemic flow receives more oxygenated blood, and in supracardiac form oxygenated pulmonary venous return is directed from the SVC towards the tricuspid valve and pulmonary artery receives more oxygenated blood than aorta. Even in absence of pulmonary venous obstruction, pulmonary vascular resistances often remain elevated after birth, which almost invariably resolves following a successful repair.

In cases where the pulmonary venous return is obstructed, pulmonary venous pressure is raised leading to pulmonary oedema. Pulmonary arterial pressure may raise to suprasystemic levels.

The main determinant of the clinical picture is thus the presence of pulmonary venous obstruction. Patients with severe pulmonary venous obstruction present in the first week of life with obvious cyanosis and breathing difficulties, similar to the patients with large VSD. Patients without severe obstruction present with heart failure at 2–3 months of age with feeding difficulties and sometimes chest infections, and cyanosis is generally not a symptom. Echocardiography is a first line diagnostic test. In complex cases MRI or cardiac CT may be used, and cardiac catheterisation is only rarely necessary, usually when there is a need for temporary procedure to improve atrial shunt by means of balloon atrial septostomy. Surgical correction is always required and in cases of obstructive TAPVC it is an emergency. Operated patients require life-long follow-up as some may develop recurrent obstruction either at the surgical repair site or at the ostiums of individual pulmonary veins.

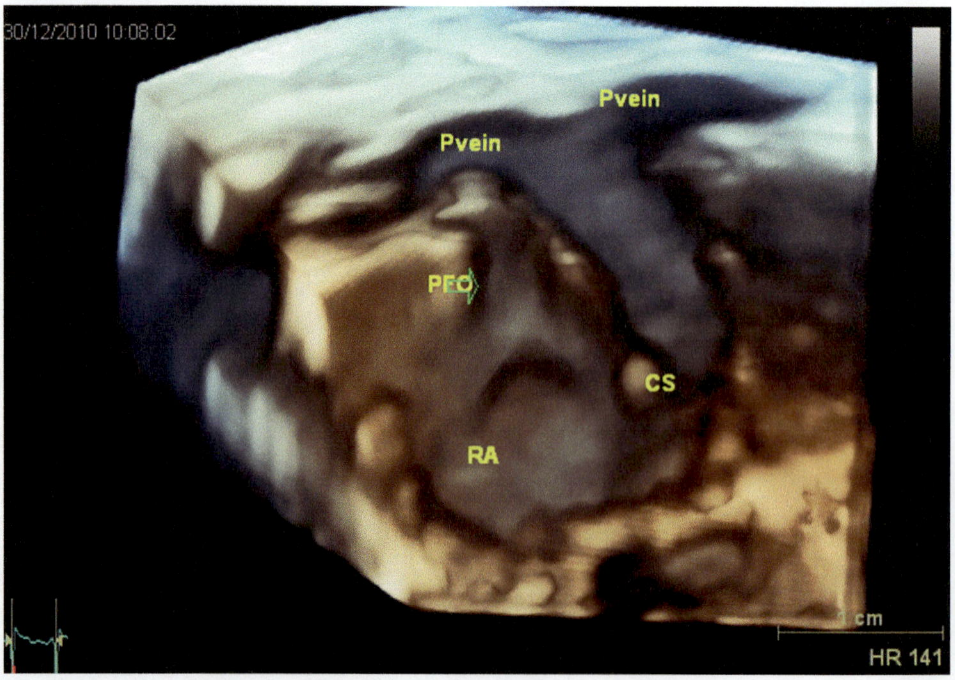

Fig. 28.17 Echocardiographic image: total anomalous pulmonary venous drainage to coronary sinus, 3D. *Pvein* pulmonary vein, *PFO* patent foramen ovale, *RA* right atrium, *CS* coronary sinus

Anomalies of the Outlet of the Heart: Congenital Semilunar Valves Malformations

Introduction and Background

Abnormalities of the left and right ventricular outflow tracts may occur at the subvalvar, valvar, supravalvar level and combination of these. They often coexist with other lesions, such as aortic coarctation or anomalies of the mitral valve in cases of aortic stenosis; pulmonary valve malformations are common in Tetralogy of Fallot. Some may be typical for certain genetic conditions, such as severely dysplastic, thickened, irregular stenotic pulmonary valves in Noonan's syndrome, or supravalvar aortic stenosis typical for Williams syndrome. This chapter will address aortic and pulmonary valvar stenosis. Pathophysiology of the obstructive lesions is similar irrespective of the level of obstruction, however treatment strategies may differ. Some of these lesions will be discussed in more detail in other parts of this chapter.

Embryology

Remodelling of the outflow tract of the developing human heart during the 5th and 6th weeks produces two vessels, each with its own arterial valve. The tissues of the outflow tract are derived from both cardiac mesenchyme and migrating cells of non-cardiogenic regions of the lateral plate and head mesenchyme. Initial septation is produced by ingrowth of mesenchyme, derived from the neural crest, into the cushions developing throughout the outflow tract. These cushions then fuse with each other, leading to separation of the intrapericardial components of the aorta and the pulmonary trunk. The proximal part of the outflow tract remains encased in a sleeve of outflow myocardium. Distal part of this sleeve, outflow cushions, produces developing aortic and pulmonary valves. Cells derived from the neural crest aid septation and the experimental removal of the neural crest from the developing embryo led to failure of septation and to formation of a persistent common arterial trunk, which will be described later in this chapter.

Epidemiologic, Genetic, Morphologic and Clinical Correlations

Congenital aortic valve stenosis accounts for 5% of all cardiac abnormalities, and if bicuspid aortic valves are included this may be the commonest congenital cardiac lesion. A bicuspid aortic valve occurs in 1%–2% of people, with a 2:1 male-to-female ratio. Familial cases of bicuspid aortic valve are not uncommon. True incidence is difficult to know as not all cases may be recognised.

A concept of the annulus, or ring, is often used to describe the semilunar valves but unlike in atrioventricular valves, the anatomic structure of the semilunar valves does not contain a distinct annulus. Semilunar valves are formed by leaflets that are hinged within the arterial root in semilunar fashion. The root itself is formed by interlocking of the aortic valvar sinuses and the supporting ventricular structures (Fig. 28.18).

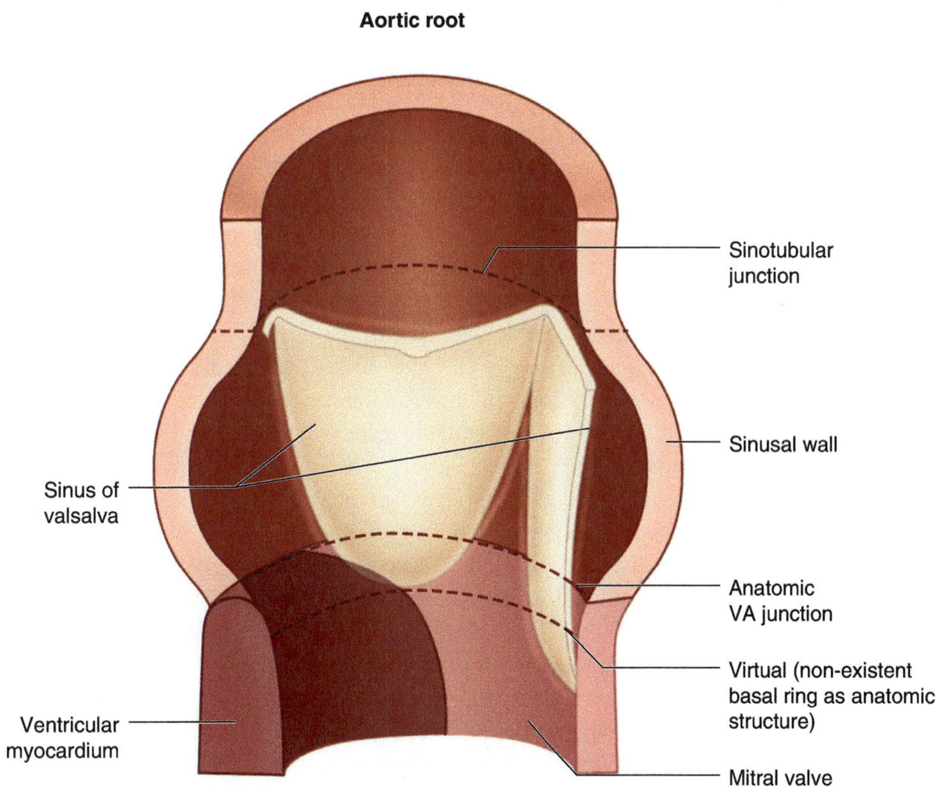

Fig. 28.18 Structure of the aortic valve

In a normal heart, aortic valve is in fibrous continuity with the mitral valve.

The stenotic semilunar valves show unicuspid, bicuspid or tricuspid patterns. However, when such valves are examined closely, even when the leaflet tissue is compartmented into less than three parts, the curtain is still suspended within three obvious sinuses. In these cases, one or more zones of apposition between leaflets fail to develop normally; leaflets are fused at the end of the zone apposition, or all along, and tethered to the sinutubular junction, resulting in a smaller than normal, funnel-like, stenotic valvar orifice. In bicuspid aortic valves, the conjoined leaflet usually represents either fusion of the two coronary leaflets or fusion of the right and non-coronary leaflets. Truly bileaflet valves are rare.

A variety of mechanisms and lesions can lead to obstruction of the subaortic and subpulmonary outflow. In subaortic obstruction, a role is often played by a fibromuscular ridge, or a diaphragm/membrane. In other cases, there is a deviation of the outlet septum resulting in malalignment of the semilunar valve in relation to the outflow tract—posterior to produce subaortic and anterior and cephalad, like in Tetralogy of Fallot, to produce subpulmonary obstruction. Asymmetric septal hypertrophy in hypertrophic cardiomyopathy, and abnormal mitral valve septal attachment in complex heart malformations can also lead to subaortic obstruction. Anomalous muscle bundles in the middle of the RV cavity can form so-called double-chambered RV, with obstruction situated deep in the cavity of the right ventricle as if dividing it into two parts. Supravalvar stenosis occurs typically above the level of the sinuses of Valsalva, incorporating sinutubular junction, and may take tubular forms. The arterial wall in these cases is very abnormal, thickened, with disorganised medial layer. Such stenosis are often associated with disorders of calcium metabolism (Williams syndrome) and are not restricted to aorta but may be encountered in aortic branches, central and peripheral pulmonary arteries.

Foetal catheter intervention can be applied to a few selected cases of critical aortic and critical pulmonary stenosis or pulmonary atresia with intact ventricular septum. The aim is not only to ensure survival but also to promote vascular and cardiac structures growth by relieving obstruction to the outflow from right and left sides of the heart and achieve biventricular circulation postnatally.

In general, clinical features of semilunar valves obstruction are a systolic ejection murmur on auscultation, hypertrophy and eventually failure of the responsible ventricle and (not universal) poststenotic dilatation of the pulmonary artery or ascending aorta.

Patients with mild aortic or pulmonary valvar stenosis are asymptomatic. In severe cases, exertional dyspnoea and fatigability may be seen; congestive heart failure is rare. Neonates with critical pulmonary stenosis are cyanosed. Whenever in order to survive, babies with aortic or pulmonary stenosis require ductal patency to be maintained, to support systemic and pulmonary blood flow, respectively, the degree of stenosis is named "critical".

In the modern era, echocardiography is the diagnostic imaging modality of choice. Severity of pulmonary stenosis is graded based on the Doppler-derived estimate of RV systolic pressure: <50% of systemic corresponds to a mild degree of stenosis, 50%–75% to moderate and >75% to severe. Severity of stenosis, except in mild forms, tends to progress with age. Turbulent high velocity flow across the stenotic aortic or pulmonary valve render the heart vulnerable and increase the background risk of infective endocarditis. For significant pulmonary valve stenosis, transcatheter balloon pulmonary valvuloplasty is the procedure of choice. Results of balloon valvuloplasty for aortic valve stenosis are not as good as those for PS and the procedure is regarded as the first step of management of symptomatic neonates and children. Surgery is required to repair or replace the aortic valve, and if significant pulmonary regurgitation results from balloon valvuloplasty with ensuing right ventricular dilatation, surgery may be required to replace pulmonary valve also.

Anomalies of the Outlet of the Heart: Tetralogy of Fallot, Pulmonary Atresia with Ventricular Septal Defect and Double Outlet Right Ventricle

Introduction and Background

Although the term "tetralogy" as described by Arthur Louis Etienne Fallot in 1888 implies four main characteristics—dextraposition of the aorta, ventricular septal defect, right ventricular outflow tract obstruction and right ventricular hypertrophy, some authors argue that all four can be resulting from a single morphological feature, namely—anterior and cephalad deviation of the septal insertion of the outlet septum (Fig. 28.19) relative to the septomarginal trabeculation. In addition, there is an anomalous arrangement of septoparietal right ventricular trabeculations so that an obstruction is produced at the entrance to the infundibulum (Fig. 28.20). The ventricular septal defect is usually very large allowing equalisation of pressure between the two ventricles (Fig. 28.20).

Embryology

The key mechanism of this malformation is malseptation of the arterial segment of the developing heart. Abnormalities occur in the formation of the endocardial cushions which normally fuse to septate the ventricular outlets. The abnormal attachment of the muscular outlet septum is sufficient to produce interventricular communication, or VSD, and the

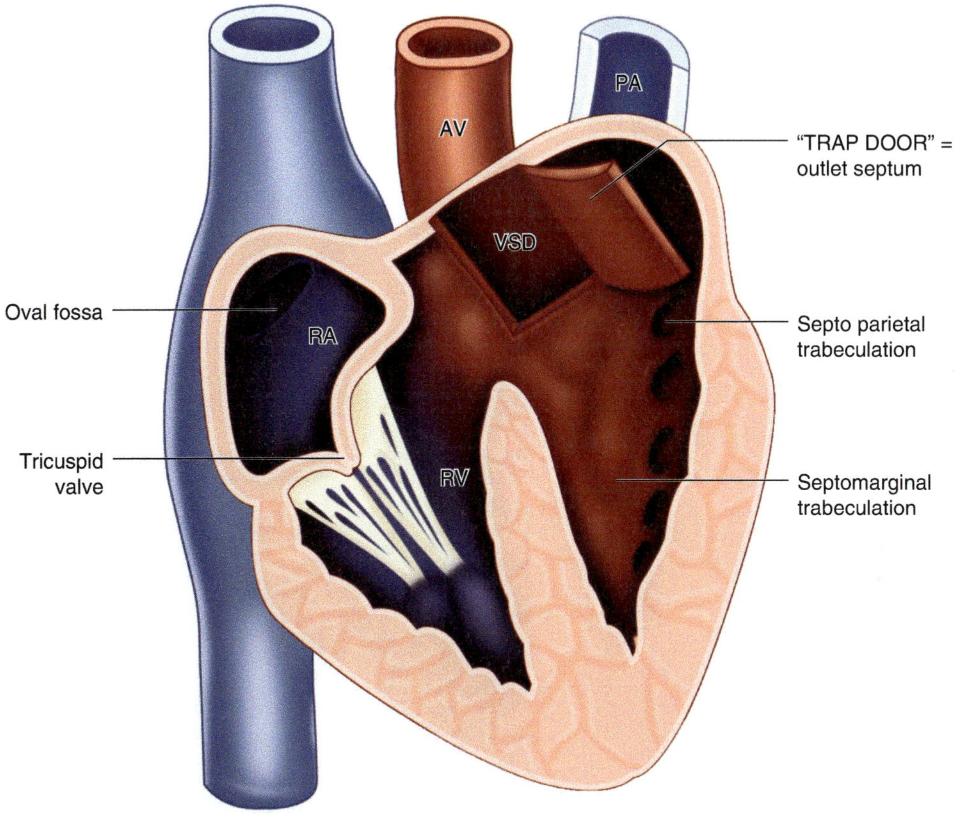

Fig. 28.19 Anatomy of the outlet septum defining Tetralogy of Fallot

overriding of the aorta, resulting in its biventricular connection. Subpulmonary muscular stenosis development requires an additional abnormality of septoparietal trabeculations. The right ventricular hypertrophy is simply a haemodynamic consequence of these anatomic lesions.

Epidemiologic, Genetic, Morphologic and Clinical Correlations

Tetralogy of Fallot is one of the commonest cyanotic heart lesions, encountered in 0.28 per 1000 live births and has no gender preponderance. The risk of recurrence in siblings is around 3%; however, precise aetiology is unknown and studies indicate polygenic model of inheritance. Tetralogy of Fallot is associated with chromosomal anomalies such as DiGeorge syndrome and Down's syndrome. The ventricular septal defect is always in perimembranous position extending to outlet septum or sometimes both into inlet and outlet septum. Aortic root is positioned further towards the right than in a normal heart (dextraposition) and aortic valve is rotated clockwise. The right ventricular outflow tract obstruction most commonly occurs in the form of muscular infundibular stenosis (over 50% of the cases), occasionally as isolated pulmonary valve stenosis, or the combination of both, and is often accompanied by the hypoplasia of pulmonary valve, main pulmonary arteries and occasionally pulmonary arteries branches. Coronary arteries anomalies and right-sided aortic arch are other common associations.

When the degree of aortic dextraposition is such that the majority of its orifice is committed to the right ventricle, the condition is named Double Outlet Right Ventricle (DORV) of Tetralogy of Fallot type (Fig. 28.21). Approximately 10% of patients with tetralogy of Fallot will have pulmonary atresia, and pulmonary blood flow will be supplied through a PDA (two-thirds) or through major aortopulmonary collaterals (MAPCA, one-third), and sometimes these two sources of pulmonary blood supply coexist. Majority of cases of pulmonary atresia is associated with hypoplasia of branch pulmonary arteries and paucity of distribution of its peripheral branches; this is more pronounced in the presence of aortopulmonary collaterals. Patients with pulmonary atresia and PDA supply to pulmonary arteries will be cyanosed at birth and require intravenous prostaglandin infusion to keep the arterial duct open.

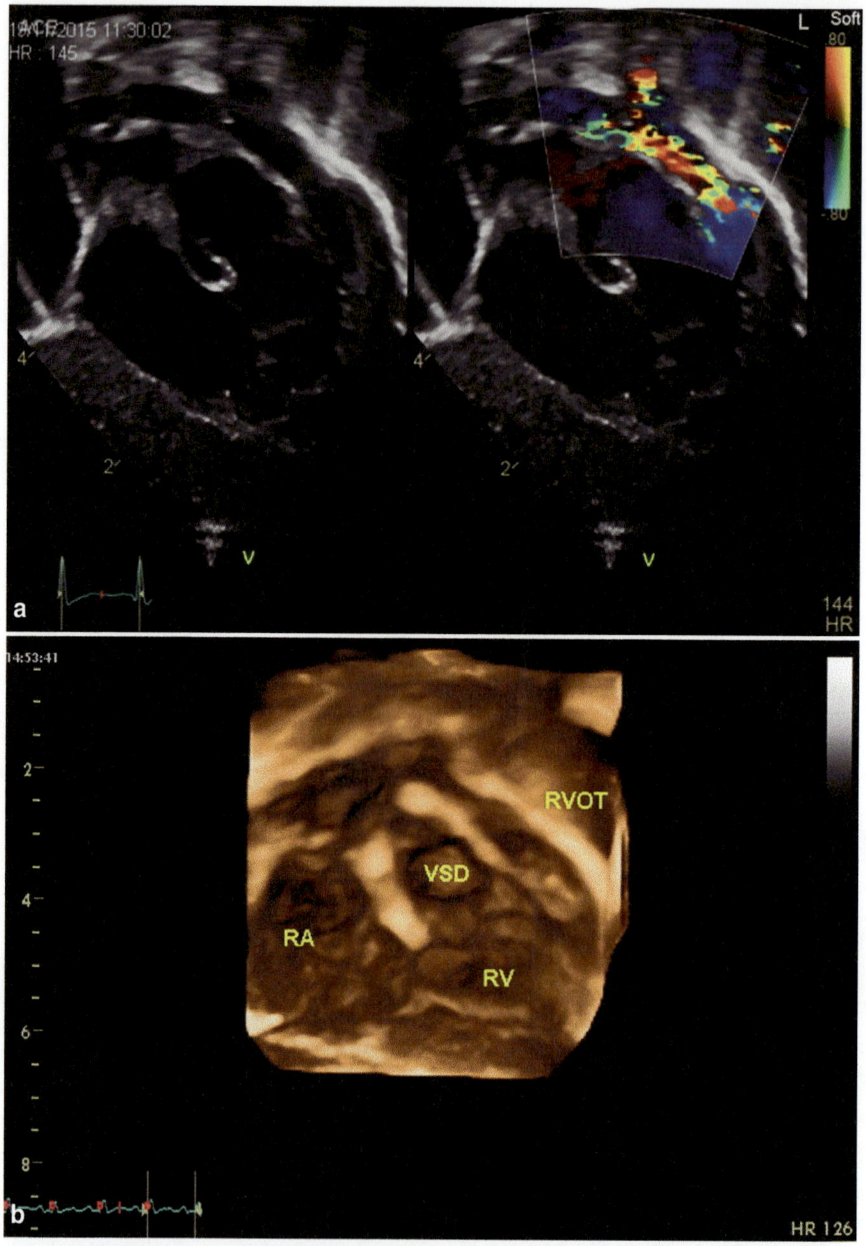

Fig. 28.20 Echocardiographic image: Tetralogy of Fallot-(**a**) 2D and colour-coded Doppler; (**b**) 3D subcostal en-face view of the interventricular septum. *RA* right atrium, *VSD* ventricular septal defect, *RV* right ventricle, *RVOT* right ventricular outflow tract

Tetralogy of Fallot lends itself to foetal diagnosis in a significant proportion of cases, up to 39% of all cases (and up to 76% of double outlet right ventricle cases) diagnosed antenatally in the UK in the recent decade.

Pathophysiology is defined by the degree of right ventricular outflow tract obstruction. As the VSD is large and pressure in both ventricles is equal, the net pulmonary blood flow may be higher than systemic if pulmonary stenosis is mild and patients will not be cyanosed or (rarely) even display signs of congestive heart failure. If pulmonary stenosis is moderate to severe, patients can present early with cyanosis and heart murmur, or may develop cyanosis gradually by 1 year of age or later.

Hypoxic spells, a clinical "hallmark" of the Tetralogy of Fallot representing a paroxysm of rapid and deep breathing with increased cyanosis usually after a prolonged bout of crying, may lead to a loss of consciousness or even death and develop in patients with severe infundibular stenosis, during the first years or occasionally months of life. Development of cyanotic spells is an indication for surgical correction. In the majority of patients, with the exception of a small group with significant branch pulmonary arteries hypoplasia or anomalous coronary arteries, corrective surgery is readily available and consists of a complete repair and is carried out between 3 and 12 months of age. Depending on the individual centre experience, a selected

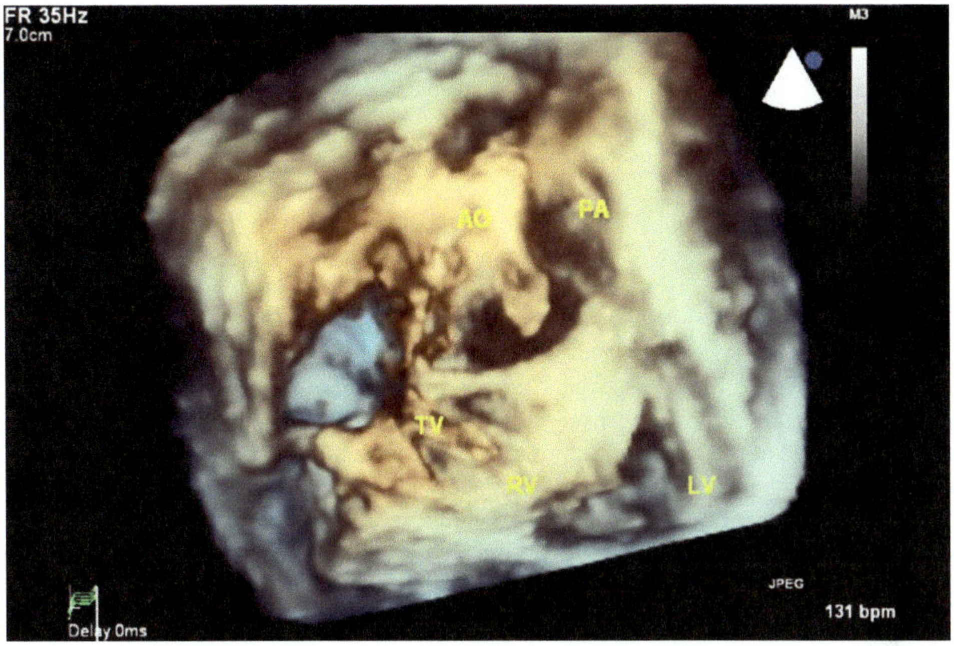

Fig. 28.21 Echocardiographic image: double outlet right ventricle, 3D subcostal en-face view of interventricular septum. *VSD* ventricular septal defect, *PA* pulmonary artery, *TV* tricuspid valve, *LV* left ventricle

group of patients not suitable for primary surgical repair may undergo a temporary palliative augmentation of the pulmonary blood flow by means of surgery (modified Blalock–Taussig interposition shunt) or transcatheter intervention (stent placement in the right ventricular outflow tract). Overall prognosis in Tetralogy of Fallot is excellent with low operative mortality and normal life expectancy. A significant proportion will require one or more reinterventions to correct residual pulmonary valve regurgitation which often results from surgical right ventricular outflow tract obstruction relief.

Anomalies of the Outlet of the Heart: Transposition of the Great Arteries

Introduction and Background

In transposition of the great arteries, aorta arises anteriorly from the right ventricle and pulmonary artery arises posteriorly from the left ventricle. It is a rare malformation accounting for about 5% of all congenital cardiac lesions but prior to the era of surgical correction it led to a disproportionately high number of cardiac deaths in infancy—up to 20%.

Embryology

The anatomy is the consequence of inappropriate separation of the arterial pole of the heart when ridges separating the arterial segment of the heart tube fuse in an abnormal straight rather than the normal spiral fashion. As a result the sixth aortic arches supposed to supply pulmonary arteries are connected to the left ventricle while the fourth aortic arches connect to the right.

Epidemiologic, Genetic, Morphologic and Clinical Correlations

There is approximately 2:1 to 3:1 male preponderance in this defect. No definite aetiological factors have been found but a higher incidence of the condition amongst infants of diabetic mothers, maternal intake of alcohol, poor nutrition and stressful events during pregnancy all were implicated as increasing the risk, while addition of folic acid to the maternal diet may result in modest reduction of risk. There is no known strong association of TGA with any of the chromosomal or genetic abnormalities. A VSD is present in 40% of cases; left ventricular outflow tract obstruction resulting in pulmonary stenosis occurs in about 5% of patients without VSD and in 30% of patients with a VSD. Over a half of patients there are no associated defects other than a PFO or a small PDA.

Transposition of the aorta and pulmonary artery results in loss of the normal sequential relationship between the systemic and pulmonary circulation and their complete separation, with hypoxaemic, or poor in oxygen content, blood circulating around the body, and hyperoxaemic, or well oxygenated, blood circulating in the pulmonary circuit. From this follows that defects permitting mixing of the blood between the two parallel circuits will be necessary to ensure survival (ASD, VSD, PDA).

Cyanosis and severe arterial hypoxaemia unresponsive to oxygen are often present from birth. Neonates with poor mixing of the two circulations require an early intervention to avoid progressive hypoxia and acidosis that may lead to a very rapid deterioration and death. In this group of patients, intravenous prostaglandin infusion is required to maintain ductal patency and a transvenous balloon atrial septostomy is required as a matter of urgency (Fig. 28.8—Echo still). In a significant minority of patients, there are anomalies of coronary arteries origins, branching pattern and epicardial course which can complicate surgical repair, and it is therefore important to ascertain the anatomical features preoperatively. Two-dimensional echocardiography provides diagnosis of the transposition and associated lesions including information on coronary arteries anatomy.

In the current era, the arterial switch operation is the surgical technique of choice. Aorta and pulmonary arteries are transsected above the valves, removed from their transposed positions and sutured onto the opposite, anatomically correct root. This procedure is accompanied by the surgical excision of the coronary arteries origins from the previously aortic and their reimplantation into the pulmonary, or neo-aortic, root. A minority of patients with associated left ventricular outflow tract obstruction may require different surgical approach. Outcomes of surgery are continuously improving and long-term prognosis of patients born with TGA without associated lesions who underwent uncomplicated arterial switch procedure is generally excellent with normal life expectancy, although life-long outpatient surveillance and follow-up imaging, and reintervention in a selected group of patients may be required.

Anomalies of the Outlet of the Heart: Common Arterial Trunk

Introduction and Background

This is a rare condition where only a single arterial trunk guarded by an abnormal semilunar valve called "truncal valve" leaves the heart and gives rise to the pulmonary, systemic and coronary circulations. A large muscular VSD is always present and associated aortic arch anomalies are very common.

Embryology

In a normally developing heart, the originally common ventricular outflow tract is separated by endocardial ridges to produce two arterial trunks and arterial valvar leaflets, along with their supporting ventricular outflow tracts. When the migration of the cells from the neural crest is perturbed, these cushions do not develop properly and may fail to fuse, thus producing common arterial trunk. It has been shown that such malformations have a strong genetic basis, and common arterial trunk has been produced when there is deficiency of SOX4, a gene normally populating the endocardial cushions of the developing outflow tracts. Arterial duct is absent in three quarters of patients with a common arterial trunk due to the presence of a larger aorto-pulmonary connection which allows the duct to disappear early in foetal life.

Epidemiologic, Genetic, Morphologic and Clinical Correlations

Collette and Edwards' classification divides this anomaly into four types (Fig. 28.22). Type IV is not a true common arterial trunk but a form of Tetralogy of Fallot with pulmonary atresia and aortic collateral supplying blood flow to the lungs.

Type B aortic arch interruption is seen in 13% of patients with common arterial trunk. Truncal valve is always abnormal, can be made of any number of cusps, and may be stenotic, regurgitant or both. Coronary artery abnormalities are common and play a role in high surgical mortality. Common arterial trunk is strongly associated with DiGeorge syndrome which is seen in 30% of patients, and an association between common arterial trunk and CHARGE syndrome is also recognised. The incidence of congenital malformations of the heart is increased to 7%–14% in offspring of patients with common arterial trunk.

The presentation is usually immediately or a few weeks after birth with signs of congestive heart failure or cardiogenic shock, and (not always) cyanosis. This is due to increase in pulmonary blood flow following the physiological postnatal reduction of pulmonary vascular resistance. A number of patients may experience myocardial ischemia due to a large "run-off" of blood from the aorta into pulmonary artery and/or presence of congenital coronary malformations, which may result in reduction of ventricular contractility or even sudden infant death. The diagnosis in the current era is often established by means of echocardiography, antenatally or during first days of life. If left untreated, patients develop pulmonary vascular disease leading to pulmonary hypertension, and most will die in infancy or early childhood. Surgical palliation with pulmonary arteries narrowing using a band (pulmonary artery banding) occasionally precedes complete repair using a valved tube called conduit substituting for the "missing" pulmonary trunk, and various forms of reconstruction of proximal branch pulmonary arteries. Early postoperative death rate is low but relatively higher than in most other congenital conditions. Life-long surveillance and several surgical operations and transcatheter interventional procedures will be required throughout childhood and in later life.

Isomerism (heterotaxia) and Anomalies of Cardiac Position

Introduction and Background

A normal heart can be abnormally located, which is called "malposition". Extreme examples are exteriorisation of the heart described as "ectopia cordis", accompanied by a gross deficiency of the fibrous pericardial sac, or union of the parts of the heart in conjoined twins. The normal scenario is when the heart is in the left hemithorax and its apex is pointing to

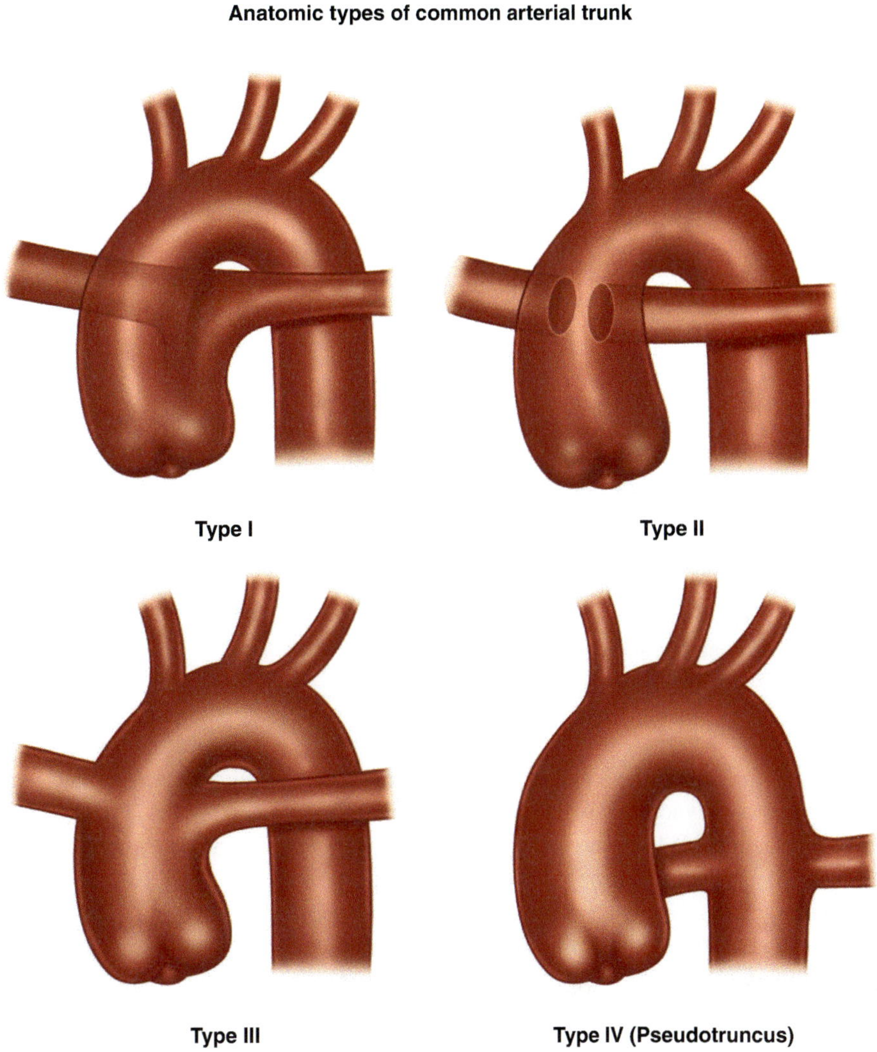

Fig. 28.22 Anatomic types of common arterial trunk (type IV—"pseudotruncus")

the left, which is called "levocardia". A complete mirror image position of the heart in the right side of the chest with its apex pointing to the right is named "dextrocardia". An indeterminate cardiac position "in the middle" without clear predisposition for the right or for the left is named "mesocardia". There may be anomalies of position of the cardiac chambers in relation to one another, such as juxtaposition of the atrial appendages, when both the appendages are to the same side of the arterial pedicle, variations of orientation of the ventricles and a particular arrangement of ventricular inlets also called "criss-cross heart".

Whenever an anomaly of the cardiac position occurs, it may be accompanied by other cardiac anomalies. They are more common if cardiac malposition is isolated, in other words not accompanied by malposition of the other organs.

The term "isomerism" describes failure of differentiation into the right- and left-sided organs, that is, "bilateral same-sidedness". In other words, right atrial isomerism (RAA) means that there are two morphologically right atriums and no morphologically left atriums at all. It is also known as Ivemark syndrome, or asplenia syndrome because the spleen is usually absent. Left atrial isomerism (LAA) means that there are two left atriums and no right atrium, and is also known as polysplenia syndrome because multiple smaller spleens are usually present. The term "heterotaxia syndromes" unifies asplenia and polysplenia syndromes. Both left and right atrial isomerisms are associated with arrhythmias and conduction defects due to the absent (in LAA) or reduplicated (in RAA) sinoatrial and atrioventricular nodes.

Embryology

(…)

Epidemiologic, Genetic, Morphologic and Clinical Correlations

Heterotaxia syndromes are rare and are encountered in 1%–2% of neonates with symptomatic congenital heart disease. Anomalies of cardiac and visceral situs are often detected during the routine antenatal screening. No conclusive evidence exists that there is a single aetiology responsible for development of atrial isomerism. Complex cardiac malformations are

Table 28.4 Cardiovascular malformations in right and left atrial isomerism

Structure	RAA (asplenia)	LAA (polysplenia)
Systemic veins	Normal IVC (35% left-sided)	Interrupted IVC (absent hepatic segment) with azygos or hemiazygos continuation (85%)
Pulmonary veins (PV)	Extracardiac TAPVC (75%), often obstructed	All PV return to left-sided atrium (50%) or right PV to right-sided and left PV to the left (50%)
Atrium and atrial septum	Bilateral right atria and sinus node, bilateral AV node, primum ASD (100%) and secundum ASD (66%)	Bilateral left atria (no sinus node), single atrium, primum ASD (60%), secundum ASD (25%)
AV valve	Common AV valve (90%)	Normal AV valve (50%), common AV valve (15%)
Ventricles	Two ventricles (50%), single/unbalanced ventricles (50%)	Balanced ventricle size almost always. VSD (65%), DORV (20%)
Great arteries	TGA and CCTGA, pulmonary stenosis (40%) or atresia (40%)	TGA (15%), pulmonary stenosis or atresia (40%)

almost always present in heterotaxia syndromes, especially with RAA (asplenia syndrome) which is associated with more severe abnormalities and worse outcomes (Table 28.4). The presence of interrupted IVC is probably the best discriminator. The coronary sinus is usually absent, and absent in all cases of RAA. A normal heart or only minimal cardiac malformations are present in up to a quarter of all patients with LAA (polysplenia syndrome) (Table 28.4).

In RAA (asplenia) bilateral three-lobed lungs with bilateral eparterial, that is, overlying branch pulmonary arteries, bronchi, and various gastrointestinal malformations such as a symmetrical midline liver and gut malrotation, are present. The stomach may be on either side. In LAA (polysplenia), bilateral bilobed lungs with bilateral hyparterial bronchi (two left lungs), symmetrical liver, absence of gallbladder and some degree of intestinal malrotation are often present.

In complex malformations there is usually a complete mixing of systemic and pulmonary venous blood in the heart. When pulmonary blood flow is reduced, as in RAA, this results in severe cyanosis. When pulmonary blood flow is increased, as in LAA, cyanosis is not as intense and congestive heart failure often develops. Heart murmurs are often audible.

Echocardiography can detect all or most of the anomalies; however, cardiac MRI or CT is usually indicated to document complex anomalies of pulmonary and systemic venous return. Without palliative surgical procedures, more than 95% of patients with RAA (asplenia) syndrome would die during the first year of life. Immune deficiency making patients with RAA (asplenia) vulnerable to infections caused by incapsulated bacteria such as *Staphylococcus pneumoniae* puts them at risk of fulminant sepsis and requires antibiotic prophylaxis and specific immunisation schedule. Tachy- and bradyarrhythmias may be significant enough to require treatment with medication and/or pacing.

Univentricular Hearts

Introduction and Background

Functionally univentricular heart results from inadequacy of one of the ventricles to support pulmonary or systemic circulation, or from the anatomic malformations where both ventricles are fully developed but the nature of associated anomalies—VSD size and position, atrioventricular valves straddling and overriding, great arteries position—make surgical restoration of the ventricular septum integrity, or septation, impossible.

Embryology

(…)

Epidemiologic, Genetic, Morphologic and Clinical Correlations

Numerous anatomic variants of functionally univentricular hearts exist (Fig. 28.23). They can be subdivided into groups based on the characteristics of the dominant chamber and described, but not limited by, the following examples:

- *Hypoplasia of the right ventricle* (tricuspid atresia, pulmonary atresia with intact ventricular septum, unbalanced atrioventricular septal defect, double inlet left ventricle, unbalanced double outlet left ventricle, severe Ebstein's malformation, Uhl's anomaly)
- *Hypoplasia of the left ventricle* (mitral stenosis/atresia usually associated with severe aortic stenosis/atresia results in hypoplastic left heart syndrome, double inlet right ventricle (very rare), unbalanced AVSD with right ventricular dominance, unbalanced double outlet right ventricle with right ventricular dominance)
- *Unseptatable ventricles*, in which most of the ventricular septum is absent but an apical muscular rim exists dividing the ventricular mass into right and left components
- *"True"* univentricular heart with a solitary ventricle of indeterminate morphological type

A different systematisation based on the atrioventricular valves connecting atrial and ventricular segments of the heart can be devised (Fig. 28.12). Any atrioventricular connection can coexist with any of the atrial arrangement and any of the ventricular morphologies. Further variations are possible at the ventriculo-arterial level:

- Both atrioventricular valves open into a common dominant ventricular chamber

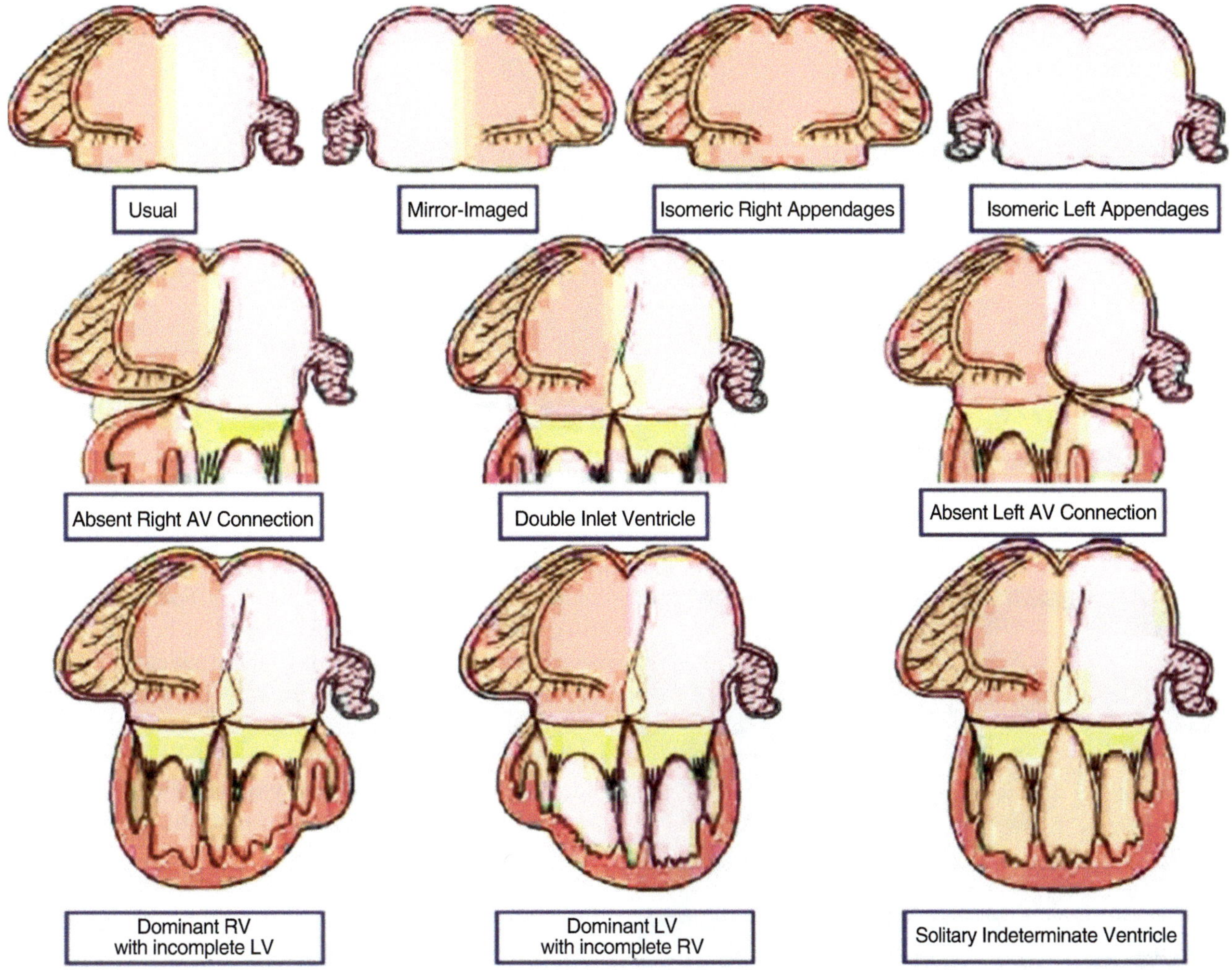

Fig. 28.23 Segmental combinations in univentricular connections (From Anderson RH, Shirali G. Sequential segmental analysis. Ann Pediatr Cardiol 2009)

  - Double inlet left ventricle (Fig. 28.24) with two separate AV valves (most common form of single ventricle)
  - Double inlet ventricle (right or left) via a common AV valve and unbalanced AVSD (Fig. 28.25)
  - Double inlet right ventricle (very rare)

- One of the atrioventricular valves is imperforate or absent
  - Tricuspid atresia or absent right atrioventricular connection (Fig. 28.26)
  - Mitral atresia or absent left atrioventricular connection (Fig. 28.27)

A rudimentary ventricle (or rudimentary outlet chamber if no atrioventricular valve is leading into it) is practically always present and connected to the dominant ventricle with a ventricular septal defect that in this case is named "bulboventricular foramen". It frequently becomes obstructive leading to the subvalvar subaortic or subpulmonary stenosis, depending on which of the great vessels arises from the rudimentary ventricular chamber. The valve of this vessel, and the vessel itself is often underdeveloped (hypoplastic) and also stenotic. There is a complete mixing of systemic and pulmonary venous blood in the ventricle, and thus oxygen saturation of the blood in aorta and pulmonary artery is identical. The saturation depends on the ratio of systemic and pulmonary blood flow. When pulmonary blood flow is reduced due to obstruction, cyanosis is marked. When there is no pulmonary obstruction and low pulmonary vascular resistance leads to a significantly increased pulmonary blood flow, patients will not be cyanosed and develop symptoms of heart failure.

Clinical course and management depends on whether pulmonary blood flow is increased or reduced. Newborn babies with severe pulmonary stenosis when pulmonary artery arises from the rudimentary chamber, or aortic stenosis, coarctation of the aorta or interrupted aortic arch when aorta arises from the rudimentary chamber, will require

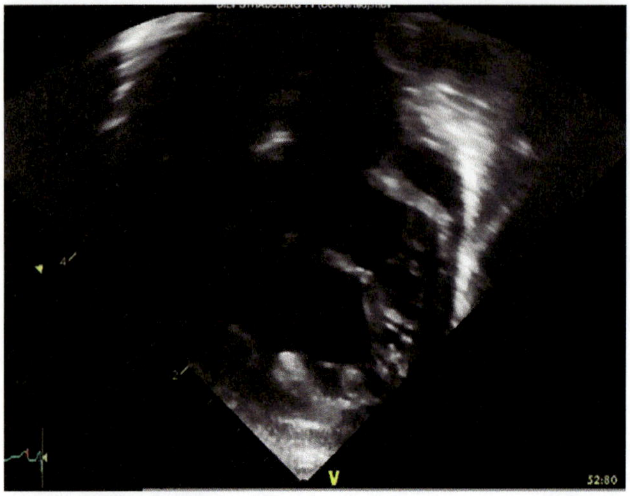

Fig. 28.24 Echocardiographic image: double inlet ventricle, apical four chambers view

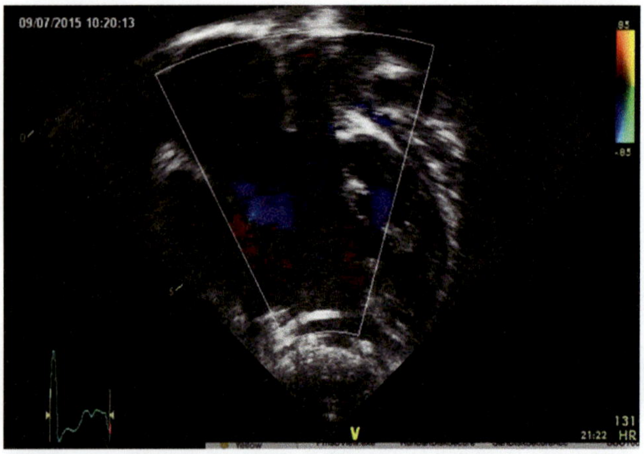

Fig. 28.25 Echocardiographic image: unbalanced AVSD (hypoplastic LV), apical four chambers view, colour Doppler

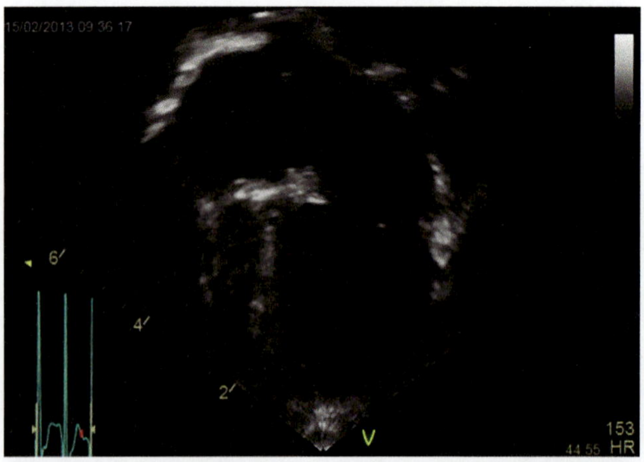

Fig. 28.26 Echocardiographic image: tricuspid atresia, hypoplastic RV, apical four chambers view

maintaining of the ductal patency with the help of intravenous prostaglandin infusion. All patients with the functionally univentricular heart will require staged univentricular

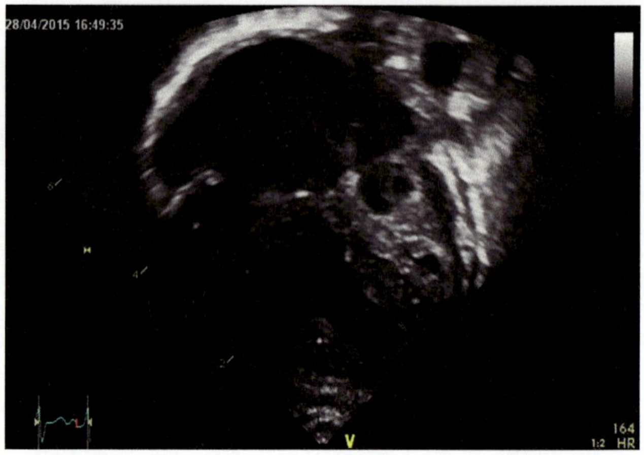

Fig. 28.27 Echocardiographic image: mitral atresia, hypoplastic left heart syndrome, apical four chambers view

surgical palliation aimed to achieve total cavo-pulmonary connection (Fontan-type procedure). In Fontan-type set-up all systemic venous return is diverted to bypass the heart directly into pulmonary arteries. Oldest patients who underwent Fontan-type procedure are in their forties, and many have good exercise tolerance and quality of life. Long-term complications of univentricular circulation include stroke and arrhythmias, less frequently encountered are protein-losing enteropathy and plastic bronchitis, and 20–30% require heart transplantation.

Recommended Readings

Anderson RH, Baker EJ, Penny DJ, Redington AN, Rigby ML, Wernovsky G. Paediatric cardiology. 3rd ed. Philadelphia, PA: Churchill Livingstone/Elsevier; 2010.

Freedom RM, Yoo S-J, Mikailian H, Williams WG. The natural and modified history of congenital heart disease. New York: Blackwell Publishing, Inc./Futura Division; 2004.

Allen HD, Gutgesell HP, Clark EB, Driscoll DJ. Moss and Adams' heart disease in infants, children, and adolescents including the fetus and young adult. 6th ed. Baltimore, MD: Lippincott Williams & Wilkins Publishers; 2000.

Rudolph AM. Congenital diseases of the heart: clinical-physiological considerations. 2nd ed. Armonk, NY: Futura Publishing Company, Inc.; 2001.

Park MK, Salamat M. Park's the pediatric cardiology handbook. 5th ed. Philadelphia, PA: Elsevier Saunders; 2016.

Casans R, Lal M, Gritsatis M. The essential revision guide to paediatric cardiology. London: Radcliffe Publishing Ltd; 2013.

Sidwell R, Thomson M. Concise paediatrics. London: Greenwich Medical Media Limited; 2000.

Frescura C, Thiene G. The new concept of univentricular heart. Frontiers in Pediatrics. 2014;2:62. https://doi.org/10.3389/fped.2014.00062.

Silvestry F, et al. Guidelines for the echocardiographic assessment of atrial septal defect and patent foramen ovale: from the American Society of Echocardiogrphy and Society for Cardiac Angiography and Interventions. J Am Soc Echocardiography. 2015;28:910–58.

Calvert PA, Rana BS, Kydd AC, Shapiro LM. Patent foramen ovale: anatomy, outcomes, and closure. Nat Rev Cardiol. 2011;8:148–60.

Anderson A, Shirali G. Sequential segmental analysis. Ann Pediatr Cardiol. 2009;2(1):24–35.

29 Embryology of the Great Vessels

Maria Ilina and Stuart Lilley

The vascular system appears in the middle of week 3 of development from fixed primitive angiogenic cells which are mesodermal in origin. The first arteries to appear are the right and left primitive aortae which appear as a continuation of the endocardial heart tubes. The primitive aortae curve dorsally in the first pharyngeal arch, around the anterior part of the foregut, and then continue dorsally as the two dorsal aortae.

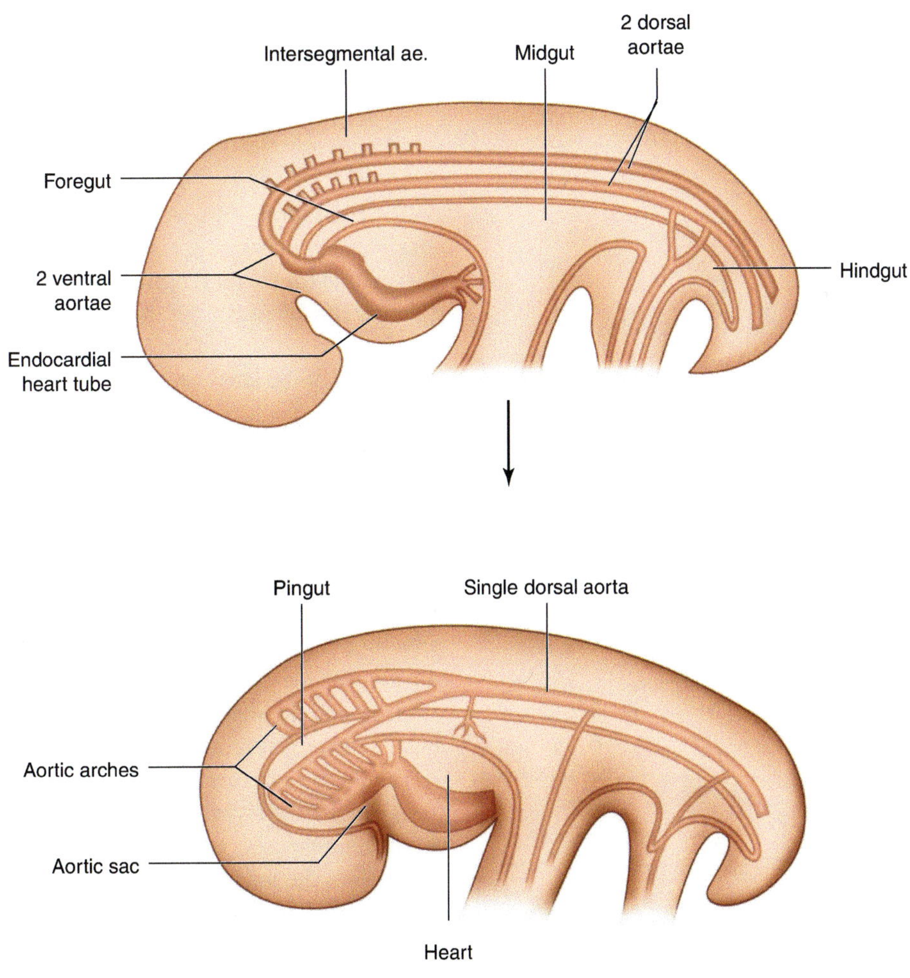

M. Ilina (✉) · S. Lilley
Queen Elizabeth University Hospital, Glasgow, UK
e-mail: Maria.Ilina@ggc.scot.nhs.uk

R. Carachi, S. H. E. Doss (eds.), *Clinical Embryology*, https://doi.org/10.1007/978-3-319-26158-4_29

The aortic sac is a dilated channel formed by the primitive aortae fusing close to the heart. This lies ventral to the pharynx and is continuous caudally with the truncus arteriosus. The two dorsal aortae pass caudally dorsal to the primitive gut. They fuse together all forming a single dorsal aorta apart from at the cranial part which remain separate and are continuous with the two ventral aortae via the first aortic arch arteries.

Aortic Arches

The aortic arches are six pairs of arteries which connect the aortic sac ventrally with the two dorsal aortae dorsally. They run in the pharyngeal arches along the side wall of the pharynx. As the pharyngeal arches begin to develop, the aortic sac sends a pair of branches to each pharyngeal arch. Each branch, known as an aortic arch artery, leaves the aortic sac and curves around the pharynx inside the corresponding pharyngeal arch to end in the dorsal aorta. The six pairs are never present at the same time as by the time the third pair develops, the first pair disappears.

Formation of the aortic arches

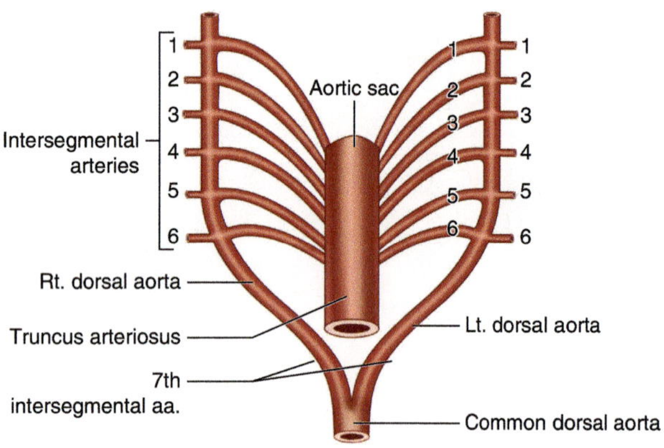

Fate of the aotic arches

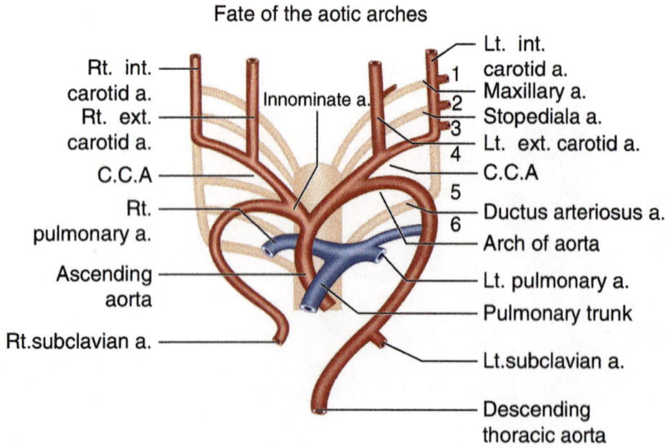

The aortic arches undergo changes in number and arrangement as follows:

Aortic arch	Right		Left	
First	Disappears except for a small part which forms the maxillary artery			
Second	Disappears except for a small part which forms the stapedial artery			
Third	Forms the common carotid artery and the proximal part of the internal carotid artery. The external carotid artery arises as a bud from the third arch			
Fourth	Forms the proximal part of the right subclavian artery		Forms the proximal part of the arch of the aorta	
Fifth	Disappears completely and very early in development			
Sixth	Each sixth arch divides into ventral (medial) and dorsal (lateral) segments			
	Ventral	**Dorsal**	**Ventral**	**Dorsal**
	Forms the right pulmonary artery which enters the right lung bud	Disappears and loses its connection with the dorsal aorta	Forms the left pulmonary artery which enters the left lung bud	Persists during intrauterine life forming the ductus arteriosus which forms a connection between the left pulmonary artery and the arch of the aorta

The right horn of the aortic sac goes on to form the brachiocephalic artery which is continuous with the right common carotid (of the third right aortic arch) and the stem of the right subclavian artery (of the fourth right aortic arch). The stem and the left horn of the aortic sac form the proximal part of the arch of the aorta.

Development of the Aortic Arch

The arch of the aorta arises from multiple sources. The proximal part (proximal to the origin of the innominate artery) arises from the stem of the aortic sac. The middle region, between the innominate and left common carotid, is from the left horn of the aortic sac, and the region distal to the left common carotid artery arises from the fourth left aortic arch and the lower part of the left dorsal aorta.

There are seven cervical intersegmental arteries arising from each of the right and left dorsal aortae. The upper six become connected by vertical anastomoses which will give rise to the second part of the vertebral artery and the deep cervical artery as well as the superior intercostal artery. The seventh intersegmental artery forms the subclavian artery.

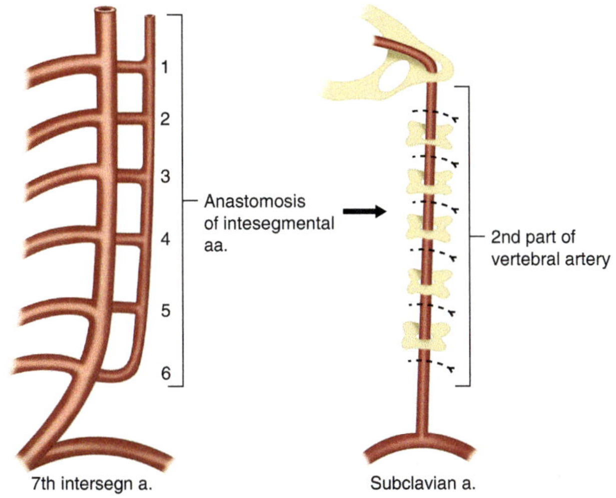

Anomalies of the Aortic Arches

1. Patent ductus arteriosus

   Normally the ductus arteriosus closes shortly after birth. Failure to close establishes a communication between the left pulmonary artery and the aortic arch. This is one of the most common congenital anomalies of the great vessels and leads to progressive hypertrophy of the left side of the heart.

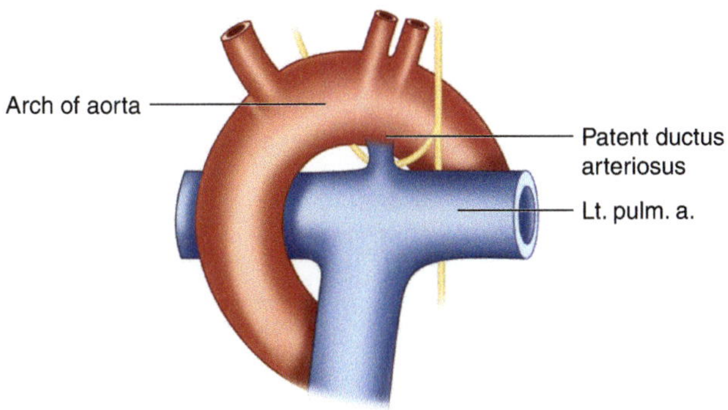

2. Coarctation of the aorta

   This condition is a congenital narrowing of the aorta distal to the origin of the left subclavian artery. It can be pre- or postductal. In the preductal type, the narrowing of the aorta is above the level of the ductus arteriosus. The ductus remains patent to maintain circulation to the lower body. With postductal, the narrowing of the aorta is distal to the level of the ductus, which closes to form the ligamentum arteriosum. In this type, the circulation to the lower part of the body is maintained by anastomosis between the arteries and the scapula and the post-intercostal arteries as well as anastomosis between the superior epigastric and inferior epigastric arteries.

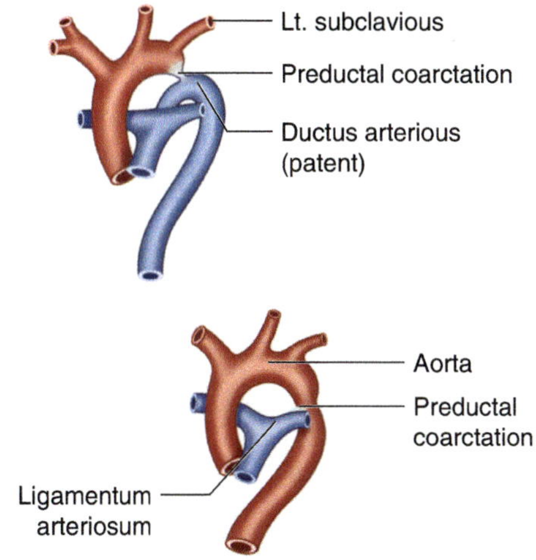

3. Right-sided aortic arch

   This anomaly results from the persistence of the distal part of the right dorsal aorta and degeneration of the distal part of the left dorsal aorta.

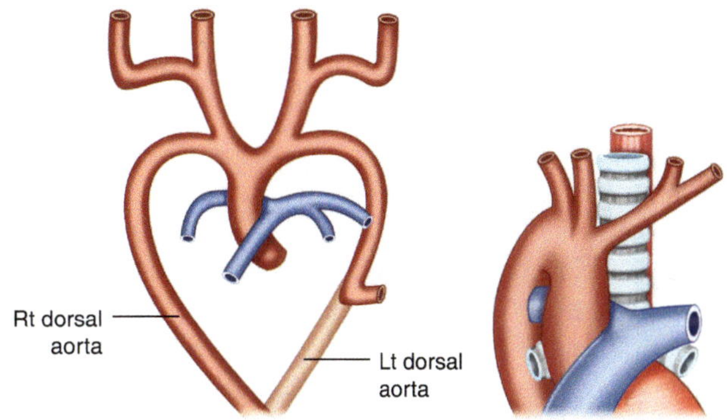

4. Double aortic arch

   Arising from the persistence of the distal part of both the right and left dorsal aortae leading to the formation of a vascular ring around the trachea and oesophagus causing difficulty in swallowing and breathing.

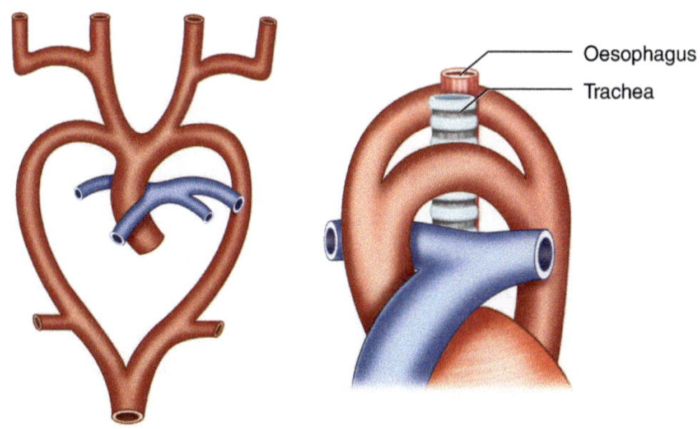

The Common Dorsal Aorta

The common dorsal aorta is formed by the union of the right and left dorsal aortae in the region extending from the fourth thoracic to the fourth lumbar somite segments.

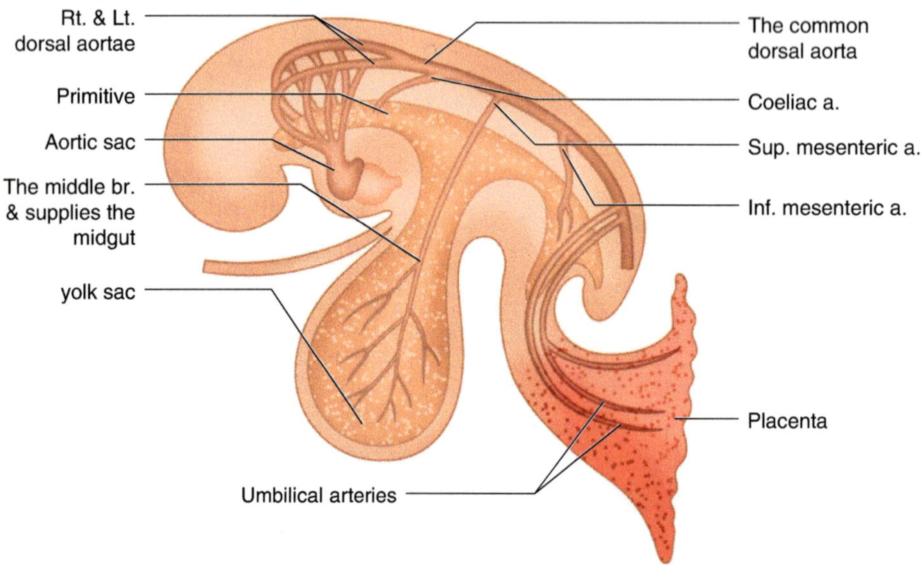

It branches into the:

1. Ventral splanchnic arteries, which arise from the ventral aspect to supply the primitive gut that includes the coeliac artery (supplies the foregut), the superior mesenteric artery (supplies the midgut) and the inferior mesenteric artery which supplies the hindgut. These arteries anastomose together both dorsally and ventrally to the gut.
2. Lateral splanchnic branches, which are paired (left and right) branches arising from the lateral aspect of the dorsal aorta to supply structures arising from the intermediate mesoderm including the middle suprarenal arteries, the renal arteries and the gonadal arteries (testicular or ovarian).
3. Intersegmental arteries, the paired branches that come from the posterolateral aspect of the dorsal aorta and pass laterally between the somites. These are represented in the adult by the posterior intercostal, subcostal and lumbar arteries. They anastomose both ventrally and dorsally, and their ventral anastomoses produce the internal thoracic, superior epigastric and inferior epigastric arteries in the adult.
4. Umbilical arteries. At first, these arise from the dorsal aorta; then they become connected to the fifth lumbar intersegmental artery by anastomosis at which time they lose their connection to the dorsal aorta. The external iliac artery arises as a branch of the fifth lumbar intersegmental artery leaving the umbilical artery attached to its distal part which becomes the internal iliac artery. The proximal part of the fifth lumbar intersegmental artery then becomes the common iliac artery.

Development of the Arteries in the Upper and Lower Limbs

The arteries of the upper limb develop from the axial artery which runs in the axis of the limb. It arises from the seventh intersegmental artery and proceeds distally as the subclavian, axillary, brachial and anterior interosseous arteries. It ends in the hand by forming a deep vascular plexus which will later become the deep palmar arch. A median artery arises as a branch from the anterior interosseous artery and runs along the median nerve to communicate with the capillary plexus in the hand. The axial artery gives off two branches in the elbow region, the ulnar artery and the radial artery.

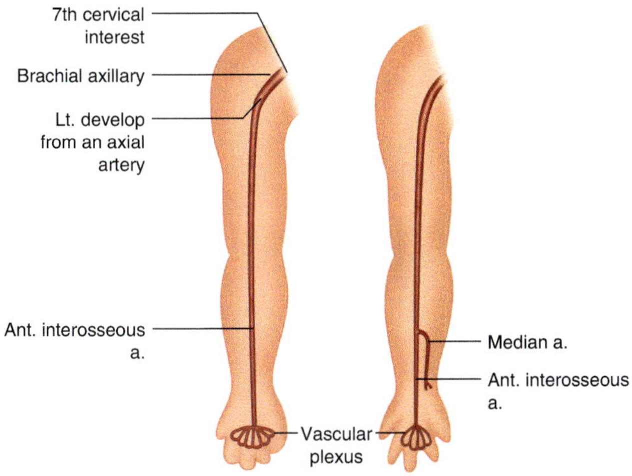

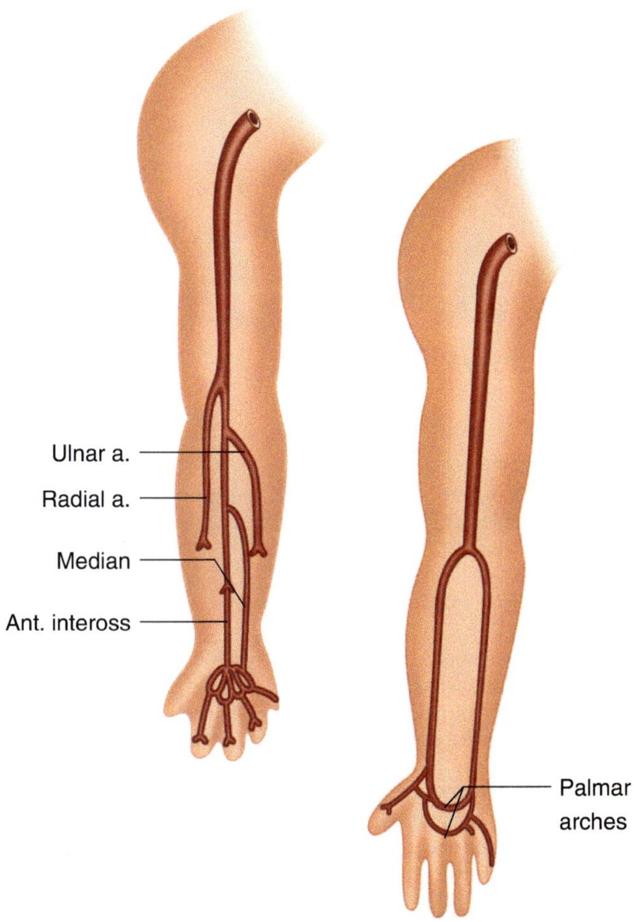

The axial artery of the lower limb is the continuation of the fifth lumbar intersegmental artery. It follows the sciatic nerve, hence its naming as the sciatic artery. It descends through the gluteal region, back of the thigh and back of the leg, deep to the popliteus and calf muscles. It ends in the sole of the foot by forming a vascular plexus. The femoral artery develops later as a continuation of the external iliac artery. It descends in the front of the thigh then curves backwards to join the axial artery in the popliteal fossa above the popliteal muscle forming the popliteal artery. The tibial arteries develop from local vascular plexuses in the front and back of the leg and become connected to the popliteal artery. The axial artery degenerates leaving remnants in the adult represented by the inferior gluteal artery, the companion artery of the sciatic nerve and the peroneal artery in the back of the leg.

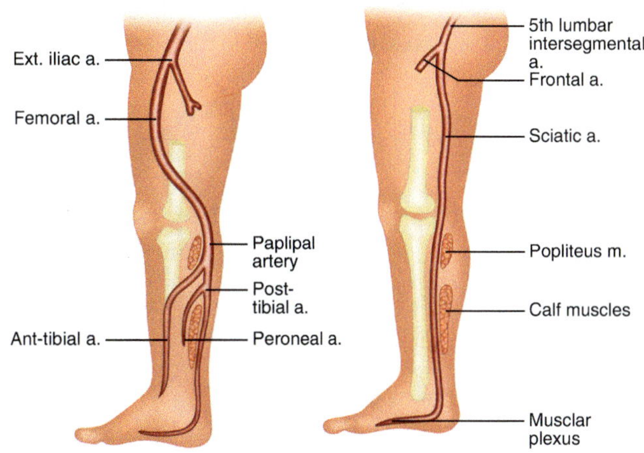

Development of the Veins

The embryo has three types of veins: vitelline, umbilical and cardinal.

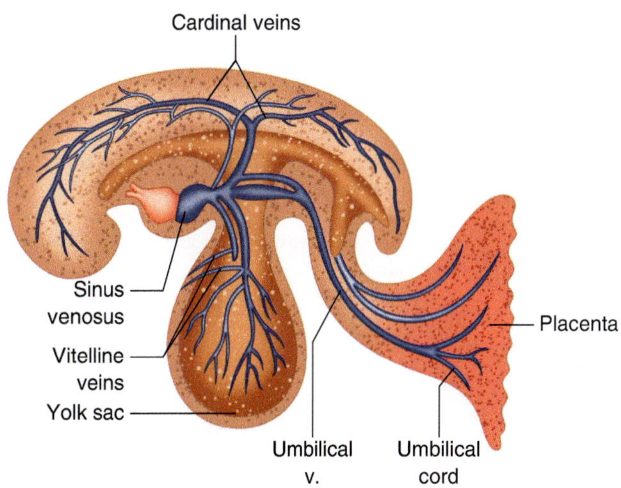

Vitelline and Umbilical Veins

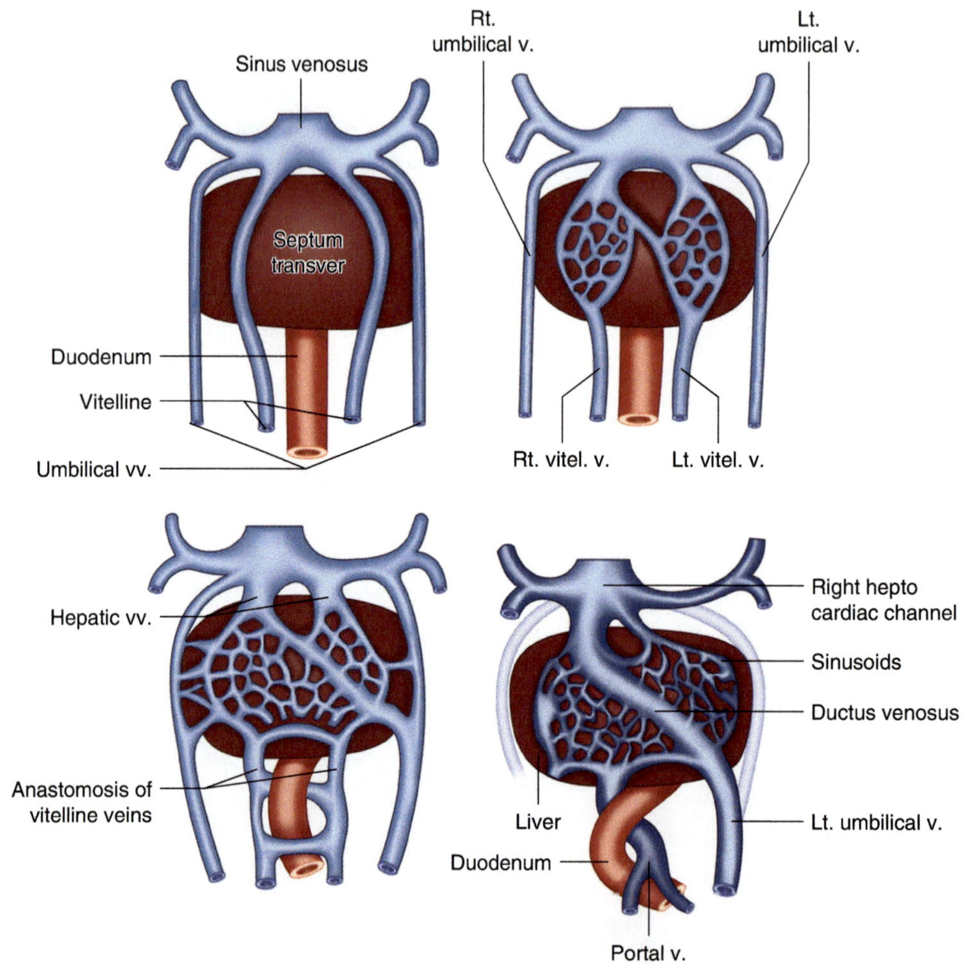

	Vitelline veins	Umbilical veins
Origin	Network of capillaries in the mesoderm of the wall of the yolk sac. Enter the body of the embryo via the yolk sac stalk	Arise from the placenta as one vein which enters the body via the umbilical cord then duplicates forming the right and left umbilical veins
Course	Pass cranially, one on each side of the duodenum, then traverse the septum transversum to end in the right and left horns of the sinus venosus	Pass cranially traversing the peripheral parts of the septum transversum to end in the right and left horns of the sinus venosus
Derivatives	1. The caudal parts become connected by three anastomotic channels (two ventral and one dorsal to the duodenum) forming a figure of 8 which undergoes partial degeneration to form the portal vein 2. The middle parts inside the septum transversum become invaded by the liver cell cords and break down into hepatic sinusoids 3. The cranial part of the left vitelline vein disappears but the right persists forming the hepatic veins	The right umbilical vein disappears completely. The left umbilical vein persists as the only vessel carrying oxygenated blood from the placenta and has the following fate: 1. The caudal part runs in the free margin of the falciform ligament 2. The middle parts inside the septum transversum are invaded by the liver cord cells to become hepatic sinusoids; as development proceeds, some sinusoids enlarge forming a large channel—the ductus venosus—which connects the proximal part of the left umbilical vein with the hepatic cardiac channel of the right vitelline vein 3. The cranial part disappears

Cardinal Veins

These are symmetrical longitudinal venous channels which appear in successive stages inside the body of the embryo to convey its venous blood. They will give rise to the main systemic veins of the adult. The cardinal veins include:

- Anterior cardinal veins
- Posterior cardinal veins
- Common cardinal veins
- Subcardinal veins
- Supracardinal veins
- Azygos line veins

Development of the Inferior Vena Cava

The inferior vena cava is formed of six segments described as follows from above downwards.

Suprahepatic	Formed by the proximal part of the right vitelline vein
Hepatic	Formed by the anastomosis between the right vitelline vein and the right subcardinal vein
Prerenal	Formed by intersubcardinal anastomosis and part of the right subcardinal vein
Renal	Formed by the right sub- and supracardinal anastomosis
Postrenal	The main part of the inferior vena cava and is formed by the right supracardinal vein
Lowest segment	Formed of the iliac anastomosis of the postcardinal veins

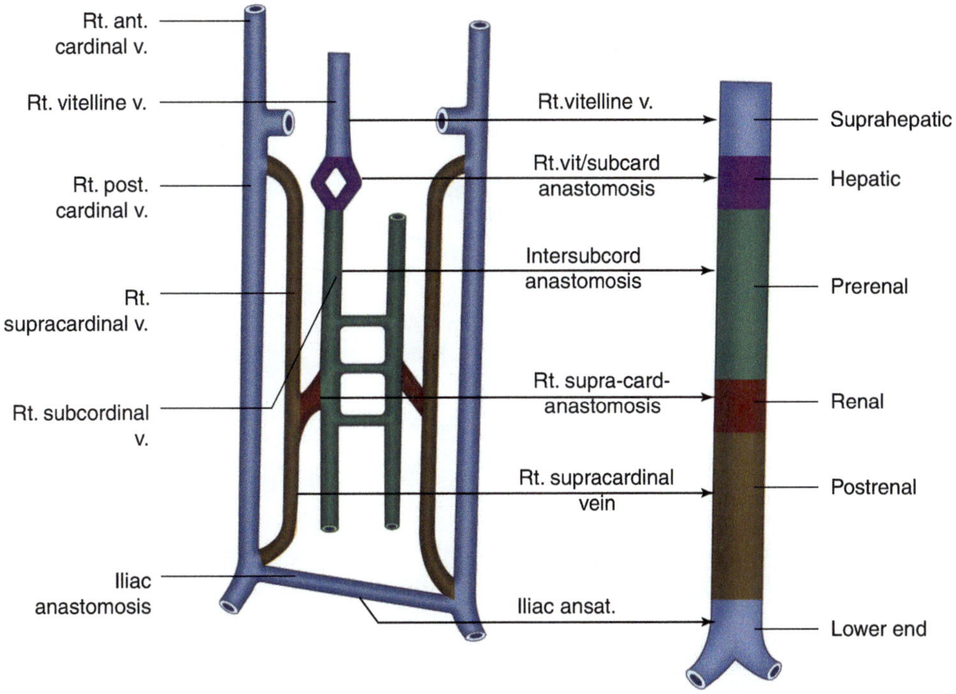

Anomalies of the Inferior Vena Cava

1. Left inferior vena cava, as the name suggests, lies on the left side of the abdominal aorta instead of the right. It results from persistence of the left supracardinal vein and disappearance of the right supracardinal vein.
2. In the case of double inferior vena cava, there is a left-sided inferior vena cava below the level of the renal veins, in addition to the normal inferior vena cava on the right side. This condition results from the persistence of both the right and left supracardinal veins.
3. Absent inferior vena cava is a rare condition in which the right subcardinal vein fails to join the hepato-cardiac channel of the right vitelline vein. The venous blood of the lower part of the body reaches the heart via the azygos vein.

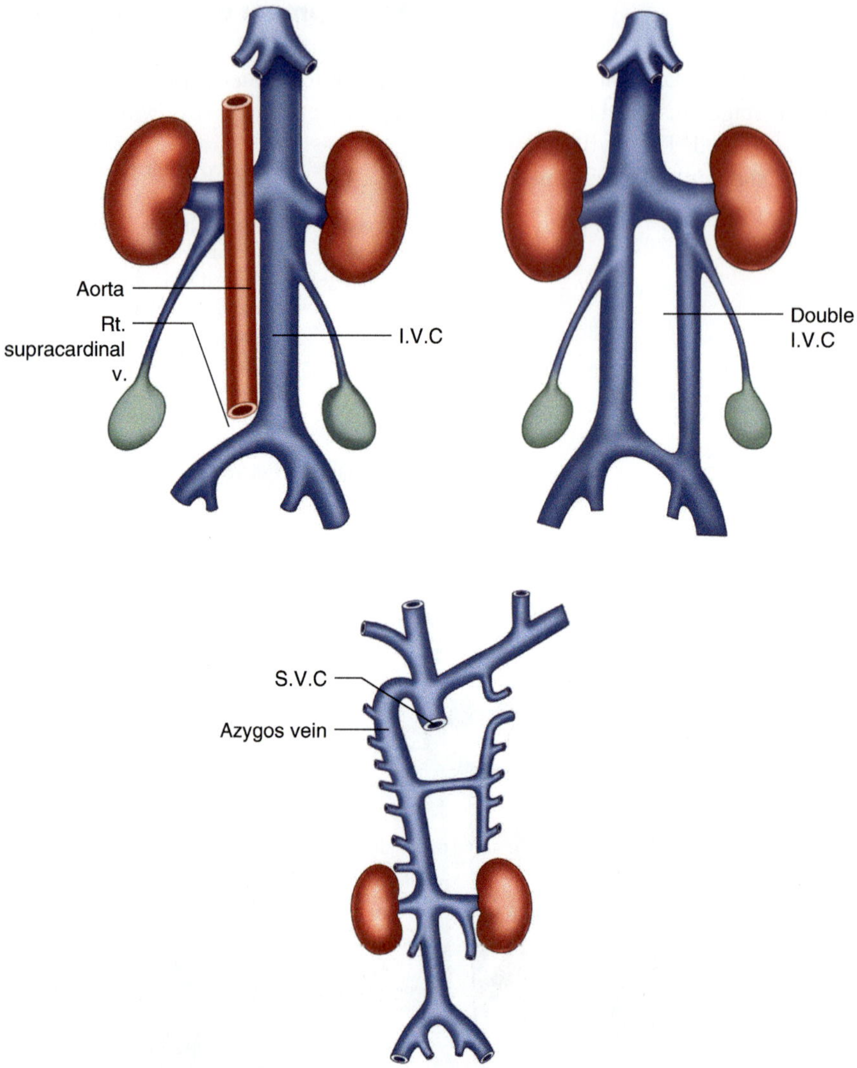

Renal Veins

The right renal vein is formed by the right metanephric vein which emerges from the metanephros to join the right subcardinal vein. The left renal vein is longer than the right and has a double origin; it forms from the left metanephric vein which joins the left subcardinal vein and from the intersubcardinal anastomoses between the two subcardinal veins.

Congenital Anomalies of the Great Vessels

Patent Arterial Duct

Introduction and Background

This common congenital lesion represents persistent postnatal patency of a normal foetal structure between the pulmonary artery and the descending aorta.

Embryology

During early foetal development, five arterial arches that are never present simultaneously link the aortic sac with the paired dorsal aortas. This will be discussed in the aortic arch abnormalities part of this chapter. The normal arterial duct develops from the dorsal portion of the left sixth arch. The arteries that feed the developing lung originate from the aortic sac which also feed the bilateral sixth arches. In the majority of cases, the right sixth arch obliterates and disappears, whilst persisting on the left side, remaining in continuity with the pulmonary arteries and pulmonary trunk.

Epidemiologic, Genetic, Morphologic and Clinical Correlations

About two-thirds of the foetal cardiac output originates from the right ventricle, with only 5–10% passing through the lungs. This means that the majority of the right ventricular output passes through the arterial duct into the descending aorta and its presence is essential for normal foetal development, allowing majority of the right ventricular output to be diverted from the "unused" and high-resistance pulmonary circulation. Premature intrauterine ductal constriction or closure may lead to the right heart failure and result in foetal hydrops. A number of teratogens including rubella, alcohol, amphetamines and certain anticonvulsants influence the duct development, the most sensitive period being between 18 and 60 days of gestation.

In the foetus and neonate, the arterial duct is a short and wide vessel of variable length connecting the pulmonary arteries to the lesser curve of the aortic arch and terminating at the point of transition from the isthmus to the descending aorta, distal to the origin of the left subclavian artery (Fig. 29.1).

Ductal patency during foetal life is maintained by the low oxygen content of the foetal blood and placenta and ductal tissue-produced cyclooxygenase-mediated products of arachidonic acid metabolism, prostaglandin and prostacyclin. These mediators cause vasodilatation through interaction with prostanoid receptors.

The duct is a muscular artery with an intima, media and adventitia, markedly differing from the adjacent pulmonary trunk and aorta, in that aortic media is composed mainly of circumferentially arranged elastic fibres, whilst ductal media consists largely of spirally arranged smooth muscle cells; the intimal layers are thicker containing abundant mucoid substance, and no collagen is seen in the media. The duct is richly innervated by the adrenergic fibres, cholinergic fibres being sparse or absent. At the time of closure, smooth muscle constricts following an abrupt increase in oxygen tension, and endothelial cells accumulate in the lumen causing increase in the ductal wall thickness. Anatomic obliteration follows functional closure, beginning with necrosis of the inner wall followed by the formation of dense fibrous tissue, and eventually the duct becomes converted into a fibrous strand named "arterial ligament", which may take several weeks to complete. Anatomical obliteration occurs by the age of 2 weeks in two-thirds of babies and by age of 1 year in almost all.

The factors responsible for persistent ductal patency beyond the first days of life include prematurity, prenatal infection and genetic factors.

Following the postnatal reduction in pulmonary vascular resistance, patent arterial duct results in left-to-right shunting from the aorta to the pulmonary arteries. The magnitude of the shunt is determined by the diameter and length of the duct and the ratio between pulmonary and systemic vascular resistances. If the arterial duct is large and shunting persists for a long time, pulmonary hypertension and pulmonary vascular disease may develop, eventually leading to Eisenmenger

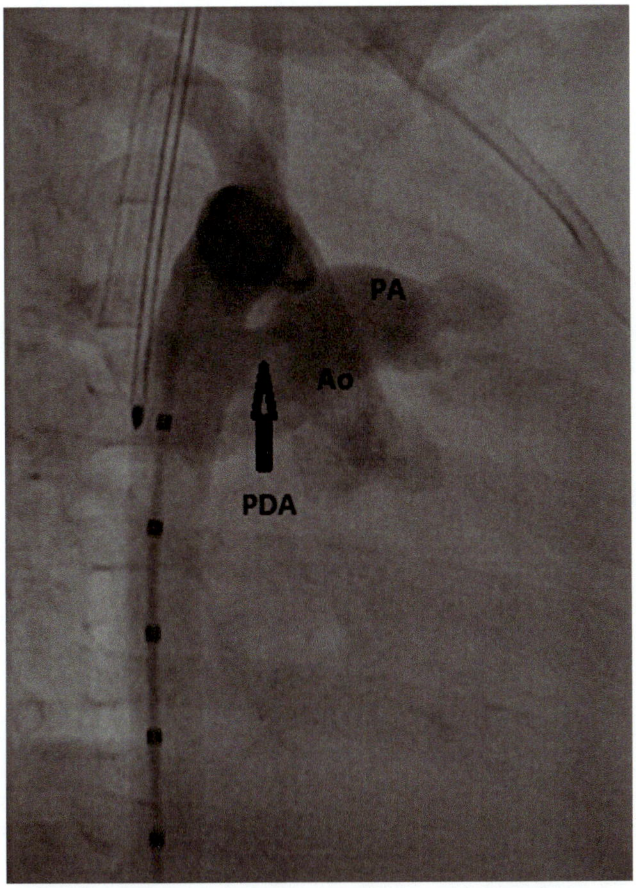

Fig. 29.1 PDA angiogram. Catheter in the aortic arch

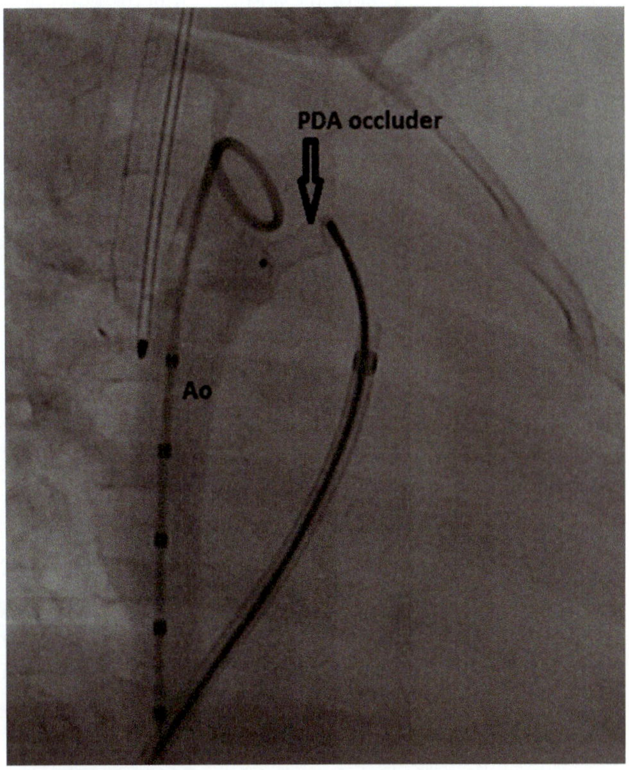

Fig. 29.2 PDA device closure. Catheters in the descending aorta and main pulmonary artery

syndrome and right-to-left shunting across the duct, and a degree of cyanosis develops. Presentation is usually with heart murmur; if the patent arterial duct is large, then signs of congestive heart failure or recurrent chest infections may develop. Patent arterial ducts require transcatheter device closure or occasionally surgical closure if transcatheter procedure is not possible for some reason.

In preterm infants, this condition is very common and encountered in over a half of infants of low and very low birth weight. It is a source of significant morbidity in this group of patients, may lead to development of congestive heart failure, contributes to respiratory disease of prematurity and increases risk of necrotising enterocolitis. Pharmacologic closure with the non-steroidal anti-inflammatory drug indomethacin may be attempted, and if medical treatment is unsuccessful or contraindicated, surgical ligation of the duct is undertaken.

Coarctation of the Aorta, Aortic Arch Anomalies and Vascular Rings

Introduction and Background

Term "aortic coarctation" indicates a narrowing at some point along the course of the aorta. The most common site of the narrowing in the context of the congenitally malformed heart is the thoracic aorta in the region of the insertion of the arterial duct. Interruption of the aortic arch is the condition when there is discontinuity between two adjacent segments of the aortic arch and it includes cases with a fibrous ligament, the remnant of the occluded segment, and thus interruption of the aorta can be described as the extreme end of the spectrum of aortic coarctation.

Due to the close proximity to the major airway and oesophagus, anomalies of the size, position and branching pattern of the aorta, arterial duct and pulmonary artery branches may cause obstruction. The terms "vascular ring", where a complete encirclement occurs, and "sling", where encirclement is incomplete, refer to a group of anomalies causing obstruction to the airways and oesophageal compression, although not all result in clinical symptoms and signs.

Embryology

There are three main aberrations in embryological development—abnormal development of the arch vessels, abnormal development of the arterial duct and changes in the ratio of flow between the pulmonary and systemic arterial pathways. In coarctation, the sling of ductal tissue forms part of the aortic wall (Fig. 29.3). It develops during the differential growth and migration of embryonic vascular structures and precursors of the subclavian artery and is present around the circumference of the aortic isthmus. The patterns of blood flow in the foetal circulation influence embryogenesis, and reduction in the volume of blood passing through the ascending aorta in foetal life leads postnatally to the development of coarctation.

Epidemiologic, Genetic, Morphologic and Clinical Correlations

Inheritance of coarctation of the aorta is multifactorial, although there are associations with some chromosomal abnormalities such as Turner syndrome, and there is a well-recognised association of 22q11 deletion (DiGeorge syndrome) and aortic arch interruption due to an abnormal migration of cells from the neural crest. The junction of the aortic arch and the arterial duct, or isthmus, is of great importance to the morphology of aortic arch anomalies. Two-thirds of the patients with aortic arch interruption between the left common carotid and subclavian arteries will have DiGeorge syndrome. This type of interruption (type B) is the commonest, accounting for over half of all

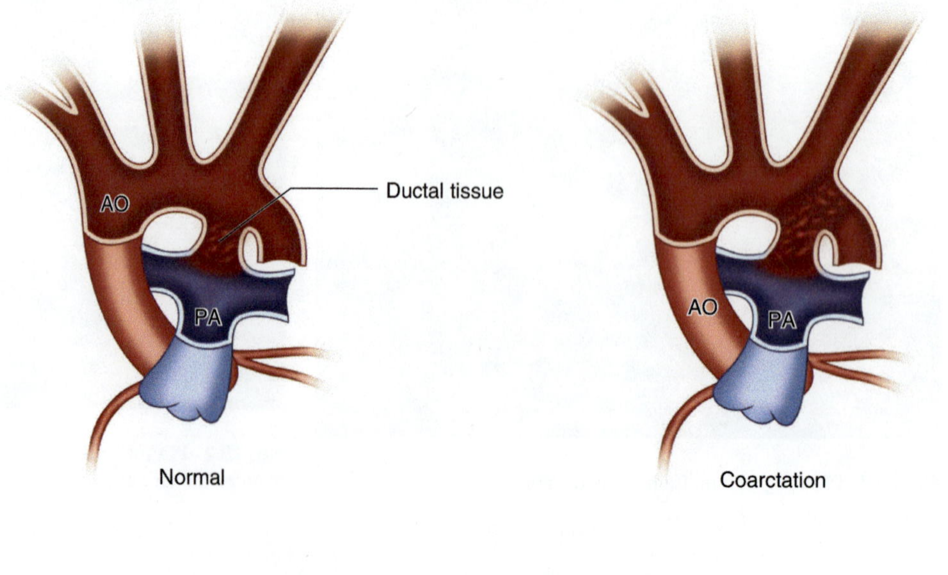

Fig. 29.3 Coarctation of the aorta

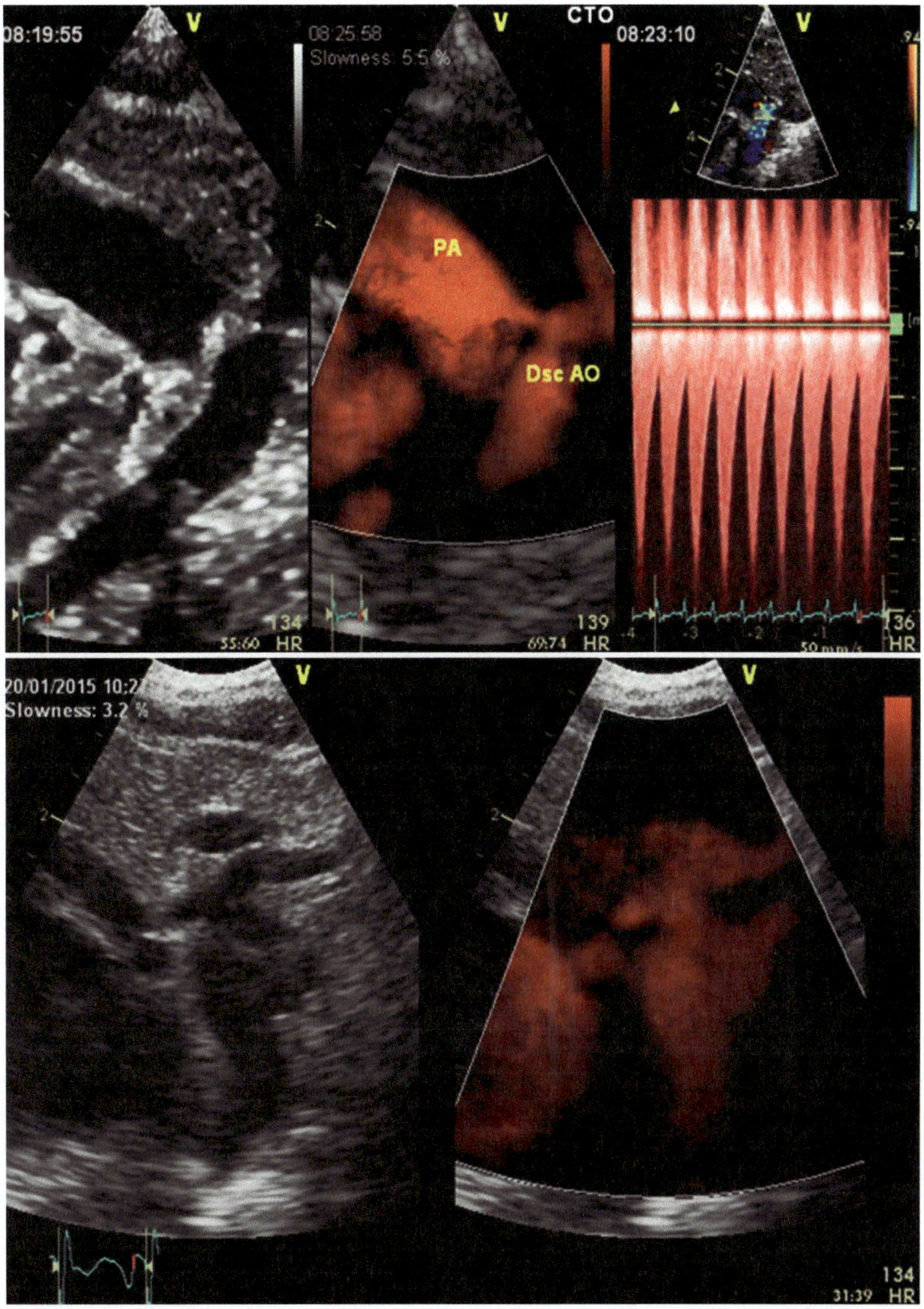

Fig. 29.4 Echo image coarctation of the aorta and PDA

cases, interruption at the isthmus (type A) for two-fifths, and interruption between the carotid arteries (type C) is very rare (Fig. 29.6a, b).

Aortic coarctation is a spectrum of lesions including variable degrees of hypoplasia of the aortic arch, obliteration of the lumen of the arch which turns into a fibrous cord between the interrupted segments that remain in anatomic continuity also called atresia of the arch and complete interruption. Foetal diagnosis of isolated aortic coarctation is challenging and often remains merely a suspicion based on the subtle imbalance of the ventricular dimensions and transverse arch size. A discrete coarctation represents a localised shelf-like lesion, often with a degree of proximal tapering of the arch towards the obstructed segment. When the arterial duct is patent, the obstruction site can be preductal, paraductal or postductal. The "shelf" is formed by ductal tissue, which completely encircles the lumen of the isthmus. When the duct closes, the ductal shelf turns into a fibrous diaphragm, often with a pinhole orifice. Postductal coarctation occurs rarely in infants in whom no improvement will be observed despite maintenance of ductal patency with prostaglandin but is a usual finding in adults.

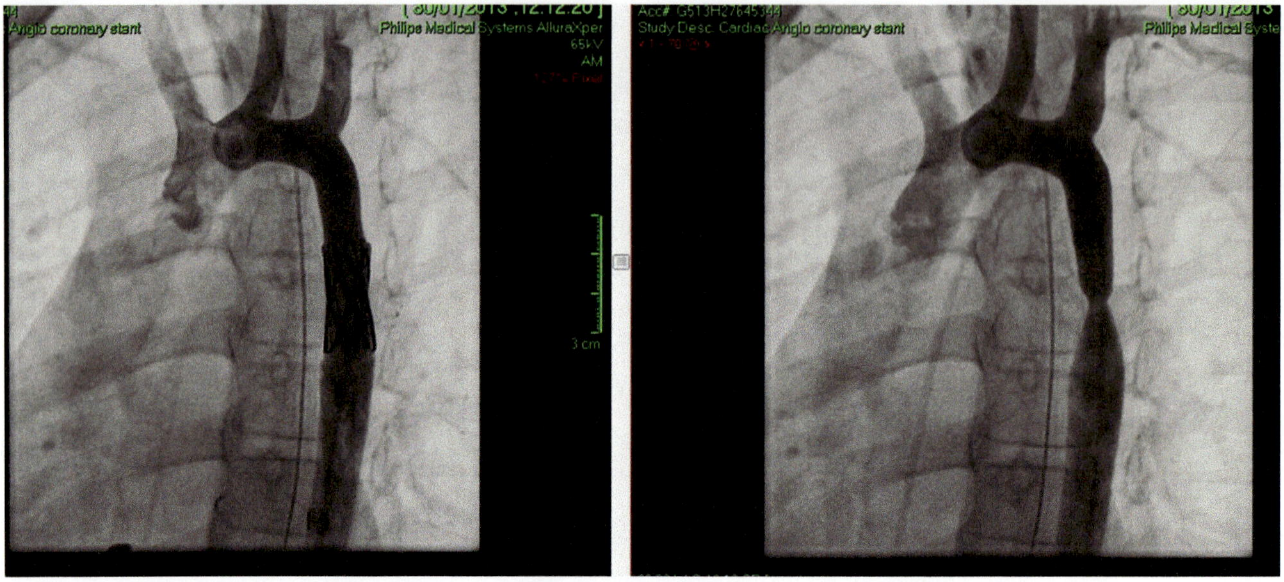

Fig. 29.5 Angiogram of the coarctation of the aorta, before and after the stent deployment

The association of aortic coarctation, hypoplasia of the isthmus, patency of the arterial duct and presence of an interatrial communication is so common as to be referred to as part of a "coarctation complex" in neonates. Other typical associated anomalies tend to increase pulmonary rather than systemic flow, such as a ventricular septal defect with posterior deviation of the outlet septum into the subaortic area leading to subaortic obstruction or perimembranous with overriding of the aortic valve. Other associations include transposition of the great arteries, double outlet right ventricle and common arterial trunk—in all these cases, aortic coarctation itself is considered an associated lesion. Another mechanism of associated lesion development is the one producing obstruction to the outflow from the left ventricle both pre- and postnatally, including valvar and subvalvar aortic stenosis, bileaflet (bicuspid) aortic valve (up to 85% of all patients with coarctation) and congenital stenosis of the mitral valve. A well-known combination consists of a parachute mitral valve, supravalvar supramitral ring, subaortic stenosis and aortic coarctation, also known as Shone's syndrome. The most extreme form of the series of obstructive lesions of the left heart including aortic coarctation is hypoplastic left heart syndrome, where mitral and aortic valves are severely hypoplastic and stenotic or atretic, and the left ventricle, ascending aorta and transverse aortic arch are often diminutive. Anomalies of the subclavian artery origin, where the anomalous artery instead of arising off the innominate artery often arises from the expanded segment of the aorta called the diverticulum of Kommerell and takes a retroesophageal course, can accompany coarctation or, more frequently, interruption. The diverticulum of Kommerell represents a remnant of the contralateral aortic arch.

Coarctation of the aorta and interruption of the aortic arch, when present with symptoms in infancy, are life-threatening conditions and require urgent intervention, mostly surgery. In a neonate, decompensation is frequently rapid following the ductal closure, and unless intervention is performed, death ensues. If the onset of obstruction is less abrupt, compensatory adaptations develop, such as left ventricular hypertrophy. Frequently, patients with coarctation go beyond infancy without detection, if obstruction is not severe enough. If coarctation remains unrepaired and patient survives to later childhood, teenage or even adulthood, collateral circulation gradually develops to bypass the obstruction site and augment perfusion to the lower body, and fibrosis and thickening of the intima may lead to obliteration of the lumen. Echocardiography is the diagnostic method of choice in infancy, and cross-sectional imaging in the form of magnetic resonance and computed tomography is frequently used to supplement the diagnostic information and provide additional assessment of the cardiac and vascular structures, confirming or ruling out associated anomalies. Primary intervention method in young children is usually surgery with balloon angioplasty and stenting reserved to the sickest infants and older patients with less severe coarctation. Lifelong follow-up is required, and arterial hypertension, late dissection and aneurysm formation are among late complications.

A number of rare vascular ring and sling varieties exist. Five major forms are most commonly encountered:

1. Double aortic arch (most common)—the right and left aortic arches completely encircle and compress the trachea and oesophagus, each giving off the common carotid and the subclavian arteries. The right aortic arch is usu-

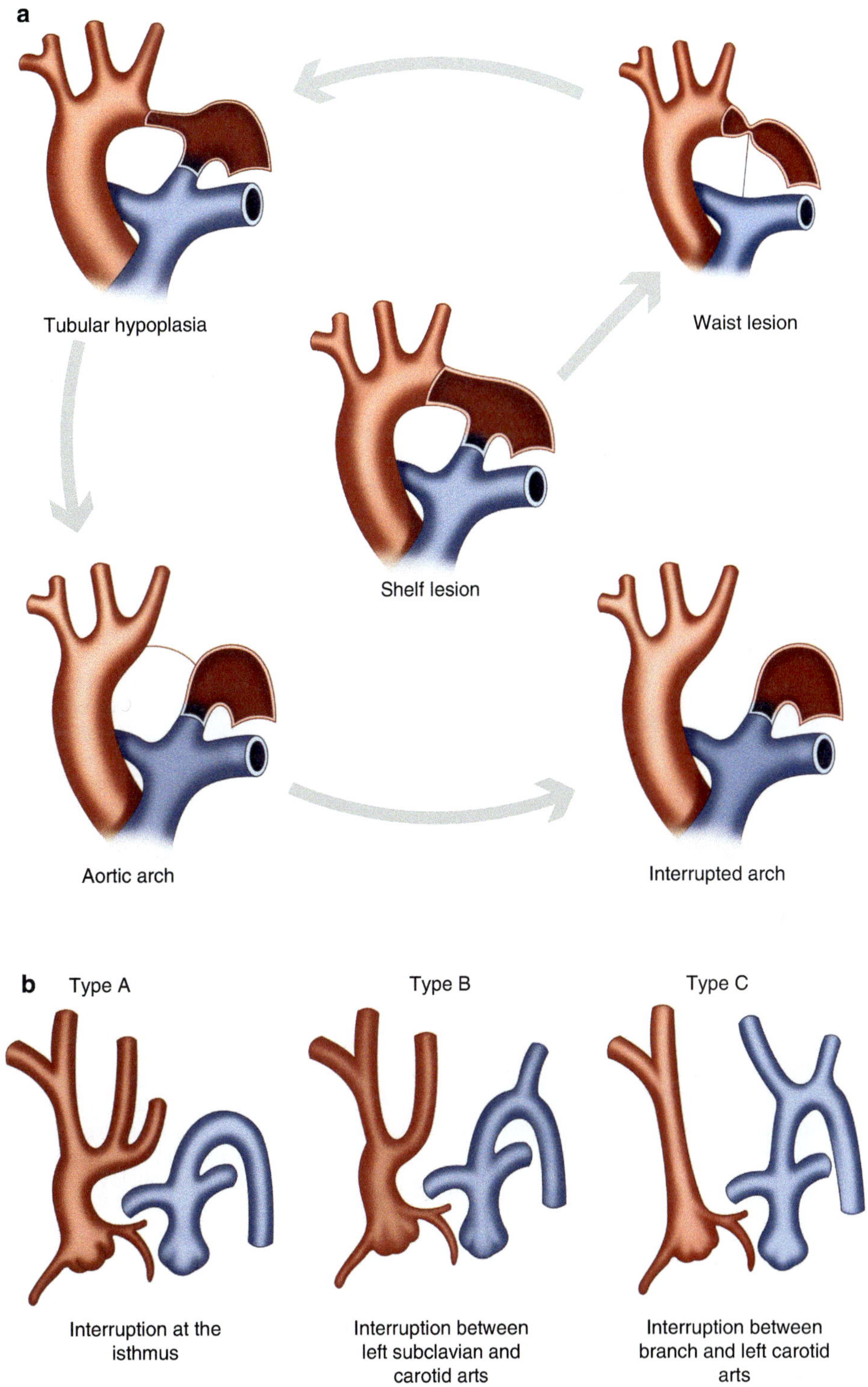

Fig. 29.6 (**a**, **b**) Interrupted aortic arch as part of the coarctation spectrum and classification of interruption

ally dominant. This condition is usually isolated but may be part of a complex congenital lesion.

2. Right aortic arch with and without the aberrant left subclavian artery and left ligamentum arteriosum connecting the subclavian artery or the descending aorta and LPA. The first form is rarely associated with intracardiac defects.
3. Anomalous innominate artery taking off too far left from the aortic arch and compresses the trachea, commonly associated with other congenital heart lesions.

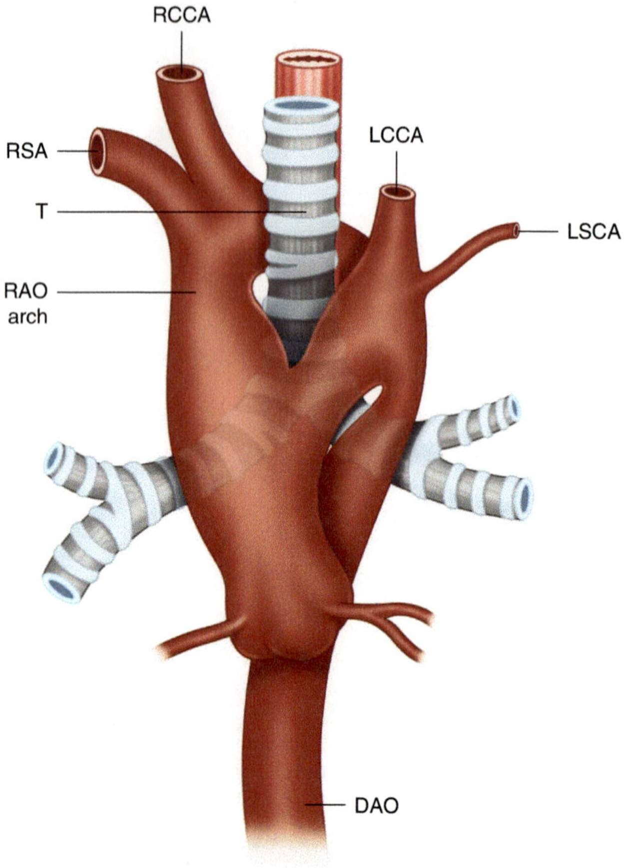

Fig. 29.7 Double aortic arch

4. Aberrant retroesophageal right subclavian artery and left aortic arch (vascular "sling") is the most common arch anomaly occurring in 0.5% of the general population, often isolated but can be associated with other lesions and very common in Down's syndrome patients with congenital heart disease.
5. LPA "sling" (anomalous origin of LPA off RPA)—rare anomaly, often associated with other forms of CHD.

Clinically these lesions manifest with respiratory distress and feeding problems of varying severity at varying ages. Diagnosis is made by means of upper GI contrast study (barium swallow) and helped by echocardiography; in the modern era, final diagnosis is confirmed by CT angiography or MRI, as they demonstrate the relationship of tracheobronchial tree, oesophagus and vascular structures. Asymptomatic patients do not require treatment; symptomatic patients undergo surgical division of the vascular ring.

Recommended Reading

Frescura C, Thiene G. The new concept of univentricular heart. Front Pediatr. 2014;2(62):1–17.

Samanek M, Slavik Z, Zborilova B, Hrobonova V, Voriskova M, Skovranek J. Prevalence, treatment, and outcome of heart disease in live-born children: a prospective analysis of 91823 live-born children. Pediatr Cardiol. 1989;10(4):205–11.

Kitchiner DJ, Jackson M, Walsh K, Peart I, Arnold R. Incidence and prognosis of congenital aortic valve stenosis in Liverpool (1960–1990). Br Heart J. 1993;69(1):71–9.

Hoffman JI, Kaplan S. The incidence of congenital heart disease. J Am Coll Cardiol. 2002;39(12):1890–900.

Bu'Lock F, Shebani S, Eltayb H, Peek G, Khare M, Mouse H. Effect of changes to UK fetal cardiac screening protocols across the spectrum of congenital heart disease requiring intervention in the first year of life. Ultrasound in Obstetrics and Gynecology, 9–12 Sep 2012, Copenhagen, Denmark.

MacDonald MJ, Hanley FL. Pulmonary atresia with ventricular septal defect and major aortopulmonary collaterals. In: Yuh DD, Vricella LA, Yang SC, Doty JR, editors. Johns Hopkins manual of cardiothoracic surgery. New York, NY: McGraw-Hill; 2007. p. 1141–8. Chapter 61.

Lupo PJ, Langlois PH, Mitchell LE. Epidemiology of Ebstein anomaly: prevalence and patterns in Texas, 1999–2005. Am J Med Genet A. 2011;155A(5):1007–14.

Anderson R. Interatrial communication [Internet]. 2007 [cited 04/01/2016]. Available from http://www.orpha.net.

Ballerini L. Truncus arteriosus [Internet]. 2005 [cited 04/01/2016]. Available from: http://www.orpha.net.

Park MK. Park's the pediatric cardiology handbook. 5th ed: Elsevier Saunders; 2016.

Fontan F, Baudet E. Surgical repair of tricuspid atresia. Thorax. 1971;26(3):240–8.

Anderson RH, Baker EJ, Penny DJ, Redington AN, Rigby ML, Wernovsky G, editors. Paediatric cardiology. 3rd ed. Philadelphia, PA: Churchill Livingstone/Elsevier; 2010.

Freedom RM, Yoo S-J, Mikailian H, Williams WG. The natural and modified history of congenital heart disease. 1st ed. New York: Blackwell publishing, Inc/Futura Division; 2004.

30 Pulmonary and Airways

Peter Carachi

Introduction

The airway and lungs are essential organs for human survival providing a mechanism for gas exchange and tissue oxygenation. In this chapter we describe how congenital anomalies of the respiratory pathway manifest. The embryological development is clearly complex with potential for multiple anomalies to occur. Our clinical decision-making may be aided by a better understanding of the presentation of the congenital anomalies and associated conditions.

Laryngo-tracheo-oesophageal Cleft

A laryngo-tracheo-oesophageal cleft (LC) is a congenital malformation involving a posterior sagittal communication between the larynx and pharynx, which may extend downward between trachea and oesophagus [1]. There is a variable incidence from 1 in 10,000 to 1 in 20,000 due to difficulty in diagnosis in mild forms of the disease. There is a high mortality in severe forms. Male incidence is slightly higher, but no geographic variation exists.

Laryngeal clefts result from failure of fusion of the posterior cricoid lamina and abnormal development of tracheo-oesophageal septum (Fig. 30.1).

The clinical presentation and age of diagnosis are dependent on the severity of the cleft. There can be presentation with respiratory distress at birth or recurrent pneumonia later in life. Disorders of the pharynx and larynx result in stridor, weak cry and copious pharyngeal secretions. In approximately 50% of cases, swallowing disorders manifest with aspiration or cyanosis during feeding or a chronic cough.

Five types of LC have been identified based on extension of the cleft, which usually correlate with the severity of symptoms (Fig. 30.2).

P. Carachi (✉)
Royal Hospital for Children, Glasgow, UK
e-mail: p.carachi@nhs.net

The classification is essential as this guides the therapeutic strategy.

- Type 0—submucosal cleft
- Type I—supraglottic interarytenoid cleft, above the vocal cord level
- Type II—cleft extending below the vocal cords into cricoid cartilage
- Type IIIa—cleft extending through cricoid cartilage but not into trachea
- Type IIIb—cleft extending through cricoid into cervical trachea
- Type IV—cleft extending into thoracic trachea, potentially down to carina

LC is associated with other congenital malformations of the respiratory and gastrointestinal tract. These include

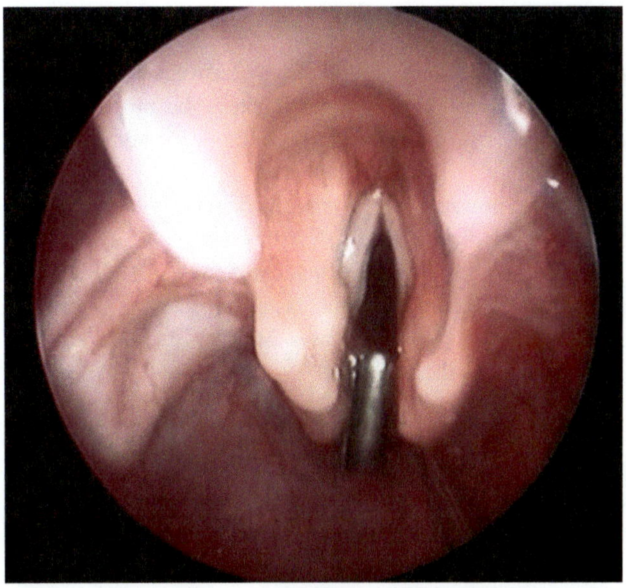

Fig. 30.1 Endoscopic view; palpation of a posterior cleft. Diagram taken from [Leboulanger et al. Laryngo-tracheo-oesophageal cleft; Orphanet Journal of Rare Diseases, 2011 6:81]

R. Carachi, S. H. E. Doss (eds.), *Clinical Embryology*, https://doi.org/10.1007/978-3-319-26158-4_30

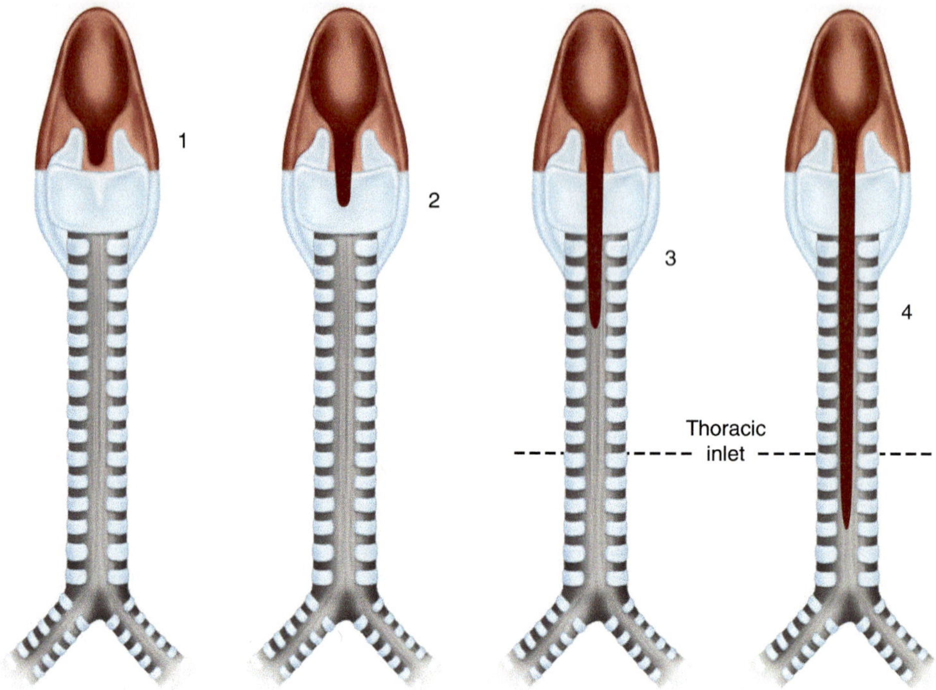

Fig. 30.2 Benjamin and Inglis' original classification. Diagram taken from Leboulanger et al. Laryngo-tracheo-oesophageal cleft; Orphanet Journal of Rare Diseases, 2011 6:81. Previously cited Benjamin B, Inglis A. Minor congenital laryngeal clefts: Diagnosis and classification; Ann Otol Rhinol Laryngol 1989, 417–426

laryngomalacia, tracheobronchial dyskinesia and gastro-oesophageal reflux disease. LC is associated with several multisystem syndromes, namely, Opitz/BBB syndrome, Pallister-Hall syndrome, VACTERL/VATER association and CHARGE syndrome. The prognosis is dependent on the severity of the cleft and other associated malformations.

Laryngomalacia

Laryngomalacia is a congenital anomaly of the larynx leading to dynamic supraglottic airway collapse. Inspiratory stridor occurs due to intermittent airway obstruction during this phase of respiration. Laryngomalacia is the most common congenital laryngeal anomaly accounting for about 60% of cases [2].

Clinical presentation is in early infancy with symptoms maximal at 6–8 months resolving by 2 years of age. Males are twice as likely to be affected than females. The aetiology is not fully understood, but gastro-oesophageal reflux is implicated in up to 80% of cases. Multiple pathophysiological mechanisms include:

- Anatomical; short aryepiglottic folds and a curled omega-shaped epiglottis predispose to dynamic airway collapse.
- Immature neural pathways leading to failure of neuromuscular coordination and subglottic hypotonia.
- Gastro-oesophageal reflux may result in mucosal oedema worsening laryngomalacia.

The clinical classification is based on the presence of stridor, resultant respiratory distress and feeding difficulties with associated failure to thrive.

Laryngeal Atresia/Stenosis

See Chap. 22.

Subglottic Haemangioma

See Chap. 22.

Tracheo-Oesophageal Fistula +/− Oesophageal Atresia

See Chap. 34.

Tracheal Agenesis

Tracheal agenesis is a rare fatal congenital abnormality in which there may be complete or partial absence of the trachea [3]. There is no communication between the larynx and the distal airways. Occasionally there is a fistula between the oesophagus and the distal airways which allows aeration of the lungs. Floyd et al. describe three subtypes of tracheal agenesis [4] (Fig. 30.3).

- Type 1: There is no upper trachea, and the distal trachea connects to the oesophagus.
- Type II: There is a common bronchus which connects right and left main bronchi to the oesophagus with an absent trachea. An isolated type II tracheal agenesis and fistula are rare, and 94% of cases are associated with other

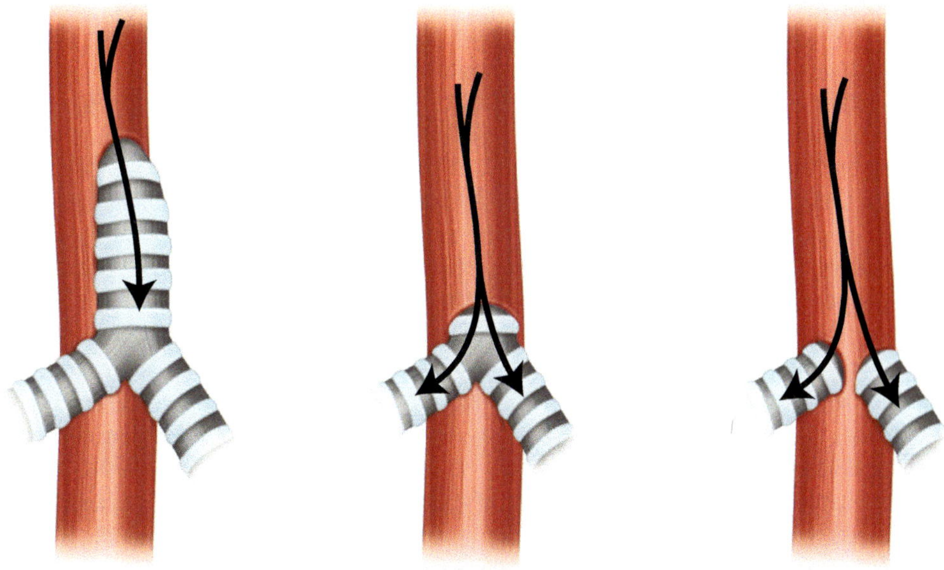

Fig. 30.3 Floyd classification (diagram taken from Diijkman KP et al. Failed resuscitation of a newborn due to congenital tracheal agenesis: a case report. Cases Journal. 2009;2:7212)

congenital abnormalities (cardiac, renal, gastrointestinal, pulmonary and CNS).

- Type III: The right and left main bronchi arise independently from oesophagus.

This condition is associated with male predominance, prematurity and polyhydramnios. The prevalence is estimated at 1 in 50,000 live births but is often associated with other multisystem congenital abnormalities. This diagnosis should be considered in the neonate with a history of maternal polyhydramnios, acute perinatal respiratory distress, cyanosis and no audible cry. Tracheal intubation proves impossible, and the condition is always fatal.

Congenital Tracheal Malformations

Congenital tracheal malformations may manifest at birth but could also present symptomatically later on in life. The most common presentation is with biphasic stridor and a prolonged expiratory phase. Tracheal malformations may be due to intrinsic causes or compression by external structures.

Tracheal Stenosis

This is a rare condition where in comparison to a tracheal web, there is more extensive pathological involvement. The area of stenosis may involve multiple tracheal rings or the entire length of the trachea. Early extubation should be aimed for in children requiring invasive ventilation to prevent oedema worsening the stenosis. Other abnormalities associated with tracheal stenosis are vascular slings, tracheo-oesophageal fistula, pulmonary hypoplasia and trisomy 21 (Fig. 30.4).

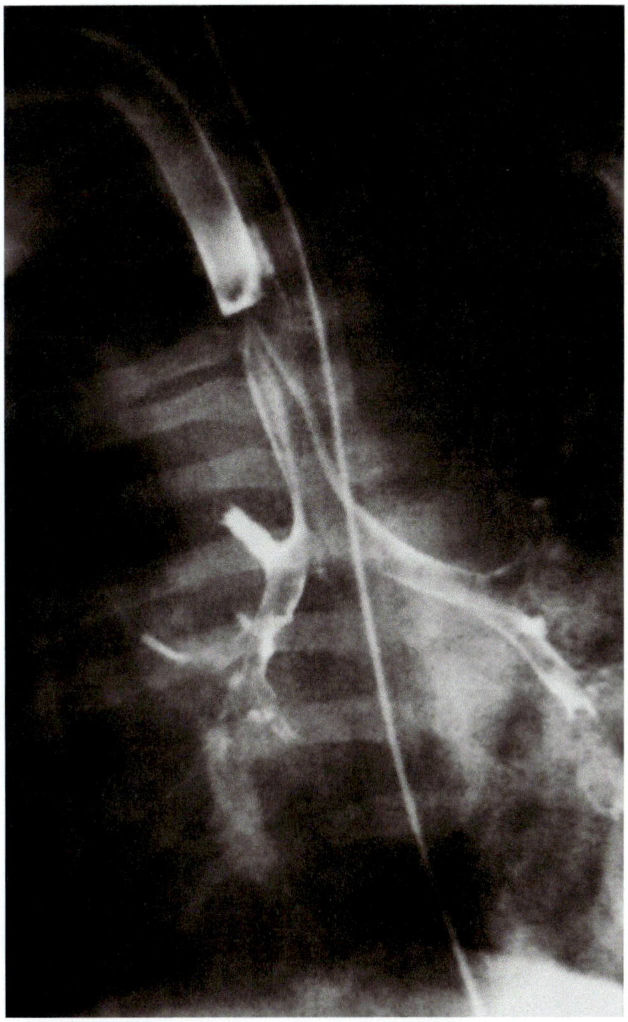

Fig. 30.4 A bronchogram demonstrating severe tracheal stenosis

Congenital Tracheomalacia

The clinical presentation is an infant with expiratory stridor, noisy breathing or upper respiratory tract infections. It can occur due to a primary congenital disorder of the tracheal rings or due to cartilaginous compression by the innominate artery or a tracheo-oesophageal fistula. Complete airway collapse occurs during heavy breathing or crying as a result of the anterior movement of the posterior tracheal wall. Tracheomalacia usually improves by the age of 18–24 months.

Bronchial Atresia

Bronchial atresia is a developmental anomaly where there is focal obliteration of the proximal segment of the bronchus. This anomaly has been caused by a traumatic event in utero. The distal bronchus may become filled with mucus and form a bronchocele. The surrounding lung tissue may become overinflated due to collateral supply. Patients are usually asymptomatic, and this is often an incidental radiological finding.

Bronchogenic Cysts

Bronchogenic cysts are a type of foregut duplication cyst (Fig. 30.5a, b). They are lined with bronchial epithelium and are most commonly located in the suprasternal notch or manubrium. They are an uncommon diagnosis in infancy, affecting symptomatic young adults due to their compressive effects.

The clinical presentation is varied from respiratory distress due to compression, recurrent infection or an incidental radiological finding. The compressive effects can lead to air trapping and a hyperlucent hemithorax. These cysts are at risk of expansion, haemorrhage or infection. Subcarinal cysts can lead to life-threatening airway compromise. Occasionally the cysts may involve the mediastinum near the hilum, the parenchyma or even extraparenchymal.

Bronchogenic cysts are diagnosed antenatally in approximately 60% of cases. They are rare lesions but account for 10% of paediatric mediastinal masses.

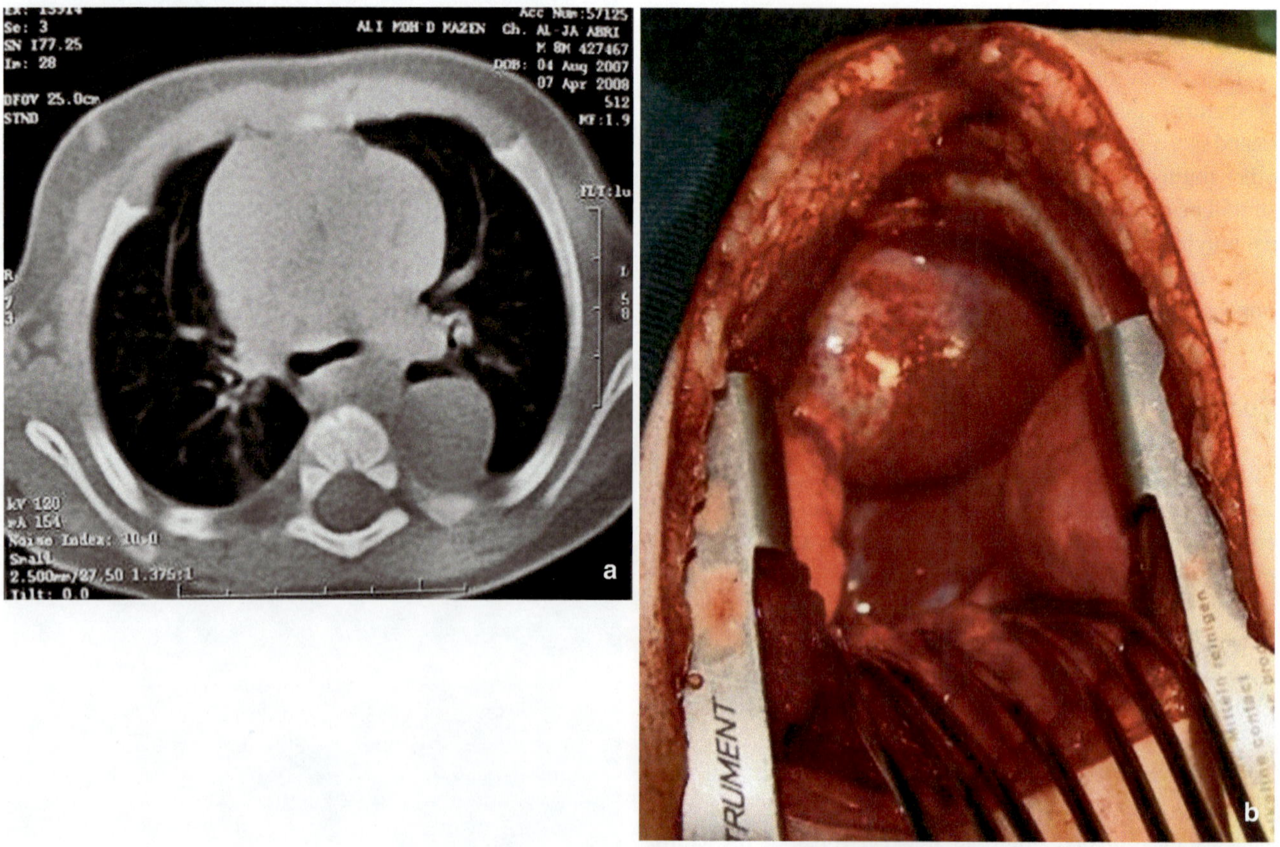

Fig. 30.5 (**a**) CT thorax showing large extraparenchymal bronchogenic cyst. (**b**) Intraoperative findings showing a large bronchogenic cyst

Hamartomas and Polyps

Hamartomas are benign disease, and approximately 90% of lesions involve the lung parenchyma. These lesions are disorganised native lung tissue which contain fat, epithelial and connective tissue.

Lung Agenesis and Pulmonary Hypoplasia

Lung agenesis is a very rare condition, and unilateral lung agenesis can be picked up incidentally on a routine chest X-ray (Fig. 30.6).

Pulmonary hypoplasia is common in the perinatal period and a significant cause of neonatal mortality. This condition refers to deficient or abnormal lung development, which can be secondary to in utero anomalies. Pathologically it is characterised by the presence of both bronchi and alveoli in an underdeveloped lobe. There is a reduction in both the size and weight of the lung.

The key factors associated with adequate lung development are:

- Adequate amniotic fluid volume
- Adequate thoracic space
- Normal respiratory movement
- Normal fluid within the pulmonary tissue

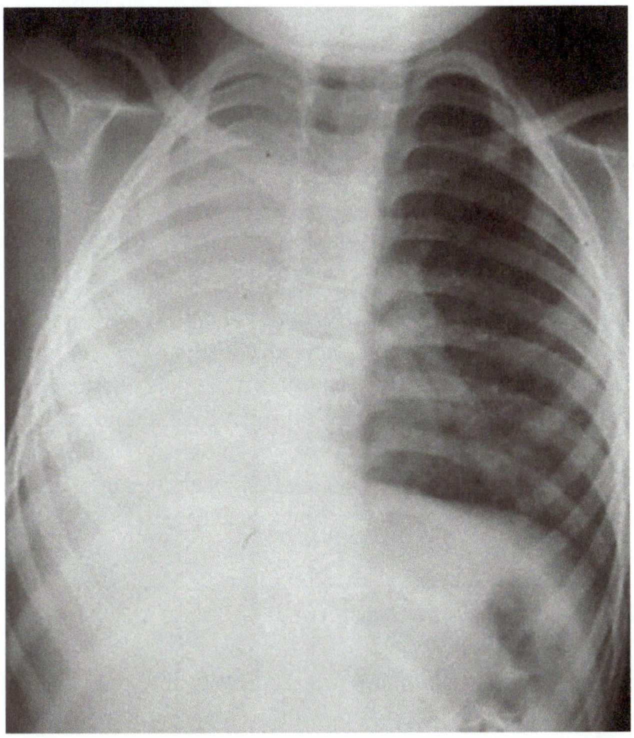

Fig. 30.6 Chest X-ray showing right lung agenesis

Lung development can be affected if any of these criteria are not met. Pulmonary hypoplasia usually occurs secondary to another congenital anomaly or a pregnancy-related complication. Primary pulmonary hypoplasia is very rare and only found at autopsy where there are structurally normal hypoplastic lungs with no identifiable cause. Pulmonary hypoplasia may be a feature of trisomy 21, associated with reduced alveolar surface area.

Intrathoracic causes of pulmonary hypoplasia include:

- Congenital diaphragmatic hernia (most common)
- Extralobar sequestration
- Mediastinal mass (e.g. teratoma)
- Reduced pulmonary (arterial) vascular perfusion
  - Congenital cardiac anomaly (Tetralogy of Fallot)
  - Unilateral absence of pulmonary artery

Extrathoracic causes are:

- Oligohydramnios and the causes
  - Potter sequence (renal anomalies)
  - Preterm premature rupture of membranes (PPROM)
- Skeletal dysplasia (narrow foetal thorax)
  - Achondrogenesis
  - Osteogenesis imperfecta
- Large intra-abdominal mass compressing the thorax
- Neuromuscular conditions associated with foetal breathing

Antenatal ultrasound may show features such as the presence of oligohydramnios and any causative anomalies. The presence of pulmonary hypoplasia is critical in determining foetal survival.

The clinical features are of immediate respiratory distress with tachypnoea, cyanosis, hypercarbia and acidosis. Other clinical signs are condition specific:

- Congenital diaphragmatic hernia: scaphoid abdomen
- Oligohydramnios: 'Potter's facies', arthrogryposis
- CNS lesions: signs suggestive of abnormal CNS function

Treatment options depend on the extent of pulmonary hypoplasia and whether targeted interventions are possible and appropriate.

Congenital Lobar Emphysema (CLE)

Congenital lobar emphysema is an abnormality, which results in progressive hyperinflation of one or two lobes of the neonatal lung (Fig. 30.7). This condition is three times more common in the male population. Clinical presentation is with respiratory distress and occurs within the neonatal

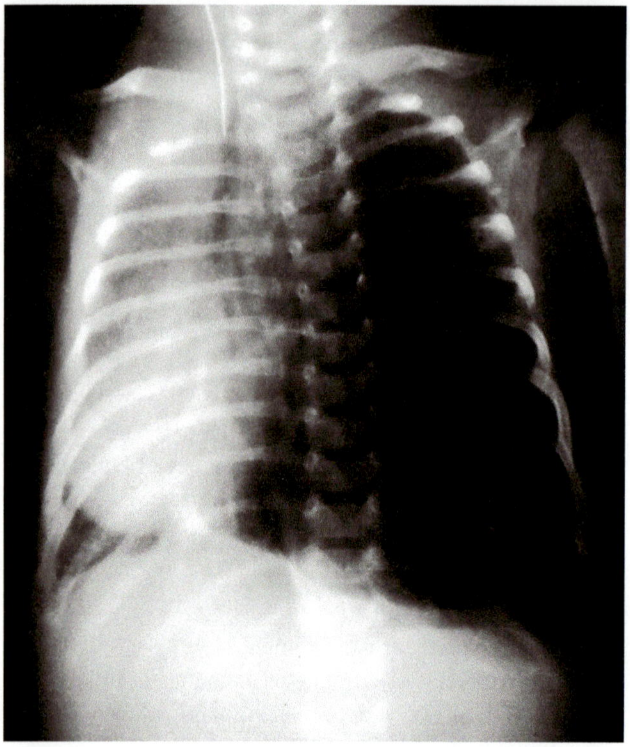

Fig. 30.7 AP chest film of intubated child with congenital lobar emphysema of the left upper lobe

period or the first 6 months of life. As a consequence of lobar hyperinflation, there may not be alveoli to participate with gas exchange.

The exact mechanism for progressive overdistension is uncertain, but obstruction, cartilage deficiency, dysplasia and immaturity have all been postulated. There are associations with congenital heart defects (namely, VSD, PDA and Tetralogy of Fallot). Progressive changes in lobar anatomy may cause depression of the diaphragm on the affected side. In severe cases overdistension may cause mediastinal compression and cardiovascular collapse. This condition is normally less dramatic, and asymptomatic lesions adopt an expectant management approach.

Congenital Pulmonary Airway Malformation

A congenital pulmonary airway malformation (CPAM) is a multicystic mass of segmental lung tissue associated with abnormal bronchial proliferation (Fig. 30.8). This is considered part of the spectrum of bronchopulmonary foregut malformations and previously known as congenital cystic adenomatoid malformation (CCAM).

This condition has a male predominance and accounts for 25% of all congenital lung lesions. The diagnosis is usually made by antenatal ultrasound. In the neonatal period, it may manifest with respiratory distress. If lesions are large, there may be pulmonary hypoplasia, and this is associated with a poor prognosis. Smaller malformations may not be diagnosed until adulthood and often present with recurrent chest infection.

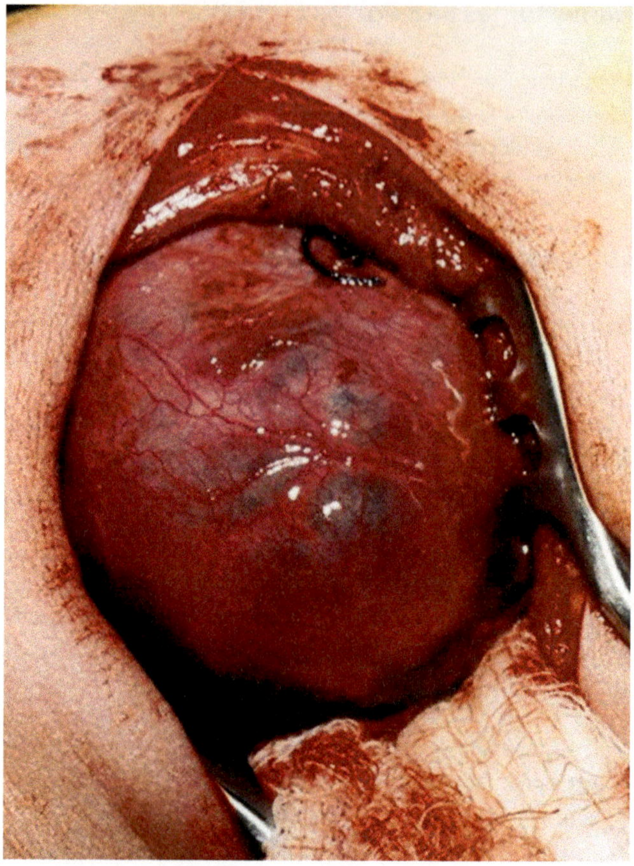

Fig. 30.8 Intraoperative findings of congenital pulmonary airway malformation (CPAM)

Pathologically there is a failure of normal bronchoalveolar development resulting in hamartomous proliferation of terminal respiratory units. Potential complications may be divided in in utero and postnatal complications. In utero complications are hydrops fetalis and pulmonary hypoplasia. Postnatal complications are recurrent pneumothorax, haemopneumothorax and risk of malignant transformation.

CPAM is associated with other pulmonary lesions (CPAM and pulmonary sequestration) and renal agenesis.

Pulmonary Sequestration

Pulmonary sequestration is where there is abnormal formation of segmental lung tissue with no connection to the arterial or bronchial tree (Fig. 30.9). The overall incidence is estimated at 0.1%. Clinical presentation is dependent on underlying pathology.

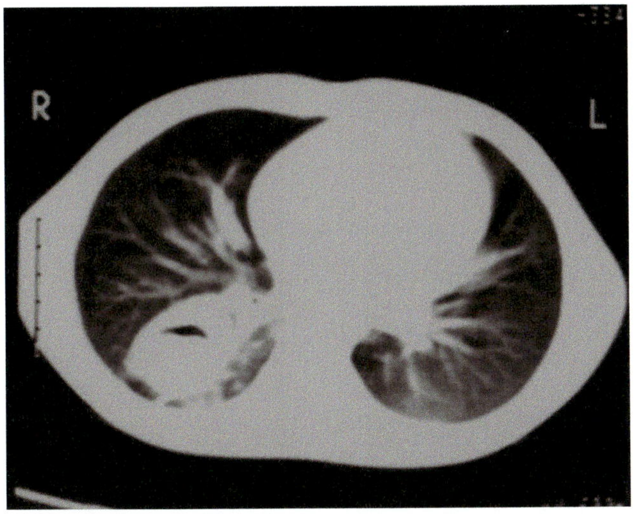

Fig. 30.9 CT thorax showing pulmonary sequestration

There are two distinct groups based on the relationship of the lung tissue to the pleura:

- Intralobar sequestration (ILS)
- Extralobar sequestration (ELS)

ILS accounts for the majority of cases of pulmonary sequestration and present later in childhood with recurrent infection. ELS presents more commonly in the neonatal period with respiratory distress, cyanosis and infection. There is a recognised predilection M: F ratio of 4:1.

The anomalous lung tissue has a systemic arterial supply from a branch of the aorta. In ILS venous drainage is via the pulmonary veins as the lung is closely connected to adjacent lung tissue with no separate pleura. ELS is separated from normal lung tissue by its own pleura and drains into systemic venous system.

Common associations with the extralobar type include CPAM/hybrid lesion, congenital heart disease and congenital diaphragmatic hernia. Associated complications include frequent respiratory tract infection and in neonates can be complicated by high output cardiac failure.

PHACES Syndrome

PHACES syndrome is the name given to a collection of symptoms commonly seen together. This is a spectrum disorder with predominance for the female sex (9:1).

This acronym stands for a collection of conditions seen together:

- Posterior fossa abnormalities
- Haemangioma
- Arterial lesions
- Cardiac abnormalities
- Eye problems
- Sternal notch or a dimple

With particular relation to the upper airway, haemangiomas affect midline structures of the lip, neck and jaw. There is also the possibility of subglottic haemangiomas. Subglottic haemangioma represents 1% of congenital anomalies of the larynx leading to life-threatening airway compromise.

Segmental haemangiomas can indicate underlying tissue problems. Mandibular haemangiomas are associated with the greatest risk of cardiac anomalies and airway lesions. Extracutaneous haemangiomas occur in 20% of patients with the subglottis being the most common site.

References

1. Leboulanger N, et al. Laryngo-tracheo-oesophageal cleft. Orphanet J Rare Dis. 2011;6:81.
2. Boardman SJ, Bailey CM. Laryngomalacia; BMJ Best Pract. Updated 01April 2016.
3. Ergun S, et al. Tracheal agenesis: A rare but fatal congenital anomaly. Mcgill J Med. 2011;13(1):10.
4. Floyd J, et al. Agenesis of the trachea. Am Rev Respir Dis. 1962;86:557–60.

The Diaphragm

31

Rania Kronfli

Development of the Diaphragm

*The diaphragm is mesodermal in origin.

*It is formed between the *eighth* and *tenth* weeks by the union of the following mesodermal structures:

1. The septum transversum
2. Two pleuro-peritoneal membranes
3. Mesoderm from the chest wall
4. Mesentery of the oesophagus
5. Mesoderm around abdominal aorta

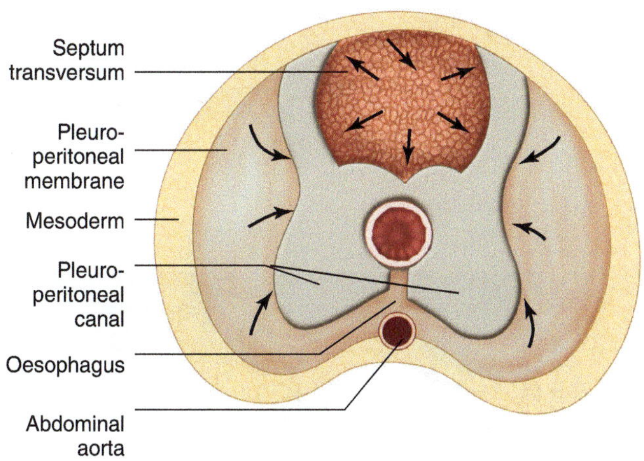

**The development of the diaphragm proceeds as follows:*

1. *Septum transversum*:

   *It is formed in the neck and descends to the lower part of the thoracic cavity.

   *It forms an incomplete septum between the thoracic and abdominal cavities since a pleuro-peritoneal canal lies posterolateral to it on each side.

   *The septum transversum expands and unites with the other components of the diaphragm to *form a large central part + the sternal and costal portions of the diaphragm.*

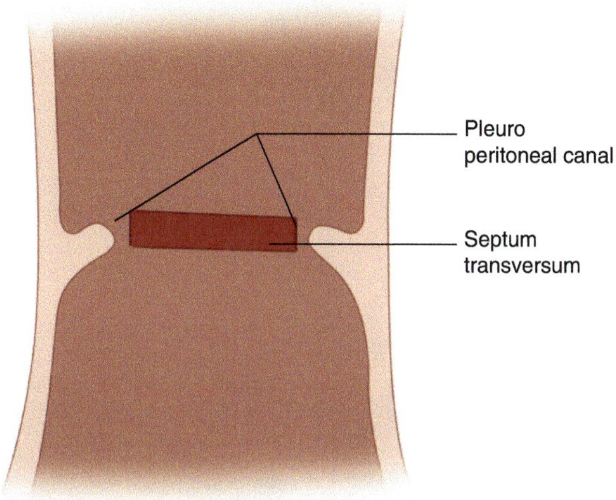

2. *The two pleuro-peritoneal membranes*:

   *These are two mesodermal folds (one on each side) which project inwards from body wall.

   *They grow medially encroaching on the pleuro-peritoneal canals until they finally fuse with the septum transversum and the mesentery of the oesophagus, thus closing the pleuro-peritoneal canals.

   *The pleuro-peritoneal membranes *form the dorsolateral parts of the diaphragm.*

3. *Mesoderm of the chest wall*:

   Grows inwards *forming the marginal part of the diaphragm* on either side of the pleuro-peritoneal membrane.

R. Kronfli (✉)
Royal Hospital for Children, Glasgow, UK
e-mail: rania.kronfli@nhs.net

R. Carachi, S. H. E. Doss (eds.), *Clinical Embryology*, https://doi.org/10.1007/978-3-319-26158-4_31

4. *Mesentery of the oesophagus*:
   Forms the dorsal median part of the diaphragm (between the oesophagus and aorta).
5. *The mesoderm around the aorta*:
   Forms the *lumbar (vertebral) part* of the diaphragm.

Congenital Anomalies of the Diaphragm

1. *Parasternal hernia of Morgagni*:
   It is due to failure of development of a small part of the diaphragm between the sternal and costal parts.
2. *Congenital hiatus hernia (oesophageal hernia)* (Fig. 31.1):
   Due to widening of the oesophageal hiatus of the diaphragm and congenital shortening of the oesophagus leading to protrusion of the stomach into the thorax.
3. *Congenital diaphragmatic hernia of Bochdalek* (Fig. 31.2):
   Due to failure of the pleuro-peritoneal membranes to close the pleuro-peritoneal canals allowing the abdominal viscera to enter the pleural cavity and compress the lungs and heart. *It is more common on the left side*.

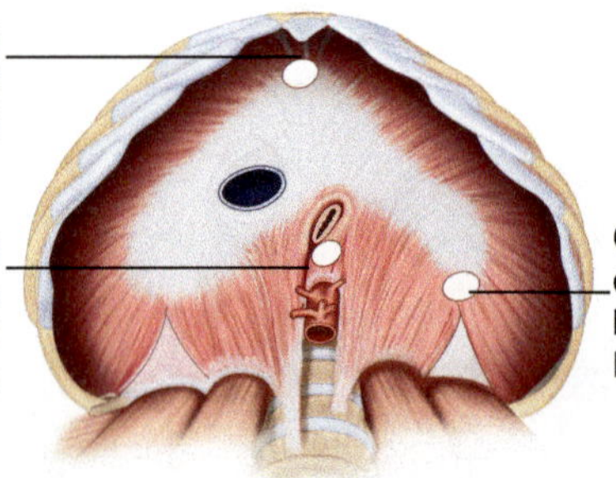

Congenital diaphragmatic hernia (CDH) is a defect in the diaphragm resulting in herniation of the abdominal viscera into the thorax. This is associated with pulmonary hypoplasia and abnormalities of lung development and subsequent pulmonary problems, namely, pulmonary hypertension.

The most common type of CDH is the Bochdalek hernia which is a posterolateral defect and occurs in 90% of cases, of which 85% are left-sided. Morgagni hernias (anterior) occur in approximately 2% and rarely present in the neonatal period. The rest comprise central defects. The size of the defect can only be classified if the patient undergoes surgery

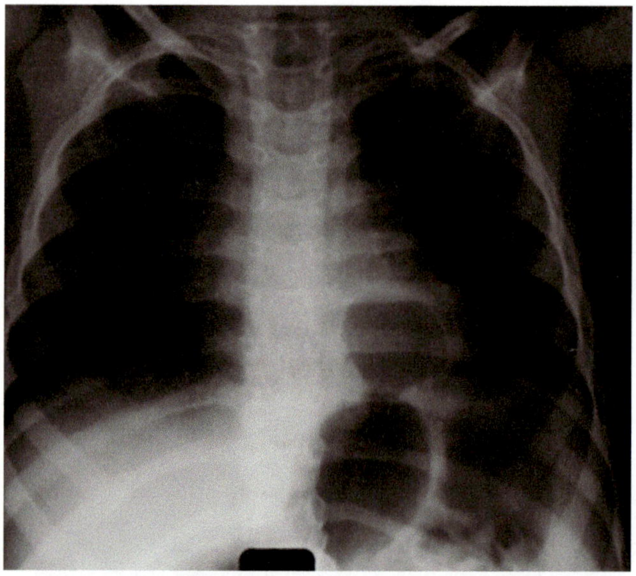

Fig. 31.1 Hiatus hernia X-ray

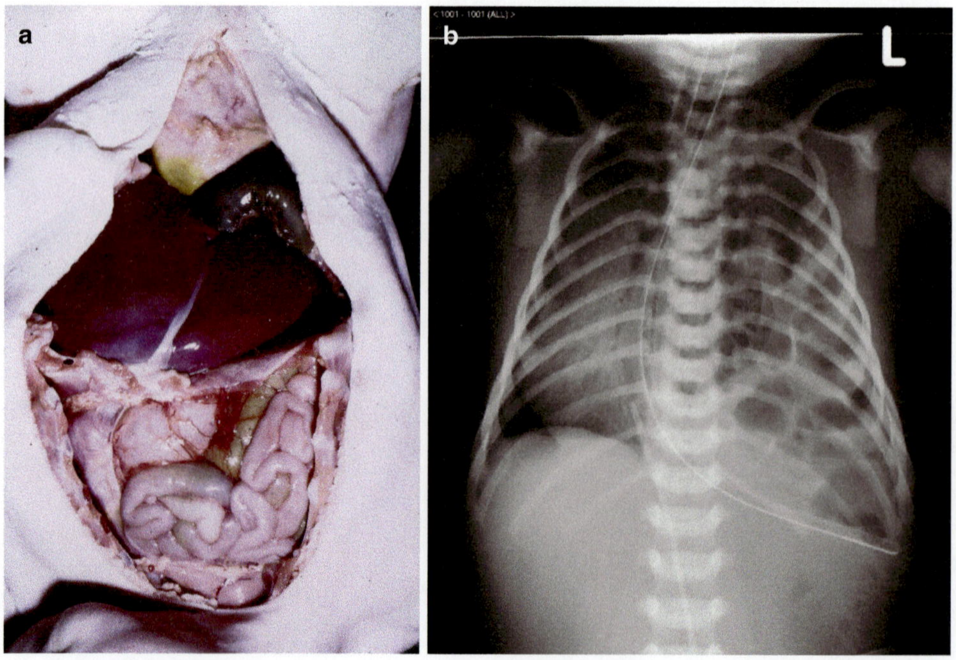

Fig. 31.2 (**a**) Congenital diaphragmatic hernia. (**b**) CDH chest X-ray

or has a post-mortem examination and has been formally classified [1].

The development of the diaphragm occurs broadly between weeks three and eight of gestation. The transverse septum descends caudally on the ventral body wall, eventually stopping at the first lumbar segment, where it folds inwards, forming the central tendon. This creates an incomplete division between the thoracic and abdominal cavities, with the primitive oesophagus, inferior vena cava and aorta in the centre. The lateral body wall folds inwards to form the pleuroperitoneal folds on either side, which migrate centrally and fuse with the oesophageal mesentery and the central tendon to close the defects. The right side closes before the left. Structural closure is complete around week eight; however, after this, myogenic precursors migrate to colonise and muscularise the diaphragm by week ten. The phrenic nerve migrates early on, from C3 to C5 with the transverse septum and then via the pleuroperitoneal folds to innervate the diaphragm by week ten.

There are two broad theories thought to explain the occurrence of CDH. One theory is that it is a primary abnormality of diaphragmatic development and occurs due to the failure of closure of the pleuroperitoneal canal [2] and/or abnormal muscularisation of the diaphragm [3], resulting in herniation of abdominal contents into the thorax and subsequent lung hypoplasia. An alternative theory is that there is a primary abnormality of lung development which subsequently impairs diaphragmatic development [4]. Finally, it has also been suggested that a single developmental anomaly results in both the diaphragmatic and pulmonary abnormalities simultaneously [5]. The use of animal models such as the nitrofen mouse model, vitamin A-deficient rodents, retinoic acid receptor knockout mice and foetal lambs with surgically created CDH has been important in studying these theories.

Irrespective of the theory, these foetuses have mediastinal shift and lung hypoplasia. In addition to the gross lung abnormalities, several microscopic lung pathologies have been noted in those with CDH. CDH lungs appear to have reduced bronchial branching, fewer alveoli, fewer pulmonary arterial branches and muscular hypertrophy in the pre-acinar and intra-acinar arterioles [6, 7], which may explain the often refractory pulmonary hypertension seen in these babies. Foetal Doppler studies have shown that foetuses with CDH have impaired pulmonary blood flow and vascularity which has led to the development of indices (e.g. the modified McGoon index) aiming to predict the degree of pulmonary vascular abnormality and thus outcome [8].

Antenatal diagnosis of CDH can be made on foetal ultrasound scan (US) as early as 14 weeks of gestation. Once identified, the foetus is monitored with US, and in some centres, magnetic resonance imaging (MRI) is performed for further attempts at prognostication.

Antenatal predictors of severity include the presence of polyhydramnios, the side of CDH, presence of 'liver up' and associated abnormalities.

Specific measurements used for attempted prognostication include the observed to expected lung to head ratio (LHR), performed on foetal US, and the observed to expected total lung volume (TLV), performed on foetal MRI. These measurements are taken by experienced foetal medicine specialists/radiologists according to specific criteria and are compared to age-matched controls in order to provide a 'corrected' value.

It has been shown that an observed to expected LHR <25% predicts a very poor postnatal survival [9]. However it is recognised that LHR is not reliably sensitive and may have a high false-positive rate. Therefore more direct methods, such as TLV on MRI, are being used and seem to offer more accurate predictive value [10].

The presence of the liver in the thoracic cavity is a poor prognostic indicator [11].

The presence of other anomalies is associated with a poorer prognosis. These are present in 50% of babies with CDH. Cardiac anomalies are most common and occur in 25%. Chromosomal anomalies do occur but are rare. Familial associations are rare. Following an antenatal diagnosis of CDH, foetal echocardiography should be performed, and amniocentesis should be offered.

As the prognosis for CDH is variable and management options are vast, antenatal counselling should ideally be performed by an experienced multidisciplinary team in a tertiary unit where referral for foetal intervention can be considered if appropriate.

Foetal endoscopic tracheal occlusion (FETO) is a controversial foetal intervention that is offered in certain regions for selective cases with poor predicted prognosis, and trials are ongoing. This involves placing a reversible balloon in the foetal trachea in the second trimester in order to encourage lung growth and reduce the mortality associated with severe pulmonary hypoplasia. Most recent evidence does suggest improved survival following the intervention in severe hypoplasia; however, considerable variation in technique remains between centres. The TOTAL trial is ongoing in Europe to determine the role of FETO in moderate lung hypoplasia [12].

CDH affects approximately 1 in 2000 to 1 in 5000 *live* births. The mortality in utero has been reported to be as high as 30% but is poorly recorded and represents a significant 'hidden' mortality.

Associations with Turner's syndrome and trisomies 13, 18, 21, 22 and 23 are seen, as well as others. Twenty per cent of prenatally diagnosed babies with CDH will have an associated chromosomal abnormality, and babies should be screened accordingly.

Animal and human studies are revealing that there does seem to be a genetic component to CDH. A gene distal to the 15q21 locus has been implicated in the normal development of the diaphragm as well as other genes such as Wilms' tumour gene 1 (WT1) and Fog2.

Cardiac anomalies are the most common associated structural anomaly associated with CDH, occurring in 15% of patients. Ventriculoseptal defects are the most common, followed by abnormalities of the aortic arch and tetralogy of Fallot. Genitourinary abnormalities are also prevalent.

Infants who have survived CDH are likely to have long-term problems. The incidence of recurrence following CDH repair varies widely. It is well recognised that patch repair has an associated rate of recurrence of up to 50%. Respiratory sequelae are common, with obstructive airways disease requiring bronchodilator ± steroid therapy occurring in up to 25% of patients over the age of 5 [13, 14]. Recurrent respiratory infections are common (up to 40%) and respiratory syncytial virus (RSV) is a problematic pathogen for which prophylaxis should be recommended [13, 14]. Interestingly, recent studies have shown VQ mismatch in children with CDH, with the biggest discrepancy being seen in children with severe CDH requiring extracorporeal life support (ECLS)/patch repair. The significance of these findings is yet to be shown. Failure to thrive is a morbidity following CDH repair that is likely to be multifactorial, with a proportion of children requiring gastrostomy insertion for nutritional supplementation. Increased work of breathing, oral aversion due to prolonged ventilation and gastro-oesophageal reflux disease (GORD) are all likely to contribute. GORD is common following CDH repair and appears more problematic if a patch is required. It is thought to occur due to a combination of abnormal position of the oesophagogastric junction, absence of a normal hiatus and increased intra-abdominal pressure following repair. Anti-reflux medications are often required, and fundoplication may be necessary but is not recommended routinely [14]. Neurodevelopmental impairment is not uncommon in survivors of CDH, with a range of domains being affected. Overall, almost 60% of children with CDH showed normal neurodevelopment, up to 25% showing mild-moderate delay and the remainder showing severe delay [15]. As expected, disease severity (including the use of ECLS) correlates with severity of neurological impairment. Longer-term studies have shown below average IQ scores in almost half of children with CDH aged 10 years. In patients with CDH, chest wall deformities (such as pectus excavatum) and scoliosis can occur [14]. These appear to be directly proportional to the size of the diaphragmatic defect and may be a consequence of the disease itself or indeed the surgery. Intervention is rarely required. It is clear that morbidity in patients with CDH is complex and occurs beyond infancy. A multidisciplinary follow-up clinic is paramount for the optimal care of these patients.

Diaphragmatic Eventration

Eventration of the diaphragm can be congenital or acquired and occurs when the diaphragm is abnormally elevated and moves paradoxically with respiration, i.e. moves up on inspiration [16].

Congenital eventuation is thought to occur due to incomplete development of the central tendon and is predominantly left-sided. Acquired diaphragmatic eventration results from damage to the phrenic nerve, which can occur during cardiac surgery or birth trauma.

Congenital diaphragmatic eventration may be difficult to distinguish from CDH as the neonate may present similarly. However pulmonary hypoplasia and pulmonary hypertension are not commonly seen with eventration. If it does not present acutely, it can present with recurrent respiratory infections, feeding difficulties or poor exercise tolerance in older children.

The diagnosis is made on ultrasound or fluoroscopy, when paradoxical movement of the diaphragm is demonstrated (Fig. 31.3).

Depending on the size and effect of the eventration, management can be conservative or operative. Conservative management is usually only feasible for small eventrations where the body can usually compensate for the paradoxical movement. Surgical repair involves plication with non-absorbable sutures and can be performed through the abdomen or the chest and can be achieved successfully with minimally invasive surgery [17].

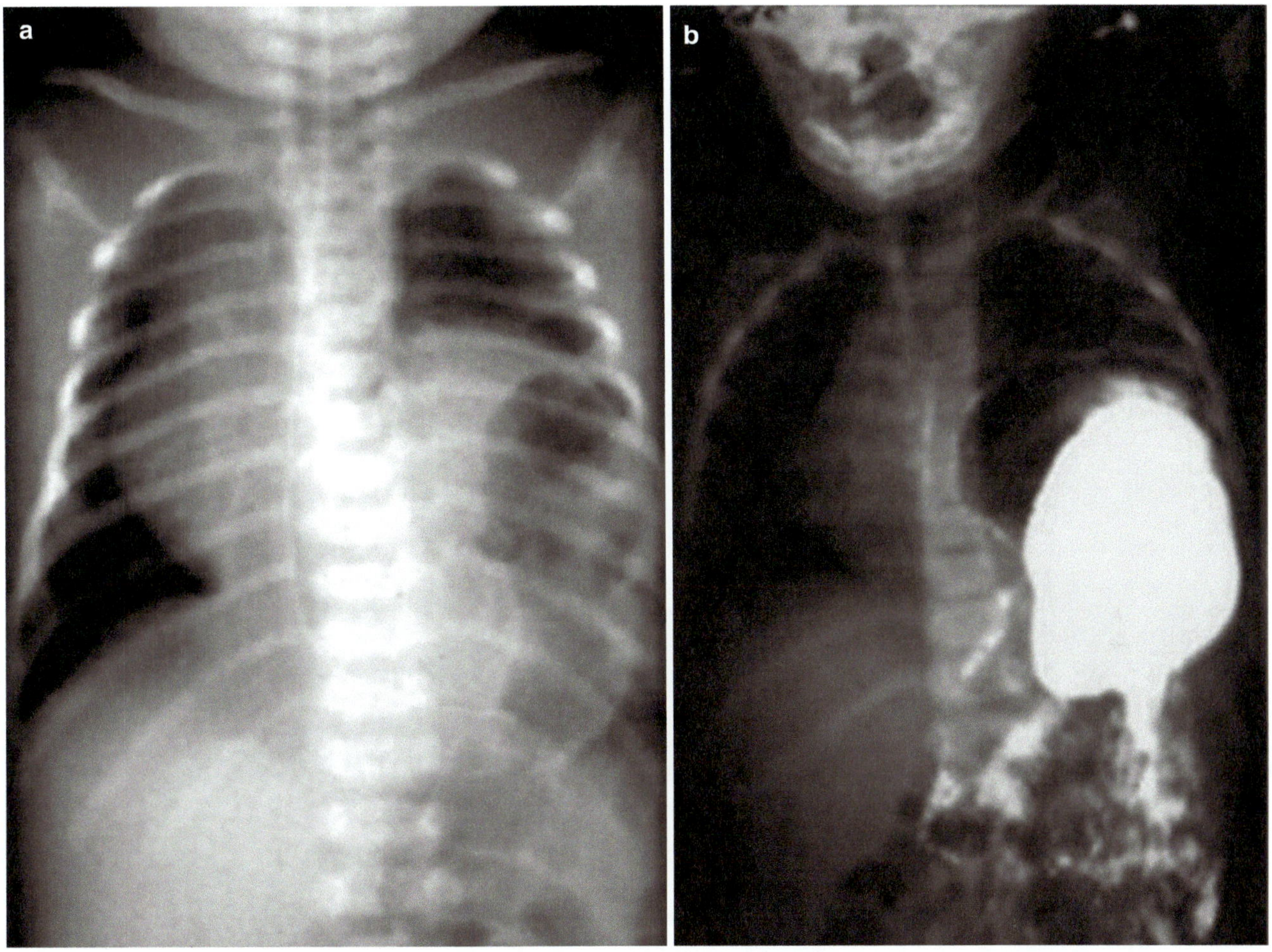

Fig. 31.3 (**a**) Eventration. (**b**) Eventration with stomach in the chest

References

1. Tsao K, Lally KP. The congenital diaphragmatic hernia study group: a voluntary international registry. Semin Pediatr Surg. 2008;17:90–7.
2. Allan DW, Greer JJ. Pathogenesis of nitrofen-induced congenital diaphragmatic hernia in fetal rats. J Appl Physiol. 1997;83:338–47.
3. Greer JJ, Allan DW, Babiuk RP, et al. Recent advances in understanding the pathogenesis of nitrogen-induced congenital diaphragmatic hernia. Pediatr Pulmonol. 2000;29:394–9.
4. Iritani I. Experimental study on embryogenesis of congenital diaphragmatic hernia. Anat Embryol (Berl). 1984;169:133–9.
5. Keijzer R, Liu J, Deimling J, et al. Dual-hit hypothesis explains pulmonary hypoplasia in the nitrofen model of congenital diaphragmatic hernia. Am J Pathol. 2000;156:1299–306.
6. Xia H, Migliazza L, Diez-Pardo JA, et al. The tracheobronchial tree is abnormal in experimental congenital diaphragmatic hernia. Pediatr Surg Int. 1999;15:184–7.
7. Jesudason EC, Connell MG, Fernig DG, et al. Early lung malformations in congenital diaphragmatic hernia. J Pediatr Surg. 2000;35:124–7.
8. Suda K, Bigras J-L, Bohn D, et al. Echocardiographic predictors of outcome in newborns with congenital diaphragmatic hernia. Pediatrics. 2000;105:1106–9.
9. Jani J, Nicolaides KH, Keller RL, et al. Observed to expected lung area to head circumference ratio in the prediction of survival in fetuses with isolated diaphragmatic hernia. Ultrasound Obstet Gynecol. 2007;30:67–71.
10. Gorincour G, Bouvenot J, Mourot MG, et al. Prenatal prognosis of congenital diaphragmatic hernia using magnetic resonance imaging measurement of fetal lung volume. Ultrasound Obstet Gynecol. 2005;26:738–44.
11. Mullassery D, Ba'ath ME, Jesudason EC, et al. Value of liver herniation in prediction of outcome in fetal congenital diaphragmatic hernia: a systematic review and meta-analysis. Ultrasound Obstet Gynecol. 2010;35:609–14.
12. www.totaltrial.eu.
13. Muratore CS, Kharasch V, Lund DP, et al. Pulmonary morbidity in 100 survivors of congenital diaphragmatic hernia monitored in a multidisciplinary clinic. J Pediatr Surg. 2001;26:133–40.
14. Bagolan P, Morini F. Long term follow up of infants with congenital diaphragmatic hernia. Semin Pediatr Surg. 2007;16:134–44.
15. Van der Cammen-van Ziip MH, Gischler SJ, Mazer P, et al. Motor function and exercise capacity in children with major anatomical congenital anomalies: an evaluation at 5 years of age. Early Hum Dev. 2010;86(8):523.
16. Wayne ER, Campbell JB, Burrington JD, et al. Eventration of the diaphragm. J Pediatr Surg. 1974;9:643–51.
17. Kozlov Y, Novozhilov V. Thoracoscopic plication of the diaphragm in infants in the first 3 months of life. J Laparoendosc Adv Surg Tech A. 2015;25:342–7.

32 Abdominal Wall Defects

Lynne A. Mcintosh

Abdominal wall defects can be divided into two groups depending on their embryological origins: gastroschisis, ectopia cordis and bladder exstrophy in one group and exomphalos in the other [1]. In practice, there is an overlap particularly with regard to bladder exstrophy, pentalogy of Cantrell and exomphalos. This chapter will focus on gastroschisis and exomphalos with bladder exstrophy being discussed elsewhere in the book.

Gastroschisis

Gastroschisis is characterised by a defect in the abdominal wall at the level of the umbilicus but not in the midline. Most commonly, the defect lies to the right of the midline with the abdominal viscera protruding freely into the amniotic cavity.

Antenatal diagnosis with ultrasonography detects around 91% [1] of cases of gastroschisis resulting in obstetric, neonatal and paediatric surgical input and monitoring. As lesions are believed to occur at gastrulation (4–6 weeks) [2], early detection of lesions is possible. In practice, diagnosis is made after the time when physiological herniation of the gut has resolved (10–12 weeks gestation). Differentiation of gastroschisis from exomphalos is possible by detection of the presence or absence of bowel coverings.

Over the years, there have been a number of embryological hypotheses which are detailed in the table below.

As can be seen from the above table, there is no clear consensus as to the underlying embryonic origin of gastroschisis. The prevalence of gastroschisis has dramatically increased over the past 30 years, with some countries experiencing a fourfold increase. This increase is not restricted to young maternal age, and some geographical areas/countries show clustering of cases [11].

The aetiology is thought to be multifactorial, with young maternal age; low socioeconomic class; smoking; vasoactive drugs such as salicylates, paracetamol and pseudoephedrine; and recreational drugs playing a role. No major genetic links have been identified.

Bowel atresias are the most common complicating factor in gastroschisis reports of incidence which vary from 6.9 to 28% [12]. Ultrasonic finding of intra-abdominal dilatation has been shown to be predictive of complex gastroschisis [13]. Other findings include extra-abdominal dilatation, gastric dilation and thickened bowel wall. The mechanisms involved are thought to be related to the cytokines and pro-inflammatory mediators within the amniotic fluid causing inflammation. With particular reference to atresias, the size of the abdominal defect and constriction of the bowel may account for their occurrence and in some cases the reduction in bowel length following necrosis. Perforation or volvulus may also complicate gastroschisis (Table 32.1).

L. A. Mcintosh (✉)
Queen Elizabeth University Hospital, Glasgow, UK
e-mail: lynneholmes@doctors.org.uk

R. Carachi, S. H. E. Doss (eds.), *Clinical Embryology*, https://doi.org/10.1007/978-3-319-26158-4_32

Table 32.1 Embryological origins of gastroschisis—seminal papers

Paper (year)	Embryological hypothesis	Related evidence (for and against)
Duhamel (1963) *Archives of Disease in Childhood* [3]	Failure of lateral folding, i.e. failure of folds to grow ventrally and fuse in midline (Fig. 32.1: body wall folding) Believed to result from early teratogenic action which prevents differentiation of the embryonic mesenchyme resulting in resorption of somatopleure and abdominal wall defect	**Miller SA J. Exp Zool (1998)** [4]: Right lateral body fold occurs after left in chick embryo models—does this explain the preponderance to right-sided defects?
Shaw (1975) *Journal of Pediatric Surgery* [5]	Rupture of amniotic membrane at the base of the umbilical cord during the time of normal 'physiological' umbilical herniation of the midgut (5–10 weeks)	**Sadler (2010)**: No evidence rupture happens at the border of the umbilical ring [2] **Sadler, Feldcamp (2008)**: Hypothesis does not explain other ventral body wall defects or the skin bridge between umbilicus and defect which is present in some cases [6]
deVries (1980) *Journal of Pediatric Surgery* [7]	Defect results from 'Disturbances to the circulation of the somatopleure at the junction with the body stalk during involution of the right umbilical vein' The entire umbilical vein and its communication with omphalomesenteric veins involute. Leaves only a small number of vessels at the junction of the somatopleure and body stalk	**Feldkamp, Sadler (2007)**: 'Umbilical veins do not supply mesenchyme in body wall or skin of paraumbilical area' therefore would not result in defect [8]
Hoyme, Higginbottom (1981) *Journal of Pediatric Surgery* [9]	Intrauterine interruption of the omphalomesenteric artery (which lies closest to the right side of the umbilical cord). Disruption leads to infarction and necrosis of the body wall	Supported by vascular nature of likely teratogens Not supported by the complex network of vessels in the area and implicated vessels do not supply skin and body wall
Stevenson (2009) *Clinical Genetics* [10]	Failure of yolk sac and related structures to become incorporated into the body stalk. Resulting in closure occurring normally but leaves a second perforation in abdominal wall and attachment of midgut to vitelline structures (Meckel's diverticulum)	Anecdotal evidence of vitelline structures and midpoint of gut always exteriorized Not every gastroschisis has Meckel's diverticulum and no yolk sac remnants. However, authors feel yolk sac remnants would have degenerated or be too small to identify

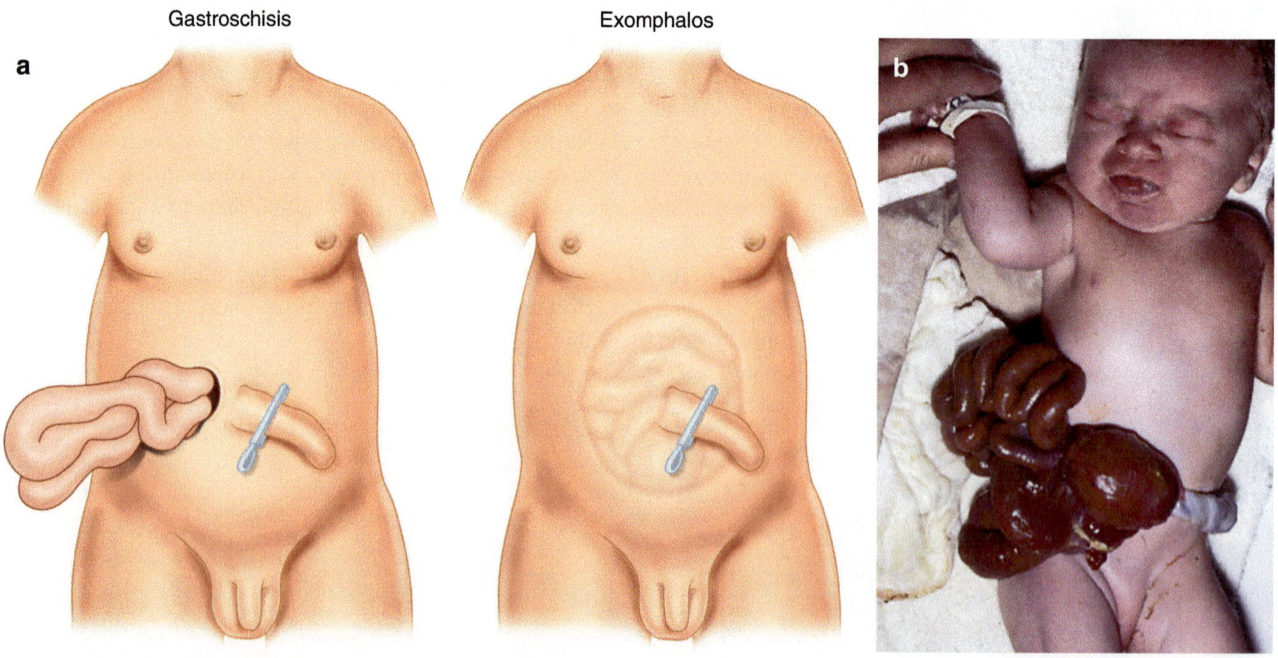

Fig. 32.1 (**a**) Abdominal wall defects. (**b**) Gastroschisis

Antenatal/Perinatal Management

Antenatal care should include detection of lesion, monitoring, counselling and mode and timing of delivery. Regular ultrasonography is required looking specifically at growth, liquor and condition of the bowel. There is no clear evidence to drive caesarean section over vaginal delivery. However, bowel damage does increase towards the end of the third trimester making early delivery an attractive option. Spontaneous preterm labour in one series occurred in 28% compared with 6% of the general population [14]. However, birth weight less than 2 kg is associated with a poorer outcome, increased time to start and full enteral feeds and ventilated days and duration of total parenteral nutrition and therefore does not support routine preterm delivery [15]. Timing and mode of delivery should therefore be made on a case-by-case basis taking into account the ultrasonographic findings. In cases of reduced liquor volume, amniotic infusion with saline or maternal oral hydration has been advocated [16].

Long-Term Outcome

Outcome is dependent on whether the condition is simple or complex. For patients with no associated abnormalities, the commonest treatment options would be either primary closure or application of preformed silo with delayed closure. The length of hospital stay is dependent on time to start enteral feeds and attain full enteral feeds but is extremely variable between patients.

Those with complex abnormalities often require multiple operations and prolonged courses of parenteral nutrition. Bacterial overgrowth in all patients can be problematic and may require regular low-dose antibiotics. A small but significant number will be affected by short bowel syndrome requiring prolonged home parenteral nutrition and the complications associated with it, with the prospect of gut lengthening surgery or transplantation in the future.

Exomphalos

Exomphalos is characterised by herniation of bowel and abdominal contents into the umbilical cord, associated with a larger than normal umbilical defect. These defects can be divided into three groups: hernia of the umbilical cord, exomphalos minor and exomphalos major. Exomphalos minor is described as herniation of bowel with a defect <4 cm, and exomphalos major is herniation of bowel and/or liver with large umbilical defect (>4 cm). There are some discrepancies in the definitions and application of these terms between surgeons with some using 5 cm, as the cutoff, and others classifying the presence of liver herniation as a major defect. The size of the defect is important in determining the likelihood of respiratory problems specifically pulmonary hypoplasia and treatment options. The prevalence is 1 in 3000–4000 [17]; this has remained unchanged for the past 50 years (Fig. 32.2).

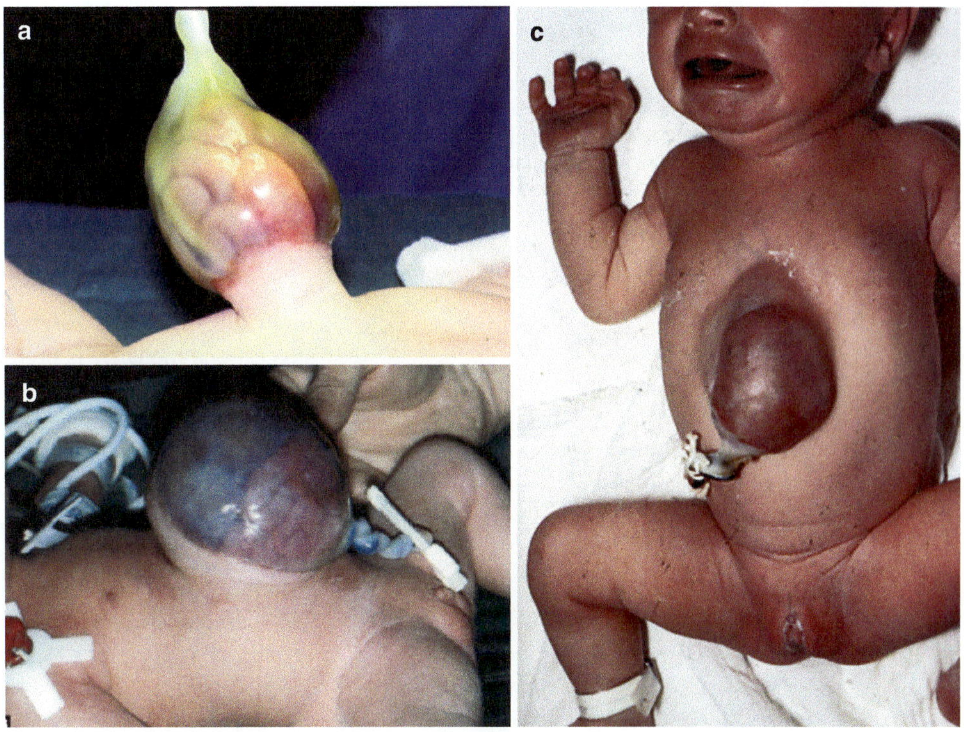

Fig. 32.2 (**a**) Hernia into the cord. (**b**) Exomphalos major. (**c**) Exomphalos supraumbilical

Exomphalos differs from gastroschisis in that the herniated bowel is not free within the amniotic cavity but contained within a three-layered sac made of peritoneum, Wharton's jelly and amnion. Also in contrast to gastroschisis, the umbilical cord is inserted directly into the sac. As previously mentioned, the sac contains herniated midgut but may also contain the liver, stomach, spleen, colon, testes or ovaries.

Embryology

Physiological herniation of the midgut occurs at week 6 of gestation and returns at week 10. This occurs as the liver grows and gut elongates and can no longer be accommodated by the abdominal cavity. On return to the abdominal cavity, the gut undergoes complex rotation and fixation to ensure that midgut volvulus does not occur (Chap. 35). For this reason, the gut in babies with exomphalos is non- or malrotated, and consideration must be given to the position of the gut when returning the bowel to the abdominal cavity. The formation of exomphalos is widely attributed to the failure of the herniated bowel loops to return to the abdominal cavity [18]. The seminal embryological papers are discussed below (Table 32.2).

Table 32.2 Embryological origin of omphalocele

Paper (year)	Problem	Associations
Duhamel (1963) *Archives of Disease in Childhood* [3]	'Exomphalos belongs to a group of ventral wall malformations that result from a disturbance of 'closing of the body of the embryo' Failure of folding at the level of the lateral folds prevents the body from closing completely, and the umbilical orifice remains widely open resulting in exomphalos	Failure of cephalic fold results in pentalogy of Cantrell/ ectopia cordis Failure of the caudal fold results in bladder exstrophy +/− exomphalos
deVries (1980) [7]	'Persistence of body stalk in the region normally occupied by the somatopleure' The surrounding tissue around the umbilicus would normally develop into fascia	Spectrum of conditions depending on the stage of development affected from body stalk anomaly to simple umbilical hernia
Sadler (2010) [2]	'Failure of loops of bowel to return to abdominal cavity after physiological herniation of the loops into the umbilical cord' This occurs independently of lateral body wall folding and closure	
Christison-Lagay (2011) [12]	'Intestine fail to return to abdominal cavity after normal embryonic herniation' In contrast to Sadler, Christison-Lagay attribute this 'to a folding defect in the abdominal wall rather than the genes involved in gut elongation and rotation'	

Associations

Unlike gastroschisis, exomphalos appears to be genetically determined and is associated with a number of chromosomal abnormalities and syndromes. No environmental factors are thought to be involved. The most common associations are trisomy 13, 18 and 21; these can occur in up to 49% of prenatally diagnosed cases [19]. Other syndromic associations include Beckwith-Wiedemann syndrome, Donnai-Barrow syndrome [12], pentalogy of Cantrell and bladder and cloacal exstrophy. In one study, isolated lesions were as low as 14% [19], respectively, for prenatally diagnosed lesions. Associated anomalies are also more commonly identified in minor exomphalos (55 vs. 36%) [20].

Antenatal/Perinatal Management

Due to the common occurrence of associated abnormalities, prenatal management focuses on identifying these lesions and providing antenatal counselling regarding likely outcome, offering karyotyping and termination of pregnancy if appropriate. It is important to ensure that the diagnosis is not made before the physiological herniation has resolved. Around 83% of exomphalos cases will be detected prenatally [1]. Antenatal counselling in exomphalos is extremely difficult due to the wide ranging associations and the varying degrees of impact these will have on outcomes. Brantberg et al. reported a study of 90 foetuses where only 23% survived past the neonatal period and only 9% were considered healthy [19].

The mode of delivery should focus on obstetric indications; similarly to gastroschisis, there is no clear evidence to suggest one modality is preferable. The size of the lesion is often considered with a preference to deliver larger lesions by caesarean section due to fear of liver rupture. There is also no indication for preterm delivery [21]. Delivery should however occur in a tertiary paediatric centre to allow access to expertise required to deal with any immediate problems. Pulmonary hypoplasia in cases of exomphalos major may require intubation and ventilation. This results from abnormal thoracic development with a narrow (bell-shaped) thorax and small lung area [12].

Table 32.3 Immediate treatment and investigations for exomphalos

Intervention	Indication
Blood glucose monitoring	Neonatal hypoglycaemia associated with Beckwith-Wiedemann syndrome
IV access and resuscitation Temperature regulation Dressing of lesion	Fluid and temperature losses are less than seen in gastroschisis but still need to be minimised and replaced
Nasogastric/orogastric tube insertion	Decompression of stomach and bowel
Cardiac monitoring and echocardiography	Identify associated cardiac abnormalities
Renal ultrasound scan	Identify associated renal anomalies

Immediate postnatal management focuses on cardiorespiratory resuscitation and identification of associated abnormalities. Table 32.3 shows early interventions required.

Long-Term Outcomes

Long-term outcome is determined by associated abnormalities in exomphalos. Exomphalos minor is universally treated by primary closure, and apart from cosmesis, long-term problems reported are only related to any associated abnormalities.

Treatment of exomphalos major varies between centres and include primary closure (with or without synthetic patch), staged closure with placement of a silo or nonoperative management with delayed closure of ventral hernia following epithelialisation of the exomphalos sac. This last method can be achieved using silver sulphadiazine dressing or similar with delayed closure being performed months or years later following serial reduction of the hernia by corset wearing.

There are a number of long-term problems related specifically to exomphalos major including pulmonary hypoplasia, gastro-oesophageal reflux disease and feeding difficulties. Gastrostomy, nasogastric or nasojejunal feeding is often required to treat reflux, feeding difficulties and resultant failure to thrive. Some patients will ultimately require fundoplication; however, in those with conservative management strategy, this will need to be delayed until definitive closure is performed. Studies suggest that reflux will resolve during childhood.

Zaccara et al. studied cardiopulmonary performance in large abdominal wall defects and found only slight reduction in FVC which was non-significant and a reduction in VO_2 max consistent with reduced physical activity suggesting that respiratory insufficiency also resolves in childhood [22].

Over 50% of patient with exomphalos (both major and minor) are unhappy with cosmetic outcome with 100% of exomphalos major patients without an umbilicus dissatisfaction. Many will undergo reconstructive surgery of their abdominal wall scars and formation of umbilicus. Cosmetic factors do not however appear to impact on quality of life [20].

Commoner Abdominal Wall Abnormalities

A much more common abnormality of the abdominal wall is divarication of the rectus muscle due to the absence of the linea alba which produces the above defect (Fig. 32.3). Other commoner abnormalities include epigastric hernia (Fig. 32.4), umbilical hernia (Figs. 32.5 and 32.6) and inguinal hernia (Fig. 32.7). No treatment is needed for this condition.

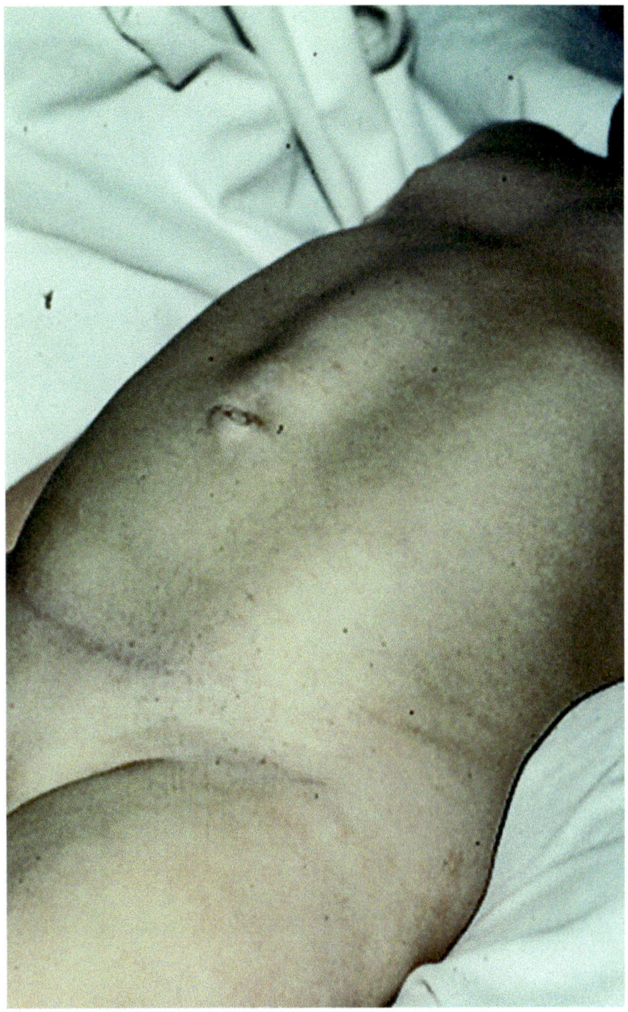

Fig. 32.3 Divarication of recti

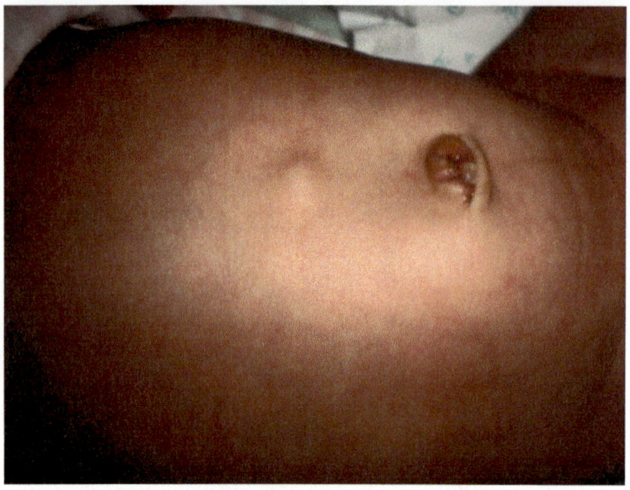

Fig. 32.4 Epigastric hernia

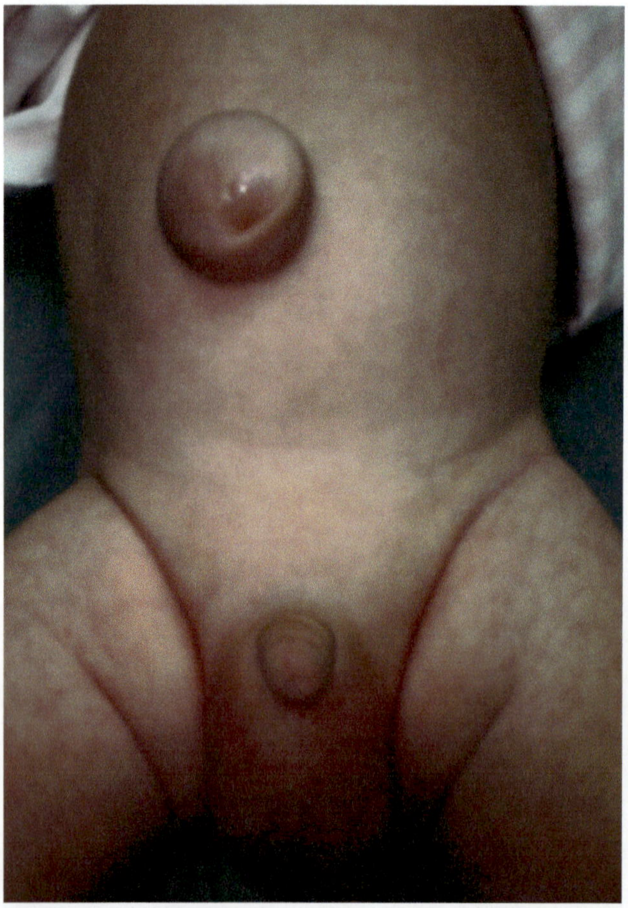

Fig. 32.5 Umbilical hernia

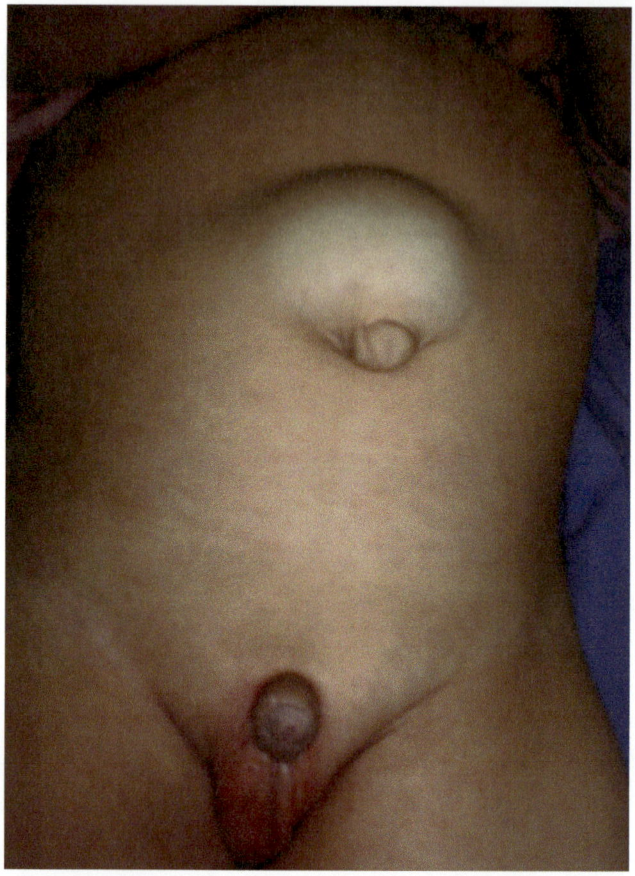

Fig. 32.6 Supraumbilical hernia

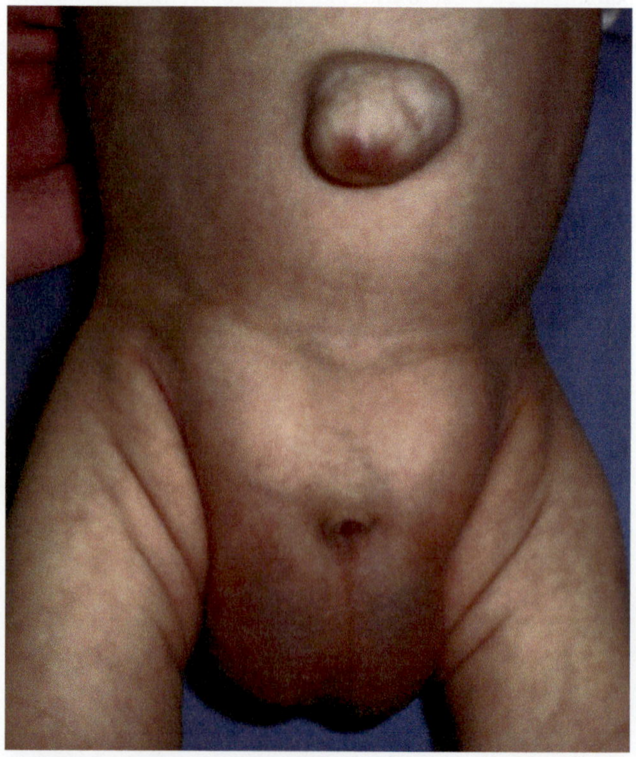

Fig. 32.7 Umbilical hernia and bilateral inguinal hernia

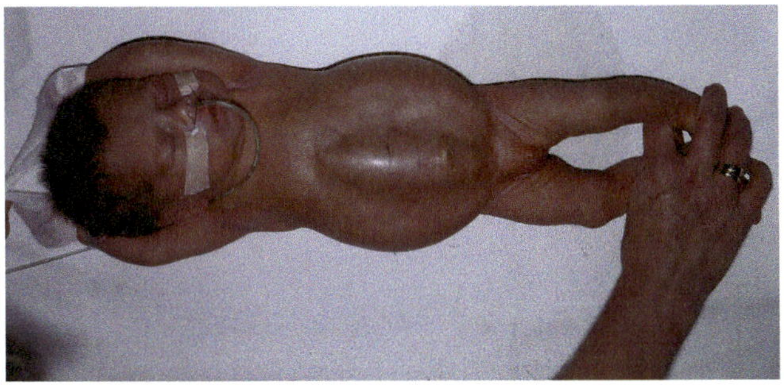

Fig. 32.8 Hypothyroidism

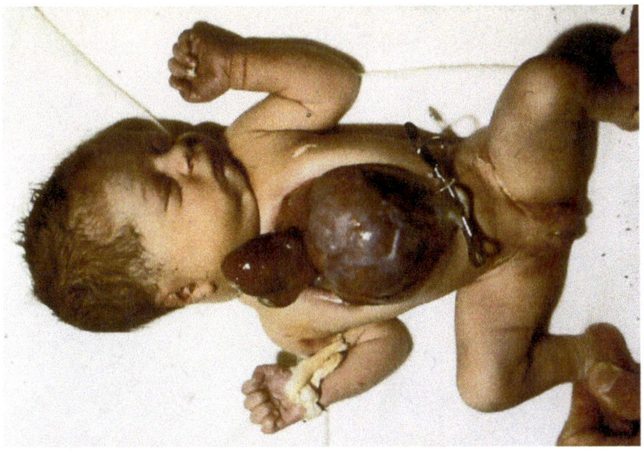

Fig. 32.9 Ectopia cordis

Rarer Abdominal Wall Abnormalities

Body stalk anomaly (also known as limb body wall complex) is a rare fatal condition where the abdominal contents herniate through a large abdominal wall defect. It is associated with a short or absent umbilical cord. Associated abnormalities include severe kyphoscoliosis, brain abnormalities and limb abnormalities. The insult occurs early in gestation at around week 4 [1].

In 1958, Cantrell and Ravitch described the combination of 'midline supraumbilical defect, defect of the lower sternum, anterior diaphragmatic deficiency, pericardial defect and congenital intra-cardiac defects' [23]. This has become known as **pentalogy of Cantrell** and is also associated with ectopia cordis, omphalocele and gastroschisis.

In **ectopia cordis**, the heart lies out with the thoracic cavity and results from the failure of lateral body wall closure in the thoracic region and often extends downwards explaining its associations as described in the pentalogy of Cantrell (Figs. 32.8 and 32.9).

OEIS complex comprises omphalocele, cloacal exstrophy, imperforate anus and spina bifida. It is thought the underlying cause may be both genetic and environmental. Lesion can be identified during the second trimester by ultrasonography [1]. The lesion is rare occurring in 1 in 250,000 [19].

References

1. Prefumo F, Izzi C. Fetal abdominal wall defects. Best Pract Res Clin Obstet Gynaecol. 2014;28:391–402.
2. Sadler TW. The embryologic origin of ventral body wall defects. Semin Pediatr Surg. 2010;19:209–14.
3. Duhamel B. Embryology of exomphalos and other ventral body wall defects. Arch Dis Child. 1963;38:142–7.
4. Miller SA. Differential proliferation in morphogenesis of lateral body wall folds. J Exp Zool. 1982;221:2005–11.
5. Shaw A. The myth of gastroschisis. J Pediatr Surg. 1975;10: 235–44.
6. Sadler TW, Feldkamp ML. The embryology of body wall closure: relevance to gastroschisis and other ventral body wall defects. Am J Med Genet C Semin Med Genet. 2008;148C:180–5.
7. deVries PA. The pathogenesis of Gastroschisis and omphalocele. J Pediatr Surg. 1980;15:245–51.
8. Feldcamp ML, Carey JC, Sadler TW. Development of gastroschisis: review of hypothesis and implications for research. Am J Med Genet. 2007;143A:639–52.
9. Hoyme HE, Higginbottom MC, Jones KL. The vascular pathogenesis: intrauterine interruption of the omphalomesenteric artery. J Pediatr Surg. 1981;98:228–31.
10. Stevenson RE, Rodgers RC, Chandler JC, et al. Escape of the yolk sac: a hypothesis to explain embryogenesis of gastroschisis. Clin Genet. 2009;75:326–33.
11. Loan M, Dolk H, Bradbury I. Increasing prevalence of gastroschisis in Europe 1980-2002: a phenomenon restricted to younger mothers? Paediatr Perinat Epidemiol. 2007;21:363–9.
12. Christison ER, Kelleher CM, Langer JC. Neonatal abdominal wall defects. Semin Fetal Neonatal Med. 2011;16:164–72.
13. Kuleva M, Khen-Dunlop N, Dumez Y. Is complex gastroschisis predictable by neonatal ultrasound? BJOG. 2011;119:102–9.
14. Lausman AY, Langer JC, Tai M, et al. Gastroschisis: what is the average gestational age of spontaneous delivery? J Pediatr Surg. 2007;42:1816–21.
15. Charlesworth P, Njere I, Alloytey J. Postnatal outcome in gastroschisis: effect of birth weight and gestational age. J Pediatr Surg. 2007;42:815–8.
16. Luton D, de Lagausie P, Guibourdenche J, et al. Effect of amnioinfusion on the outcome of prenatally diagnosed gastroschisis. Fetal Diagn Ther. 1999;14(3):152–5.
17. Gamba P, Midrio P. Prenatal diagnosis, newborn management and long-term outcomes. Semin Pediatr Surg. 2014;23:283–90.
18. Holcomb GW III, Murphy JP. Ashcraft's pediatric surgery (5th Ed). Philadelphia: Saunders Elsevier.
19. Brantberg A, Blaas HGK, et al. Characteristics and outcome in 90 cases of fetal exomphalos. Ultrasound Obstet Gynecol. 2005;26:527–37.
20. Van Eijck FC, Hoogeveen YL, van Weel C, et al. Minor and giant omphalocele: long-term outcomes and quality of life. J Pediatr Surg. 2009;44:1355–9.
21. Segal SY, Marder SJ, Parry S, et al. Fetal abdominal wall defects and mode of delivery: a systematic review. Obstet Gynecol. 2001;98(5):867–73.
22. Zaccara A, Lacobelli BD, Calzolari A, et al. Cardiopulmonary performances in young children and adolescents born with large abdominal wall defects. J Pediatr Surg. 2003;38:478–81.
23. Engum SA. Embryology, sternal clefts, ectopia cordis and Cantrell's pentalogy. Semin Pediatr Surg. 2008;17:154–60.

Hernia and Hydrocele

33

Aileen Rooney

The testis develops high on the posterior abdominal wall behind the peritoneum. A fibromuscular band, the gubernaculum, extends from the lower pole of the testis to the scrotum and traverses the muscles of the anterior abdominal wall in the inguinal canal to reach the scrotum where it becomes attached. The processus vaginalis, which is an evagination of the peritoneal sac, will accompany the gubernaculum to reach the scrotum.

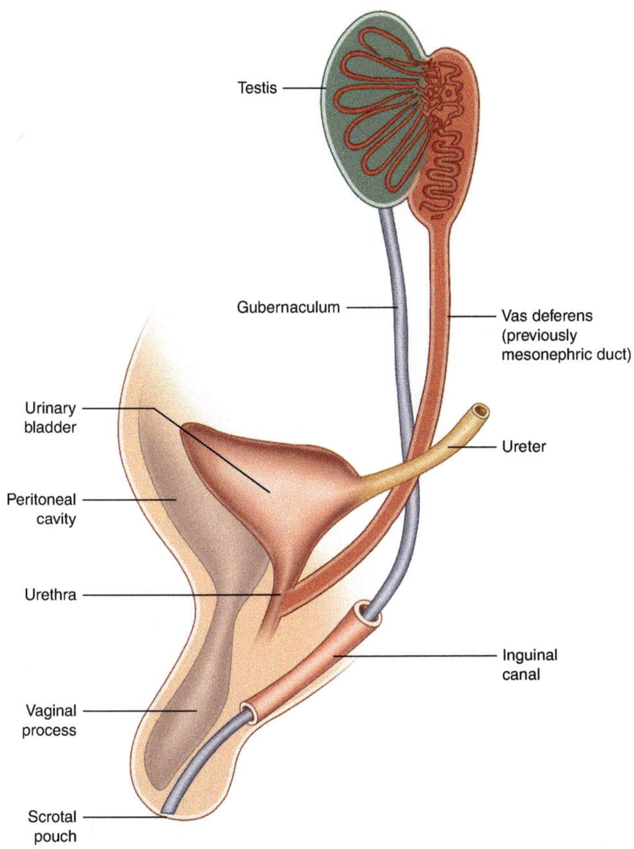

Hormonal action and increased intra-abdominal pressure will help shorten the gubernaculum and dragging down of the testis. By the third month, the testis should lie in the iliac fossa; by the seventh month, it should traverse the inguinal canal reaching the superficial ring by the eighth month; and by the ninth month, it should enter the scrotal sac. The remainder of the gubernaculum disappears completely, and the part of the processus vaginalis inside the scrotum will form a serous cavity for the testis called the tunica vaginalis. The proximal part of the processus vaginalis becomes obliterated.

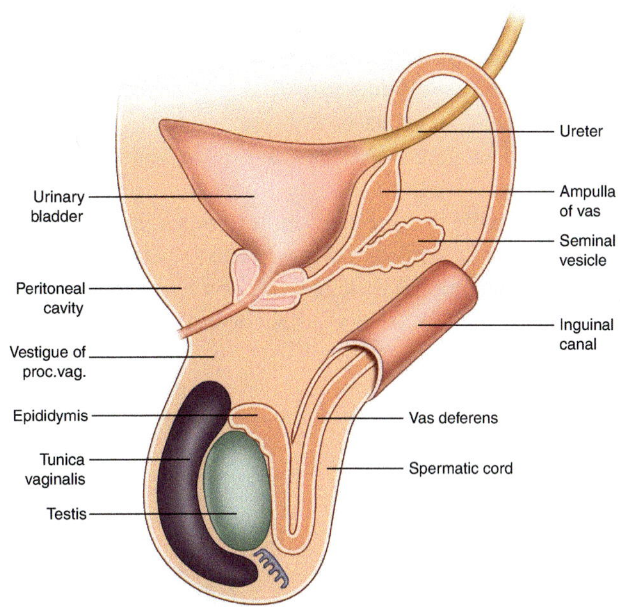

Hernias and Hydroceles

Introduction

Hernias and hydroceles are amongst the most common presentations to paediatric surgeons. Inguinal hernias alone

A. Rooney (✉)
Department of Neonatal and Paediatric Surgery, Royal Hospital for Children, Glasgow, UK

R. Carachi, S. H. E. Doss (eds.), *Clinical Embryology*, https://doi.org/10.1007/978-3-319-26158-4_33

occur in 1–5% of all term infants and up to 25% of preterm infants, and all will require surgical repair [1]. Unlike in adults where hernias are mainly the result of acquired weakness of surrounding tissue, most paediatric hernias are intrinsically linked to processes that occur during embryological and newborn development.

Inguinal Hernia

Embryological Considerations and Pathophysiology

The processus vaginalis is an evagination of the peritoneum through the internal ring into the inguinal canal [2]. It can first be identified at 3 months of gestation. In the paediatric population, almost all inguinal hernias result from failure of the processus vaginalis to obliterate. This results in peritoneal fluid or intra-abdominal organs being able to pass through the canal, resulting in a hydrocele or indirect inguinal hernia (Fig. 33.1).

The testes begin developing retroperitoneally high up on the posterior abdominal wall (medial to the mesonephros) at approximately 5 weeks of gestation. A fibromuscular band (the gubernaculum) extends from the lower pole of the testis down through the inguinal canal to the scrotum. The processus vaginalis herniates through the layers of the lower abdominal wall, along the pouch formed by the gubernaculum [3]. The intra-abdominal testes descend to reach the internal inguinal ring at around 7 months of gestation and pass down through the processus vaginalis to reach the scrotum around term. The part of the processus vaginalis which remains in the scrotum forms the tunica vaginalis. The part which lies above the testicle should become obliterated, so that the connection between the peritoneal cavity and the scrotum is terminated. If the processus vaginalis remains patent, it allows peritoneal fluid to track down and the development of a hydrocele. If it is open more widely, it can allow bowel or other intra-abdominal organs to pass through and the development of an indirect inguinal hernia.

The process that results in closure of the processus vaginalis appears to follow or be linked to the descent of the testes, since there is a high incidence of patency of the processus vaginalis associated with undescended testes [2]. The factors controlling the inguinoscrotal phase of testicular descent and closure of the patent processus vaginalis (PPV) are not yet fully understood. Foetal androgens and mechanical factors resulting from increased abdominal pressure appear to play a significant role. It has more recently been suggested that calcitonin gene-related peptide (CGRP) released from the genitofemoral nerve also contributes [4].

In females the canal of Nuck corresponds to the processus vaginalis, and when patent it communicates with the labia majora. The ovaries descend only to the pelvis, but the gubernaculum passes through the inguinal canal. The upper part of the gubernaculum becomes the ovarian ligament, and the lower part becomes the round ligament. This extends between the ovary and the labia majora. The canal of Nuck normally closes at approximately 7 months of gestation.

There is a window of time during which the processus vaginalis can close, the exact boundaries of which are not known. There is a high incidence of PPV at birth, estimated to be between 40 and 60%. The incidence is highest in premature infants, which is not unexpected given the role of the PPV in testicular descent (the final stages of which occur between the seventh and ninth months of foetal development). Processus vaginalis patency falls substantially within the first 6 months of life, and very few will close after age 3 to 5. It is reported that approximately 5% remain patent at autopsy [5].

Incidence and Associations

The prevalence of inguinal hernias at different ages correlates with the incidence of PPV at the same age and the natural history of closure of the processus vaginalis. The

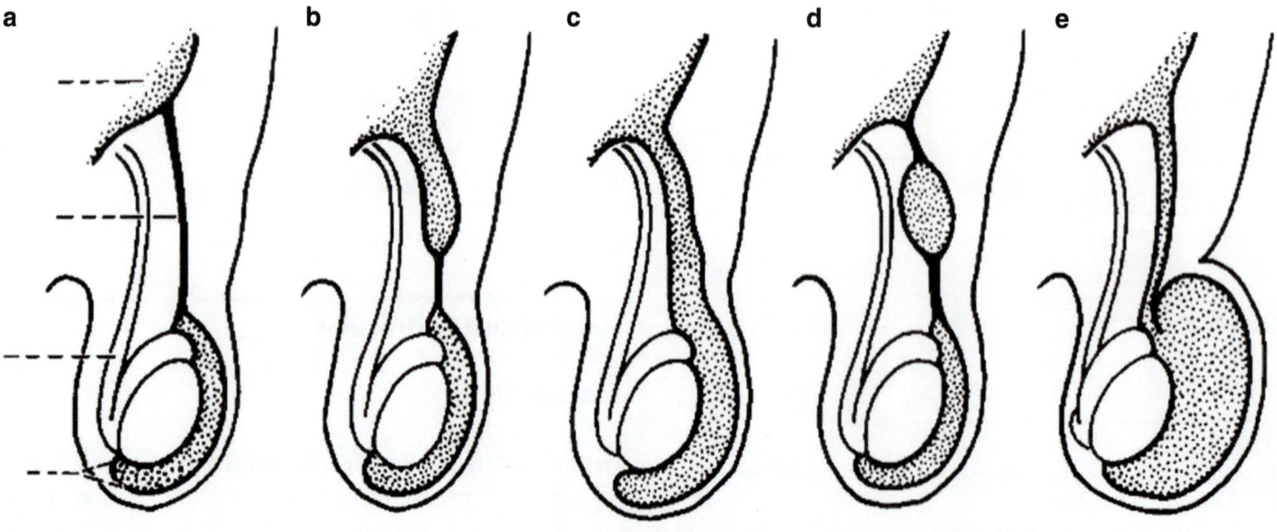

Fig. 33.1 (**a**) Normal, (**b**) inguinal hernia, (**c**) inguinoscrotal hernia, (**d**) hydrocele of the cord, (**e**) communicating hydrocele

incidence of inguinal hernia is approximately 0.8–4.4% in term babies and 10–30% in preterm infants. Hernias in premature infants are also more likely to be bilateral. Inguinal hernias most commonly present within the first 6 months up to a year, and one-third of children undergoing inguinal herniotomy are less than 6 months of age [1]. There is a positive family history in just over 11.5% of cases, and rates amongst twins are 10.6% in males and 4.1% in females [6].

Sex

There is a much higher incidence of inguinal hernias in males. The male to female ratio is between 3 and 10:1. In premature infants, the male to female ratio is more even. The markedly higher incidence in males is thought to be due to passage of the testes through the PPV (resulting in the channel being more widely distended) and that it begins closing later than in females.

Side

Sixty per cent of inguinal hernias occur on the right side, 30% are left sided, and 10% are bilateral at presentation. This has been attributed to the later descent of the right testis and subsequent later closure of the right processus vaginalis, but this does not explain the same occurrence in females.

Intersex:

Seventy-six per cent of patients with complete androgen insensitivity syndrome (CAIS) present with an inguinal hernia [7]. It is therefore important to inspect the gonad in the hernia sac of a phenotypically normal female to exclude presentation of an intra-abdominal testis. If doubt exists, samples can be sent from the gonad, and karyotyping can be performed.

Risk Factors for Development of an Inguinal Hernia

The presence of a patent processus vaginalis (PPV) is an essential component in the development of an inguinal hernia but most PPVs will not develop into a clinically apparent hernia. There are several other factors which confer an increased risk of developing an inguinal hernia (Table 33.1).

Clinical Presentation

Inguinal hernias can present electively or as an emergency. The vast majority will present with a painless swelling in the groin with or without extension into the scrotum or the labia majora. This is often noticed by parents or by physicians during a routine examination. The swelling will often appear to increase in size when the infant or child coughs or cries due to increased abdominal pressure. This is commonly mistaken for distress from the hernia itself as they occur in tandem. Most inguinal hernias are largely asymptomatic and reduce spontaneously (particularly when lying supine) with gentle pressure over the site. The hernia will often contain intra-abdominal viscera, most commonly loops of bowel or the ipsilateral ovary in females.

The major risk is incarceration, potentially leading to strangulation and peritonitis. The risk is highest in infants, particularly within the first 6 months. Clinically, the child presents upset or inconsolable with a tense, swelling in the groin. Twelve to seventeen per cent will present with an incarcerated hernia; most of these (approximately 90%) will be able to be reduced in the emergency department [5]. If the hernia is able to be reduced, the patient is admitted for observation and will be fixed urgently as an inpatient within 24–48 h (once the accompanying swelling has reduced). Incarceration of an ovary does not always impair the blood supply; if the patient is asymptomatic with a non-tender hernia, the repair can be performed on an urgent outpatient basis rather than as an emergency case.

If the hernia is irreducible, there may be accompanying signs of bowel obstruction (abdominal pain and distention, vomiting and absolute constipation), and this may be evident on abdominal X-ray. The hernia is tense and tender, and there can be overlying skin discoloration. If it is not able to be reduced, there are signs of vascular compromise of the trapped contents (strangulation) or frank peritonitis with accompanying signs of shock which will require operative management as an emergency. The progression from strangulation to perforation can happen rapidly, within a few hours of incarceration, and may be evident at first presentation. All parents should be made aware of the potential for the hernia to become incarcerated, and they should be advised to present as an emergency if this occurs (Fig. 33.2).

Overview of Management and Areas of Ongoing Debate

Inguinal hernias will not resolve spontaneously and thus require operative management. The procedure is most commonly performed under general anaesthesia, but it can be performed under spinal anaesthetic in preterm infants. Most surgeons will elect to repair hernias soon after diagnosis, but the timing of repair often depends on the age of the child and the relative risk of incarceration. Those under a year of age, but particularly less than 6 months, should be fixed as soon as possible. For preterm infants in neonatal intensive care, repair of the hernia is normally undertaken prior to discharge or when the infant is greater than 2 kg (depending on patient co-morbidities and individual surgeon's preference).

The principle of repair is high ligation of the hernia sac. This can be performed either open or laparoscopically. This is largely centre dependent, and no method has yet conclusively been shown to be better for unilateral inguinal hernias in terms of risk of complications, length of stay or time to resumption of normal activity [8].

Exploration of the contralateral side at the time of the primary procedure is another area of ongoing debate. There has been a shift of opinion over the past 20 years, with surgeons in the past favouring routine exploration with or without repair of the contralateral side. Most of the findings that have emerged from more recent studies have meant that on balance most surgeons currently would not routinely explore the other side. The potential benefits are that approximately

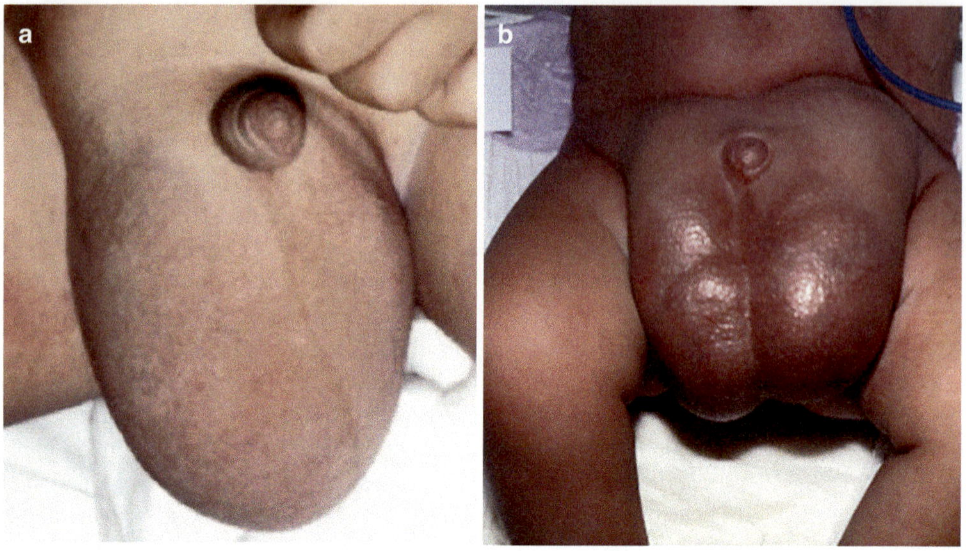

Fig. 33.2 (**a**) Inguinal hernia. (**b**) Bilateral inguinal hernia

Table 33.1 Risk factors for development of an inguinal hernia

General: male sex, prematurity, low birth weight
Cryptorchidism
Increased abdominal pressure: ventriculoperitoneal shunts, ascites, peritoneal dialysis
Abdominal wall defects: gastroschisis, exomphalos
Respiratory: cystic fibrosis and chronic lung disease
Connective tissue disorders: Marfan and Ehlers-Danlos syndromes, Hunter and Hurler syndromes

10% of all patients will go on to develop a metachronous inguinal hernia, and in these patients a further procedure with a second anaesthetic could have been avoided. The negative aspect of routine exploration is the potential complications (particularly the risk of damage to the vas and vessels) conferred by performing an unnecessary procedure on the 90% of patients who would never have required a herniotomy on the contralateral side. The use of laparoscopy has added another dimension to this in that it is often possible to visualise the contralateral side and to determine if there is a PPV.

Direct Hernias

Direct inguinal hernias are uncommon in the paediatric population. They occur when abdominal viscera pass through an area of weakness in the posterior wall of the inguinal canal (Hesselbach's triangle) instead of through the inguinal canal itself as in the case of an indirect hernia. They can be identified at the initial examination but are more commonly identified at the time of repair, when the defect is noted to be medial to the inferior epigastric vessels. Direct hernias can also be noted when a patient represents with what is thought to be a recurrence of an indirect inguinal hernia. It may be that it was a direct defect that was missed at the time of the original procedure. Additional care should therefore be taken if an obvious hernia sac is not visualised during a primary inguinal hernia repair to check that there is not a direct defect.

Femoral Hernias

Femoral hernias occur when there is herniation through the femoral ring into the femoral canal. They are also rare in children, accounting for less than 0.5% of all groin hernias, but approximately 10–20% are found to be bilateral. They are classically difficult to diagnose. Clinically they can be detected by eliciting a cough impulse when a hand is place 1 finger breadth medial to the femoral artery. They are described as occurring being below and lateral to the pubic tubercle. The risk of strangulation is higher than that of inguinal hernias, and there is an increased risk of recurrence post repair [9] (Fig. 33.3).

Congenital Hydroceles

There are several types of hydrocele as shown in diagram A (embryology section). The underlying pathophysiology is the same as for indirect inguinal hernias. The processus vaginalis remains patent; in the formation of hydroceles, this allows peritoneal fluid to track down and surround the testicle between the layers of the tunica vaginalis. There are two main types of congenital hydrocele: communicating and non-communicating. In communicating hydroceles, the processus vaginalis is open and allows fluid to continue to flow between the peritoneal cavity and the tunica vaginalis. The history and clinical features are normally of a soft, painless, swelling around the testis which can have a bluish appearance and which transilluminates. Its appearance may coincide with a viral illness, which has caused an increased amount of free intra-abdominal fluid. Parents may also give

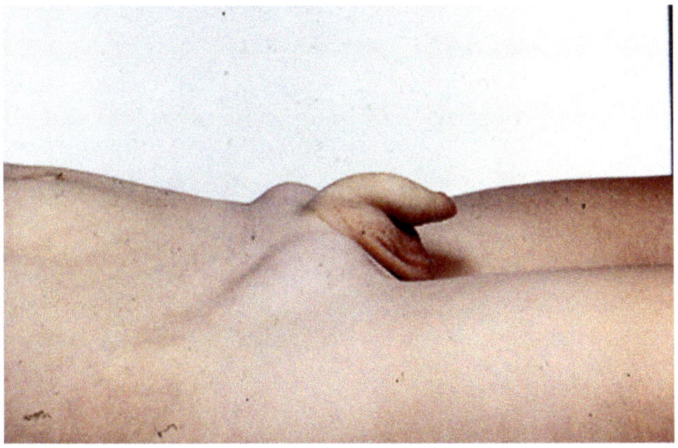

Fig. 33.3 Femoral hernia

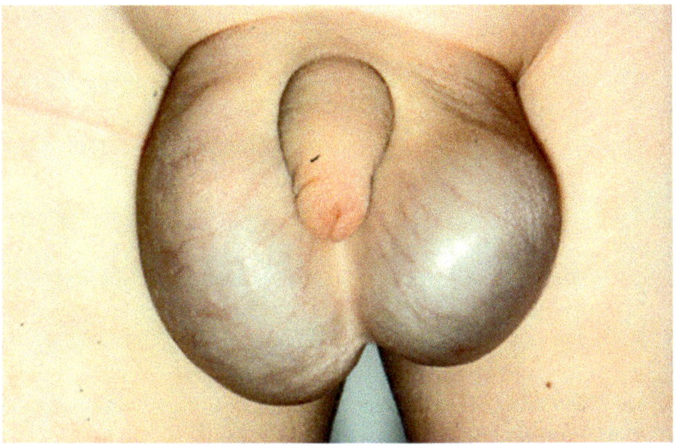

Fig. 33.4 Bilateral hydrocele

a history of a fluctuation in size over the course of a day, being more noticeable at night due to the effects of gravity, and smaller first thing in the morning as the fluid is reabsorbed when the patient is lying flat overnight (Fig. 33.4).

Non-communicating hydroceles occur when (a) the upper part of the processus vaginalis closes but fluid remains distal to this section and (b) the processus vaginalis closes above and below, leaving a pocket of fluid (a hydrocele of the cord). A hydrocele of the cord can be mistaken for an incarcerated indirect inguinal hernia, but in most cases a section of the spermatic cord is able to be palpated above the swelling which confirms it to be a hydrocele of the cord.

Hydroceles are common; most will resolve without operative intervention before the age of 2 years, and there is evidence that some can still resolve after this point. In our hospital, operative intervention is not advocated for asymptomatic hydroceles below about age 4 to allow maximal time for potential spontaneous resolution.

Epigastric and Supraumbilical Hernias

Epigastric hernias occur in the midline between the xiphoid process and the umbilicus. They are relatively common in children and occur through a defect in the linea alba. If located just above the umbilicus, they can be referred to as supraumbilical hernias and can be difficult to distinguish from an umbilical hernia. The exact pathogenesis is not known, but two main theories have been proposed. Moschowitz's hypothesis was that epigastric hernias are congenital and occur at sites where small vessels perforate the linea alba. This theory is still accepted, but it is thought that it cannot be responsible for of all epigastric hernias since it is rare to find a perforating vessel at the time of surgical repair. The more recent theory, proposed by Askar, is that epigastric hernias are acquired defects that occur when small tears develop as a consequence of forceful contraction of fibres which originate at the diaphragm and insert into the linea alba [10].

Epigastric hernias are often small and asymptomatic. The 'bulge' noted most commonly contains preperitoneal fat only (occasionally a small piece of omentum may also have been pulled up) and thus does not present the same danger if the herniated contents become incarcerated. Epigastric hernias will not resolve spontaneously, and thus if they are symptomatic (abdominal wall pain), or progressively enlarging, they require surgical repair.

Spigelian

Spigelian hernias are rare in adults and are even less common in the paediatric population. They occur between the rectus abdominis muscles medially and the semilunar line laterally, around the level of the arcuate line. There is often no visible swelling as the hernia occurs between the muscles of the abdominal wall. They have a relatively high risk of strangulation as the defect is often narrow.

References

1. Singal et al. http://www.pediatricurologybook.com/inguinal_hernia.html.
2. Coran AG, Adzick NS, Krummel TM, et al. Pediatric surgery. 7th ed. vol 1. Ch76. Philadelphia: Elsevier Saunders; 2012.
3. Pansky B. Review of medical embryology. Chapter 9. Section 103. Alameda: Embryome Sciences; 1982.
4. Hutson JM, Hasthorpe S, Heyns CF. Anatomical and functional aspects of testicular descent and cryptorchidism. Endocr Rev. 2013;18(2):259.
5. Holcomb GW III, Murphy JD. Ashcraft's pediatric surgery. 5th ed. Ch 51. Philadelphia: Elsevier Saunders; 2010.
6. Kingsworth A, Le Blanc KA, et al. Management of abdominal hernias. 4th Ed. Ch 11. Berlin: Springer; 2013.
7. Viner RM, Teoh Y, Williams DM, et al. Androgen insensitivity syndrome: a survey of diagnostic procedures and management in the UK. Arch Dis Child. 1997;77(4):305.
8. Yang C, Zhang H, Pu J, et al. Laparoscopic vs open herniorrhaphy in the management of pediatric inguinal hernia: a systemic review and meta-analysis. J Pediatr Surg. 2011;46(9):1824–34.
9. Gallegos NC, Dawson J, Jarvis M, et al. Risk of strangulation in groin hernias. Br J Surg. 1991;78(10):1171–3.
10. Ponten JE, Somers KY, Nienhuijs SW. Pathogenesis of the epigastric hernia. Hernia. 2012;16(6):627–33.

Gastrointestinal Tract 1: Foregut

34

Gregor M. Walker

Introduction

Foregut abnormalities discussed in this chapter are related to the oesophagus and stomach and will be considered in descending anatomical order. Given the common embryological origins, there is a close association with the tracheobronchial tree, but abnormalities of this system will be discussed in another chapter. Although the transition between foregut and midgut is the second part of the duodenum, pathology related to this part of the intestinal tract will be discussed in the midgut chapter. The timing of presentation of abnormalities of the foregut relates to the impact on normal functional dynamics of the oesophagus and stomach, ranging from antenatal diagnosis, often due to failure of the foetus to swallow/absorb liquor, to presentation in later childhood.

Oesophageal Atresia and Tracheo-oesophageal Fistula

Although the outcome of oesophageal atresia (OA) has gradually improved since the first report of a successful operative repair over 70 years ago [1], there is still considerable patient morbidity related to post-operative complications, impaired oesophageal function and associated anomalies that can cause problems throughout childhood and into adult life [2].

The incidence of OA in two recently published, multicentre surveillance studies was between 1 in 2300–5700 live births [3, 4]. There are a few classifications of OA, although all describe the anatomy in relation to the presence and site(s) of associated tracheo-oesophageal fistula(e) (TOF) (Fig. 34.1). The most common arrangement affecting around 85% of patients has a thickened, dilated, blind-ending upper oesophagus ending around the level of the second to fourth thoracic vertebral bodies and a thin lower oesophagus that forms a communication between the stomach and the trachea, usually at, or just above, the level of the carina (Gross type C, Vogt type IIIB). In a small number of patients (<1% of total), in addition to a distal fistula, there will be a fistula from the upper oesophageal pouch, most commonly located around the level of the thoracic inlet (Gross type D, Vogt type IIIC). "Pure" oesophageal atresia, where there are blind-ending upper and lower oesophageal pouches with no fistulae, affects around 8–10% of patients (Gross type A, Vogt type II). Some patients (~4% of total) with no distal fistula will have a fistula from the blind-ending upper oesophageal pouch, located at the level of the thoracic inlet (Gross type B, Vogt type IIIA). As demonstrated by the relative proportions of the described types, the likelihood of a proximal pouch fistula can be predicted from the presence of a distal fistula—with up to one third of apparent "pure" oesophageal atresia actually having an upper pouch connection. Around 4% of the total cases will have an isolated tracheo-oesophageal fistula (at the thoracic inlet) despite having no oesophageal discontinuity (Gross type E, Vogt type II).

The primitive foregut separates into two structures at around 4–5 gestational weeks with ingrowth of two lateral epithelial folds that eventually fuse in the midline to form the oesophagus dorsally and trachea ventrally. Much of the evidence supporting partial failure of this separation leading to the pathogenesis of oesophageal atresia has come from the Adriamycin rat model which has also highlighted abnormalities in various signalling molecules and transcription factors [5].

Most cases are sporadic, but there is a slight male preponderance [3, 4]. The recurrence risk among parents with one affected child is 0.5–2% but increases to 20% with additional affected siblings. Various environmental factors have been implicated such as prolonged use of contraceptive pills, methimazole and thalidomide exposure in pregnancy, foetal alcohol syndrome and maternal conditions such as diabetes and phenylketonuria [2].

G. M. Walker (✉)
Department of Paediatric Surgery, Royal Hospital for Children, Glasgow, UK
e-mail: Gregor.Walker@ggc.scot.nhs.uk

R. Carachi, S. H. E. Doss (eds.), *Clinical Embryology*, https://doi.org/10.1007/978-3-319-26158-4_34

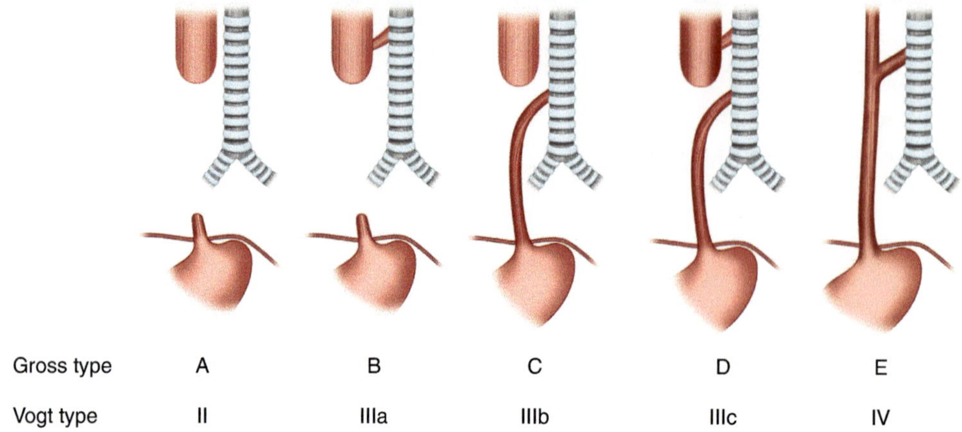

Fig. 34.1 Classification of oesophageal atresia and/or tracheo-oesophageal fistula

OA can be suspected antenatally, often due to polyhydramnios, and occasionally through identification of a dilated upper oesophageal pouch, although pure OA (Gross type A) is more likely to be identified on antenatal scan [3]. Postnatally, patients will present with inability to swallow saliva or milk, with bubbling of saliva at the mouth. Aspiration of saliva/milk into the trachea will result in coughing or respiratory distress. The diagnosis is confirmed by failure to pass a nasogastric tube into the stomach. Attempted passage usually results in failure to advance beyond 12 cm from the nose, and a plain radiograph will confirm that the tube is arrested/coiled in the dilated upper oesophageal pouch. Once identified (antenatally or postnatally), it is important to recognise that approximately 50% of patients will have another anomaly and 30% will have multiple anomalies. Perhaps the most widely known non-random association is the VACTERL association which is found in around 10–13% of patients with OA. This is diagnosed when patients have abnormalities of at least three of the following systems: *V*ertebral, *A*norectal, *C*ardiac, *T*racheo-*oE*sophageal, *R*enal and *L*imb. Antenatal and postnatal investigations should specifically look at each of these systems. The most likely additional anomaly is cardiac with around one third of patients having abnormalities identified on echocardiography, most commonly tetralogy of Fallot, atrial septal defect, ventricular septal defect and transposition of the great arteries [3, 4]. Plain X-rays should also include views of the abdomen to identify if air has passed into the intra-abdominal intestine, confirming the presence of a distal TOF (Fig. 34.2).

The gas pattern on abdominal radiography should also identify if there is associated duodenal atresia (Fig. 34.3), which occurs in around 5% of cases and presents a more complex surgical management problem.

Chromosomal anomalies are found in up to 10% of cases (Table 34.1). Aneuploidy is the most common chromosomal anomaly with trisomy 18 (Edwards syndrome) being more frequently encountered than trisomy 21. Autosomal dominant single-gene syndromes include anophthalmia-oesophageal-genital (AEG) syndrome, CHARGE syndrome, Feingold syndrome and Pallister-Hall syndrome. Fanconi anaemia is the only autosomal recessive syndrome, and the X-linked Opitz G/BBB syndrome is rarely encountered.

There is no antenatal intervention to treat OA, although in a large prospective European study, around 27% of pregnancies where OA was identified antenatally were terminated due to severe chromosomal or multiple anomalies [4]. The survival for newborns with this condition has gradually improved, and now, in the absence of significant cardiac or chromosomal anomalies, the expected survival should be >95%. Long-term morbidity can be related to post-operative complications, gastro-oesophageal reflux, disordered oesophageal motility, tracheomalacia, musculoskeletal anomalies related to neonatal thoracotomy and issues related to the associated malformations [2]. Parents of children with this condition in the UK have found the TOFS charity to be particularly useful is providing information and support (www.tofs.org.uk).

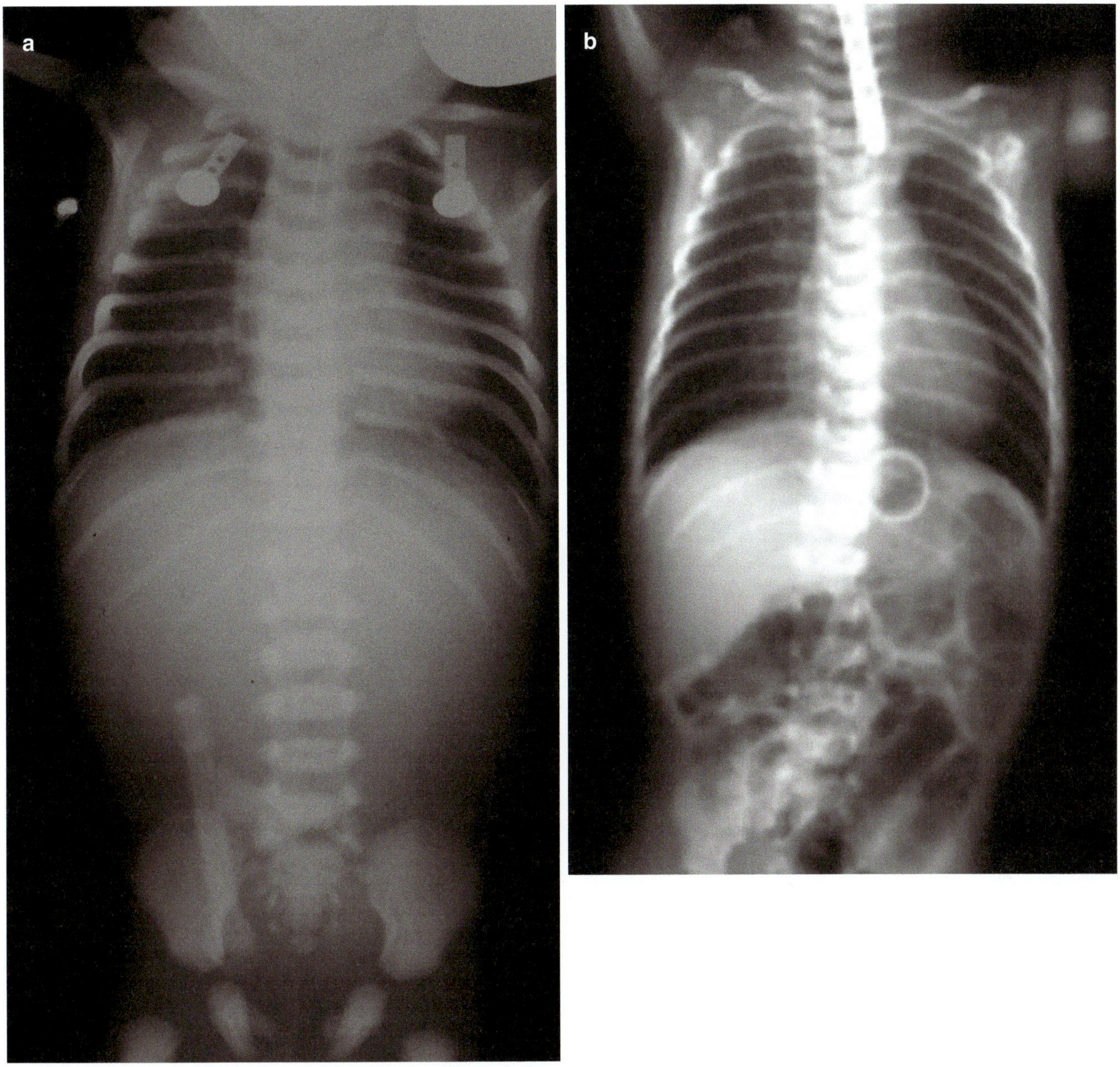

Fig. 34.2 (**a**) Pure oesophageal atresia. (**b**) Oesophageal atresia with bougie in upper pouch

Congenital Oesophageal Stenosis

Congenital oesophageal stenosis is a fixed intrinsic narrowing typically of the distal oesophagus with an incidence of 1 in 25,000–50,000 live births. Around one third of cases are seen in associated with OA, but the remaining cases are isolated [6]. It is rarely diagnosed in the newborn period and often presents later in infancy, after the introduction of solid food with vomiting, dysphagia and failure to thrive. There are three histological types: ectopic tracheobronchial remnants in the oesophageal wall, segmental fibromuscular hypertrophy of the muscle and submucosal layers and a membranous diaphragm or stenosis [7].

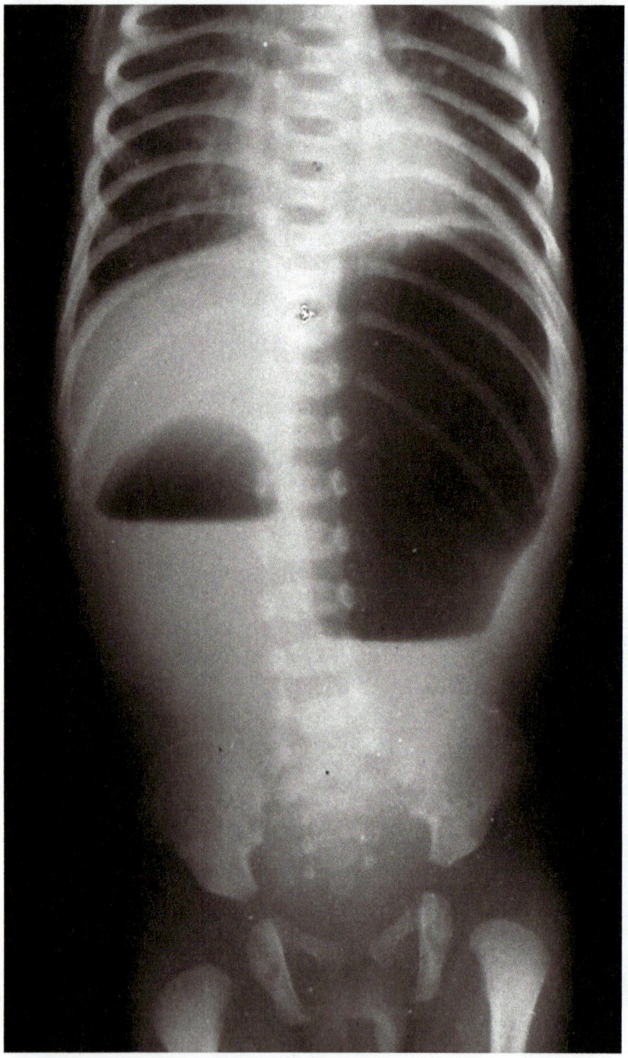

Fig. 34.3 Duodenal atresia double bubble

Diagnosis is usually confirmed by a contrast oesophagogram which may show an abrupt or tapered stenosis and can be complimented by endoscopy. A fibrotic stricture associated with reflux oesophagitis is the main differential diagnosis, but a pH study will clarify this. In patients with ectopic tracheobronchial remnants, non-surgical options such as dilatation or bouginage are unlikely to be successful, and surgical resection is recommended [7, 8]. Many patients, however, have persistent dysphagia despite any form of treatment [7].

Foregut Duplications

Duplications can be associated with the entire length of the gastrointestinal tract and oesophageal duplications account for around 15% of the total, with an incidence of around 1 in 8000 live births. Although they can occur anywhere along the oesophagus, the majority are situated on the right side of the thoracic oesophagus [6]. Most are cystic duplications, approximately 50% are lined with gastric mucosa and communication with the oesophageal lumen is uncommon. A small number of patients will have a thoracoabdominal cyst, with caudal extension from the posterior mediastinum, through the diaphragmatic hiatus into the abdominal cavity. These tubular duplications have a higher proportion of heterotopic gastric mucosa, and the infra-diaphragmatic portion can connect to the stomach, duodenum, pancreas or jejunum [9].

Associated thoracic vertebral anomalies are present in 20% of thoracic oesophageal duplications and up to 88% of thoracoabdominal cysts, giving rise to the "split notochord theory" (Fig. 34.4). The notochord, which is present from the end of the third gestational week, results from infolding of the

Table 34.1 Syndromic conditions associated with OA/TOF

Syndrome	Clinical characteristics in newborn	Prevalence of association in OA/TOF
Trisomy 18	Microcephaly/prominent occiput, choroid plexus cysts, low-set abnormal ears, micrognathia, cleft lip/cleft palate, ocular hypertelorism, short sternum, renal tract anomalies, cardiac anomalies, exomphalos, oesophageal atresia, clenched hands, underdeveloped thumbs, absent radius, webbing of second/third toes, rocker bottom feet	~6% (including stillbirths/ terminations)
Trisomy 21	Flat occiput, hypotonia, upward slanting eyes with epicanthic folds, large tongue, small low-set ears, short webbed neck, cardiac anomalies, OA/duodenal atresia, single palmar and sole crease, short fingers, sandal gap	~2%
Anophthalmia-oesophageal-genital syndrome	Anophthalmia/microphthalmia, OA/TOF, urogenital anomalies	
CHARGE syndrome	**C**oloboma of the eye, **H**eart anomalies, choanal **A**tresia, mental **R**etardation, **G**enital anomalies, **E**ar anomalies (up to 20% have OA/TOF)	1%
Feingold syndrome	Oesophageal/duodenal atresia, microcephaly, syndactyly, cardiac defects	Rare
Pallister-Hall syndrome	Hypothalamic hamartoma, central and postaxial polydactyly, imperforate anus, renal anomalies, bifid epiglottis, tracheal clefts, OA/TOF (rare)	Rare
Fanconi anaemia	Bone marrow failure, abnormal skin pigmentation, radial defects, eye anomalies, renal anomalies, cardiac defects, abnormal ears, central nervous system anomalies, hearing loss, gastrointestinal anomalies including OA/TOF	Rare
X-linked Opitz G/BBB	Cleft lip, laryngeal cleft, cardiac defects, hypospadias, agenesis of the corpus callosum, OA/TOF	Rare

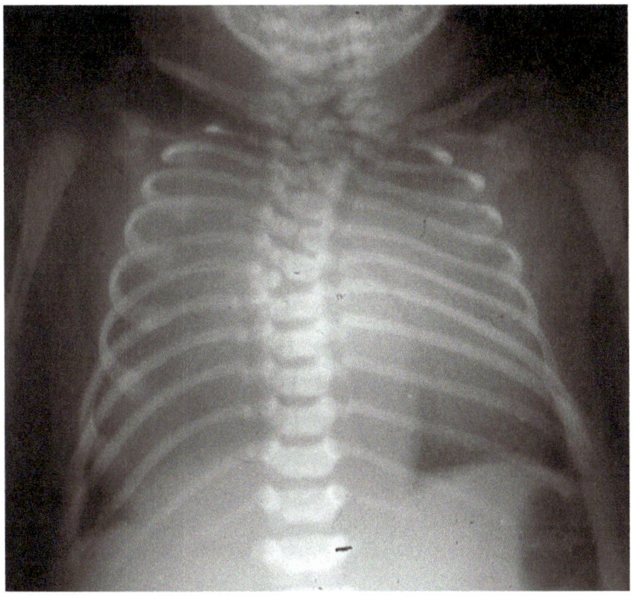

Fig. 34.4 X-ray demonstrating thoracic vertebral anomalies in association with an oesophageal duplication

notochordal plate and forms the longitudinal midline axis around which the vertebral bodies are organised. If the notochord splits, this will allow the endodermal gut to herniate through the gap and may result in a cyst or fistula which interferes with anterior fusion of the vertebral mesoderm [10].

Patients tend to present with symptoms due to local compressive effects, such as respiratory symptoms, vomiting or dysphagia, although some may be diagnosed incidentally on X-ray. Diagnosis can be made by a contrast swallow, although cross-sectional imaging such as MRI can identify any potential extension into the neural canal. Even in apparently isolated oesophageal duplication cysts, abdominal imaging is recommended as around 25% were associated with a metachronous intra-abdominal duplication cyst [9]. Because the duplication may increase in size with time and compress surrounding structures, operative resection is the treatment of choice.

Gastric duplications that account for around 8% of all duplications are rare anomalies that generally occur along the greater curvature [9]. They are most commonly cystic duplications, and communication with the gastric lumen is rare. With routine antenatal ultrasound screening, foetal diagnosis is possible. Associated malformations are not common, and there appears to be a female predilection. Unless there are signs of gastric outlet obstruction or gastrointestinal bleeding, normal gastric feeding is safe after delivery, and surgical excision can be delayed until beyond the newborn period [11]. Patients can present in infancy with vomiting and an epigastric mass and ultrasound scan is generally diagnostic, demonstrating discrete cyst with a trilaminar "gut signature" (Fig. 34.5). Important differential diagnoses include choledochal cysts and pancreatic pseudocyst, and abdominal CT scan or an upper gastrointestinal contrast study may be required to clarify the anatomy. All gastric duplications should be excised as there is a risk of gastrointestinal bleeding. Resection is usually possible without entering the stomach lumen, although larger duplications may require a partial gastrectomy. There have been case reports of malignancies arising in duplication cysts, and, although these generally do not present until after the fourth decade of life, gastric duplications have been associated with younger malignant transformation, in teenage years [12].

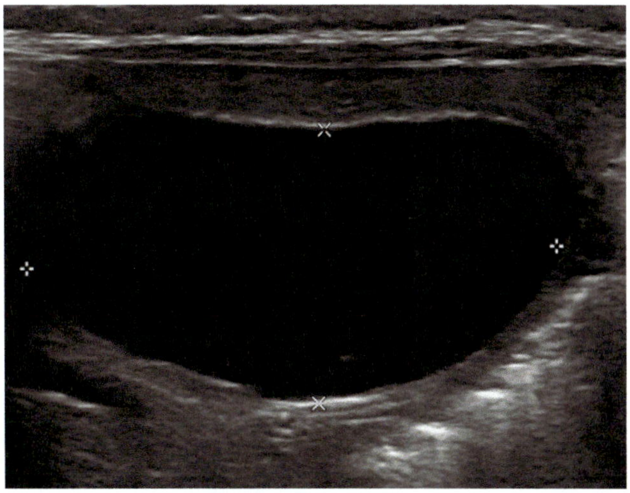

Fig. 34.5 Ultrasound image of gastric duplication cyst showing "trilaminar" wall. The white crosses on the image represent measures of dimension

Gastric Volvulus

During embryological development of the stomach, the four gastric ligaments form attachment adjacent structures: the gastrocolic, gastrohepatic, gastrophrenic and gastrosplenic ligaments. The ligaments, in additional to the pylorus and the gastro-oesophageal junction, anchor the stomach. Failure of these normal attachments through agenesis, elongation or disruption, results in an incompletely fixed stomach which is at risk of volvulus, termed "primary" gastric volvulus [13]. Other disorders can result in "secondary" gastric volvulus, such as abnormalities leading to acute gastric distension and abnormalities leading to displacement of the stomach (congenital diaphragmatic hernia, diaphragmatic eventration, paraoesophageal hernia) or agenesis/anomalies of adjacent structures (asplenia, hypoplasia of left lobe of the liver).

Gastric volvulus is described as two different entities, based on the axis of gastric rotation (Fig. 34.6). Organoaxial gastric volvulus is rotation around the long axis of the stomach (Fig. 34.6a, b), but the relative anatomical positions of the pylo-

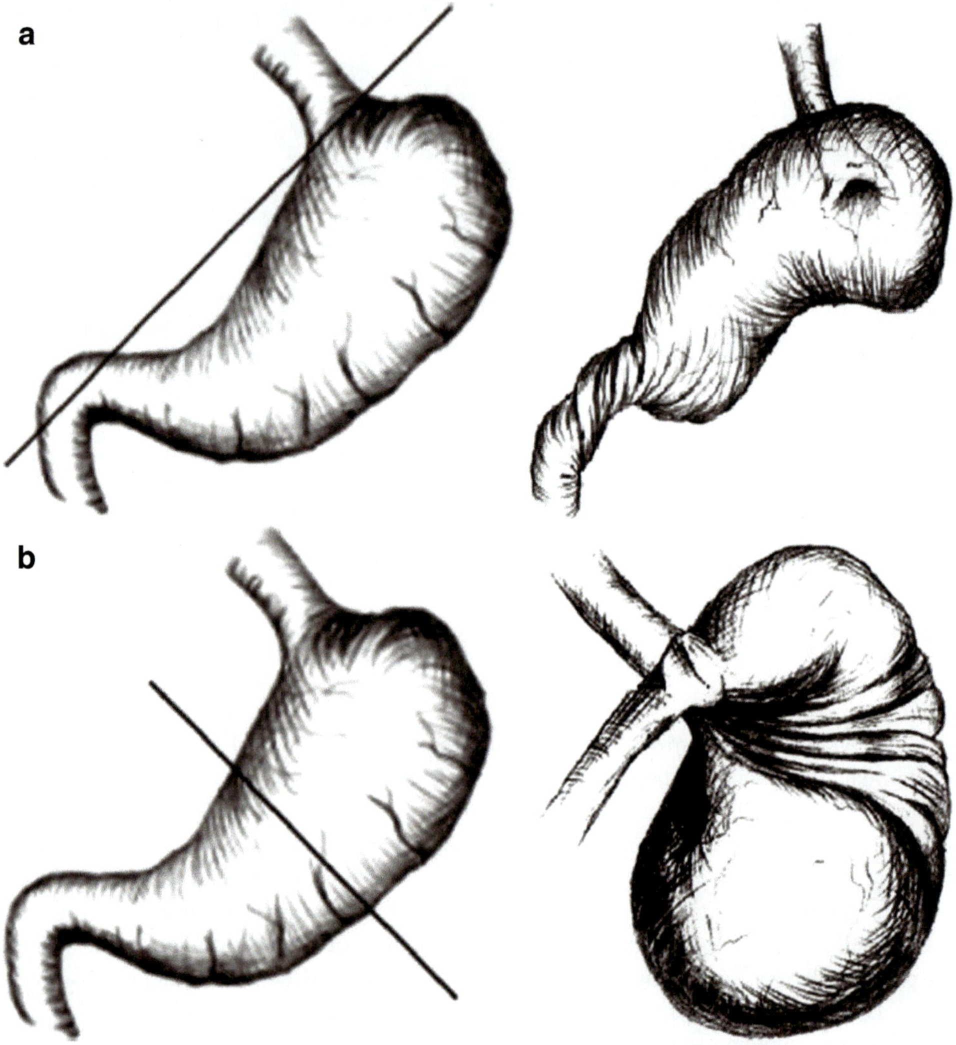

Fig. 34.6 (**a**) Organoaxial axis of the stomach, and the position of organoaxial volvulus. (**b**) Mesenteroaxial axis of the stomach, and the position of mesenteroaxial volvulus

rus and gastro-oesophageal junctions do not change significantly. Mesenteroaxial gastric volvulus is rotation about the gastric short axis, transecting the greater and lesser curvatures (Fig. 34.6c, d), and the pylorus is displaced superiorly.

Around half of the cases present as an emergency with acute abdominal pain, intractable retching and the inability to pass a nasogastric tube into the stomach lumen, the "classic" triad of Borchardt. The average age of presentation is 2.5 years, and it affects equal numbers of males and females [14]. The majority of children presenting acutely in this fashion have an associated anomaly (secondary gastric volvulus), and the organoaxial variety is slightly more common (around 55%) [13].

Those patients presenting with chronic symptoms are generally younger, the majority are of primary aetiology, and around 85% are the organoaxial variety [13]. Non-bilious vomiting is the most common presenting symptom, but feeding issues, growth failure and gastro-oesophageal reflux are also seen frequently.

Diagnosis is confirmed by plain abdominal radiography, although the anatomy in the chronic form of the condition can be highlighted by contrast studies. Treatment consists of patient resuscitation, nasogastric decompression and surgical correction with fixation of antrum and fundus. Unless there is irreversible ischaemia of the stomach, a full recovery can be expected.

Congenital Pyloric Obstruction

Pyloric atresia is a rare disease with an incidence of 1 in 100,000 live births [14]. It can be diagnosed antenatally with polyhydramnios and a large stomach [15] but more often presents in the newborn period but non-bilious vomiting due to gastric outlet obstruction.

Around 18–40% of cases of pyloric atresia are associated with epidermolysis bullosa (EB), a group of hereditary severe bullous skin disorders divided into three types based on skin histology: EB simplex, junctional EB and dystrophic EB. All three types have been reported in association with pyloric atresia, but junctional EB is the most frequently reported [16]. The combination of these conditions is also called the congenital pyloric atresia epidermolysis bullosa syndrome (CPA-EB syndrome), or Carmi syndrome. This clinical entity may result from a mutation in one of the integrin genes, with the absence of alpha 6 integrin being reported [17]. The other important associations are with aplasia cutis congenita and multiple intestinal atresias. Some cases can be associated with both EB and aplasia cutis congenita (Bart syndrome).

Similar to other intestinal atresia, pyloric atresia can take the form of a diaphragm (or web), a cord-like connection between atretic ends, a complete gap between ends or multiple intestinal atresias. In reported series, the most common type is the pyloric diaphragm [17], but all types can be associated with EB. Although the pathogenesis of pyloric atresia may have an ischaemic aetiology in line with other forms of intestinal atresia, the development of pyloric atresia in the setting of EB is thought to be secondary to separation of the mucosal lining from insufficient hemi-desmosome function leading to progressive fibrosis of the pyloric channel [15].

The diagnosis is suggested on plain X-ray but can be confirmed with a contrast study. Treatment is through surgical correction, often with a Billroth type I gastroduodenostomy, although some pyloric preserving techniques have been described for the diaphragm-type of anomaly. Unless associated with EB, the outcome is excellent.

Antral web is where there is a membranous occlusion of the gastric outlet before the pylorus [14, 18]. Aetiology is unknown but is thought to be congenital as a result of an inflammatory process. As there may be a perforation within the web, the gastric outlet obstruction is often incomplete, and presentation may be in older children or adults. The condition may be difficult to distinguish from hypertrophic pyloric stenosis in infants or other forms of inflammatory pyloric stenosis in older children and adults. Plain X-rays are often normal, and diagnosis will be achieved by upper gastrointestinal contrast series or endoscopy [14]. Treatment is through surgical ablation via laparotomy, laparoscopy or endoscopic techniques, and the outcome is generally excellent.

References

1. Haight C. Congenital atresia of the esophagus with tracheoesophageal fistula : reconstruction of esophageal continuity by primary anastomosis. Ann Surg. 1944;120(4):623–52.
2. Bax K. Esophageal atresia and tracheoesophageal malformations. In: Holcomb GW, Murphy JP, Ostlie DJ, editors. Ashcraft's pediatric surgery. 5th ed. Philadelphia: Saunders; 2010. p. 345–61.
3. Burge DM, Shah K, Spark P, Shenker N, Pierce M, Kurinczuk JJ, et al. Contemporary management and outcomes for infants born with oesophageal atresia. Br J Surg. 2013;100(4):515–21.
4. Pedersen RN, Calzolari E, Husby S, Garne E, EUROCAT Working group. Oesophageal atresia: prevalence, prenatal diagnosis and associated anomalies in 23 European regions. Arch Dis Child. 2012;97(3):227–32.
5. Jacobs IJ, Que J. Genetic and cellular mechanisms of the formation of esophageal atresia and tracheoesophageal fistula. Dis Esophagus. 2013;26(4):356–8.
6. Chandler N, Colombani PM. The esophagus. In: Holcomb GW, Murphy JP, Ostlie DJ, editors. Ashcraft's pediatric surgery. 5th ed. Philidelphia: Saunders; 2010. p. 330–44.
7. Michaud L, Coutenier F, Podevin G, Bonnard A, Becmeur F, Khen-Dunlop N, et al. Characteristics and management of congenital esophageal stenosis: findings from a multicenter study. Orphanet J Rare Dis. 2013;8:186–90.
8. Takamizawa S, Tsugawa C, Mouri N, Satoh S, Kanegawa K, Nishijima E, Muraji T. Congenital esophageal stenosis: therapeutic strategy based on etiology. J Pediatr Surg. 2002;37(2):197–201.
9. Keckler SJ, Holcomb GW. Alimentary tract duplications. In: Holcomb GW, Murphy JP, Ostlie DJ, editors. Ashcraft's pediatric surgery. 5th ed. Philidelphia: Saunders; 2010. p. 517–25.
10. Carachi R, Azmy A. Foregut duplications. Pediatr Surg Int. 2002;18(5–6):371–4.
11. Foley PT, Sithasanan N, McEwing R, Lipsett J, Ford WD, Furness M. Enteric duplications presenting as antenatally detected abdominal cysts: is delayed resection appropriate? J Pediatr Surg. 2003;38(12):1810–3.
12. Ma H, Xiao W, Li J, Li Y. Clinical and pathological analysis of malignancies arising from alimentary tract duplications. Surg Oncol. 2012;21(4):324–30.
13. Cribbs RK, Gow KW, Wulkan ML. Gastric volvulus in infants and children. Pediatrics. 2008;122(3):e752–62.
14. Koontz CS, Wulkan M. Lesions of the stomach. In: Holcomb GW, Murphy JP, Ostlie DJ, editors. Ashcraft's pediatric surgery. 5th ed. Saunders: Philidelphia; 2010. p. 391–9.
15. De Jenlis Sicot B, Deruelle P, Kacet N, Vaillant C, Subtil D. Prenatal findings in epidermolysis bullosa with pyloric atresia in a family not known to be at risk. Ultrasound Obstet Gynecol. 2005;25(6):607–9.
16. Short SS, Grant CN, Merianos D, Haydel D, Ford HR. A case of congenital pyloric atresia with dystrophic epidermolysis bullosa. Pediatr Surg Int. 2014;30(6):681–4.
17. Al-Salem AH, Abdulla MR, Kothari MR, Naga MI. Congenital pyloric atresia, presentation, management, and outcome: a report of 20 cases. J Pediatr Surg. 2014;49(7):1078–82.
18. Nam SH, Koo SH, Chung ML, Jung YJ, Lim YJ. Congenital antral web in premature baby. Pediatr Gastroenterol Hepatol Nutr. 2013;16(1):49–52.

Gastrointestinal Tract II: Midgut

35

Marie Steven

The midgut extends from the anterior intestinal portal to the posterior intestinal portal. The derivatives of the midgut are the lower half of the duodenum, the jejunum and the ileum, as well as the caecum, appendix, ascending colon and the right two-thirds of the transverse colon.

At the beginning of week 5, the part of the primitive gut that extends between the anterior and posterior intestinal portals forms a u-shaped loop which has a cranial limb, apex and a caudal limb, which shows a caecal swelling.

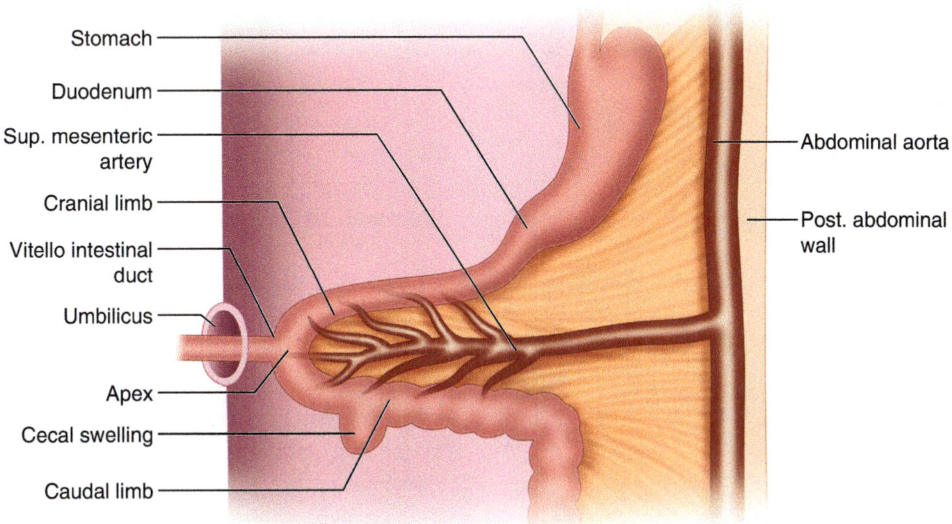

Part of the u-shaped loop proximal to the caecal swelling will form the lower half of the duodenum, the jejunum and the ileum. The apex of the loop is connected to the yolk sac via the vitello-intestinal duct. The caecal swelling will eventually develop into the caecum and appendix, and the part distal to the caecal swelling will develop into the ascending colon and the right two-thirds of the transverse colon.

At the 6th week, the intestinal loop elongates rapidly forming many loops which leave the abdominal cavity and enter the umbilical cord. This physiological umbilical hernia will remain until the end of week 10.

M. Steven (✉)
Department of Surgical Paediatrics, Royal Hospital for Children, Glasgow, UK
e-mail: mairisteven@doctors.org.uk

R. Carachi, S. H. E. Doss (eds.), *Clinical Embryology*, https://doi.org/10.1007/978-3-319-26158-4_35

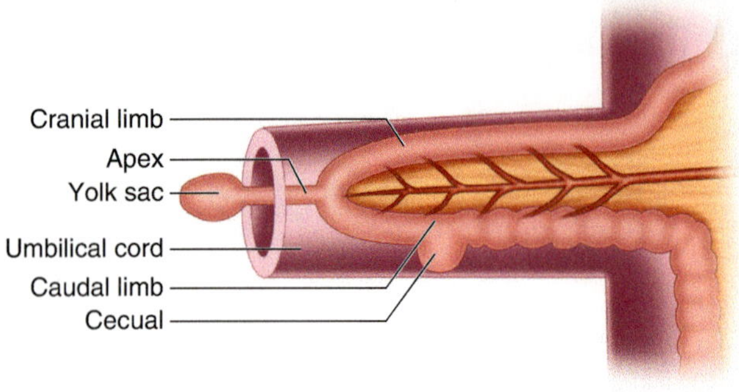

At the same time as the elongation of the loop, the intestines are also rotating by 270° in an anticlockwise direction around a longitudinal axis formed by the superior mesenteric artery.

As a result of this rotation, the upper part of the small intestine comes to lie behind the large intestine.

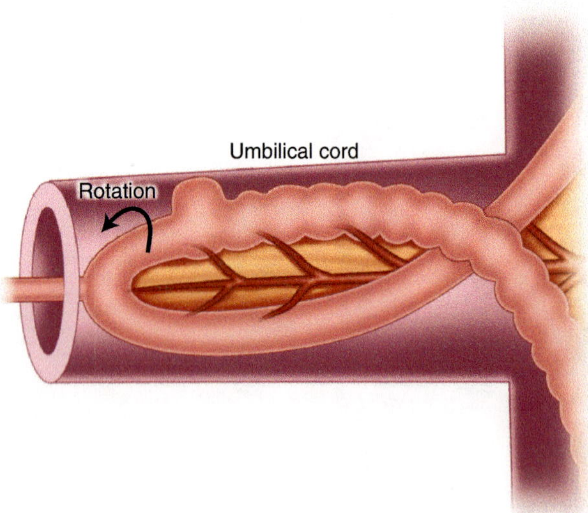

By week 10, the abdominal cavity enlarges, and the herniated loops begin to return to the abdominal cavity.

- *The upper part of the jejunum is the first part to re-enter the abdomen and lies on the left side.*
- *The loops returning next gradually lie more and more to the right side.*
- *The caecal swelling is last to return and will lie below the liver.*

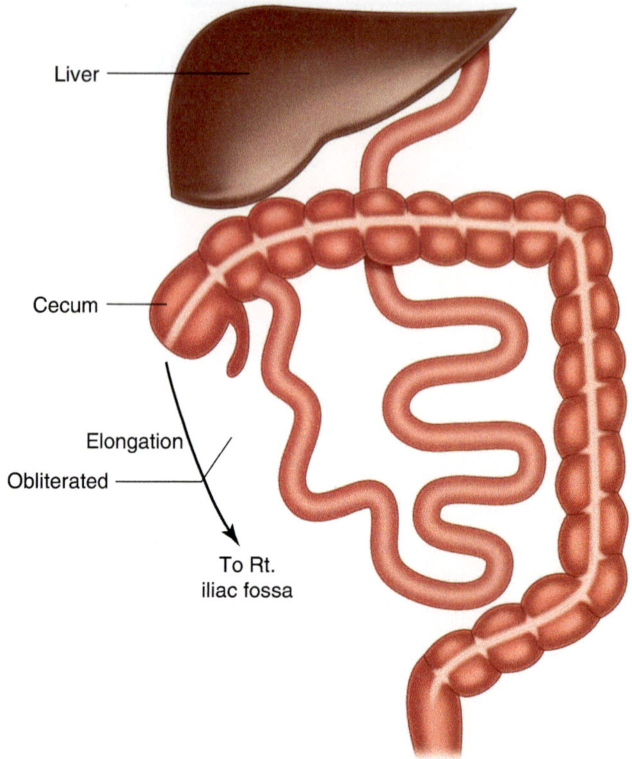

As development proceeds, the caecal swelling elongates downwards to reach the right iliac fossa, forming the right colic flexure and ascending colon. Finally, the vitello-intestinal duct disappears.

Introduction

Midgut abnormalities encompass different pathologies that often present in neonates. It is important to mention that abdominal wall pathologies and umbilico- or vitello-intestinal pathologies are in discussed in another chapter. This chapter will include duodenal and small bowel atresias and rotational abnormalities and associated syndromes.

Atresias

Atresia of the intestine is a malformation where there is a narrowing or absence of a segment of the intestine. In 1900 Tandler et al. described the theory of "failure of recanalisation" of the solid tube of bowel, but this now thought to be

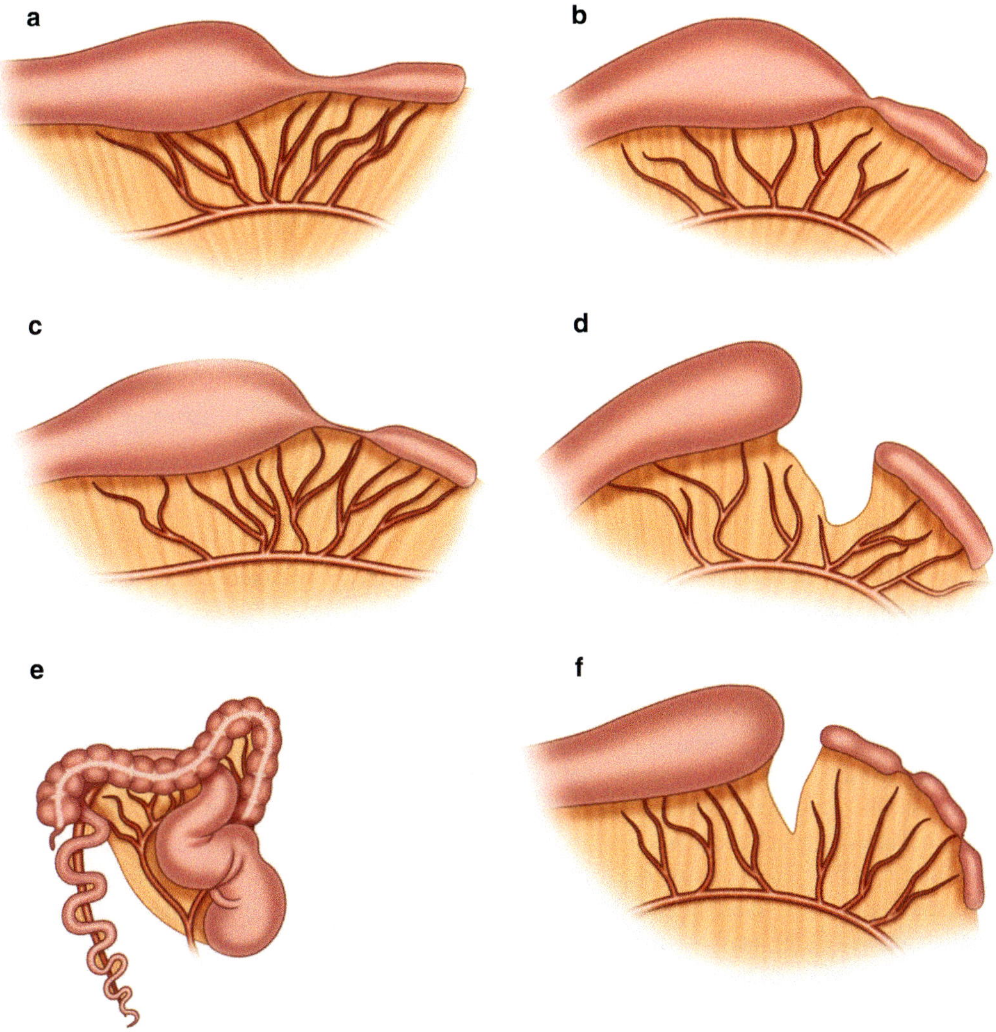

Fig. 35.1 Classification of intestinal atresia. (**a**) type 1 atresia with a intraluminal membrane, (**b**) type 2 atresia with a cord between the two segments, (**c**) type 3 atresia with complete disconnection but no mesenteric defect, (**d**) type 4 atresia with complete disconnection and a mesenteric defect, (**e**) an apple peel type atresia and lastly (**f**) multiple atresias

hold true only for duodenal atresia, whereas atresias in the rest of the bowel are thought to arise from impedance to the blood supply of the bowel, and this was famously demonstrated by Louw and Barnard in 1955 with their work inducing mesenteric vascular accidents in puppies [1, 2]. They also showed that in utero intestinal volvulus, intussusception, internal hernia or closing gastroschisis could lead to intestinal atresia. Atresias were originally classified into four types, i.e. type 1 or A is an intraluminal membrane or diaphragm, type 2 or B is a cord, and type 3 or C is a completely disconnected segment of bowel without a mesenteric defect and type 4 or D with a mesenteric defect. In 1979 Grosfield et al. added a further two types of the apple peel type and those that have multiple atresias [3] (Fig. 35.1).

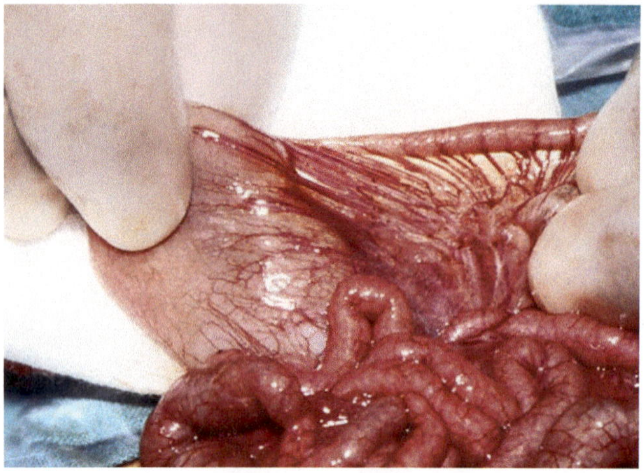

Fig. 35.2 Small bowel atresia at operation

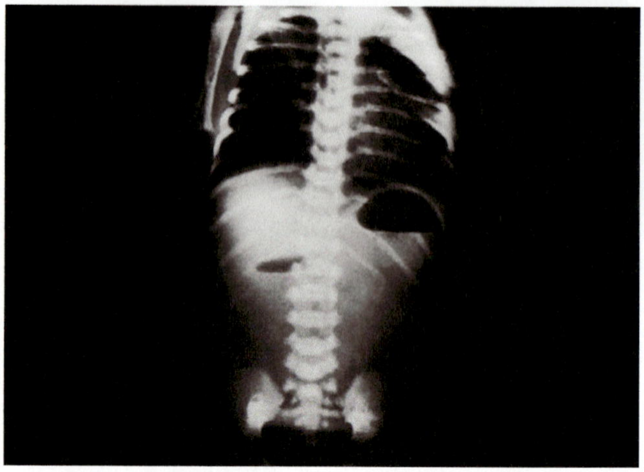

Fig. 35.3 Plain abdominal radiograph showing a "double bubble"

Intestinal atresia or stenosis can occur anywhere along the GI tract, and the anatomic location of the obstruction determines the clinical presentation. The most common site of intestinal atresia is the small intestine (jejunum and ileum). The incidence of jejunal and ileal atresia ranges from one in 1500 to 5000 births. It can be rarely related to other conditions such as Hirschsprung's disease, cystic fibrosis, malrotation, Down syndrome, anorectal malformations, congenital heart disease and other atresias which are found in up to one-third of patients. Diagnosis may be suggested antenatally or postnatally; there may be bilious vomiting and features of obstruction. Plain abdominal X-ray may show dilated loops of bowel with no gas in the pelvis or even meconium calcification. A contrast enema may be performed if meconium disease is suspected or if the diagnosis is not clear. Management is surgical and at laparotomy the bowel is inspected fully and the level of obstruction found (Figs. 35.2 and 35.3).

Further atresias should be sought and depending on the presence of these and on the calibre of the bowel, resection and primary anastomosis is performed. Multiple atresias or apple peel may lead to extensive resection and short bowel syndrome (SBS). Indeed a bowel length of less than 40 cm from DJ flexure requires long-term parenteral nutrition and then may result in associated liver disease and can lead to significant morbidity and mortality in these patients [4]. Bowel lengthening procedures including the serial transverse enteroplasty or "STEP" procedure and the Bianchi technique have been used with some success, but SBS remains an incredibly challenging condition to manage [5, 6]. Colonic atresia is the rarest of all the atresias with an incidence of 1 in 20,000. It presents with more distal obstruction and has a higher chance of perforation than other atresias. It may be related to Hirschsprung's disease. Following prompt diagnosis, however, the outcome is generally excellent [7, 8].

Duodenal atresia (DA) is often considered a separate entity to other intestinal atresias. It is thought to result from a failure of recanalisation other than a vascular insult as in more distal atresias. Duodenal atresia occurs in 1 in 20,000 to 40,000 births. Approximately 30% of infants with duodenal atresia have Down syndrome and over half an associated anomaly such as a cardiac anomaly. The atresia is usually post-ampullary (80%), and so the child often presents with bilious vomiting. It may be suspected antenatally with the appearance on ultrasound of a classic "double bubble". Further suspicion is raised if the foetus is known or suspected to be trisomy 21. The most common type of duodenal atresia is a type 1 (90%), i.e. an intraluminal membrane or diaphragm. An annular pancreas is another possible cause of this type of obstruction. Around one in four patients with DA has another anomaly of their gastrointestinal tract. If not suspected antenatally, the neonate presents with vomiting often bilious and distension of the upper abdomen. Plain abdominal X-ray shows a "double bubble". Management is surgical and is usually in the form of a diamond duodenoduodenostomy. This is now being performed laparoscopically in some centres [9]. A trans-anastomotic tube may or may not be passed depending on the surgeon's preference as some believes this leaves to achievement of enteral feeds [10]. Enteral feeds are gradually increased. Early complications include perforation, wound infection or anastomotic leak. Late complications are often related to the abnormal and dysmotile proximal duodenum and include delayed gastric emptying, GORD, adhesive obstruction and blind loop syndrome.

Rotational Deformities (Figs. 35.4 and 35.5)

Malrotation of the gut results when the intestinal rotation and fixation that should occur between the 4th and 12th weeks of gestation does not. Normal rotation 270° anticlockwise around the superior mesenteric artery should result in the DJ

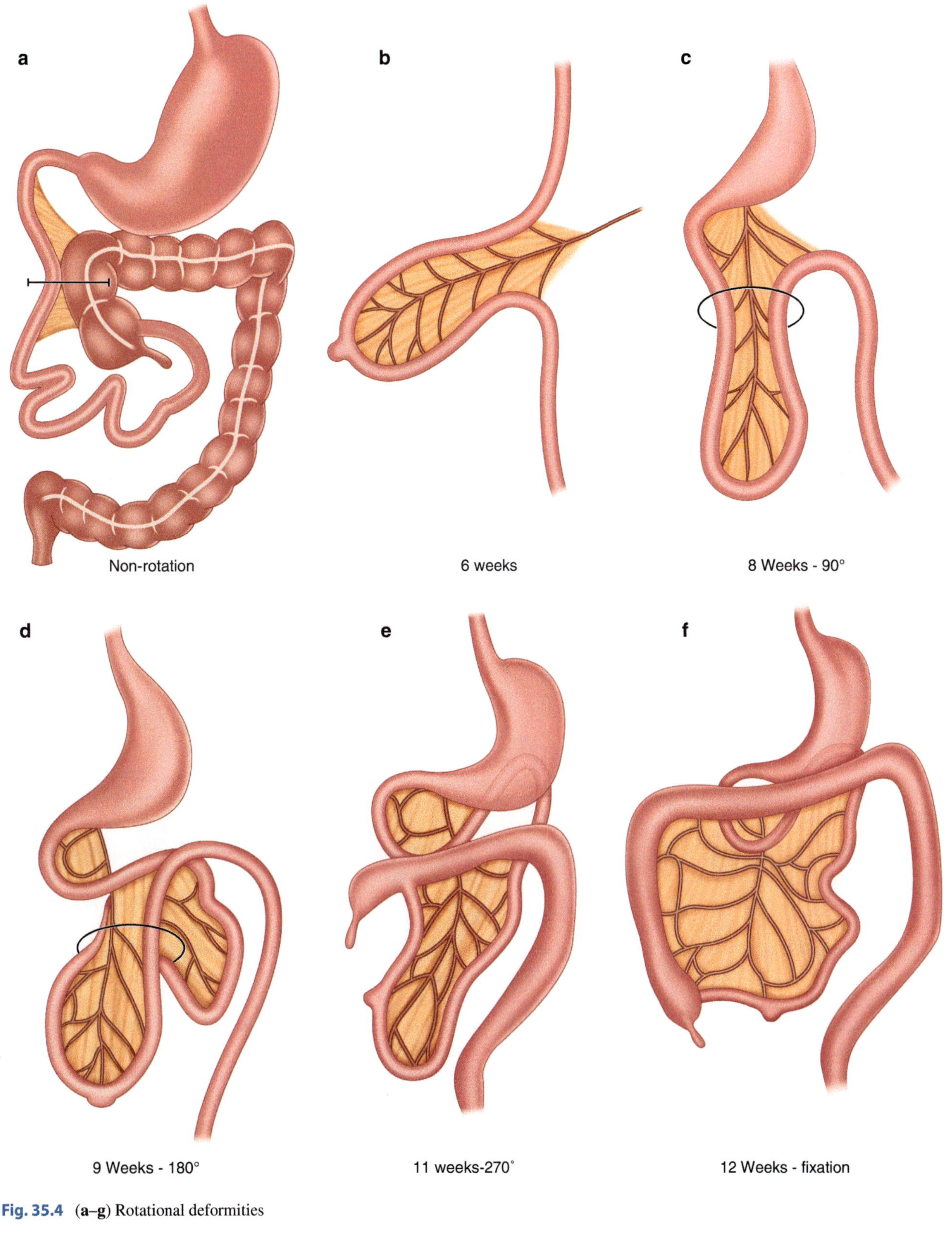

Fig. 35.4 (**a–g**) Rotational deformities

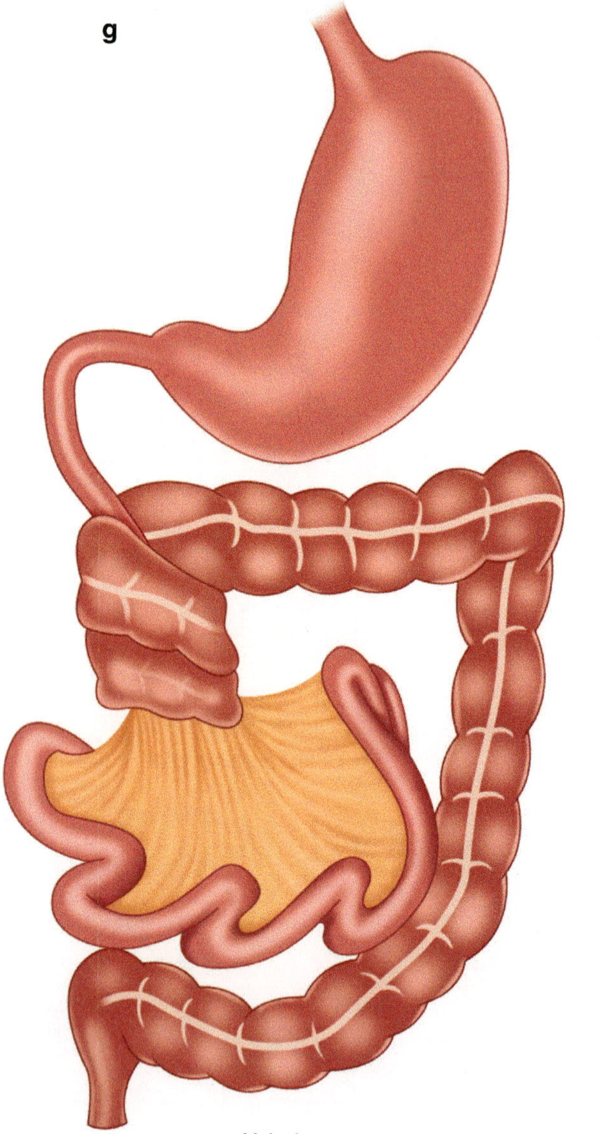

Fig. 35.4 (continued)

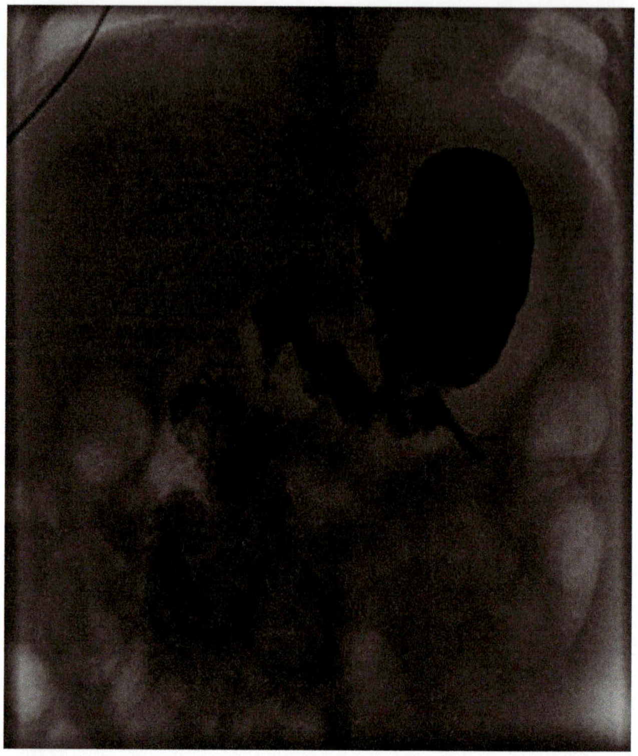

Fig. 35.5 Upper GI contrast showing malrotation and volvulus

flexure lying to the left of the midline, the caecum lying in the right iliac fossa and the transverse colon anterior to the small bowel mesentery. The commonest abnormality with malrotation is that the caecum lies close to DJ flexure, resulting in an abnormally narrow midgut mesentery which is liable to twist, i.e. leading to volvulus. This is a surgical emergency. Any neonate with bilious vomiting has malrotation with volvulus until proven otherwise, i.e. the child needs an urgent upper GI contrast. Unfortunately, it has been shown that not all neonates with bilious vomiting are referred to a paediatric surgeon and that there is a real discrepancy in the public and medical professions understanding of what colour bile is [11, 12]. It is important to reiterate that bile is green and any green vomit needs to be acted on expeditiously. If the child is malrotated, then they should proceed to laparotomy. The bowel is then inspected and any volvulus reduced. A Ladd's procedure is then performed, and in some centres this is now performed laparoscopically although this is not the author's practice [13]. The steps in a Ladd's procedure involve straightening or kocherising the duodenum, widening of the small bowel mesentery and then returning the entire bowel to the peritoneal cavity and placing the large bowel on the left and the small bowel on the right. This is not the anatomically correct position but the best arrangement to prevent any future volvulus. Appendicectomy traditionally is also performed, although some surgeons prefer to leave the appendix in situ and warn the parents that the appendix now resides in the left upper quadrant, should the child develop acute appendicitis in later life.

There are many syndromes related to malrotation. These include heterotaxy syndromes; familial intestinal malrotation; Ivemark syndrome [14]; Rubinstein-Taybi syndrome [15]; prune belly syndrome; Marfan syndrome; trisomies 13, 18 and 21; Cantrell syndrome; and Cornelia de Lange syndrome. There is a current debate as to whether some of these children especially those with heterotaxy should be screened for malrotation and whether this should be treated surgically [16, 17]. Symptomatic children should obviously be investigated urgently and operated on if they are malrotated, but asymptomatic children are more of a surgical quandary especially as many of these children often have

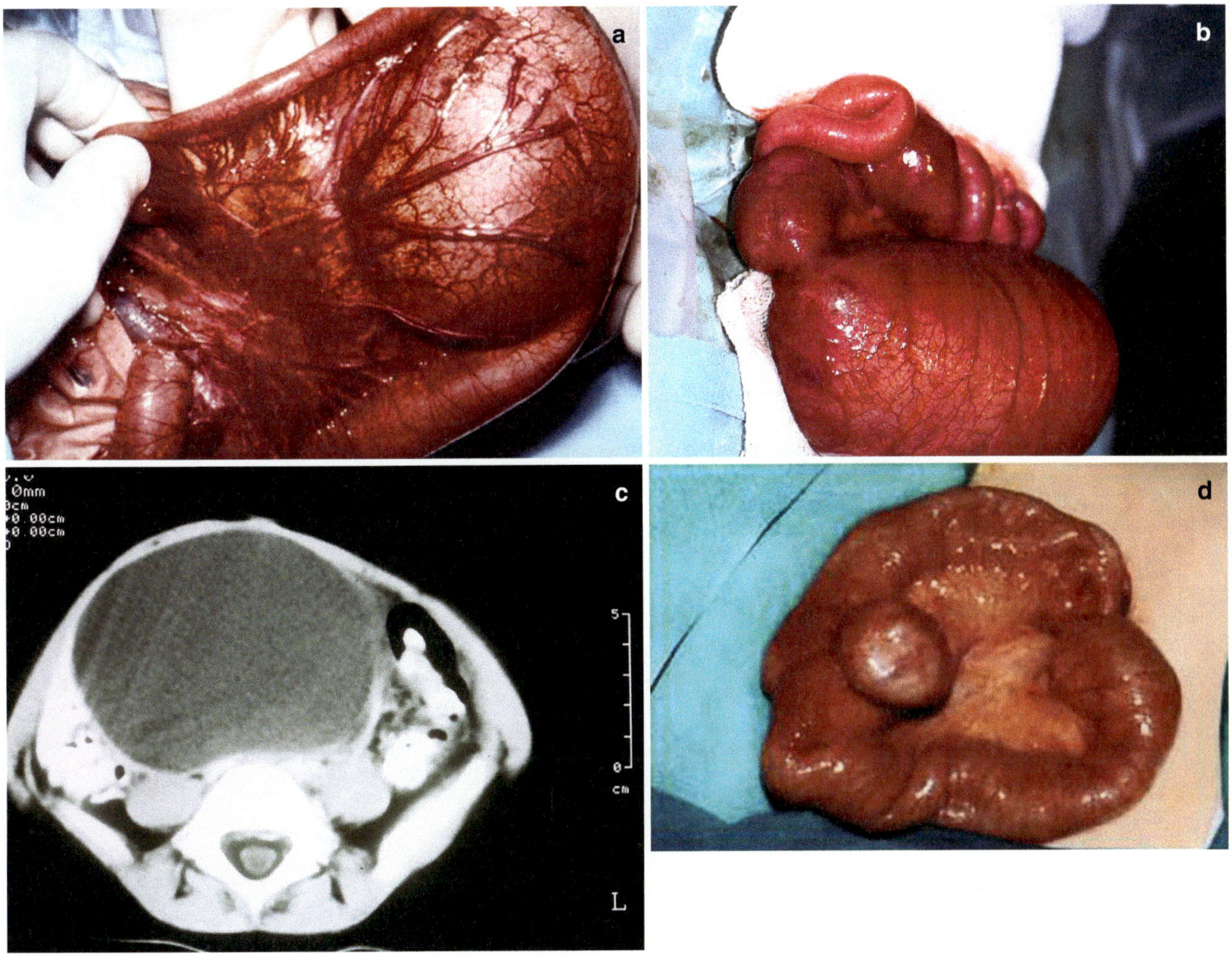

Fig. 35.6 (**a**) Intraoperative appearance of a small bowel duplication. (**b**) Duplication cyst small intestine. (**c**) Omental cyst on CT scan. (**d**) Meckel's diverticulum

other comorbidities making them an increased anaesthetic risk. Certain specific genetic abnormalities have also recently been shown to lead to intestinal malrotation such as inactivating heterozygous mutations in the forkhead transcription factor *FOXF1* [18].

Alimentary Duplications (Fig. 35.6)

Duplications can occur anywhere along the gastrointestinal tract, e.g. thoracic, gastric, pyloric and duodenal, but most cystic duplications in the abdomen are small bowel in origin.

In 1989 Holcomb et al. reported that they commonly arise on the mesenteric border and the duplication shares a common blood supply with that of the neighbouring intestine [19]. They are often suspected antenatally but may present in the neonate with a mass and have been reported to a higher incidence of associated congenital anomalies [20]. The differential diagnosis must include choledochal and ovarian cysts, and the presence of a typical bowel signature on ultrasound should confirm the diagnosis. In the older child, they may present with vague abdominal pain and can be wrongly diagnosed as acute appendicitis. Other complications or sequelae of small bowel duplication include intussusception, perforation or volvulus. Surgical options include resection and primary anastomosis, or if this is not thought to be feasible or safe, then marsupialisation or mucosectomy can be performed.

References

1. Grosfeld JL. Jejunoileal atresia and stenosis. In: O'Neill JA, Rowe MI, Grosfeld JL, Fonkalsrud EW, Coran AG, editors. Pediatric surgery. 5th ed. Louis: Mosby; 1998. p. 1145–58.
2. Dave S, Gupta DK. Jejunoileal atresia. In: Gupta DK, editor. Text book of neonatal surgery. 1st ed. New Delhi: Modern Publishers; 2000. p. 181–91.

3. Grosfeld JL, Ballantine TV, Shoemaker R. Operative management of intestinal atresia and stenosis based on pathologic findings. J Pediatr Surg. 1979 Jun;14(3):368–75.
4. Dalla Vecchia L, Grosfield J, West K, Rescorla F, Scherer L, Engum S. Intestinal atresia and stenosis. Arch Surg. 1998;133:490–7.
5. Kim HB, Lee PW, Garza J, et al. Serial transverse enteroplasty for short bowel syndrome: a case report. J Pediatr Surg. 2003;38:881–5.
6. Bianchi A. Intestinal loop lengthening – a technique for increasing small intestinal length. J Pediatr Surg. 1980;15:145–51.
7. AC W, Sabharwal AJ, MacKinlay GA, Munro FD. Congenital colonic atresia: should primary anastomosis always be the goal? Pediatr Surg Int. 2003;19(1-2):14–7.
8. Baglaj M, Carachi R, MacCormack B. Colonic atresia: a clinicopathological insight into its etiology. Eur J Pediatr Surg. 2010;20(2):102–5. https://doi.org/10.1055/s-0029-1242735.
9. Parmentier B, Peycelon M, Muller CO, El Ghoneimi A, Bonnard A. Laparoscopic management of congenital duodenal atresia or stenosis: a single-center early experience. J Pediatr Surg. 2015;50(11):1833–6. pii: S0022-3468(15)00361-9. https://doi.org/10.1016/j.jpedsurg.2015.05.007.
10. E A, Larsson M, Finkel Y, Karpe B. Transanastomotic feeding tube after an operation for duodenal atresia. Eur J Pediatr Surg. 2002;12(3):159–62.
11. Walker GM, Raine PA. Bilious vomiting in the newborn: how often is further investigation undertaken? J Pediatr Surg. 2007;42(4):714–6.
12. Walker G, Neilson A, Young D, Raine P. Colour of bile vomiting in intestinal obstruction in the newborn: questionnaire study. BMJ. 2006;332:1363. https://doi.org/10.1136/bmj.38859.614352.55.
13. Hagendoorn J, Vieira-Travassos D, van der Zee D. Laparoscopic treatment of intestinal malrotation in neonates and infants: retrospective study. Surg Endosc. 2011;25(1):217–20. https://doi.org/10.1007/s00464-010-1162-3.
14. Stankiewicz P, Sen P, Bhatt SS, Storer M, Xia Z, Bejjani BA, Ou Z, Wiszniewska J, Driscoll DJ, Bolivar J, Bauer M, Zackai EH, McDonald-McGinn D, Nowaczyk MM, Murray M, Shaikh TH, Martin V, Tyreman M, Simonic I, Willatt L, Paterson J, Mehta S, Rajan D, Fitzgerald T, Gribble S, Prigmore E, Patel A, Hrusca A, Rachisan AL, Lucian B, Oprita S, Manole S, Cainap S. Ivemark syndrome-a rare entity with specific anatomical features. Rev Med Chil. 2015;143(3):383–6.
15. Stevens CA. Intestinal malrotation in Rubinstein-Taybi syndrome. Am J Med Genet A. 2015;167A(10):2399–401. https://doi.org/10.1002/ajmg.a.37167. [Epub ahead of print]
16. M C, Borenstein SH, Hornberger L, Langer JC. Heterotaxia syndrome: the role of screening for intestinal rotation abnormalities. Arch Dis Child. 2005;90(8):813–5.
17. Borenstein SH, Langer JC. Heterotaxia syndromes and their abdominal manifestations. Curr Opin Pediatr. 2006;18(3):294–7.
18. Shaffer LG, Carter NP, Cheung SW, Langston C, Shaw-Smith C. Genomic and genic deletions of the FOX gene cluster on 16q24.1 and inactivating mutations of FOXF1 cause alveolar capillary dysplasia and other malformations. Am J Hum Genet. 2009;84:780–91.
19. Holcomb GW, Gheissari A, O'Neill JA, Shorter NA, Bishop HC. Surgical management of alimentary tract duplications. Ann Surg. 1989;209(2):167–74.
20. Y G, Skába R, Kalousová J, Rousková B, Hrîbal Z, Snajdauf J. Alimentary tract duplications in children: high incidence of associated anomalies. Eur J Pediatr Surg. 2004;14(2):79–84.

Gastrointestinal Tract III: Hindgut

36

Tim J. Bradnock

Development of Hindgut

Formation of the Cloaca

*The lower end of the hindgut dilates to form an expanded part called the cloaca (endodermal).

*The cloaca is connected to the umbilicus by the allantois and closed below by the cloacal membrane which is bilaminar, i.e., formed two layers:

(a) Outer ectodermal layer
(b) Inner endodermal layer

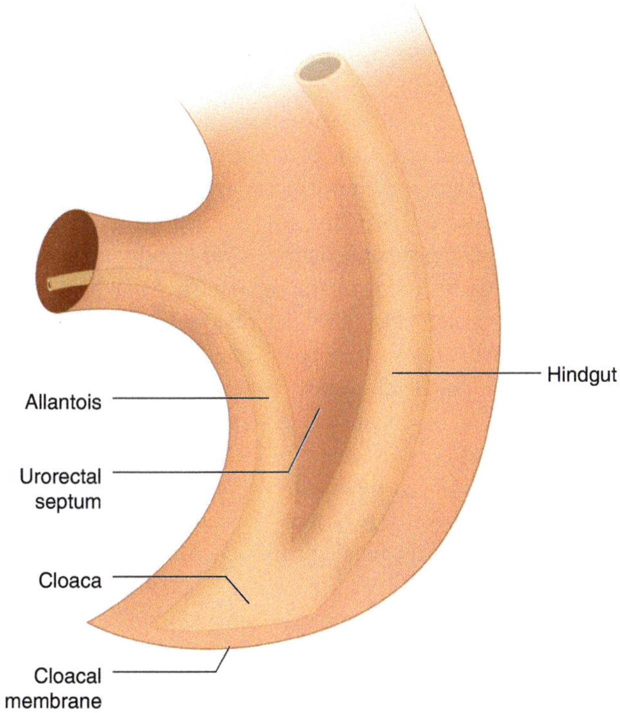

Division of the Cloaca into Two Parts

*The tissue between the hindgut and allantois forms the uro-rectal septum which grows caudally towards the cloacal membrane dividing the cloaca into:

(a) An anterior part called the *primitive urogenital sinus*
(b) A posterior part called the *anorectal canal*

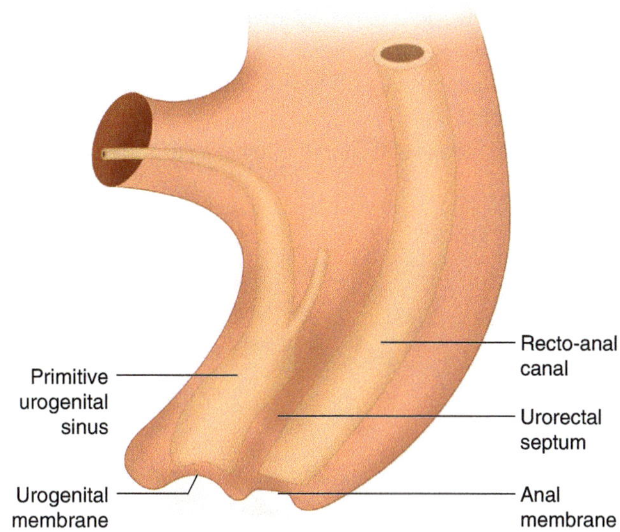

*At the same time the cloacal membrane is divided into urogenital membrane anteriorly and anal membrane posteriorly.

T. J. Bradnock (✉)
The Royal Hospital for Children, Glasgow, UK
e-mail: Timothy.Bradnock@ggc.scot.nhs.uk

R. Carachi, S. H. E. Doss (eds.), *Clinical Embryology*, https://doi.org/10.1007/978-3-319-26158-4_36

Formation of the Proctodeum

An ectodermal depression called the proctodeum is formed opposite the lower end of the recto-anal canal and is separated from it by the anal membrane.

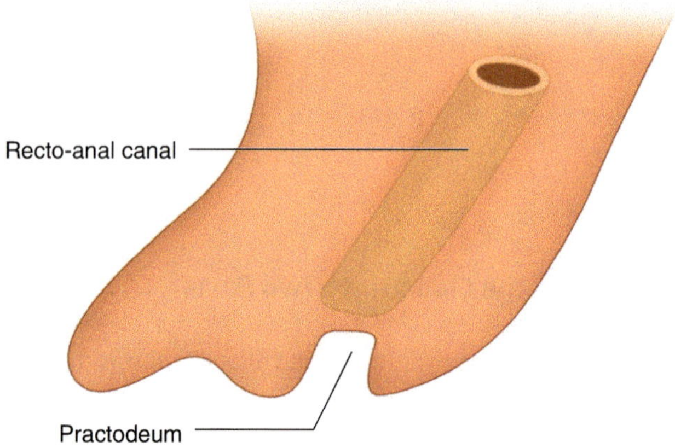

Rupture of the Anal Membrane and Formation of Anal Canal

*The anal membrane finally ruptures and the proctodeum becomes continuous with the recto-anal canal.

*The recto-anal canal will form the rectum + the upper ½ of the anal canal.

*The proctodeum will form the lower ½ of the anal canal.

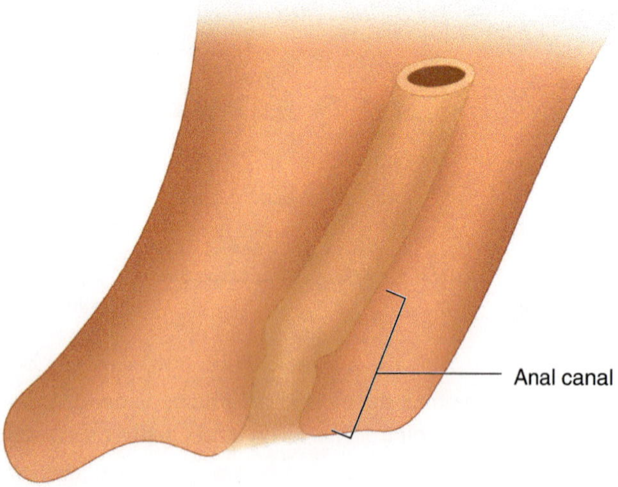

Introduction

Congenital abnormalities of the hindgut can be considered in terms of aberrations of function, anatomy or both. Functional anomalies affecting the hindgut are exemplified by Hirschsprung's disease. In addition, a wide spectrum of quantitative or qualitative gastrointestinal neuromuscular abnormalities have been characterized histopathologically, but there is little consensus regarding whether the observed histopathological abnormalities are the cause of the underlying motility disorder or simply a morphological consequence of a chronic and profound disturbance in gastrointestinal function. Despite recent attempts to standardize the histopathological diagnosis [1] and classification [2] of these anomalies, there remains little consensus regarding whether they represent true clinical entities. The principal congenital anatomical abnormalities affecting the hindgut are alimentary duplications and atresias. The spectrum of anorectal malformations and cloacal anomalies are the most studied of these conditions and are covered in a separate chapter.

Although the embryological transition from the midgut to the hindgut lies at a point two-thirds of the way along the transverse colon, for pragmatic reasons, congenital anomalies affecting any part of the colon or rectum will be considered here.

Congenital Motility Disorders of the Hindgut

Hirschsprung's Disease

Hirschsprung's disease (HD) is the clinical manifestation of an embryological failure of enteric neural crest cells (ENCCs) to appropriately colonize the distal intestine during foetal development [3]. In histopathological terms, this results in an absence of intrinsic parasympathetic ganglia (aganglionosis) in the submucosal and myenteric plexuses of the distal colon and, in most cases, the presence of thickened extrinsic parasympathetic nerve trunks in the myenteric plexus, submucosa and mucosa [4]. Clinically, a functional distal intestinal obstruction ensues, secondary to an inability to achieve co-ordinated propagation of smooth muscle peristalsis and a failure of the distal, aganglionic colon to relax (Fig. 36.1).

The overwhelming majority of infants with isolated HD have normal antenatal scans, although other associated anomalies may be identified in syndromic cases. Foetal bowel dilatation is not a feature of HD, although hyperechogenic bowel has been reported in a few cases, but usually only after retrospective review of the antenatal imaging following a postnatal diagnosis of HD [5]. No antenatal interventions have been described. Over 90% of affected individuals present during the neonatal period [6], classically with the triad of delayed passage of meconium, abdominal distension and bilious vomiting. Around 98.5% of normal term infants pass meconium within 24h of delivery, and the remainder by 48h [7]. Classical teaching, currently incorporated into national guidance on the management of paediatric constipation [8], contends that infants with HD exhibit delayed passage of meconium beyond 48h, but two recent studies [6, 9] suggest that up to 40% of infants with HD,

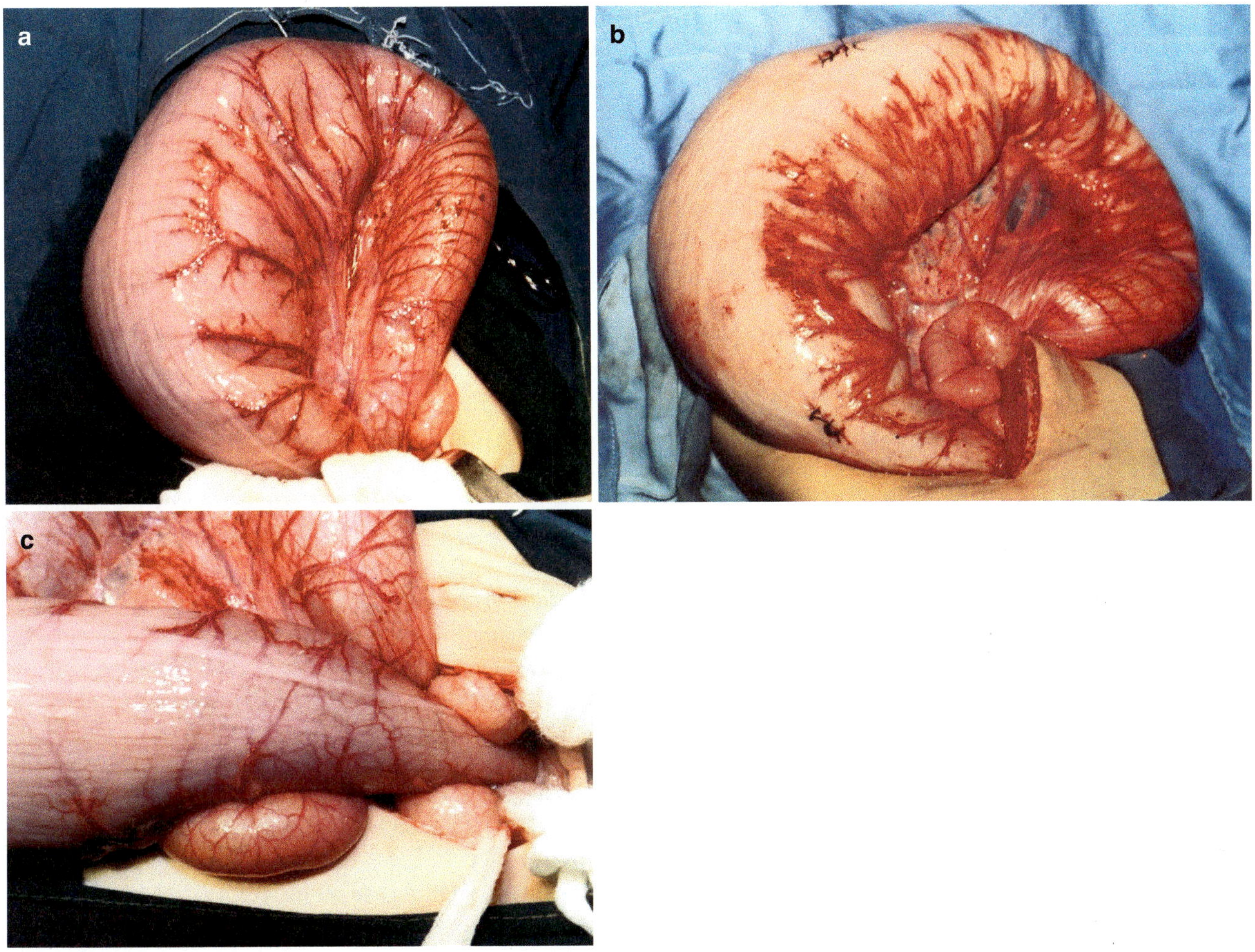

Fig. 36.1 (**a**) Hirschsprung's disease colon. (**b**, **c**) Distal colonic transition zone

including preterm infants [9], will also pass meconium within 48h of birth. HD is four times more common in males than females [10], although normalization of the sex ratio in long-segment and total colonic forms has been reported [11].

Since the distal extent of aganglionosis always lies at the anal canal, HD is classified by the proximal extent of the aganglionic segment. 75–85% of patients have 'classical' HD, with aganglionosis confined to the rectum and sigmoid [12]. In total colonic HD, which affects 9% of patients with HD [13], the aganglionic segment typically comprises the entire colon and a short but variable segment of the ileum. The term 'long-segment HD' is used inconsistently, with some authors using it to describe aganglionosis 'extending to the splenic flexure or transverse colon' [12] and others using it to describe aganglionosis extending into the descending colon [14]. This may in part explain the wide variation in the proportion of long-segment cases reported in large series, accounting for anything between 3.9 and 23.7% of cases [12]. In its most severe form, the aganglionosis extends throughout the entire intestine, a condition called total intestinal aganglionosis (TIA).

Normal Enteric Nervous System Development

To understand the underlying pathogenesis of HD, one must begin by examining normal enteric nervous system (ENS) development. The human ENS is exclusively derived from the neural crest and contains around 500 million neurons and 2 billion supporting glia, distributed along the entire bowel length in two interconnected layers [3]. The myenteric plexus runs between the longitudinal and circular muscular layers of the bowel wall and is principally responsible for affecting the co-ordinated motor activity of enteric smooth muscle required for normal motility. The submucosal plexus lies deep to the mucosa and is primarily concerned with sensing the luminal environment, regulating blood flow and controlling epithelial function [15].

Normal development of the ENS relies on the ability of neural crest cells to migrate along the entire gastrointestinal tract, proliferating as they go; before differentiating into at least 14 different neuronal subtypes, as well as glial cells; and successfully colonizing the bowel wall, where they must then survive and become functionally active

[16]. Neural crest cells leave the vagal neural crest and undergo craniocaudal migration to populate the entire length of the developing bowel, entering the foregut by week 4 of gestation and the terminal hindgut by week 7 [17]. There appears to be a time-critical element to the migration of NCCs, as the colonic population of enteric nervous system cells are principally derived from the trans-mesenteric migration of NCC from the future small bowel to the future colon, which occurs at a point in development when these structures lie in close proximity [18]. In addition to the craniocaudal migration of vagal NC cells, the sacral neural crest additionally contributes to the innervation of the colon as far as the caecum [19]. The myenteric plexus is the first to form, followed by the submucous plexus, which is populated by neuroblasts migrating across the circular muscle layer of the bowel wall [20]. Proliferation and maturation of ENCCs begins during migration and continues up to 1 month after birth [15]. A key element of ENS maturation is the formation of ganglia which occurs later, between weeks 14 and 19 [21]. The precise mechanisms that drive clustering of nitrergic neurons into ganglia remain unclear.

Multipotent neural crest cells (NCCs) have varying differentiation potential dependent upon their position within the neural crest. For example, NCCs from the cranial end of the neural crest give rise to the cranial nerves; head mesenchyme including the cartilage, bone and connective tissue; the inner ear sensory hair cells; and melanocytes [15]. Other NCCs migrate extensively and give rise to the adrenal medulla and the autonomic nervous system. The term neurocristopathy is used to define the heterogeneous group of clinical conditions that result from aberrant NCC migration and differentiation. HD is a neurocristopathy that occurs either as a result of the premature arrest of the craniocaudal migration of vagal NCCs or due to a failure of proliferation, survival or maturation of these cells. The rare occurrence of zonal aganglionosis or skip lesions supports the theory that local factors in the bowel wall may promote segmental apoptotic loss of neuroblasts that have successfully completed migration [22]. The extensive and varied end-organ colonization with cells derived from a common neural crest origin helps explain the wide range of associated anomalies found in the major neurocristopathies associated with HD.

A Brief Overview of the Developmental Genetics of HD

HD is a non-Mendelian, oligogenic disease, with at least 16 known genes implicated in the pathogenesis of HD at the time of writing [15]. It is thought that these known mutations account for about 50% of the familial cases and 20% of the sporadic cases of HD [23]. All genes contain coding and non-coding sequences. It is now recognized that mutations in the non-coding regions of genes, such as RET, confer much of the susceptibility to HD [23]. The penetrance of the HD is low, but is also sex and extent of aganglionosis dependent, with a higher rate of penetrance in males and infants with long-segment or total colonic aganglionosis [14]. HD has variable expression with the same genetic mutation implicated in giving rise to distinctly different phenotypes and lengths of aganglionosis [24]. The genetic mutations associated with HD occur in the so-called susceptibility genes, in that they increase an individual's risk of HD, but do not in themselves predict the occurrence of HD [24].

There are two main signalling pathways implicated in the pathogenesis of HD. The Rearranged during Transfection (RET) gene and its four ligands, of which glial-derived neurotrophic factor (GDNF) is the most important, play a critical role in ENS development [4]. RET is the major susceptibility gene for HD accounting for 50% of familial and 15% of sporadic cases [14] and is also implicated in multiple endocrine neoplasia type 2 (MEN2A) and familial medullary thyroid carcinoma (FMTC). The other main pathways comprise the endothelin receptor type B (EDNRB) gene and its ligand endothelin 3 [15]. Both pathways play a critical role in the survival, proliferation, differentiation and migration of NCCs. The loss of an adequate pool of precursor NCCs during early development has been implicated in the pathogenesis of aganglionosis [23]. Recent studies suggest that RET may play a central role in maintaining an adequate pool of precursor NCCs for intestinal invasion and that the variable penetrance of RET may be explained through this effect on cell numbers [23]. A more detailed discussion of this developmental genetics of HD is beyond the scope of this chapter, and the reader is directed to the following excellent reviews of the subject [4, 14, 15].

Associated Anomalies

Overview

The traditional view is that HD occurs as an isolated anomaly in around 70% of cases with 12% having a chromosomal abnormality and a further 18% of cases having at least one additional congenital anomaly [14]. However, a recent study combining prospective, cross-sectional methodology with a diagnostic algorithm that included urinary tract and cerebral ultrasound scans, a cardiology assessment with echocardiography, audiometry, ENT and ophthalmology assessment with additional specialist evaluations based on clinical features identified 112 associated anomalies in 61/106 (58%) infants [25]. The predominant anomalies identified were ophthalmic anomalies (typically refraction anomalies) in 43.4%, visual impairment in 9.4%, congenital anomalies of the kidney and urinary tract (CAKUT) in 20.7%, congenital heart disease in 4.7%, hearing impairment or deafness in 4.7%, CNS anomalies in 2.3%, chromosomal or syndromic anomalies in 8.5% and other anomalies in 12%. The overall percentage anomaly rate was consistent across the sexes and by length of aganglionosis. Only 28/61 (46%) of these infants had their anomaly identified prior to the study, suggesting under-reporting of anomalies in studies using retrospective methodology. Pini-Prato et al. propose a diagnostic algorithm, which is dichotomized by the presence or absence of an underlying syndrome [25].

Congenital Anomalies of the Kidney and Urinary Tract (CAKUT)

The link between CAKUT and HD is based on the common genetic origins of the ENS and urinary tract, in particular because of the importance of the RET-GDNF signalling pathway [26]. Many phenotypes have been described including renal agenesis/ dysplasia, multicystic dysplastic kidneys (MCDK); pelviureteric junction obstruction (PUJO); hydronephrosis; vesicoureteric reflux; nonobstructed, nonrefluxing primary megaureter; duplex collecting system; posterior urethral valves; and horseshoe kidney [26]. A recent systematic review of 5693 patients with HD found a 3.6% incidence of CAKUT. This suggests a considerable degree of under-reporting compared to the prospective Pini-Prato study, which found a 20.7% rate of association [25]. All infants with HD, particularly those with syndromic HD, should undergo screening for urological anomalies, and this should include urinary tract ultrasound.

Cardiac Anomalies

Normal cardiac outflow septation shares a common neural crest origin with the ENS [14]. Retrospective studies report that around 5% of infants with HD have associated cardiac defects, most commonly septation defects (atrial, ventricular or atrioventricular) and conotruncal developmental defects [27]. A recent prospective study has suggested that overall as many as 8% of HD infants and 3.8% of non-syndromic infants with HD have a cardiac anomaly and that 4.5% of all infants with HD require cardiac surgery [27]. Based on these results, the authors recommend echocardiographic assessment of *all* infants with HD.

Cardiac anomalies occur in 20–80% of syndromic HD cases. The strongest association is seen in Down syndrome and Mowat-Wilson syndrome, where 70 and 24% of infants, respectively, have an associated cardiac anomaly [28].

Central Nervous System (CNS) Anomalies

The CNS and ENS develop in the same spatio-temporal timeframe and are governed by the same neural growth factors [29]. A number of syndromes associated with HD exhibit neurological impairment, and the association between anencephaly and aganglionosis is absolute [29]. More subtle cognitive disturbances and other issues such as attention deficit disorder and epileptic seizures have also been linked to HD [29]. As a result, Moore recommends that all children with HD should have a screening neurological examination and age-appropriate developmental assessment as part of their long-term follow-up [29]. Routine MRI brain or EEG are not indicated but may be considered in selected cases.

Syndromic Associations of HD

It is important to recognize that many of the syndromes associated with HD are neurocristopathies, with a shared developmental neural crest origin. Examples of neurocristopathies include multiple endocrine neoplasia type 2A and congenital central hypoventilation syndrome, which is associated with HD in 20% of cases (Haddad syndrome) [14]. The most common syndrome associated with HD is Down syndrome (DS). Infants with DS have a 40 times increased risk of HD. 7.3% of infants with HD have DS and 2.6% of infants with DS have HD [30]. Higher rates of post-operative enterocolitis and soiling have been reported in infants with DS associated HD [30]. The other principal syndromic associations of HD are summarized in Table 36.1. For further reading, the reader is directed to the excellent overview of the syndromic associations of HD provided by Amiel et al. [14].

The association between MEN2A and HD requires special mention, mainly because of the risk of early and aggressive medullary thyroid carcinoma in infants with MEN2A. Both may arise due to mutations in the RET proto-oncogene. However, their co-occurrence is intuitively unexpected since MEN2A relies on 'gain of function' in RET and HD occurs due to 'loss of function' in RET [31]. Around 7% of infants with MEN2A have HD [31], and around 3% of infants with HD have MEN2A [14].

Table 36.1 Major susceptibility genes in isolated and syndromic HD

Syndrome	Related gene (s)	Gene product	Ch.	Clinical features
Neuroblastoma	PHOX2B	Paired-like homeobox 2b transcription factor	4p12	Most common solid tumour of infancy. Common NCC origin with HD and occasional association
Shah Waardenburg syndrome type 4 (WS4)	EDNRB EDN3 SOX10	Endothelin-B receptor Endothelin-3 SRY-related HMG box gene 10 transcription factor	13q22 20q13 22q13.1	Genetically heterogeneous (EDNRB, EDN3, SOX10) Pigmentary anomalies (white forelock, patchy hypopigmentation and iris hypoplasia), sensorineural deafness. If SOX10 mutation—also seizures, ataxia and central/ peripheral demyelination
Congenital central hypoventilation (CCHS)	PHOX2B	Paired-like homeobox 2b transcription factor	4p12	Rare. Failure of autonomic regulation of respiration Predisposition to NCC-derived tumours. Haddad syndrome is associated with HD and CCHS (20% of all CCHS)
Multiple endocrine neoplasia 2A (MEN2A)	RET/GFRA	Receptor tyrosine kinase/ GDNF family receptor alpha-1	10q11.2	Medullary thyroid carcinoma, phaeochromocytoma, parathyroid hyperplasia
McKusick-Kaufman	MKKS	McKusick-Kaufman/ Bardet-Biedl syndromes putative chaperonin protein	20p12	Hydrometrocolpos, glanular hypospadias and prominent scrotal raphe, postaxial polydactyly, congenital heart defect
RET mutation	RET	Receptor tyrosine kinase	10q11.2	Coding and non-coding region mutations. Implicated in 50% of familial and 15% of sporadic cases
Bardet-Biedl	MKKS	McKusick-Kaufman/ Bardet-Biedl syndromes putative chaperonin protein	20p12	Progressive pigmentary retinopathy, truncal obesity, male hypogenitalism, female genitourinary anomalies, mild mental retardation, postaxial polydactyly
Mowat-Wilson	ZFHX1B (SIP1)	Zinc-finger homeobox 1B (or SMAD interacting protein-1) transcription factor	2q22	Classic facial gestalt (square face, hypertelorism, broad nasal bridge, large flaring eyebrows, uplifted earlobes) hypospadias + renal anomalies, epilepsy, microcephaly, agenesis of corpus callosum and severe mental retardation
Smith-Lemli-Opitz	DHCR7	7-dehydro-cholesterol reductase	11q12-q13	Growth restriction (foetal and postnatal), microcephaly, severe mental retardation, hypospadias, second and third toe syndactyly, facial dysmorphism
BRESHEK	MBTPS2	Membrane-bound transcription factor peptidase site 2	Xp22.12-p22.11	**B**rain abnormalities, **r**etardation, **e**ctodermal dysplasia, **s**keletal malformation, **HD**, **e**ar/eye anomalies, **k**idney dysplasia. <10 cases. All male
Goldberg-Shprintzen	KIAA1279	Kinesin family member 1 binding protein	10q22.1	Mental retardation, polymicrogyria, microcephaly, hypotonia, marfanoid habitus, coloboma, facial dysmorphism (hypertelorism, prominent nose and sparse hair)
HD with limb anomalies	–	–	–	Rare. Combinations of polydactyly, brachydactyly or nail hypoplasia ± cardiac anomaly ± dysmorphism
Down syndrome	–	–	–	Trisomy 21. Accounts for 7% of HD cases. 2.5% of DS infants have HD. Associated with duodenal atresia, complex atrioventricular septation anomalies, imperforate anus
Cartilage-hair hypoplasia syndrome	RMRP	RNA component of mitochondrial RNA processing endoribonuclease	9p13	Short limb dwarfism, metaphyseal dysplasia, fine blond hair and immunodeficiency. HD in ~10% of cases

Adapted from Kapur RP [4], Puri [12], Amiel [14] and Gariepy CE [24]

Genetic Counselling

The overall familial recurrence rate for isolated HD is between 4 [14] and 7.6%, with the highest risk of recurrence seen in infants with total colonic aganglionosis (TCA), where up to one in five cases is recurrent [13]. The relative risk for siblings of an infant with isolated HD lies somewhere between 200 and 380. For isolated non-syndromic HD, the risk of recurrent HD in future offspring can be calculated based on the sex and extent of aganglionosis of the proband, as well as the sex of the sibling. For example, if the proband is a female with long-segment disease, male and female siblings, respectively, have a 33 and 9% chance of having HD. In contrast, for a male proband with short-segment HD, the chances of a male and female sibling, respectively, having HD are 5 and 1% [14]. Genetic counselling should be offered to all parents of an infant with HD. This allows the possibility of systematic examination for associated syndromes and anomalies as well as mutational analysis for specific RET gene mutations which would predispose the infant to an increased risk of multiple endocrine neoplasia (MEN) type 2A (exons 10 and 11 of the RET gene) [31].

Outcomes

The main arbiter of outcome in HD is long-term bowel function after surgery. Successful definitive surgery (pull-through procedure) for HD relies on accurate histopathological determination of the transition zone, mobilization and excision of the affected colon, a safe rectal dissection and some form of colo-anal or colorectal anastomosis. The main trends in the surgical management of HD in the UK and Ireland have been for increasing the use of minimally invasive techniques. Currently, the predominant approach to infants with right-sided disease is a staged, open Duhamel procedure, whilst the majority of infants with left-sided disease undergo a single-stage, laparoscopic-assisted or purely transanal, endorectal pull-through [32].

A full discussion of outcomes is beyond the scope of this chapter. There is some evidence that HD represents a field change abnormality with persistent but subtle abnormalities in the residual (non-resected) ENS and in the protective mucin layers that coat the bowel wall. Deficiencies in mucosal immune function and alterations in the normal gut microbiome may predispose to enterocolitis [33], which affects 11–50% of infants with HD [34]. Outcomes may be related to the type of surgery but are not worse in single-stage compared to multistage operations [35]. Key determinants of outcome are meticulous surgical technique, including avoidance of overstretching the anal sphincters and not anastomosing the colon too close to the dentate line, as these mistakes will render the patient incontinent; the extent of aganglionosis; and the presence of associated syndromes and other co-morbidities. A detailed review of outcomes is presented in the thoroughly excellent book by Puri and Holschneider (see further reading). Although improvement in continence may occur with time after surgery [36], up to 60% of infants with rectosigmoid HD and 86% of those with TCA have some degree of incontinence at between 7 and 17 years after definitive surgery [37]. Mortality has declined from 30% to around 3–4% in recent years, although the odds ratio of death remains around 2.5 times higher in infants with DS and HD [30].

Registries

The Hirschsprung Disease Research Collaborative (HDRC), a multicentre research study led by Dr. Aravinda Chakravarti at the McKusick-Nathans Institute of Genetic Medicine at Johns Hopkins University (https://hdrcstudy.org/documents), are seeking to enrol 3000 patients in a genetic study examining the role of genetic mutations in the pathogenesis of HD. This major undertaking will provide valuable insight into the genotype-phenotype relationship in HD and establish a large biobank for future studies. We will await the findings of this study with great interest.

Future Therapies

The inconsistent long-term outcomes after surgery for HD have prompted clinicians and scientists to explore novel therapies to reconstitute the defective ENS. The field of ENS stem cell therapy shows great promise. This work is predicated on the premise that ENS stem cells can be isolated, optimized and transplanted into the distal colon of infants and children with HD with restoration of normal ENS function. An excellent recent white paper by the leaders in the field summarizes the state of the art and the obstacles that must be overcome for this promising work to become a clinically relevant therapy [3].

Allied Disorders of Hirschsprung's Disease

Background

The concept of 'pseudo-Hirschsprung's' disease was proposed by Ravitch in 1958 [38], who used the term to describe a group of conditions which behaved in clinical terms like HD, but which lacked the classic pathological and radiological hallmarks of that condition. Since then, the terminology used to describe and classify these conditions has continued to evolve and prompted much debate. The terms 'allied disorders' or 'variants of HD' are commonly used. Puri proposed a diagnostic algorithm to standardize the classification of these conditions, based on an initial suction rectal biopsy with haematoxylin and eosin and acetylcholinesterase staining, followed by anorectal manometry, an assessment of the response to standard medical therapy and, in selected cases, a full-thickness rectal biopsy with the application of additional histological, immunohistochemical and electron microscopic techniques (Table 36.2) [39].

Controversies

However, the status of these conditions and their acceptance has differed over time and between centres. Compared to the objective clarity with which a diagnosis of HD can be made, attempting to diagnose a heterogeneous group of conditions

Table 36.2 Proposed variants of HD (adapted from Friedmacher and Puri) [40]

Intestinal neuronal dysplasia (IND)
Intestinal ganglioneuromatosis
Isolated hypoganglionosis (congenital and acquired)
Immature ganglia
Absence of the argyrophil plexus
Internal anal sphincter achalasia (IASA)
Megacystis microcolon intestinal hypoperistalsis syndrome (MMIHS)

characterized by either *quantitative* or *qualitative* neuronal abnormalities of the ENS is extremely difficult [40]. To further confound the issue, poorly understood histopathological abnormalities have been described and then given 'disease status', without a full understanding of the normal values of the cells on which a diagnosis depends or an appreciation that these normative values differ by colonic site and patient age [41]. Furthermore, different laboratories have used different reference ranges and staining algorithms in the assessment of these conditions. The clinico-pathological correlation and understanding of the natural history of these putative conditions is poor, and attempts to frame subtle histopathological abnormalities as the architects of clinically significant motility disorders are speculative [3]. In an attempt to provide normative control values for ganglion cell density and the number of ganglion cells per ganglion, Coerdt et al. [41] examined age-matched cadaveric controls. They found that both the density of ganglia and the number of ganglion cells per ganglia fall with age [41]. By applying one consensus conference criterion [42] for a diagnosis of IND type B (more than 10% of observed ganglia must be giant ganglia comprising more than seven ganglion cells), all of the normal specimens they examined would have been diagnosed as IND type B, since 20% of neonatal and 11% of adult ganglia were 'giant' [41].

To try and overcome these problems, an international working group (the Gastro 2009 IWG) of experts from the fields of pathology and paediatric and adult gastroenterology and surgeons with a special interest in neurogastroenterology, came together to classify neuromuscular GI disorders into what has become known as the London classification [2]. They have also defined the techniques for histopathological assessment, interpretation and reporting of intestinal neuromuscular disorders [1]. Subsequent work by the group suggested that the clinical utilization of inaccurate terminology and inconsistent diagnostic criteria remained common, and this prompted an excellent position paper, which summarizes the important information from the group's previous papers and places this in an accessible form for the practising clinician [40].

Ultrashort Segment HD

Ultrashort-segment HD or internal anal sphincter achalasia has been inconsistently defined with the required length of aganglionic segment varying between 2 and 10 cm [43]. The possibility of an accurate diagnosis is further confounded by the fact that the normal length of aganglionosis above the dentate line has not been defined. The other widely held diagnostic criterion of an absence of the normal recto-anal inhibitory reflex (RAIR) on anorectal manometry has also been questioned, based on the risk of a false-positive result using a 'one-sized balloon fits all approach' in infants with a mega-rectosigmoid [44]. The IWG conclude that the existence of ultrashort-segment HD remains debatable [40]. Current putative but controversial treatments for this condition include botulinum toxin A injection to the internal anal sphincter or myectomy.

Intestinal Neuronal Dysplasia

The term IND was first coined by Meier-Ruge in 1971 [45]. Since then, the 'diagnostic' histopathological criteria for IND type B have changed frequently and have been applied inconsistently, leading to doubt about its existence [40]. The current pathological diagnostic criteria proposed by the IWG rely on the finding of giant ganglia (comprising >eight neurons per ganglia) in more than 20% of at least 25 submucosal ganglia, in a patient older than 1 year of age [2]. These changes can occur in isolation or in the context of HD. It is thought that these findings may represent a developmental stage, since they are more common in preterm infants and improve with time, both clinically and in histopathological terms [40]. For these reasons, even authors who firmly believe in this clinical entity, advocate conservative management of the constipation with laxatives and enemas and avoidance of major resectional surgery [39]. Martucciello attempted to classify IND type B into two types—rectocolonic, which was the most common form, and diffuse. Associated anomalies, mainly of the GI tract, were identified in 80% of diffuse cases and 20% of rectocolonic cases [46].

Intestinal Ganglioneuromatosis

Ganglioneuromatosis is a feature of MEN2B, an autosomal dominant cancer predisposition syndrome, associated with gain-of-function mutations in RET. In this condition, the excessive neuronal tissue (ganglioneuromatosis) is found throughout the GI tract from the mouth to the anus and can be biopsied to confirm the diagnosis [40]. GI dysmotility may be the first recognizable feature, and this should prompt a genetic review looking for the specific activating mutation in RET, which confers the risk of medullary thyroid carcinoma [47].

Intestinal Hypoganglionosis

Hypoganglionosis is a controversial entity, which has been reported as an isolated phenomenon or as part of the transition zone of infants with HD. There is no accepted lower limit of normal ganglion cell density, and for this reason the IWG recommends that each reference centre must establish its own normal ranges for children and that the diagnosis should only be made by an expert pathologist based on the analysis of a donut of bowel measuring at least 1 cm [40]. An important confounding variable for this condition, whose diagnosis relies on quantifying a single component of the enteric nervous system, is that even if the absolute number of ganglion cells remains constant with age (which it appears they do not) [41], the density of these cells will be reduced by any condition, such as chronic idiopathic constipation, that promotes bowel wall distension [44].

Megacystis Microcolon Intestinal Hypoperistalsis Syndrome

Megacystis microcolon intestinal hypoperistalsis syndrome (MMIHS) is a rare, almost uniformly fatal congenital condition, characterized by the finding of gross abdominal distension due to a hugely dilated but non-obstructed foetal/neonatal urinary bladder, in the presence of a microcolon with decreased or even absent intestinal peristalsis [48]. The cause of MMIHS was initially thought to be neurological, but it has now been shown to be the result of mutations in the smooth muscle actin (ACTG2) gene [49]. A systematic review found an overall survival rate of 19%, with survival to 24 years of age reported with home TPN or multi-visceral transplantation [48].

Congenital Anatomical Abnormalities of the Hindgut

Duplications of the Hindgut

Based on neonatal autopsy studies, the overall incidence of enteric duplications is thought to be around 1 in 4500 [50]. Colorectal duplications comprise about one fifth of all intestinal duplications (15% colonic and 6% rectal) [51]. There is no clear gender predilection. Intestinal duplications may be diagnosed on a routine foetal anomaly ultrasound scan or antenatal foetal MRI from 20-week gestation [52]. The differential diagnoses of a foetal intra-abdominal cyst include mesenteric, ovarian or choledochal cysts and anterior myelomeningocoele. Enteric duplications rarely cause problems antenatally or require foetal intervention and do not routinely need to be delivered in a tertiary centre [52]. It has been suggested that the antenatal diagnosis of enteric duplication cysts allows for earlier postnatal investigations and treatment before the onset of symptoms or complications [53], although the natural history of asymptomatic lesions has not yet been fully characterized to validate this contention. 75% of the lesions that become symptomatic will do so by the age of 2 years, with symptoms dependent on both the location of the lesion and the lining of the duplication [54]. In general the lining of the duplication will match the mucosa of the adjacent, normal bowel. Around 30% of duplications are lined by ectopic tissue [51], most commonly gastric mucosa, although exocrine or endocrine pancreatic tissue has also been described. Acidic secretions from the ectopic gastric mucosa may cause peptic ulceration of the adjacent bowel with significant gastrointestinal bleeding. Compared to other sites, the occurrence of ectopic gastric mucosa is less common in hindgut duplications [54], occurring with a frequency of around 10%. Most colonic duplications occur in the caecum, are cystic in nature and lie on the antimesenteric bowel wall [51]. Tubular colonic duplications are extremely rare and highly variable [55]. A complete colorectal duplication (affecting the entire colon and rectum) may be associated with duplication of other pelvic viscera including the bladder, uterus, vagina and anus or with an abnormal fistulous opening on the perineum or within the vestibule or an anorectal atresia of either or both lumens [56].

Enteric duplications are generally thought to occur between weeks 4 and 8 of embryonal development [52]. Several theories have been proposed to explain this disordered morphogenesis, including aberrant recanalization; partial twinning, which may explain best the co-occurrence of hindgut and genitourinary duplications (Lewis 1961); and Bentley's split notochord theory, which may explain best the association of enteric duplications and spinal anomalies [51]. The caudal duplication syndrome describes the most severe form of this anomaly, where multiple structures derived from the notochord and cloaca are duplicated to varying extent [57]. To date, no single theory has provided a convincing developmental explanation for the wide spectrum of anatomical aberrations and associated anomalies observed, and the exact aetiology should be considered unresolved. No genetic predisposition for intestinal duplications has been identified, and there are no known syndromic associations.

In the context of hindgut duplications, multiple associated anomalies have been described. 80% of tubular colonic duplications have an associated anomaly, most commonly bladder or external genital duplication, and 50% of cases have a fistulous opening, either onto the perineum or the genitourinary tract [55, 58]. Screening for other anomalies should be tailored to the individual and can be guided by the perineal anatomy [55] but may include vertebral X-rays or an MRI spine in the presence of a suspected spinal anomaly; urinary tract ultrasound, micturating cysto-genito-urethrogram, with cystovaginoscopy if genitourinary abnormalities are suspected; and contrast studies to confirm normal rotation and delineate the anatomical arrangement of the duplicated hindgut.

Enteric duplications may become symptomatic due to distension of the lesion causing pain, intestinal obstruction related to a mass effect or with complications related to the presence of an ectopic mucosa, such as perforation or bleeding. All of these cases are managed surgically. Although the management of asymptomatic lesions is less uniform, many authors advocate elective surgery in the first 6 months of life, to prevent the occurrence of complications and to remove a reported risk of malignant transformation [59]. Cystic hindgut duplications can usually be enucleated or resected without difficulty. Tubular duplications can be extremely long, affecting the entire colon and rectum. In this case, the lesion may be managed with simple laxatives, provided the duplication is open distally. In lesions with a closed distal end, this may be opened and allowed to drain in to the native bowel via an end-to-end or side-to-side anastomosis. A conservative approach in this selected group may be reasonable as the risk of ectopic gastric mucosa is low and a colorectal carcinoma has not been described in these lesions once opened distally [51]. A full discussion of the various surgical approaches to the more complex hindgut duplication anomalies is beyond the scope of this chapter, but any reconstruction should be discussed in the context of a multidisciplinary team of surgeons, urologists, radiologists, and neonatologists. Regardless of the approach, it is important to recognize that the duplication and adjacent native bowel share a common blood supply and this must be preserved during resection and that these infants may suffer additional morbidity related to associated urogenital abnormalities.

Rectal duplications are rare, usually presacral, lying posterior to the native rectum, and may communicate with the perineum via a fistulous connection (Fig. 36.2). They may cause constipation due to a mass effect and can mimic HD [55]. Much information regarding the internal anatomy of these lesions can be gleaned by careful perineal inspection, and a classification system for these anomalies has been proposed based on the status of the colon proper and the duplicated colon in terms of their respective association with a normal or imperforate anus or a fistula [60]. Surgical approaches to rectal duplications include transanal marsupialization with division of the septum between the rectum and cyst or complete resection via a posterior sagittal approach, usually with colostomy cover [51, 54].

True anal canal duplications are second anal orifices which lie posterior to the true anus and end blindly without connection to the rectum and with the normal histological features of an anus [61]. They occur far more commonly in females, and around a third to a half have associated anomalies including cleft lip or palate, intestinal malrotation, sacral dysgenesis, presacral mass including teratoma, dermoid cyst or anterior meningocele [61], all of which may occur as part of Currarino's triad. Two theories have been proposed to account for anal duplications. The first proposes that anal duplications arise due to duplication of the dorsal cloaca in the early developmental stage [62]. The second proposes that this is a late embryonic defect, due to excessive elongation of the dorsal cloacal membrane, which then recanalizes to form a second, blind-ending anal canal, after development of the external anal sphincter [63]. The diagnosis can be confirmed by fistulogram, with an MRI pelvis recommended to exclude associated presacral anomalies. Most authors report excision via a posterior or perineal approach [61].

Colonic Atresia

The incidence of colonic atresia is thought to be between 1 in 10,000 and 1 in 66,000 live births [64], accounting for 1.8–15% of all intestinal atresias [65]. Distal intestinal atresias may be suspected on antenatal ultrasound due to gross bowel

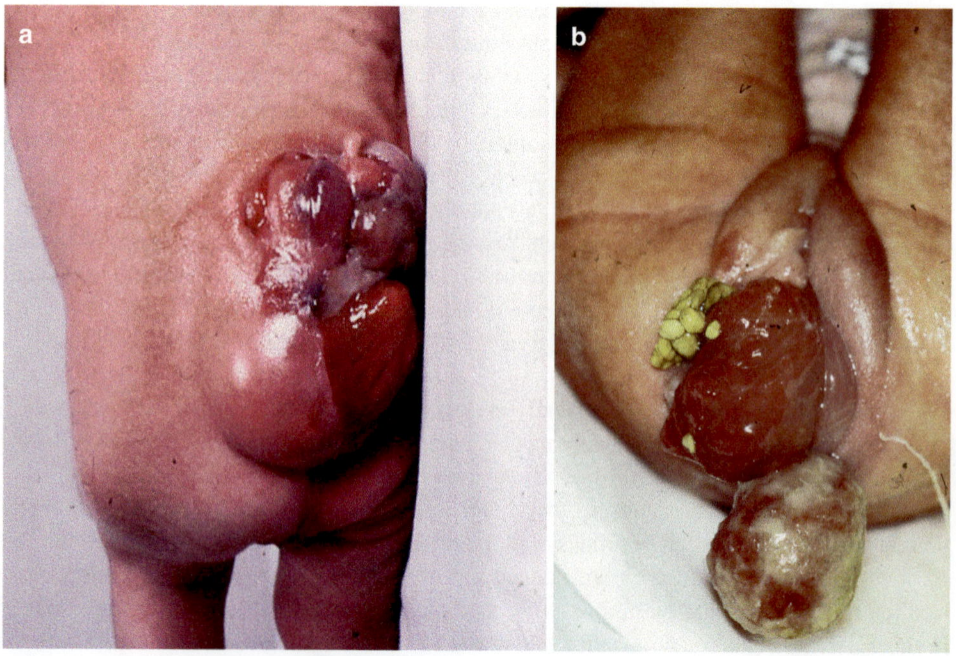

Fig. 36.2 (**a**) Posterior enteric duplication. (**b**) Rectal duplication

dilatation, but polyhydramnios is not normally seen, as amniotic fluid reabsorption takes place in the proximal small bowel. The majority of colonic atresia cases are diagnosed postnatally, presenting with progressive abdominal distension and bilious vomiting [66]. Progression to perforation may be rapid, particularly in the 40% of cases with a competent ileo-caecal valve. A 40-year review of the literature identified 224 cases of colonic atresia [67]. 122 of the 208 cases where the location of the atresia was known, had a right-sided colonic atresia [67]. Colonic atresias can be classified according to the modified classification scheme proposed by Grosfeld in 1979 [68]. This is described in more detail in the midgut chapter.

The classical aetiological pathway associated with small intestinal atresia is thought to be an acute vascular compromise due to in-utero volvulus, intussusception or internal herniation. These factors are often used to explain the occurrence of colonic atresias, despite the fact that colonic intussusception and volvulus are extremely rare occurrences. This realization prompted Baglaj and Carachi to review 30 cases of colonic atresia at their centres [65]. They identified two groups with distinct clinical features. 13 (43%) cases were associated with an anterior abdominal wall defect (2 umbilical cord hernias and 11 cases of gastroschisis). 11 of this group were born prematurely, and 1 in 3 of these infants had involvement of the small and large bowel in the atresia. Rather than being 'associated' with colonic atresia, it would appear that the constricting effect of the gastroschisis defect on the returning caeco-colic loop is the primary cause of the atresia in these cases. This may also explain the extensive small and large bowel involvement often seen in atresias associated with an anterior abdominal wall defect, with the extreme end of this spectrum being the high jejunal atresia and distal colonic atresia observed in a closing gastroschisis [65]. The remaining 17 (57%) infants had an isolated colonic atresia, which was confined to the right colon (caecum, ascending and proximal transverse colon) in 15 of the 17 cases. Only four of this group were born prematurely. The high prevalence of right-sided colonic atresias in this group led the authors to speculate that the final part of the returning caeco-colic loop may be the most vulnerable to compression by the closing umbilical ring, analogous to the situation with a gastroschisis defect. Over 80% of the right-sided atresias were type 3a, which is in agreement with other large series [67], and may support the theory of a two-point compression by the gastroschisis defect or umbilical ring. In contrast, the majority of colonic atresias beyond the splenic flexure vascular watershed area are type 1 atresias, supporting an alternative pathogenetic mechanism [69].

Associated anomalies are identified in 30–47.3% of cases of colonic atresia [67, 70]. The overwhelming majority of these are anterior abdominal wall defects or intestinal anomalies including malrotation and other intestinal atresias [67, 71]. Other reported orthopaedic anomalies include syndactyly, polydactyly, the absence of the radius and clubfoot [70]. An association between micro-ophthalmia and facial asymmetry with colonic atresia has recently been described, including in the context of Goldenhar syndrome [72].

A potential association between colonic atresia and Hirschsprung's disease has been hypothesized, with some series reporting that up to 0.9% of infants with HD have an intestinal atresia [73], which is usually jejunoileal. Several theories have been put forward to justify the apparent association between colonic atresia and HD [70]. The first purports that HD occurs because the colonic atresia acts as a 'mechanical barrier' to migrating NCCs. The presence of ganglionic bowel distal to the majority of colonic atresias would not support this theory. The second theory suggests that HD causes the colonic atresia as a result of segmental volvulus of a dilated, meconium-filled proximal loop of the colon, with subsequent ischaemic loss of intestinal continuity [74]. Others have hypothesized a common genetic origin to both conditions [75]. Regardless of the pathogenetic mechanisms or whether this represents a true association or the co-occurrence of two unrelated anomalies, the importance of recognizing concurrent HD lies in the fact that a primary colo-colonic anastomosis is destined to fail and require a delayed enterostomy in the presence of undiagnosed HD. The search for predictors of HD associated with CA has led some authors to suggest that the presence of a coiled, non-fixed and foreshortened colon in the pelvis of an infant with colonic atresia is highly suggestive of concurrent HD [76]. In the light of this, the pertinent questions for the surgeon to consider are: should a diverting enterostomy be performed with subsequent rectal biopsy to confirm normal ganglionosis prior to restoration of continuity? Alternatively, should a rectal biopsy with intraoperative confirmation of normal ganglionosis be performed up front, if the intention is to perform a primary anastomosis?

There is little consistency in surgical practice for colonic atresia, perhaps reflecting the rarity of the condition. Despite this, certain operative principles hold true, regardless of approach. First, given the high frequency of associated intestinal anomalies, careful inspection of the remaining intestine should always be made to confirm normal midgut rotation and to exclude additional atresias or a suspicion of functional obstruction due to HD. The end of the proximal atretic bowel is usually grossly dilated and adynamic, with associated abnormalities in vascularity and innervation, and should be back-resected during the original procedure, provided the total bowel length is adequate [77] (Fig. 36.3). Most surgeons reserve primary anastomosis for carefully selected cases, where distal luminal patency and functional integrity has been confirmed, but there are reports of a higher incidence

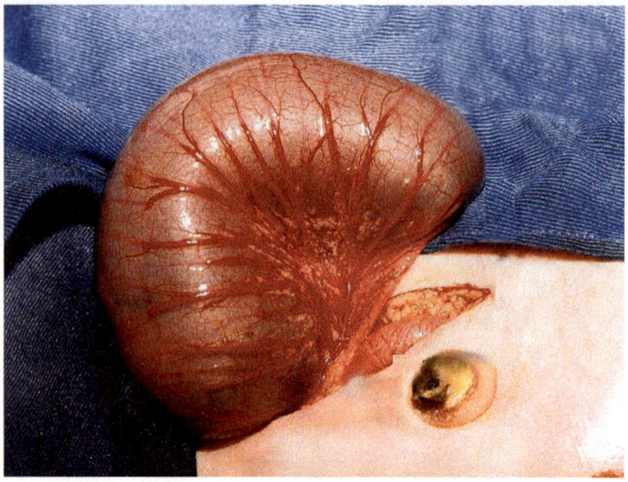

Fig. 36.3 Blind ending colon in Colonic atresia

of complications using this approach, usually as a result of missed distal pathology [78]. In most instances, an enterostomy is preferred, with confirmation of distal ganglionosis prior to stoma closure. The long-term outlook is generally excellent, with prognosis related to residual small intestinal length, associated anomalies and the occurrence of surgical complications. Infants with colonic atresia and gastroschisis pose a significant management challenge and have a worse outcome [72]. Although a 40-year systematic review found an overall mortality rate of 25% [67], recent series with no mortality have been reported.

References

1. Knowles CH, De Giorgio R, Kapur RP, et al. Gastrointestinal neuromuscular pathology: guidelines for histological techniques and reporting on behalf of the Gastro 2009 International Working Group. Acta Neuropathol. 2009;118:271–301.
2. Knowles CH, De Giorgio R, Kapur RP, et al. The London classification of gastrointestinal neuromuscular pathology: report on behalf of the Gastro 2009 International Working Group. Gut. 2010;59:882–7.
3. Burns AJ, Goldstein AM, Newgreen DF, et al. White paper on guidelines concerning enteric nervous system stem cell therapy for enteric neuropathies. Dev Biol. 2016;417(2):229–51. https://doi.org/10.1016/j.ydbio.2016.04.001.
4. Kapur RP. Practical pathology and genetics of Hirschsprung's disease. Semin Pediatr Surg. 2009;18:212–23.
5. Jakobson-Setton A, Weissmann-Brenner A, Achiron R, et al. Retrospective analysis of prenatal ultrasound of children with Hirschsprung disease. Prenat Diagn. 2015;35:699–702.
6. Singh SJ, Croaker GDH, Manglick P, et al. Hirschsprung's disease: the Australian Paediatric surveillance Unit's experience. Pediatr Surg Int. 2003;19:247–50.
7. Clark DA. Times of first void and first stool in 500 newborns. Pediatrics. 1977;60:457–9.
8. National Institute for Health and Clinical Excellence (NICE) (2010) Nice guidelines [CG99]. Constipation in children and young people: diagnosis and management. http://www.nice.org.uk/cg99. Accessed 6 Mar 2016.
9. Downey EC, Hughes E, Putnam AR, et al. Hirschsprung disease in the premature newborn: a population based study and 40-year single center experience. J Pediatr Surg. 2015;50:123–5.
10. Badner JA, Sieber WK, Garver KL, et al. A genetic study of Hirschsprung disease. Am J Hum Genet. 1990;46:568–80.
11. Ryan ET, Ecker JL, Christakis NA, et al. Hirschsprung's disease: associated abnormalities and demography. J Pediatr Surg. 1992;27:76–81.
12. Puri P, Montedonico S. Hirschsprung's disease: clinical features. In: Holschneider AM, Puri P, editors. Hirschsprung's disease and allied disorders. 3rd ed. Berlin: Springer; 2008.
13. Laughlin DM, Friedmacher F, Puri P. Total colonic aganglionosis: a systematic review and meta-analysis of long term clinical outcome. Pediatr Surg Int. 2012;28:773–9.
14. Amiel J, Sproat-Emison E, Garcia-Barcelo M, et al. Hirschsprung disease, associated syndromes and genetics: a review. J Med Genet. 2008;45:1–14.
15. Bergeron K-F, Silversides DW, Pilon N. The developmental genetics of Hirschsprung's disease. Clin Genet. 2013;83:15–22.
16. Burns AJ, Roberts RR, Bornstein JC, Young HM. Development of the enteric nervous system and its role in intestinal motility during fetal and early postnatal stages. Semin Pediatr Surg. 2009;18:196–205.
17. Wallace AS, Burns AJ. Development of the enteric nervous system, smooth muscle and interstitial cells of Cajal in the human gastrointestinal tract. Cell Tissue Res. 2005;319:367–82.
18. Nishiyama C, Uesaka T, Manabe T, et al. Trans-mesenteric neural crest cells are the principal source of the colonic enteric nervous system. Nat Neurosci. 2012;15:1211–8.
19. Burns AJ, Le Duoarin NM. The sacral neural crest contributes neurons and glia to the post-umbilical gut: spatiotemporal analysis of the development of the enteric nervous system. Development. 1998;125:4335–47.
20. Rolle U, Memeth L, Puri P. Nitrergic innervation of the normal gut and in motility disorders of childhood. J Pediatr Surg. 2002;36:551–67.
21. Roman V, Bagyanszki M, Krecsmarik M, et al. Spatial pattern analysis of nitrergic neurons in the developing myenteric plexus of the human fetal intestine. Cytometry A. 2004;57:108–12.
22. Moore SW, Sidler D, Schubert PA. Segmental aganglionosis (zonal aganglionosis or "skip" lesions) in Hirschsprungs disease: a report of 2 unusual cases. Pediatr Surg Int. 2013;29:495–500.
23. Emison ES, Garcia-Barcelo M, Grice EA, et al. Differential contributions of rare and common, coding and noncoding ret mutations to multifactorial Hirschsprung disease liability. Am J Hum Genet. 2010;87:60–74.
24. Gariepy CE. Genetic basis of Hirschsprung disease: implications in clinical practice. Mol Genet Metab. 2003;80:66–73.
25. Pini Prato A, Rossi V, Mosconi M, et al. A prospective observational study of associated anomalies in Hirschsprung's disease. Orphanet J Rare Dis. 2013;8:184. https://doi.org/10.1186/1750-1172-8-184.
26. Hoffman AD, Duess JW, Puri P. Congenital anomalies of the kidney and urinary tract (CAKUT) associated with Hirschsprung's disease: a systematic review. Pediatr Surg Int. 2014;30:757–61.
27. Tuo G, Pini Prato A, Derchi M, et al. Hirschsprung's disease and associated congenital heart defects: a prospective observational study from a single institution. Front Pediatr. 2014;2:99. https://doi.org/10.3389/fped.2014.00099.
28. Duess JW, Puri P. Syndromic Hirschsprung's disease and associated congenital heart disease: a systematic review. Pediatr Surg Int. 2015; https://doi.org/10.1007/s00383-015-3744-6.
29. Moore SW. Hirschsprung's disease and the brain. Pediatr Surg Int. 2011;27:347–52.

30. Friedmacher F, Puri P. Hirschsprung's disease associated with down syndrome: a meta-analysis of incidence, functional outcomes and mortality. Pediatr Surg Int. 2013;29:937–46.
31. Wells SA, Asa SL, Dralle H, et al. Revised American Thyroid Association guidelines for the management of medullary thyroid carcinoma. Thyroid. 2015;25:567–610.
32. Bradnock TJ, Walker GM. Evolution in the management of Hirschsprung's disease in the UK and Ireland. A national survey of practice revisited. Ann R Coll Surg. 2011;93:34–8.
33. Gosain A. Established and emerging concepts in Hirschsprung's-associated enterocolitis. Pediatr Surg Int. 2016;32:313–20.
34. Frykman PK, Short SS. Hirschsprung-associated enterocolitis: prevention and therapy. Semin Pediatr Surg. 2012;21:328–35.
35. Pierro A, Fasoli L, Kiely EM, et al. Staged pull-through for rectosigmoid Hirschsprung's disease is not safer than primary pull-through. J Pediatr Surg. 1997;32:505–9.
36. Rescorla FJ, Morrison AM, Engles D, et al. Hirschsprung's disease. Evaluation of mortality and long-term function in 260 cases. Arch Surg. 1992;127:934–42.
37. Ludman L, Spitz L, Tsuji H, et al. Hirschsprung's disease: functional and psychological follow up comparing total colonic and rectosigmoid aganglionosis. Arch Dis Child. 2002;86:348–51.
38. Ravitch MM. Pseudo Hirschsprung's disease. Ann Surg. 1958;147:781–95.
39. Friedmacher F, Puri P. Classification and diagnostic criteria of variants of Hirschsprung's disease. Pediatr Surg Int. 2013;29:855–72.
40. Schappi MG, Staiano A, Milla PJ, et al. A practical guide for the diagnosis of primary enteric nervous system disorders. JPGN. 2013;57:677–86.
41. Coerdt W, Michel J-S, Rippin G, et al. Quantitative morphometric analysis of the submucous plexus in age-related control groups. Virchows Arch. 2004;444:239–46.
42. Borchard F, Meier-Ruge WA, Wiebecke B, et al. Innervationsstorungen des Dickdarms: Klassifikation und Diagnostik. Pathologe. 1991;12:171–4.
43. Holschneider AM, Kunst M. Anal sphincter achalasia and ultra-short Hirschsprung's disease. In: Holschneider AM, Puri P, editors. Hirschsprung's disease and allied disorders. 3rd ed. Berlin: Springer; 2008.
44. Pena P, Bischoff A. Idiopathic constipation and other motility disorders. In: Pena P, Bischoff A, editors. Surgical treatment of colorectal problems in children. 1st ed. Switzerland: Springer; 2015.
45. Meier-Ruge W. Uber ein erkrankungsbild des colon mit Hirschsprung-symptomatik. Ver Dtsch Ges Pathol. 1971;55: 506–10.
46. Martucciello G, Torre M, Pini Prato A, et al. Associated anomalies in intestinal neuronal dysplasia. J Pediatr Surg. 2002;37: 219–23.
47. Waguespack SG, Rich TA, Perrier ND, et al. Management of medullary thyroid carcinoma and MEN2 syndromes in childhood. Nature reviews. Endocrinology. 2011;7:596–607.
48. Gosemann J, Puri P. Megacystis microcolon intestinal hypoperistalsis syndrome: systematic review of outcome. Pediatr Surg Int. 2011;27:1041–6.
49. Wangler MF, Gonzaga-Jauregui C, Gambin T, et al. Heterozygous de novo and inherited mutations in the smooth muscle actin (ACTG2) gene underlie megacystis-microcolon-intestinal hypoperistalsis syndrome. PLoS Genet. 2014;10:e1004258. https://doi.org/10.1371/journal.pgen.1004258.
50. Potter EL. Pathology of the fetus and newborn. Chicago (Ill): Year Book Medical Publishers; 1961.
51. Shew SB, Holcomb GWIII. Alimentary tract duplications. In: Ashcraft KW, Holcomb III GW, Murphy JP, editors. Pediatric surgery. 4th ed. Philadelphia: Elsevier Saunders; 2005.
52. Laje P, Flake AW, Adzick NS. Prenatal diagnosis and postnatal resection of intraabdominal enteric duplications. J Pediatr Surg. 2010;45:1554–8.
53. Puligandla PS, Nguyen LT, St-Vil D, et al. Gastrointestinal duplications. J Pediatr Surg. 2003;38:740–4.
54. Holcomb GW III, Gheissari A, O'Neill JA Jr, et al. Surgical management of alimentary tract duplications. Ann Surg. 1989;209:167–74.
55. Yousefzadeh DK, Bickers GH, Jackson JH Jr. Tubular colonic duplication- review of 1876-1981 literature. Pediatr Radiol. 1983;13:65–71.
56. Kaur N, Nagpal K, Sodhi P, et al. Hindgut duplication—case report and literature review. Pediatr Surg Int. 2004;20:640–2.
57. Dominguez R, Rott J, Castillo M, et al. Caudal duplication syndrome. Am J Dis Child. 1993;147:1048–52.
58. Azmy AF. Complete duplication of the hindgut and lower urinary tract with diphallus. J Pediatr Surg. 1990;25:647–96.
59. Adair HM, Trowell JE. Squamous cell carcinoma arising in a duplication of the small bowel. J Pathol. 1981;133:25–31.
60. Craigie RJ, Abbaraju JS, Ba'ath ME, et al. Anorectal malformation with tubular hindgut duplication. J Pediatr Surg. 2006;41:31–4.
61. Kratz JR, Deshpande V, Ryan DP, et al. Anal canal duplication associated with presacral cyst. J Pediatr Surg. 2008;43:1749–52.
62. Hamada Y, Sato M, Hioki K. Anal canal duplication in childhood. Pediatr Surg Int. 1996;11:577–9.
63. Choi SO, Park WH. Anal canal duplication in infants. J Pediatr Surg. 2003;38:708–12.
64. Adams SD, Stanton MP. Malrotation and intestinal atresias. Early Hum Dev. 2014;90:921–5.
65. Baglaj M, Carachi R, MacCormack B. Colonic atresia: a clinicopathological insight into its etiology. Eur J Pediatr Surg. 2010;20:102–5.
66. Cox SG, Numanoglu A, Millar AJW, et al. Colonic atresia: spectrum of presentation and pitfalls in management. A review of 14 cases. Pediatr Surg Int. 2005;21:813–8.
67. Etensel B, Temir G, Karkiner A, et al. Atresia of the colon. Pediatr Surg Int. 2005;40:1258–68.
68. Grosfeld J, Ballantine T, Shoemaker R. Operative management of intestinal atresia and stenosis based on pathologic findings. J Pediatr Surg. 1979;14:368–75.
69. Boles ET, Vassy LE, Ralston M. Atresia of the colon. J Pediatr Surg. 1988;23:852.
70. Lauwers P, Moens E, Wustenberghs K, et al. Association of colonic atresia and Hirschsprung's disease in the newborn: report of a new case and review of the literature. Pediatr Surg Int. 2006;22:277–81.
71. Kronfli R, Bradnock TJ, Sabharwal A. Gastroschisis and atresia: a 26-year review. Pediatr Surg Int. 2010;26:891–4.
72. Haxhija EQ, Schalamon J, Hollwarth ME. Management of isolated and associated colonic atresia. Pediatr Surg Int. 2011;27:411–41.
73. Ikeda K, Goto S. Additional anomalies in Hirschsprung's disease: an analysis based on the nationwide survey in Japan. Z Kinderchir. 1986;41:279–81.
74. Hyde GA, De Lorimier AA. Colon atresia and Hirschsprung's disease. Surgery. 1968;64:976–8.
75. Croaker GDH, Harvey JG, Cass DT. Hirschsprung's disease, colonic atresia, and absent hand: a new triad. J Pediatr Surg. 1997;32:1368–70.
76. Fishman SJ, Islam S, Buonomo C, et al. Nonfixation of an atretic colon predicts Hirschsprung's disease. J Pediatr Surg. 2001;36:202–4.
77. Louw JH. Congenital intestinal atresia and stenosis in the newborn. Observations on its pathogenesis and treatment. Ann R Coll Surg Engl. 1959;25:209.
78. Karnak I, Ciftci AO, Senocak ME, et al. Colonic atresia: surgical management and outcome. Pediatr Surg Int. 2001;17:631–5.

Further Reading

Overview of Hirschsprung's Disease and Allied Disorders

Holschneider AM, Puri P. Hirschsprung's disease and allied disorders. 3rd ed. Berlin: Springer; 2008.

Development of the Enteric Nervous System

Burns AJ, Roberts RR, Bornstein JC, Young HM. Development of the enteric nervous system and its role in intestinal motility during fetal and early postnatal stages. Sem Ped Surg. 2009;18:196–205.

Genetics of Hirschsprung's Disease

Amiel J, Sproat-Emison E, Garcia-Barcelo M, et al. Hirschsprung disease, associated syndromes and genetics: a review. J Med Genet. 2008;45:1–14.

Bergeron K-F, Silversides DW, Pilon N. The developmental genetics of Hirschsprung's disease. Clin Genet. 2013;83:15–22.

Kapur RJ. Practical pathology and genetics of Hirschsprung's disease. Sem Ped Surg. 2009;18:212–23.

Future Therapies—Stem Cell Transplantation

Burns AJ, Goldstein AM, Newgreen DF et al. White paper on guidelines concerning enteric nervous system stem cell therapy for enteric neuropathies. Dev Biol. 2016. Doi:https://doi.org/10.1016/j.ydbio.2016.04.001

The Umbilicus

37

Paul Cullis

The Umbilical Cord

******Definition***:** A bundle of vessels enclosed inside a tubular sheath of amnion which extends from the placenta to the abdominal wall of the foetus.

******Development***:**

Formation of the Primitive Umbilical Ring

- As a result of folding, the embryonic disc bulges into the amniotic cavity and the amnio-ectodermal junction is carried onto the ventral aspect of the embryo.
- The line of reflection between the amnion and ectoderm acquires an oval outline and is called the ***primitive umbilical ring.***

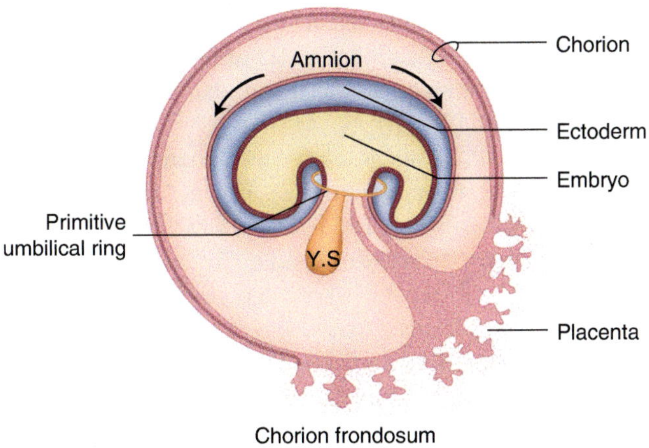

Formation of the Primitive Umbilical Cord

- By the fifth week, the primitive umbilical ring constricts to form a tubular sheath which encloses the body stalk, the yolk sac and its vessels and part of the allantois. This tubular sheath is called the ***primitive umbilical cord.***

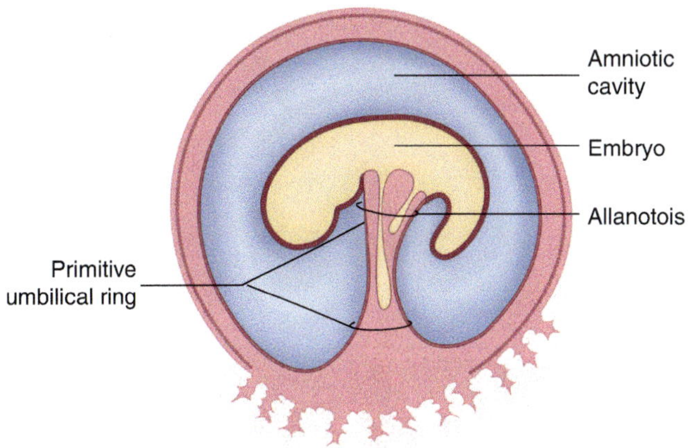

Formation of the Definitive Umbilical Cord

The umbilical cord elongates and its constituent structures undergo changes as follows:

1. The *extra embryonic mesoderm* of the body stalk changes to a mucoid substance called "**Wharton's jelly**" (forms the main bulk of the cord).
2. The remnants of *extra embryonic coelom* inside the cord gradually disappear.
3. The *yolk sac* becomes obliterated together with the vitello-intestinal duct connecting the yolk sac to the midgut.

P. Cullis (✉)
Queen Elizabeth University Hospital, Glasgow, UK

R. Carachi, S. H. E. Doss (eds.), *Clinical Embryology*, https://doi.org/10.1007/978-3-319-26158-4_37

4. The distal part of the *allantois* becomes obliterated while its vessels persist and elongate to form the umbilical vessels.
5. At the sixth week, a part of the midgut loop enters the umbilical cord (*physiological hernia*) and then returns again to the abdominal cavity after the 10th week.

Anatomical Features of Definitive U-cord

****Shape***: soft tortuous cord having a smooth surface (covered by amnion).

****Length***: 50–60 cm long and 1 cm in diameter.

****Structure***: it is formed of outer covering of amnion endorsing:

(a) Gelatinous ground substance (Wharton's jelly)
(b) Three vessels: two umbilical arteries and one umbilical vein
(c) Remnant of the allantois

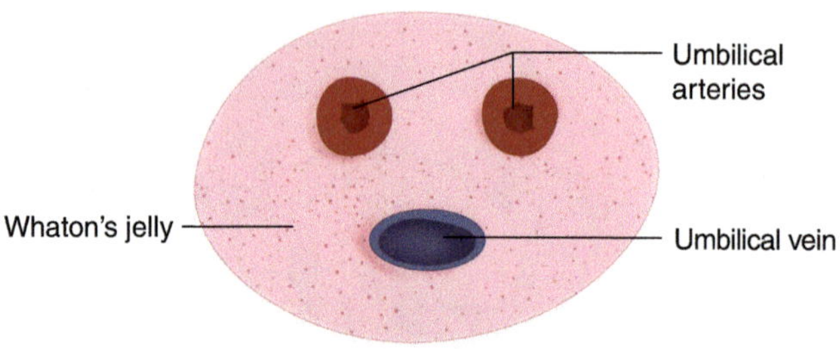

Introduction

No structure confers more importance to the foetus during development than the umbilicus, which, by means of its containing vessels, provides the necessary nutrition and waste disposal needed for growth. After birth however it retires and spends its life as a frivolous oddity whose little value is aesthetic. That is under normal circumstances. The umbilicus and its remnants can present to midwives, doctors, surgeons and family physicians at any stage in life but particularly during childhood.

Embryological Considerations

Development of the umbilicus begins in the fourth week of gestation. The embryo at this stage has two stalks: a ventral yolk sac stalk which contains the vitelline duct and vessels, and a caudal connecting stalk which contains the umbilical vessels and allantois. The latter stalk becomes increasingly ventral as the embryo undergoes cephalocaudal flexion, such it fuses with the yolk sac stalk, forming the primitive umbilical cord. Therefore, the early umbilical cord comprises the allantois, umbilical vessels, vitelline duct and vessels, within a mesenchymal core. This is covered by a layer of epithelial tissue continuous with the amnion and the embryo's outermost lining [1].

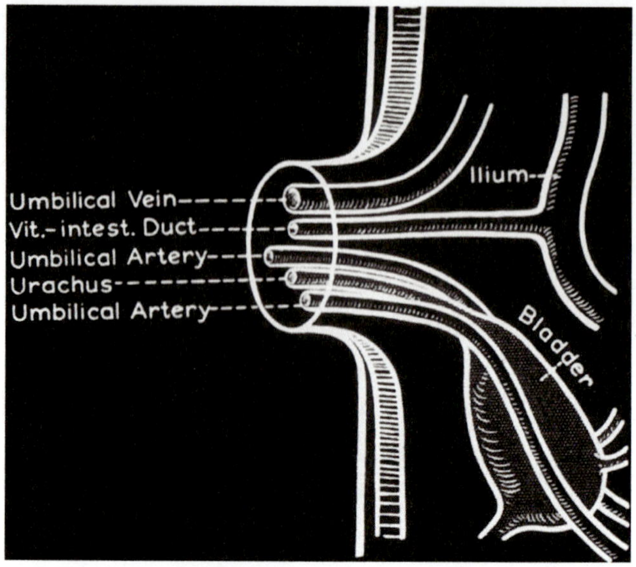

Fig. 37.1 Contents of the umbilical cord

Between the sixth and tenth weeks, rapid growth of the primitive gastrointestinal tract results in a 'physiological hernia'. That is to say the midgut leaves the peritoneal cavity and lies within the umbilical cord. The contents eventually return to lie within the peritoneal cavity. The umbilical ring gradually contracts, such that at birth, it has closed or is near closure. This was thought to be due to lateral infolding and flexion of the ventral abdominal wall, but this concept has been challenged [2, 3].

Involution of the allantois, which is continuous with the urinary bladder, forms the urachus. This normally involutes to form the median umbilical ligament which lies in the preperitoneal space, the space of Retzius. By the end of week 5, the yolk sac stalk separates from the gastrointestinal tract; however, in 1.2%, this persists as a true ileal diverticulum, known as a Meckel's diverticulum [1, 4].

At full gestation, the umbilical cord contains the umbilical vessels (two arteries and a single vein) alone within a smooth mesenchyme called Wharton's jelly. A single umbilical artery is present in 1 in 180 births and is associated with other defects in almost a third, particularly cardiac, renal and chromosomal anomalies. Average umbilical cord length is 55 cm. Long cords may be associated with true knots, entanglements and prolapse. True knots occur in around 1% of cases whereby the foetus passes through a loop of cord. Usually this carries little significance, but if the knot is created early in pregnancy, and tightens with ongoing gestation, subsequent fetocide can result from impaired blood supply. Conversely, short cords may indicate decreased foetal movement due to foetal construction as in oligohydramnios, for example, or foetal anomalies such as limb defects. Short umbilical cords can occasionally pull the placenta from its attachment during delivery [1, 5].

After birth, it takes 3 days to 10 weeks under normal circumstances for umbilical cord separation with an average of 2 weeks. Delayed separation is weakly associated with underlying immunologic abnormalities, that is to say, there are many normal babies whose cord fails to separate by 2 weeks after birth [6].

Umbilical Hernia

It is thought that there exist three subtypes of umbilical hernia: the omphalocele, the infantile umbilical hernia and the acquired umbilical hernia of adulthood. Failure of the physiological hernia to resolve results in the omphalocele, described in detail elsewhere, along with its associated anomalies. The infantile hernia represents incomplete contraction of the fibromuscular umbilical ring. Unlike the omphalocele, this defect is covered with the skin. The natural history of the umbilical hernia is a progressive reduction in the magnitude of the defect and eventual resolution in most cases [7]. These more commonly affect children of Afro-Caribbean descent and preterm or low birthweight neonates (with as many as three quarters of extremely low birthweight neonates affected) [8], but there are many other associations (Table A). In Caucasian populations studied, 10–15% of neonates are found to have umbilical hernias, whilst figures in African populations suggest that herniation is the norm under 6 years of age.

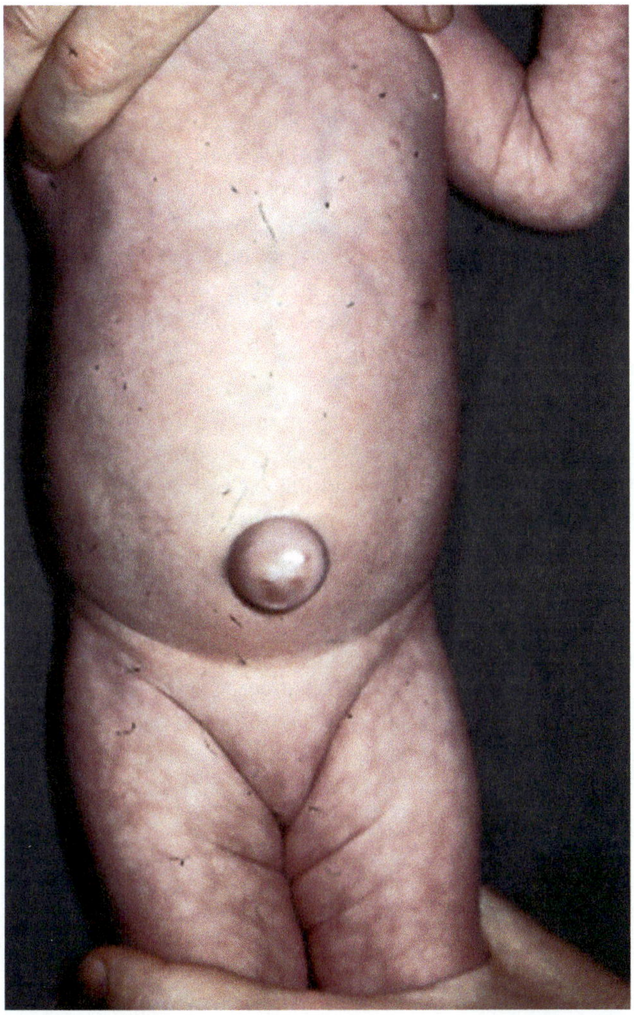

Fig. 37.2 Umbilical hernia

Table 37.1 Factors associated with an increased incidence of umbilical hernia

Preterm
Low birthweight
Afro-Caribbean descent
Down syndrome
Beckwith-Wiedemann syndrome
Marfan's syndrome
Ascites
Ventriculo-peritoneal shunts
Obesity

The hernia may appear to increase in size, which exacerbates parental concern. This apparent increase in size does not indicate that the hernia defect is enlarging but that more abdominal contents are herniating. This, in itself, is of no great consequence. The magnitude of the defect is more important. The smaller the defect, the more likely it is to close spontaneously. Unless obstruction or strangulation occurs, the umbilical hernia is asymptomatic. Risk of incarceration in a large historical series was 1 in 1500 cases [9]; therefore, traditionally umbilical hernias were considered 'safe'. Nevertheless, there have been several published series which suggest that risk of incarceration is much higher than previously thought. Rupture is rare but has been reported [10, 11].

As with hernias in other bodily sites, obstruction is typically heralded by the onset of abdominal colic and vomiting, which may be bilious. A strangulated hernia is seen as a tender, erythematous and irreducible lump at the site of the umbilicus.

In the developed world, it is a routine practice to repair those hernias which persist to beyond the fourth or fifth birthday and defects >2 cm in diameter. A more conservative approach is followed in Africa, generally, when operative intervention is undertaken if the patient becomes symptomatic. A single large West African study, however, has suggested that spontaneous closure may occur until the age of 14 years, at least in that population [12].

In some cases, the umbilicus has a proboscoid appearance. Underlying this, there is no hernial defect, and in fact, the appearance is caused simply by excess skin and subcutaneous fat. Nevertheless, the finding is associated with several rare syndromes: Aarskog-Scott, Reiger's and Robinow's [3].

Umbilical Granulomas and Polyps

Umbilical granulomas are not true congenital defects. They represent non-epithelialized granulation tissue and inflammation of the umbilicus and appear as wet, round, pink and often pedunculated nodules, ranging from 1 mm to 1 cm in size [5]. Conversely, the umbilical polyp is usually a vitelline duct remnant (occasionally urachal), appearing as a shiny red and smooth lump at the base of the umbilicus. Histologically it contains small bowel mucosa to a variable degree if derived from the vitelline duct. Whilst it can usually be differentiated from an umbilical granuloma by its slightly larger and scarlet red appearance, it is sometimes identified when the lump thought to be a granuloma fails to respond adequately to topical silver nitrate application, a common treatment for granulomas [3].

Urachal Remnants

Failure of obliteration of the urachal tract which connects the dome of the bladder to the umbilical ring results in various pathologies. Prevalence amongst the paediatric population has been estimated at approximately 1 in 7500 for patent urachal and 1 in 5000 for urachal cysts based on autopsy findings. Obliteration of proximal and distal tract results in urachal cyst formation. Most often these are subclinical, and it is only when superimposed infection intervenes that their presence is suspected. In fact, infective complications are the most common means of presentation of urachal remnants. *Staphylococcus aureus* is the most frequently identified microbe. A tender suprapubic mass with skin erythema is typical. If obliteration occurs at only one end, either an umbilical sinus or bladder diverticulum forms. The latter is the least common variant of urachal remnants. Drainage of fluid with characteristics of urine from the umbilicus is suggestive of a patent urachal tract or umbilical sinus. There may be associated skin excoriation. Plain radiograph after instillation of contrast to the abnormality can be useful. Of note, the patent urachal connection may be associated with distal urinary tract obstruction, as in posterior urethral valves, in which case it is protective. Adequate evaluation of the urinary tract is therefore important before surgical correction. Finally, bladder diverticula are usually subclinical anomalies and are identified incidentally during urinary tract investigations [3, 5, 13, 14].

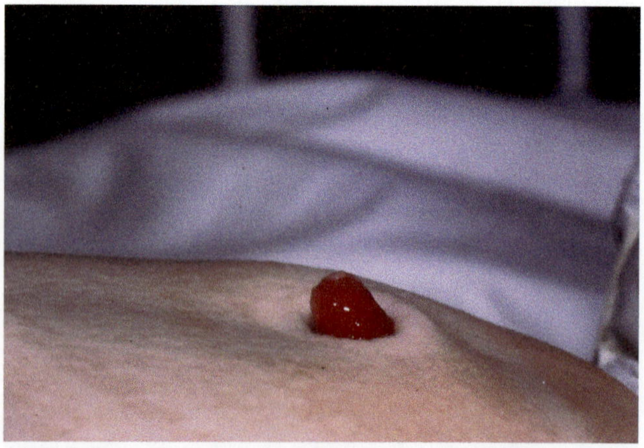

Fig. 37.3 Umbilical Polyp

Vitelline Duct Remnants

Incomplete separation and involution of the vitelline duct, which normally occurs by the ninth week, result in an array of congenital anomalies. The most common variant is the simple Meckel's diverticulum, a proximal remnant of the duct, around 60 cm from the ileocaecal valve proximally, which in most individuals is no cause for concern. In fact, this anomaly occurs in 1.2% of the population; however, in only 4% of these does it cause issue and does so most often within the first 2 years of life. Ectopic gastric and pancreatic mucosa may be identified in a third. The former may cause gastrointestinal tract bleeding as a result of corrosive ulceration of local ileal mucosa. Less commonly, diverticulitis occurs where the lumen is obstructed resulting in a similar series of events culminating in gangrene and perforation as seen in acute appendicitis. Small bowel intussusception is a further means of presentation whereby the diverticulum acts as a lead point. Rarely, Meckel's diverticula can present as Littre's hernia or malignancy within the ectopic tissue. If full patency of the duct exists at birth, the alarming appearance of faecal drainage will occur. Partial regression of the vitelline duct may result in a variety of presentations, including sinuses, polyps, cysts, fibrous bands and a persistent vitelline artery. The latter two may cause small bowel obstruction and volvulus [3, 4, 15–17].

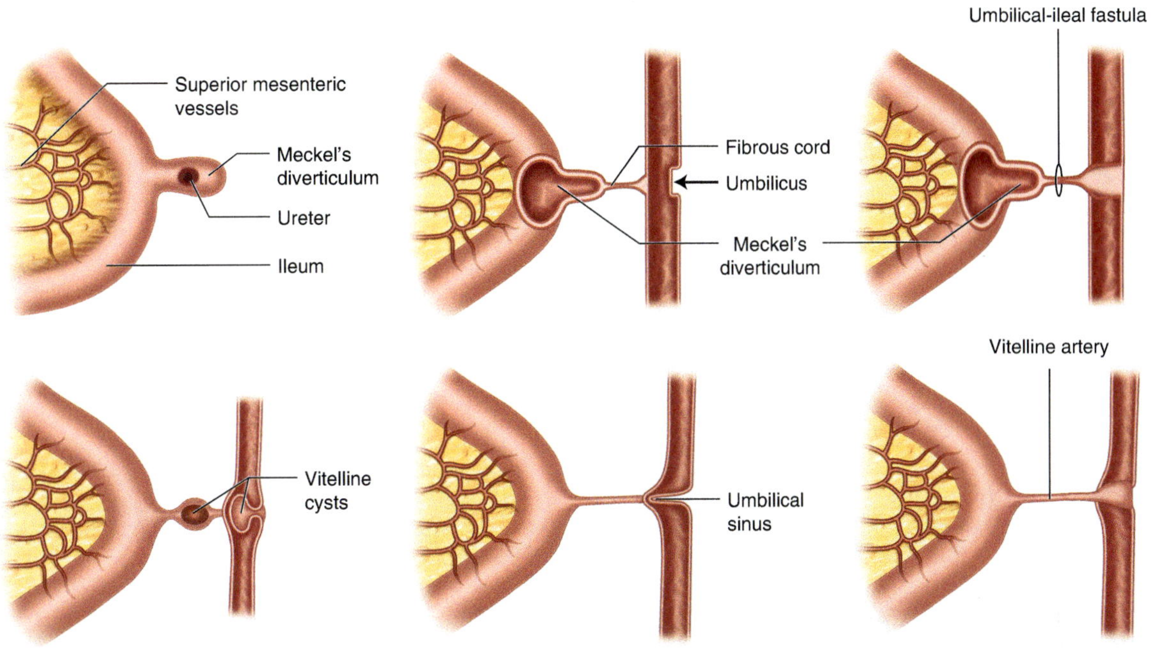

Fig. 37.4 The variable presentations of vitello-intestinal duct remnants

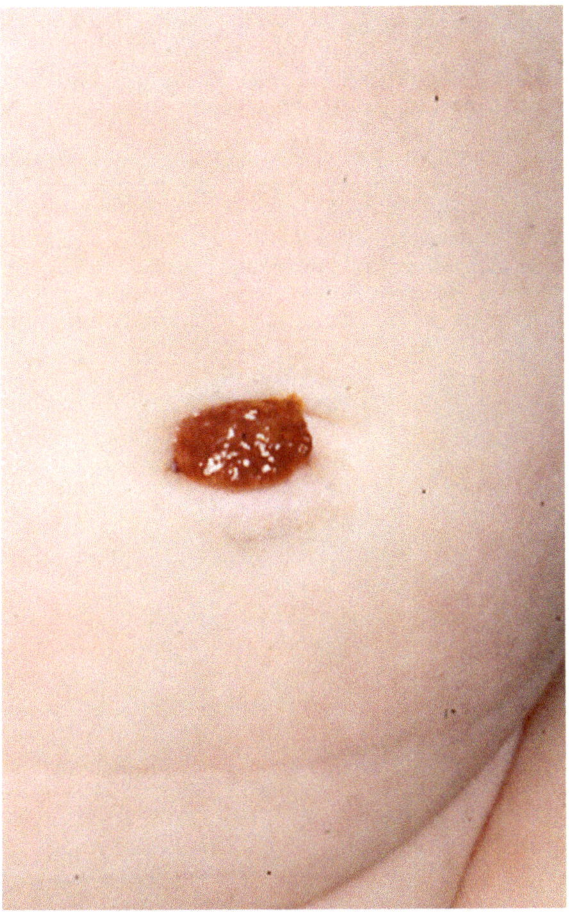

Fig. 37.5 Vitello-intestinal remnant

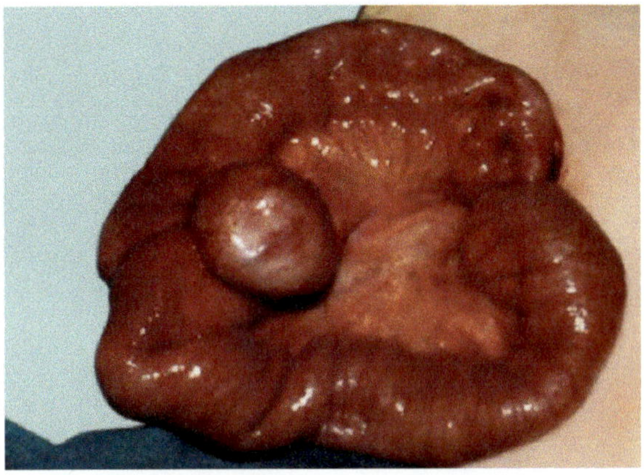

Fig. 37.6 Meckel's diverticulum

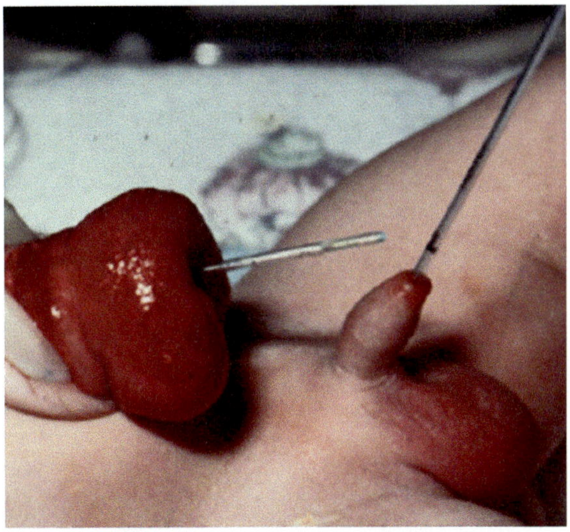

Fig. 37.7 Urachus

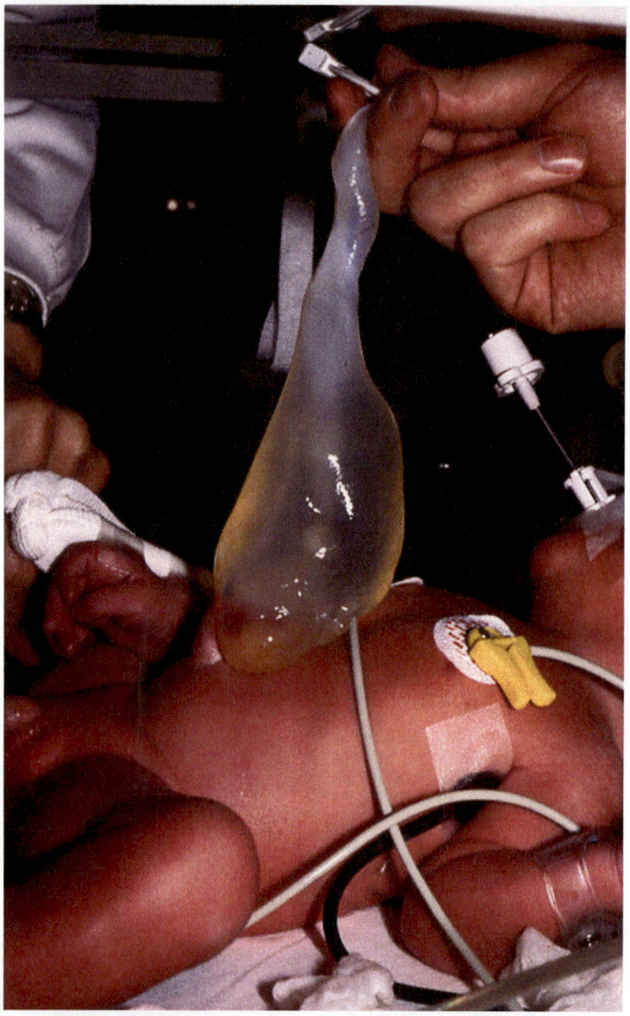

Fig. 37.8 Umbilical cord

References

1. Pansky B. Review of medical embryology (Ch 34–35). Alameda: Embryome Sciences; 1982.
2. Holcomb GW III, Murphy JD. Ashcraft's pediatric surgery (Ch 49). 5th ed. Philadelphia: Elsevier Saunders; 2010.
3. Mattei P. Fundamentals of pediatric surgery (Ch 70). 1st ed. Philadelphia: Springer; 2011.
4. Zani A, Eaton S, Rees CM, et al. Incidentally detected Meckel diverticulum: to resect or not to resect? Ann Surg. 2008;247:276–81.
5. Pomeranz A. Anomalies, abnormalities and care of the umbilicus. Pediatr Clin N Am. 2004;51:819–27.
6. Coran AG, Adzick NS, Krummel ™ et al. Pediatric surgery (Ch 74). 7th ed. Vol. 2. Philadelphia: Elsevier Saunders; 2012.
7. Muschaweck U. Umbilical and epigastric hernia repair. Surg Clin N Am. 2003;83:1207–21.
8. Vohr BR, Rosenfield AG, Oh W. Umbilical hernia in the low-birthweight infant (less than 1,500 gm). J Pediatr. 1977;90(5):807–8.
9. Mestel AL, Burns H. Incarcerated and strangulated umbilical hernias in infants and children. Clin Pediatr. 1963;2:368–70.
10. Lassaletta L, Fonkalsrud EW, Tovar JA, Dudgeon D, Asch MJ. The management of umbilical hernia in infancy and childhood. J Pediatr Surg. 1975;10:405–9.
11. Fall I, Sanou A, Ngom G, et al. Strangulated umbilical hernias in children. Pediatr Surg Int. 2006;22:233–5.
12. Meier DE, OlaOlorun DA, Omodele RA, et al. Incidence of umbilical hernia in African children: redefinition of "normal" and reevaluation of indications for repair. World J Surg. 2001;25:645–8.
13. Snyder CL. Current management of umbilical abnormalities and related anomalies. Semin Pediatr Surg. 2007;16:41–9.
14. McCollum MO, Macneily AE, Blair HK. Surgical implications of urachal remnants, presentation and management. J Pediatr Surg. 2003;38:798–803.
15. Rouma BS, Lakhoo K. Vitelline duct anomalies. In: Paediatric surgery: a comprehensive text for africa (Ch 64). Seattle: Global HELP; 2010.
16. St-Vil D, Brandt ML, Panic S, et al. Meckel's diverticulum in children: a 20-year review. J Pediatr Surg. 1991;26:1289–92.
17. Menezes M, Tareen F, Saeed A, et al. Symptomatic Meckel's diverticulum in children: a 16-year review. Pediatr Surg Int. 2008;24:575–7.

The Liver and Gallbladder

38

Mark Davenport

Development of Liver and Gallbladder

Biliary Atresia

***The liver and gallbladder arise from the endoderm of the foregut as follows:**

1. In the middle of the third week, a ***liver bud*** (hepatic diverticulum) develops from the endoderm of the lower and of foregut.

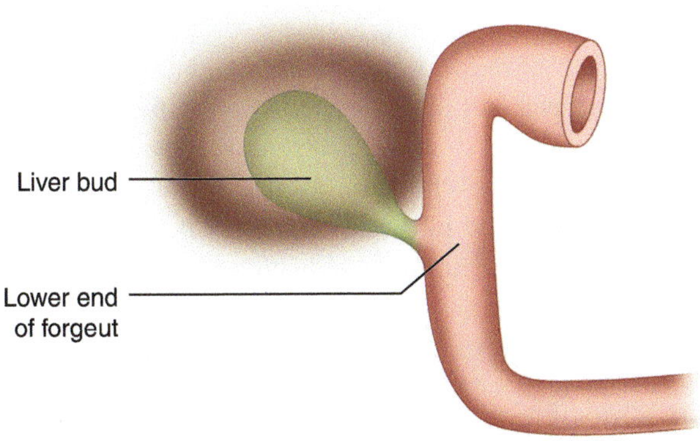

2. ***The liver bud divides into two parts*:**
   (a) ***Pars hepatica*:** is the large cranial part which will give the future liver
   (b) ***Pars cystica*:** is the small caudal part which will give the future gallbladder

   The original stalk of the liver bud will form the common bile duct.

M. Davenport (✉)
Department of Paediatric Surgery, King's College Hospital, London, UK

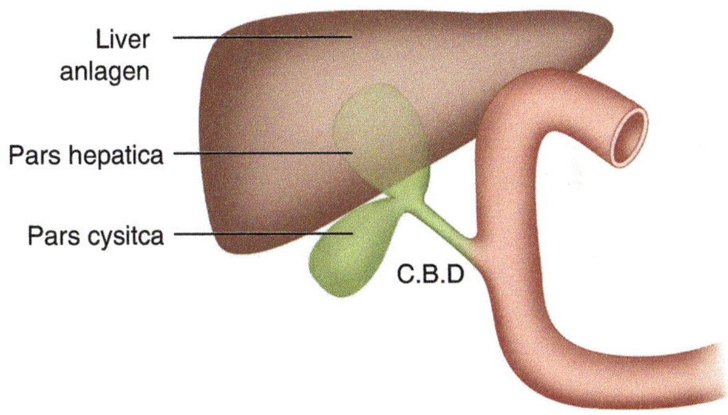

3. The pars hepatica invades the *septum transversum* and developing liver anlagen with the distal part assuming a *funnel-shaped* appearance.

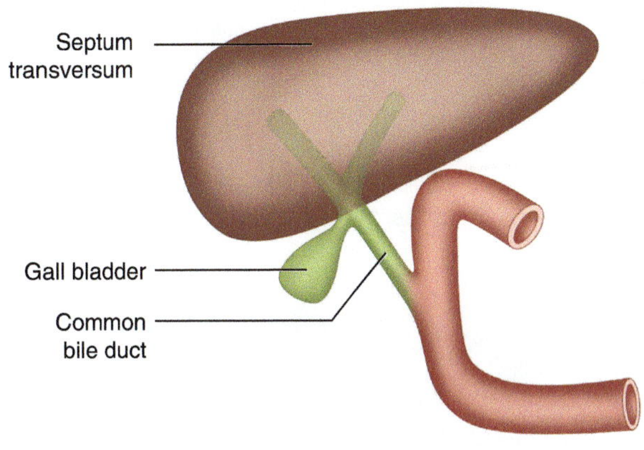

4. Hepatoblasts within the liver anlagen differentiate into future columns of hepatocytes broken up by the ingrowing vitelline (portal) venous network as *liver sinusoids*. The branching network of biliary epithelial cells tubularises around this venous ingrowth as the intrinsic bile duct system in-continuity with the extrinsic bile duct at the porta hepatis.

R. Carachi, S. H. E. Doss (eds.), *Clinical Embryology*, https://doi.org/10.1007/978-3-319-26158-4_38

5. The mesoderm of the septum transversum gives rise to the *fibrous tissue stroma and the capsule of the liver.*

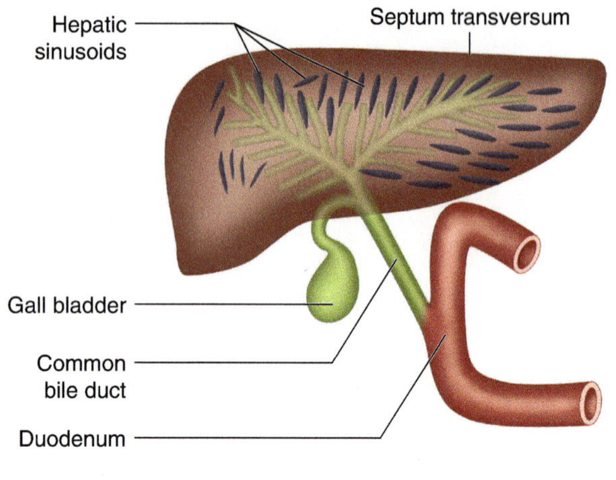

6. *The ligaments of the liver* develop from the septum transversum as follows:

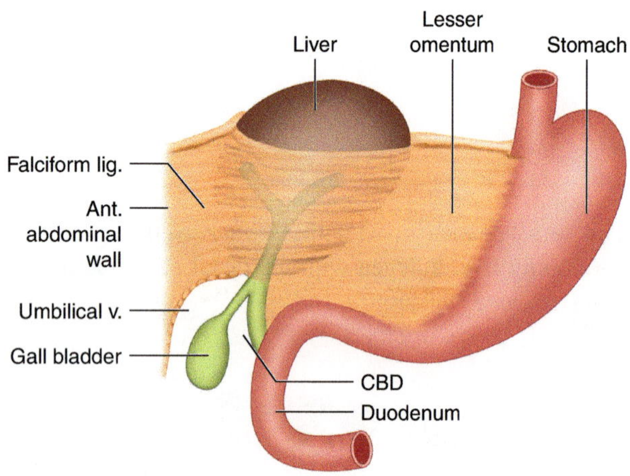

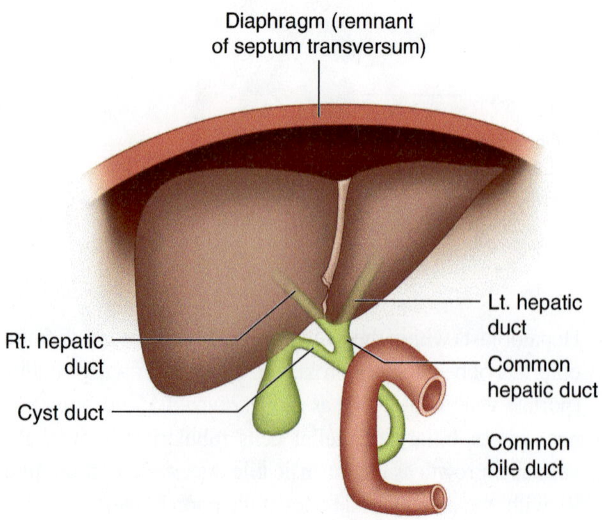

   (a) The mesoderm of septum transversum between the liver and the ant-abdominal wall becomes stretched and farms the ***falciform ligament*** and the umbilical v. originally traversing the septum lying in the free margin of the falciform lig.
   (b) The mesoderm of the septum transversum lying between the liver and the stomach becomes stretched to form the ***lesser omentum***.
   (c) The ***liver separates from the septum transversum*** (the remnant of which will form part of the diaphragm). Only a small area (the bare area of liver) remains in contact with the septum transversum.
7. ***The ducts*** of the liver develop as follows:
   (a) The ***right and left hepatic ducts*** develop from the stems of the right and left branches of the pars hepatica.
   (b) The original stalk of the liver bud elongates to form the ***common bile duct*** which opens at first into the ant-wall of the duodenum.
   (c) Later, as a result of rotation of the stomach the opening of the C.B.D. migrates to the posteromedial aspect of the second part of duodenum.

Introduction

Biliary atresia can be characterised as an obliterative cholangiopathy, but whether it is actually congenital or not can be the subject of fierce debate. So some variants (e.g. cystic biliary atresia and the biliary atresia splenic malformation syndrome) are clearly in utero developmental biliary problems, while in other infants (e.g. those with CMV IgM + ve- associated BA), the damage may actually be an inflammatory immune-mediated process initiated after birth [1, 2].

Background

It is a rare condition estimated to occur in 1 in 17,000 infants in the UK [3]. However, it is much more common in the East and specifically Taiwan and Japan (1 in 5–10,000) [4].

Broadly, the condition can be divided into four main groups depending on clinical characteristics. These are:

[*]*Congenital anomalies*:

(1) Congenital atresia of the common bile duct with distension of gallbladder and biliary passages.

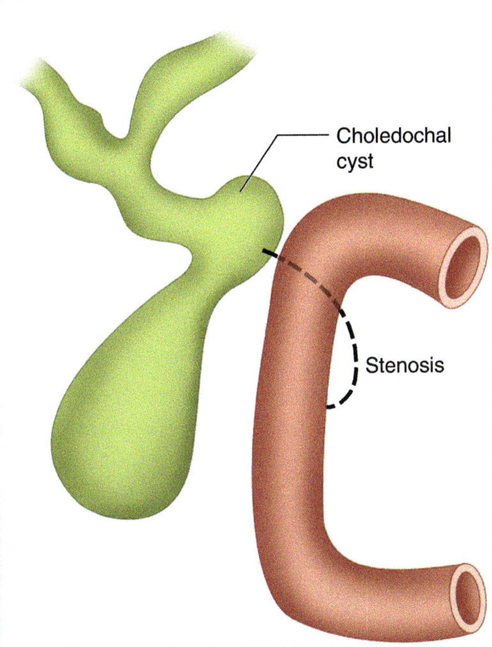

(2) Duplication of gallbladder which may be partial or complete.

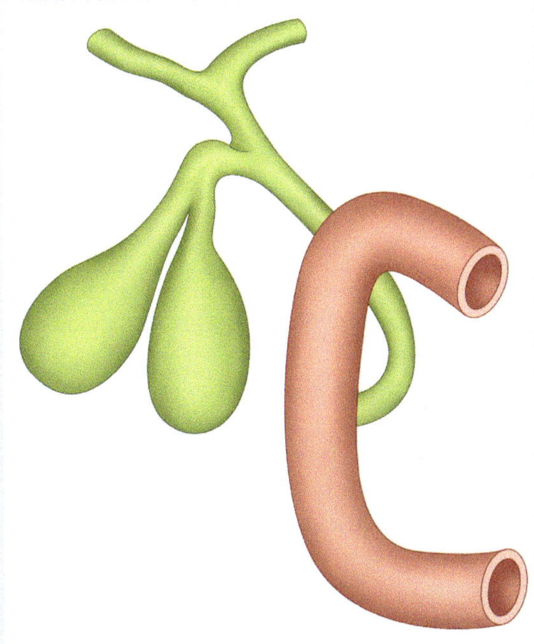

BA Splenic Malformation (BASM) Syndrome (Box 38.1)

Box 38.1 Biliary atresia splenic malformation syndrome

Components

- Biliary atresia (99%)
- Preduodenal portal vein (40%)
- Situs inversus (45%)
- Malrotation (40%)
- Polysplenia/asplenia (100%)
- Absence of vena cava (40%)
- Cardiac anomalies (50%) and malrotation (40%)

Incidence: 1 in 100,000 live births

Origin

- Possible genetic mutation in *CFC-1*
- Part of diabetic embryopathy spectrum (also includes transposition of great vessels, sacral agenesis)
- Key developmental stages in all of above 20–40 days' gestation

These infants (usually female) are characterised by a constellation of unusual anomalies including polysplenia (sometimes asplenia), vascular anomalies (preduodenal portal vein, absence of the cava), situs inversus and cardiac anomalies [5]. In these, there is a primary failure of extrahepatic bile duct development—the gallbladder is invariably atrophic and the CBD absent.

Cystic Biliary Atresia

This can be defined as cystic change in an otherwise obliterated biliary tract and is seen in about 5–10% of most large series [6]. The cyst itself may contain bile (~20%) or mucus implying onset after establishment of continuity between intra- and extrahepatic bile ducts (10–12 weeks' gestation) (Caponcelli et al., 2008). Around half of these are picked up antenatally on the maternal ultrasound scan.

CMV-IgM +ve-Associated Biliary Atresia

These are a group of infants who have IgM antibodies to cytomegalovirus and occur in about 10% of European series [2]. Clinically, they were older at diagnosis and are more jaundiced with larger spleens than comparable IgM-ve BA infants. Their extrahepatic duct appearance is more inflammatory with prominent nodes.

It appears likely that these have had abnormal exposure of bile duct-related antigens by early transient exposure to CMV. This has triggered an abnormal immune response led by T cells and NK cells damaging bile ducts and inducing fibrosis. Other viruses may be involved, but clinical evidence is lacking.

Isolated BA

This actually forms the largest clinical group (70–80%), but there are no distinguishing features to indicate actual aetiology.

All infants with BA will have conjugated jaundice, pale stools and dark urine. Some may also have a bleeding tendency postnatally because of lack of the fat-soluble vitamin K. Most come to surgery which aims at restoration of bile flow and salvage of their native liver. The median age for this in the UK is now below 50 days of age.

It is potentially a fatal disease if untreated; however, current results show that a clearance of jaundice rate of 55–60% is possible with experienced teams which translates to an overall survival of 90% and a native liver survival rate of about 45% at 10 years. Still BA remains the commonest indication for liver transplantation in childhood.

Development of the Bile Ducts and Liver

The bile ducts and liver develop as an endodermal bud from the distal foregut starting around the 20th day of gestation projecting within the ventral mesogastrium and into the mesenchyme of the **septum transversum**. Ingrowing hepatoblasts also have an intimate relationship with endothelial cell lined primitive vascular sinusoids.

The biliary tract itself is derived in two distinct ways, each with a different schedule. So, the extrahepatic bile duct is derived directly from foregut endoderm, while the intrahepatic ducts are derived later from **hepatoblasts** differentiating within the liver primordium. The original endodermal bud assumes a funnel-shaped appearance, always with a lumen, and a separating gallbladder visible by day 45. The lining **cholangiocytes** are of foregut origin, expressing transcription factors common to the pancreas and duodenum (e.g. *PDX-1, PROX-1, HNF-6*). By 28 days, the liver anlagen is populated by **hepatoblasts** and **haematopoietic cells** initially derived from yolk sac with the former arranged in plates, initially 3–4 cells thick, lining the vascular sinusoidal network.

Intrahepatic bile ducts only appear distinctly from about 49 days. Hepatoblasts will differentiate into hepatocytes or biliary epithelial cells (BEC) under the influence of Notch and canonical Wnt signalling. The so-called ***ductal (or limiting) plate*** is the first distinct primitive intrahepatic duct system and consists of a branch of the infiltrating portal venous network surrounded by a layer of mesenchyme and then a cylindrical double cell layer of darkly staining BECs. These then through a process of selection and deletion form a tabularised BEC lined network of interconnected bile ducts enveloped into the mesenchyme extended progressively from hilum to periphery.

Clearly, the extra- and intrahepatic ducts have to join and link up successfully to allow continuity with the interface believed to be at the porta hepatis [7]. One cause of what we later recognise as biliary atresia maybe failure to achieve this interface. Bile from hepatocytes begins to be secreted from their canalicular membrane of the hepatocyte from about 11 to 12 weeks.

Choledochal Malformation and the Common Channel (Fig. 38.1)

Introduction

This is rather an umbrella term for a number of unrelated conditions but characterised by dilatation of some part of the bile ducts and illustrates a working classification dividing it into five groups based on which part appears most affected. Type 1 is further divided according to which shape the extrahepatic dilatation best resembles—a cyst (i.e. spherical, "bladder"-shaped with a clear cut-off point at top and bottom) or fusiform (transition from normal to abnormal is gradual). The overall dimension of a Type 1f is smaller than a Type 1c.

Types 2 and 3 are rarely seen in practice and are illustrated for completeness. Type 4 includes both intra- and extrahepatic dilatation, while Type 5 can be simply an isolated, usually intrahepatic biliary cyst or part of the Caroli syndrome (Box 38.2).

Background

The incidence of choledochal malformation is not known with any precision, but again there is a marked geographical variation with most large surgical series being reported from China and Japan. It is increasing in incidence, but this is probably illusory as with increasingly liberal use of ultrasound in the diagnostic approach to children's illness more subtle biliary pathology is being picked up.

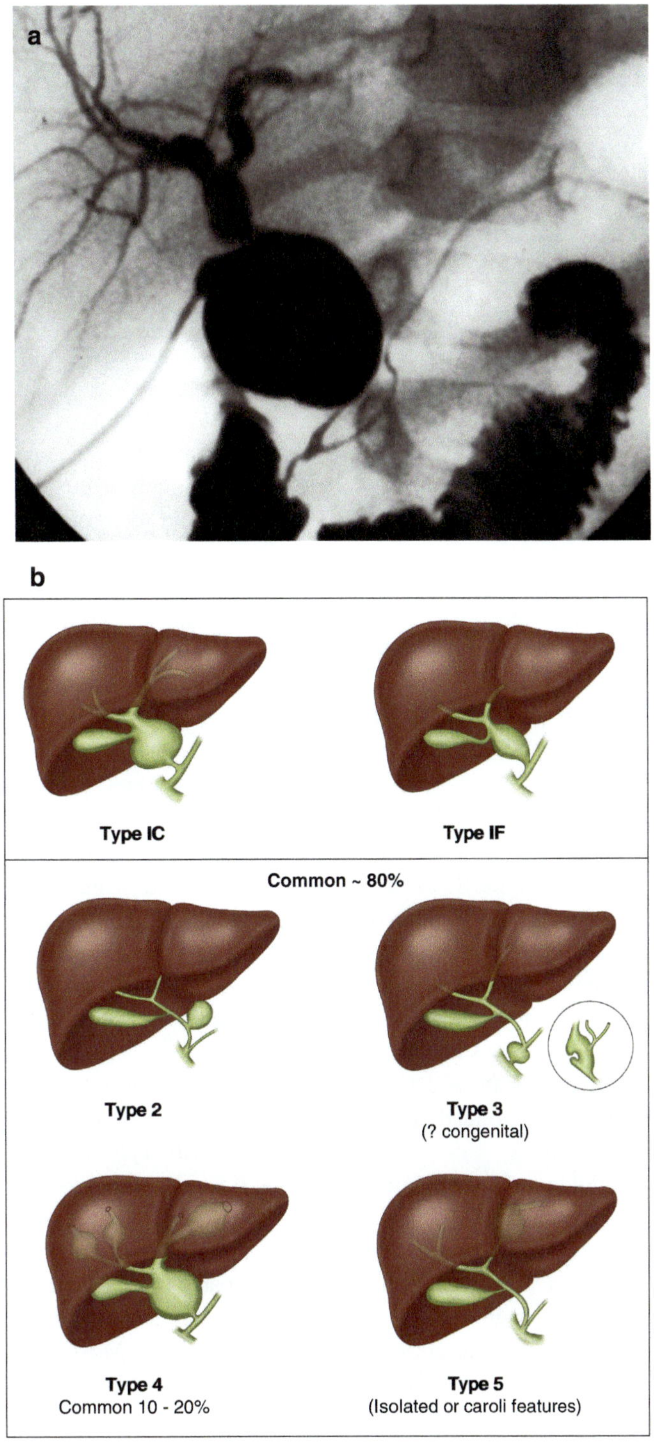

Fig. 38.1 (**a**) Cystic choledochal malformation (Type 1C), (**b**) Kings college hospital classification for choledochal malformation

Box 38.2 Caroli syndrome (if additional extrahepatic anomalies present), Caroli disease usually refers to isolated intrahepatic dilatation

Described by Jacques Caroli, a French gastroenterologist in 1958

Components

- $M = F$
- Segmental cystic dilatation of intrahepatic bile ducts
- Polycystic kidney disease (usually autosomal recessive PCKD)
- Risk of malignancy
- Liver fibrosis—portal hypertension

Incidence: Unknown

Origin

- Usually autosomal recessive genetic mutation in *PKHD1* gene (codes for fibrocystin)
- Key developmental stages, 50–70 days' gestation. Persistence of ductal plate malformation is a typical finding on liver histology

The aetiology of the CM varies with the type. So in Caroli syndrome, the intrahepatic biliary dilatation is genetic in origin as is probably the accompanying liver fibrosis. There is a degree of failure of maturation of the intrahepatic bile ducts, though this does not usually manifest as neonatal jaundice, and indeed many patients present much later as adults. The rare Type 2 diverticulum of the bile duct may be a simple congenital lesion but could represent a constrained perforation of the bile duct with subsequent healing.

There is still doubt about the aetiology of the commoner Types 1c and 1 f. It is possible to detect the cystic CM antenatally at the time of the fetal anomaly scan (20 weeks' gestation). Although they may be confused with cystic biliary atresia, these are different pathologies. They don't cause any problems during fetal life, although some can present in the neonatal period with obstructive jaundice. However, unlike cystic BA, the intrahepatic bile ducts will retain a normal tree-like branching pattern and dilate. So far, all antenatally detected CM have been of Type 1c. Although this doesn't exclude a congenital aetiology for Type 1f as they are smaller in size and degree of dilatation, it perhaps makes it less likely.

Distal biliary obstruction appears to be a likely aetiological factor in the commonest types of CM. This leads to increasing sustained intrabiliary pressure and hence dilatation [8]. There is a gradation of increasing pressures from Type 1f, through 1c to Type 4. What actually causes the obstruction is less clear. In some of those coming to early surgery because of neonatal jaundice, there is a very fine connection with the pancreatic duct, which is presumably developmental.

All the common types of CM (1c,1f, 4) are associated with the so-called **common channel**. This is where the junction of bile and pancreatic ducts occurs outside of the ampulla of Vater. Hence, there is the possibility of free intermixing of those respective organs' juices. In older children, the presence of bile in the pancreatic duct seems to be a cause of recurrent pancreatitis, while pancreatic juice in the bile duct was formerly believed to have some causative effect on biliary dilatation (Babbitt hypothesis).

The ventral pancreas and its duct are outgrowths of the main common bile duct, proximal to the gallbladder. It swings around to the back of duodenum to meld with the dorsal pancreatic anlage at about sixth week of gestation. Here, parenchyma and the two pancreatic ducts combine and meld. The final stage is absorption of the common pancreatobiliary channel into the wall of the duodenum and the creation of the ampulla as a way of separation.

Clinical Features

About 10–15% of CM now presented with an abnormal antenatal ultrasound scan, of which about 50% will actually become jaundiced during the neonatal period [9]. Otherwise, there are two principle modes of presentation with jaundice due to obstruction perhaps from 2 to 3 years of age. This is usually painless. The other mode is with recurrent abdominal pain, some with actual pancreatitis, and this usually occurs later on in childhood. Although there is a major degree of overlap, those with Type1c tend to present with the former and those with Type 1f, the latter. Ultrasound and now MR scans are the predominant methods of achieving a correct diagnosis, while surgical excision is the mainstay of treatment. The principle of surgery is to excise the dilatation extrahepatic part of the bile duct down to the junction with the pancreatic duct, leaving a transected common hepatic duct to anastomose with. Most centres use a Roux loop fashioned from the proximal jejunum to restore biliary continuity.

Congenital Anomalies of the Gallbladder

The gallbladder is initially a tubular outpouching from the main ventral bile duct and is definable in the embryo from about 30 days. It becomes a saccular structure from about the 11th week in gestation. Most congenital anomalies will actually be clinically silent, but a number are recognised and may cause symptoms.

Agenesis of the Gallbladder

True isolated agenesis (i.e. not associated with biliary atresia) is perhaps surprisingly common with an estimated incidence of 1 in 6300 [10]. These cases should be entirely asymptomatic.

Once again it is possible to detect the absence of a gallbladder on the maternal ultrasound scan [11]. At this stage, this might have implications on other associated pathologies. So it can be a finding in biliary atresia, cystic fibrosis and even some chromosomal abnormalities such as trisomy 18.

Left-Sided Gallbladder

This is a rare anomaly and is defined as a gallbladder to the left of the falciform ligament. Usually, the cystic duct then passes below the ligamentum teres to enter the common hepatic duct on its right side—as normal.

The condition can be entirely isolated and will be seen therefore as an incidental finding. Sometimes it is seen in combination with key developmental problems such as situs inversus, preduodenal portal vein, interrupted inferior vena cava and anomalous intrahepatic branching of the portal vein.

Double and Triple Gallbladder

The estimated incidence of duplication of the gallbladder is about 1 in 4000 and occurs because of bifurcation of the developing gallbladder during the fifth week of gestation [12]. Actual separation can occur at various levels with division of the gallbladder but leaving a single cystic duct.

At least one of the gallbladder elements will have some functional impairment and therefore predispose to stone formation, but this usually occurs during adult rather than childhood. Isolated case reports exist of the even less common triplicated gallbladders [13].

The so-called "Phrygian Cap" appearance was described as a radiological finding in 1935 [14] and is a variation with kinking of the fundus, sometimes simulating a mass or stone within. It again is surprisingly common and estimated at 4% but has no real pathological significance.

Multiseptate Gallbladder

A multiseptate gallbladder is divided into smaller compartments by thin-walled membranes lined with columnar epithelium and was first recognised as an entity in 1963 by Simon and Tandon [15]. Externally, they can look normal but often have a spongy bosselated feel which when cut open shows a honeycomb appearance. There is communication with each septae via small perforating holes. A muscular layer within each septa joining the outer muscular layer of the gallbladder may or may not be present.

Multiseptate gallbladder can be seen in association with other ductal anomalies, gallbladder ectopy, choledochal malformations and anomalous pancreaticobiliary junction. Most clinical reports suggest that they can be a cause of right upper quadrant pain and discomfort, but increasing use of ultrasound will identify many of these incidentally. Nevertheless removal is probably advised [16].

A number of hypotheses have been put forward to explain a multiseptate gallbladder but none with any real evidence. Early authors suggested that the initial embryological solid gallbladder fails to vacuolise appropriately or that there was variation in kinking and clefting of gallbladder during development. However, more recently and actually based on observational studies in embryos and fetuses, the whole concept of a "solid stage" in any part of the extrahepatic part of the biliary tree has been refuted [17, 18].

Congenital Vascular Anomalies of the Portal Venous System

Introduction

The fully developed vascular anatomy of the liver has arisen by a whole series of complex pathways. As a result, a number of different structural anomalies are recognised with many being the cause of symptoms and pathologies. Most of these concern the development of the portal venous system rather than from the later ingrowth of the hepatic arteries. Indeed although anatomical variation in the hepatic arteries is common, with replaced and accessory arteries actual functional issues are not described.

Congenital Portosystemic Shunts

Background

The first clinical description of a congenital portocaval shunt (CPS) was by John Abernethy, a surgeon practising at St Bartholomew's Hospital, London. He described the post-mortem findings of a 10-month-old girl with multiple congenital anomalies and included a description of the portal vein terminating in the IVC at the level of the renal veins rather than entering the liver. Sometimes this is still referred to as an Abernethy malformation in his honour [19].

There are two broad categories of CPS): those where the junction is end-to-side and therefore there is no intrahepatic portal venous system (Type 1) and those where there is a more side-to-side arrangement with a degree of intrahepatic venous preservation (Type 2).

Both types must be clearly distinguished from a persistence of a **ductus venosus**. This is a normal structure which connects the left portal vein with the left or middle hepatic vein and allows bypass of the liver sinusoids by oxygenated blood returning via the umbilical vein from the placenta. Following clamping of the cord at birth, intraportal pressure falls, changes occur in the muscularis of the ductus and it functionally closes during the first week. In a small proportion of infants, this does not happen and can be detected on ultrasound. In contrast with the other CPS), most examples of a patent ductus venosus will close spontaneously in the first year of life.

Clinical Features

The clinical presentation of CPS) is extremely variable [20]. It may be entirely asymptomatic and be an incidental finding on prenatal (10%) or postnatal ultrasound. Some are detected as part of a neonatal screening programmes for galactosaemia (10–20%) or as part of an investigation strategy for neonatal jaundice.

Perhaps more importantly the two main areas of actual pathological symptoms are due to:

1. Abnormal arterialisation of the liver (due to venous deprivation) causing parenchymal neoplastic change. These are usually benign tumours such a focal nodular hyperplasia but can be malign as in hepatoblastoma.
2. Bypass of the normal metabolic processes of the liver parenchyma. Thus "unfiltered" mesenteric blood is emptied directly into the cava and systemic venous system. The degree of this can be reflected in elevated plasma ammonia levels. In adults, but rarely children, this can cause overt encephalopathy. However, there is increasing evidence that even in children there can be behavioural and developmental issues.

Treatment is aimed at surgical (occasionally radiological) closure of the CPS. This should be possible in all those with Type 2 lesions but may be impossible in those with Type 1 CPS. For these then liver transplantation is the only possibility, and that would depend on their symptoms or the pathological state of the liver.

Developmental Origins

Three pairs of major veins exist in the fifth week of embryologic development:

1. **Vitelline veins**—these surround the developing midgut arising originally from the yolk sac and terminate in the sinus venosus. Coming towards the liver anlage, they have interconnecting branches resembling as stepladder which undergoes from the 8 to 12th week of gestation remodelling to form a single but now S-shaped portal vein around the duodenum (Fig. 38.2). A preduodenal portal vein may result from failure of part of the vitelline system to regress.
2. **Umbilical veins**—the right umbilical vein disappears by the eighth week of gestation with the left enveloped by the liver and connected to the developing portal venous system.
3. **Posterior cardinal veins**—these carry blood back from the lower half of the body along the posterior body wall. These ultimately become the paired azygous and hemiazygous venous system which empty into the superior vena cava. They are superseded in function (from the sixth week gestation) by subcardinal and supracardinal veins draining the kidney and gonads. The **inferior vena cava** is a conglomerate of these later venous channels which connect with the hepatic venous confluence and right side of the heart (Fig. 38.3).

The veins intimately associated with the developing liver such as the terminal part of the right vitelline vein and subcardinal vein are in close approximation, and presumably CPS shunts arise from aberrant interlinkage. They don't represent however any normal stage in venous development.

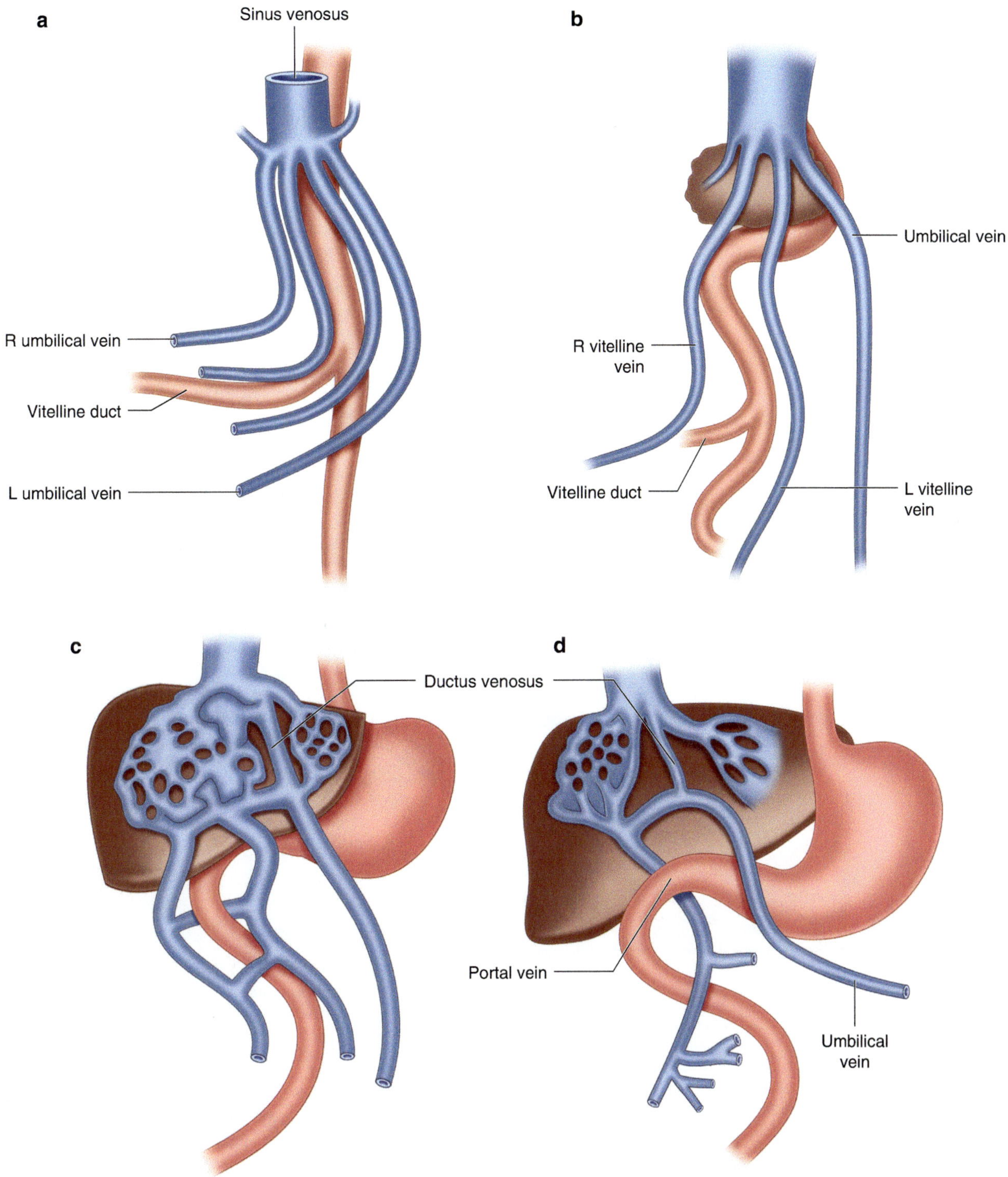

Fig. 38.2 Development of the portal venous system

Fig. 38.3 Development of the inferior vena cava

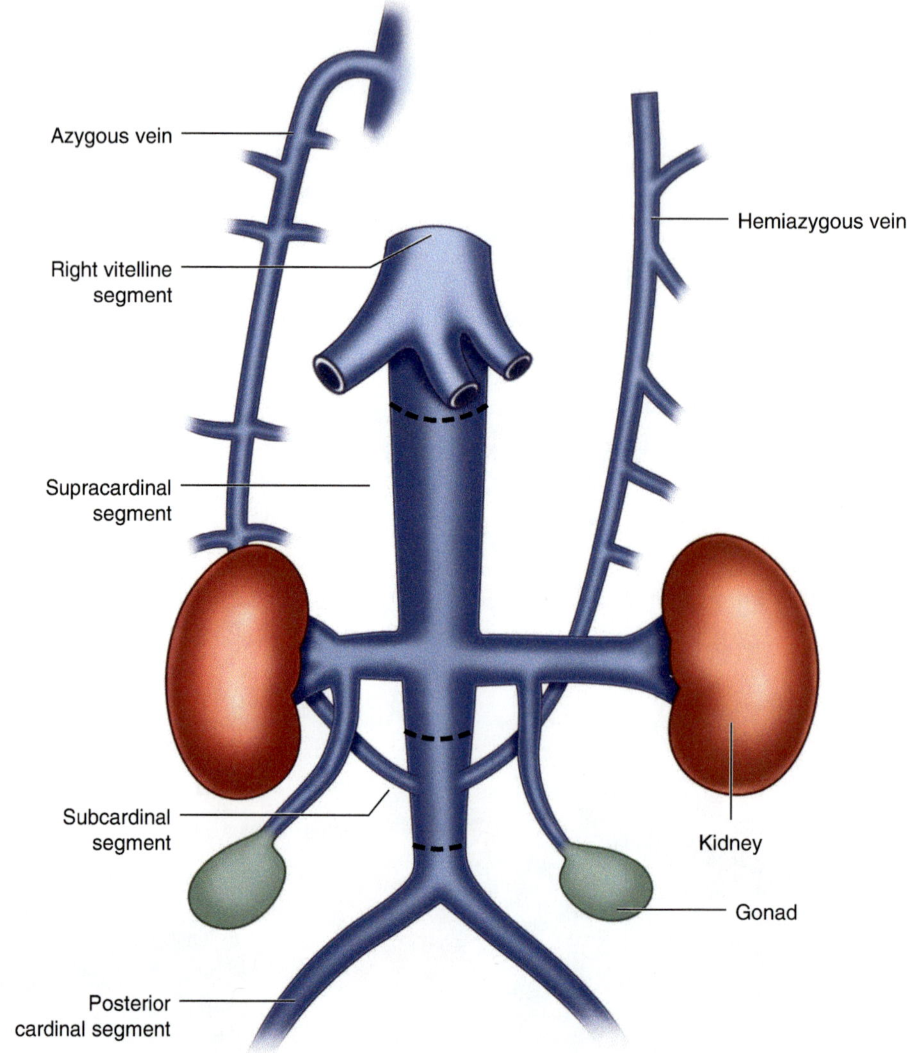

References

1. Hartley JL, Davenport M, Kelly DA. Biliary atresia. Lancet. 2009;374(9702):1704–13.
2. Zani A, Quaglia A, Hadzić N, Zuckerman M, Davenport M. Cytomegalovirus-associated biliary atresia: an aetiological and prognostic subgroup. J Pediatr Surg. 2015;50(10):1739–45. https://doi.org/10.1016/j.jpedsurg.2015.03.001.
3. Livesey E, Cortina Borja M, Sharif K, Alizai N, McClean P, Kelly D, Hadzic N, Davenport M. Epidemiology of biliary atresia in England and Wales (1999-2006). Arch Dis Child Fetal Neonatal Ed. 2009;94:F451–5.
4. Hsiao CH, Chang MH, Chen HL, Lee HC, et al. Universal screening for biliary atresia using an infant stool color card in Taiwan. Hepatology. 2008;47:1233–40.
5. Davenport M, Tizzard SA, Underhill J, Mieli-Vergani G, Portmann B, Hadzić N. The biliary atresia splenic malformation syndrome: a 28-year single-center retrospective study. J Pediatr. 2006;149:393–400.
6. Caponcelli E, Knisely AS, Davenport M. Cystic biliary atresia: an etiologic and prognostic subgroup. J Pediatr Surg. 2008;43:1619–24.
7. Tan CE, Driver M, Howard ER, Moscoso GJ. Extrahepatic biliary atresia: a first-trimester event? Clues from light microscopy and immunohistochemistry. J Pediatr Surg. 1994; 29:808–14.
8. Turowski C, Knisely AS, Davenport M. Role of pressure and pancreatic reflux in the aetiology of choledochal malformation. Br J Surg. 2011;98:1319–26.
9. Redkar R, Davenport M, Howard ER. Antenatal diagnosis of congenital anomalies of the biliary tract. J Pediatr Surg. 1998;33:700–4.
10. Bennion RS, Thompson JE Jr, Tompkins RK. Agenesis of the gallbladder without extrahepatic biliary atresia. Arch Surg. 1988;123:1257–60.
11. Shen O, Rabinowitz R, Yagel S, Gal M. Absent gallbladder on fetal ultrasound: prenatal findings and postnatal outcome. Ultrasound Obstet Gynecol. 2011;37:673–7.
12. Boyden EA. The accessory gallbladder, an embryological and comparative study of aberrant biliary vesicles occurring in man and the domestic mammals. Am J Anat. 1926;38:117.
13. Schroeder C, Draper KR. Laparoscopic cholecystectomy for triple gallbladder. Surg Endosc. 2003;17:1322. Epub 2003 Jun 13

14. Boyden EA. The phrygian cap in cholecystography: a congenital anomaly of the gallbladder. Am J Radiol. 1935;33:589.
15. Simon M, Tandon BN. Multiseptate gallbladder. A case report. Radiology. 1963;80:84–6.
16. Geremia P, Tomà P, Martinoli C, Camerini G, Derchi LE. Multiseptate gallbladder: clinical and ultrasonographic follow-up for 12 years. J Pediatr Surg. 2013;48:E25–8.
17. Vijayan V, Tan CE. Development of the human intrahepatic biliary system. Ann Acad Med Singap. 1999;28:105–8.
18. Tan CE, Moscoso GJ. The developing human biliary system at the porta hepatis level between 29 days and 8 weeks of gestation: a way to understanding biliary atresia. Part 1. Pathol Int. 1994;44:587–99.
19. Howard ER, Davenport M. Congenital extrahepatic portocaval shunts--the Abernethy malformation. J Pediatr Surg. 1997;32:494–7.
20. Bernard O, Franchi-Abella S, Branchereau S, Pariente D, Gauthier F, Jacquemin E. Congenital portosystemic shunts in children: recognition, evaluation, and management. Semin Liver Dis. 2012;32:273–87.

Further Reading

Schoenwolf GC, Bleyl SB, Brauer PR, Francis-West P. Larsen's Human Embryology. Philadelphia: Churchill Livingstone; 2015.

39 The Pancreas

Suzanne McMahon

Development of Pancreas (Figs. 39.1, 39.2 and 39.3)

1. The pancreas develops in the fourth week from two endodermal buds:

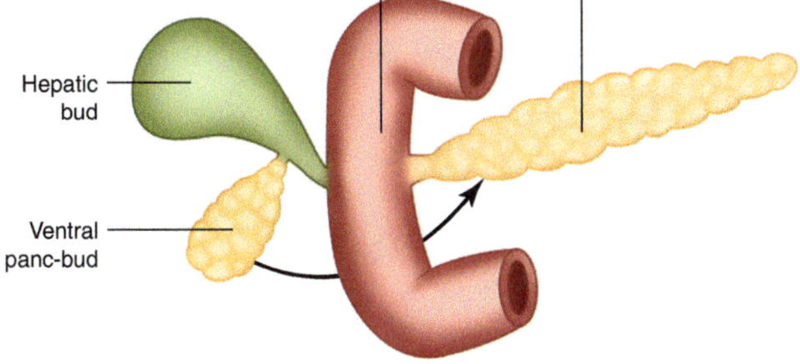

   (a) ***Dorsal pancreatic bud*** which arises from the endoderm of the dorsal wall of the duodenum slightly above the liver bud and extends dorsally and upwards into the mesoduodenum.
   (b) ***Ventral pancreatic bud*** which arises from the ant. wall of the stem of the hepatic bud.
2. In the seventh week, the ventral bud migrates dorsally to lie just below and behind the dorsal bud. The two buds fuse together to form the pancreas. The ventral bud forms the greater part of the head + the uncinate process while the dorsal bud forms the remainder of the pancreas.

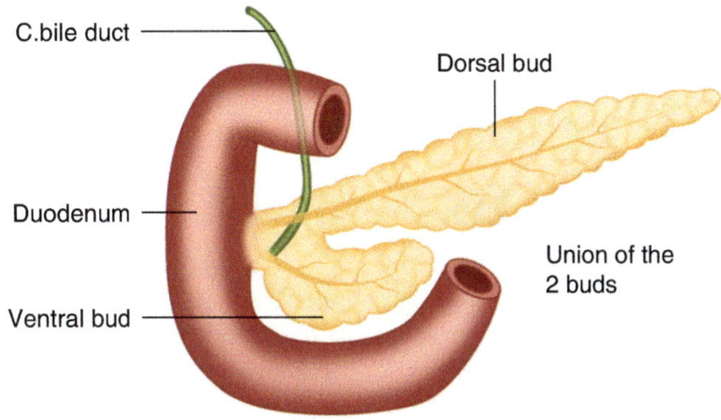

3. Later on, ***the ducts of the dorsal and ventral pancreatic buds join*** each other to form the main pancreatic duct which is joined by the common bile duct to form the ampulla of Vater opening into the duodenum. The part of the dorsal pancreatic passing to the duodenum is obliterated but occasionally it remains patent forming the accessory pancreatic duct.

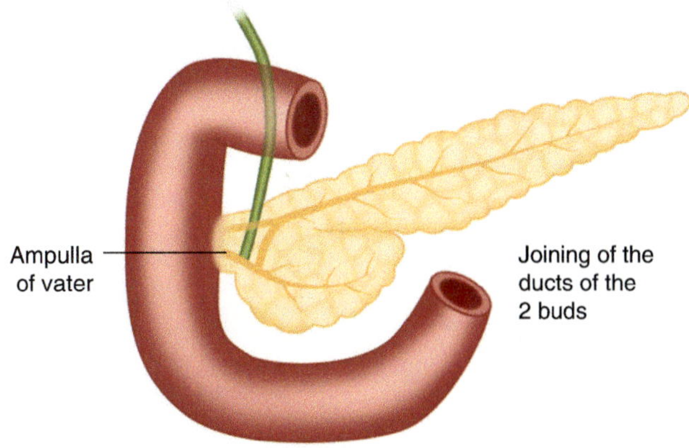

S. McMahon (✉)
Queen Elizabeth University Hospital, Glasgow University, Glasgow, UK

R. Carachi, S. H. E. Doss (eds.), *Clinical Embryology*, https://doi.org/10.1007/978-3-319-26158-4_39

4. ***The islets of Langerhans*** appear in the third month but insulin secretion begins about the fifth month.

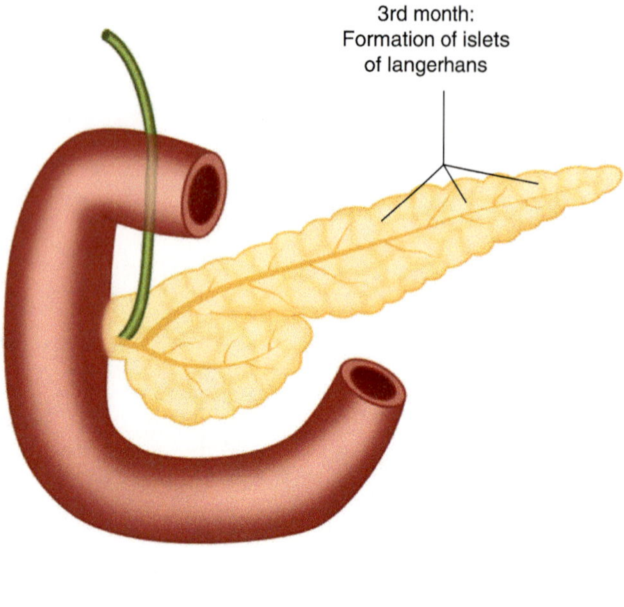

The pancreas is derived from the Greek word meaning '*pan*', 'all,' and '*Kreas*', 'flesh'. It is an important endocrine and exocrine organ sited in the retroperitoneum of the epigastrium. The pancreas is derived from the foregut and so is endodermal in origin. Arising from the endoderm of the duodenum, the pancreas initially starts off as two buds (the ventral and dorsal buds) which subsequently fuse to form the organ as we know it.

Pancreas organogenesis begins at 4 weeks' postconception. Initial tissue develops under the influence of pancreas and duodenal homeobox gene 1 (Pdx1). This is a pancreas promoting transcription factor expressed in the stomach and preduodenal endoderm and promotes dorsal and ventral pancreatic bud formation. Indeed, homozygous defects of Pdx 1 in mice have resulted in pancreatic agenesis.

The two 'buds' of pancreatic tissue are expanded by the activation and deactivation of Sonic hedgehog signalling in both the pancreatic parenchyma and the adjacent duodenal tissue. A ventral bud of endodermal tissue is formed in the ventral mesentery (connecting the foregut to the ventral abdominal wall). It takes position initially lying adjacent to the liver bud. At the same time, a larger cluster of endoderm forms the dorsal pancreatic bud within the dorsal mesentery (connecting the foregut with the posterior abdominal wall), more specifically the mesoduodenum. This is placed in a slightly more cephalad position relative to the ventral pancreatic bud and liver bud. The two pancreatic buds develop independently with their own respective ducts draining into the wall of the duodenum as the ventral and dorsal ducts.

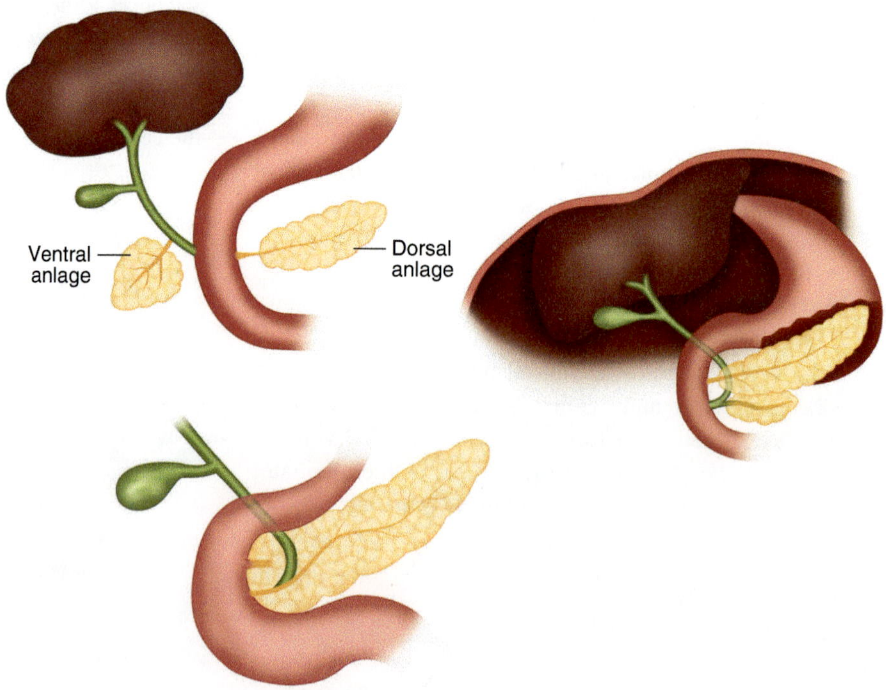

Fig. 39.1 Development of pancreas

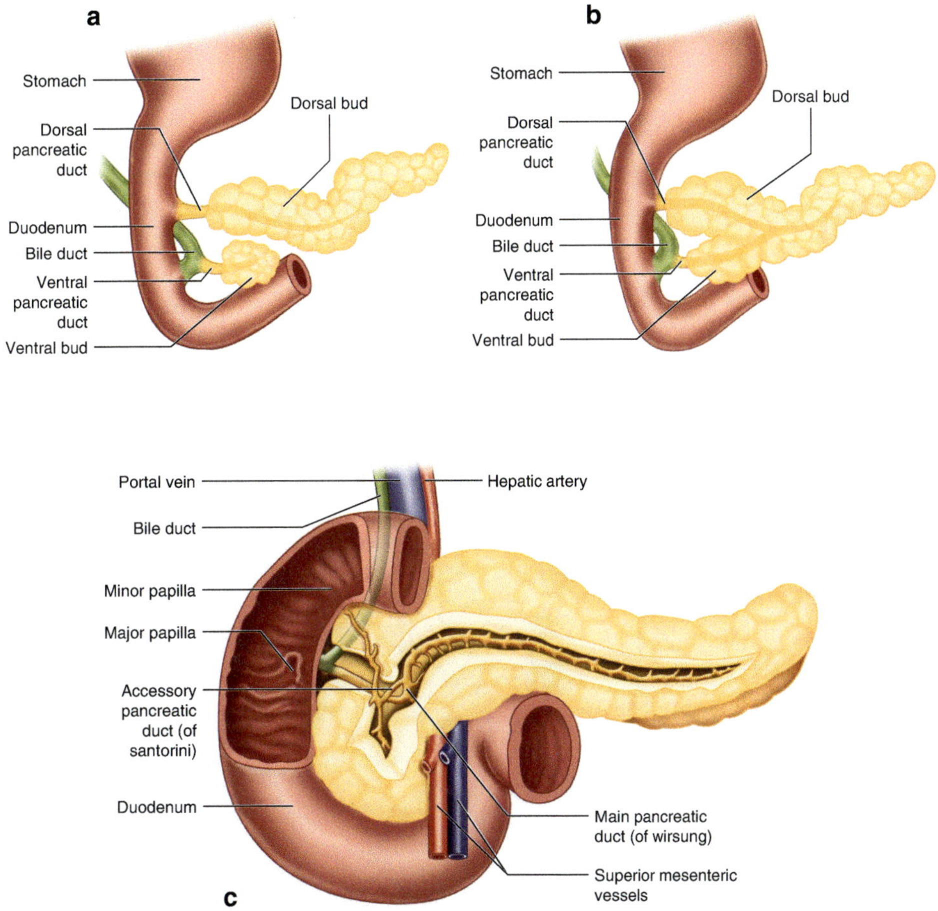

Fig. 39.2 Development of pancreas

At 6 weeks, the duodenum moves rightward and the ventral bud rotates posteriorly, coming to lie just inferior and posterior to the dorsal bud. By the end of the sixth week, the two buds lie adjacent and have created the shape of the anatomical pancreas, with the ventral bud forming the uncinate process and posterior head of the pancreas and the dorsal bud forming the anterior head, neck, body and tail of the pancreas. Initially, these two buds are separate and drained by separate ducts into the duodenum (the ventral and dorsal ducts). At around 7 weeks, the two parenchymal buds fuse, and after the seventh week, the ductal systems fuse. The distal part of the larger dorsal duct fuses with the entire smaller ventral bud duct to form the main pancreatic duct (of Wirsung). This ductal arrangement is joined by the common bile duct draining the liver and ends in the ampulla Vater entering the medial duodenal wall at D2 as the major duodenal papilla. The segment of the dorsal duct that is proximal to the fusion with the ventral duct is either obliterated or persists as an accessory pancreatic duct (of Santorini) draining directly into the duodenum as the minor duodenal papilla.

Initially, the posterior rotation of the ventral bud causes the superior mesenteric vessels to become encased by pancreatic tissue, but with time the vessels are seen to emerge from the substance of the pancreas.

It is during the sixth week that the pancreas buds are seen to develop acini and ducts. The islets of Langerhans can be seen from 3 months' gestation but do not begin to perform their endocrine function of insulin secretion until about the fifth month.

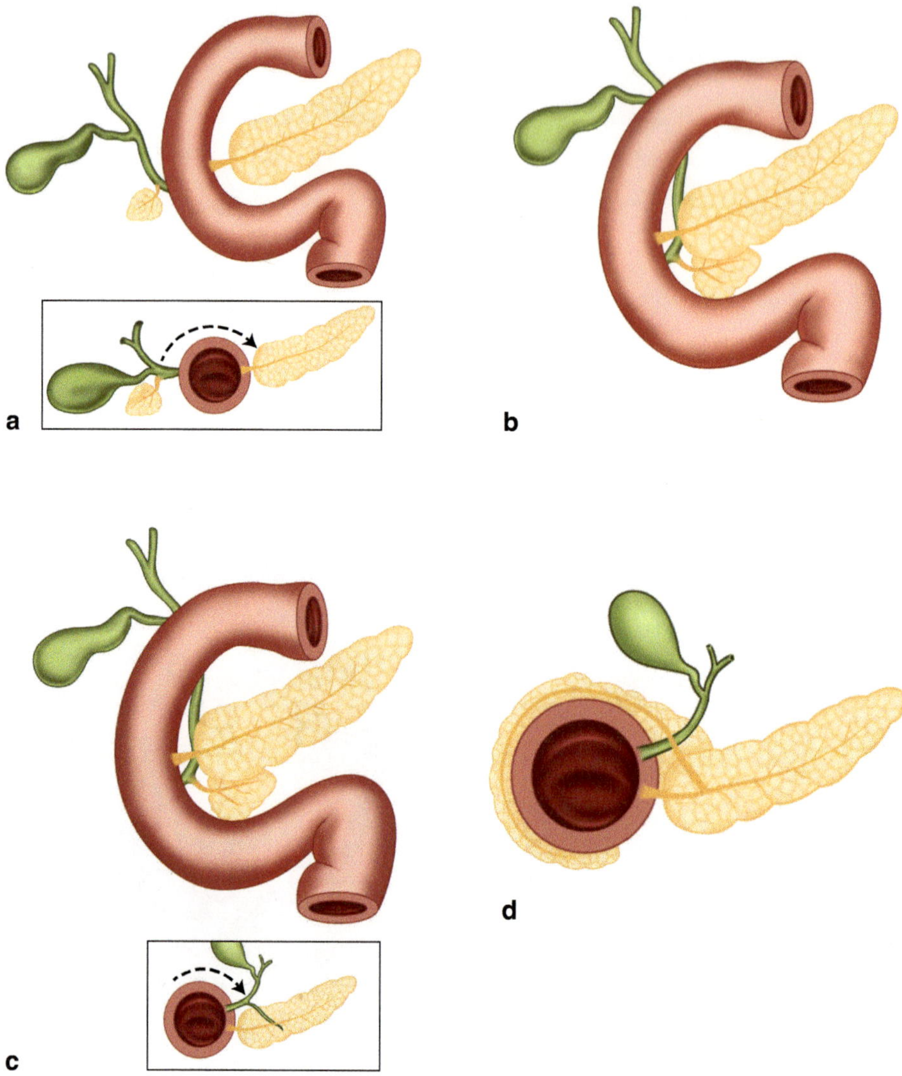

Fig. 39.3 (**a–c**) Normal pancreatic development D Lecco's theory of pancreatic development

Congenital Malformations

Anomalies of the pancreas are not uncommon and vary significantly in their presentation.

Agenesis and Hypoplasia

Complete pancreatic agenesis occurs in the absence of formation of the pancreatic buds in the fourth week. Its incidence is unknown as it results in severe intrauterine growth retardation at an early stage and subsequently intrauterine death. As mentioned previously, this has been observed in mice with a targeted homozygous Pdx1 mutation, although this remains undetermined in humans.

Partial agenesis is compatible with life and most commonly affects the dorsal pancreatic bud. Dorsal pancreatic agenesis can be an isolated phenomenon; however, it has also been associated with heterotaxia syndromes. It can be complete, resulting in the absence of the anterior head, neck, body and tail of the pancreas as well as the absence of the duct of Santorini and the minor duodenal papilla, or it can be partial, which will result in loss of a variable amount of tissue originating from the dorsal pancreatic bud but with definite preservation of the duct of Santorini and the minor papilla. Patients with partial dorsal pancreatic bud agenesis may be completely asymptomatic depending on the volume and function of remaining pancreatic parenchyma. However, as significant portion of the tail of the pancreas is populated with islets of Langerhans, patients deficient in 'tail' tissue often present with diabetes mellitus.

Accessory Pancreas

Ectopic or accessory pancreatic tissue (heterotopic pancreas) is defined as pancreatic tissue with no anatomic or vascular connection to the normal pancreas. The incidence has been reported to be between 2 and 15% of the population. It can be found anywhere in the GI tract, although is most commonly found in the distribution of the foregut, between the distal duodenum and the primary intestinal loop. Most often it is found in the mucosa of the stomach, the wall of the duodenum or proximal jejunum or within 6% of Meckel's diverticulum, which can be confirmed on histological examination of an excised specimen. Around 50% are reported to be within the stomach and the duodenum, although rarely, ectopic pancreatic tissue has been found in the gallbladder, appendix, colon and mesentery. Most areas of ectopic pancreatic tissue are detected incidentally and are asymptomatic; however, patients can present with epigastric pain, dyspeptic symptoms, haemorrhage, intussusception or obstruction. Ectopic pancreatic tissue has the potential to become inflamed or malignant in the same way the 'normal' pancreas can; however, adenocarcinoma of ectopic pancreatic tissue is rare with only around 30 published cases reported.

Annular Pancreas

On occasion, erroneous rotation of the pancreas during the sixth week will occur resulting in an *annular pancreas*. This is thought to occur at least in part due to Sonic hedgehog signalling dysfunction.

Annual pancreas describes a congenital anomaly in which there is a ring or strip of pancreatic tissue originating from the ventral pancreatic bud surrounding the second part of the duodenum, which causes the obstruction. This encircling tissue may contain a major pancreatic duct that curves posteriorly around the second part of the duodenum to enter the main pancreatic duct (of Wirsung) close to the ampulla of Vater. The strip of encircling pancreatic tissue is histopathologically identical to 'normal' pancreatic tissue and is therefore susceptible to the same pathological processes.

Annular pancreas can be complete or incomplete and affects 1/1000 to 1/20000 population. The circular encasement of the duodenum in pancreatic tissue was first described by Tiedemann (1818) and was later named by Ecker (1862) who reported 'a strip of glandular substance lying across the descending portion of the duodenum'.

At least 50% of patients with annular pancreas present as neonates usually on the first or second day of life with duodenal obstruction and/or bilious vomiting and are often included in reference with duodenal atresia as its diagnosis and management are the same. In fact, up to 70% of infants with annular pancreas have associated congenital anomalies, including duodenal stenosis/atresia, Down syndrome, tracheoesophageal fistula and various congenital cardiac defects.

Patients who present later in life have an incomplete annular pancreas and often present with symptoms of pain and or pancreatitis rather than obstructive symptoms. Associated congenital anomalies are much less frequent in adults with annular pancreas, occurring in only 16%, although it is strongly associated with pancreas divisum in adults, occurring in up to one third of patients.

Overexpression of the ventral-specific gene transmembrane 4 superfamily member 3 (tm4sf3) has also been associated with annular formation, and cases of familial annular pancreas have been described, implying a further heritable underlying role of genetics.

Many theories have been postulated as the cause of the annular pancreas, but no clear consensus has been reached. *Tieken* hypothesised in 1901 that the creation of the annular pancreas was secondary to hypertrophy of both the ventral and dorsal buds which resulted in them meeting and fusing anteriorly over the descending duodenum. *Lecco*, like many who came after him in 1910, suggested that the ventral bud became fused at its base with the wall of the duodenum. When duodenal rotation follows, the pancreatic tissue, taken with it, then encircling the duodenum.

Around the same time, Baldwin postulated the initial presence of a bilobed ventral bud, with the left bud developing to eventually encircle the duodenum.

Other theories include the development of annular pancreas secondary to the fusion of accessory pancreatic tissue from the wall of the duodenum (*Erimoglu*), and indeed that annular pancreas is a result of failed duodenal development rather than the other way around (*Elliot*).

Variations in ductal anatomy in patients with annular pancreas have prompted hypotheses that perhaps the embryogenesis is multifactorial in origin, but generally most support is given to the theories of Baldwin and Lecco.

Pancreas Divisum and Accessory Ducts

Around the seventh week, fusion of the two buds and subsequently their ductal systems occurs. However, failure of fusion at this stage is not an uncommon occurrence. **Pancreas divisum** is the failure of fusion of the two pancreatic buds in the presence of normal rotation. It is estimated that this occurs in between 3 and 22% of the population and is more common in Caucasians than Asian populations. It can be complete or incomplete. Failure of fusion means that secretions from the head, neck, body and tail which develop from the dorsal bud are forced to drain through the minor duodenal papilla and only secretions from the smaller uncinate process, derived from the ventral bud, will drain through the major duodenal papilla. Subsequently, impaired drainage of the original dorsal bud (head, neck body and tail) can leave the patient predisposed to pancreatitis. Because no fusion occurs in complete pancreas divisum, the resulting duct systems are described from their embryonic origin as the dorsal and ventral pancreatic ducts.

Pancreas divisum is the most common congenital variation of pancreatic development. The significance of pancreas divisum is disputed. It can be asymptomatic, and a significant proportion of the population are found to have this anomaly at autopsy. However, the rate is increased in those investigated with chronic or acute recurrent pancreatitis.

Incomplete fusion resulting in fusion of parenchyma without complete fusion of ductal systems gives a number of possible anatomical configurations. Persistence of the dorsal pancreatic bud duct as the minor duodenal papillae/accessory pancreatic duct creates a bifid drainage of the pancreas. Although the bifid configuration with dominant duct of Wirsung drainage is most common (60%), a rudimentary, non-draining duct of Santorini (30%), or dominant duct of Santorini without divisum (1%), also occurs. Rarely, at ERCP, the duct of Santorini is seen to take a curved course before it fuses with the duct of Wirsung. This is known as *Ansa Pancreatica* and can be distinguished from annular pancreas as it neither crosses the duodenum, nor has a direct ampulla entering it.

Syndromes Affecting the Pancreas

As mentioned previously, ***heterotaxia syndromes or isomerisms*** are known to affect the embryology and subsequent anatomy of the pancreas of the developing foetus, along with many other organs and systems including respiratory, cardiac, gastrointestinal, hepatobiliary, central nervous and urology systems. Partial agenesis is compatible with life and most commonly affects the dorsal pancreatic bud, leaving a truncated pancreas with only a head and part of a body of pancreas.

Beckwith-Wiedemann syndrome (BWS) is a genetic overgrowth disorder affecting 1/12000 live births. It can be hereditary but is more commonly a sporadic (85%) mutation on chromosome 11. Cases are affected varyingly with a constellation of potential manifestations. Classical features include macroglossia, macrosomia, exomphalos, naevus flammeus and neonatal hypoglycaemia, although it is unusual that a patient would have all of these. Other systems can also be affected and are associated with anomalies of the multiple pancreas.

Neonatal hypoglycaemia is common but usually transient in patients with BWS. Rarely, it will persist beyond the neonatal period requiring further treatment. It is thought to be due to hyperinsulinism.

Beckwith-Wiedemann syndrome can be associated with cystic dysplasia of the pancreas. Pancreatic hypertrophy and cystic dysplasia of the pancreas are caused by ductal proliferation, virtual absence of normal exocrine tissue and an increase in endocrine tissue, namely, giant islet-like structures formed by smaller subunits, although the architecture still reflects that of the normal pancreas with non-B cells surrounding B cells.

Finally, amongst other embryological tumours of infancy, patients with BWS are at increase risk of developing pancreaticoblastoma. Although this is rarely congenital, increased risk of various solid organ tumours, especially in the first 4 years of life, mandates regular screening with ultrasound and serology testing (*alpha-fetoprotein*). This is advised until 8 years old when the risk falls.

Recommended Reading

Etienne D, John A, Menias CO, Ward R, Tubbs RS, Loukas M. Annular pancreas: a review of its molecular embryology, genetic basis and clinical considerations. Ann Anat. 2012;194(5):422–8.

Mitchell B and Sharma R. Embryology: an illustrated colour text. 2nd edn.

Sadler TW. Langman's medical embryology. 11th edn.

Stefan Y, Bordi C, Grasso S, Orci L. Beckwith-Wiedemann syndrome: a quantitative, immunohistochemical study of pancreatic islet cell populations. Diabetologia. 1985;28(12):914–9.

Türkvatan A, Erden A, Akif Türkog˘lu M, Yener O. Congenital variants and anomalies of the pancreas and pancreatic duct: imaging by magnetic resonance cholangiopancreaticography and multidetector computed tomography. Korean J Radiol. 2013;14(6):905–13.

White J, Robert Z, Gest T, Beale E. Pancreas divisum: a common developmental variant that deserves attention in preclinical medical education. Clin Anat. 2014;27:1038–45.

Spleen

40

Mohamed Abdel-Latif and Mohamed Sameh Shalaby

Introduction

Congenital anomalies of the spleen include persistent lobulation, accessory spleen, polysplenia, asplenia, splenogonadal fusion, and wandering spleen. These anomalies can be solitary or part of a syndrome.

Although lobular spleen and accessory spleens are relatively frequent, other congenital anomalies of the spleen are rare. Anomalies of the spleen are difficult to diagnose due to unfamiliarity to clinicians and radiologists; however, caregivers should keep in mind the differential diagnosis and management of different pathologies.

Lobular Spleen

Embryological splenic lobules usually disappear before birth. Persistence of splenic lobules results in lobular spleen, which is a normal variant. Typically, lobules of the spleen are present along its medial part; however, they may extend anterior to the upper pole of the left kidney. Lobular spleen has no clinical importance, but radiologists should discriminate it from splenic laceration in patients with abdominal trauma and from masses arising from the left upper kidney.

M. Abdel-Latif
Helwan University, Cairo, Egypt

Ain Shams University Hospitals, Cairo, Egypt

M. S. Shalaby (✉)
Ain Shams University Hospitals, Cairo, Egypt

Bristol Royal Children's Hospital, Bristol, UK
e-mail: mshalaby@doctors.org.uk

Accessory Spleens (Splenunculi)

Introduction

Splenunculi are common congenital anomaly of the spleen, present in 10–30% of the population. They are either single (splenunculus) or multiple (splenunculi).

Site

Splenic hilum is the most common site (75%). Other sites are along the splenic vessels, the gastrosplenic ligament, the lienorenal ligament, and the tail of pancreas, within the wall of the stomach or the bowel, in the greater omentum, in the mesentery, or anywhere in the abdomen or the pelvis [1].

Etiology

Splenunculi arise from failure of fusion of one or more parts of the splenic anlage.

Clinical Significance

- The histology is identical to the main spleen. Furthermore, any pathological condition that affects the main spleen will affect the splenunculi.
- Although the majority of splenunculi are asymptomatic and accidentally discovered, their reporting is of clinical significance in certain instances.
- Splenunculi can be mistaken for enlarged lymph nodes or tumors in the adrenal gland, tail of pancreas, stomach, and intestine. Diagnosing splenunculi in the tail of pancreas represents a challenge because of their imaging similarity to pancreatic neoplasms, so diagnosis is usually pathologically after unnecessary exploration and removal of the splenunculi.
- Inability to localize splenunculi before elective splenectomy due to hematological or immunological disorders will result in their hypertrophy and recurrence of symptoms.

R. Carachi, S. H. E. Doss (eds.), *Clinical Embryology*, https://doi.org/10.1007/978-3-319-26158-4_40

Diagnosis

- Ultrasound (US), CT, or MRI can be used to diagnose splenunculi. Splenunculus appears as homogenous round mass (less than 4 cm in diameter) that has the same echogenicity of the main spleen.
- Technetium-99 m colloid scintigraphy is highly sensitive and specific to the splenic and hepatic tissues and hence can help in diagnosing splenunculi present in rare sites.

Asplenia

Introduction

Asplenia is the congenital absence of the spleen. It occurs in context of a syndrome or isolated.

Syndromic Asplenia (Ivemark Syndrome)

- Syndromic asplenia is characterized by abnormal arrangement of thoracic and abdominal organs and vessels (situs ambiguous or heterotaxia).
- Males are more affected than females.
- Congenital heart disease is present in 99–100% of affected individuals. That is why up to 95% of cases die in the first year of life [2].

Isolated Asplenia

Isolated asplenia is rare and is either familial or sporadic.

Heterotaxia with Asplenia (Table 40.1)

Heterotaxia results from defective left-right axis of development (***lateralization defect***).

Anatomical Arrangement [3]

- Hepatobiliary: Most cases are characterized by the central position of the liver and gallbladder.
- Pancreas: Midline in position and is truncated as it is formed from the head alone, which lies between the superior mesenteric artery and vein.
- Intestinal rotation: Abnormalities include non-rotation, incomplete rotation, or reversed rotation.
- Heart: Two-chambered (cor biloculare) due to failure of development of the interatrial and interventricular septa. The two atria of the heart and the two lungs are morphologically right and known as bilateral right-sidedness (right isomerism). In some cases, the cardiac apex is contralateral to the stomach. There is pulmonary stenosis or atresia.
- Great vessels: The abdominal aorta and IVC are ipsilateral in some cases, while in others the aorta is present slightly to the right and IVC slightly to the left. Rarely, there is an interrupted IVC with continuation of the azygos or hemiazygos. Other anomalies of the great vessels like right-sided aortic arch, transposition of great vessels, and truncus arteriosus are present in few cases.

Table 40.1 Differences between asplenia and polysplenia patients

	Asplenia	Polysplenia
Sex	• More common in males	• More common in females
Isomerism	• Right isomerism (bilateral right-sidedness)	• Left isomerism (bilateral left-sidedness)
Spleen	• Absent or rudimentary	• Multiple with variable sizes and numbers • Left- or right-sided but always keep their normal relation to the stomach
Hepatobiliary	• Central position of the liver and gallbladder	• Mostly central liver but may lie in the right or left side • Biliary atresia may be present
Pancreas	• Midline in position • Truncated (formed from head alone)	• Midline or to the right • Short truncated
Intestinal rotational abnormalities	• Common	• Common
Heart	• Two-chambered (cor biloculare) • The cardiac apex may be contralateral to the stomach • Pulmonary stenosis or atresia • Congenital heart disease in 99–100% of syndromic cases	• Levocardia in most cases • The cardiac apex may be contralateral to the stomach • Some patients have normal hearts or minor cardiac anomalies
Great vessels	• Interrupted IVC with continuation of the azygos or hemiazygos: rare • Anomalies include right-sided aortic arch, transposition of great vessels, and truncus arteriosus	• Interrupted IVC with continuation of the azygos or hemiazygos: 65–80% • Pre-duodenal portal vein: common

Clinical Significance

- The spleen stands as a blood filter and plays a fundamental role in immunity against the encapsulated bacteria.
- Asplenia is usually associated with hematologic abnormalities in the presence of Howell-Jolly bodies, target cells, siderocytosis, leukocytosis, and transient normoblastemia. In addition, the osmotic fragility decreases.
- Pneumococcal sepsis is often the first faced sign of such immune defect.

- The absence of the spleen on US and the presence of Howell-Jolly bodies on blood smears diagnose isolated cases.
- Once diagnosed, the use of the pneumococcal vaccine and the appropriate prophylactic antibiotic may prevent sepsis and save the lives of affected cases.

Polysplenia

Introduction

- Polysplenia syndrome is a condition characterized by the presence of multiple spleens in association with situs ambiguous. There is no pathognomonic set of anomalies, but instead a wide range of anomalies is present.
- It is more common in females.

Heterotaxia with Polysplenia (Table 40.1)

The defect in left-right axis of development results in bilateral left-sidedness (left isomerism).

Anatomical Arrangement [4]

- Spleen: Multiple discrete spleens of variable sizes and numbers are present in the majority of patients. Multiple spleens are left sided or right sided, but they always keep their normal relation to the stomach.
- Hepatobiliary: The liver and gallbladder are mostly central in position. In addition, they may lie normally in the right side or in the left side. The central liver is often symmetrical in shape. The biliary system may be normal or anomalous. Biliary atresia may be present.
- Pancreas: A short truncated pancreas is usually present. Its position is to the right of the midline or in the midline.
- Intestinal rotational abnormalities are common in such patients. The stomach may be right sided or left sided.
- Heart: Levocardia is present in most of cases. In about half of cases, the cardiac apex is contralateral to the stomach. Some patients have normal hearts or minor cardiac anomalies, while others have severe anomalies.
- Great vessels: About 65–80% of cases have interruption of the IVC with continuation of the azygos or hemiazygos veins. The intrahepatic IVC may be present to the right or left of the midline. In few cases, the intrahepatic IVC is duplicated; the right one is interrupted with azygos continuation, whereas the left one drains into the right atrium. The abdominal aorta is mostly present to the left of the midline. Commonly, there is a pre-duodenal portal vein.
- Renal: Some cases suffer unilateral renal agenesis, hypoplastic kidney, or duplex system.

Biliary Atresia Splenic Malformation (BASM) Complex

Introduction

- Biliary atresia (BA) is a rare neonatal disease characterized by progressive obstructive cholangiopathy affecting the biliary tree leading to cirrhosis and liver failure. The incidence of BA is higher in East Asian countries with a slight female predominance.
- There are four clinical forms of BA [5]: Isolated BA, syndromic BA, cystic BA, and CMV-associated BA.
- BASM is a subgroup of syndromic BA. It is more common in Europe and the USA as it accounts for about 10% of BA cases with a marked female predominance (2:1).

Box 40.1 Associated anomalies with BASM [5, 6]

1. Hepatobiliary: The liver may be right sided, left sided, central, and symmetrical. The extrahepatic biliary remnant is scanty with complete absence of the distal common bile duct in 45% of cases.
2. Splenic anomalies (100%): Polysplenia (90%), double spleen, and asplenia.
3. Absent inferior vena cava (70%).
4. Abnormal portal vein (60%): Pre-duodenal portal vein or absent portal vein with mesenterico-hemiazygos shunt and arterialized liver.
5. Intestinal malrotation (60%).
6. Cardiac anomalies (45%): Important in prognosis. They vary in severity including hypoplastic left heart, teratology of Fallot, left atrial isomerism, septal defects, and patent ductus arteriosus.
7. Situs inversus abdominis (37%).
8. Pancreatic anomalies (10%): Annular pancreas, absent tail, and anterior position.
9. Other infrequent anomalies: Tracheoesophageal fistula, duodenal atresia, absent kidney, and unilateral pulmonary hypoplasia.

Etiology

There is an association between BASM and maternal diabetes or thyrotoxicosis. Bile duct pathology and other developmental anomalies usually occur at the same time during the embryonic phase of organ development at the fifth week.

Clinical Significance

- Histologically, there is no difference in liver parenchyma between syndromic and non-syndromic forms of BA.

- All cases of BASM should undergo screening for cardiac anomalies. Significant cardiac defects that affect cardiopulmonary function should be corrected first. The appropriate time of surgery in BASM cases is the same as in non-syndromic cases.
- The placement of Roux-en-Y loop is challenging in those patients due to difficult orientation caused by altered anatomy and gut rotation abnormalities.
- The overall prognosis is worse than non-syndromic forms, which may be attributed to associated anomalies.

Splenogonadal Fusion (SGF)

Introduction

- Splenogonadal fusion is a rare congenital anomaly characterized by abnormal connection between the spleen and the gonad or between the spleen and derivatives of the mesonephros.
- This anomaly usually affects males (95%) and mostly on the left side (98%).
- The incidence in females may be underestimated. This is attributed to the intraperitoneal position of the female gonads and the lack of symptoms. In females, the ectopic spleen is found adjacent to the ovary or the mesovarium.

Types

- There are two types of SGF, continuous and discontinuous.

Continuous SGF (60%)

- There is a direct anatomical connection between the main spleen and the accessory one through a cord of splenic or fibrous tissue, which may contain beads of splenic tissue [1].
- Cryptorchidism is the most common associated abnormality affecting 44% of cases, more on the left side but may be bilateral or on the right side.
- Limb defects (*splenogonadal fusion-limb defect syndrome, SGFLD*) and micrognathia always associate this type of SGF.
- Other less frequent anomalies are cleft palate, cardiac defects, hypoplastic lung, diaphragmatic hernia, spina bifida, hypospadias, and anal anomalies [7].

Discontinuous SGF (40%)

- The discontinuous type lacks the attachment between both spleens [1]. The ectopic spleen is encapsulated and is attached to the testes or the epididymis (Fig. 40.1).
- Only 14% of cases with the discontinuous type have cryptorchidism [7].

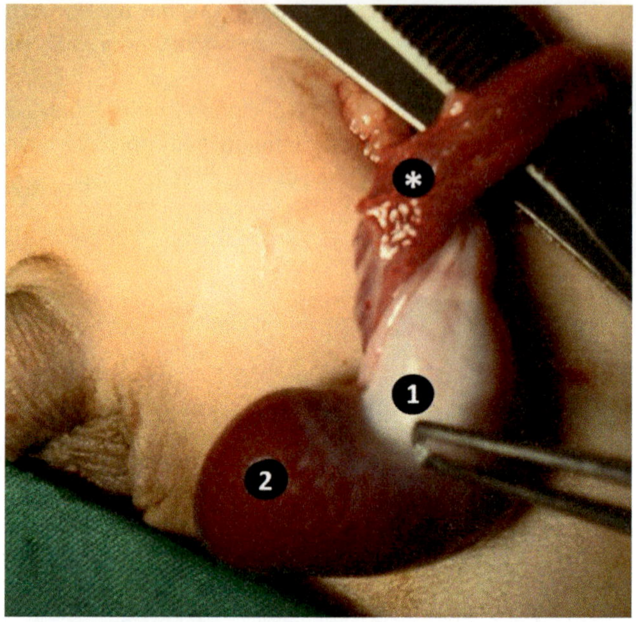

Fig. 40.1 Left discontinuous SGF with the forceps pointing at the junction between the left testes (1) and the accessory spleen (2). The spermatic cord (*) was separated through an inguinal exploration for left testicular mass

Etiology

- The etiology of SGF is still uncertain. The proposed theory for continuous type suggests that fusion occurs between the fifth and eight weeks of gestation. Gut rotation brings the splenic clusters in close proximity to the left urogenital fold containing the gonadal mesoderm. Unknown defect causes fusion between these two organs.
- The timing of events can explain the associated anomalies because the limb buds and the Meckel's cartilage (the mold of the bony mandible) develop during the same period. The former theory cannot explain the discontinuous type, which may represent a rare variant of accessory spleen [8].
- Cases with right-sided SGF are explained by direct fusion if there is situs inversus or by contralateral retroperitoneal migration of splenic cells along the dorsal mesentery.
- Loss of the connection between the main and accessory spleens results in the discontinuous type [7].

Clinical Significance

- Most patients present with a scrotal mass, 10% of scrotal masses, close to the upper pole of the testes. Other cases are discovered incidentally during hernia, hydrocele, or undescended testes operations. Rare presentations include bowel obstruction, traumatic rupture, or association with intra-abdominal testicular tumor.

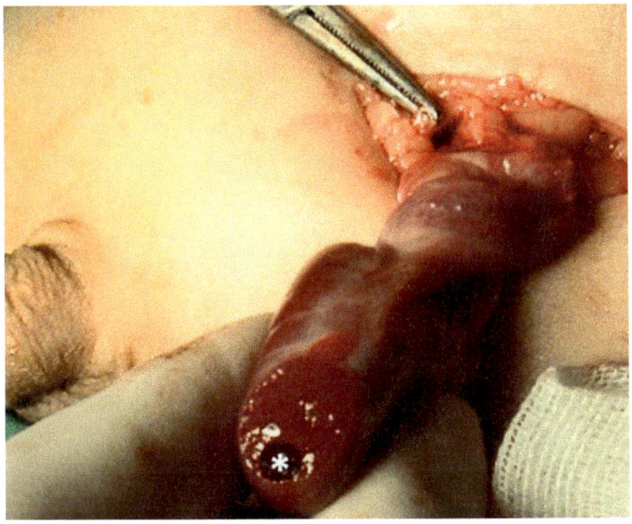

Fig. 40.2 Biopsy (*) of the splenic tissue in cases of SGF may help in confirming the diagnosis and avoid unnecessary orchiectomy

- It is very difficult to differentiate SGF from paratesticular lymphomas or rhabdomyosarcomas.
- Cases may present at any age; however, 80% of cases are below 30 years (50% are children). Most of the patients are diagnosed by postoperative pathological examination. A high index of suspicion is mandatory, and intraoperative biopsy (Fig. 40.2) may help in confirming the diagnosis of splenic tissue and avoiding unnecessary orchiectomy.

Management

- Imaging by US +/− MRI is usually performed for investigating the mass.
- Excision of the ectopic spleen with sparing the gonads is the rule; however, we can preserve both in cases of difficult separation.

Splenogonadal Fusion-Limb Defect Syndrome (SGFLD) [9]

- SGFLD is a heterogeneous syndrome that should be recognized from SGF without limb defects.
- SGFLD is characterized by the association of:
  1. SGF (83% continuous type).
  2. Limb anomalies: There is no relation between the type of SGF and the severity of limb affection. The main limb defects consist of amelia or rudimentary limbs.
  3. Orofacial anomalies: (mandibular hypoplasia in 70% of cases, cleft palate or bifid uvula, and microglossia).
- This syndrome represents a developmental field defect that occurs early in development during blastogenesis. The earlier the insult takes effect, the greater the number of anomalies presents. The embryological events that lead to this syndrome are unclear. Most of the cases are stillborn or death within the first year of life.

Wandering Spleen

Introduction

- Normally, the gastrosplenic and lienorenal ligaments keep the spleen in position.
- Wandering spleen is a rare entity, defined as the migration of the spleen from its normal location in the left hypochondrium. The length of the splenic pedicle determines the range of the splenic movement. Wandering spleen is either acquired or congenital.
- Acquired cases are more common in females in the childbearing period due to laxity of the supporting ligaments secondary to hormonal changes of pregnancy and multiparity-associated abdominal laxity.
- The congenital form is due to the absence or underdevelopment of the supporting ligaments.
- Children constitute about one third of all cases. Both males and females are affected, usually before the age of 10 years. Males predominate in the first year of life, whereas females dominate after that [10].

Etiology

- Children with prune belly syndrome are possibly at risk of splenic hypermobility.
- Embryologically, the two layers of the dorsal mesogastrium host the spleen and the body and tail of pancreas. The developing spleen bulges through the left layer of the dorsal mesogastrium. The mesentery dorsal to the spleen fuses with the peritoneum of the posterior wall and forms the lienorenal ligament, whereas the mesentery ventral to it forms the gastrosplenic ligament (Fig. 40.3).
- Increased splenic mobility is probably due to incomplete fusion of the dorsal mesogastrium posteriorly, so that the lienorenal ligament is missed. The gastrosplenic ligament alone is not sufficient to fix the spleen. In some cases, both the lienorenal and gastrosplenic ligaments are missing.

Presentations

- Cases with wandering spleen may be asymptomatic or present with acute, chronic, or intermittent abdominal pain secondary to torsion of its pedicle.
- Some cases present with a mobile abdominal mass. The movement of the mass is painful in all directions except toward the left hypochondrium.
- The most common presenting symptom in the pediatric age is acute abdominal pain; however, abdominal mass is the most common presentation in those less than 1 year old.

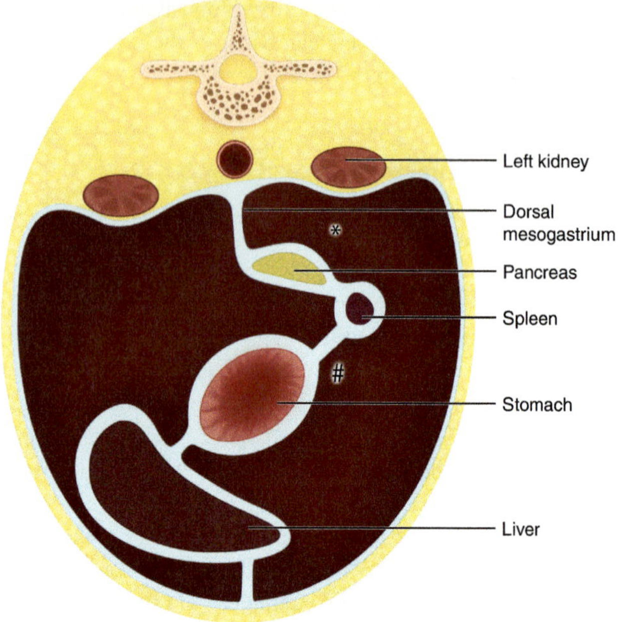

Fig. 40.3 Normal embryological development of the spleen showing the supporting lienorenal (*) and gastrosplenic (#) ligaments

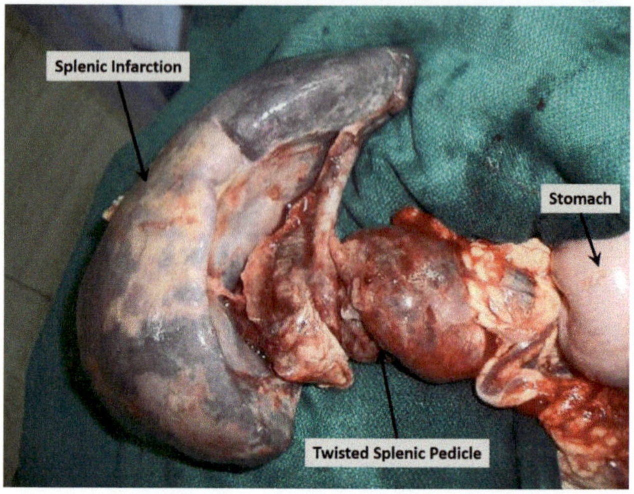

Fig. 40.4 Splenic infarction secondary to torsion of wandering spleen

Clinical Significance

- Loss of the splenic supporting ligaments predisposes the wandering spleen to torsion of its pedicle and hence its infarction. The torsion may be acute, chronic, or intermittent.
- Acute splenic torsion represents an acute surgical emergency. It may lead to acute abdominal pain and localized peritonitis secondary to splenic ischemia and infarction (Fig. 40.4). It can also cause intestinal obstruction and pancreatic necrosis as the pancreatic tail is included in the proximal splenic pedicle.
- Chronic or intermittent torsion is associated with hypersplenism and causes attacks of vague abdominal pain, anorexia, and vomiting. The resultant splenic vein occlusion causes venous congestion and splenomegaly.
- Other complications include gastric volvulus, gastric and duodenal compression by an overlying spleen, and mass effect caused by an impacted spleen within the pelvis.

Diagnosis

- Clinical diagnosis without imaging studies is almost impossible. US and CT are the commonly used radiological modalities.
- US is a reliable, inexpensive, and noninvasive modality for diagnosing wandering spleen. Diagnosis is based on visualization of the spleen in an abnormal position. Doppler scan can assess the splenic vascularity.
- CT findings (Fig. 40.5) are more informative and include the presence of the spleen away from its anatomical site, the whirl sign of the splenic pedicle denoting splenic torsion, and attenuation of the spleen indicating infarction.
- Liver-spleen isotope scan is another modality to diagnose wandering spleen.

Management

- The management of wandering spleen is surgical. Splenopexy is the treatment of choice for viable spleen (Fig. 40.6). Splenectomy is indicated in cases of splenic infarction. Conservative management is not recommended, as most of the cases with acute presentation have no relevant past history, i.e., they were asymptomatic.

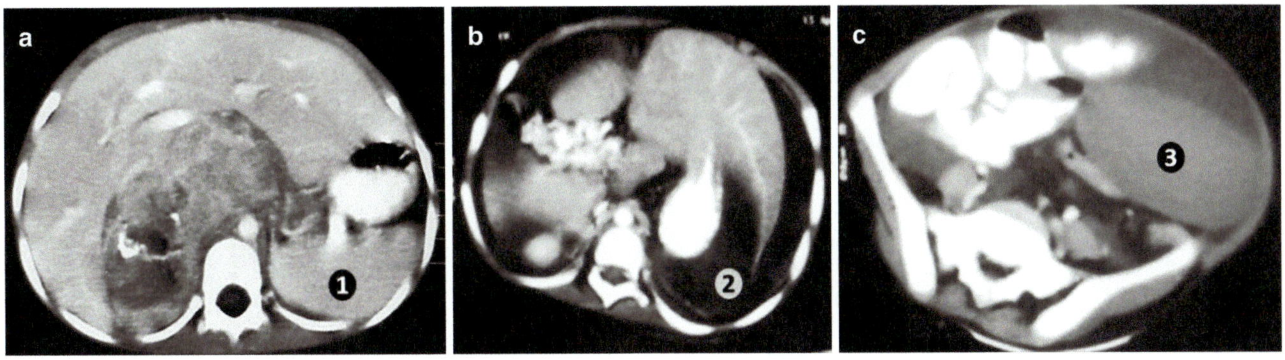

Fig. 40.5 CT with oral and IV contrast is useful in diagnosing wandering spleen. (a) Normal CT showing spleen (1) in its normal position. (**b** and **c**) Show CT imaging of a patient with wandering spleen. The expected splenic site (2) is empty, with the wandering spleen (3) in an ectopic position

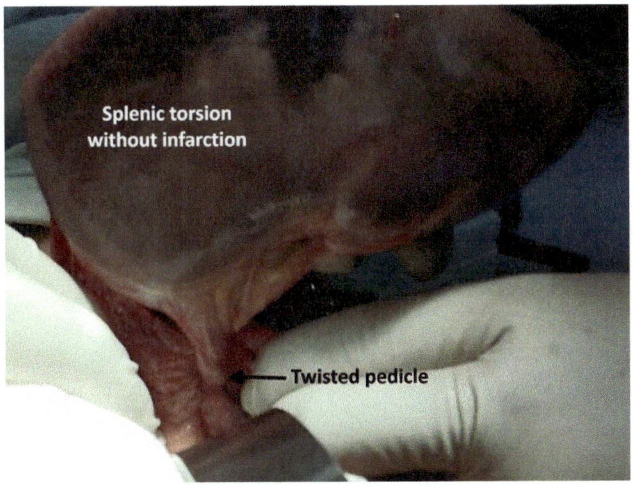

Fig. 40.6 Intraoperative picture of torted wandering spleen. The spleen is still viable with no infarction and hence splenopexy is the preferred treatment

References

1. Varga I, Galfiova P, Adamkov M, et al. Congenital anomalies of the spleen from an embryological point of view. Med Sci Monit. 2009;15:RA269–76.
2. Van Mierop LHS, Gessner IH, Schiebler GL. Asplenia and polysplenia syndromes. Birth Def. 1972;8:74–82.
3. Aguilar MJ, Crane JT, Stephens HB. Syndrome of congenital absence of the spleen with associated cardiovascular and gastroenteric anomalies. Circulation. 1956;14:520–31.
4. Fulcher AS, Turner MA. Abdominal manifestations of situs anomalies in adults. Radiographics. 2002;22:1439–56.
5. Hartley JL, Davenport M, Kelly DA. Biliary atresia. Lancet. 2009;374:1704–13.
6. Davenport M, Tizzard SA, Underhill J, et al. The biliary atresia splenic malformation syndrome: a 28-year single-center retrospective study. J Pediatr. 2006;149:393–400.
7. Cortes D, Thorup JM, Visfeldt J. The pathogenesis of cryptorchidism and splenogonadal fusion: a new hypothesis. Br J Urol. 1996;77:285–90.
8. Le Roux PJ, Heddle RM. Splenogonadal fusion: is the accepted classification system accurate? BJU Int. 2000;85:114–5.
9. Bonneau D, Roume J, Gonzalez M, et al. Splenogonadal fusion limb defect syndrome. Report of five new cases and review. Am J Med Genet. 1999;86:347–58.
10. Brown CV, Virgilio GR, Vazquez WD. Wandering spleen and its complications in children: a case series and review of the literature. J Pediatr Surg. 2003;38:1676–9.

Suggested Readings

Carragher AM. One hundred years of splenogonadal fusion. Urology. 1990;35:471–5.

Anorectal Malformations and Cloacal Anomalies

41

Constantinos A. Hajivassiliou

Embryology

The cloaca is the dilated endodermal region of the lower hindgut. It is connected to the umbilicus by the allantois and closed below by the bilaminar cloacal membrane.

Development of Hindgut

Formation of the Cloaca

*The lower end of the hindgut dilates to form an expanded part called the cloaca (endodermal).

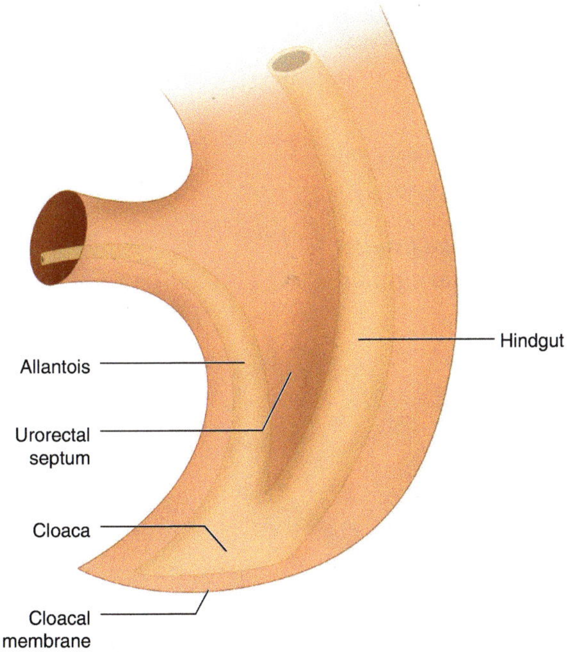

*The cloaca is connected to the umbilicus by the allantois and closed below by the cloacal membrane which is bilaminar, i.e., formed two layers:

(a) Outer ectodermal layer
(b) Inner endodermal layer

Division of the Cloaca into Two Parts

*The tissue between the hindgut and allantois forms the uro-rectal septum which grows caudally towards the cloacal membrane dividing the cloaca into:

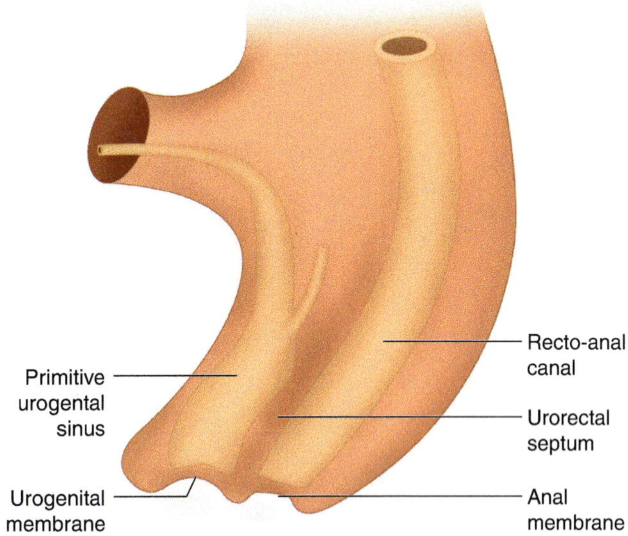

(a) An anterior part called the *primitive urogenital sinus.*
(b) A posterior part called the *anorectal canal.*

*At the same time the cloacal membrane is divided into urogenital membrane anteriorly and anal membrane posteriorly.

C. A. Hajivassiliou (✉)
Department of Paediatric Surgery, Royal Hospital for Sick Children and University of Glasgow, Glasgow, UK
e-mail: constantinos.hajivassiliou@glasgow.ac.uk

R. Carachi, S. H. E. Doss (eds.), *Clinical Embryology*, https://doi.org/10.1007/978-3-319-26158-4_41

Formation of the Proctodeum

An ectodermal depression called the proctodeum is formed opposite the lower end of the recto-anal canal and is separated from it by the anal membrane.

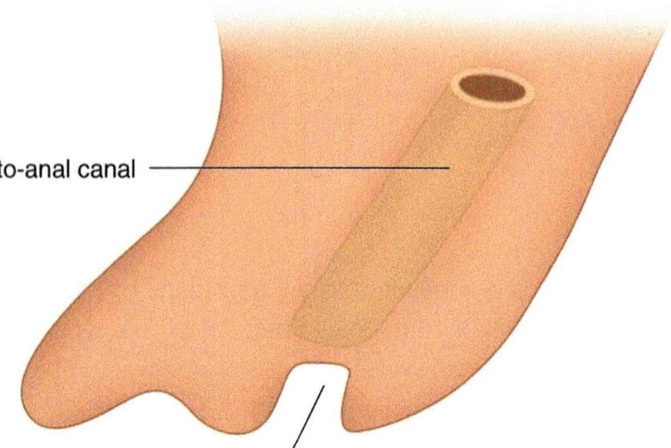

Rupture of the Anal Membrane and Formation of Anal Canal

*The anal membrane finally ruptures and the proctodeum becomes continuous with the recto-anal canal.

*The recto-anal canal will form the rectum + the upper 1/2 of the anal canal.

*The proctodeum will form the lower 1/2 of the anal canal.

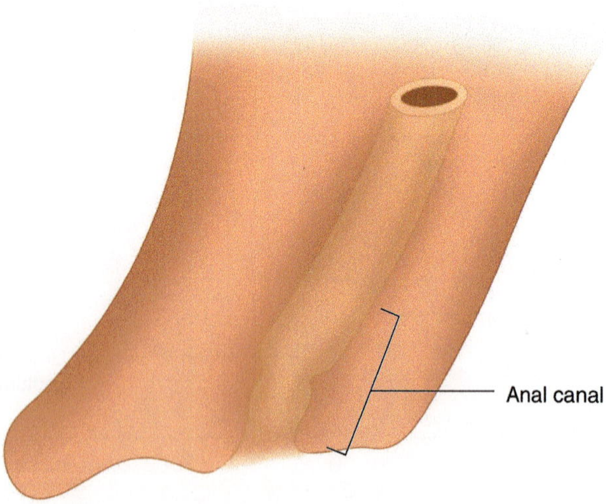

Derivatives of the Hindgut

1. Left 1/3 of transverse colon, (2) descending colon, (3) sigmoid colon, (4) rectum, (5) upper 1/2 of the anal canal

Congenital Anomalies of the Hindgut

1. ***Imperforate anus*:** due to failure of rupture of the anal membrane (between the endodermal and ectodermal parts of the anal canal)

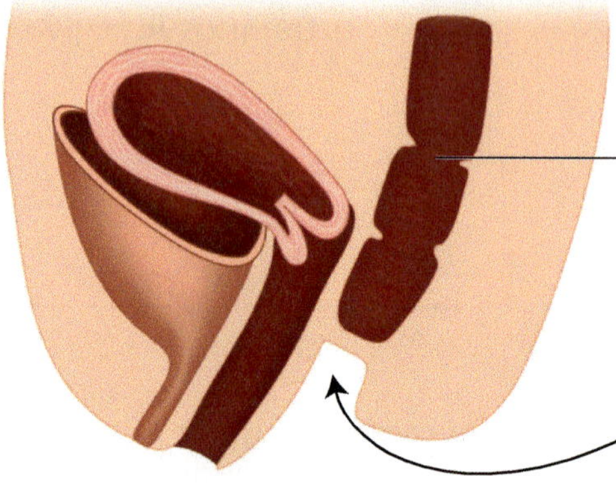

Imperforate anus

2. ***Atresia of rectum***: Due to failure of development of the proctodeum, the rectum ends blindly and is separated from the surface by a mass of connective tissue.

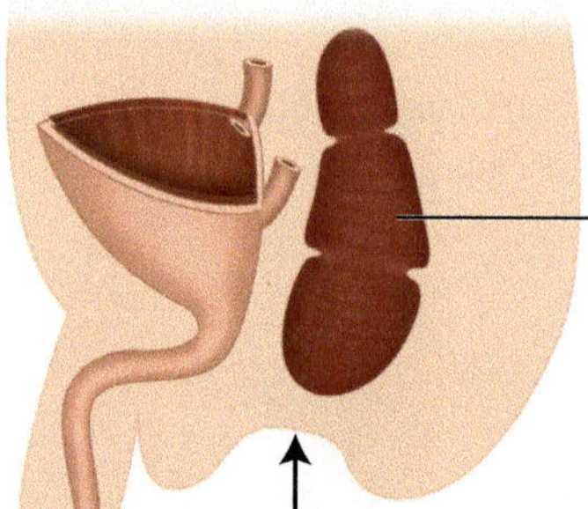

Artresia of rectum

3. ***Rectovaginal fistula***
4. ***Recto-urinary fistula***

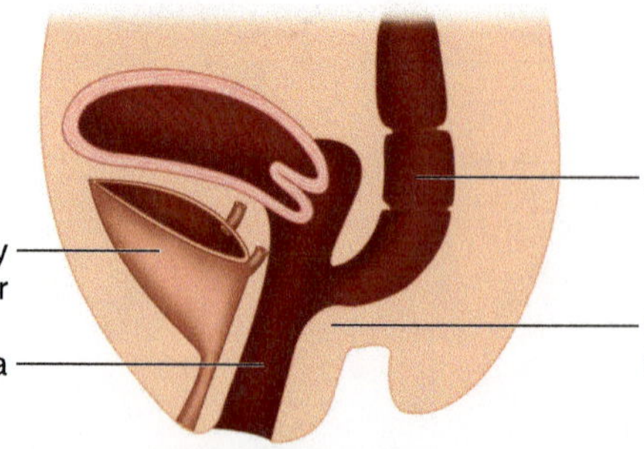

Recto vaginal fistula

These are seen frequently with imperforate anus and caused by incomplete division of the cloaca into two parts so the rectum remains connected to the urinary bladder, urethra, or vagina.

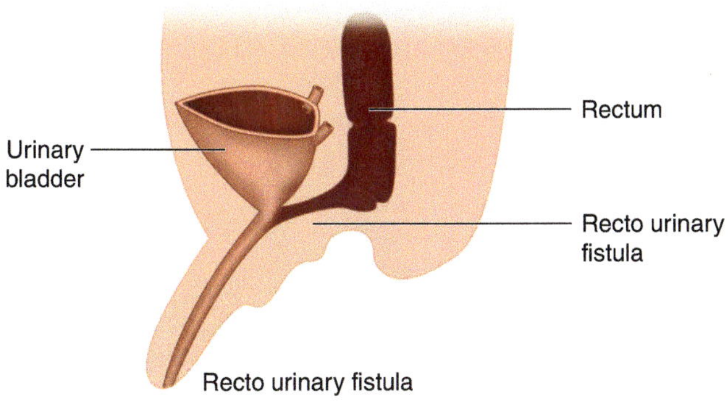

The tissue between the hindgut and allantois forms the urorectal septum which grows caudally towards the cloacal membrane dividing the cloaca into and anterior primitive urogenital sinus and a posterior part, the anorectal canal. At the same time, the cloacal membrane is divided into the urogenital membrane anteriorly and the anal membrane posteriorly.

The proctodeum, an ectodermal depression, is formed opposite the lower end of the recto-anal canal and is separated from it by the anal membrane. The anal membrane finally ruptures, and the proctodeum becomes continuous with the recto-anal canal. The canal will form the rectum and the upper half of the anal canal, with the proctodeum forming the lower half of the anal canal.

Anorectal Malformations

Introduction

Anorectal malformations (ARMs) include a wide range of anatomical and functional abnormalities of the perineal structures of varying degrees of severity. The common factor is an abnormal position of the termination of the GI tract, which either opens aberrantly in the perineum, communicates abnormally with another viscus or is narrow or blind ending.

Since early descriptions of this condition [1], there have been numerous attempts to review, classify [2] and explain and unify their pathogenesis, [3–16], still with incomplete success [5]. Although their pathogenesis is unclear and many aspects of our knowledge are still disputed, there is a clear contribution of both genetic [5, 6] and non-genetic [17] or teratogenic [5] factors. Although abnormalities can be isolated [9, 16, 18], the fact that they are usually part of well-recognised patterns and associations (Triad, VACTERL, OEIS) [8, 11, 19–21] points to strong unifying—genetic and/or environmental—pathogenetic factors.

There is a clear correlation between the severity of the anomaly and the likelihood of perineal dysfunction, which can lead to incontinence. Consequently, understanding of the pathogenesis is intricately interdependent not only in choosing the most appropriate management but also in developing new methods of treatment. Although nonfatal, the commonly associated incontinence is a social and functional disaster. The principal management aim is to establish the best possible level of faecal and urinary continence [22, 23] in addition to ensuring all other associated conditions and malformations are managed appropriately.

Congenital Malformations of the Rectum and Anus

In the 5-week embryo, the urogenital sinus and hindgut empty into a common cavity—the cloaca—separated from the exterior by the cloacal membrane. By the downgrowth of a mesodermic fold (urorectal septum), the separation of the urogenital tract from the rectum is normally completed by the seventh week. For a time, there is a small opening between the rectum and the urogenital tract known as the cloacal duct. The urorectal septum also divides the cloacal membrane into the urogenital membrane anteriorly and the anal membrane posteriorly, and these become the external openings. A small invagination—the proctodeum—develops in the region of the future anus, and the proctodeum and rectum join by the rupture of the anal membrane during the eighth week. Rectal and anal anomalies and associated malformations are due to arrests in development in the seventh and eighth weeks of foetal life. A range of anomalies occur in the anorectal region, and these will be discussed individually, but the subdivisions are not always so clear in practice.

Anal Stenosis

The anus is in its normal position, but the anal canal shows a degree of stenosis and is less distensible than normal. This condition may be missed at birth and can present weeks, months or even years later with chronic constipation and faecal soiling. There is delay in the passage of meconium, and subsequently the infant has to press excessively to defaecate through the narrow canal ribbon-like stools. This condition will be missed completely if rectal examination is omitted. Repeated dilatation is necessary, with the size of dilators being gradually increased until the little finger can be inserted. The infant may then be sent home and the mother given a supply of finger cots and instructed to continue dilatation for several months. Failure to continue dilatation may lead to retention of faeces and subsequently to rectal inertia, a secondary megarectum and overflow incontinence. This is very difficult to treat. Rarely, some form of anoplasty may be necessary.

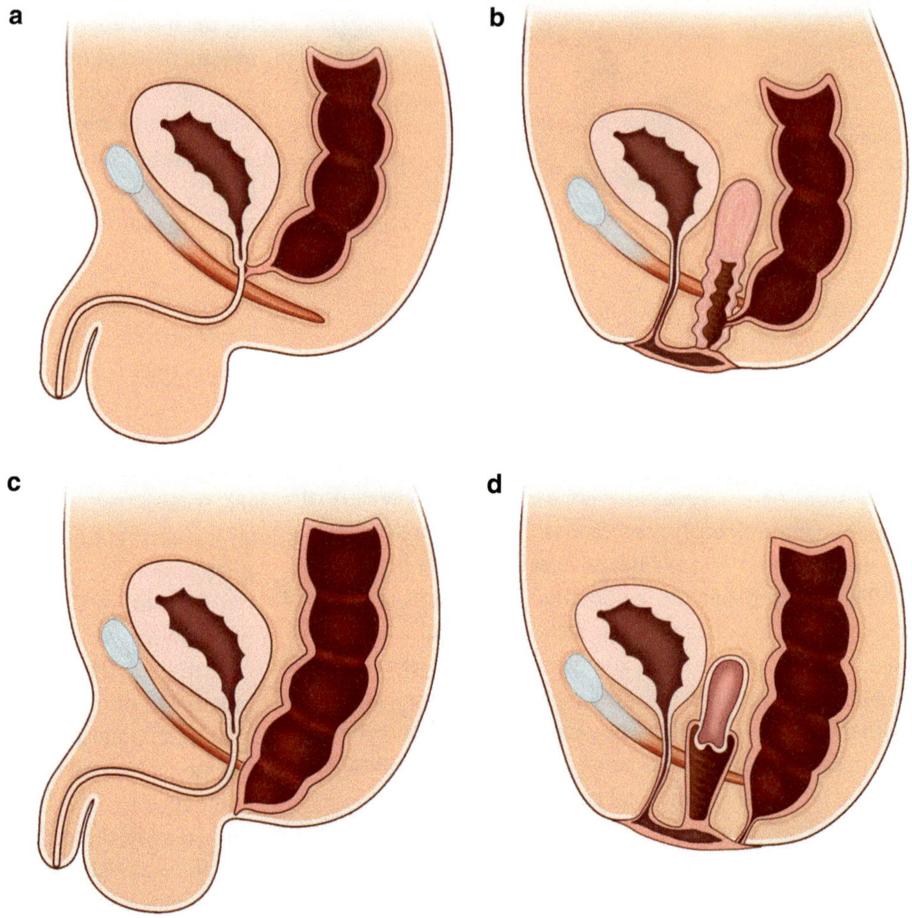

Fig. 41.1 Anorectal anomalies: Diagrammatic representation of high (**a**, **b**) and low (**c**, **d**) lesions in the male (**a**, **c**) and female (**b**, **d**)

Anal Membrane. A membrane occludes a normally situated anal canal. Meconium may be seen shining through the membrane which bulges when the infant cries. The membrane should be divided/excised and the patient supervised over the first few weeks to ensure no stenosis develops.

Covered Anus. This is a more common anomaly and is often described as a low anorectal anomaly as the rectum comes down through the pelvic floor before coming to the malformed part. There is no evidence of the anus in the normal position. In a baby boy, a raised ridge of skin or a narrow tract filled with meconium runs forward in the perineal raphe as far as the posterior aspect of the scrotum where a speck of meconium may be seen at the abnormal orifice—the rectoperineal fistula. Treatment is by anoplasty with postoperative follow-up to ensure that stenosis of the new anus does not develop.

Rectal Atresia. The anus looks apparently normal, but the rectal pouch ends blindly in the hollow of the sacrum. This is a rare anomaly and results in failure to pass meconium, the baby presenting with low intestinal obstruction. Diagnosis is made on perianal examination. Treatment varies from division of a septum to a more major pull-through procedure.

Clinical Presentation of Anorectal Anomalies. Anorectal anomalies are a spectrum of malformation from the high to low lesions, the definition of high or low being whether the bowel terminates above the pelvic floor muscles in the high anomalies or in low lesions whether the bowel comes through the pelvic floor muscles normally and then terminates by a fistula to the perineum. The latter are more common. High lesions are frequently one of a number of anomalies in the baby, and in only one third is it an isolated defect of low anomalies where two thirds are otherwise normal. There are also less frequent intermediate anomalies where the bowel comes into the pelvic floor muscles and terminates as a fistula to the bulbous urethra in the male and to the vulva in the female. All these lesions should be detected on the initial neonatal examination and the baby referred for investigation and diagnosis of the specific type at a specialist neonatal surgical unit.

Because of the different pelvic anatomies of the male and female, the anomalies vary with sex in high lesions. The boys have a fistula from the rectum to the urethra or bladder, and this may result in meconium being passed per urethra. In females, the genital tract is interposed between the alimentary and urinary tracts so that the fistulous communication is from the rectum to the vagina. This may be high in the vagina and not visible, but is more often low and may be seen. The low lesions have usually a fistula to the perineum. In the male, this may be close to the scrotum or even onto the median raphe of the scrotum, whereas in the females it is to the perineum or introitus. The uncommon intermediate types usually communicate with the bulbous urethra in the male and the vestibule in the female.

Investigations of the infant with an imperforate anus should commence with a thorough general examination because of the frequency of other anomalies. Ultrasound examination of the urinary tract and to the perineum should determine whether the infant has a high or a low lesion. At 24 h of age, once adequate time has allowed air to pass through the bowel to the rectum, X-rays of the abdomen and pelvis, in particular a lateral film with the baby inverted, are taken. If air is present below a line drawn from the pubis to the coccyx, the baby has a low lesion and if above, a high lesion.

Treatment for the high or intermediate varieties in the male is by colostomy and subsequently division of fistula and anoplasty usually through a posterior sagittal approach. The colostomy is then closed. In females, the fistula is usually larger allowing the baby to spontaneously decompress and pass meconium and faeces, and the posterior sagittal anoplasty may be done without colostomy.

With low lesions, perineal operations are performed to establish the anus in the normal site.

The results of the high and low groups are significantly different. With the high group, long-term continence results are much less satisfactory than in the low group where most achieve satisfactory continence. The associated anomalies also significantly affect the outcome, e.g. infants with partial sacral agenesis have poor continence, and some may require a permanent colostomy.

Background

The spectrum of ARMs (loosely referred to as "anorectal atresia") is a rare condition estimated to occur in approximately 1 in 5000 newborns in the UK. In addition to well-recognised patterns of associations/syndromes, it is becoming increasingly evident that there are rarer forms coexisting with less obvious correlations (Table 41.1).

Voiding dysfunction often coexists with ARM, both with [37] and without [38, 39] spinal cord malformations. Continence depends on the coordinated contraction and relaxation of the perineal musculature under both voluntary and involuntary controls. The anatomy and physiology of continence is well researched and documented, the mainstay of which is involuntary tonic activity to inhibit faecal/urinary effluent at rest, complemented by voluntary control when socially dictated. Anatomically, the conti-

Table 41.1 Rare/less well-known associations with anorectal malformations

Condition	Incidence	Other associations	Comments	Ref.
ARM without fistula	5%	Presacral mass		[24, 25]
ARM without fistula		Accessory labioscrotal folds, perineal lipoma, accessory scrotum, prune belly syndrome		[26–28]
ARM with oesophageal atresia	8.3%	Patients with OA have: • Worse sacral ratio Higher incidence of: • Tethered cord • Cardiac • Renal • Duodenal • Vertebra • Extremity • Tracheal • Developmental defects	• Prostatic fistula most common in males (45.9%) • Cloaca most common in females (57.9%) Patients with OA had worse bowel (47 vs. 67%) and urinary control (56.6 vs. 79.4%)	[29]
H-type urethroanal fistula		Oesophageal H-type fistula		[18]
ARM with ileal atresia				[30]
ARM with sacrococcygeal teratoma				[25, 31, 32]
ARM with Prune belly syndrome				[19, 27, 33]
ARM with segmental colonic dilatation				[34]
ARM with Hirschsprung's disease	2.3–3.4%	High incidence of other conditions/syndromes		[35, 36]

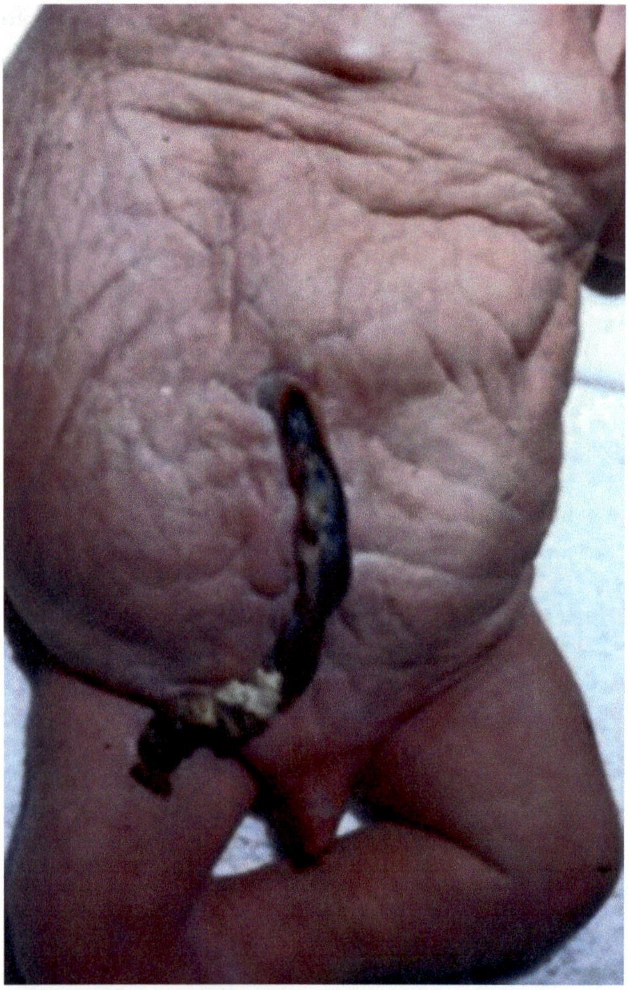

Fig. 41.2 Prune belly syndrome

nence apparatus consists of the neurological control pathways (brain, spinal cord, somatic and autonomic afferent and efferent pathways), smooth and striated muscle sphincters and biomechanical factors (anorectal angle, stool consistency), all in conjunction with normal intestinal motility.

Development of the Anorectal Apparatus

The descriptive aspects and timeline of human embryology have been extensively documented as relating to healthy and abnormal development. There are also increasing number of animated resources which provide a simulated 3D, dynamic depiction of different aspects of human development, including the hindgut/anorectum [40].

Early theories to explain congenital aberrations of anorectal anatomy [8, 9] suggested that two independent mechanisms subdivide the endodermal cloaca into the anterior urogenital sinus and posterior rectum, one craniocaudal and one lateromedial. The first process indents the cloaca from above from the junction of the allantois and the hindgut and is brought about by growth, rotation and folding processes ending at the level of Müller's hillock. The second process completes the subdivision from that point down to the cloacal membrane and takes place by lateral growth. It was postulated that the relative absence of one or other mechanisms also explains the level of the defect; rectal anomalies occur due to failure of subdivision of the cloaca into urinary and alimentary canals, whereas anal deformities occur after this subdivision is complete and are caused by defects of the anal pit and perineum [8, 9, 11–13, 41].

In the female, the patterns of abnormalities also depend on the superimposition of the developing Müllerian ducts on the undivided cloaca. This is why it was originally thought that a true rectovesical fistula could not occur in the female; this has been demonstrated in the presence of a didelphic uterus (*personal observation*).

Recent animal studies and the advent of scanning electron microscopy have shed more light on and finally elucidated most embryologic events which lead to abnormal hindgut development: the previously held "segmentation theories" were challenged by Kluth [5, 6], who suggests that—although the process does start during the early embryonic stages—the development of the septum is much more passive than active. The cloacal membrane is too short/deficient dorsally, leading to a partly absent dorsal cloaca; as a result a recto-urethral fistula results by the continuing abutment/communication of the hindgut to the urogenital sinus. The finding that the embryonic cloaca does not exhibit similarities to any form of anorectal malformations in the neonate (even to "cloacas" in females) [5, 42] suggests that there is still much more to be discovered in this field. Further answers will lie in the realms of molecular biology, molecular genetics and epigenetics.

Classification

There have been numerous ways to classify these conditions proposed by working groups of specialists [2]. Broadly, the condition in both females and males can be divided into groups depending on the anatomical level of the defect, which correlates with their specific potential to attain continence with appropriate management: the lower the defect, the higher the chance of attaining normal function.

ARMs Include (Modified from Pena [22]):

Males

- Low defects: cutaneous fistula, anal stenosis, anal membrane and "bucket handle" malformations
- Recto-urethral fistula
  - Bulbar
  - Prostatic

- Rectovesical fistula
- Imperforate anus without fistula
- Rectal atresia
- Rectal stenosis

Some Clinical Examples of ARMs in the Male

Females

- Cutaneous (perineal) fistula
- Vestibular fistula
- Vaginal fistula
- Imperforate anus without fistula
- Rectal atresia
- Rectal stenosis
- Rectovesical fistula (only possible with didelphic uterus [9], Figure 16—and *personal observation*)
- Persistent cloaca

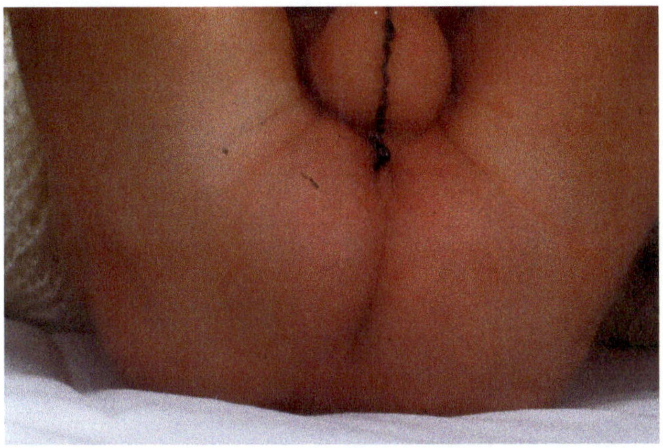

Fig. 41.3 Low imperforate anus with meconium in a cutaneous fistula

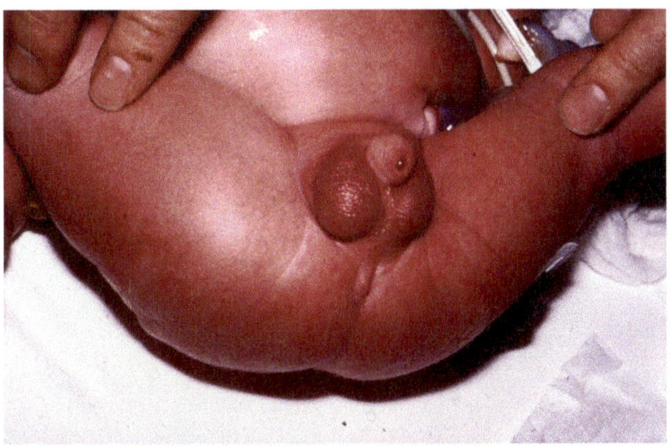

Fig. 41.4 High imperforate anus with meconium in the urethra

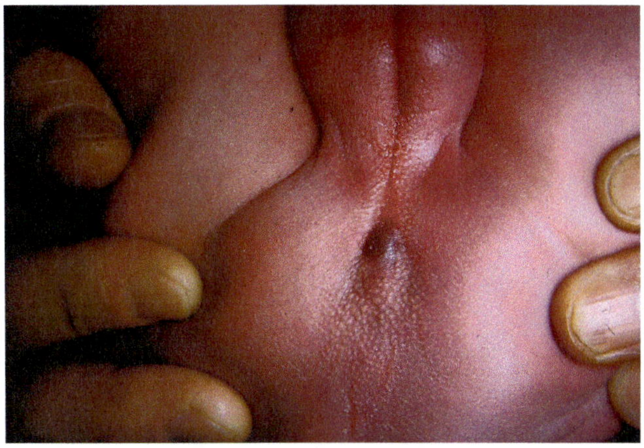

Fig. 41.5 Low imperforate anus with meconium behind anal membrane

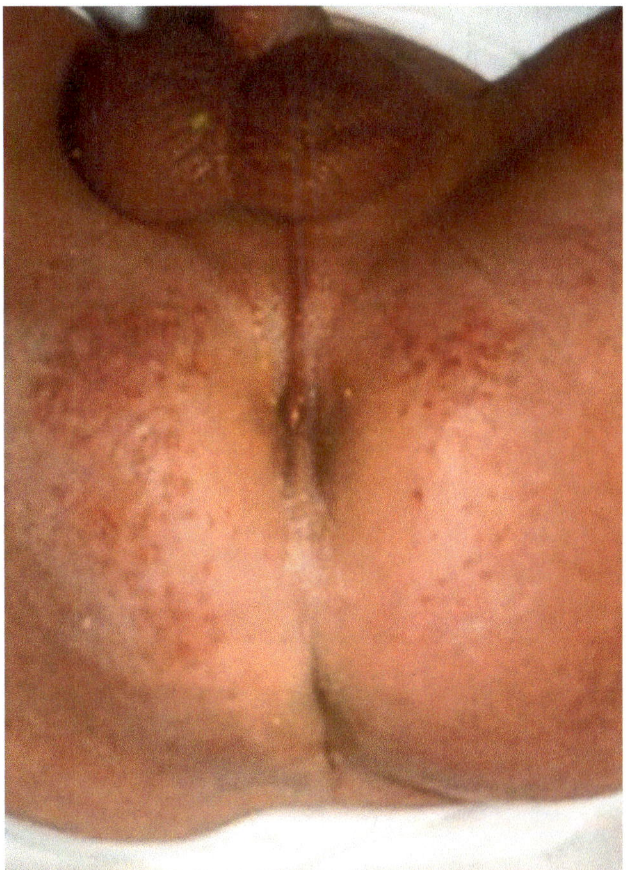

Fig. 41.6 Low imperforate anus, with bifid scrotum

Some Clinical Examples of ARMs in the Female

Complex Defects

Groups or isolated cases of unusual heterogeneous defects (e.g. bladder exstrophy and perineal exstrophy).

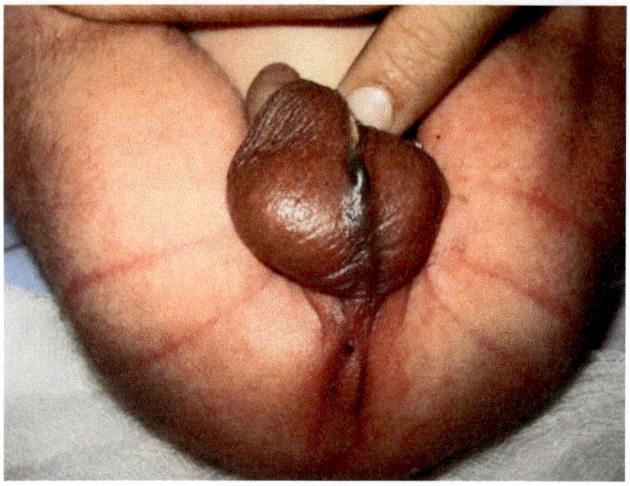

Fig. 41.7 Meconium tracking in fistula in a low anomaly

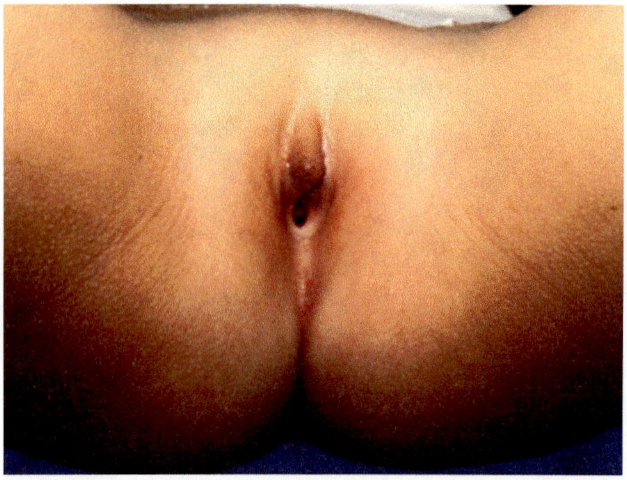

Fig. 41.8 Imperforate anus

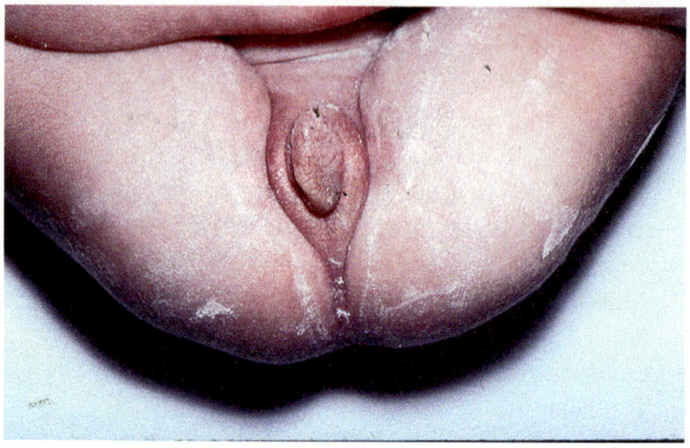

Fig. 41.9 Anorectal anomaly in a neonate

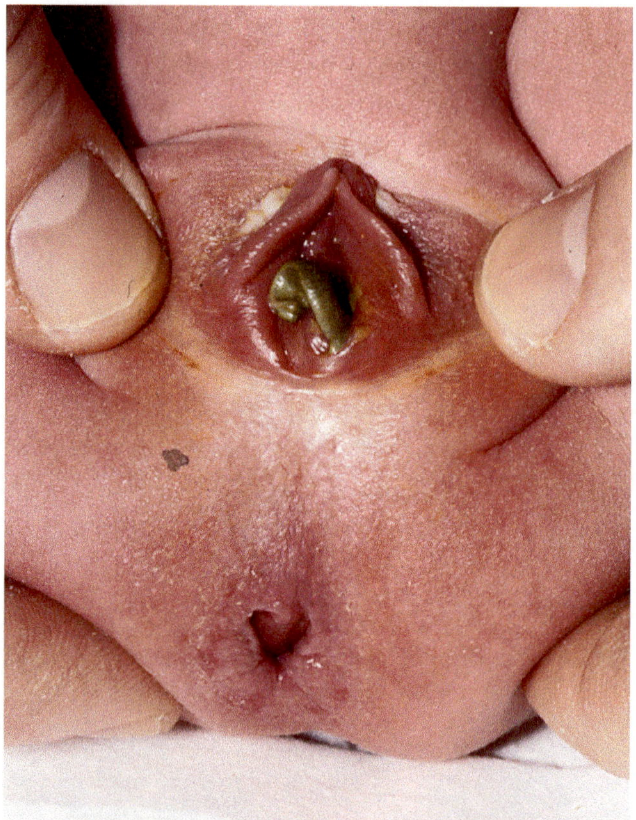

Fig. 41.10 Rectovaginal fistula with an anal opening

Some Other Anomalies of Female Genitalia

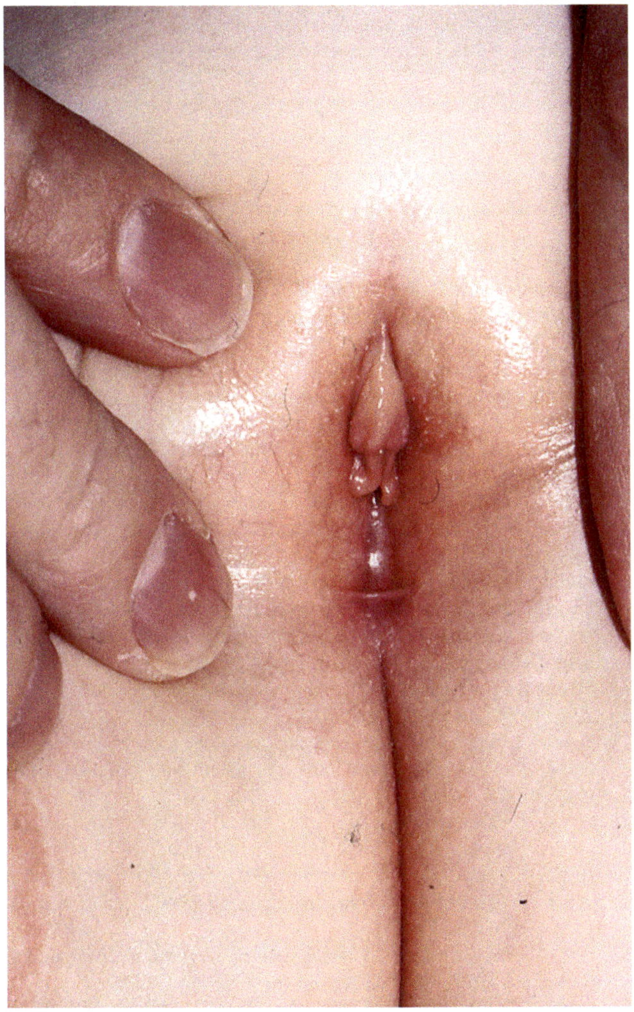

Fig. 41.11 Adherent labia

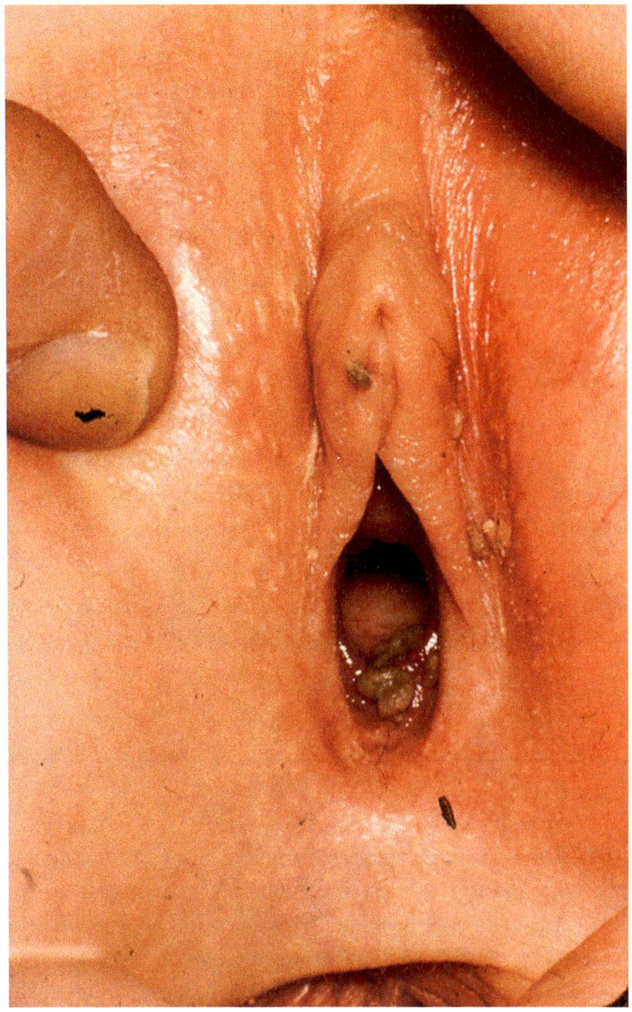

Fig. 41.12 Vaginal cyst

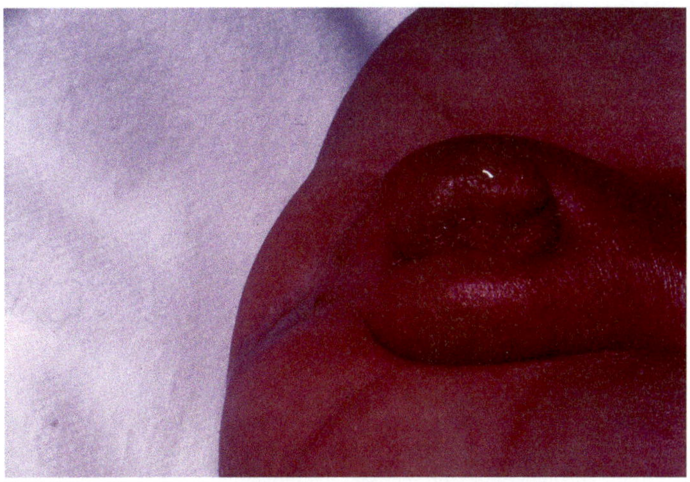

Fig. 41.13 Ambiguous genitalia

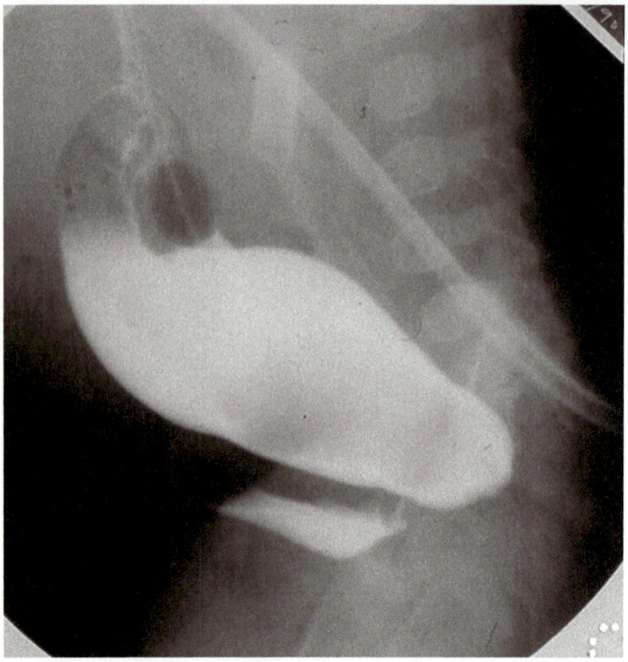

Fig. 41.14 X-ray of hydrocolpos

Summary

The aetiology of anorectal malformations remains unknown, but both genetic and environmental factors appear to be involved, and our understanding is continuously evolving. The majority of cases conform to well-recognised patterns and associations involving the distal perineum, spinal cord, GI tract and other mainly mesodermal structures. More recently, new associations are being recognised suggesting that hitherto unrecognised pathogenetic factors are involved.

The anatomical classification of these conditions correlates with prognosis. Understanding of the pathogenesis of ARMs is intricately interdependent not only in choosing the most appropriate management but also in developing new methods of treatment.

References

1. Costa J. Anorectal atresia with rectovaginal fistula; case report and review of literature. Pediatr Prat. 1950;21:117–36.
2. Stephens FDS, Smith ED. Classification, identification, and assessment of surgical treatment of anorectal anomalies. Pediatr Surg Int. 1986;1:200–5.
3. Dujovny N, Quiros RM, Saclarides TJ. Anorectal anatomy and embryology. Surg Oncol Clin N Am. 2004;13:277–93.
4. Gonzalez Utrilla J, Martinez Urrutia MJ. Surgical treatment of anorectal atresia. New contributions. Cir Pediatr. 1988;1:66–8.
5. Kluth D. Embryology of anorectal malformations. Semin Pediatr Surg. 2010;19:201–8.
6. Kluth D, Lambrecht W. Current concepts in the embryology of anorectal malformations. Semin Pediatr Surg. 1997;6:180–6.
7. Smith ED, Stephens FD. High, intermediate, and low anomalies in the male. Birth Defects Orig Artic Ser. 1988;24:17–72.
8. Stephens FD. Congenital imperforated rectum, recto-urethral and recto-vaginal fistulae. Aust N Z J Surg. 1953;22:161–72.
9. Stephens FD. Malformations of the anus. Aust N Z J Surg. 1953;23:9–24.
10. Stephens FD. Congenital malformations of the rectum and anus in female children. Aust N Z J Surg. 1961;31:90–104.
11. Stephens FD. The female anus, perineum and vestibule. Embryogenesis and deformities. Aust N Z J Obstet Gynaecol. 1968;8:55–73.
12. Stephens FD. Embryologic and functional aspects of "imperforate anus". Surg Clin North Am. 1970;50:919–27.
13. Stephens FD. Embryology of the cloaca and embryogenesis of anorectal malformations. Birth Defects Orig Artic Ser. 1988;24:177–209.
14. Stephens FD. Potential anatomic sphincters of anorectal malformations in males. Birth Defects Orig Artic Ser. 1988;24:155–61.
15. Stephens FD. Rectal ectasia: primary and secondary associated with anorectal anomalies. Birth Defects Orig Artic Ser. 1988;24:99–104.
16. Stephens FD. Wingspread anomalies, rarities, and super rarities of the anorectum and cloaca. Birth Defects Orig Artic Ser. 1988;24:581–5.
17. Zwink N, Rissmann A, Potzsch S, Reutter H, Jenetzky E, Consortium, C. U.-N. Parental risk factors of anorectal malformations: analysis with a regional population-based control group. Birth Defects Res A Clin Mol Teratol. 2016;106(2):133–41.
18. Stephens FD, Donnellan WL. "H-type" urethroanal fistula. J Pediatr Surg. 1977;12:95–102.
19. Nunn IN, Stephens FD. The triad syndrome: a composite anomaly of the abdominal wall, urinary system and testes. J Urol. 1961;86:782–94.
20. Pinsky L. The syndromology of anorectal malformation (atresia, stenosis, ectopia). Am J Med Genet. 1978;1:461–74.
21. Tarry WF, Duckett JW, Stephens FD. The Mayer-Rokitansky syndrome: pathogenesis, classification and management. J Urol. 1986;136:648–52.
22. Pena A. Atlas of surgical management of anorectal malformations, New York. Berlin, Heidelberg: Springer-Verlag; 1990.
23. Smith ED, Stephens FD, Holschneider AM, Nixon HH, Pena A. Operations to improve continence after previous surgery. Birth Defects Orig Artic Ser. 1988;24:447–79.
24. Bischoff A, Frischer J, Dickie BH, Pena A. Anorectal malformation without fistula: a defect with unique characteristics. Pediatr Surg Int. 2014;30:763–6.
25. Hamrick M, Eradi B, Bischoff A, Louden E, Pena A, Levitt M. Rectal atresia and stenosis: unique anorectal malformations. J Pediatr Surg. 2012;47:1280–4.
26. Chu SM, Ming YC, Chao HC, Luo CC. An accessory labioscrotal fold associated with anorectal malformation in female neonates. J Pediatr Surg. 2009;44:E17–9.
27. Shoshany G, Gottfied E, Bar-Maor JA. Accessory scrotum and anorectal malformation associated with "pseudo" prune belly in a neonate. J Perinatol. 1996;16:224–6.
28. Ozcan C, Celik A, Erdener A. Accessory perineal scrotum associated with anorectal malformation. BJU Int. 1999;83:729–30.
29. Fernandez E, Bischoff A, Dickie BH, Frischer J, Hall J, Pena A. Esophageal atresia in patients with anorectal malformations. Pediatr Surg Int. 2014;30:767–71.
30. Asabe K, Handa N. Anorectal malformation with ileal atresia. Pediatr Surg Int. 1997;12(4):302.
31. Shija JK. Anorectal malformation presenting as Hirschsprung's disease: a case report. East Afr Med J. 1995;72:130–1.
32. NG WT, NG TK, Cheng PW. Sacrococcygeal teratoma and anorectal malformation. Aust N Z J Surg. 1997;67:218–20.

33. Mahajan JK, Ojha S, Rao KL. Prune-belly syndrome with anorectal malformation. Eur J Pediatr Surg. 2004;14:351–4.
34. Mathur P, Mogra N, Surana SS, Bordia S. Congenital segmental dilatation of the colon with anorectal malformation. J Pediatr Surg. 2004;39:e18–20.
35. Hofmann AD, Puri P. Association of Hirschsprung's disease and anorectal malformation: a systematic review. Pediatr Surg Int. 2013;29:913–7.
36. Wetherill C, Sutcliffe J. Hirschsprung disease and anorectal malformation. Early Hum Dev. 2014;90:927–32.
37. Borg H, Holmdahl G, Olsson I, Wiklund LM, Sillen U. Impact of spinal cord malformation on bladder function in children with anorectal malformations. J Pediatr Surg. 2009;44:1778–85.
38. Borg H, Holmdahl G, Doroszkievicz M, Sillen U. Longitudinal study of lower urinary tract function in children with anorectal malformation. Eur J Pediatr Surg. 2014;24:492–9.
39. Senel E, Akbiyik F, Atayurt H, Tiryaki HT. Urological problems or fecal continence during long-term follow-up of patients with anorectal malformation. Pediatr Surg Int. 2010;26:683–9.
40. Duke-University. Simbryo GI development animation. Durham NC: Duke University; 2015. https://web.duke.edu/anatomy/embryology/embryology.html
41. N'guessan G, Stephens FD. Covered anus with anocutaneous fistula: the muscular sphincters. J Pediatr Surg. 1986;21:33–5.
42. Kluth D, Fiegel HC, Metzger R. Embryology of the hindgut. Semin Pediatr Surg. 2011;20:152–60.

The Kidneys and Ureters

42

Boma Lee and Martyn Flett

The kidneys, the ureters and the trigone of the urinary bladder all develop from the intermediate mesoderm. The human embryo develops three embryonic kidney structures during development, the pronephros, mesonephros and metanephros.

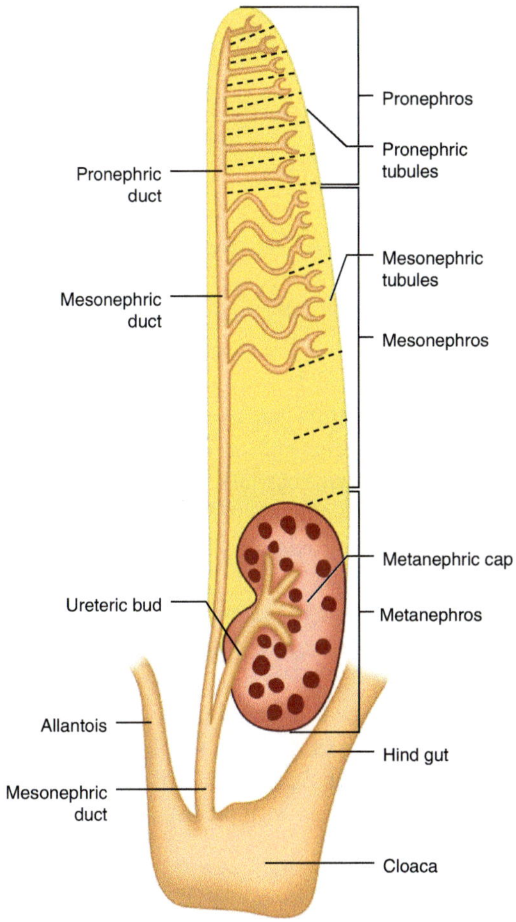

Pronephros

This is the first and simplest kidney system which appears in the human embryo. It develops in the cervical region of the segmented intermediate mesoderm and consists of 7–10 excretory pronephric tubules. A longitudinal collecting duct, the pronephric duct, extends downwards and opens into the cloaca. The pronephros has no function in the human embryo, and the tubules degenerate completely by the end of the fourth week. The pronephric duct persists to become the mesonephric duct.

Mesonephros

As the pronephros degenerates, the mesonephros begins to form in the region of the thoracic and upper lumbar region of the intermediate mesoderm. Each segment of the mesoderm develops several mesonephric excretory tubules, each of which are s-shaped with a medial end invaginated by a glomerulus and a lateral end opening into a longitudinal mesonephric duct. The majority of these tubules disappear by the end of the second month, while a few persist near the site of the developing gonads. The fate of the mesonephric tubules and duct is summarised below (Table 42.1).

Table 42.1 Development of the kidney and ureters

	Fate in the male embryo	Fate in the female embryo
Mesonephric duct	Upper tubules degenerate Middle and lower tubules form the efferent ductules of the testis and the head of the epididymis	Degenerates, a few rudimentary tubules may remain which contribute to the mesovarium
Mesonephric tubule	Forms the body and tail of the epididymis, the ductus deferens, the ejaculatory duct, the seminal vesicle, the ureteric bud and the trigone of the bladder	Gives rise to the ureteric bud, trigone of the bladder and Gartner's duct, which is a rudimentary structure

B. Lee (✉) · M. Flett
Queen Elizabeth University Hospital, Glasgow, UK
e-mail: Boma.adikibi@nhs.net

R. Carachi, S. H. E. Doss (eds.), *Clinical Embryology*, https://doi.org/10.1007/978-3-319-26158-4_42

Metanephros

The metanephros is the last to develop and forms the permanent kidney in the human. It develops in the sacral region and forms from two sources. The nephrons develop from the metanephric cap, which is the lowest part of intermediate mesoderm. The collecting tubules and ureter develop from the ureteric bud which arises as a diverticulum from the mesonephric duct.

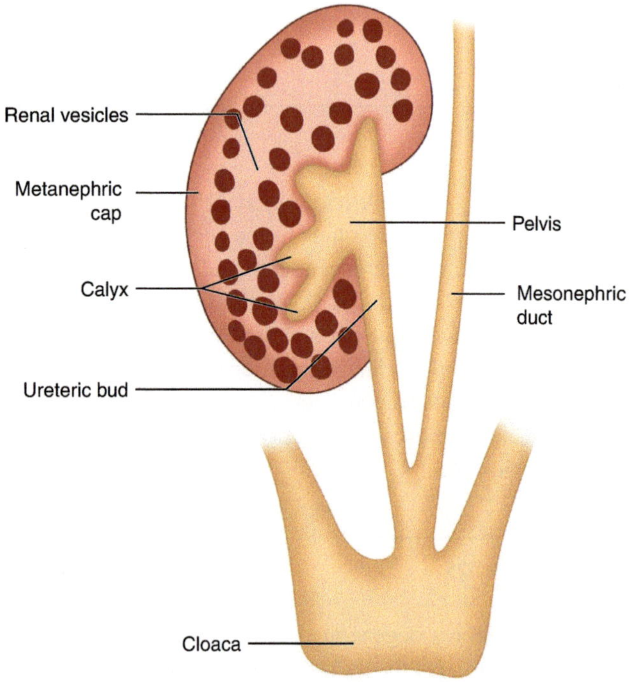

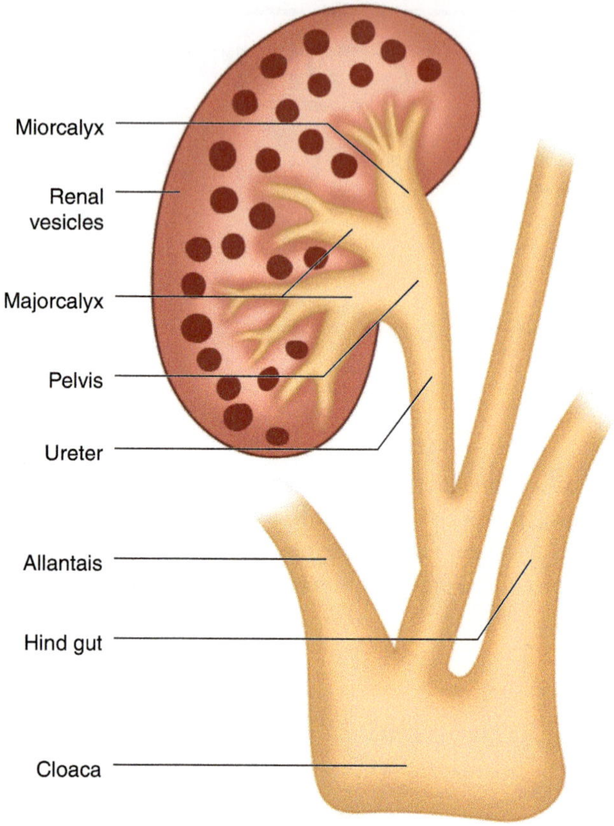

Development of the Collecting Tubules and Ureter

The ureteric bud forms as an outgrowth from the dorsomedial part of the mesonephric duct close to its entrance into the cloaca. The bud grows cranially, penetrating the metanephric cap. The upper end of the ureteric bud enlarges to form the primitive pelvis of the ureter which divides into two to three major calyces. Each major calyx then divides into minor calyces and collecting tubules which become connected to the nephrons of the metanephric cap.

Development of the Nephron

The metanephric cap develops from the caudal part of the intermediate mesoderm and divides into spherical masses or renal vesicles. Each vesicle surrounds the free end of a collecting tubule and forms an s-shaped tubule—the future nephron. One end of the nephron is invaginated by a glomerulus forming Bowman's capsule, while the other end of the nephron joins the collecting duct. The nephron elongates and forms the distal and proximal convoluted tubules and the loop of Henlé.

Early in development, the kidney is a lobulated organ, but as development proceeds, the grooves disappear leaving a smooth surface. In the early stages, the kidney lies in the pelvic region, and then it migrates up towards its adult position. At first, it receives a blood supply from the median sacral, common iliac arteries and lower part of the abdominal aorta. As the kidney ascends, it receives blood supply from the aorta only.

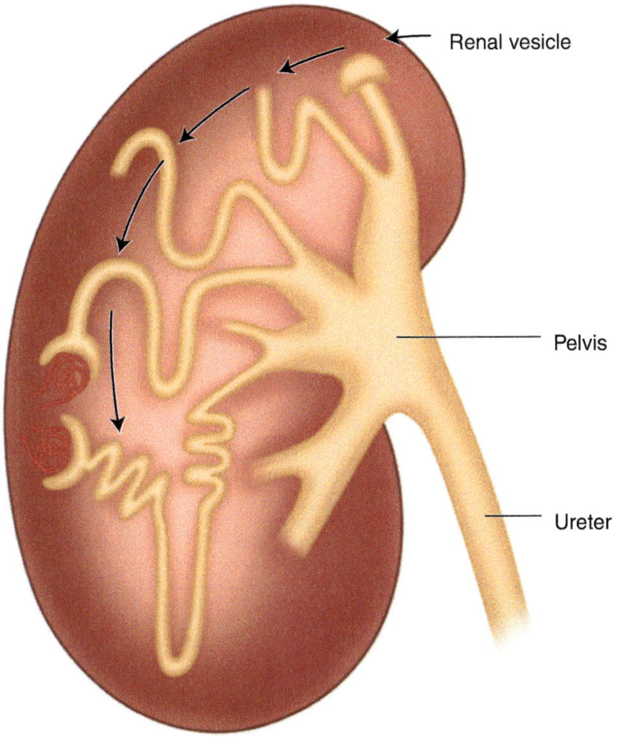

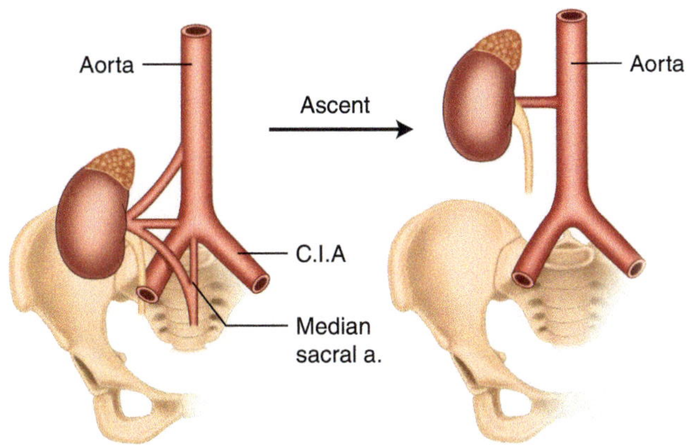

From an embryological viewpoint, there are three critical moments in the development of the upper urinary tract:

1. The development of the ureteric bud from the mesonephric duct at the end of the fifth week
2. The joining of the ureteric bud and the mesonephros in week 6
3. The ascent of the kidneys during the sixth and seventh week

An abnormality occurring in the first two stages can result in agenesis, hypoplasia or duplication anomalies. A deviance from the normal sequence in the third stage can result in fusion abnormalities or an ectopic position.

Kidney

Anomalies in Number

Unilateral Renal Agenesis

Recent increases in prenatal ultrasonography have led to increased detection of unilateral agenesis (Fig. 42.1). There is also evidence to suggest that a proportion of unilateral renal agenesis is actually dysplastic or multicystic dysplastic kidneys that have involuted antenatally [1].

Epidemiology

Autopsy and clinical studies report a range of the incidence of unilateral renal agenesis of between 1 in 500–2900 [2, 3]. A systematic review with 33 papers reporting on incidence between 1910 and 2012 calculated the incidence at 1 in 2000 [4].The literature suggests a male predominance (1.5:1) with the left side more frequently affected [3]. Familial studies have reported siblings and monozygotic twins more commonly affected with data suggesting autosomal dominant inheritance with 50–90% penetrance [5].

Pathophysiology

The embryological aberration is thought to be failure of the ureteric bud to induce the metanephric blastema. This could be due to failure of the ureteric bud or Wolffian duct to form or failure of the ureteric bud to reach or invade the mesonephros. An abnormality in the mesonephros is felt to be less likely because the ipsilateral gonad is rarely abnormal. These probably occur in the fourth or fifth week when the ureteric bud forms and the Wolffian duct derivatives are forming.

Associated syndromes

- Kallmann syndrome
- Turner syndrome
- Poland syndrome
- Klippel-Feil syndrome
- Frazer syndrome
- Branchio-oto-renal syndrome
- DiGeorge anomaly
- Townes-Brock syndrome
- VACTERL (20–30%)
- MURCS syndrome (Mullerian hypoplasia/aplasia, renal agenesis and cervicothoracic somite dysplasia)

Urinary tract anomalies

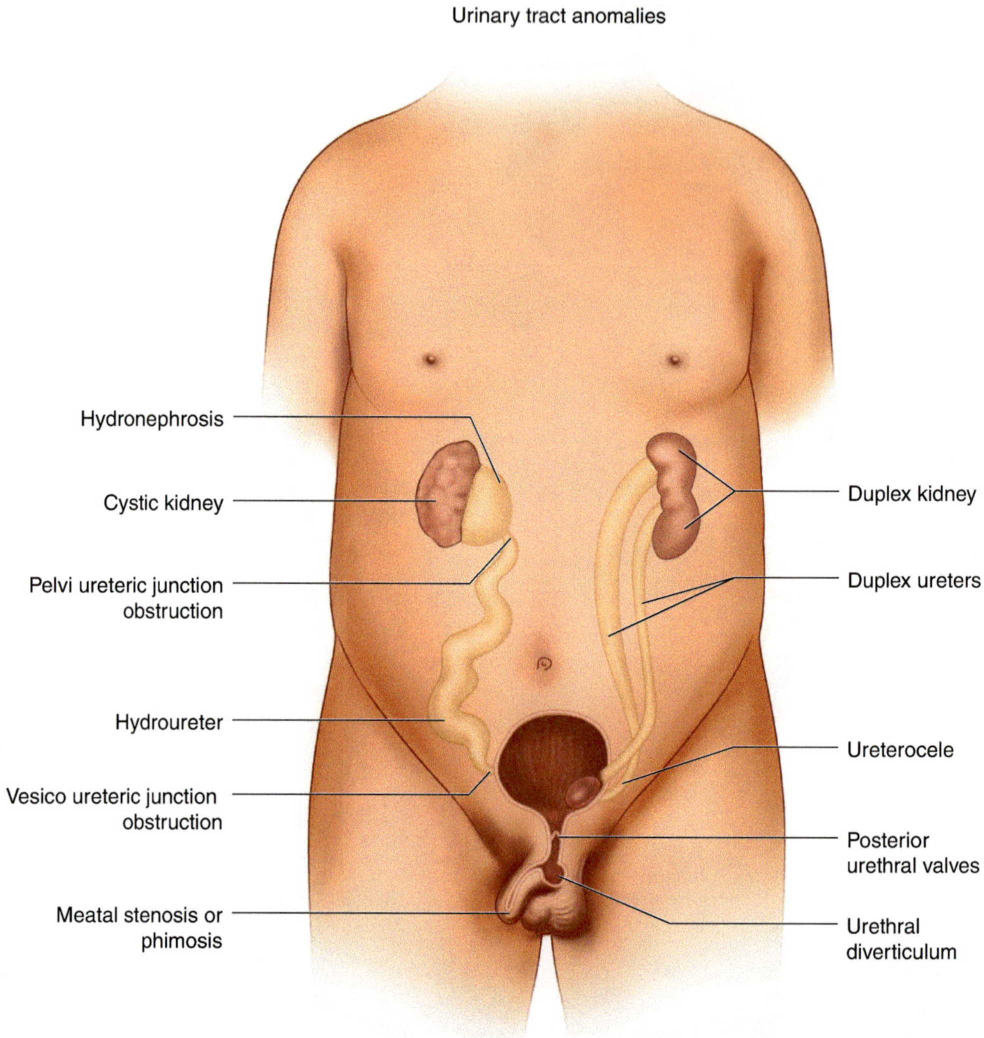

Ultrasound screening should be performed in the above conditions and should be considered in a child who has anomalies of two systems, i.e. VSD and undescended testes.

Associated Anomalies

The ipsilateral ureter is absent in the majority of cases, with complete absence in 60% of cases [6]. This is associated with "failure of the ipsilateral trigone to develop". However more recent work has questioned whether the trigone, of urogenital sinus origin, does develop but with no ureteric tunnel is indistinguishable from the surrounding detrusor.

Excluding renal ectopia or malrotation, abnormalities of the contralateral kidney are uncommon [7]. The contralateral ureter, however, will frequently be found to have functional or anatomical changes [8]. Studies have shown a rate of VUR of 30% which is often high grade and PUJO and VUJO in 11 and 7% of cases, respectively [9].

In males abnormalities that are associated with unilateral renal agenesis occur in 10–15%. The testes and head of epididymis, both of mesonephric duct origin, are usually present; however, all structures proximal to this, of Wolffian duct origin, are absent in 50% of cases. Abnormalities include absence of the body and tail of the epididymis, vas, seminal vesicles and ejaculatory ducts, seminal vesicle cysts and cystic dysplasia of the rete testes. The proportion of males with an absent vas that have a unilateral renal agenesis has been reported as high as 85% [10] with reports of unilateral agenesis of the kidney and bilateral absence of the vas. The diagnosis should be sought in children who have an absent vas or epididymal abnormalities found at orchidopexy or inguinal herniotomy.

In females Mullerian duct formation is inhibited by defective Wolffian duct resulting in a number of anomalies such as unicornuate or bicornuate uterus, double or septate uterus

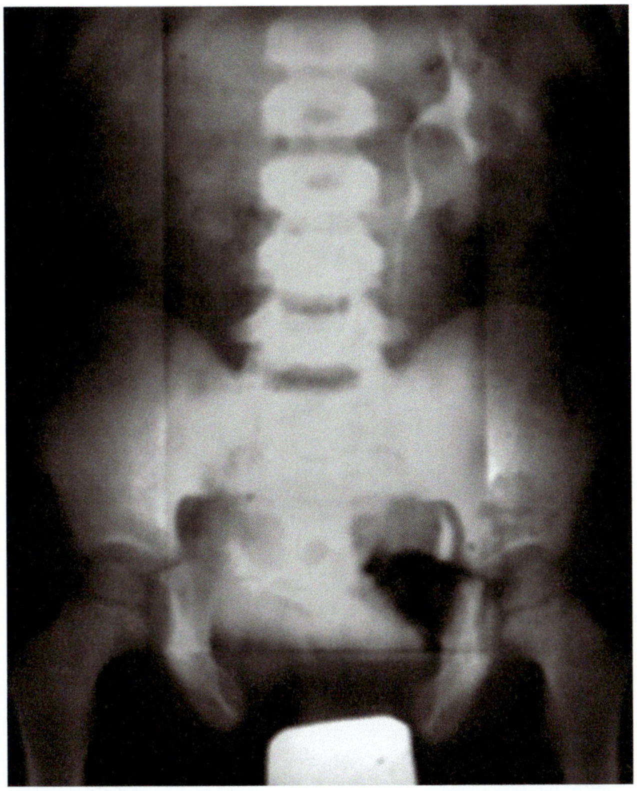

Fig. 42.1 Renal agenesis with pubic diastasis in a patient with exstrophy of the bladder

and partial or complete vaginal atresia. Women with Mullerian anomalies should be screened for renal anomalies, in one cohort of women with a unicornuate uterus and rudimentary horn, 38% had associated unilateral renal agenesis (URA) [11]. The proportion of females with URA and a Mullerian anomaly ranges from 11 to 25% [4, 12]. Two syndromes featuring Mullerian anomalies are OHVIRA syndrome (Obstructed HemiVagina and Ipsilateral Renal Anomaly) [13] and type 2 Mayer-Rokitansky-Kuster-Hauser syndrome (Table 42.2) [14].

Anomalies of other systems are frequently found. A systematic review looking at associated anomalies extracted data from 16 studies with 709 patients of which 31% had extrarenal anomalies. The most common are gastrointestinal anomalies such as imperforate anus and oesophageal abnormalities comprising 16%. Cardiovascular anomalies such as septal and valvular lesions made up 14%. Musculoskeletal problems were less common making up 13% such as vertebral or phalangeal anomalies [14]. Preauricular pits and minor ear tags are also associated with renal agenesis.

Long-Term Outcome

The risk from trauma was examined by the American Academy of Pediatrics [15]. They recommended that there should be no restriction on non-contact sports. Parental judgement should be utilised for contact sports. The relative risk of head injury is fivefold higher in activities associated with higher risk of renal injury, and hence an element of perspective is required. The sports associated with a higher grade of renal injury are bicycling, sledding, downhill skiing, snowboarding and equestrian sports. Protective clothing can be used as required and dependent on the individual. The risk for children in a road traffic accident is higher than injury from sports, and parents should be aware of this and take necessary safety precautions.

In the systematic review, there were 11 studies with 437 patients looking at long-term renal outcome with unilateral renal agenesis. The pooled mean GFR was normal; however, hypertension, microalbuminaemia and renal function impairment were found in 16, 21 and 10%, respectively [4]. Another study found a high frequency of end-stage renal disease at 30 years of age [16]. They found the risk for the need of dialysis was significantly higher for patients with a solitary kidney compared to renal hypoplasia, multicystic or horseshoe kidneys and was independent of other risk factors. This suggests that subclinical defects of the solitary kidney may be the reason for a poorer prognosis in this group compared with other causes of congenital anomalies of the kidney and urinary tract.

There is evidence to support that differences in ureteric bud branching lead to a reduction in final nephron number. Individuals with a lower nephron number have been shown to be more susceptible to long-term problems. This may be applicable to unilateral agenesis.

The solitary kidney undergoes hypertrophy and hyperfiltration in the long term; this may have an adverse effect depending on the starting number of nephrons. There have been reports of glomerulosclerosis in patients with URA confirming the pathological stigmata of hyperfiltration. Studies have also shown a higher rate of pregnancy-related problems such as hypertension, pre-eclampsia and proteinuria in patients with URA [12].

Lifelong surveillance is required with annual blood pressure and urinalysis for protein to serve as markers for early renal deterioration.

Bilateral Renal Agenesis

Bilateral renal agenesis was first identified in the seventeenth century. However it was Edith Potter in 1946 who described the pathognomic features of bilateral renal agenesis in 20 cases at foetal autopsy. Her name is now used eponymously to describe this constellation of features [17].

Epidemiology

The incidence is estimated at 1 in 4000 births with a 2.5:1 male predominance. A risk factor for the anomaly is increas-

Table 42.2 Associated syndromes and anomalies of the kidney

Condition	Associated syndromes	Associated anomalies
Kidney		
Unilateral renal agenesis	Turner syndrome Poland syndrome Frazer syndrome Branchio-oto-renal syndrome DiGeorge syndrome Townes-Brock syndrome VACTERL MURCS syndrome OHVIRA Mayer-Rokitansky-Kuster-Hauser syndrome	Absent ipsilateral ureter Renal ectopia Renal malrotation VUR PUJO VUJO Wolffian duct anomalies Mullerian duct anomalies Preauricular pits/minor ear tags GI- imperforate anus, oesophageal anomalies CVS anomalies Musculoskeletal anomalies
Bilateral renal agenesis	Frazer syndrome Klinefelter syndrome Kallmann syndrome	Potter facies Sirenomelia Lumbar meningocele Hydrocephalus Cryptophthalmos Cardiac lesions GI anomalies in 60% including oesophageal, intestinal atresia and imperforate anus
Supernumerary kidneys	None	Ectopic ureter Coarctation of aorta Duplication of urethra Vaginal atresia Horseshoe kidney
Ectopic kidney	None	Renal malrotation PUJO VUJO Renal dysplasia (contralateral) Genital anomalies Skeletal Cardiac
Horseshoe kidney	Turners syndrome Trisomy 21 Neural tube Townes-Brock syndrome SALL1 transcription factor defect	Hypospadias Cryptorchidism Bicornate uterus/septate vagina Duplication ureter VUR PUJO Calculi UTI Skeletal/cardiovascular (VSD)
Multicystic dysplastic kidney	None	Segmental horseshoe kidney and moieties of duplex Contralateral VUR most grade III or less PUJO VUJO
Ureter		
Duplex kidney	None	Renal scarring Hydronephrosis Ectopic ureteroceles
Pelviureteric junction obstruction	VACTERL	Contralateral PUJO Renal dysplasia/multicystic Unilateral renal agenesis Duplex (lower>upper) Horseshoe Ectopic VUR
Vesicoureteric junction obstruction	None	None

ing maternal age. There is a genetic element to the disease which is not fully understood. When the pedigree of an index case was screened, unilateral renal agenesis was found in 4.5% [18] and bilateral agenesis in 3.5% [5]. In view of these risks, it is reasonable to screen siblings and first-degree relatives of infants born with any form of renal agenesis [19].

Associated Syndromes:

- Oesophageal atresia
- Cryptophthalmos
- Frazer syndrome
- Klinefelter syndrome
- Kallmann syndrome

Associated Anomalies

The characteristic facies is well recognised, and these infants are often growth restricted with low centile birth weights and measurements. Most babies will have a pathognomic skin-fold beginning over each eye and extending onto the cheek. The nose is blunted with low-set ears and a prominent depression between the lower lip and chin. The skin is generally loose and their hands appear disproportionately large. The lower limbs are bowed and clubbed with excessive flexion at the knees and hips.

Associated anomalies include sirenomelia, lumbar meningocele and GI problems in 60% especially imperforate anus.

The diagnosis is frequently made antenatally, with scans showing no discernible renal tissue or bladder and severe oligohydramnios after 14–16 weeks. Other features that assist with the diagnosis include small-volume lungs and chest diameter and "pancake" adrenal glands. Termination is an option at this stage when the diagnosis is certain. Foetal MRI can be a useful adjunct to confirm the diagnosis.

Forty percent of affected babies are stillborn, with the rest surviving less than 48 hours due to pulmonary insufficiency secondary to hypoplasia.

Supernumerary Kidney

Supernumerary kidneys are additional discrete kidneys that are either completely separate or joined to the ipsilateral kidney by loose areolar tissue only. They have their own renal capsule, collecting system and blood supply but are usually smaller than the orthotopic kidney.

Epidemiology

The incidence of this condition is unknown as it is extremely uncommon. Since it was first described in 1656, there have been about 100 cases reported. There is equal sex distribution with cases, but the left side is affected more frequently [20]. There are a few cases of bilateral supernumerary kidneys [21]. The majority of supernumerary kidneys are located in a caudal position to the dominant kidney which sits in an orthotopic position. In half of cases, the ureters converge, and in the remaining half, there are two separate ureters entering the bladder separately. In the majority of cases, the ureteric insertion of the kidneys will follow the Weigert-Meyer rule (see below).

Pathophysiology

The embryological events for this anomaly appear to start with two ureteric buds or branching of a solitary bud. The metanephros also needs to be separate and separate entirely when induced to differentiate by the two or bifid ureteric buds [20].

Another theory outlined by Geisinger in 1937 states that the kidneys developed separately due to fragmentation secondary to linear infarction, and the separate fragments only develop when induced by two ureteric buds [22].

The development of the ureteric bud from the Wolffian duct is stimulated by glial cell line-derived neurotrophic factor (GDNF) secreted by the metanephros. Another intercellular signalling system including Slit-family proteins (SLIT2) or its receptor Roundabout receptor (ROBO2) is crucial to produce a solitary ureteric bud in the correct position. Mutant mice lacking either SLIT2 or ROBO2 develop supernumerary ureteric buds [23]. This may form part of the underlying problem in humans, but this has not been confirmed to date.

Associated Syndromes

There are no specific genetic or syndromic associations.

Associated Anomalies

The two normal orthotopic kidneys are normal.

Associations reported:

- Ectopic ureter
- Coarctation of the aorta
- Duplication of urethra
- Vaginal atresia
- Horseshoe kidney [24]

Long-Term Outcomes

In general the condition is asymptomatic in childhood and diagnosed after the third decade with symptoms of infection or obstruction. There are reports of malignancy associated with supernumerary kidneys. Both Wilms tumour and clear-cell tumour have been reported [25]. If the diagnosis is made, then surveillance is warranted as these patients do have associated morbidity. However a large proportion of patients are asymptomatic with 25% only detected at autopsy [26].

Abnormalities of Ascent

Ectopic Kidneys

Ectopic kidneys are a heterogeneous group where the kidney has ascended in an abnormal manner resulting in an aberrant position. The commonest variant is a pelvic kidney. Other types include iliac, abdominal, thoracic and cephalad and crossed with or without fusion.

Cephalad are seen with omphalocele, the herniated liver results in the kidney ascending higher as there is no physical barrier. They are usually found beneath the diaphragm at the level of the tenth thoracic vertebrae. Thoracic ectopic kidneys (<5%) are seen at autopsy with a frequency of 1 in 13,000 and can be partially or completely above the diaphragm, residing in the posterior mediastinum within the foramen of Bochdalek. They are usually found as an incidental finding on chest imaging.

Epidemiology

Autopsy studies have recorded an incidence of 1 in 500 to 1 in 1200. As the condition is asymptomatic in the majority of cases, the observed incidence is significantly lower and reported to be 1 in 10,000 [27]. In autopsy studies males and females are affected at equal frequency. In clinical studies more females are observed, and this is attributed to the more frequent uroradiology performed for females. The condition is observed bilaterally in 10% of cases.

The ascent of the kidneys is normally completed by the eighth week of gestation; therefore, it is proposed that failure of this process occurs between 5 and 8 weeks. The two embryological theories are either an abnormality of the ureteric bud development or of the metanephric tissue to induce appropriate ascent. The aetiology for this is still debated with different reports attributing it to genetic, maternal or teratogens and a vascular barrier caused by persistence of foetal blood supply to the kidney. To date there are no convincing data to support a unifying cause.

Associated Syndromes

There are no specific genetic or syndromic associations.

Associated Anomalies

The ectopic kidney is classically smaller than normal and may retain foetal lobulations giving it an irregular appearance. Bilateral ectopia may be present in 10% of cases. The renal axis is often abnormal with failure of normal rotation and persistent anterior lie. In a retrospective review of 82 ectopic kidneys, 56% were found to be hydronephrotic. Around half were due to anatomical obstruction with 70% at the pelviureteric junction obstruction (PUJO) and 30% at the vesicoureteric junction obstruction (VUJO). In the remainder the hydronephrosis was more functional with a pseudo PUJO related to vesicoureteric reflux (VUR) or malrotation.

Hydronephrosis may also occur in 25% of non-ectopic contralateral kidneys [27].

VUR was seen in 20, 30 and 71% of crossed fused ectopia, simple ectopia and bilateral ectopia, respectively. In a significant proportion of cases (85% with unilateral simple renal ectopia), the reflux was to the orthotopic normal kidney. The majority of the ectopic kidneys have reduced differential function on DMSA [28]. Other genitourinary anomalies are observed in 14% with the most common being contralateral renal dysplasia, cryptorchidism and hypospadias.

The ureter may be slightly tortuous but usually has an orthotopic orifice into the bladder. The vascular supply is variable and dependant of the final position. The blood supply may be from the aorta, aortic bifurcation, iliac vessels or inferior mesenteric artery.

Genital anomalies are seen in both sexes with an overall rate of between 15 and 45%. In females the rate is 20–66% and includes cloaca, uterine and vaginal anomalies. In one study with cloacal patients, the incidence of ectopic kidney was 14% [29]. In males the rate of genital anomalies is between 10 and 20% and includes undescended testes, duplication of the urethra and hypospadias [30].

Abnormalities in other systems are seen in 21% with skeletal and cardiac malformations the commonest [31].

Rarely are there associated adrenal anomalies which can be absence or malposition.

Long-Term Outcome

Generally the ectopic kidney has no increased risk of disease except hydronephrosis and calculi formation [32]. This has been attributed to the malrotated position and resultant impaired drainage and in some cases an anomalous vascular supply.

There have been reports of renovascular hypertension in patients with ectopic kidney, but no definite causal relationship has been proven [33].

There is no increase in malignancy reported; historically cases of solitary ectopic kidney were excised due to concern regarding pelvic malignancy, but with modern imaging this should no longer occur.

The ectopic kidney is theoretically at increased risk of injury trauma due to less protection in some positions although literature supporting this is sparse.

The ectopic kidney frequently will have a relative decreased function on DMSA with one study showing mean function of 38% with an interquartile range of 33–43%. While it does seem likely that ectopic kidneys do have reduced function, there are technical issues with the renogram studies where the pelvis may shield the emissions and lead to a false reduced count; this should be taken into consideration when planning management of these patients. The global renal function was normal, and with a median follow-up of 7.7 years, there was no trend to a reduction in function. Whether this change is sustained, postpuberty remains to be seen with further long-term studies [33].

The high rate of VUR and relative reduced function with a propensity to drainage problems and stones are all potential risk factors for loss of function with time. These children therefore require long-term follow-up and surveillance.

Abnormalities of Fusion

Horseshoe Kidney

Epidemiology

The incidence of horseshoe kidney is 1 in 400 live births. It is the commonest fusion anomaly of the kidneys (Fig. 42.2). The male to female ratio is 2:1 [34]. In autopsy studies there is a higher prevalence in younger individuals and in association with other anomalies. This suggests that it is commonly linked to other conditions that are not compatible with life [35]. There are studies showing increased occurrence in twins and siblings supporting a genetic predisposition with low penetrance.

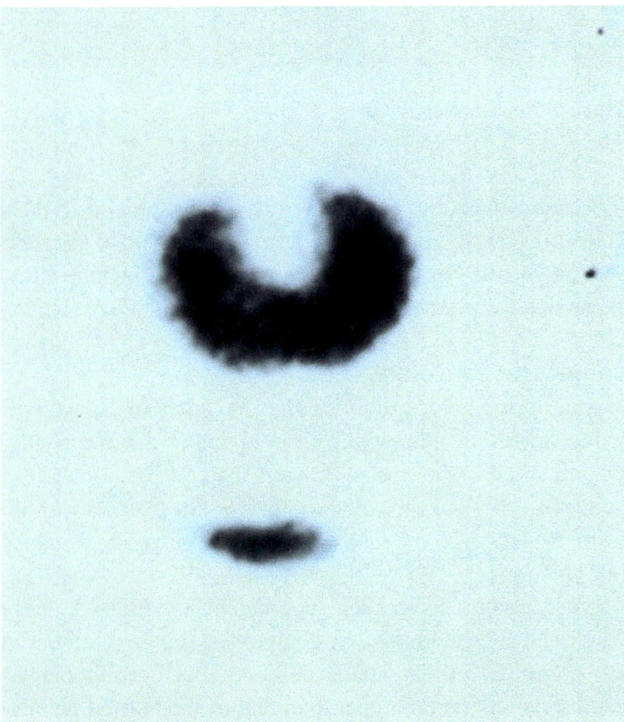

Fig. 42.2 Isotope scan of horseshoe kidney

The critical time in gestation is after ureteric bud, and metanephric blastemal association before significant migration has taken place. A case report in 1931 described a horseshoe kidney in a 6-week foetus and postulated that the fusion occurred at 4.5 weeks [36]. Embryologists have proposed a mechanism where the kidneys are medially displaced and come into contact (frequently the lower poles); fusion then occurs. Vascular variations such as different position of the umbilical artery or common iliacs at this critical stage may be enough to cause this shift and subsequent fusion. Other theories suggest that abnormal formation of the tail of the embryo or other pelvic organs may be the precursor to fusion.

From a genetic perspective, knockout studies of the sonic hedgehog gene, which is highly expressed on the notochord and floor plate, resulted in kidney fusion. This may be the causal link with the increased incidence of horseshoe kidney and vertebral and neural tube defects [37].

Associated syndromes

- Turners syndrome 60%
- Trisomy 18 20%
- Townes-Brock syndrome

Associated Anomalies

Neural tube defects are seen in up to 3% of patients with a horseshoe kidney [38]. They found horseshoe kidneys were associated with lesions between T9 and L1. Renal agenesis and duplications were seen in patients with lesions between T5 and T8 and sacral lesions, respectively.

Renal associated anomalies such as VUR, PUJO, duplicated ureters and calculi are seen with higher frequency than the normal population.

In a study examining the outcomes of 52 children with horseshoe kidneys (median age of 3.9 years), 52% had urological anomalies, 32% had VUR with approximately half having renal parenchymal damage and a further 24% had PUJO [39].

Other studies have shown rates of VUR of up 26% and PUJO 13–32% [40–42]; however, interpretation of these studies is difficult as the imaging was not uniform across the cohort (only 50% of patients has MCUGs and none underwent isotope studies). The study with the lowest rate of VUR had the oldest cohort of patients [41]. This would support other literature showing the VUR resolution rate is similar to the normal population with 75% rate of resolution [39].

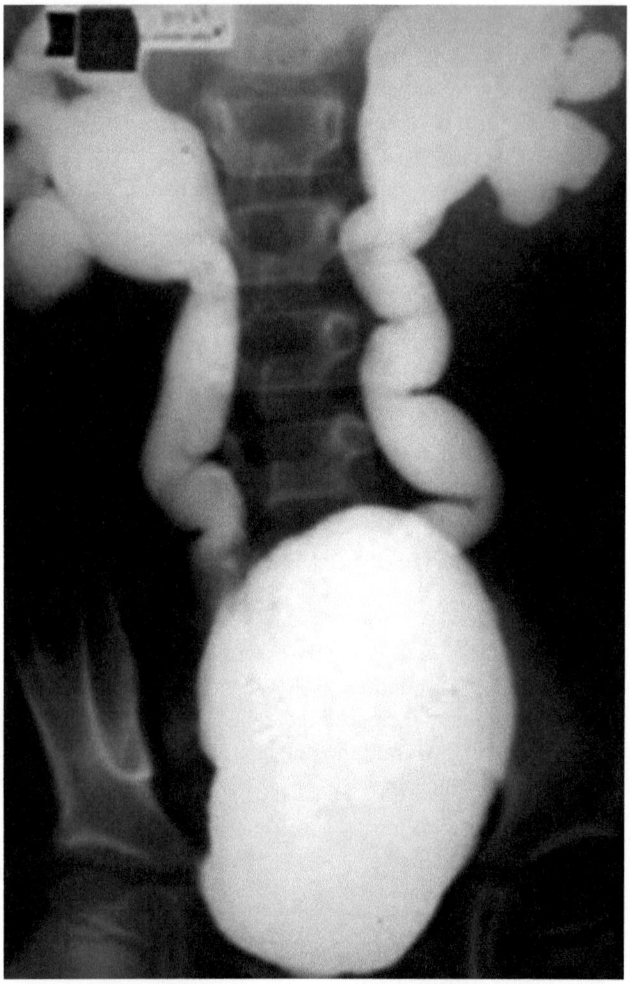

Bilateral reflux grade V

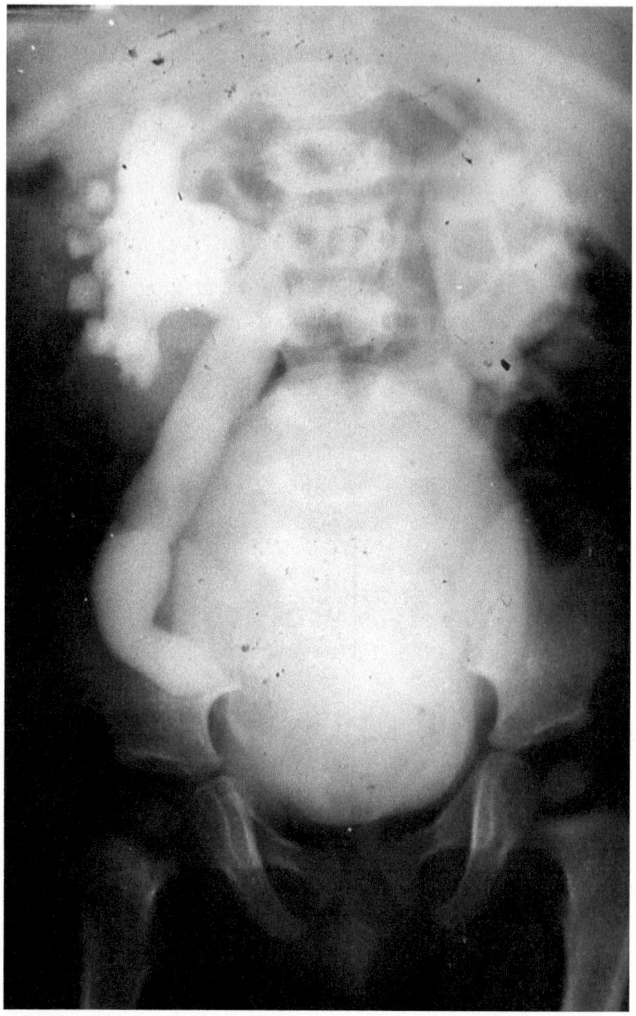

Megacystis-megaureter reflux

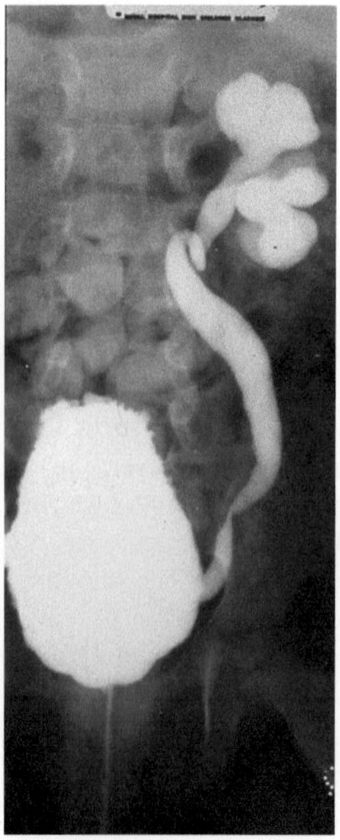

Cystogram reflux

Genital anomalies are seen with hypospadias and cryptorchidism occurring in 4%. Bicornuate uterus/septate vagina is seen in 7% of series. Other system anomalies include ARM skeletal and cardiovascular anomalies particularly VSD.

Long-Term Outcome

A long-term study examining the outcomes of 51 patients with a horseshoe kidney [43] showed that 60% were asymptomatic after 10 years of follow-up. The rate of infection or pain was 13% and the calculi rate 17%. In patients who do form stones, ECSWL and PCNL achieve good clearance rates [44, 45].

Malignancies have been reported in horseshoe kidneys with renal cell carcinoma the commonest reported tumour in this group (50%). The incidence of renal cell carcinoma has been shown to be no higher than that of the normal population. However, the incidence of Wilms tumour has been reported to be higher in horseshoe kidney. The National Wilms Tumour Study Group reported 41 cases of horseshoe kidney in 8617 patients affected by Wilms tumour, an inci-

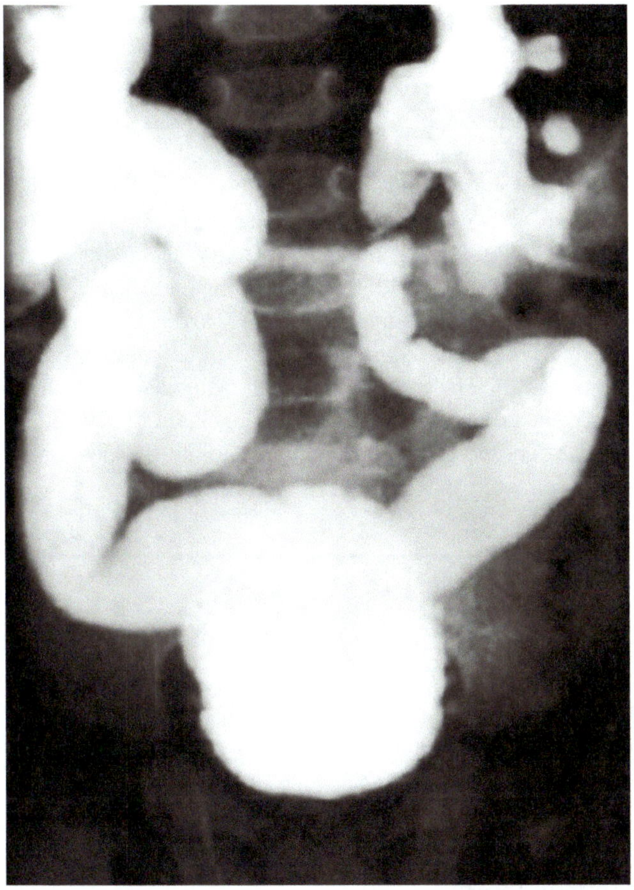

Bilateral grade V reflux

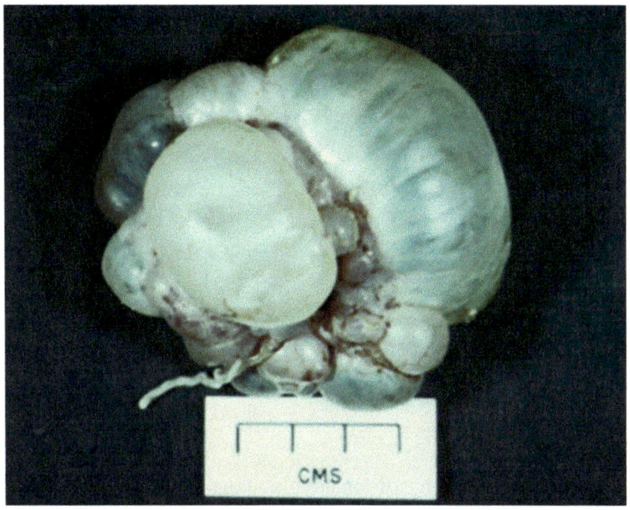

Multicystic dysplastic kidney

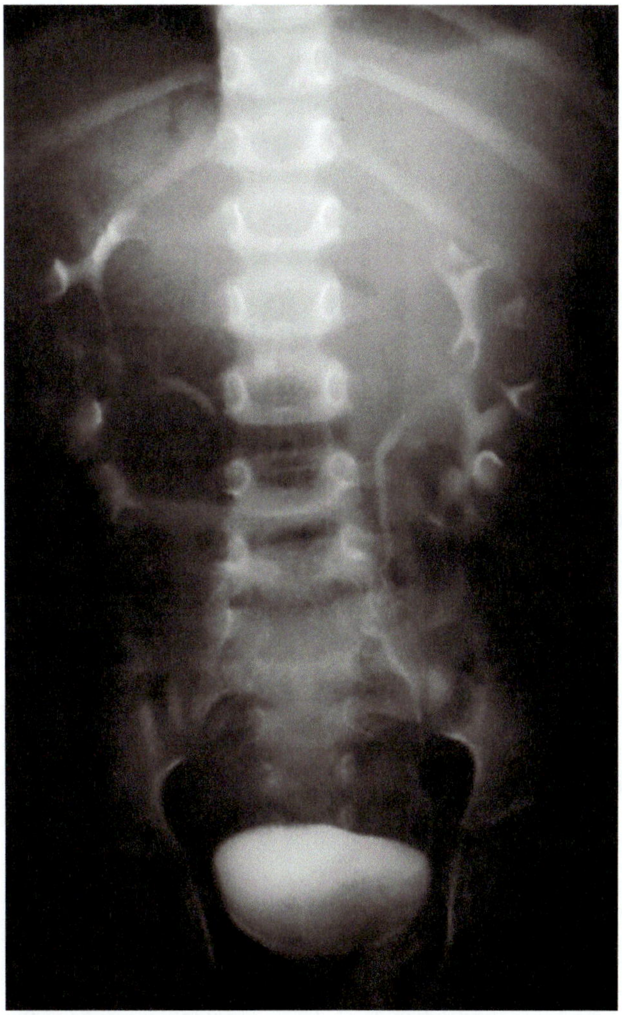

Polycystic kidneys

dence of 0.48%. This is estimated at a twofold increased rate compared to the population risk. The tumours are predominantly left sided and of favourable histology. The affected children have similar morbidity compared to children without horseshoe kidneys [46, 47].

Parenchymal Anomalies

Multicystic Dysplastic Kidney

Multicystic dysplastic kidneys (MCDKs) are characterised by multiple cysts of variable size that classically do not communicate, minimal or no normal parenchyma and an atretic ureter. It is the second commonest cause of a palpable abdominal mass in a neonate.

Epidemiology

The incidence is reported to between 1 in 2, 2000 and 1 in 4300 live births [48, 49]. There is a male predominance with a ratio of 1.5:1 for unilateral disease with the left kidney being more commonly affected [49]. In contrast bilateral cases are more common in females 2:1 and are associated with poor outcome similar to bilateral renal agenesis.

Pathophysiology

There are two types of MCDK, infundibulopelvic and hydronephrotic types. The former is more common and due to atresia of the pelvis and ureter, and the latter is due to ureteric atresia alone. Both have hypoplastic vessels.

The embryological events that result in a MCDK are not fully understood; however, there are two proposed mechanisms. The first is obstruction secondary to atresia of the ureter. Animal models replicating early obstruction of the kidney have resulted in dysplasia but not multicystic dysplasia [50]. The left-sided bias in MCDKs is thought to support this as other obstructive conditions are also known to be more common on the left (PUJO and obstructing megaureter).

The second hypothesis proposes an abnormal interaction between the ureteric bud and the metanephric mesenchyme resulting in multicystic dysplasia [51]. Proponents highlight known genes implicated in ureteric bud development which are also associated with renal dysplasia.

There are reports of maternal infections and medications being associated with development of the condition and rare familial reports [52–54].

Associated Syndromes

None

Associated Anomalies

Associated anomalies have been reported to occur in one in three patients with MCDK [49]. The commonest anomaly is contralateral VUR with a rate of 19.7% in a meta-analysis with over 3500 patients. The VUR was generally low grade with the majority grade III or less [49, 55]. Other anomalies were PUJO (4.6%), ureterocele (1.3%) and horseshoe kidney (0.6%, 4 × the normal population rate of 0.15%).

Long-Term Outcome

There has been debate regarding the management of MCDK for decades with a shift to more conservative management over the years. The potential sources for debate are the rate of spontaneous involution advocating conservative management versus the risk of hypertension and malignancy requiring active treatment.

A review of 105 papers with MCDK attempted to tease out the answers to some of these controversies; the authors found a complete involution rate of 20% at 3 years and 50% by 5 years [56].

From the viewpoint of hypertension, an incidence of 5.4 per 1000 cases was estimated by Narchi which is lower than the general population [57]. In Wacksman's review for the multicystic kidney registry, there were no cases of hypertension in 441 patients [58]. However other small series have suggested a link between MCDK and hypertension; however, nephrectomy failed to resolve the hypertension in two thirds of patients making a causative association difficult to confirm [59]. Other researchers have demonstrated that contralateral renal disease may be the most important predictive factor [60].

The risk of malignancy was historically the underlying concern for a more aggressive management approach. Two independent reporters utilising different techniques have found similar results with Beckwith citing the risk of a Wilms tumour in a MCDK kidney as 1 in 2000 and Noe calculating the nephrectomies required to prevent one case of Wilms tumour as 2000 [61, 62]. All reported malignancies occurred pre 4 years of age, with none in an involuted MCDK and a similar good prognosis in accordance to other patients with Wilms tumour.

Ureter

Duplex Systems

This condition covers a wide range of anatomical abnormality with a very variable potential for pathology. Duplication may be splitting of the pelvis up to full doubled system with separate pelvis and ureters (Figs. 42.3 and 42.4).

Epidemiology

The incidence of ureteric duplication in autopsy studies is 1 in 125 patients (0.8%) and is the most common ureteric anomaly [63]. The right and left ureters are affected with equal frequency, and unilateral cases are six times more common than bilateral cases. There are more females affected with a ratio of approximately 1.7:1. The most common clinical manifestation is a urinary tract infection. A cohort of 774 children under the age of 6 years at the time of first UTI found an 8% incidence of children with duplications [63].

It is postulated that the gene responsible is an autosomal dominant gene with incomplete penetrance. In Atwell's study where they performed IVPs in siblings and parents of index cases, the incidence of duplex systems increased to one in nine [64].

Pathophysiology

Incomplete ureteral duplication occurs when a ureteral bud bifurcates soon after its origin from the mesonephric duct. If this occurs after the ureteral bud has penetrated the metanephros, the result is a bifid pelvis. However if this occurs before penetration, varying degrees of incomplete duplication with single common ureter entering the bladder may occur. Complete duplication requires two separate ureteric buds to develop.

The lower pole ureter develops from a ureteric bud that starts at a lower position on the mesonephric duct and there-

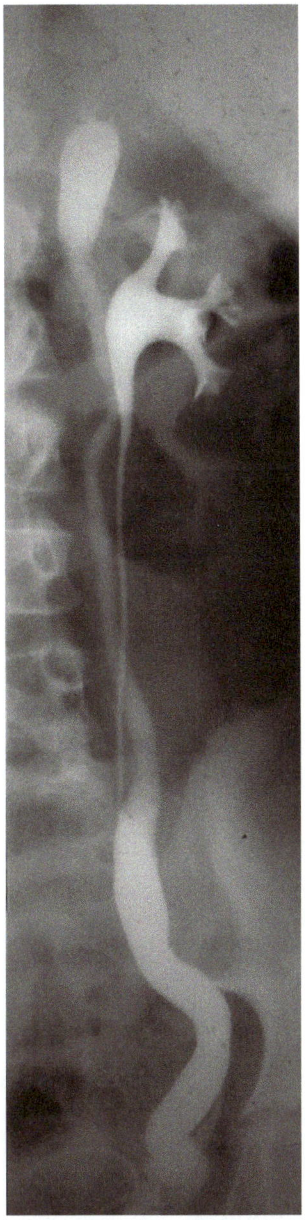

Fig. 42.3 Duplex system with ectopic ureter

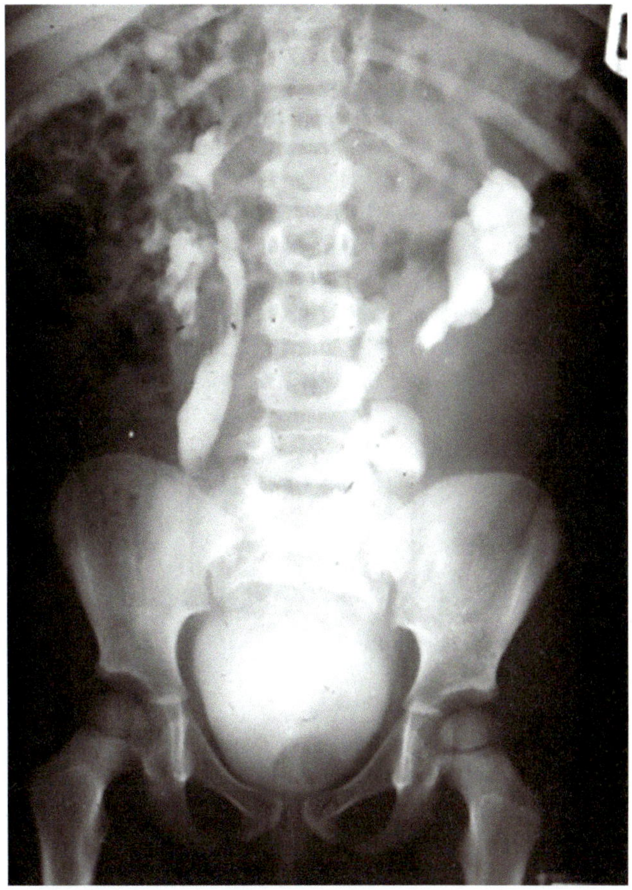

Fig. 42.4 Duplex kidney

fore arrives earlier at the urogenital sinus. Following UG migration and development of the trigone, the orifice ends in a more lateral and cranial position where there is weakened trigonal support and a predisposition to reflux due to the short intramural tunnel.

Conversely the upper pole ureter has a ureteric bud that starts at a higher position on the mesonephric duct and arrives later at the urogenital sinus. This results in a final ureteric orifice position that is caudal and medial. If the starting position is substantially higher, it may not be possible to gain incorporation into the bladder at all, and the ureteric orifice may be in the urethra or mesonephric remnants. In the male the site of insertion may be the epididymis, vas or seminal vesicle. In females Gartner duct is the distal mesonephric remnant; it traverses from the broad ligament of the uterus along the lateral wall of the vagina to end in the hymen. The ureteric orifice can enter anywhere along this with rupture of the duct into the vagina.

This differential migration of the ureters is known at the Meyer Weigert law which states in complete duplication the medial and caudal ureteric orifice is the ureter to the upper pole [51] (Stephens and Mackie 1975).

There is still debate regarding the embryology of ureteroceles, which is in part due to the wide variety in their morphology. One theory is the persistence of Chwalle's membrane, which is a two-cell layer membrane present at the time of ureteric bud emergence. In normal development this should regress, but where there is persistence, the ureteral-meatal orifice will be stenotic resulting in an ureterocele [65]. This can explain cases where there is a stenotic orifice or where there is evidence of muscular hypertrophy of the ureteric wall. Another theory is that the distal ureteric segment undergoes the same expansion forces that change the urogenital sinus into the bladder [66]. In cases where there is delay in the formation of the ureteric bud lumen, this expansion results in an ureterocele. However, as all caudal ectopic ureters do not form ureteroceles, this cannot be a unifying embryological explanation.

Associated Syndromes

None

Associated Anomalies

Duplex systems are associated with a number of urological pathologies. Many of the pathologies affecting a single system kidney will be affecting the lower pole (PUJO and VUR). The upper pole problems are commonly secondary to anomalies of insertion such as ectopia and ureteroceles leading to obstruction. VUR is the most common associated renal pathology with a high rate in children presenting with a UTI. The lower grades have a significant resolution rate but take longer with a 50% resolution taking 84 months compared with 54 months for simplex systems [67]. The rate of success with STING is also lower with a success rate of 50% compared with 73% from a meta-analysis of over 8000 renal units [68].

Pelviureteric Junction Obstruction

This is the commonest cause of antenatal hydronephrosis responsible for 50% of cases of urological significance [69].A study conducted examining the outcome of antenatal hydronephrosis found that 1 in 60 pregnancies are affected [70](Thomas). The condition is caused by impaired flow of urine from the renal pelvis to the ureter. The aetiology can be divided into intrinsic and extrinsic causes. The commonest cause is an adynamic section impairing urinary flow, with polyps and valves being less frequent in paediatrics. Extrinsic causes include anomalous crossing vessels or high takeoff of the ureter, kinks or adhesions.

Epidemiology

The incidence of PUJO is 1 in 1000–2000 livebirths. The condition is more common in males with a 2–3 times increased incidence. The left side is more commonly affected with a 2:1 ratio. The condition can be bilateral in 10–20% of cases; this is usually found in infants less than 6 months. Increasingly this condition is diagnosed antenatally with the UK introduction of universal anomaly screening at around 16 week.

Pathophysiology

The metanephric mesenchyme stimulates the ureteric bud to form from the posterior aspect of the mesonephric duct and proceed to the metanephros in the fourth week. The ureteral bud grows in a cranial direction and penetrates the metanephros. Reciprocal induction then begins with the tissues stimulating each other to develop. The metanephros develops into specialised nephrons, and the ureteric bud divides to form the collecting system. At 6 weeks the mid-ureter is solid and recanalisation occurs in both directions [71]. The failure to complete this process completely may be the underlying mechanism for the PUJO. Another proposal is that the smooth muscular differentiation of the ureter occurs in a retrograde manner from the bladder to the kidney. If this process stops early, this would result in the adynamic area in the pelviureteric area and a functional obstruction.

Usually urine produced by the kidney can pass via the ureter to the bladder by the 10th–12th week. The kidney continues to mature, and by the 18th week of gestation, the majority of the amniotic fluid is foetal urine.

There is less debate regarding the embryological origin of the crossing vessels compared to their role in causing impairment of flow. As the kidney ascends and rotates during development, it acquires and sheds its blood supply from branches of the aorta. During this process if a vessel does not involute, it may cross the ureter leading to kinking and obstruction. This extrinsic compression may be the cause of the hydronephrosis, but some authors suggest that the intrinsic changes often seen are more important—there remains no clear consensus.

Hydronephrosis discovered at antenatal screening is now the commonest mode of diagnosis in countries where anomaly screening is universal. It enables affected infants to be diagnosed before symptoms commence. Natural history studies have shown that hydronephrosis of more than 20 mm is predictive of the need for surgery postnatally. Foetal intervention is not appropriate in unilateral disease and seldom considered even in a bilateral situation. Patients are generally followed postnatally with serial scans to monitor for progression following delivery; further scanning is undertaken and functional studies to try and identify kidneys that require treatment.

Generally the condition occurs sporadically although linkage to the HLA region on chromosome 6 has been found in some families.

Associated Syndromes

None

Associated Anomalies

Half of children with PUJO will have associated anomalies. The commonest is a contralateral PUJO. Renal dysplasia and multicystic kidney may also be present. In 5% of patients, there may be contralateral renal agenesis. The condition can be associated with duplex system with the lower moiety generally being affected (although upper moiety and combined cases are recognised). Other renal associations include ectopic kidney, horseshoe kidney and VUR which is normally of low grade.

Long-Term Outcome

The condition appears to have a low risk of long-term problems. With lesser degrees of dilation, the chance of surgery is low with the anticipation that the dilation will resolve spontaneously with growth. Those who do require treatment are usually offered pyeloplasty (open or increasingly minimally invasive techniques). The overall results are good with low incidence of renal injury, recurrence or other complication. However, special consideration should be given where the dilation is rapidly progressive, in the younger age group, and where there is significantly reduced differential function as this group does not respond as well to treatment.

Vesicoureteric Junction Obstruction

The abnormalities of the vesicoureteric junction were originally classified into obstructed, refluxing and nonrefluxing and also subdivided into primary and secondary [72]. Subsequently the nomenclature for subclassification was changed to what is generally still used, obstructed, refluxing, neither or both obstructed and refluxing [73].

Epidemiology

Congenital megaureters are more common in males with a ratio of 2:1, and the left side is more commonly affected; the incidence is variably reported around 1 in 4000 children [74].

Pathophysiology

The knowledge regarding the normal development of the ureter comes from animal studies. The sequence of smooth muscle differentiation is not fully understood but animal studies assist in shedding some light on the subject. A study looking at bladder specimens from piglets as foetus, newborns and 6 months found that the bladder detrusor muscle developed first followed by the extravesical and then intravesical ureter. The intravesical portion did not develop fully until after birth. The smooth muscle showed an increase in bulk between the neonatal and later sections [75]. This is similar to human studies which show that smooth muscle cells of the extramural ureter are first visible between 17 and 22 weeks of gestation, and ureteral muscle development proceeds in a cranio-caudal direction [76]. Tanagho also observed that the distal ureter is the last portion to develop the muscular coat and early differentiation is of the circular layer [77]. The change from the characteristic foetal circular muscular wall to the full-term neonate with a double muscle layer may take up to 2 years and may help to explain the transient dysfunction observed in the VUJ during this time. The possible theoretical aetiologies for VUJO include a thick muscular cuff around the distal portion of the ureter [78], increased deposition of collagen fibres [79] and dysregulation in the neuromodulatory control of the vesicoureteric junction [80]. A study of 18 obstructed ureters of patients up to 2 years of age demonstrated increased myocyte apoptosis, reduced vascular and neural elements and increased collagen fibres [81]. Another group comparing refluxing and obstructed ureters in 14 patients found disparity between the number of smooth muscle cells, myocyte apoptosis indices and number of C-kit-positive Cajal-like cells.

Another possible aetiology is the persistence of Chwalle's membrane. In 1927 Chwalle demonstrated the closure of the embryonic ureteral membrane by an epithelial membrane during formation at the 6 week stage. The membrane usually would disintegrate establishing patency by the eighth week. Persistence of this membrane has been suggested by some to the aetiology of ureteral obstruction and has been seen in rats with pantothenic acid deficiency [65, 82]. This membrane has also been implicated in the formation of ureteroceles (see above).

The prenatal detection of a dilated ureter is difficult without associated hydronephrosis. As a consequence megaureters without significant hydronephrosis are less likely to be picked up antenatally. A UK study of 101 patients with hydroureteronephrosis found only 20% had hydroureter detected prenatally [83].

There is no evidence from the SFU consensus statement correlating the prenatal finding of a dilated ureter with postnatal outcome [84].

Associated Syndromes

None.

Associated Anomalies

None.

The risk of UTI is higher for this cohort of patients compared to PUJO and tends to occur in the first 6 months of life [85]. Thirty-five percent of patients with megaureter required admission for UTI [86], and the incidence of UTI in untreated infants is 0.94 per year. The use of antibiotic prophylaxis reduced this by 83% in the first 6 months and 55% in the first year.

Long-Term Outcome

There has been a move in recent times from an active surgical intervention to more conservative management of these patients [87]. A group of 74 patients with primary obstructed megaureter observed for a median of 10 years found that 73% did not require surgery but cautioned regarding long-term follow-up as some of these patients can have changes much later in life.

There is one case report of a 14 year old boy who had reoccurrence of the megaureter and presented 6 years after discharge with decreased function [88]. Another series of 55 patients who represented in their third decade, 20 patients had renal calculi, 5 had bilateral megaureters with chronic renal impairment and 2 subsequently died due to this [89].

This is a poorly understood heterogeneous group but lifelong follow-up may be appropriate even where the megaureter seems to improve during the paediatric period.

References

1. Hitchcock R, Burge DM. Renal agenesis: an acquired condition. J Pediatr Surg. 1994;29:454.
2. Barakat AJ, Drougas JG. Occurrence of congenital abnormalities of kidney and urinary tract in 13,775 autopsies. Urology. 1991;38(4):347–50.
3. Parikh CR, McCall D, Engelman C, Schrier RW. Congenital renal agenesis: case-control analysis of birth characteristics. Am J Kidney Dis. 2002;39(4):689–94.
4. Westland R, Schreuder MF, Ket JCF, van Wijk JAE. Unilateral renal agenesis: a systematic review on associated anomalies and renal injury. Nephrol Dial Transplant. 2013;28(7):1844–55.
5. McPherson E, Carey J, Kramer A, et al. Dominantly inherited renal adysplasia. Am J Med Genet. 26:8631987.
6. Ashley DJB, Mostofi FK. Renal agenesis and dysgenesis. J Urol. 1960;83:211.
7. Longo VJ, Thomson GJ. Congenital solitary kidney. J Urol. 1952;68:63–8.
8. Chow JS, Benson CB, Lebowitz RL. The clinical significance of an empty renal fossa on prenatal sonography. J Ultrasound Med. 2005;24(8):1049–54.
9. Cascio S, Paran S, Puri P. Associated urological anomalies in children with unilateral renal agenesis. J Urol. 2002;162:2566–8.
10. Donohue RE, Fauver HE. Unilateral absence of the vas deferens. JAMA. 1989;261:1180.
11. Heinonen PK. Unicornuate uterus and rudimentary horn. Fertil Steril. 1997;68:224–30.
12. Heinonen PK. Gestational hypertension and preeclampsia associated with unilateral renal agenesis in women with uterine malformations. Eur J Obstet Gynecol Reprod Biol. 2004;114:39–43.
13. Smith NA, Laufer MR. Obstructed hemivagina and ipsilateral renal anomaly (OHVIRA) syndrome: management and follow-up. Fertil Steril. 2007;87:918–22.
14. Guerrier D, Mouchel T, Pasquier L, Pellerin I. The Mayer-Rokitansky-Küster-Hauser syndrome (congenital absence of uterus and vagina)—phenotypic manifestations and genetic approaches. J Negat Results Biomed. 2006;5:1.
15. Psooy K. Sports and the solitary kidney: what parents of a young child with a solitary kidney should know. Can Urol Assoc J. 2009;3:67–8.
16. Sanna-Cherchi S, Ravani P, Corbani V, et al. Renal outcome in patients with congenital anomalies of the kidney and urinary tract. Kidney Int. 2009;76:528–33.
17. Potter EL. Facial characteristics in infants with bilateral renal agenesis. Am J Obstet Gynecol. 1946;51:885, cited in Chapter 17 The kidney and ureter. Embryology for Surgeons. Williams and Wilkins. Ed. Skandalis and Gray 2nd Edition 594-670.
18. Roodhooft AM, Birnholz JC, Holmes LD. Familial nature of congenital absence and severe dysgenesis of both kidneys. N Engl J Med. 1984;310:1341.
19. McPherson E. Renal anomalies in families of individuals with congenital solitary kidney. Genet Med. 2007;9(5):298–302.
20. N'Guessan G, Stephens FD. Supernumerary kidney. J Urol. 1983;130:–649.
21. Oto A, Kerimoglu U, Eskicorapci C, et al. Bilateral supernumerary kidney: imaging findings. JBR-BTR. 2002;85(6):300–3.
22. Geisinger JF. Supernumerary kidney. J Urol. 1937;38:331, cited in Chapter 17 The kidney and ureter. Embryology for Surgeons. Williams and Wilkins. Ed. Skandalis and Gray 2nd Edition 594-670.
23. Grieshammer U, Le M, Plump AS, et al. SLIT2 mediated ROBO2 signaling restricts kidney induction to a single site. Dev Cell. 2004;6:709–17.
24. Mustafa M, et al. Bilateral supernumerary kidneys in conjunction with horseshoe anomaly. Saudi J Kidney Dis Transpl. 2012;23(6):1243.
25. Upsdell SM. Supernumerary kidney. Br J Urol. 1989;64(6):650.
26. Carlson HE. Supernumerary kidney: a summary of fifty-one reported cases. J Urol. 1950;64:221, cited in Chapter 17 The kidney and ureter. Embryology for Surgeons. Williams and Wilkins. Ed. Skandalis and Gray 2nd Edition 594-670.
27. Gleason PE, Kelalis PP, Husmann DA, Kramer SA. Hydronephrosis in renal ectopia: incidence, etiology and significance. J Urol. 1994;151:1660.
28. Guarino N, Tadini B, Camardi P, et al. The incidence of associated urological abnormalities in children with renal ectopic. J Urol. 2004;172(4 Pt. 2):1757–9.
29. Warne SA, Wilcox DT, Ledermann SE, Ransley PG. Renal outcome in patients with cloaca. J Urol. 2002;167(6):2548–51; discussion 2551
30. Thompson DP, Lynn HB. Genital anomalies associated with solitary kidney. Mayo Clin Proc. 1966;41(8):538–48, cited in Chapter 17 The kidney and ureter. Embryology for Surgeons. Williams and Wilkins. Ed. Skandalis and Gray 2nd Edition 594-670.
31. Downs RA, Lane LW, Burns E. Solitary pelvic kidney: its clinical implications. Urology. 1973;1:51.
32. Benchekroun A, Kasmaoui EH, Jira H, et al. Pathological pelvic kidney. Ann Urol. 2002;36:231–5.
33. Van den Bosch CMA, van Wijk JAE, Beckers GMA, van der Horst HJR, Schreuder MF, Bökenkamp A. Urological and Nephrological findings of renal Ectopia. J Urol. 2010;183(4):1574–8.
34. Weizer AZ, Silverstein AD, Auge BK, Delvecchio FC, Raj G, Albala DM, et al. Determining the incidence of horseshoe kidney from radiographic data at a single institution. J Urol. 2003;170(5):1722–6.
35. Scott JES. Fetal, perinatal, and infant death with congenital renal anomaly. Arch Dis Child. 2002;87(2):114–7.
36. Boyden EA. Description of a horseshoe kidney associated with left inferior vena cava and disc-shaped suprarenal glands, together with a note on the occurrence of horseshoe kidneys in human embryos. Anat Rec. 1931;51:187, cited in Chapter 17 The kidney and ureter. Embryology for Surgeons. Williams and Wilkins. Ed. Skandalis and Gray 2nd Edition 594-670
37. Tripathi I, Guo Q, Wang Y, et al. Midline signaling regulates kidney positioning but not nephrogenesis through Shh. Dev Biol. 2010;340:518–27.
38. Whitaker RH, Hunt GM. Incidence and distribution of renal anomalies in patients with neural tube defects. Eur Urol. 1987;13(5):322–3.
39. Cascio S, Sweeney B, Granata C, Piaggio G, Jasonni V, Puri P. Vesicoureteral reflux and ureteropelvic junction obstruction in children with horseshoe kidney: treatment and outcome. J Urol. 2002;167(6):2566–8.
40. Segura JW, Kelalis PP, Burke EC. Horseshoe kidney in children. J Urol. 1972;108(2):333–6.
41. Pitts WR, Muecke EC. Horseshoe kidneys: a 40-year experience. J Urol. 1975;113(6):743–6.
42. Wilson C, Azmy AF. Horseshoe kidney in children. Br J Urol. 1986;58:361.
43. Glenn JF. Analysis of 51 patients with horseshoe kidneys. N Engl J Med. 1959;261:684–7, cited in Chapter 17 The kidney and ureter. Embryology for Surgeons. Williams and Wilkins. Ed. Skandalis and Gray 2nd Edition 594-670
44. Küpeli B, Isen K, Biri H, Sinik Z, Alkibay T, Karaoğlan U, et al. Extracorporeal shockwave lithotripsy in anomalous kidneys. J Endourol. 1999;13(5):349–52.
45. Raj GV, Auge BK, Weizer AZ, et al. Percutaneous management of calculi within horseshoe kidneys. J Urol. 2003;170:48–51.
46. Mesrobian HG, Kelalis PP, Hrabovsky E, Othersen HB, deLorimier A, Nesmith B. Wilms tumor in horseshoe kidneys: a report from the national Wilms tumor study. J Urol. 1985;133(6):1002–3.
47. Neville H, Ritchey ML, Shamberger RC, Haase G, Perlman S, Yoshioka T. The occurrence of Wilms tumor in horseshoe kidneys:

a report from the National Wilms Tumor Study Group (NWTSG). J Pediatr Surg. 2002;37(8):1134–7.
48. Mallik M, Watson AR. Antenatally detected urinary tract abnormalities: more detection but less action. Pediatr Nephrol. 2008;23(6):897–904.
49. Schreuder MF, Westland R, van Wijk JAE. Unilateral multicystic dysplastic kidney: a meta-analysis of observational studies on the incidence, associated urinary tract malformations and the contralateral kidney. Nephrol Dial Transplant. 2009;24(6):1810–8.
50. Beck AD. The effect of intra-uterine urinary obstruction upon the development of the fetal kidney. J Urol. 1971;105(6):784–9.
51. Mackie GG, Stephens FD. Duplex kidneys: a correlation of renal dysplasia with position of the ureteral orifice. J Urol. 1975; 114:274.
52. Watanabe T, Yamazaki A, Kurabayashi T, Hanaoka J. Familial multicystic dysplastic kidney. Pediatr Nephrol. 2005;20(8):1200.
53. Petrikovsky BM, Lipson SM, Kaplan MH. Viral studies on amniotic fluid from fetuses with and without abnormalities detected by prenatal sonography. J Reprod Med. 2003;48(4):230–2.
54. Carta M, Cimador M, Giuffrè M, Sergio M, Di Pace MR, De Grazia E, et al. Unilateral multicystic dysplastic kidney in infants exposed to antiepileptic drugs during pregnancy. Pediatr Nephrol. 2007;22(7):1054–7.
55. Miller DC, Rumohr JA, Dunn RL, Bloom DA, Park JM. What is the fate of the refluxing contralateral kidney in children with multicystic dysplastic kidney? J Urol. 2004;172(4 Pt 2):1630–4.
56. Cambio AJ, Evans CP, Kurzrock EA. Non-surgical management of multicystic dysplastic kidney. BJU Int. 2008;101(7):804–8.
57. Narchi H. Risk of Wilms' tumour with multicystic kidney disease: a systematic review. Arch Dis Child. 2005;90(2):147–9.
58. Wacksman J, Phipps L. Report of the multicystic kidney registry: preliminary findings. J Urol. 1993;150(6):1870–2.
59. Husmann DA. Renal dysplasia: the risks and consequences of leaving dysplastic tissue in situ. Urology. 1998;52(4):533–6.
60. Seeman T, John U, Bláhová K, Vondrichová H, Janda J, Misselwitz J. Ambulatory blood pressure monitoring in children with unilateral multicystic dysplastic kidney. Eur J Pediatr. 2001;160(2):78–83.
61. Beckwith JB. The John Lattimer lecture. Wilms tumor and other renal tumors of childhood: an update. J Urol. 1986;136(1 Pt 2):320–4.
62. Noe HN, Marshall JH, Edwards OP. Nodular renal blastema in the multicystic kidney. J Urol. 1989;142(2 Pt 2):486–8, discussion 489
63. Siomou E, Papadopoulou F, Kollios KD, Photopoulos A, Evagelidou E, Androulakakis P, et al. Duplex collecting system diagnosed during the first 6 years of life after a first urinary tract infection: a study of 63 children. J Urol. 2006;175(2):678–81, discussion 681–2
64. Atwell JD, Cook PL, Howell CJ, Hyde I, Parker BC. Familial incidence of bifid and double ureters. Arch Dis Child. 1974;49(5): 390–3.
65. Chwalle R. The process of formation of cystic dilatations of the vesical end of the ureter and of diverticula at the ureteral ostium. Urol Cutan Rev. 1927;31:499, cited in Chapter 17 The kidney and ureter. Embryology for Surgeons. Williams and Wilkins. Ed. Skandalis and Gray 2nd Edition 594-670
66. Tanagho EA. Embryologic basis for lower ureteral anomalies: a hypothesis. Urology. 1976;7:451–64.
67. Afshar K, Papanikolaou F, Malek R, Bagli D, Pippi-Salle JL, Khoury A. Vesicoureteral reflux and complete ureteral duplication. Conservative or surgical management? J Urol. 2005;173(5): 1725–7.
68. Elder JS, Diaz M, Caldamone AA, Cendron M, Greenfield S, Hurwitz R, et al. Endoscopic therapy for vesicoureteral reflux: a meta-analysis. I. Reflux resolution and urinary tract infection. J Urol. 2006;175(2):716–22.
69. Dhillon HK. Prenatally diagnosed hydronephrosis: the great Ormond street experience. Br J Urol. 1998;81(2):39–44.
70. Thomas DF. Fetal uropathy. Br J Urol. 1990;66(3):225–31.
71. Alcaraz A, Vinaixa F, Tejedo-Mateu A, Forés MM, Gotzens V, Mestres CA, et al. Obstruction and recanalization of the ureter during embryonic development. J Urol. 1991;145(2):410–6.
72. Smith ED. Report of Working Party to establish the international nomenclature for the large ureter. In: Bergsma D, Duckett JW, eds. Birth defects. Original articles series. 13th edn. 1977 p. 3–8.
73. King LR. Megaloureter: definition, diagnosis and management. J Urol. 1980;123(2):222–3.
74. Incidence megeureter.
75. Pirker ME, Rolle U, Shinkai T, Shinkai M, Puri P. Prenatal and postnatal neuromuscular development of the ureterovesical junction. J Urol. 2007;177(4):1546–51.
76. Matsuno T, Tokunaka S, Koyanagi T. Muscular development in the urinary tract. J Urol. 1984;132(1):148–52.
77. Tanagho EA. Intrauterine fetal ureteral obstruction. J Urol. 1973;109(2):196–203.
78. Dixon JS, Jen PY, Yeung CK, et al. The structure and autonomic innervation of the vesico-ureteric junction in cases of primary ureteric reflux. Br J Urol. 1998;81(1):146–51.
79. Lee BR, Silver RI, Partin AW, Epstein JI, Gearhart JP. A quantitative histologic analysis of collagen subtypes: the primary obstructed and refluxing megaureter of childhood. Urology. 1998;51(5):820–3.
80. Hertle L, Nawrath H. In vitro studies on human primary obstructed megaureters. J Urol. 1985;133(5):884–7.
81. Payabvash S, Kajbafzadeh AM, Tavangar SM, Monajemzadeh M, Sadeghi Z. Myocyte apoptosis in primary obstructive megaureters: the role of decreased vascular and neural supply. J Urol. 2007;178(1):259–64.
82. Vermooten V. A new etiology for certain types of the dilated ureters in children. J Urol. 1939;41:455–63, cited in Chapter 17 The kidney and ureter. Embryology for Surgeons. Williams and Wilkins. Ed. Skandalis and Gray 2nd Edition 594-670
83. Liu HY, Dhillon HK, Yeung CK, Diamond DA, Duffy PG, Ransley PG. Clinical outcome and management of prenatally diagnosed primary megaureters. J Urol. 1994;152(2
84. Nguyen HT, Herndon CDA, Cooper C, Gatti J, Kirsch A, Kokorowski P, et al. The Society for Fetal Urology consensus statement on the evaluation and management of antenatal hydronephrosis. J Pediatr Urol. 2010;6(3):212–31.
85. Song SH, Lee SB, Park YS, Kim KS. Is antibiotic prophylaxis necessary in infants with obstructive hydronephrosis? J Urol. 2007;177(3):1098–101.
86. Gimpel C, Masioniene L, Djakovic N, Schenk J-P, Haberkorn U, Tönshoff B, et al. Complications and long-term outcome of primary obstructive megaureter in childhood. Pediatr Nephrol. 2010;25(9):1679–86.
87. Farrugia M-K, Hitchcock R, Radford A, Burki T, Robb A, Murphy F. British Association of Paediatric Urologists consensus statement on the management of the primary obstructive megaureter. J Pediatr Urol. 2014;10(1):26–33.
88. Shukla AR, Cooper J, Patel RP, et al. Prenatally detected primary megaureter: a role for extended followup. J Urol. 2005;173(4):1353–6.
89. Hemal AK, Ansari MS, Doddamani D, Gupta NP. Symptomatic and complicated adult and adolescent primary obstructive megaureter--indications for surgery: analysis, outcome, and follow-up. Urology. 2003;61(4):703–7.

The Bladder and Urachus

43

Emily Broadis

Development of the Urinary Bladder and Urethra

1. The lower end of the hindgut expands to form a dilatation called the ***cloaca*** (endodermal). The **cloaca** is connected to the umbilicus by a duct called the ***allantois*** and is closed below by a membrane called cloacal membrane which is bilaminar (inner endoderm and outer ectoderm).

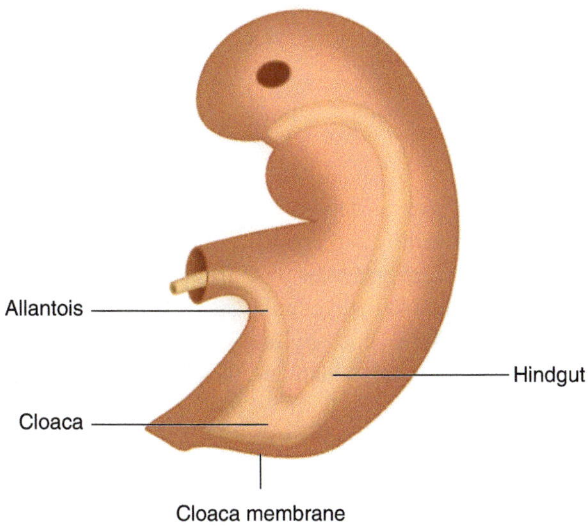

The cloaca is pierced on each side by the *mesonephric duct*.

2. A mesodermal tissue called the ***uro-rectal septum*** develops in the angle between the allantois and the hindgut. It grows caudally, cutting through the cloaca, until it fuses with the cloacal membrane.

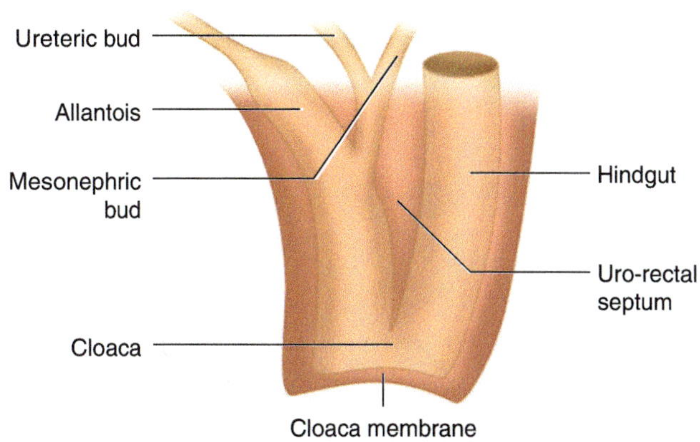

3. The **cloaca becomes divided** by this septum into two parts:
   (a) Posterior part: the ***anorectal canal***, which develops into the rectum and upper part of anal canal (**see p. 41**)
   (b) Anterior part: the ***primitive urogenital sinus***, which develops into the urinary bladder and urethra

   The *cloacal membrane also becomes divided* into two parts:

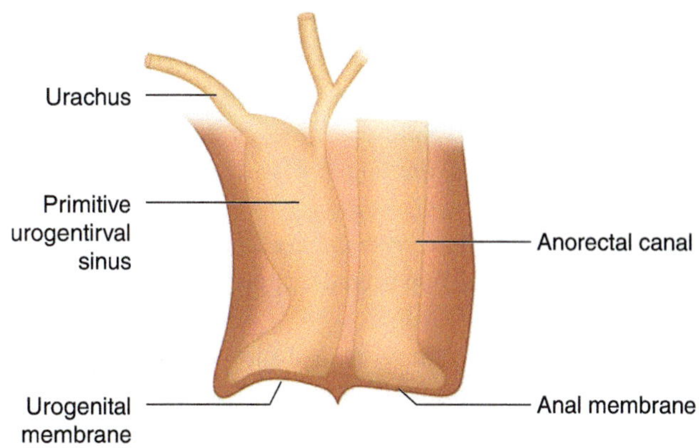

   (a) Urogenital membrane anteriorly
   (b) Anal membrane posteriorly

E. Broadis (✉)
NHS Ayrshire and Arran, University Hospital Ayr, Ayr, UK
e-mail: EBroadis@doctors.org.uk

R. Carachi, S. H. E. Doss (eds.), *Clinical Embryology*, https://doi.org/10.1007/978-3-319-26158-4_43

4. A *constriction* appears in the urogenital sinus at the site of the entrance of mesonephric duct dividing it into two parts:

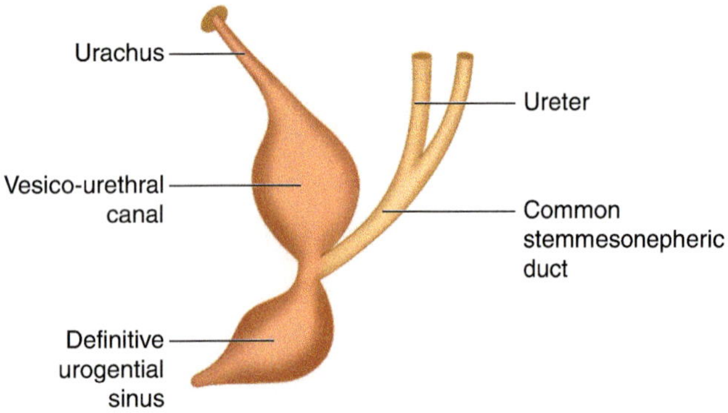

   (a) An upper part called ***vesico-urethral canal***
   (b) A lower part called ***definitive urogenital sinus***
5. Further development of these two parts differs in the male and the female embryo as follows.

Development of Urinary Bladder in the Male

(a) The **vesico-urethral canal** develops into the *urinary bladder* and the *upper 1/2 of the prostatic urethra*.

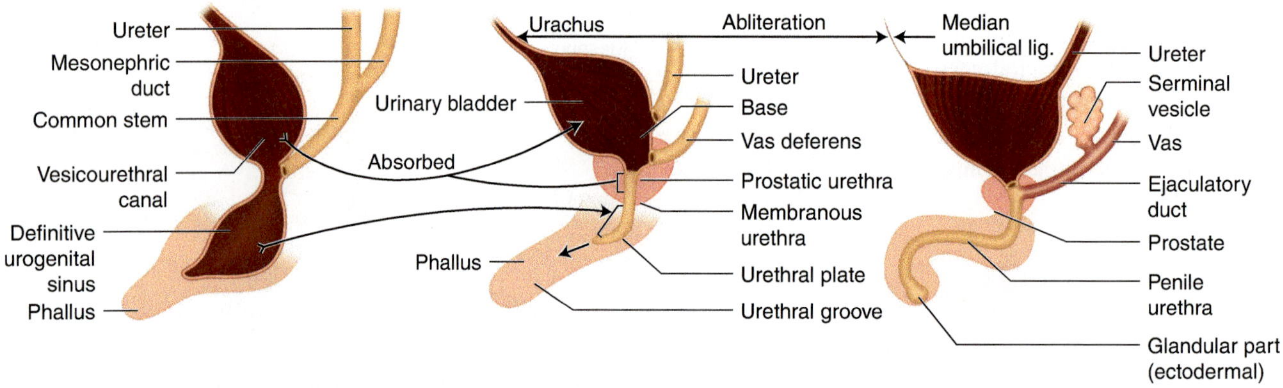

(b) The ***common stem*** of the mesonephric duct and the ureter becomes absorbed into the urinary bladder forming its *trigone* (mesodermal) and thus the ureter and the ejaculatory duct open separately into the urinary bladder.
(c) With differential growth of the posterior bladder wall, the opening of the ureter moves upwards to the postero-superior angle of the bladder while the vas deferens (formerly the mesonephric duct) moves downwards to the prostatic urethra, limiting the part derived from the vesico-urethral canal.
(d) The ***lower 1/2 of the prostatic urethra and the membranous urethra*** develop from the upper (pelvic) part of the definitive urogenital sinus.
(e) The *penile urethra* develops as follows:
   - A **urethral plate** (endodermal) extends from the lower (phallic) part of the definitive urogenital sinus on the undersurface of the phallus (primitive penis) surrounded by two genital (urethral) folds.
   - The two urethral folds unite from behind, moving forwards around the urethral plate to form the penile urethra which is lined by endoderm except its terminal (glandular) part which develops from the ectoderm.

N.B

1. ***The seminal vesicle*** develops as a diverticulum from the lower end of the vas (mesonephric duct). The part distal to it becomes the ejaculatory duct.
2. ***The prostate gland*:** develops as multiple (15–20) outgrowth (buds) from the lining of the prostatic urethra which becomes canalized to form the alveoli and ducts of the gland. The connective tissue and capsule are derived from the surrounding mesoderm.

Development of Urinary Bladder and Urethra in the Female

1. The ***vesico-urethral canal*** develops into the urinary bladder and the whole urethra.

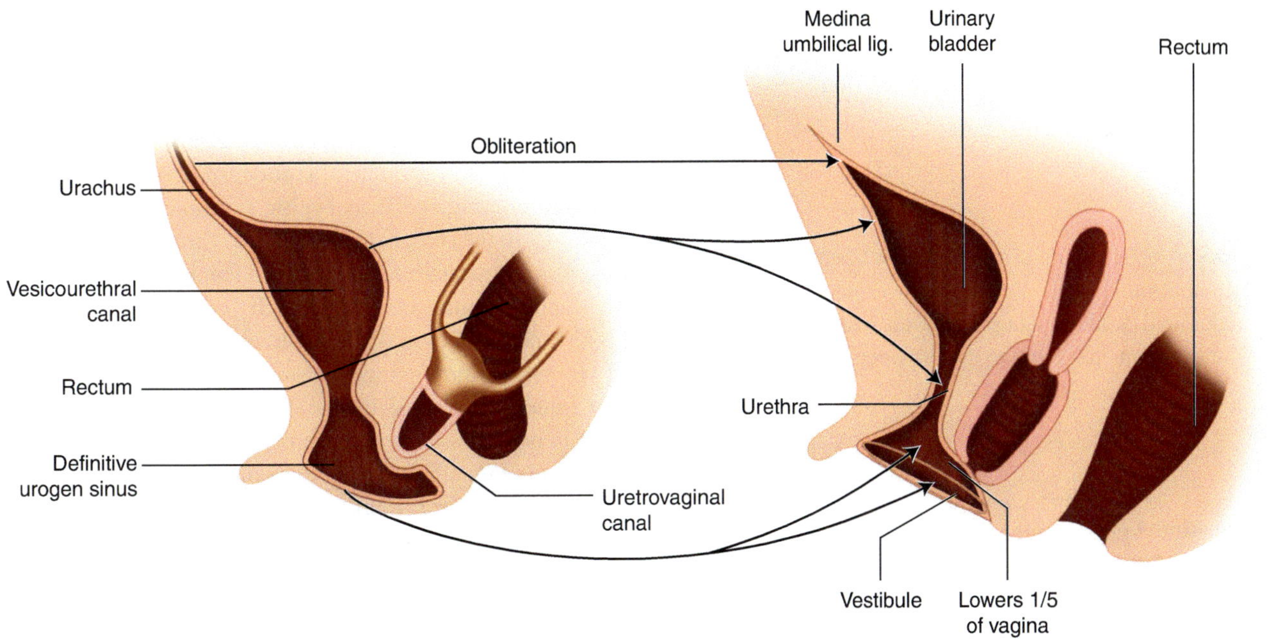

2. The ***definitive urogenital sinus*** develops into the lower $\frac{1}{5}$ of the vagina and its vestibule.

N.B

The upper $\frac{4}{5}$ of the vagina develops from the uterovaginal canal (see page 55). The union between the upper $\frac{4}{5}$ and the lower $\frac{1}{5}$ of vagina is demarcated by the **hymen**.

The Urachus

It is a tubular structure extending from the apex of the urinary bladder to the umbilicus (the remnant of the allantois). It becomes obliterated after birth and forms the median umbilical lig.

Congenital Anomalies of the Bladder and Urethra

1. *Urachal fistula*: due to failure of obliteration of the urachus after birth leading to dribbling of urine from the umbilicus.

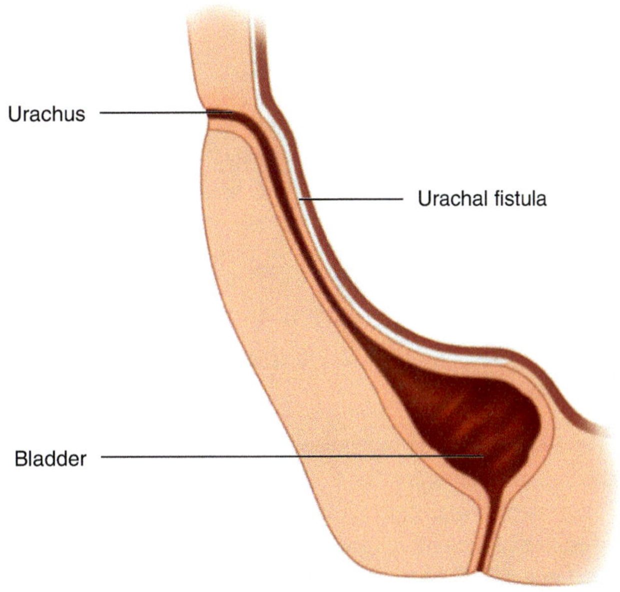

2. *Urachal cyst*: due to incomplete obliteration of the urachus an intra-abdominal cyst may develop from this unobliterated part.

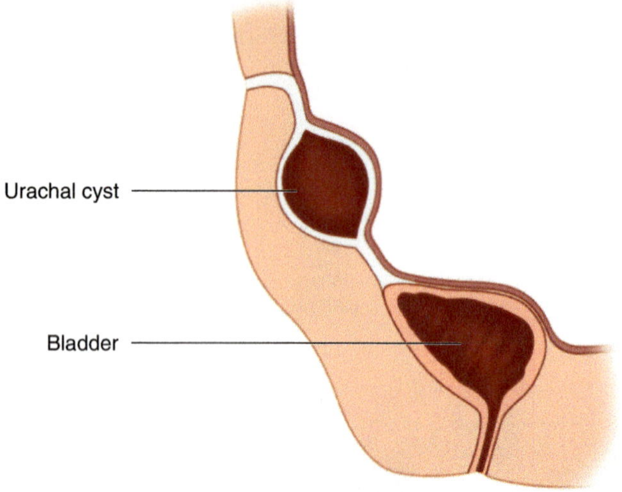

3. *Ectopia vesicae*: the urinary bladder opens into the anterior abdominal wall where the lower part of the anterior abdominal wall is occupied by an oval area of bladder mucosa on which the two ureters open.

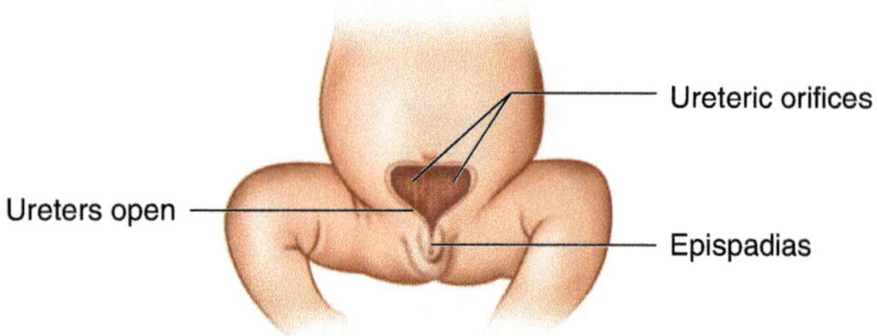

4. *Hypospadias*: the external urethral meatus appears on the under surface of the penis.

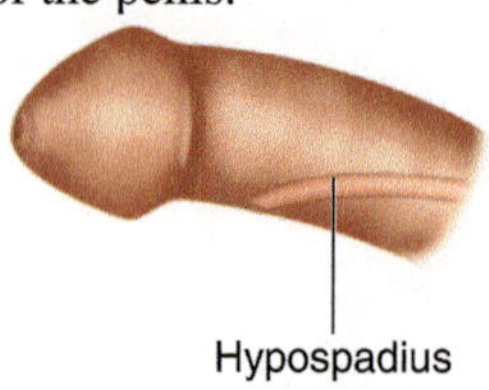

5. *Epispadias*: the external urethral meatus appears on the dorsal surface of the penis.

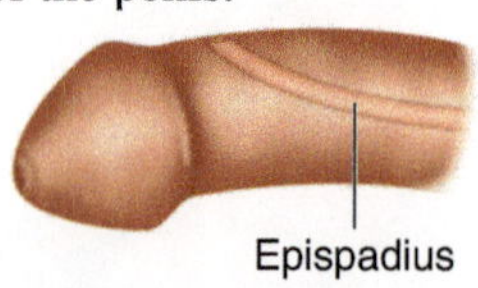

In 1930, Begg [1] recognised the importance of the urachus in embryonic urinary tract drainage and, in an attempt to increase knowledge of its structure and development, published one of the only papers specifically focusing on human foetal urachal development.

The urachus is the fibrous remnant of the allantois, the canal which drains the urinary bladder of the foetus. It is a three-layered tubular structure, with an innermost layer of transitional epithelium, a middle connective tissue layer and an outermost muscular layer continuous with the detrusor muscle. It lies in the space of Retzius, the extraperitoneal space between the pubic symphysis and the urinary bladder. Its dorsal surface is covered with peritoneum, whilst the extraperitoneal surface is covered with transversalis fascia.

The lumen of the urachus may remain patent throughout life, but often it is obliterated by an accumulation of epithelial cells that have desquamated from the walls of its canal. After lumen obliteration, it becomes the median umbilical ligament, connecting the apex of the bladder with the umbilicus.

During development of the urinary tract, an allantois-urachus-vesical communication is described [2]. The extra-abdominal component is called the allantois; the urachus makes up the intra-abdominal section. At around 5 weeks of gestation, the mesonephros begins to produce urine, which passes down the mesonephric duct into the cloaca. It is possible that the patent urachus acts as a temporary urinary outlet, until urethral development is complete at around 12–13 weeks of gestation. Pazos et al. [3] demonstrated that all foetuses without congenital malformations had the urachal lumen obliterated by the 17th week postconception. After lumen closure an absence of transitional epithelium, a decrease in smooth muscle fibres and an increase in type I collagen were noted. Female foetuses were shown to have more connective tissue, but otherwise the urachal structure was similar to males.

Urachal anomalies are infrequent, with an incidence of 1:5000 [3] and an estimated prevalence of around 1.09% [4]. They are more common in males [5] and usually detected at birth. Various anomalies have been described and classified into urachal remnants, urachal cysts, urachal sinus and vesicourachal diverticuli.

Urachal remnants are the most common and make up around 89% of urachal anomalies [4]. Urachal cysts are thought to arise during elongation of the urachus, when it may tear and cause epithelial cells from the centre to separate and degenerate [1]. They are most common in the distal third of the urachus.

A urachal sinus can be further classified into an incomplete tract and a complete tract, also known as a patent urachus, which is thought to account for between 1 and 15% of urachal anomalies [6]. It is here that knowledge of the blood supply to the urachus becomes important. A branch of the superior vesical artery, usually the left, passes up the ventral surface of the urachus and is closely applied to it. A patent urachus is thought to occur when there is increased vascularisation of the first part of the cord, which prevents the onset of dry gangrene, and separation of the cord does not take place (See Fig. 43.1).

Vesicourachal diverticuli are rare outpouchings of the bladder at its insertion of the urachus. As mentioned previously, the dorsal surface of the urachus is covered with serous peritoneum, but it becomes deficient as it passes

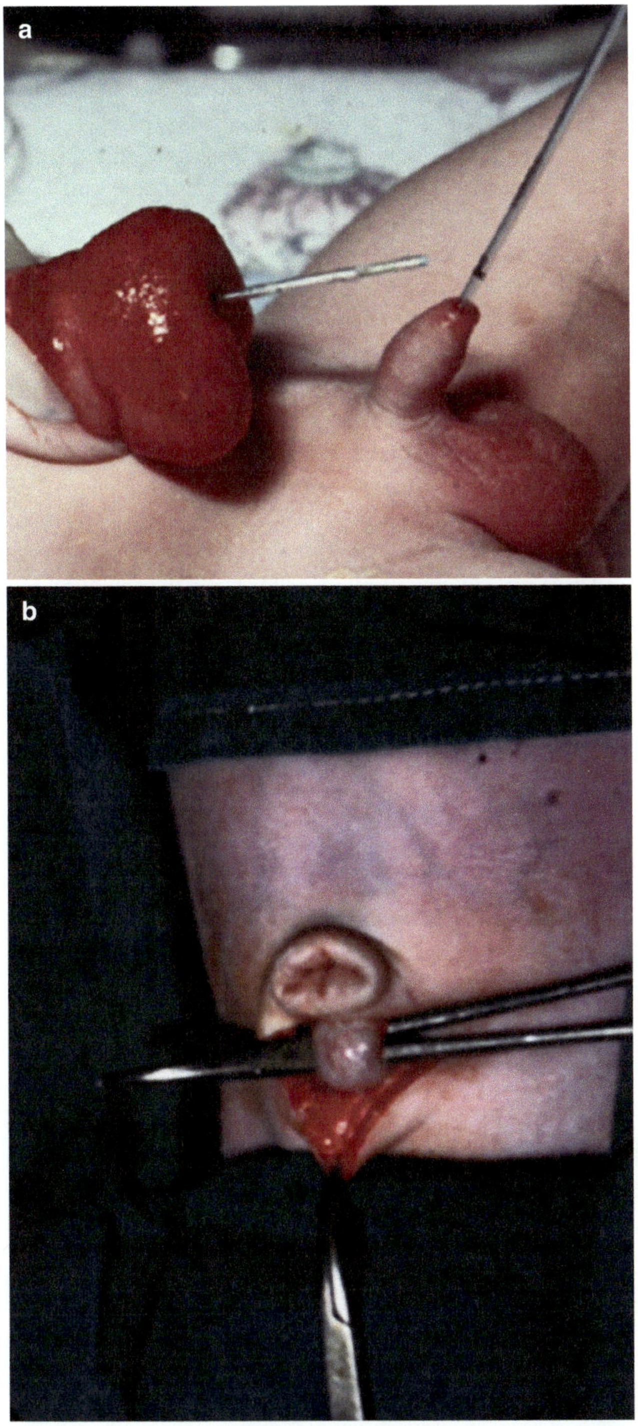

Fig. 43.1 (**a**) Urachus. (**b**) Urachus at operation

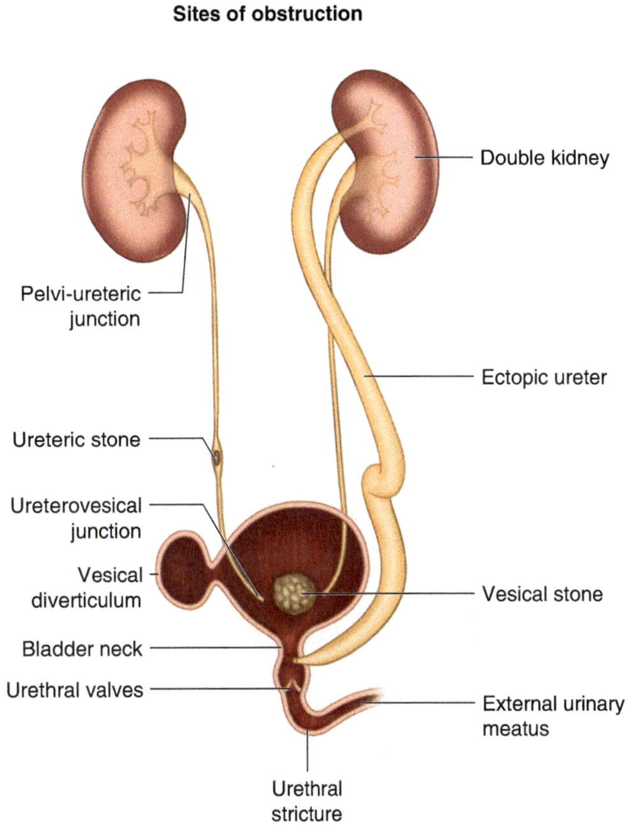

Fig. 43.2 Sites of obstruction

intramurally for about 1 cm in the wall of the bladder. The bladder has a thick adventitial coat, which does not cover the urachus, leading to a weak point at which diverticula are able to form (See Fig. 43.2).

Complications of urachal anomalies include urinary tract infections; omphalitis; leakage of urine, or mucus discharge; and rarely tumours (benign teratomas [1] and adenocarcinoma [4]). Symptomatic lesions can be excised via a curvilinear infraumbilical incision in infants or a slightly lower transverse incision in older children, taking care to excise all mucosa at the umbilicus and the entirety of the tract, along with a small cuff of bladder at its insertion [7]. However, children with asymptomatic lesions do not appear to benefit from prophylactic excision. Gleason et al. [4] estimated that the risk of malignancy later in life is remote, and the number needed to treat (NNT) is 5721 to prevent a single case of urachal adenocarcinoma.

Urachal anomalies can be associated with other genitourinary anomalies including crossed fused renal ectopia, vesicoureteric reflux and hypospadias [3].

In prune belly syndrome, around half of patients have a patent urachus [8]. However, in bladder outlet obstruction due to posterior urethral valves, urachal patency is uncommon [9].

Bladder Duplication

Duplication of the bladder is very rare, with as few as 50 cases reported in the literature [10]. In 1961 Abrahamson [11] attempted to both classify the anomaly and put forward possible embryological theories for why it occurs.

Duplication of the bladder can be described as incomplete or complete. It is classified as complete when there are two bladders lying side by side, with normal mucosal and muscular layers, and separated by a peritoneal fold and loose areolar tissue. Each bladder drains the ipsilateral kidney and empties through its own urethra. The axis of the dividing septum is used to classify complete bladder duplication as either sagittal or coronal (also known as frontal). Sagittal duplication is more common than coronal duplication.

Incomplete bladder duplications also demonstrate two bladders lying side by side separated by a peritoneal fold and loose areolar tissue, but although each bladder drains the ipsilateral kidney and ureter, distally the bladders fuse and empty via a single urethra.

Complete and incomplete septums have also been described, whereby the bladder is divided into two chambers, which can be of equal or unequal size. The septum may be thick and muscular or consists only of mucosa. Each chamber drains the ipsilateral kidney, and there is a single urethra. In the complete septum anomaly, one of the chambers is blocked from emptying, resulting in an obstructed ureter on that side. With incomplete septum, the attachment is insufficient at the lower end, resulting in a crescent-shaped inferior border allowing both chambers to drain via the urethra. This anomaly is also classified by the axis of the dividing septum. In coronal (or frontal) the septum divides the bladder into antero-superior and postero-inferior chambers, with the ureters draining into the postero-inferior chamber.

Other, much rarer bladder anomalies also described by Abrahamson are multiseptate bladder, hourglass bladder, trigonal curtains and bladder neck valves.

Complications which may arise from duplications of the bladder include urinary tract infections, hydronephrosis and vesicoureteric reflux.

Bladder duplication may be an isolated anomaly, but associated anomalies should always be sought. In duplication with coronal septum, urological anomalies are frequent and non-urogenital anomalies are less common.

Around 40% of associated anomalies involve the gastrointestinal tract: anal stenosis, imperforate anus and fistulas from the bladder, urethra, rectum or vagina [10]. Genitourinary anomalies such as duplication of external genitalia, vagina or urethra account for around one third of anomalies, and skeletal anomalies such as spina bifida, meningocele and complete duplication of the lower lumbar vertebral column make up around 10%.

There is uncertainty about how bladder duplications can arise [12]. Abrahamson [11] has put forward two explanations. The first is that there is excessive constriction between the urogenital and vesicourethral portions of the ventral cloaca. The second is that a supernumerary cloacal septum indents the epithelium of the urinary bladder. Since this publication, no further theories have been suggested.

Bladder Exstrophy

Bladder exstrophy forms part of the spectrum of anomalies known as exstrophy-epispadias complex (EEC), a rare anterior midline defect affecting the external genitalia, pelvis, urinary tract and infraumbilical abdominal wall (See Figs. 43.3 and 43.4).

In classic bladder exstrophy, the bladder is situated in the centre of the abdomen between two separated pubic rami, with the mucosal surface of the posterior wall lying open and continuous with the skin. The ureters drain to the inferior section.

The incidence of bladder exstrophy has been estimated at around 1 in 40,000 live births [13], which equates to around 14–18 babies per year born in the United Kingdom.

Previously it was thought to be far more common in males than females [14, 15]; however recent studies have shown a male-to-female ratio of around 1:1 [16].

There are no known teratogens [17], and the condition has a multifactorial inheritance. There is a 400 times increase in relative risk for the offspring of affected individuals [14], and monozygotic twins have been found to have a high concordance rate (62%) compared to dizygotic twins (11%) [18].

The body cavities are formed late in the 3rd week of gestation by the infolding of the mesodermal layer. The intermediate layer of mesoderm invaginates to form the urogenital system, whilst the lateral plate mesoderm goes on to form the gastrointestinal tract.

During the seventh and eighth weeks of gestation, the folding of the mesodermal layer is complete and the mesodermal myotomes are fused in the midline. A mesodermal plate develops between the thoracic cavity and the yolk sac, known as the septum transversum, which, together with migration of the ventromedial upper abdominal mesodermal folds, forms the diaphragm.

The cloaca is separated into the urogenital sinus and the anorectal canal by the urorectal septum, a ridge of mesodermal

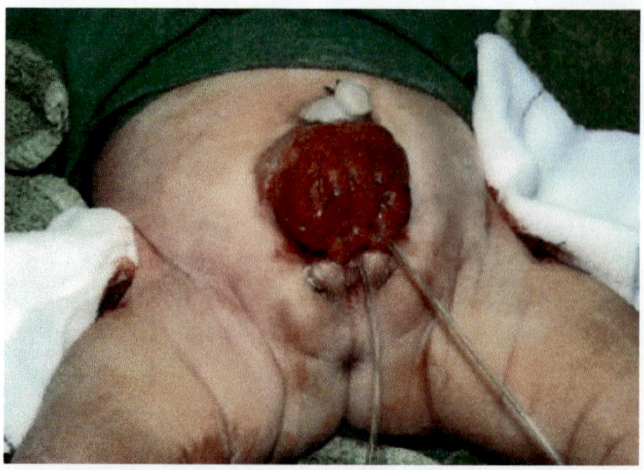

Fig. 43.3 Bladder exstrophy

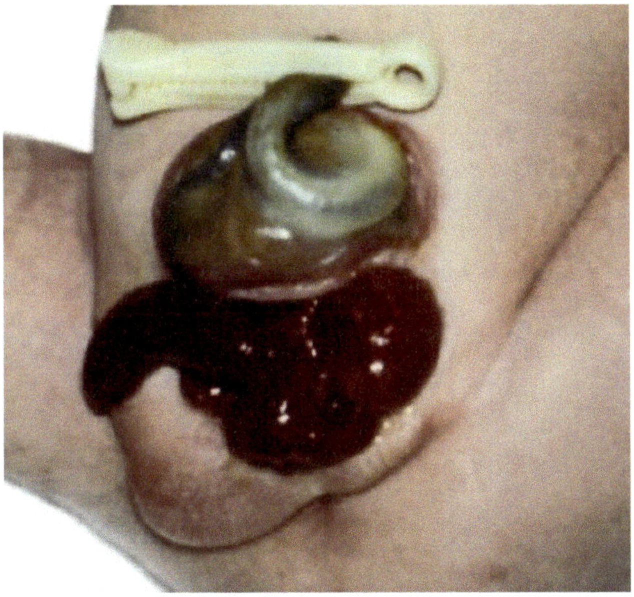

Fig. 43.4 Cloacal exstrophy

myotomes. Migration of the mesodermal myotomes around the cloacal membrane leads to the formation of the anterior wall of the bladder and its overlying abdominal wall.

It is the disruption of these layers that is thought to lead to the development of EEC, the severity of which is determined by the timing of the disruption. As early as the fourth week of gestation, if there is inadequate migration of myotomes, then there is a resulting lack of abdominal wall muscle, fascia and skin of the anterior abdominal wall.

Although three independent theories have been proposed to explain the occurrence of bladder exstrophy, it is likely that a combination of them all is responsible.

The first theory is that the cloacal membrane ruptures prematurely and prevents normal migration of the mesodermal layer into the midline. This was initially described by Muecke in 1964 [19] in chick embryos and then backed up by Thomalla et al. [20] who also noted the link between the timing of disruption and the severity of resulting exstrophy.

The second theory is known as the wedge hypothesis [21]. It describes mechanical obstruction to the lateral mesodermal layer migration, thereby preventing it from fully reaching the midline and being able to fuse properly.

Finally, the third theory is that the disruption occurs at a cellular level, as EEC has been found to occur in a mouse model with the absence of p63 [22].

Antenatal diagnosis of bladder exstrophy is possible but difficult [23]. From 15 weeks of gestation, high-resolution ultrasound is able to visualise the foetal bladder. In the absence of bladder filling, other signs suggestive of bladder exstrophy should be looked for. These include a lower abdominal mass, a low-set umbilicus, a wide pubic ramus, abnormal widening of the iliac crests and small genitalia, which may also be in a higher-than-normal position [13, 23].

Bladder exstrophy results in a raised maternal serum alpha-fetoprotein (MSAFP), and in the presence of an unexplained MSAFP, the foetal bladder should always be sought and documented.

In recent years, foetal MRI is gaining popularity in diagnosing cases that are unclear, as it has the advantage of enhanced soft tissue resolution, which may allow a more clear definition of the anatomy and also gender determination [24–26].

Karyotyping has also been found a useful adjunct to aid prenatal counselling.

There are no patient support groups in the United Kingdom; however, parents can be directed towards a website which enables contact between patients with similar diagnosis [27] The Association for Bladder Exstrophy Community [28] in America also provides helpful information about the condition.

References

1. Begg RC. The Urachus: its anatomy, histology and development. J Anat. 1930;64(2):170–83.
2. Pal K, Ashri H, Al-Ghazal FA. Allantoic cyst and patent urachus. Indian J Pediatr. 2009;76:221–3.
3. Pazos HMF, Costa WS, Sampaio FJB, Favorito LA. Structural and ontogenetic study of the urachus in human fetuses. Cells Tissues Organs. 2010;191:422–30.
4. Gleason JM, Bowlin PR, Bagli DJ, et al. A comprehensive review of pediatric urachal anomalies and predictive analysis for adult urachal adenocarcinoma. J Urol. 2015;193:632–6.
5. Copp HL, Wong IY, Krishnan C, et al. Clinical presentation and urachal remnant pathology: implications for treatment. J Urol. 2009;182:1921–4.
6. Cuda SP, Vanasupa BP, Sutherland RS. Nonoperative management of a patent urachus. Urology. 2005;66(6):1320.
7. Snyder CL. Current management of umbilical abnormalities and related anomalies. Semin Pediatr Surg. 2007;16:41–9.
8. Lattimer JK. Congenital deficiency of the abdominal musculature and associated genitourinary anomalies: a report of 22 cases. J Urol. 1958;79:343–52.
9. Kaefer M, Keating MA, Adams MC, Rink RC. Posterior urethral valves, pressure pop-offs and bladder function. J Urol. 1995;154:708–11.
10. Berrocal T, Novak S, Arjonilla A, Gutiorrez J, Prieto C. Complete duplication of bladder and urethra in the coronal plane in a girl: case report and review of the literature. Pediatr Radiol. 1999;29:171–3.
11. Abrahamson J. Double bladder and related anomalies: clinical and embryological aspects and a case report. Br J Urol. 1961;33:195–8.
12. Gajbhiye V, Nath S, Gosh P, Chatterjee A, Haldar D, Das SK. Complete duplication of the urinary bladder: an extremely rare congenital anomaly. Urol Ann. 2015;7(1):91–3.
13. Goyal A, Fishwick J, Hurrell R, Cervellione RM, Dickson AP. Antenatal diagnosis of bladder/cloacal exstrophy: challenges and possible solutions. J Pediatr Urol. 2012;8:140–4.
14. Shapiro E, Lepor H, Jeffs RD. The inheritance of the exstrophy-epispadias complex. J Urol. 1984;132:308–10.

15. Ives E, Coffey R, Carter CO. A family study of bladder exstrophy. J Med Genet. 1980;17:139–41.
16. Nelson CP, Dunn RL, Wei JT. Contemporary epidemiology of bladder exstrophy in the United States. J Urol. 2005;173:1728–31.
17. Stec AA. Embryology and bony and pelvic floor anatomy in the bladder exstrophy epispadias complex. Semin Pediatr Surg. 2011;20:66–70.
18. Reutter H, Qi L, Gearhart JP, Boemers T, Ebert AK, Rösch W, Ludwig M, Boyadjev SA. Concordance analyses of twins with bladder exstrophy-epispadias complex suggest genetic etiology. Am J Med Genet A. 2007;143A:2751–6.
19. Muecke EC. The role of the cloacal membrane in exstrophy: the first successful experimental study. J Urol. 1964;92:659–67.
20. Thomalla JV, Rudolph RA, Rink RC, et al. Induction of cloacal exstrophy in the chick embryo using the CO_2 laser. J Urol. 1985;134:991–5.
21. Manner J, Kluth D. The morphogenesis of the exstrophy-epispadias complex: a new concept based on observations made in early embryonic cases of cloacal exstrophy. Anat Embryol. 2005;210:51–7.
22. Cheng W, Jacobs WB, Zhang JJ, et al. DeltaNp63 plays an anti-apoptotic role in ventral bladder development. Development. 2006;133:4783–92.
23. Robinson JN, Abuhamad AZ. Abdominal wall and umbilical cord anomalies. Clin Perinatol. 2000;27(4):947–78.
24. Hsieh K, O'Loughlin MT, Ferrer FA. Bladder exstrophy and phenotypic gender determination on fetal magnetic resonance imaging. Urology. 2005;65:998–9.
25. Celentano C, Caulo M, Cardellini T, Matarrelli B, Lisi G, Chiesa LP, et al. Bladder exstrophy and prenatal diagnosis using 3-D ultrasound and fetal MRI. Ultrasound Obstet Gynecol. 2008;31: 597–609.
26. Herman-Sucharska I, Urbanik A, Mamak-Balaga A, Grzyb A, Szafirska M, Tomaszczyk J. MR in the fetal genitourinary system abnormalities. Ultrasound Obstet Gynecol. 2008;31: 597–609.
27. www.cafamily.org.uk.
28. www.bladderexstrophy.com.

Normal Development of the Penis and Urethra

44

Ahmed T. Hadidi

Introduction

The embryology of the penis and urogenital region has been controversial and confusing. This is related to the fact that in literature, observations have been extrapolated to the human from many different species including mice, rats, red squirrels, pigs and dogs. Current theories suggest that the penile urethra (columnar epithelium) forms as a result of fusion of the medial edges of the endodermal urethral folds like closing a zip in the midline. The ectodermal edges of the urethral groove fuse to form the median raphe. The glanular urethra (stratified squamous epithelium) is supposed to develop due to an ingrowth of ectoderm from the tip of the glans or due to endodermal cellular differentiation.

If the urethra develops through simple fusion (according to the current classic theories), why do we see chordee? How can torsion occur? Why do we see variable width and depth of the urethral plate? Why is there a wide range of corpus spongiosum hypoplasia and dysplasia in hypospadias? Why is there a variable degree of glans clefting (i.e. grooving)? How can simple fusion explain the presence of ano-cutaneous tract on the ventral penis in patients with imperforate anus?

The description in this chapter is a simplified summary of the chapter on development of the penis and urethra [1], and it is based on the study of histological sections of human embryos and foetuses by Hadidi [2] and van der Putte [3] and intrauterine photos and other histological studies available in literature.

The Cloaca

At about 6.5 weeks, the distal part of the hindgut widens into a *chamber.* This chamber receives the allantois anteriorly and the gut posteriorly. It continues distally as a narrow blind-ended tube called the *tail gut* (Fig. 44.1a). The fusion of the *mesonephric ducts* with the lateral wall of this chamber marks the transition of this chamber into *the cloaca.* This endodermal chamber will form the most caudal part of the definitive gut tube in the developing embryo (Fig. 44.1b).

During the 6.5–8-week period, the **cloaca** divides internally into *urogenital* and *anal* compartments. Externally, the **cloaca eminence** develops and forms the *genital tubercle* and *urogenital folds.*

The simple chamber-like cloaca becomes a U-shaped structure due to an indentation, including the coelom, that grows cranio-caudally and divides the cloaca into an anterior *urogenital* and a posterior *anal* compartments. The communication between both compartments becomes gradually narrower. Meanwhile, the *tail gut* disintegrates and disappears completely (Fig. 44.1c).

Development and Fate of the Cloacal Membrane

The *cloacal membrane* is the membrane that forms the floor and limits the cloacal cavity ventrally in the midline. Due to rapid proliferation of the infraumbilical mesenchyme in the midline, the cloacal cavity is displaced caudally. The orientation of the **cloacal membrane** gradually changes from being vertical facing anteriorly to become horizontal (Fig. 44.1b and c).

The cloacal membrane is now a longitudinal slitlike membrane that extends from the genital tubercle anteriorly to a small postanal swellings posteriorly. It is opposed side to side in the sagittal plane by the urogenital folds and becomes thickened (four layers) ventrally to form *the urogenital plate.* The anterior part forms a small **glans plate**.

Posteriorly, the cloacal membrane gradually becomes thinner and disintegrates to open into the amniotic cavity. It thus provides initially a single opening, *the cloacal orifice*, for both urogenital and anal compartments (Fig. 44.1c). This means that, for a short while, the orifices of the urogenital sinus and anal compartment open separately connected by a

A. T. Hadidi (✉)
Department of Pediatric Surgery, Emma and Sana Offenbach Hospitals, Seligenstadt, Germany

R. Carachi, S. H. E. Doss (eds.), *Clinical Embryology*, https://doi.org/10.1007/978-3-319-26158-4_44

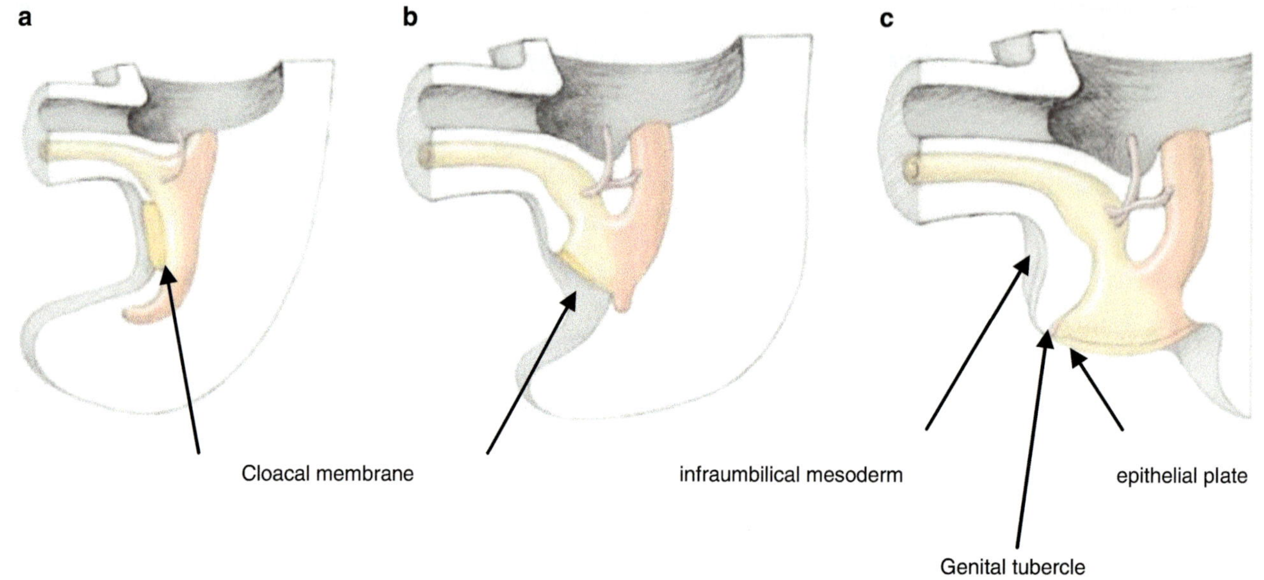

Fig. 44.1 Development of the cloaca, urogenital sinus and genital tubercle. The distal end of the hindgut continues distally as the tail gut. It is called the cloaca after receiving the mesonephric gut. The tail gut disappears completely. (**a**) The cloaca is the terminal end of the hindgut, and the cloacal membrane is vertical. (**b**) Due to folding, the tail and the allantois are incorporated into the embryo contour with the formation of the urogenital sinus. The cloacal membrane is oblique. (**c**) Further infraumbilical mesenchymal proliferation and migration results in the appearance of the genital tubercle, and the cloacal membrane becomes horizontal (From Hadidi; with permission)

short *cloacal groove* which before long is incorporated into the surface of the perineum. Rapid proliferation of the urogenital compartment follows which results in both urogenital and anal orifices being visible on the surface.

Development of the Urogenital and Labioscrotal Folds

During the formation of the cloaca, mesenchyme migrates from the primitive streak and proliferates around the cloacal membrane to form a slightly elevated **cloacal eminence**. The anterior part of the eminence proliferates anteriorly to form the **genital tubercle**. The lateral part forms the **urogenital folds** on either side of the cloacal membrane and the posterior part forms **anal folds**.

In the meantime, another pair of elevations, **the labioscrotal swellings**, becomes visible on each side of (lateral to) the urogenital folds. These swellings later form the scrotal swellings in the male and the labia majora in the female.

Development of the Urogenital Plate

The urogenital plate (which is the anterior thickened part of the cloacal membrane) becomes more superficial due to rapid growth of the ventral part of the cloacal eminence. Later, dehiscence in the central part enlarges the cavity and widens the already existing urogenital orifice.

Until the end of the ninth week, it is impossible to differentiate between the two sexes.

Development of the Male Penile (Spongy) Urethra

Proximal Penile Urethra

In 8-week gestation of embryos, the urogenital compartment opens in the perineum close to the anal orifice and becomes the urogenital sinus. Disintegration of the central part of the urogenital plate results in widening of the urological orifice to extend from the perineum, near the anal orifice until the base of the glans (Fig. 44.2).

Early in development, the urethral orifice is a longitudinal slit or oval opening that extends from the perineum, near the anal orifice posteriorly to the base of the glans anteriorly, and is bound laterally by the urogenital folds. Rapid growth of the urogenital sinus surrounding erectile and fascial primordial results in a lengthening of the penile urethra and the formation of the penile shaft (Fig. 44.3). This growth is more pronounced dorsally resulting in "physiological temporary chordee".

This is followed by differential rapid proliferation of the ventral aspect of the genital tubercle, more than the dorsal (cranial) aspect. This corrects the temporary physiological chordee and causes gradual shift of the posterior edge of the urethral opening anteriorly. Consequently, the urethral opening becomes gradually smaller, until the posterior edge reaches the base of the glans (Fig. 44.3).

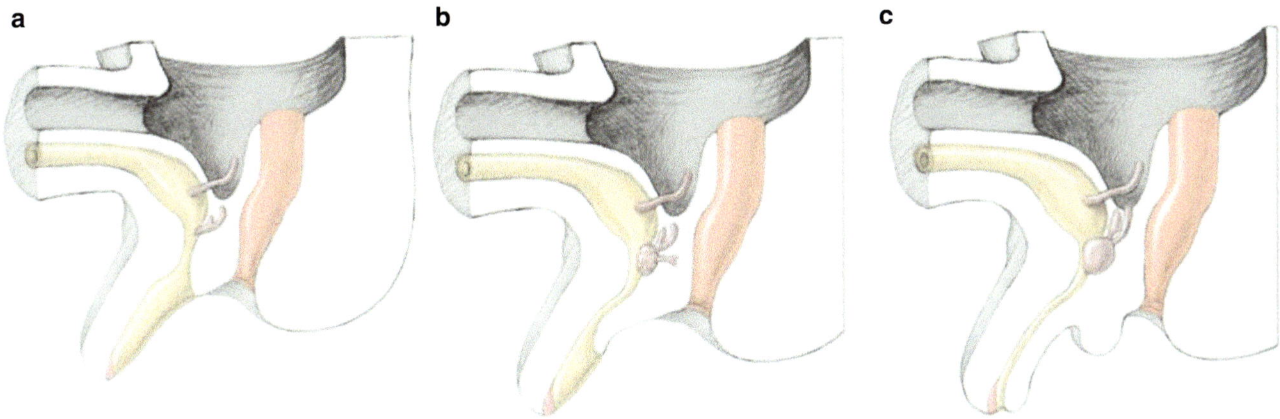

Fig. 44.2 Development of the penile urethra. (**a**) The cloacal membrane continues to disintegrate to form the cloacal groove. (**b**) Gradually, both urogenital and anal orifices reach the surface. (**c**) The posterior edge of the urogenital orifice gradually migrates forward to reach the coronal sulcus. (**d**) The glanular urethra has a different origin from the glans plate (From Hadidi with permission)

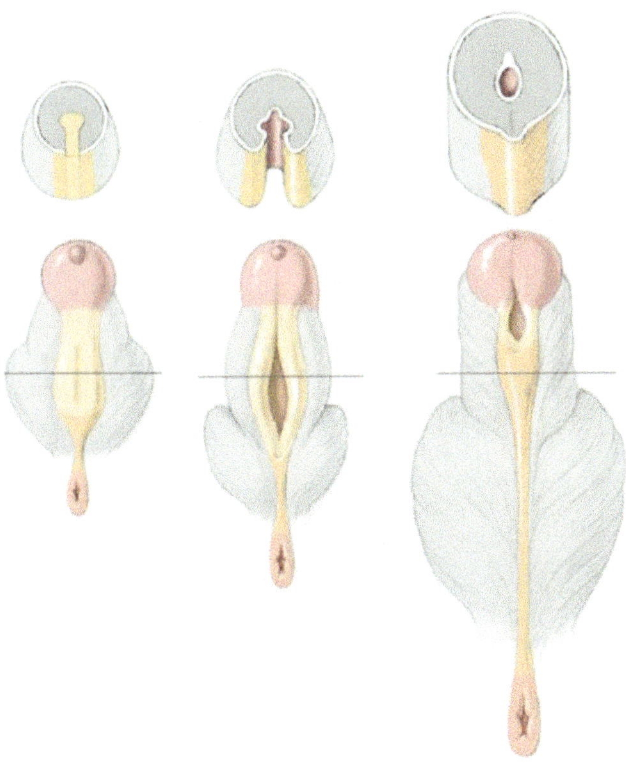

Fig. 44.3 Stages of spongy urethra development. The development of the spongy urethra is influenced by the distal migration of the ventral penile mesenchyme and passes through three stages (from Hadidi with permission). *Stage 1*: Solid urethral plate. The urethral plate begins as a solid outgrowth from the anterior wall of the urogenital sinus. *Stage 2*: Deep urethral groove. There is gradual grooving of the urethral plate due to degeneration of the caudal cells resulting in the formation of the urethral groove. *Stage 3*: Fusion of the urethral edges. Differential growth and migration of the ventral penile mesenchyme brings the edges of the urethral groove together causing proliferation, adherence and apoptosis of the urethral groove edges. This results in fusion of the edges of the urethral groove to form the floor of the penile urethra (From Hadidi; personal communication)

Glanular Urethra

Anterior to the urethral orifice is the solid glans plate. Thereafter, there is canalisation of the solid glans plate. This is followed by disintegration of the floor of the glanular canal to form the wide navicular fossa and *the glans groove*. The developing prepuce grows over the glans and the glanular groove forms the floor of the glanular urethra and the frenulum (Fig. 44.4).

To summarise, the urethral orifice starts as a longitudinal slit or oval opening that extends from the perineum, near the anal orifice posteriorly to the base of the glans anteriorly, and is bounded laterally by the urethral labia (or urogenital folds). Gradual shift of the posterior boundary of the urethra anteriorly stimulates the closure of the urethral groove forming the penile urethra.

The roof of the glanular urethra including fossa navicularis is formed by canalisation and disintegration of the floor of the glanular canal. The floor of the glanular urethra is formed by the migration of the developing prepuce. The lining of the proximal penile urethra is endodermal (columnar epithelium) originating from the cloaca. The lining of the roof and lateral wall of the glanular urethra is endodermal (from the cloaca). The lining of the floor of the glanular urethra is ectodermal (stratified squamous epithelium) originating from the ectodermal prepuce.

Development of the Genital Tubercle and Its Derivatives

As the infraumbilical mesenchyme migrates towards the cloacal membrane, it fuses anteriorly and proliferates to form the cloacal eminence with its prominent genital tubercle.

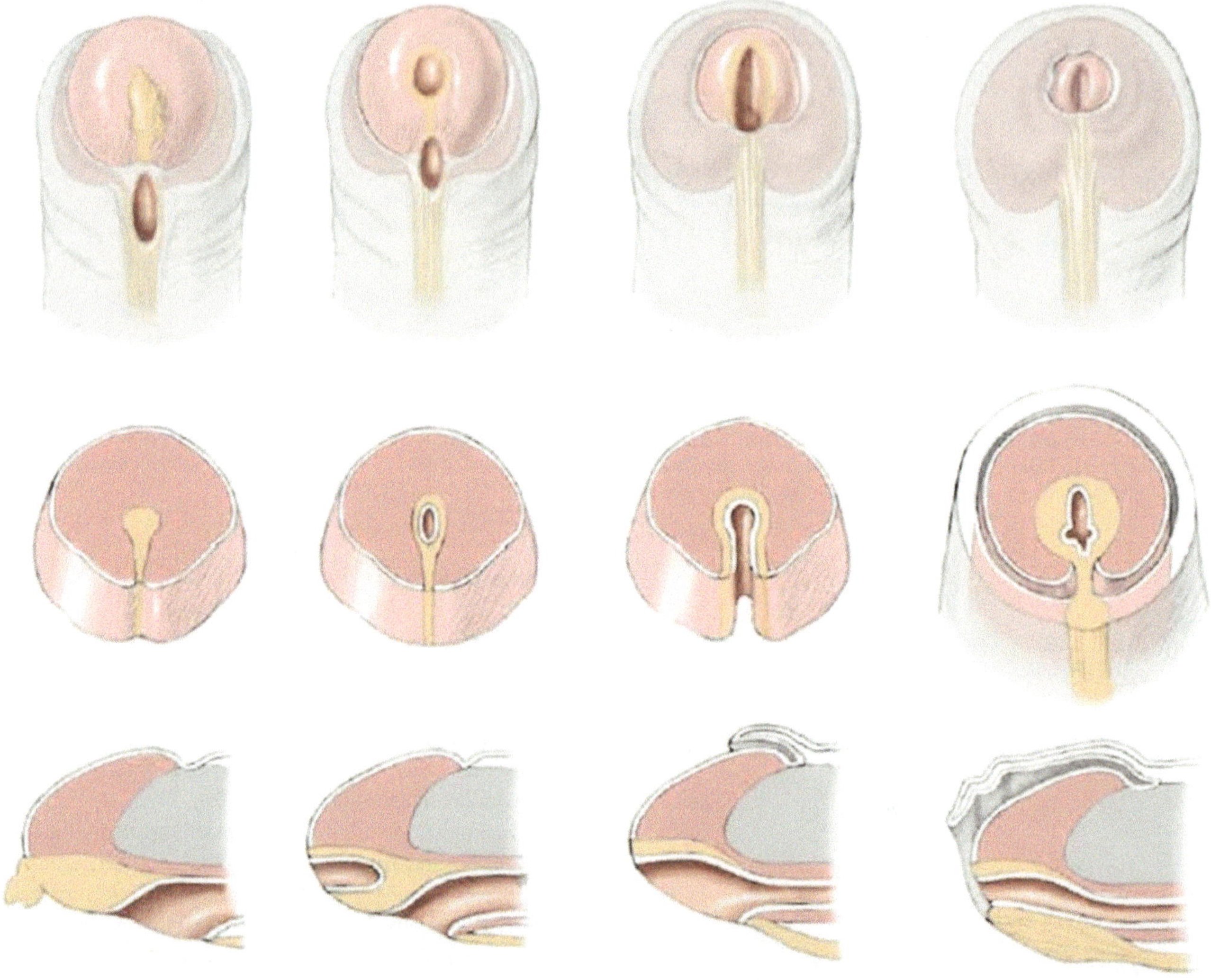

Fig. 44.4 Stages of development of the glanular urethra. Complete development of the glanular urethra is influenced by distal migration of ventral mesenchyme and normal development of the prepuce and passes through four stages (from Hadidi with permission). *Stage 1*: Solid epithelial plate. There is a solid epithelial plate that reaches the tip of the glans penis. *Stage 2*: Blind central glanular canal. Degeneration of the central cells of the epithelial plate results in formation of blind central glanular canal. Stage 3: Deep glanular groove. Degeneration of the floor of the central glanular canal results in the formation of a deep glanular groove and fossa navicularis. *Stage 4*: Complete glanular urethra. The floor of the glanular urethra develops from the frenulum and the glanulo-preputial lamella that forms the prepuce (From Hadidi, with permission)

Differences in density and structure appear in this mesenchyme to form a dense *fascial stroma* and a *fibrovascular zone*. The fibrovascular zone gives rise to a series of erectile structures: glans, corpora cavernosa and bulbi of the corpus spongiosum (Fig. 44.5).

The Glans

The glans forms the main component of the primitive genital tubercle. It is a cap-like structure formed of dense small cells. The glans is thick in the centre and becomes gradually thinner towards the periphery. It envelops the distal parts of the corpora cavernosa. The undersurface of the glans contains the solid glans plate in the midline (Fig. 44.6). The glans is gradually separated from the rest of the shaft by the rapidly growing glanulo-preputial lamella forming eventually the prepuce. The glans itself does not form a sheath. It is linked to the corpus spongiosum and corpora cavernosa by the sheaths of these three corpora and by numerous vascular anastomoses.

The Corpora Cavernosa

The two corpora cavernosa develop from the fibrovascular zone proximal to the glans. Proximally, they form the deviating *crura* of the penis. Distally, they continue to grow, sepa-

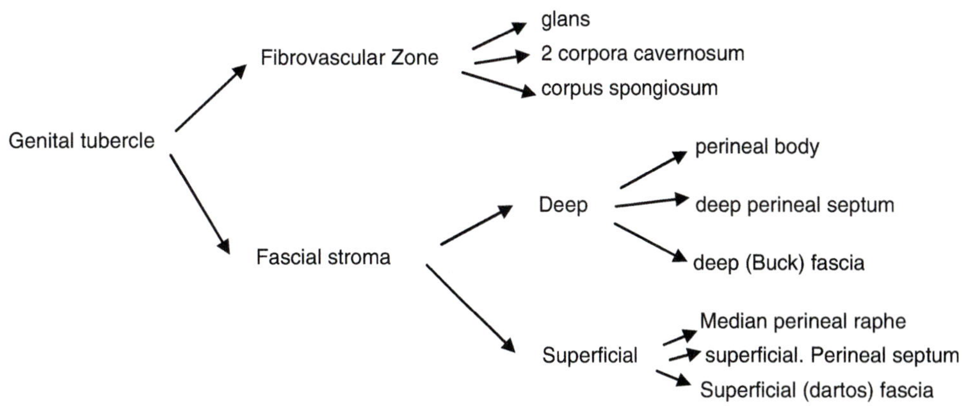

Fig. 44.5 The derivatives of the primitive genital tubercle

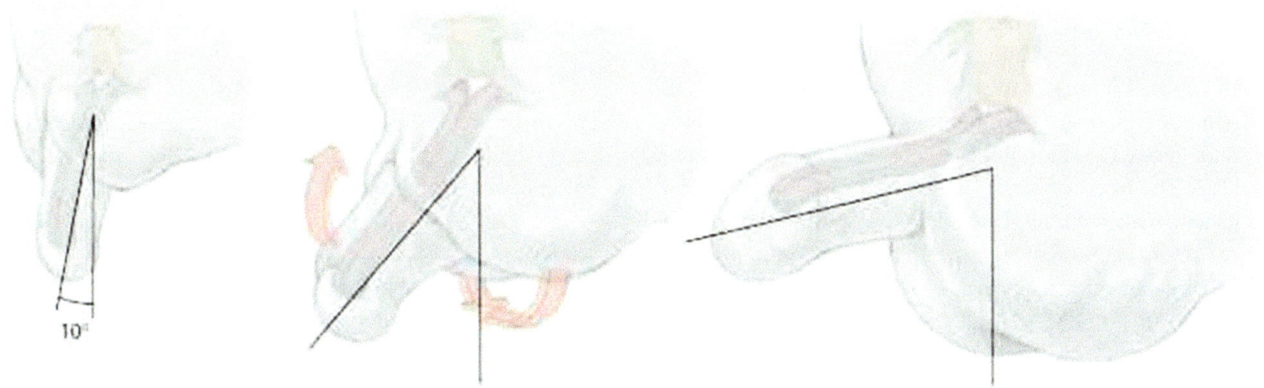

Fig. 44.6 The primitive genital tubercle makes 10° angle with the body axis. Differential growth of the ventral fascia changes the angle to 90°, and the scrotal folds migrate caudally to meet the median raphe in the midline while maintaining a fixed relation to the symphysis pubis From Hadidi with permission)

rated only by a sheath. They are formed of dense small cells (like the glans) that differentiate later into a *central* fibrovascular core and a *peripheral* tunica albuginea.

The corpora cavernosa gradually become large trunks with a thick double-layered tunica albuginea on the outside and dense vascular network centrally. This vascular network is predominantly venous which forms large communicating channels and trabeculae. The arterial element of this network is supplied by the dorsal arteries of the penis.

The Corpus Spongiosum

The corpus spongiosum develops from the deepest part of the fibrovascular zone in the form of two oval fibrovascular structures (bulbi spongiosi) on both sides of the primitive spongy urethra and underneath the crura of the corpora cavernosa. They are bordered by the lamina propria medially and the bulbospongiosus muscle laterally.

The two bulbi approach each other in the midline posterior to the urethra being separated only by a thin deep perineal septum to form the *bulb of the penis*. The dense bulbar tissue gradually differentiates into a loose connective tissue rich in prominent arteries. These prominent arteries are later surrounded by venous plexuses that form large cavities or sinuses.

The distal part of the bulb grows along with the proliferating spongy urethra between the lamina propria and the deep dorsal stroma distal to the bulbospongiosus muscle. The longitudinal venous plexuses of the lamina propria become gradually integrated into the distal corpus spongiosum and may even replace the bulbar component of the corpus spongiosum near the urethral orifice distally.

During development, the corpus spongiosum receives an extra vascular supply from plexuses between the urethra and corpora cavernosa. Part of these plexuses persist in the male as branches of the central arteries of the corpora cavernosa. These branches perforate the tunica albuginea and supply most of the distal part of the corpus spongiosum.

Development of the Scrotum

During the cloacal period, the infraumbilical mesenchymal cells migrate and proliferate around the cloacal membrane to form the **genital tubercle** *anteriorly***.** They form the **urogenital folds** (or labia) on either side of the cloacal membrane. Later, another pair of elevations, **the labioscrotal swellings,** becomes visible on each side of (lateral to) the urogenital folds (Fig. 44.6).

Early in development, the genital tubercle (the future penis) projects caudally making about 10° angle with the longitudinal axis of the body, and the labioscrotal folds lie laterally to the genital tubercle and urogenital folds (or labia). The urethral orifice is an oval, long slitlike with the posterior boundary close to anal orifice at the perineum and the anterior boundary at the base of the glans. The urethral orifice is bounded laterally by the urogenital folds. Deep stromal tissue and dartos fascia occupy most of the scrotal swellings (Fig. 44.6).

With later development, three mechanisms occur simultaneously: (1) *The corpus spongiosum, perineal septum and raphe* and other ventral penile structures proliferate much more than the dorsal structures of the penis, i.e., the corpora cavernosa. This increases the space between the anus and the posterior boundary of the urethral orifice which migrates gradually along the shaft of the penis to reach the base of the glans. (2) The *penis rotates* cranially to project anteriorly so that its axis makes 90° angle with the longitudinal axis of the body. *The two scrotal swellings* move closer to the midline and occupy part of this space.

It is worth noticing that all through the developmental process, the scrotal swellings maintain a fixed distance from the symphysis pubis. This strongly suggests that it is the penis that rotates forward (anteriorly) and moves forward rather than the scrotal swellings moving backwards (posteriorly).

Development of the Prepuce and Frenulum

The prepuce develops from two components at the junction between the glans and the shaft penis (the corona): (a) a *major* component from the ***shaft*** (dartos fascia and superficial stroma) and (b) a *minor* component from the ***glans*** stroma. Between the glans stroma of the glans itself and the preputial glans component grows the solid glanulo-preputial lamella outwards in a concerted growth. That growth is outward because the position of the deepest part of the lamella remains the same in relation to the underlying nerves to the glans and corpora cavernosa (Fig. 44.7).

The formation of the median ventral part of the prepuce is due to rapid forward (distal) proliferation of the median ventral raphe and synchronously the medial ventral part of the growing prepuce more than the dorsal part of the prepuce. The formation of the frenulum is due to an expansion of the corona of the glans and glanulo-preputial lamella proximal to the urethral orifice. As a result, the median ventral raphe and septum form the floor of the glanular urethra and remain as the frenulum inside the medial ventral part of the prepuce which covers the ventral surface of the glans (Fig. 44.7).

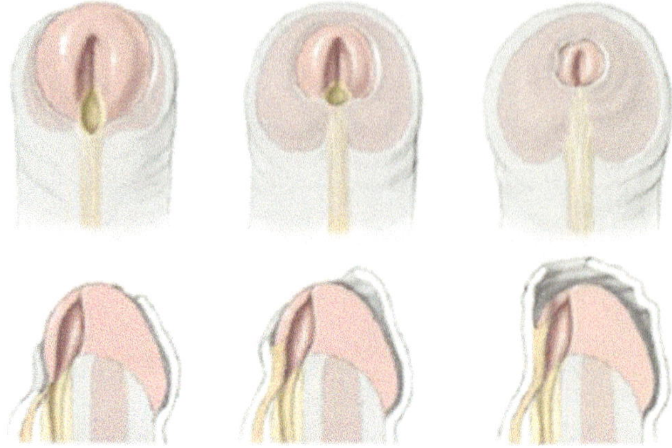

Fig. 44.7 Development of the lateral part of the prepuce from the preputial lamella. The median part and frenulum from the median raphe and penile septum (From Hadidi with permission)

This process ends beyond the tip of the penis. Cornification of the epithelium of the lamella during the last trimester of gestation forms the preputial sac.

Urethral Development: The Migration Hypothesis (Fig. 44.8)

Aetiology of Hypospadias: The Disorganisation Hypothesis (Figs. 44.9, 44.10 and 44.11)

The wide spectrum of hypospadias is most likely due to *disorganisation* and arrested distal migration of the ventral penile mesenchyme into ventral stroma and fascia. Arrested migration of the ventral mesenchyme will keep the urethral edges apart, preventing fusion of the edges of the urethral groove and resulting in *hypospadias.*

Disorganisation and arrested distal migration of the ventral stroma and fascia will also result in curvature of the normally growing corpora cavernosa which are firmly attached to the underdeveloped urethra. Such firm attachment maintains the early developmental position of the penis causing *chordee*. The degree of disorganisation and inadequate growth will determine the severity of clinical presentation:

The median raphe begins normally at the perineum and migrates distally. Disorganisation and arrested distal migration of the ventral mesenchyme will result in bifurcation of the median raphe proximal to the urethral orifice into two branches that continue forward to form the lateral edges of the deficient prepuce.

If one branch of the median raphe grows inadequately and is shorter than the other branch, this will result in *torsion* and rotation of the glans towards the longer branch.

In the condition known as megameatus intact prepuce (MIP), there is normal development of the preputial-glanular lamella, but it fails to fuse with the deep glanular groove and the floor of the glanular urethra as well as the frenulum that are not developed.

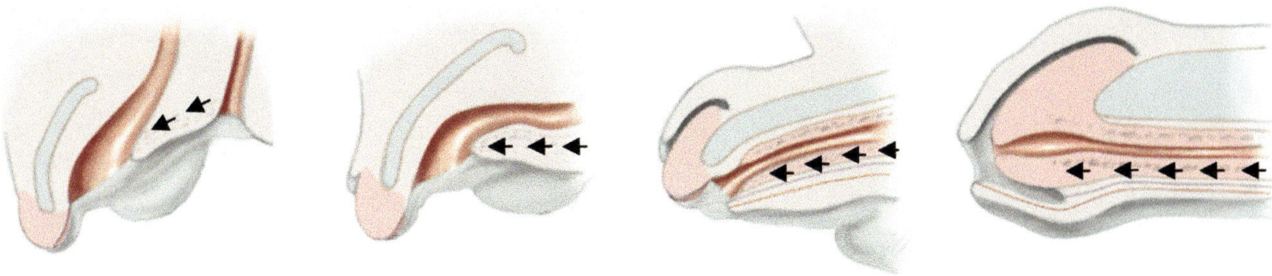

Fig. 44.8 The migration hypothesis showing the crucial role of ventral penile mesenchyme in the development of the penis, urethra and penile fascia. Differential growth and distal migration of the ventral fascia brings the edges of the urethral groove together which fuse to form the penile urethra. The corpus spongiosum grows around the developing urethra. The prepuce plays an important role in the development of the floor of the glanular urethra. Superficial stroma forms dartos fascia which continues to grow into the prepuce. For illustration purposes, only the ventral dartos fascia is shown in the figure (From Hadidi; with permission)

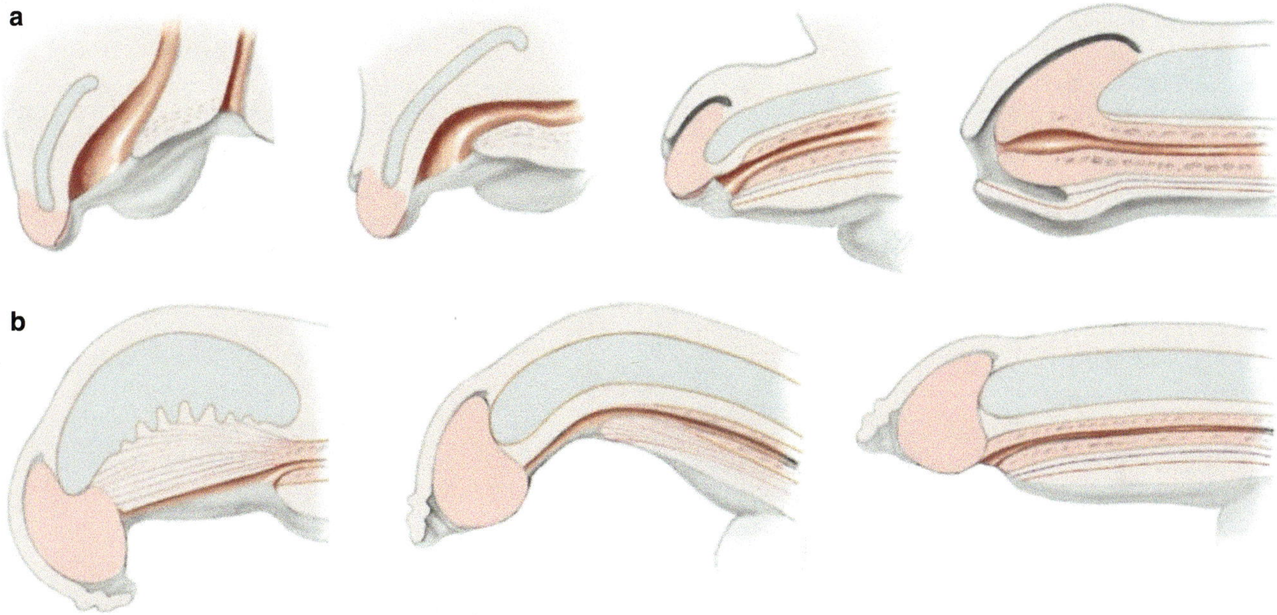

Fig. 44.9 (a) Steps of normal development of the penis and urethra. (b) Disorganisation and arrested growth of the ventral fascia results in different forms of hypospadias and different degrees of chordee (From Hadidi with permission)

Posterior Urethral Valves (PUV)

Posterior urethral obstruction may be diagnosed in utero when the dilated posterior urethra and dilated bladder with hydroureter and hydronephrosis are detected on ultrasound examination. Otherwise there is often a delay in the diagnosis because of failure to suspect the condition. This is especially so in infants in nappies where dribbling and straining to pass urine or a poor stream may easily be overlooked.

Clinical Features

The infant has difficulty in voiding urine from the earliest days of life, and the act of micturition will show diminution in the volume and the force of the urinary stream. There may be only a gentle dribble from the meatus. Palpation of the abdomen reveals a distended bladder, and this needs to be confirmed by ultrasound examination. If the urinary tract becomes infected early, the infant may show signs of uraemia, vomiting and failure to thrive. The first abnormality to be noticed may be gross enlargement of the abdomen due to the large non-emptying bladder. The older boy may present with persistent wetting due to overflow incontinence.

Management

Full clinical assessment of the infant or child, examination of the urine and biochemical assessment of renal function are

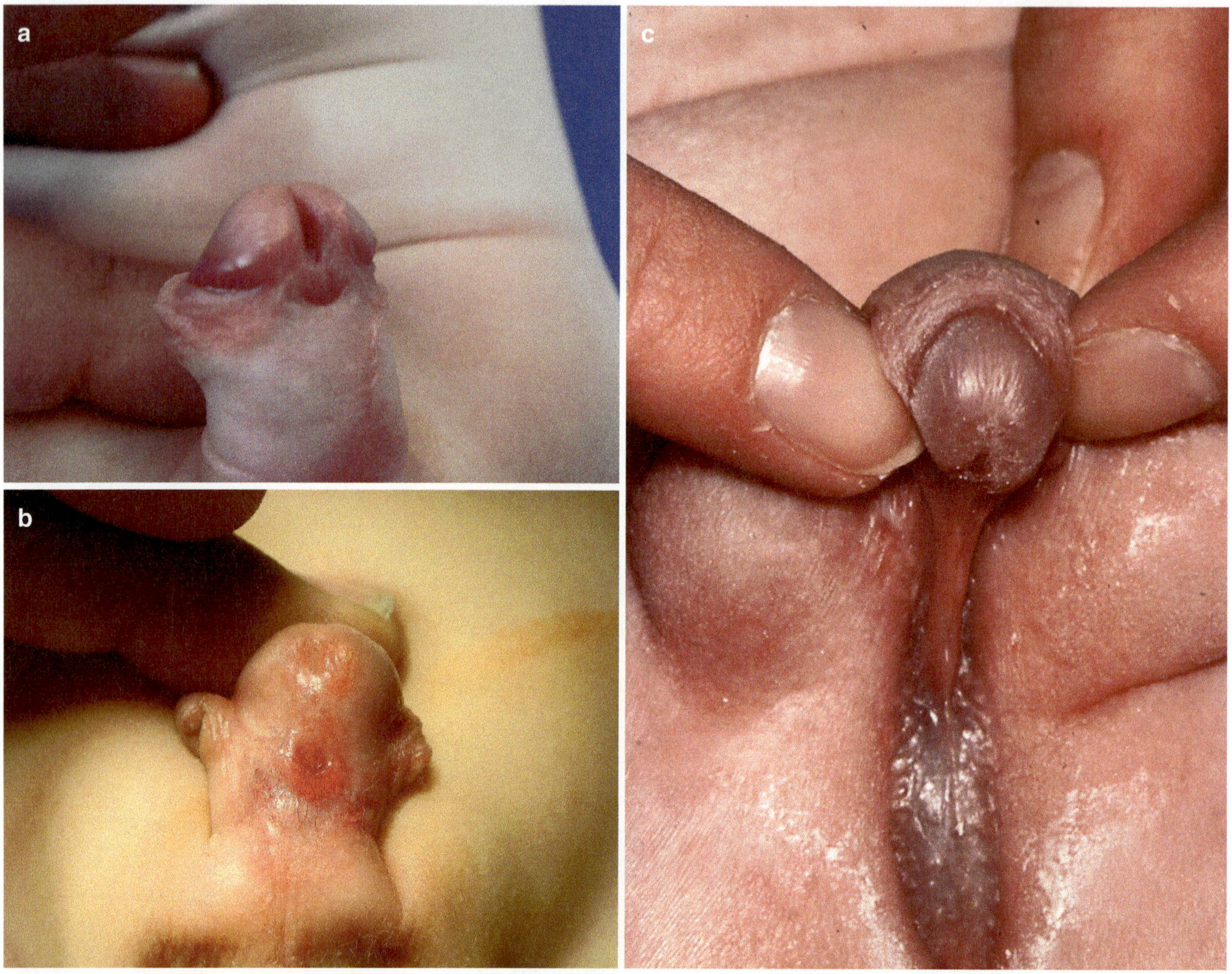

Fig. 44.10 (**a**) Coronal hypospadias. (**b**) Mid-shaft hypospadias. (**c**) Perineal hypospadias with chordee

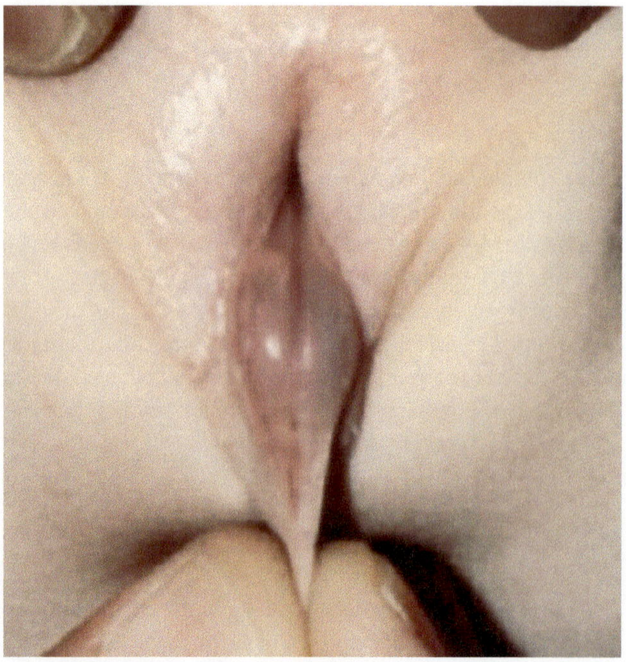

Fig. 44.11 Epispadias

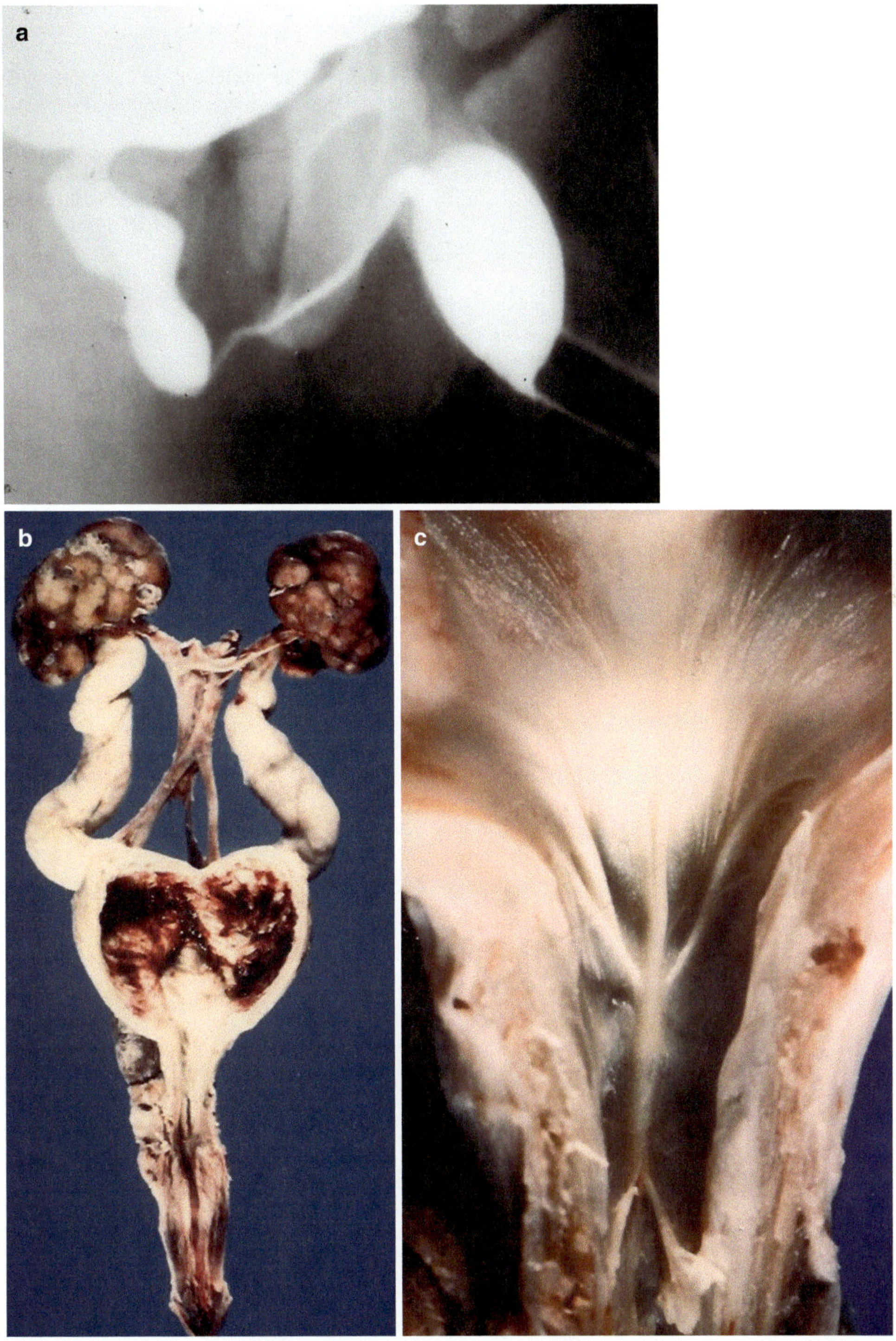

Fig. 44.12 (**a**) Posterior urethral valves x-ray with contrast. (**b**) Pathology of PUV. (**c**) Pathology close-up of the valves

necessary. Ultrasound examination should include posterior urethra, bladder, ureters and kidneys. When posterior urethral obstruction is suspected, a small catheter is passed and the bladder partially decompressed. A cystourethrogram should be performed and films taken in the lateral position during micturition or bladder expression after the catheter removal (Fig. 44.12). If the kidneys are enlarged and tense, bilateral nephrostomies or ureterostomies may be preferable to cystostomy. When the child's general condition has improved, the posterior urethral obstruction is usually corrected by diathermy of the valves viewed through an operating cystourethroscope. Long-term outlook for these children is generally poor when there is reflux because of severe renal impairment. Unlike many conditions the earlier the age of diagnosis the worse the prognosis as the more severe cases present earlier and the mild ones may not present for some years.

References

1. Hadidi AT. Development of the penis and urethra. In: Hadidi AT, editor. Hypospadias surgery, an illustrated guide. Berlin-Heidelberg: Springer-Verlag; 2004.
2. Hadidi AT. Development of the human male urethra: a histochemical study on human embryos. J Pediatr Surg. 2014;49:1146–52.
3. van der Putte SCJ. The development of the perineum in the human. Adv Anat Embryol Cell Biol. 2005;177:1–135.

45 Disorders of Sex Development (DSD)

John Hutson

Normal Embryology

• Day 19	Trilaminar embryo Notochord in midline induces neural tube Paraxial mesoderm forms somites Intermediate mesoderm forms urogenital ridges Lateral plate mesoderm forms body wall
• 3–5 weeks	Primordial germ cells form in caudal yolk sac near allantois Migrate around hindgut and into ambisexual gonad (anteromedial to mesonephros)
• 7–8 weeks	Onset of sexual differentiation triggers by SRY expression in male gonad Mesonephros regresses, leaving developing gonad on mesentery (mesorchium) Sertoli cells form and surround germ cells to form testicular cords
• 8–12 weeks	Testosterone from Leydig cells stimulates Wolffian (mesonephric) ducts to form epididymis vas deferens and seminal vesicles (Fig. 45.1) Anti-Müllerian hormone (AMH) from Sertoli cells causes regression of Müllerian ducts Lack of AMH (and possibly oestrogen) allows Müllerian ducts to develop into the fallopian tubes, uterus and vagina (Fig. 45.2). Initially the Müllerian ducts reach the urogenital sinus at the Müllerian tubercle, which triggers the development of the lower vaginal plate. The vaginal plates are solid initially but cavitate to make a vaginal cavity as the end of the vaginal plate slides caudally to open separately into the vestibule. Breakdown of the end of the vaginal plate (i.e. the hymen) occurs perinatally (Fig. 45.2) Dihydrotestosterone causes masculinisation of external genitalia (Fig. 45.3) Lack of androgen in females (and oestrogen) causes feminisation (Fig. 45.4)
• 12–40 weeks	Androgens responding to pituitary axis (gonadotrophins) cause external genitalia (esp penis) to enlarge
• 10–15 weeks	Insulin-like hormone 3 from Leydig cells causes swelling reaction in gubernaculum for first phase of testicular descent (transabdominal phase)
• 25–35 weeks	Androgens cause inguinoscrotal phase of testicular descent (gubernacular migration, elongation of processus vaginalis inside gubernaculum, closure of proximal processus vaginalis)
• 8–12 weeks	Masculinisation of internal and external genitalia and canalisation and fusion of urethral plate to form male urethra Inner genital folds fuse to form skin of penile prepuce and shaft Outer genital folds fuse to scrotum Distal urogenital sinus regresses to leave utriculus masculinis Wolffian ducts absorbed into back of proximal urogenital sinus to form trigone of bladder and verumontanum

Wolffian and Müllerian ducts are masculinised by testosterone (+AMH) acting in 'exocrine' manner by secretion down Wolffian ducts.

External genitalia need expression of enzyme (5α reductase) to convert testosterone to DHT, which binds × 5–10-fold more tightly to androgen receptors (DHT needed at 8–12 weeks because testis is still too small to generate adequate endocrine levels of T in the blood) (Fig. 45.5).

J. Hutson (✉)
Department of Paediatrics, University of Melbourne, Melbourne, Australia

Department of Urology, The Royal Children's Hospital, Melbourne, Australia

Murdoch Children's Research Institute, Melbourne, Australia
e-mail: john.hutson@rch.org.au

R. Carachi, S. H. E. Doss (eds.), *Clinical Embryology*, https://doi.org/10.1007/978-3-319-26158-4_45

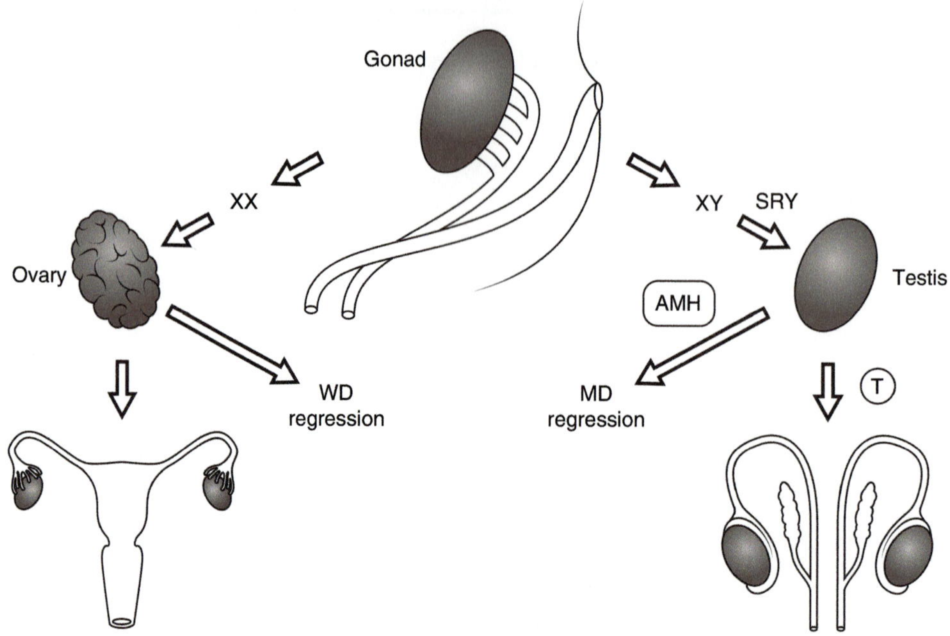

Fig. 45.1 Basic schema of sexual differentiation after 8 weeks' gestation. If the undifferentiated gonad is exposed to SRY expression (from Y chromosome), the gonad becomes a testis, which makes androgen and anti-Müllerian hormone (AMH). Androgens cause the Wolffian duct to form into epididymis, vas deferens and seminal vesicle. AMH causes regression of the Müllerian ducts. If the gonad forms an ovary (XX chromosomes and no SRY expression), the Müllerian ducts form primitive fallopian tubes, uterus and upper vagina. The lower vagina persists with oestrogen and absence of androgens

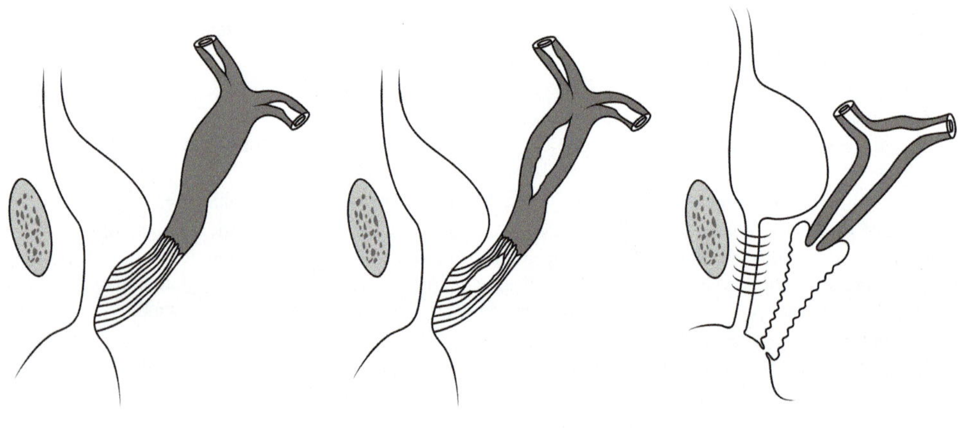

Fig. 45.2 Schema showing development of lower female genital tract from Müllerian ducts which caudally are solid cords that fuse to make uterovaginal primordium after reaching the urogenital sinus at the Müllerian tubercle. The vaginal plate cavitates (in the absence of androgens) as its connection to the urogenital sinus slides caudally to join separately from the urethra in the introitus. The solid Müllerian cords cavitate to form the uterine lumen, cervical canal and vaginal vault.

Abnormal Embryology in DSD

• Failed sex determination	Faulty expression of SRY, absent Y chromosome (e.g. Turner syndrome)	
• Complete gonadal dysgenesis (Swyer syndrome)	46, XY DSD Abnormal SRY action	
• Failure of normal testis development	Symmetrical gonadal dysgenesis Asymmetrical gonadal dysgenesis	Secondary to genes controlling testis development Secondary to mosaic sex chromosomes, e.g. XO/XY
• Failure of placental hormone function (required at 8–15 weeks)	46,XY gonadal dysgenesis	
• Failure of hypothalamic Pituitary axis (required at 15–40 weeks)	46, XY DSD with micropenis and UDT	
• Failure of INSL3 action	Intra-abdominal testes	
• Failure of AMH action	Persistent Müllerian duct syndrome	
• Failure of testosterone action	Enzyme defects (e.g. 17 ketosteroid reductase deficiency) AR receptor mutations, androgen insensitivity	
• Failure of regional morphogenesis	Bladder exstrophy/cloacal exstrophy Penile agenesis Anorectal malformations Ectopic labia/scrotum	

DSD may be caused by a large number of different genetic, hormonal and morphological defects that disturb the process of sexual differentiation.

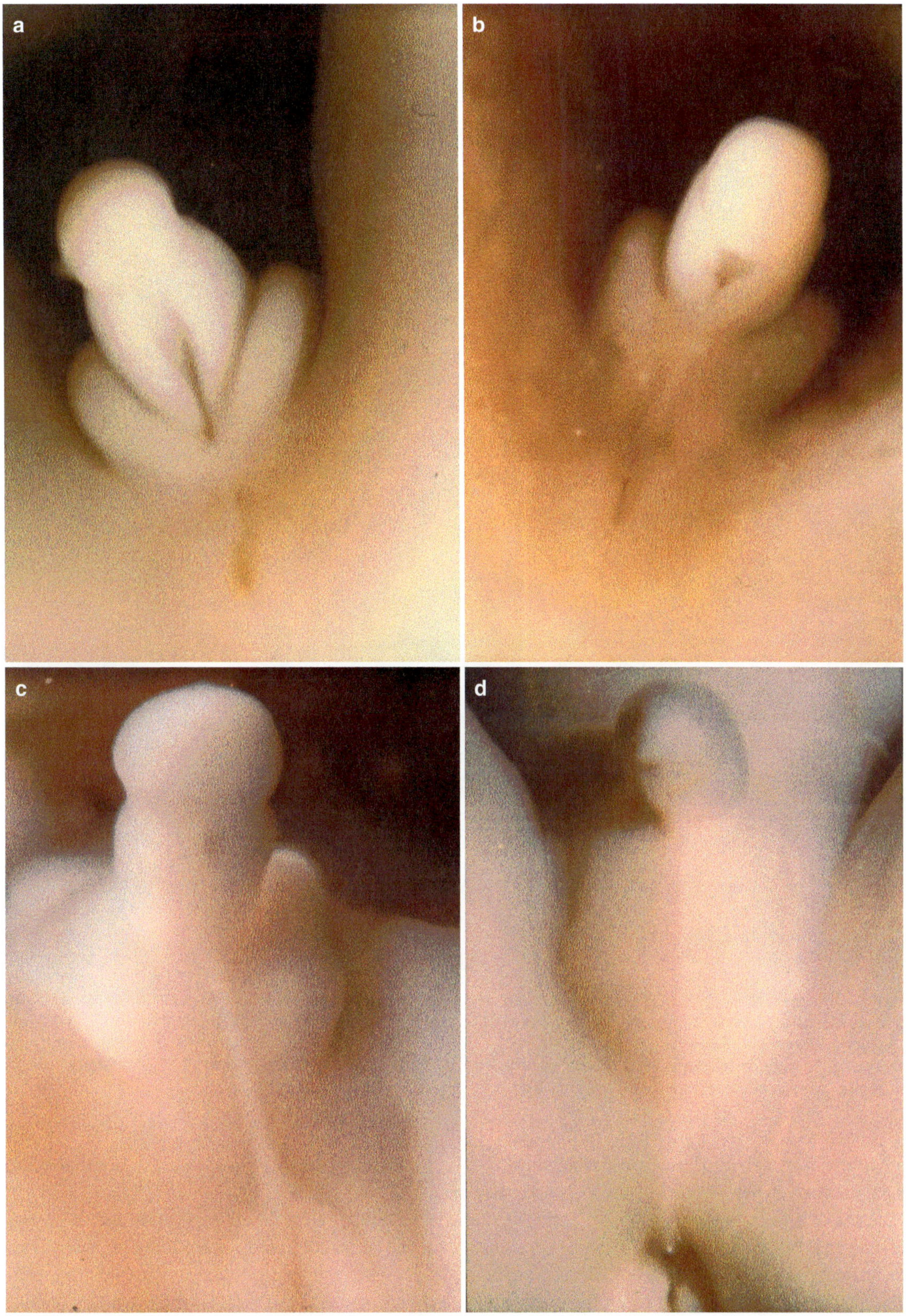

Fig. 45.3 (**a–d**) Development of the external genitalia in males between 8 and 12 weeks' gestation: (**a**) Ambisexual stage at 8–9 weeks. (**b**) Partially fused urogenital folds and canalisation of urethral plate at 9 weeks. (**c**) Urogenital fold fusion and genital tubercle enlarging at 10 weeks. The future glans is visible and the urethra opens at the future coronal groove. (**d**) By week 12 the foreskin is forming to cover the glans and the urethral plate in the glans in canalising to form (unseen) the glanular urethra (Reproduced with permission from England 1983; Clarnette et al. 1998)

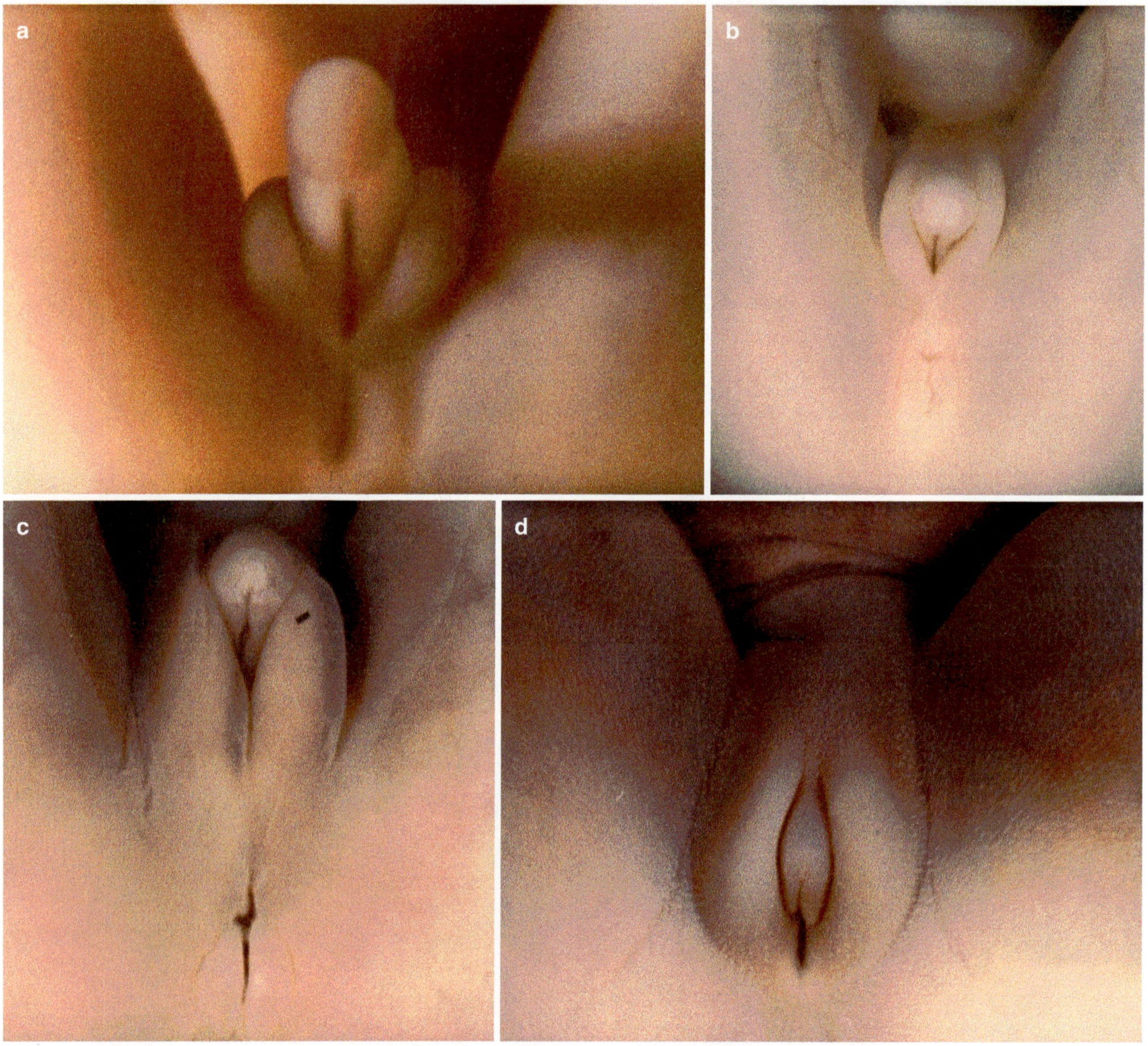

Fig. 45.4 (**a–d**) Development of the external genitalia in females between 8 and 20 weeks. (**a**) Ambisexual stage at 8–9 weeks. (**b**) The genital tubercle grows slower than the foetus at 9 weeks, making it look progressively smaller. In addition, apoptosis of the endodermal urethral plate on the ventral surface of the genital tubercle causes it to bend ventrally, as seen at 13 weeks. The genital folds remain unfused to show the developing vestibule. (**c**) By week 17 the folding of genital tubercle shows that is now has the appearance of a clitoris. (**d**) By week 20 the external anatomy is fully formed as female (Reproduced with permission from England, 1983)

Abnormalities of sex determination occur when there is a chromosomal anomaly, when either the X or Y chromosome is absent, such as 45X (Turner syndrome). Expression of the SRY gene (the sex-determining region on the Y chromosome) may be faulty in some rare forms of DSD, such as XY females, or it may be ectopically located on the X chromosome or an autosome in XX males. Defects in the 'downstream' signalling pathway triggered by SRY expression are the likely cause for 46,XY complete gonadal dysgenesis (Swyer syndrome), where a phenotypic female presents in adolescence with absent pubertal development, and on investigation is found to have streak (i.e. nonfunctional) gonads and XY karyotype.

Failure of gonadal differentiation into a normal testis is caused by both chromosomal anomalies, such as mosaicism, (e.g. XO/XY mixed gonadal dysgenesis), as well as in the specific genes involved in the complex chain reaction required to make a testis. In mixed gonadal dysgenesis with mixed chromosomal DSD, the external genitalia are commonly asymmetrical, with one side more masculine, with a descended testis into a hemiscrotum, while the other side more feminine with no gonadal descent (Fig. 45.6).

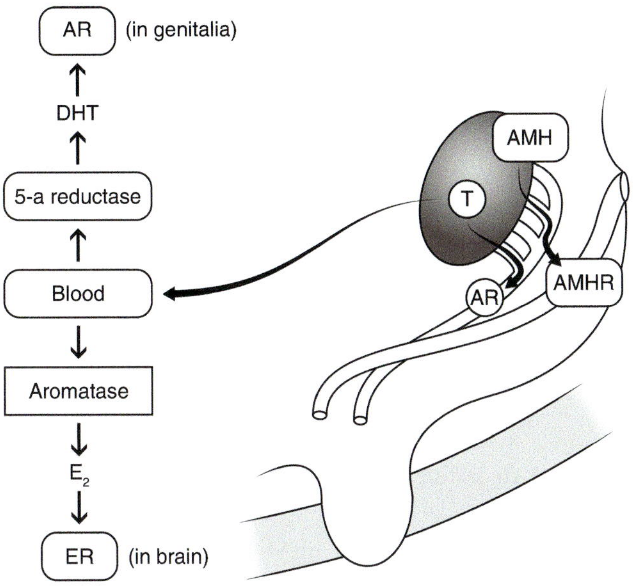

Fig. 45.5 Schema of urogenital ridge at 8–10 weeks, where testosterone produced in the developing testis is secreted through the rete testis and down the Wolffian duct (WD) to reach the androgen receptors (AR) in the WD, triggering its preservation and differentiation into epididymis, vas deferens and seminal vesicle. The AMH also is secreted down the WD and then diffuses into the adjacent MD to trigger regression. Testosterone entering the blood stream (in low concentrations as the testis is still tiny) reaches the external genitalia, where the enzyme 5α-reductase converts it into dihydrotestosterone (DHT). DHT binds ×5–10 times more tightly to AR than testosterone itself, effectively producing a five–tenfold increase in androgen levels for external genital masculinisation. Some testosterone is converted by aromatase to oestrogen, which binds to oestrogen receptors (especially in the central nervous (?) system, to trigger masculinisation of some neural pathways)

When placental function is deficient, there is commonly poor foetal growth, leading to intrauterine growth retardation (IUGR), and if the placental dysfunction is affecting the production of chorionic gonadotrophin (hCG) between 8 and 15 weeks' gestation, then a premature infant may present with IUGR and 46,XY gonadal dysgenesis (Fig. 45.7).

However, postnatally no specific hormonal anomaly can be identified in the testis, as the pituitary axis in the infant is normal and the testicular function has recovered.

The hypothalamic-pituitary-gonadal axis is perturbed by mutations affecting hypothalamic and pituitary function and presents with 46,XY DSD with a micropenis and cryptorchidism (Fig. 45.8). Depending on the precise defect in the hypothalamic function, there may be other

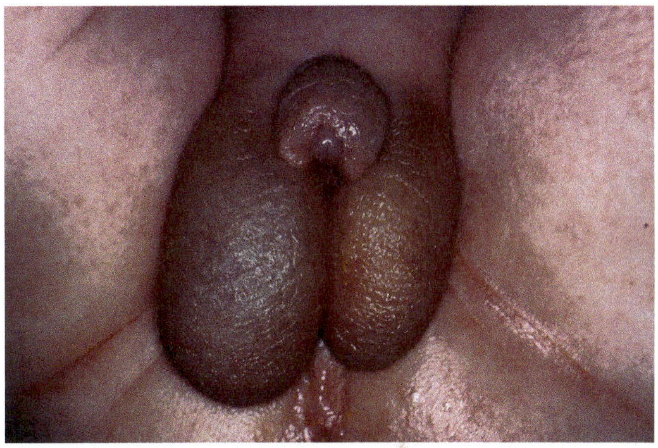

Fig. 45.7 An example of 46,XY DSD caused by inadequate placental stimulation (by hCG) of the developing testis, with secondary gonadal dysgenesis (in the cases, mostly delayed development, which eventually catches up, so that postnatally there is no obvious endocrinopathy

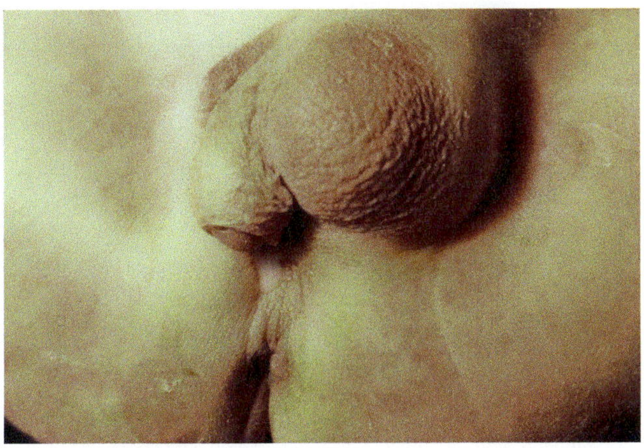

Fig. 45.6 Asymmetrical genital development is common in chromosomal DSD, with mosaicism (in this case 45,X/46,XY mixed gonadal dysgenesis). The gonad has developed well enough on one side to control its own descent, but on the contralateral side, there is a nonfunctional streak gonad with no descent. Evidence of a retained MD on the side of the streak gonad is manifested by the vaginal mucus extruding from the urogenital sinus

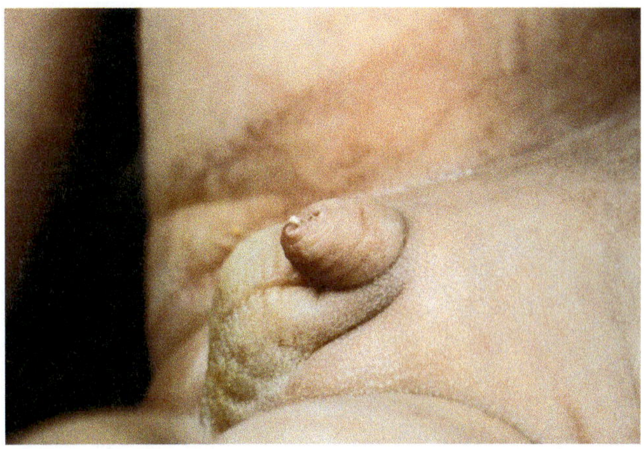

Fig. 45.8 Fully formed male genitalia except for micropenis and undescended testes, as inguinoscrotal phase of descent and penile growth requires androgens between 15 and 40 weeks. In this case the child has congenital hypopituitarism, preventing normal function of the foetal hypothalamic-pituitary-gonadal axis

clues to central nervous system abnormalities, such as nystagmus, which if present at birth often indicates blindness secondary to anomalies affecting the pituitary stalk and the optic chiasm.

When the testis has formed normally, but this is a defect in production or action of the hormones (T, INSL3 or AMH), there is undescended testis. As INSL3 controls the gubernacular swelling reaction to hold the testis near the inguinal region during foetal growth of the abdominal and pelvic cavity, INSL3 gene mutations or defects in the INSL3 receptor (relaxin family receptor 2, RXFP2) cause interruption of the transabdominal phase of testicular descent. In this case the undescended testis is impalpable, as it is intra-abdominal.

Defects in AMH action (AMH gene or AMH receptor type II) lead to the persistent Müllerian duct syndrome (PMDS), where the Müllerian ducts persist as infantile fallopian tubes and uterus (as there is no oestrogen), in a boy with normal male external genitalia but with undescended testes (Fig. 45.9). In 80% of cases, both testes are intra-abdominal, with the gubernacular cords >10 cm long rather than <½ cm, so that despite normal gubernacular migration to the scrotum and a patent PV, the testes remain intra-abdominal. In about 20% of cases, one of the testes has 'descended' into the open PV, but because of the abnormally long gubernacular cord (and fallopian tube preventing closure of PV), it can be easily displaced back into the abdominal cavity. The contralateral testis is usually found hanging loosely within the peritoneum on its gubernacular cord. This phenotype is called 'hernia uteri inguinalis', as the inguinal hernia contains the fallopian tube and fundus of the uterus. In ~10% of cases of PMDS, both testes have 'descended' into the same PV, producing transverse testicular ectopia.

The vanishing testis is common in PMDS, which is thought to be secondary to the excessive mobility of the testes with the extremely long gubernacular cords, which predisposes to gonadal torsion and ischaemic atrophy of the testis.

When there is a defect in testosterone action, with defects in production or action, there is cryptorchidism as well as abnormal external genital development. In complete androgen insensitivity syndrome (CAIS), where the androgen receptor has a mutation preventing normal androgen signalling, the testes are usually palpable (and visible, because they are enlarged) in the groin in a phenotypic female (Fig. 45.10). Investigation shows that the

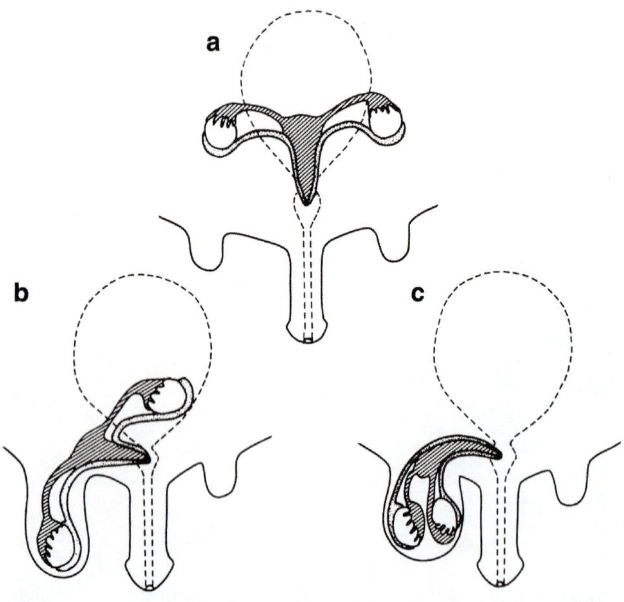

Fig. 45.9 Schema showing the three common presentations of persistent Müllerian duct syndrome (PMDS): (**a**) bilateral intra-abdominal testes with retained uterus (infantile, as no oestrogen in a male) and tubes (~70%); (**b**) hernia uteri inguinalis, a hernia containing a testis and uterus) (20%); (**c**) transverse testicular ectopia (10%), where both testes (and uterus and tubes) are in the same inguinal hernia

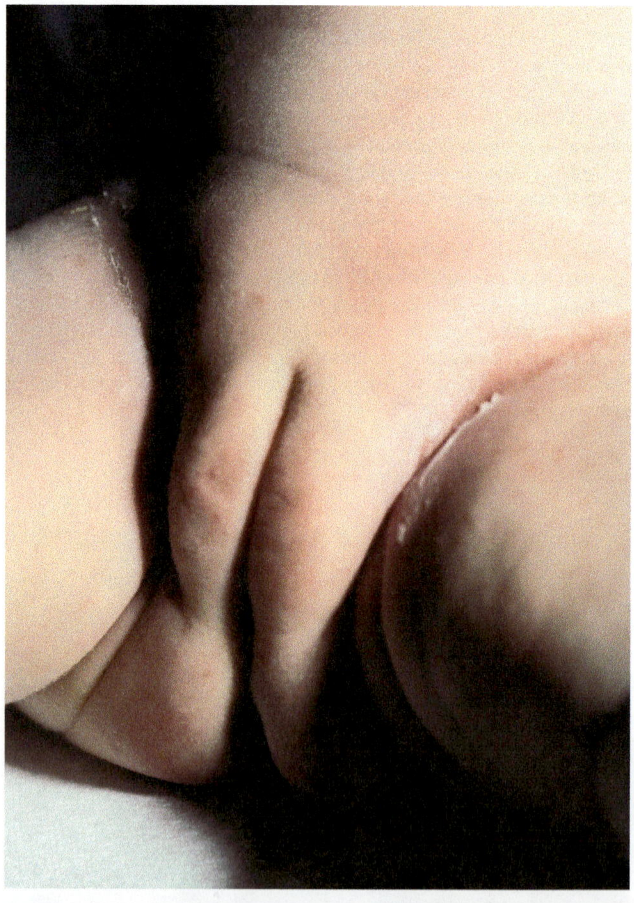

Fig. 45.10 Complete androgen insensitivity syndrome (CAIS), caused by a mutated, nonfunctional androgen receptor. The infant has 46,XY DSD with a female phenotype and visible groin lumps because testicular descent is arrested at the start of the inguinoscrotal phase in the absence of normal androgen signalling. The gonads are testes, which predict normal MD regression, so there will be no uterus, but absent androgen signalling allows the vagina to develop. These inguinal testes are larger than normal because lack of a normal androgen interferes with negative feedback on the hypothalamus, so gonadotrophins remain high

genotype is XY, and because the rest of testicular development is normal, the Müllerian ducts have regressed (normal AMH function) and the testis has completed normal transabdominal descent to the groin (normal INSL3 function). Because the Müllerian ducts have regressed (but after inducing a vaginal plate), the vagina is present but slightly shorter than normal, as there is no uterus and cervix and the vault of the vagina is absent. Despite the slightly shorter vagina, at puberty the vagina grows and enlarges with sexual function.

Another anomaly in androgen production occurs when there is an enzyme defect in the production of cortisone and aldosterone in the foetal adrenal gland. The build-up of the intermediate metabolites triggers activation of an alternative pathway to make androgen, which is usually not functional until adolescence (to cause axillary and pubic hair in girls). If the foetus is a female, it exposes her to variable levels of androgens, causing the commonest DSD, congenital adrenal hyperplasia (CAH). Lack of cortisone (± aldosterone) production prevents negative feedback on the hypothalamic-pituitary axis, leading to elevated ACTH (and melanin-stimulating hormone), which produces adrenal hypertrophy and pigmentation (often obvious in the genitalia and palms). The excess androgens lead to masculinisation of the external genitalia in females, producing ambiguous genitalia (Fig. 45.11). The genetic defect is often in $P450_{C17}$ (17-hydroxylase), which is on chromosome 10q2.43. Both sexes are affected (autosomal recessive), but abnormal adrenal androgens only cause a genital anomaly in girls, but if not recognised at birth (which is the norm in boys), infants with CAH usually present with a salt-losing adrenal crisis causing vomiting and circulatory collapse at 2–3 weeks of age.

When there is a morphological anomaly outside the spectrum between normal female and male, the DSD is caused by a regional anomaly of perineal development, rather than a hormonal cause, what I call the nonhormonal DSD (Fig. 45.12). Examples are numerous and include anorectal malformation, bladder exstrophy and cloacal exstrophy, as well as rarer anomalies such as penile agenesis.

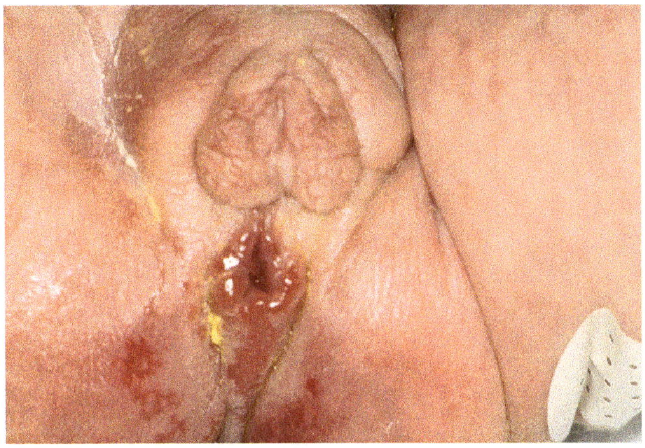

Fig. 45.12 An example of a nonhormonal DSD, where the labia minora and clitoral hood are enlarged, but the clitoris and introitus are the normal size. There is anterior ectopic anus, and pulling open the introitus revealed a double vagina

Ectopia of the labia or scrotum is usually not caused by specific genetic anomalies but is the result of foetal deformity where the heel of one of the feet compresses the perineum. This occurs when the early foetus is compressed into an abnormal position by an anomaly of the amniotic membrane, or otherwise unrelated orthopaedic anomalies. The extrinsic pressure of the heel can displace the genital folds, causing ectopic hemiscrotum and labia, and also cause exstrophy of the testis (prolapse through a 'pressure sore' in the scrotum) and primary urethral fistula (urethral development is normal, but secondary pressure necrosis of corpus spongiosum is producing a proximal urethra-cutaneous fistula) (Fig. 45.13).

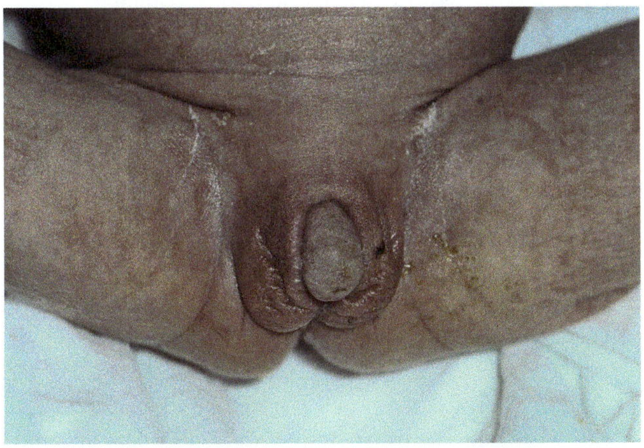

Fig. 45.11 Congenital adrenal hyperplasia (CAH) in a female infant with masculinisation of the external genitalia but no descended gonads (as the ovaries are normally sited in the abdomen)

Clinical Assessment of the Newborn with a DSD

Once the normal embryology is understood, the underlying cause of the DSD can often be predicted by the clinical examination, using a set of simple rules, as outlined below.

1. As testes descend and ovaries do not, a palpable gonad will be a testis or at least have enough testicular tissue to control its own descent. The presence of palpable gonad predicts that genome contains an SRY gene, so the chromosomes are likely to contain a Y chromosome.
2. As testicular descent is tightly linked with Müllerian duct regression (as AMH is predicted to prevent elongation of the gubernacular cord into a round ligament), a palpable

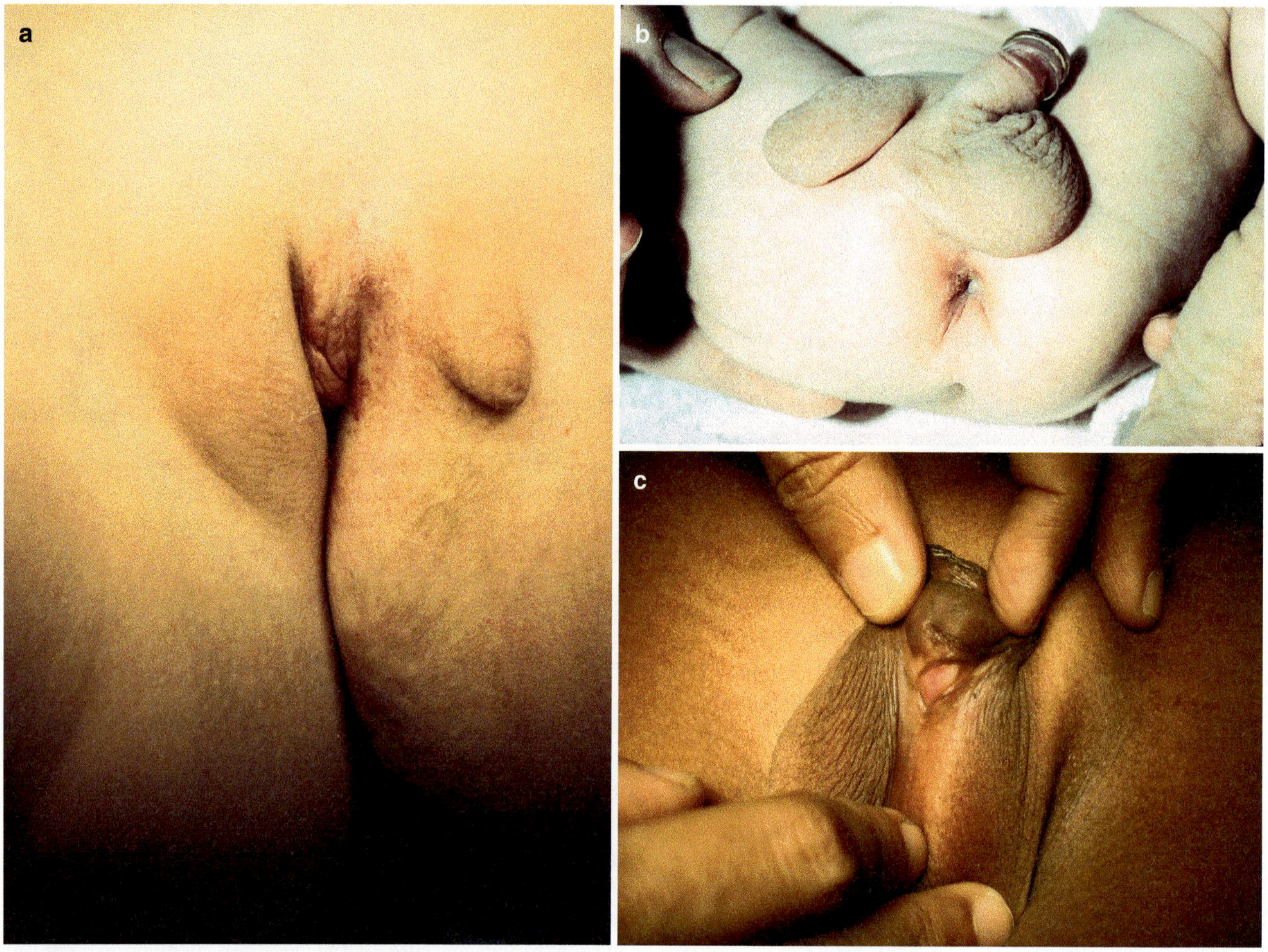

Fig. 45.13 (**a–c**) Examples of nonhormonal DSD caused by extrinsic compression of the perineum by the heel of one of the foetus' feet. (**a**) Ectopic labium; (**b**) ectopic scrotum; (**c**) urethral fistula

gonad predicts regression of the ipsilateral Müllerian duct. Therefore there will be no fallopian tube on that side.

3. Where the gonad is in the hemiscrotum, we can predict that the ipsilateral epididymis and vas deferens are present, as inguinoscrotal testicular descent is controlled by androgens.
4. If physical exam (rectal exam) and/or pelvic ultrasound scan demonstrate a uterus, this predicts absent AMH action on the ipsilateral side.
5. Where the external genitalia are asymmetrical, with one descended gonad (it must be a testis) and one undescended gonad, the ipsilateral hormonal regulation of the genital ducts (by exocrine secretion of the testosterone/AMH down the ipsilateral Wolffian duct) predicts that there will be no Müllerian duct but a vas on the side of the descended gonad. On the contralateral side (where the gonad is impalpable and undescended), there is likely to be a Müllerian duct but no vas.
6. Because the masculinisation of the external genitalia (and regression of the vaginal plate) is directly proportional to the amount of androgen, the level of effective androgen function during early development can be predicted from the degree of masculinisation. In addition, the degree of masculinisation predicts the degree of vaginal regression: when there is incomplete scrotal fusion, then the vaginal remnant is larger.
7. Because growth of the external genitalia (after formation of the penis and scrotum at 8–12 weeks) required ongoing androgen, micropenis predicts a hypothalamic or pituitary anomaly. This is because early virilisation is controlled by placental hCG, while after mid-gestation (~15 weeks) the foetal hypothalamic-pituitary axis takes over control of testicular androgen production.
8. Because growth of the genital tubercle is androgen dependent, any enlargement of the erectile tissue of the clitoris indicates abnormal exposure of the female foetus to androgen. The usual cause for this is congenital adrenal hyperplasia (where the mutation is not so severe that the genitalia are obviously ambiguous at birth).

9. Because hormonal causes of DSD disturb genial development towards male or female (but everything else is normal, when other anomalies are present outside this narrow male-female spectrum), we can predict that the cause is a nonhormonal DSD, and endocrine assessment should be completely normal.
10. Because the external genitalia respond like a 'bioassay for androgen', DSD with deficient androgen action cause abnormality of all genital structures: genital tubercle, urethra and genital folds and gonadal descent. Therefore in 'hypospadias' it is simple to exclude serous DSD by checking whether there are two descended testes in a fused scrotum. By contrast, in penoscrotal 'hypospadias' you should assume that the infant has a DSD and needs neonatal investigation rather than elective referral to urology clinic for 'simple hypospadias' surgery.

Recommended Reading

Bailez MM. Intersex disorders (Chapter 91). In: Puri P and Höllwarth M, editors. Pediatric surgery diagnosis and management. Springer.

Gatti JM. Disorders of sexual differentiation (Chapter 63). In: Murphy H, Ostlie, editors. Ashcraft's pediatric surgery. 6th ed. New York: Elsevier Publishing.

Hutson J, Grover S, O'Connell M, Bouty A, Hanna C, editors. Disorders/differences of sex development. 2nd ed: Springer; in press.

Öcal G. Current concepts in disorders of sexual development. J Clin Res Pediatr Endocrinol. 2011;3(3):105–14.

Sharp R. Intersex (Chapter 94). In: Prem Puri, editors. Newborn surgery. 2nd ed. Arnold.

Witchel SF. Disorders of sex development. Best Pract Res Clin Obstet Gynaiecol. 2017.

Yatsenko SA, Witchel SF. Genetic approach to ambiguous genitalia and disorders of sex development: what clinicians need to know. Semin Perinatol. 2017 Jun;41(4):232–43.

The Testis

46

John Hutson

Normal Embryology

• 7–8 weeks	Testis forms from ambisexual gonad in male
	Mesonephros regresses, leaving testis on mesentery (mesorchium)
	Mesorchium with cranial ligamentous thickening: cranial suspensory ligament; caudal ligamentous thickening; genitoinguinal ligament or gubernaculum
• 8–12 weeks	INSL3 from Leydig cells causes swelling reaction in gubernaculum, making it short and fat
	Testosterone from Leydig cells causes regression of cranial ligament
	Bulky gubernaculum and loss of cranial ligament allows testis to be held near future inguinal ring as foetus grows (Fig. 46.1a and b)
• 15–25 weeks	It is thought that during this period androgens masculinise the genitofemoral nerve, which supplies gubernaculum and scrotum
• 25–35 weeks	Testosterone regulates outgrowth of gubernaculum and growth of processus vaginalis inside gubernaculum. Gubernacular outgrowth is possible because of overlying mammary line in inguinal skin. Testis is still attached to the inside of processus vaginalis by gubernacular cord
	Gubernaculum elongates towards scrotum like embryonic limb bud, pulling testis with it. Migration regulated by neurotransmitter (calcitonin gene-related peptide CGRP) in animal models and possibly humans (Fig. 46.1c)
	Inguinal fat pad remodels with enzymes dissolving extracellular matrix to allow gubernacular migration
• 35–45 weeks	Bulky gubernacular mesenchyme resorbs and scrotal fat pad remodels to allow gubernacular cord and processus vaginalis to adhere to inside of scrotum
	Proximal processus vaginalis obliterates in response to CGRP from genitofemoral nerve, leaving testis inside satellite peritoneal cavity in hemiscrotum, tunica vaginalis (Fig. 46.2)

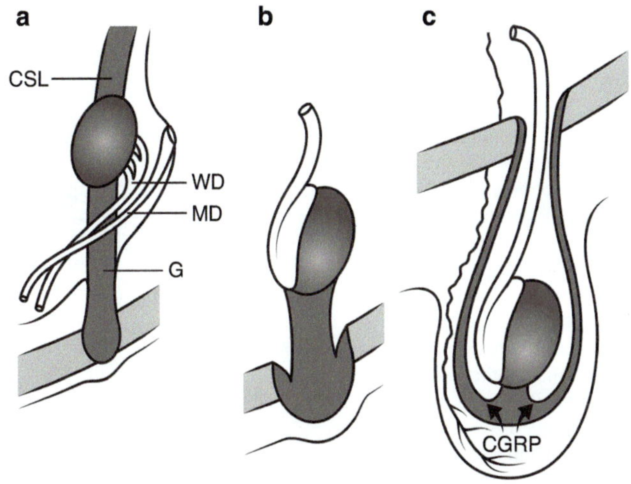

Fig. 46.1 The two stages of testicular descent: the transabdominal phase and the inguinoscrotal phase. (**a**) The initial gonadal position at the onset of sexual differentiation at about 8 weeks' gestation. The developing testis on the anteromedial surface of the urogenital ridge, with the mesonephric (Wolffian duct, WD) and paramesonephric (Müllerian duct, MD) ducts. The cranial suspensory ligament (CSL) is a thickening in the mesogenitale, while the caudal genitoinguinal ligament, or gubernaculum (G), is a thickening connecting the urogenital ridge to the anterior abdominal wall muscles. (**b**) In the transabdominal phase which is complete by 15 weeks, the gubernaculum swells and remains short, holding the testis near the future internal inguinal ring. The peritoneum over the gubernaculum begins to invade the gubernaculum to create the processus vaginalis (PV), separating the gubernaculum into a distal bulb, inner cord (attaching to gonad and developing epididymis) and outer layer, where cremaster muscle forms. (**c**) Beginning at about 25 weeks' gestation, the gubernaculum protrudes out of the abdominal wall like an embryonic limb bud and migrates to the scrotum, controlled by a chemotactic gradient of CGRP from the genitofemoral nerve

J. Hutson (✉)
Department of Paediatrics, University of Melbourne, Melbourne, Australia

Department of Urology, The Royal Children's Hospital, Melbourne, Australia

Murdoch Children's Research Institute, Melbourne, Australia
e-mail: john.hutson@rch.org.au

R. Carachi, S. H. E. Doss (eds.), *Clinical Embryology*, https://doi.org/10.1007/978-3-319-26158-4_46

Abnormal Embryology

- Failure of INSL3 action
  - Intra-abdominal testis
- Morphological faults in urogenital ridge
  - Absent testis (agenesis)
  - Splenogonadal fusion (left side)
- Morphological faults in gubernaculum
  - Misplaced gubernaculum
  - (?) Spigelian hernia with UDT
  - Gubernaculum torn or unattached
    UDT in gastroschisis
    Transverse testicular ectopia
    Prune belly syndrome
  - Failure of PV closure (inguinal hernia)
  - Failure of PV adhesion to scrotum (vanishing testis after torsion)
- Failure of testosterone action
  - UDT in 46, XY DSD, e.g. Klinefelter syndrome (47, XXY DSD)
  - Gonadal dysgenesis
  - Complete/partial androgen insensitivity

Undescended Testis

As testicular descent involves a complex sequential action between the hormones controlling the process and major anatomical remodelling, undescended testis (Fig. 46.3) is caused by a wide range of hormonal and morphological defects (Table 46.1).

It is thought that abnormalities in androgen action are the commonest cause for undescended testis (UDT), but the exact site(s) of the defect remain obscure. One plausible explanation for this is that there is deficiency of placental human chorionic gonadotrophin (hCG) because of an anomaly in placental function, which is not seen on postnatal endocrine investigations. Another possible cause is environmental toxins, which interfere with androgen function, such as endocrine disruptors (synthetic plastics that have weak oestrogenic function) and cigarette smoke. All the other causes of deficient androgen action (e.g. pituitary anomalies, gonadal dysgenesis, synthesis defects and non-functional androgen receptors) are rare.

Another group of causes that are likely to cause UDT are those where there is a morphological anomaly, particularly

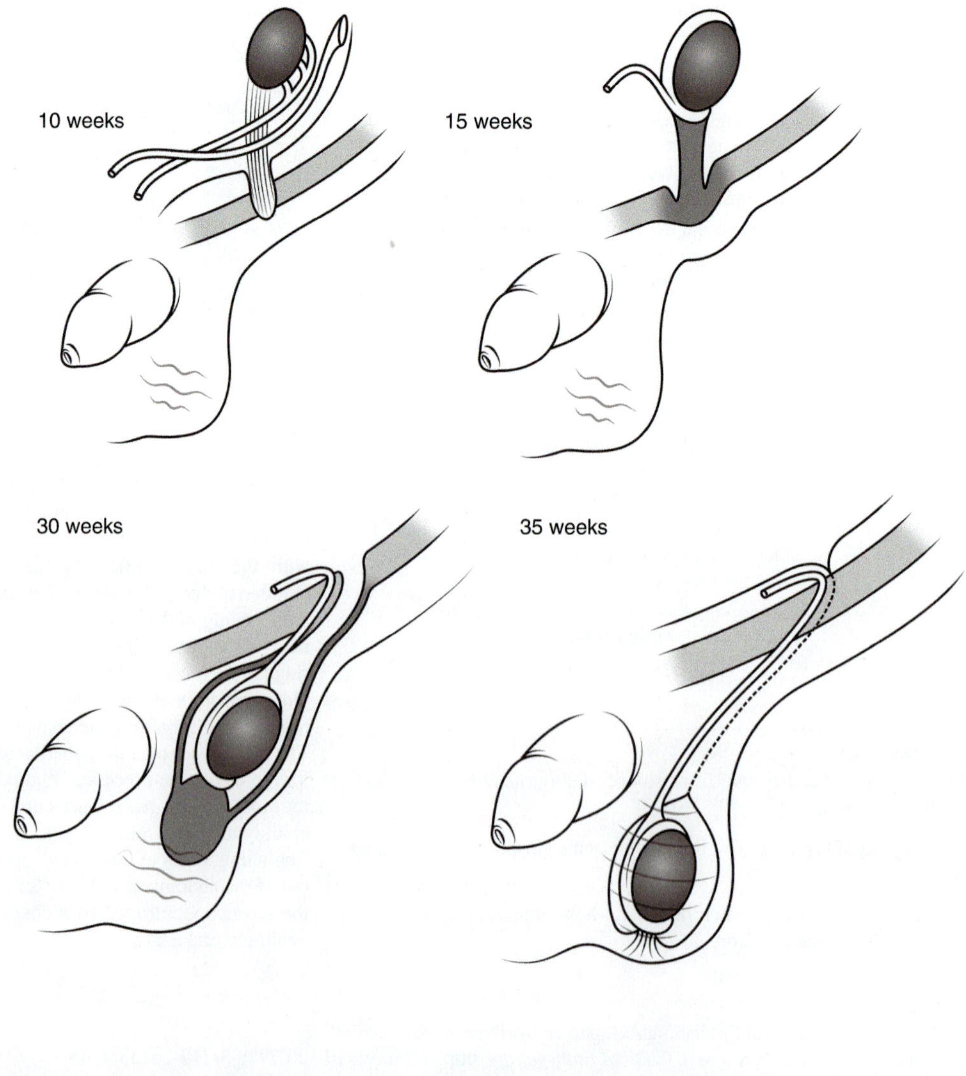

Fig. 46.2 A schema showing the phases of descent in relation to the inguinoscrotal region. At 10 weeks the testis has just formed, and this is at the start of the transabdominal phase. By 15 weeks the transabdominal phase is complete. At 30 weeks the gubernaculum is halfway through the inguinoscrotal migration with the testis inside the PV. At 35 weeks and beyond, the gubernaculum and testis have reached the scrotum, and the two final processes are occurring: fibrous attachment of the tunica vaginalis to the inside of the scrotum and obliteration of the proximal PV, leaving the testis inside the tunica vaginalis

Table 46.1 Predicted causes of cryptorchidism

Abnormal androgen action	Placental gonadotrophin deficiency Pituitary gonadotrophin deficiency Gonadal dysgenesis Defects of enzymes in androgen synthesis (rare) Defects of androgen receptors (rare)
Mechanical anomalies	Defects in gubernacular migration (? common) Prune belly syndrome, where bladder blocks inguinal canal (rare) Posterior urethral valve with bladder blocking inguinal canal (rare) Abdominal wall defects (low abdominal pressure +/or rupture of gubernaculum) (rare) Connective tissue disorders (blocks migration)
Neurological anomalies	Myelomeningocele (with abnormal genitofemoral nerve, GFN) Defects in GFN morphology/function Defects in CGRP production/action
Chromosomal anomalies	Many chromosomal defects and malformation syndromes have undescended testes (for multiple reasons, mostly unknown)
Acquired anomalies	Cerebral palsy (causes acquired UDT by spasm of cremaster muscle) Ascending testes (caused by persisting fibrous remnant of processus vaginalis)

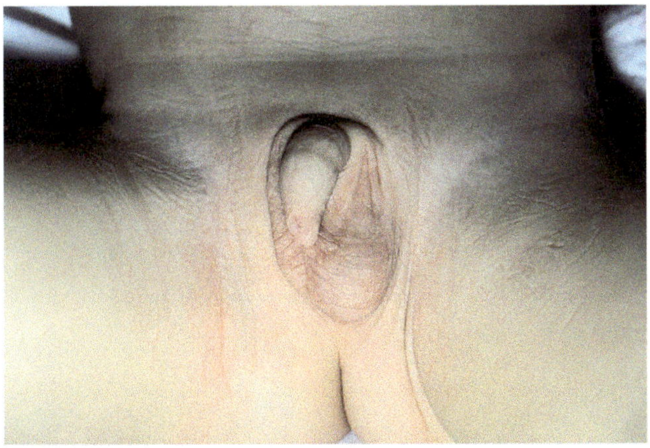

Fig. 46.3 Undescended right testis

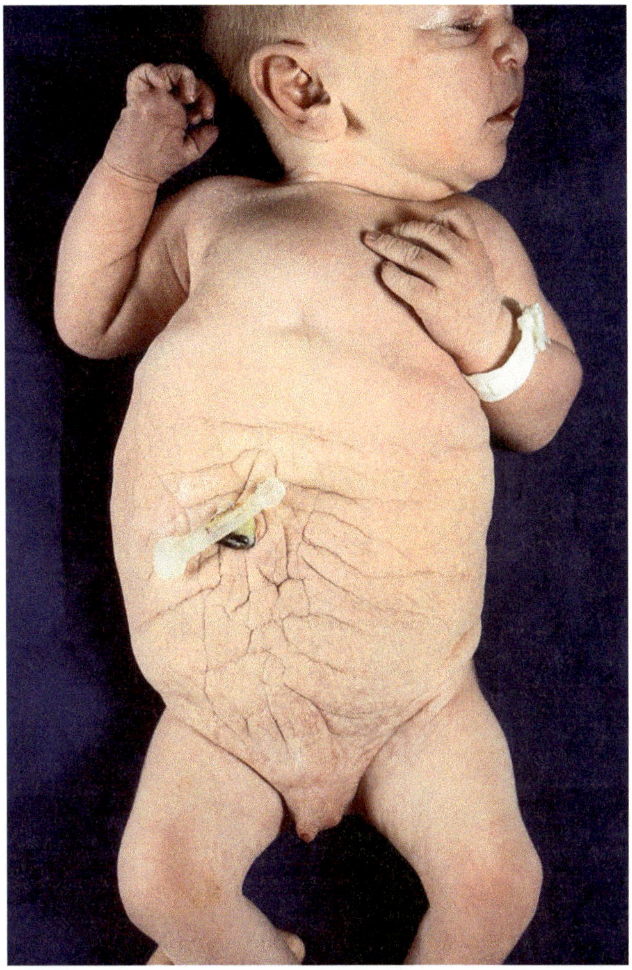

Fig. 46.4 Prune belly syndrome, where urethral obstruction between 10 and 20 weeks causes massive enlargement of the bladder, which leads to ischaemic atrophy of the lower anterior abdominal wall muscles. The urethral obstruction resolves around 20 weeks, leaving the wrinkled abdominal wall. The huge bladder obstructs inguinal canal preventing the testis from transiting to the external ring, leaving the testes intraabdominal

the very complex remodelling involved in gubernacular migration from the inguinal canal to the scrotum. This is not an area that has been investigated extensively so far, as the precise mechanism of how androgen modifies the genitofemoral nerve (GFN), and how the nerve regulates the migrating gubernaculum is not fully understood. In addition, in multiple anomaly syndromes, cryptorchidism is a frequent problem, highlighting the fact that there must be multiple steps in the mechanism of descent that can be disrupted.

In both prune belly syndrome and posterior urethral valve, the bladder outlet is obstructed which may prevent normal development of an inguinal canal by the enlarged bladder displacing the peritoneum away from the internal inguinal ring (Fig. 46.4).

Gross neurological anomalies, such as myelomeningocele, have a high rate of cryptorchidism, especially where the anomaly affects the high lumbar spinal cord, which is the origin of the genitofemoral nerve (GFN). More localised anomalies of the nerve itself, such as mislocation of the genital branch from the scrotum to the adjacent perineum, are thought to be the cause of the ectopic perineal testis (Fig. 46.5). Functional anomalies of neurotransmitters' function in the GFN are likely to also cause cryptorchidism in humans, similar to the trans-scrotal (TS) rat, where the GFN contains excess CGRP, which is thought to perturb the chemotactic gradient controlling inguinoscrotal migration.

Morphological anomalies at the time when the urogenital ridge develops may lead to agenesis of the testis, secondary to failure of the gonad to form. A more common putative

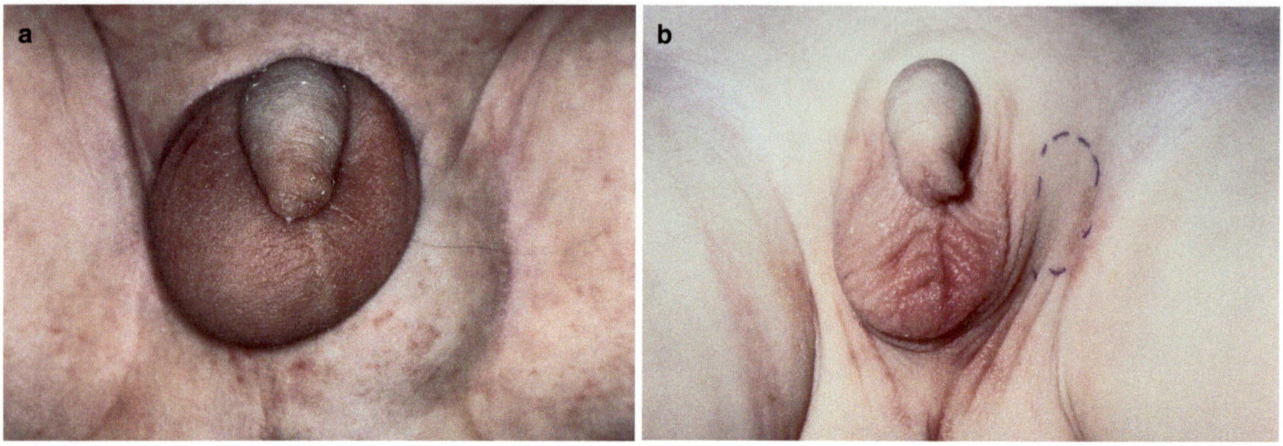

Fig. 46.5 (**a**) Perineal testis and (**b**) Inguinofemoral testis. In both circumstances, the gubernacular migration is adequate, but the direction of migration is defective

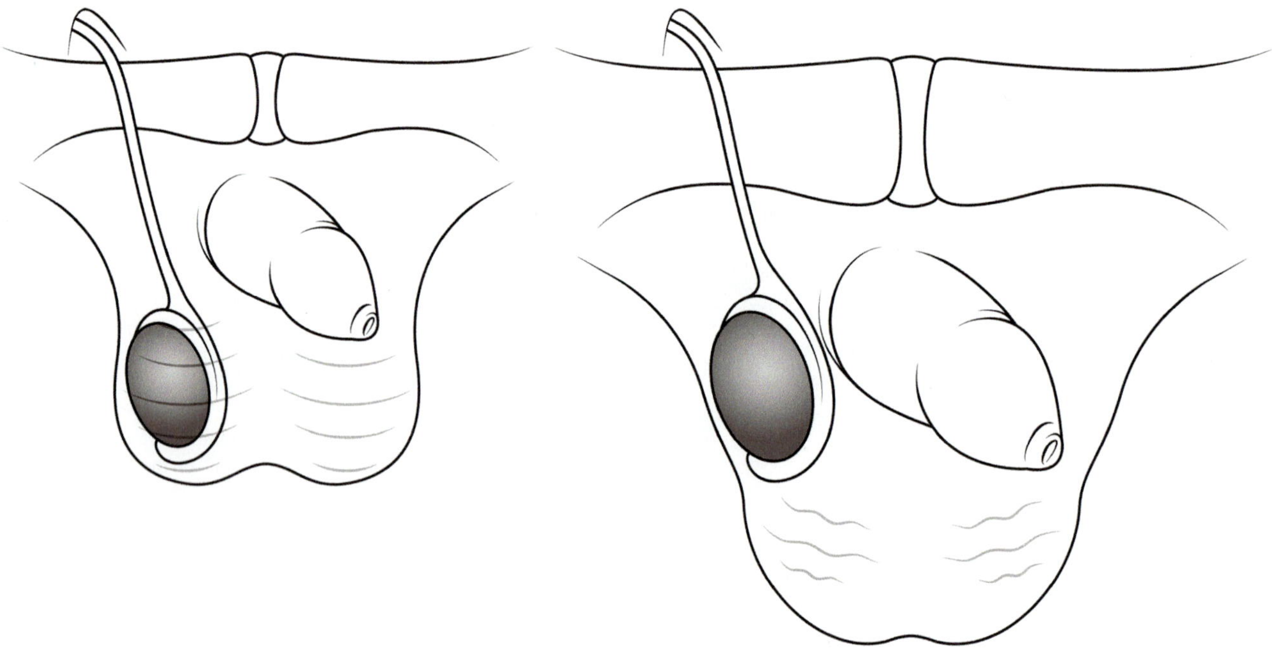

Fig. 46.6 A schema showing the likely cause of ascending, acquired undescended testis. At birth the distance between external inguinal ring and midscrotum is 4–5 cm, but by 10 years of age, this has increased to 8–10 cm. So-called ascending testes occur when the spermatic cord is unable to elongate because it contains a fibrous remnant of the PV (that should have undergone apoptosis and disappeared). The fibrous PV remnant cannot grow and prevents growth of the vas and vessels. Severe retractile testes are likely to be testes that are developing into ascending testes (i.e. they are in a 'grey zone' between normal and abnormal testicular position)

cause of absent testis, however, is ischaemic injury to the testis during descent (presumed result of torsion of the spermatic cord as the gubernaculum is migrating or just before it becomes adherent to the scrotum on completion of descent). More profound morphological causes of cryptorchidism may be related to adherence between the urogenital ridge on the left with the primordium of the spleen, leading to splenogonadal fusion. Rarely the defect in the urogenital ridge affects the caudal attachment of the gubernaculum to the anterior abdominal wall that leads to the 'inguinal canal' forming in the femoral canal or in a Spigelian hernia.

Common morphological anomalies of the gubernaculum and its enclosed peritoneal diverticulum, the processus vaginalis (PV), include failure of the PV to close at the completion of testicular descent, so the infant has an inguinal hernia or hydrocele.

Another common anomaly, not yet proven, is the possibility that acquired, ascending testes result in failure of the obliterated PV to undergo complete apoptosis, which leaves a fibrous remnant in the spermatic cord that does not elongate with age. This leaves the spermatic cord at the same length it was at birth, so that with postnatal growth of the boy, the testis gets 'left behind' as the scrotum becomes further from the groin (in millimetres) (Fig. 46.6).

In boys with cerebral palsy with upper motor neurone abnormality affecting the adductor muscles of the thigh (and the cremaster muscle), failure of normal elongation of the spermatic cord is caused by the abnormally high tone in the cremaster muscle.

Recommended Readings

1. Gurney JK, McGlynn KA, Stanley J, Merriman T, Signal V, Shaw C, Edwards R, Richiardi L, Hutson J, Sarfati D. Risk factors for cryptorchidism. Nat Rev. Urol. 2017;14(9):534–48.
2. Hutson JM, Thorup JM, Beasley SW. Descent of the testis. 2. Springer International Publishing; 2016.
3. Hutson JM, Balic A, Nation T, Southwell B. Cryptorchidism. Semin Pediatr Surg. 2010;19(3):215–24.
4. Hutson JM, Hasthorpe S. Abnormalities of testicular descent. Cell Tissue Res. 2005;322(1):155–8.
5. Hughes IA, Acerini CL. Factors controlling testis descent. Eur J Endocrinol. 2008;159(Suppl 1):S75–82.
6. Hutson JM, Lopez- Marambio FA. The possible role of AMH in shortening of the gubernacular cord in testicular descent: a reappraisal of the evidence. J Pediatr Surg. 2017;52(10):1656–60.
7. Punwani VV, Wong JSY, Lai CYH, Chia JCY, Hutson JM. Testicular ectopia: why does it happen and what do we do? J Pediatr Surg. 2017;52(11):1842–7.
8. Hutson JM, Baker M, Terada M, Zhou B, Paxton G. Hormonal control of testicular descent and the cause of cryptorchidism. Reprod Fertil Dev. 1994;6(2):151–6.

47 Congenital Abnormalities of the Testis and Epididymis

Prabhu Sekaran

Development of The Male Genital Ducts (Epididymis, vas Deferens, Seminal Vesicle and Ejaculatory Duct)

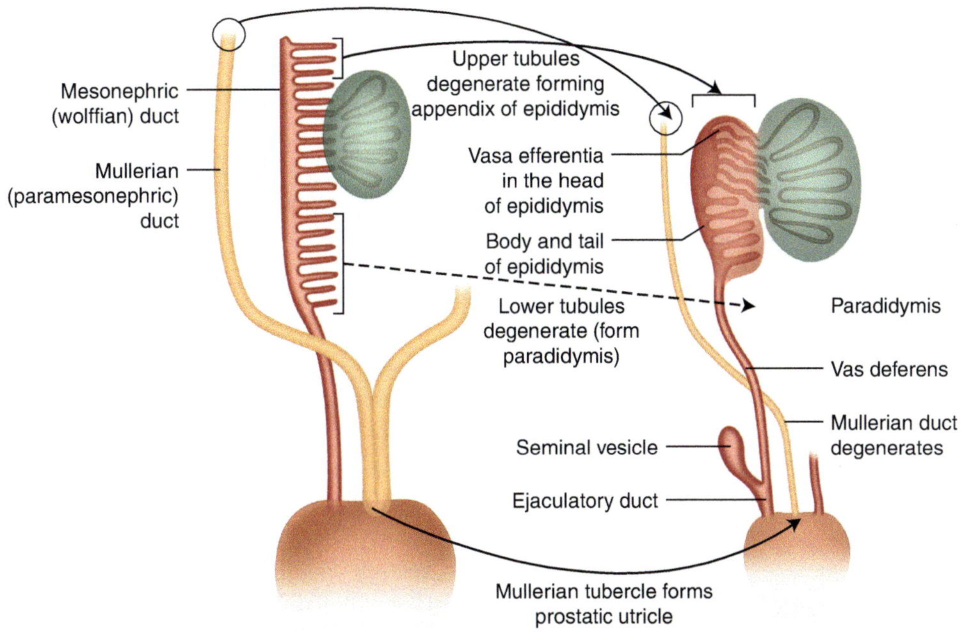

***The male genital ducts develop from the mesonefros as follows:**

1. ***The excretory mesonephric tubules*** adjoining the developing testis (6–12 tubules) will form the vasa efferentia (head of the epididymis) and become connected to the rete testis. The remaining mesonephric tubules above these vasa efferentia will degenerate forming appendix of epididymis while those below the testis will degenerate forming the paradidymis.

2. ***The mesonephric (Wolffian) duct***: differentiates into the following:
   - (a) The part below vasa efferentia becomes highly convoluted to form body and tail of epididymis.
   - (b) The part next to the tail of epididymis will acquire a very thick muscular wall and becomes the vas deferens.
   - (c) The lower part of the mesonephric duct gives out a pouch which forms the seminal vesicle.
   - (d) The lower end of the mesonephric duct below seminal vesicle will become the ejaculatory duct.
3. ***The Mullerian (paramesonephric) duct*** degenerates completely in the male except:
   - (a) Its upper end which gives the appendix of testis
   - (b) Its lower end which gives the prostatic utricle

P. Sekaran (✉)
Royal Hospital for Sick Children, Edinburgh, UK
e-mail: Prabhu.sekaran@nhs.net

R. Carachi, S. H. E. Doss (eds.), *Clinical Embryology*, https://doi.org/10.1007/978-3-319-26158-4_47

Testis

Testicular abnormalities such as anorchism and polyorchism are rare. Anorchism often presents as an impalpable testis, which may represent either as intra-abdominal testis or true testicular agenesis leading to anorchia. Diagnostic laparoscopy will determine if an intra-abdominal testis is present or whether there is anorchia. In anorchism often there is a blind-ending vas deferens and gonadal vessels. A vascular abnormality affecting the gonadal vessels leads to ischaemia and testicular atrophy and thus anorchia. This often occurs in utero during embryological development of the gonad.

Polyorchism is another rare entity and the current literature consists of 150 case reports. Polyorchidism is usually asymptomatic, but some patients may present with scrotal pain, scrotal swelling, hydrocele, epididymo-orchitis or testicular torsion. Clinical examination usually reveals that the testes are normally descended inside the scrotum. Polyorchidism is commonly associated with inguinal hernia (24%), undescended testis (22%) and microlithiasis. Triorchidism is the most common variant of this condition, where there are three testes with the accessory testis being situated on the left (in 65% of the cases). In 5% of patients with polyorchidism, there is bilateral involvement with four testes. The embryological aetiology of this condition is unclear, but it may arise due to division of the primordial germ cells at the genital ridge before the eighth week of development thus leading to an extra gonad. The accessory testis may have its own vas deferens and thus have reproductive potential or share a common vas deferens with its partner testis.

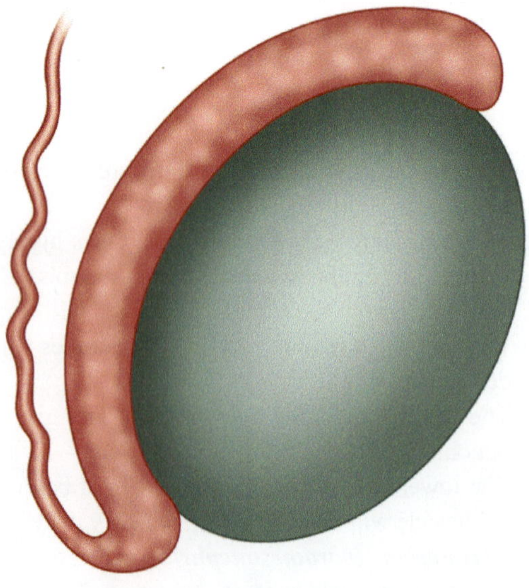

Fig. 47.1 Normal testis

The position of the testis within the scrotum may also be abnormal leading to epididymal abnormalities. In anterior inversion the relative position of the testis and the epididymis is reversed so that the epididymis lies in front of the testis. Polar inversion occurs when the epididymis—instead of having its epididymal head at the superior pole of the testis—is actually found at the base of the testis.

Appendages

There are five testicular appendages as shown in diagram 1. The testicular appendage is a remnant of the Mullerian duct (paramesonephric duct), whereas the other appendages are remnants of the Wolffian duct (mesonephric duct).

Appendages can undergo torsion and present with scrotal pain, and more than 80% of cases present between the ages of 7 and 14. On examination the scrotum will be tender, erythematous and oedematous. Differentiation between testicular torsion and torsion of an appendage may be difficult to determine on clinical examination. However, a necrotic appendage may be visible as a blue dot under the scrotal skin thus aiding diagnosis (Fig. 47.3).

Vas Deferens

Abnormalities of the vas deferens are often encountered during herniotomy. The vas deferens may be absent or hypoplastic. The surgeon may be concerned that they have unwittingly damaged the vas during dissection. Aplasia and hypoplasia of the vas deferens are associated with renal agenesis and cystic fibrosis. In cystic fibrosis the underlying pathogenesis causes tubular obstruction of the vas leading hypoplasia or absence and thus infertility. Maldevelopment of the renal tract during metanephrogenesis may lead to renal agenesis. This may occur due to failure of reciprocal induction of the metanephric mesenchyme and the ureteric bud. As gonadal and Wolffian duct development occurs in the same region as metanephric development, any local maldevelopment of the renal tract may also lead to maldevelopment of the vas deferens leading to hypoplasia or aplasia.

Cysts

The most common paratesticular mass is an epididymal cyst, which is also referred to in the literature as a spermatocele. These cysts are located at the head of the epididymis and are usually asymptomatic. These cysts arise as a result of tubular obstruction after trauma or epididymo-orchitis. The resultant

Fig. 47.2 (**a**) Inverted testis, (**b**) horizontal testis, (**c**) double testis

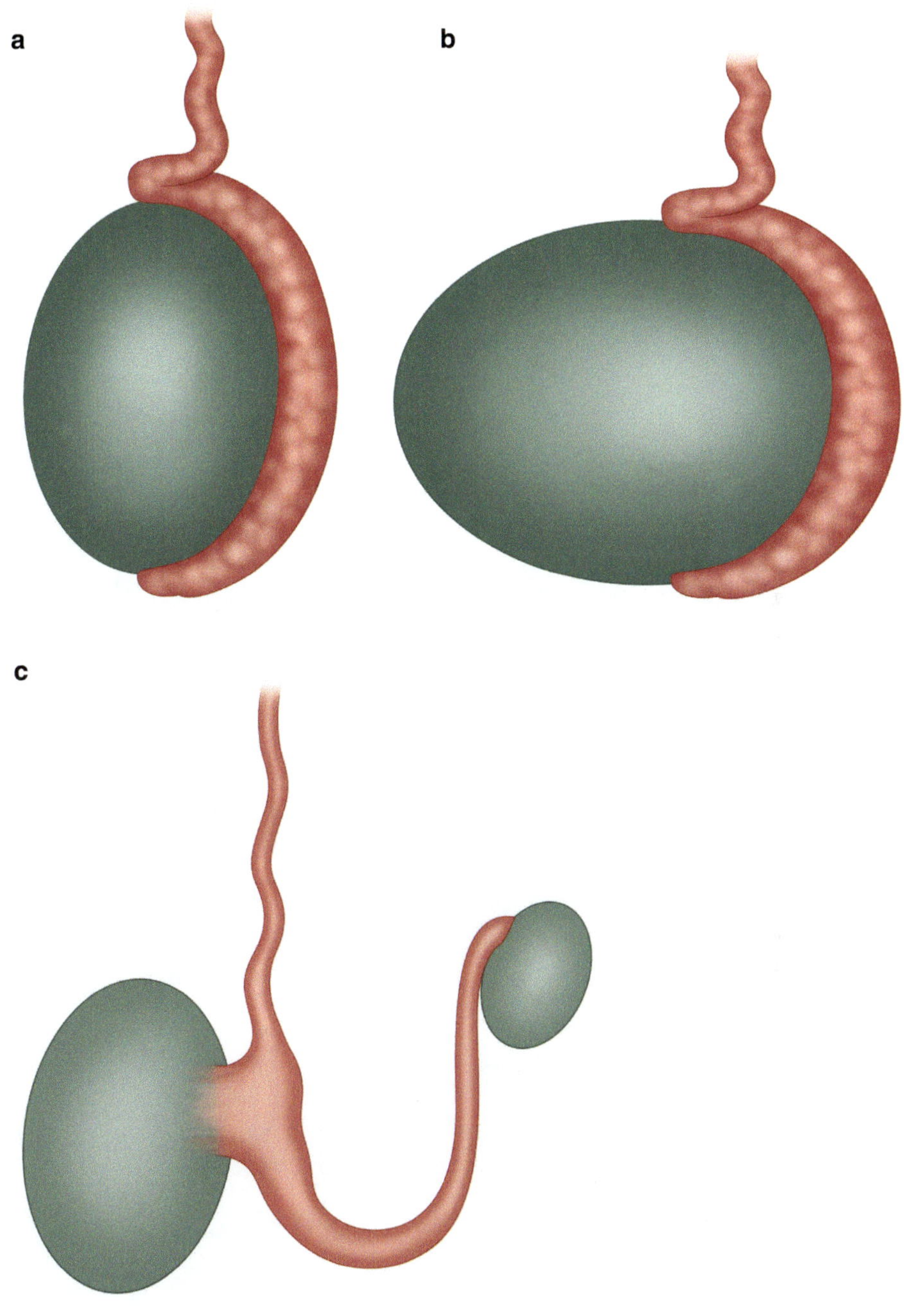

cyst may be filled with spermatozoa hence the synonym “spermatocele”. Cysts need to be differentiated from other paratesticular masses such as a rhabdomyosarcoma or testicular masses such as a germ cell tumour. Ultrasonography is usually discriminatory in diagnosis. If the patient has symptoms, then surgical excision of epididymal cysts is curative.

Testicular epidermoid cysts are benign cysts, which are found within the parenchyma of the testis and account for 1% of all testicular masses. They usually present as painless testicular masses. They are composed of layers of keratinous debris lined with keratinising squamous epithelium.

Calcification and Microcalcification of the Testis

Testicular microlithiasis is usually an incidental finding in boys between 10 and 50 years of age. The incidence is unknown, but in paediatric autopsy specimens, reports have

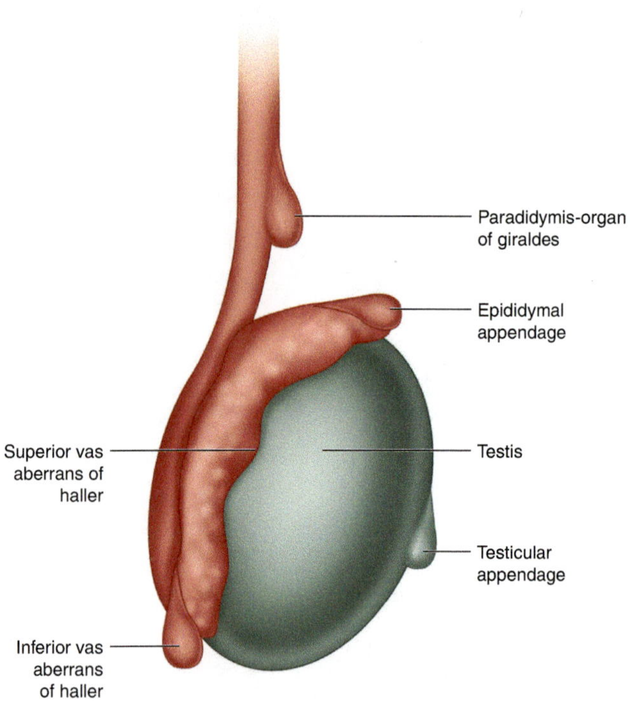

Fig. 47.3 Appendages of the testis, epididymis and vas

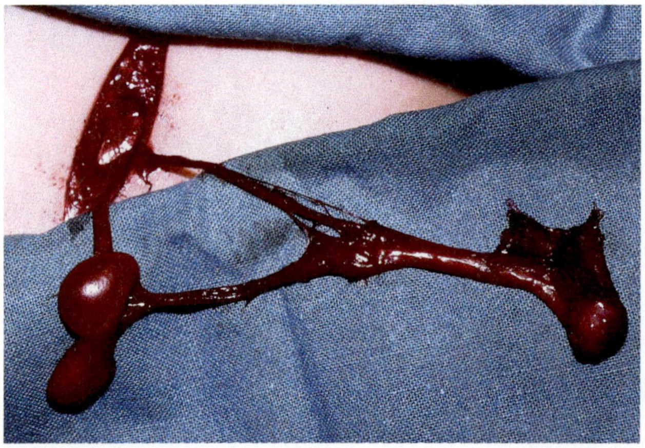

Fig. 47.4 Double testis

suggested the occurrence is from <1 to 5%. Most patients are asymptomatic, but some boys may present with chronic orchialgia. Testicular microcalcification has been associated with various testicular pathologies including germ cell malignancy, varicocele and cryptorchidism. Due to the association with malignancy, careful clinical examination is recommended with serum tumour marker levels (b-HCG, AFP) and imaging of the testes. Ultrasound of the affected testis will demonstrate hyperechoic foci of 1–2 mm in diameter in both testes. The surrounding peritesticular structures are normal. Careful follow-up of patients is suggested; however if there is any suggestion of malignancy, then a biopsy is recommended.

Fig. 47.5 Albrecht von Haller

Fig. 47.6 Joachim A. C. C. Giraldés

Historical Biography

Giovanni Battista Morgagni (1682–1771) was an Italian anatomist and pathologist whose works helped make pathological anatomy an exact science. After graduating in 1701 at Bologna with degrees in philosophy and medicine, Morgagni acted as prosector to A.M. Valsalva and then succeeded him in his position as anatomy demonstrator. In 1710 he moved to Padua, to become Professor of Medicine, and in 1715 he became Professor of Anatomy.

In 1761 Morgagni published his greatest work, *De Sedibus et Causis Morborum per Anatomen Indagatis* (*The Seats and Causes of Diseases Investigated by Anatomy*), which marked him as a founder of morbid anatomy. The work contains records of 640 dissections. The testicular appendage is named after him as well as an eponymous diaphragmatic hernia (Morgagni hernia—retrosternal diaphragmatic hernia).

Albrecht von Haller (1708–1777) was a Swiss biologist and is referred to as the father of experimental physiology. At the University of Göttingen (1736–1953), where he served as professor of medicine, anatomy, surgery and botany, Haller wrote his seminal work *Elementa Physiologiae Corporis Humani* (8 vol., 1757–1966; *Physiological Elements of the Human Body*).

Haller was the first to recognize the mechanism of respiration and the autonomous function of the heart; he discovered that bile helps to digest fats, and he wrote original descriptions of embryonic development. He also summarized anatomical studies of the genital organs, the brain and the cardiovascular system.

Joachim A. C. C. Giraldés (1808–1875) was a Paris surgeon from Portugal who died of a wound acquired during an autopsy.

The Limbs

48

Laura Burton and Robert Carachi

Development of the Limbs

1. ***Development of limb buds*:**
   (a) ***Time*:** They appear at the beginning of the fifth week.
   (b) ***Site*:**-The U.L. bud develops opposite the segments C4 to T2.
   The L.L. bud develops the lower four lumbar and upper 3 sacral segments
   (c) ***Shape*:** Each limb bud is flattened, having two borders:
   - Preaxial (cranial) border: marked by the thumb or big toe.
   - Postaxial (caudal) border: marked by the little finger or little toe.

   (d) ***Structure*:** Each limb bud is composed of mass of mesoderm covered by ectoderm. The mesoderm is derived from (a) lat-plate mesoderm and (b) migrating myotomes

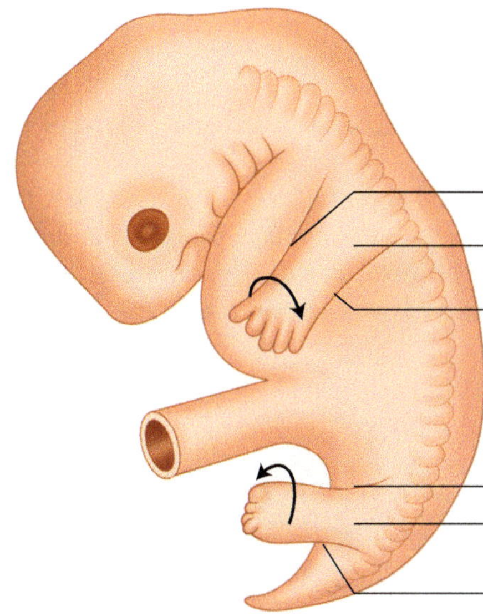

2. ***Differentiation of the mesoderm into bones and muslces*:**
   (a) The central (axial) mesoderm forms cartilaginous skeleton which will ossify into bones.
   (b) The surrounding mesoderm (derived from myotomes) gives the muscles as follows:
   - The ventral mesodermal mass develops into the flexors and adductor muscles and derives its nerve supply from the ant-divisions of the ventral 1ry rami of spinal nerves.
   - The dorsal mesodermal mass develops into the extensor and abductor muscles and are supplied by the post-divisions of the ventral 1ry rami of spinal nerves.

L. Burton
Queen Elizabeth University Hospital, Glasgow, UK

R. Carachi (✉)
University of Glasgow, Glasgow, UK
e-mail: Robert.Carachi@glasgow.ac.uk

R. Carachi, S. H. E. Doss (eds.), *Clinical Embryology*, https://doi.org/10.1007/978-3-319-26158-4_48

3. ***Segmentation of the limb bud*****:**
   Creases appear in each bud dividing it into three segments
   arm, forearm and hand (in the U.L.)
   thigh, leg and foot (in the L.L.)

4. ***Rotation of the limb buds*****:** occurs in the seventh and eighth weeks as follows:
   (a) The U.L. is adducted and rotated laterally and thus the preaxial border and thumb become directed laterally.
   (b) The L.L. is adducted and rotated medially and thus the preaxial border & the big toe become directed medially.

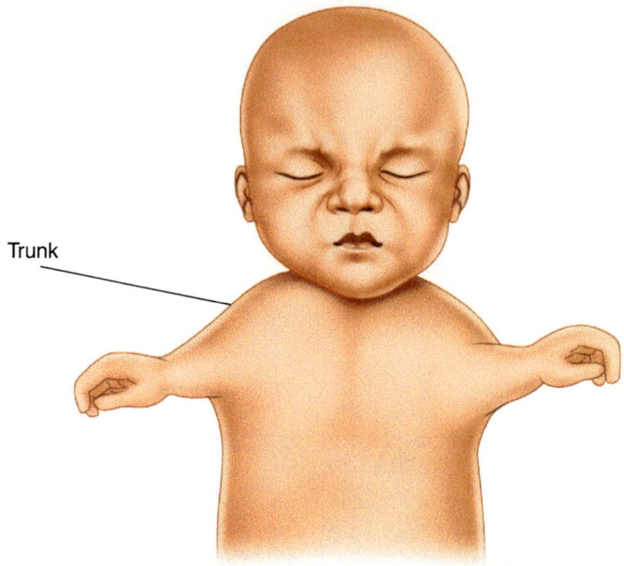

Congenital Anomalies of the Limbs

1. ***Amelia*****:** complete failure of development of one (or more) limb
2. ***Focomelia*****:** the hand or the foot is attached directly to the trunk due to absence of the proximal part of the limb
3. ***Polydactyly*****:** the presence of extra fingers or toes
4. ***Syndactyly*****:** fusion of two fingers or toes

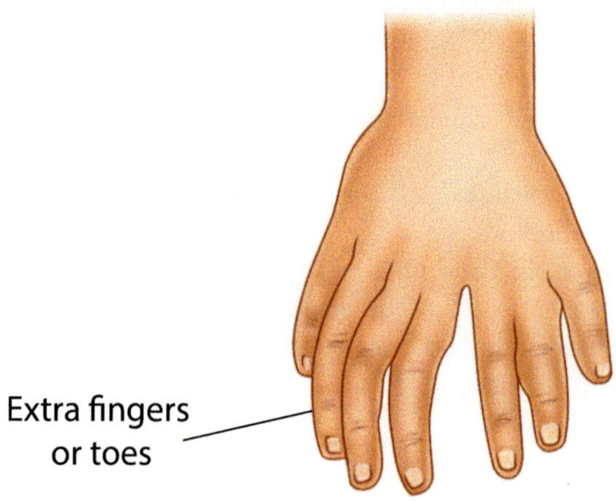

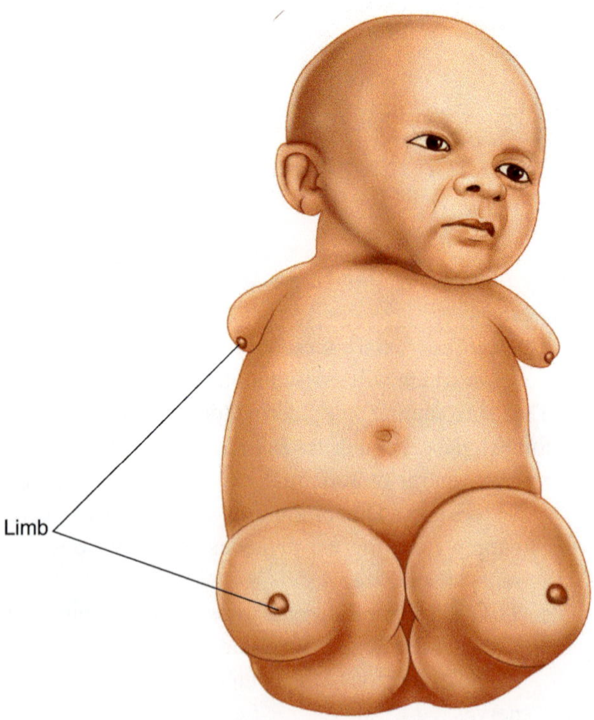

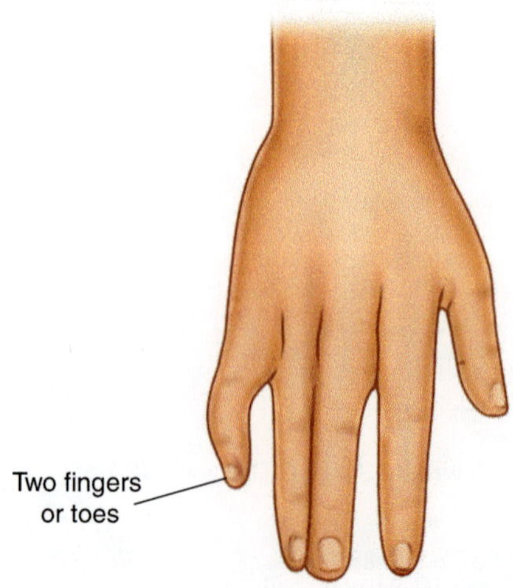

Limb Development and Growth

At the end of week 4 of development, the limb buds first become visible. The upper limb buds appear first as ridges in the ventrolateral body wall opposite segments C4 to T2 followed by the appearance of small bulges opposite the lower four lumbar and upper three sacral segments which will form the lower limb. The limb buds are comprised of a mesenchymal core covered by a layer of cuboidal ectoderm. This mesenchyme is derived from the parietal layer of lateral plate mesoderm, as well as a contribution from the migrating myotome (Fig. 48.1).

The ectoderm at the distal edge of the developing limb begins to thicken to form the apical ectodermal ridge (AER) which forms an inductive relationship with the mesenchymal core. The AER appears between the dorsal and ventral limb ectoderm and is a key signalling centre, without which limbs fail to develop normally. Limb outgrowth is driven by the interactions between the mesenchyme and ectoderm. The cells directly adjacent to the AER, known as the progress zone, remain undifferentiated, whilst mitosis is stimulated. As growth progresses, cells furthest from the AER begin to differentiate.

By the sixth week, the terminal portion of each limb bud becomes flattened and paddle shaped and become distinguishable as either a hand plates or footplate. Borders appear—firstly the preaxial or cranial border, marked by the thumb or big toe, and the postaxial or caudal border, marked by the little finger or little toe. Development of the limb proceeds in a proximodistal direction. Creases appear in each bud which divide limbs into three structural segments

- Upper limb bud—upper arm, forearm and hand
- Lower limb bud—thigh, leg and foot

The proximal segments differentiate first, with the distal segments remaining in an undifferentiated state due to the influence of FGFs and Wnts.

Cell death is an important mechanism in development. In the AER, apoptosis in specific regions of the developing hand plates and footplates enables separation of the structure into digits. Between the forming digits, the AER regresses, but the distal tips with the surviving AER continue to outgrow. Separation should be complete by day 56.

Rotation of the Limb Buds

During week 7, limb rotation occurs and the developing limbs rotate to different degrees and in opposite directions, first from the coronal to the parasagittal plane then along the long axis of the embryo. Initially, the flexor surface of the limb is ventral with the extensor surface dorsal. The upper limb is adducted and rotated 90° laterally; thus the preaxial border and thumb becomes directed laterally with the elbows pointing backwards. The lower limb is adducted and rotated 90° medially resulting in the postaxial border and big toe to become directed medially with the knees facing forwards.

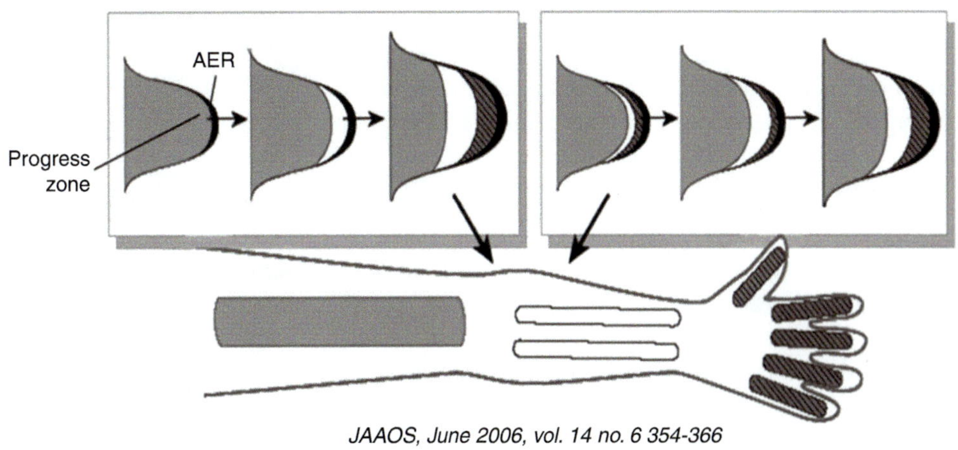

Fig. 48.1 Limb buds

Stages of Limb Bone and Muscle Development

Formation of the skeleton can first be observed as mesoderm in the developing limb buds becomes condensed in the proximal part of the limb. The ectoderm in the limb bud inhibits cartilage differentiation in the region below the ectoderm. Chondrocyte differentiation from mesenchymal stem cells begins, and differentiation occurs in a highly ordered proximodistal sequence with digit formation occurring from the fifth to the first digit. Hyaline cartilage models are laid down to establish the cartilaginous skeleton which will eventually ossify into bone. The surrounding mesoderm derived from myotomes gives the muscles as follows:

1. Ventral mesodermal mass develops into the flexor and adductor muscles and derives its nerve supply from the anterior divisions of the ventral rami of spinal nerves.
2. Dorsal mesodermal mass develops into the extensor and abductor muscles and is supplied by the posterior divisions of the ventral rami of spinal nerves.

Congenital Abnormalities of the Limbs

Limb anomalies are common and vary in severity. Due to the visibility of the defects, they are among the most commonly reported congenital defects. Minor defects are common, with major limb malformations relatively rare. Limb malformations are often associated with craniofacial, cardiac and/or genitourinary defects. The majority of congenital abnormalities of the limb tend to occur during the period of rapid limb development between weeks 6 and 0.8. Causes tend to be attributed to chromosomal abnormalities and environmental factors, e.g. teratogenic insult, or a combination of both. In addition, mechanical factors such as reduced amniotic fluid can also contribute to deformities.

Types of Deficiencies

Types of deficiencies can be classified in many ways. The most widely accepted worldwide classification now defines anomalies into seven categories (Table 48.1) which will be discussed below.

Table 48.1 Classification of limb abnormalities

Group I	• Failure of formation
Group II	• Failure of differentiation
Group III	• Duplication
Group IV	• Overgrowth
Group V	• Undergrowth
Group VI	• Constriction band syndromes
Group VII	• Generalised anomalies and syndromes

IFSSH International Federation of Societies for Surgery of the Hand

Partial or complete failure of development of one or more limbs is an example of failure of formation. Severity varies; sometimes the long bones are absent with the hand or foot directly attached to the trunk due to the absence of proximal limb structures, e.g. phocomelia, or all limb structures are absent in one or more of the extremities as seen in cases of amelia. Amelia is often associated with other malformations including cleft lip/palate. Failure of formation is rare and mainly heredity but can be teratogen induced as seen with infants born after thalidomide exposure in utero.

Sirenomelia, where the lower limbs are partially or completely fused, is due to failure of differentiation. Affected infants may have one foot, no feet or both feet, which may be rotated externally. The sacrum is often partially or completely absent, and the disorder is often fatal. Syndactyly, fusion of fingers or toes, is similarly the result of failed differentiation and absence of apoptosis in the fused area.

Duplication

Polydactyly—presence of extra fingers or toes

Overgrowth and undergrowth

Hemihypertrophy

Constriction Band Syndrome

General Anomalies and Syndromes

Club foot

Congenital hip dislocation

Advances in limb embryogenesis and the molecular regulation of development are improving our ability to detect and manage limb abnormalities (Table 48.2). Currently, only a small proportion have been mapped to specific chromosomal locations, but further identification of genetic mutations will

Table 48.2 Principles of management

Failure of formation	• Prosthesis, microvascular transfer of phalanges.
Failure of differentiation	• Separation.
Duplication	• Excision.
Overgrowth	• Epiphysiodesis. total physeal resection, debulking. partial amoutation.
Undergrowth	• Pollicization (thumb), web space deepening, metacarpal transposition, phalangeal transfer, bone graft.
Constriction band syndromes	• Prosthesis (amputation), excision of groove and functionless tissue (ring constrictions).

improve genetic testing and diagnosis of such disorders. Improved ultrasound examination of the developing foetus will advance in utero detection and surgery and will hopefully prove to be a major step forwards in the diagnosis and treatment of limb conditions in the near future.

Clinical Embryology of the Limbs

The development of the limbs is a complex process which begins at the end of week 4. It can be simply split into five sections.

1. Development of limb buds
2. Differentiation of the mesoderm into bones, muscles and joints
3. Differentiation of the ectoderm into the digits
4. Segmentation of the limb bud
5. Rotation of the limb buds

An error or problem at any of these stages can lead to serious congenital malformation of the limbs.

Development of Limb Buds

The limb buds appear on the ventrolateral body wall at the end of week 4. They have their basic structure by 8 weeks. The upper limb bud develops opposite the C4 to T2 spinal segments, while the lower limb bud develops opposite the L1 to S3 segments. They are initially flat structures with an upper and lower border marked by the thumb or big toe and the little finger or little toe, respectively. Each limb bud is composed of a mass of mesoderm covered by ectoderm. The mesoderm is derived from (a) lateral plate mesoderm and (b) migrating myotomes.

Differentiation of the Mesoderm and Ectoderm into Bones, Muscles and Joints

Morphogenesis of lower limbs is 1–2 days behind that of upper limbs. The central mesoderm forms a cartilaginous skeleton which will ossify into bones. Primary endochondral ossification centres appear in the diaphysis of long bones at week 12 of development and promote growth towards ends of the bone. At birth the diaphysis is completely ossified with the epiphyses forming new cartilaginous secondary ossification centres. It does this via epiphyseal plates, temporary cartilaginous plates between the diaphysis and epiphysis of the bones. Endochondral ossification occurs on both sides of plate though long bones have one plate at both ends while smaller bones only have one plate at one end. The epiphyseal plate lengthens bones but disappears when the bone achieves full length allowing the epiphyses to fuse with the diaphysis. They epiphyseal plate can be seen on X-rays of children and should not be confused with a fracture.

Hyaline cartilage models appear at ends of bones by week 6 of development formed by chondrocytes from the mesoderm. Cartilaginous condensations and arrest of chondrogenesis create a joint interzone. In the joint interzone, there is an increased cell number and density. Cell death in this interzone creates the joint cavity, while cellular differentiation on the periphery creates the joint capsule.

The surrounding mesoderm (derived from myotomes) develops into the muscles. The ventral mesodermal mass develops into the flexor and adductor muscles as the dorsal mesodermal mass differentiates into the extensor and abductor muscles.

Differentiation of the Ectoderm into the Digits

The epithelial ectoderm begins to thicken at the distal border of the limb bud to form the apical ectodermal ridge (AER). The AER promotes the growth and development of the limbs in the proximodistal axis to form the fingers and toes. It does this by inducing surrounding mesenchyme to form the progress zone (PZ) of undifferentiated, rapidly proliferating cells. Hand plates and footplates appear in week 6 of development as flattened terminal portions of limb buds. Cell death (apoptosis) in AER divides the ridge into five parts with mesenchymal condensation forming cartilaginous digital rays which continue to grow. At the beginning of week 8, the digits of the hand are short and webbed. However, at the end of week 8 due to further apoptosis, there are distinct regions in the limbs, with long fingers and distinct toes.

Segmentation of the Limb Bud

Creases appear in each limb bud dividing them into three segments. The arm, forearm and hand in the upper limb and the thigh, leg and foot in the lower limb. First one constriction separates the hand plates and footplates from the proximal segment in week 6 of development, and a second one divides the proximal portion into two segments.

Rotation of the Limb Buds

Rotation of the limb buds occurs in the seventh and eighth weeks of development. The upper limb is adducted and rotated laterally with the thumb directed laterally, while the lower limb is adducted and rotated medially with the big toe directed medially.

Congenital Anomalies of the Limbs

Congenital anomalies of the limbs affect approximately 6:10000 live births. More specifically, upper limb anomalies affect 3.4: 10000 live births, and lower limb anomalies affect 1.1: 10000 live births. They are often associated with craniofacial, cardiac and/or genitourinary defects. Antenatal screening can often detect these abnormalities allowing referral for treatment, specialist delivery or termination according to the parents' wishes and medical opinion (Table 48.3).

Risk factors for congenital limb anomalies can be intrinsic or extrinsic as shown below:

The most sensitive period for teratogen-induced limb defects is weeks 4–5 of development. Examples include thalidomide, an anti-emetic once used in pregnancies from 1957 to 1962 which caused a syndrome of limb defects, intestinal atresia and cardiac anomalies. Use of thalidomide has returned for treatment of AIDS and cancer. Other teratogen-induced limb defects include warfarin, phenytoin, valproic acid, alcohol and cocaine (Figs. 48.2 and 48.3).

There is an international classification system for congenital anomalies of the limbs [1, 2] which is divided into the following seven groups:

1. Failure of formation
2. Failure of differentiation
3. Duplication
4. Overgrowth
5. Undergrowth
6. Constriction band syndrome
7. Generalised anomalies and syndromes

Table 48.3 Risk factors for congenital limb anomalies

Intrinsic	Extrinsic
• Chromosomal abnormalities	Teratogens
– Inherited	Nutrient deficiency (e.g. folate)
– Sporadic	Infections (VACTERL)
• Constriction bands	Failed abortion
	Removal of IUD

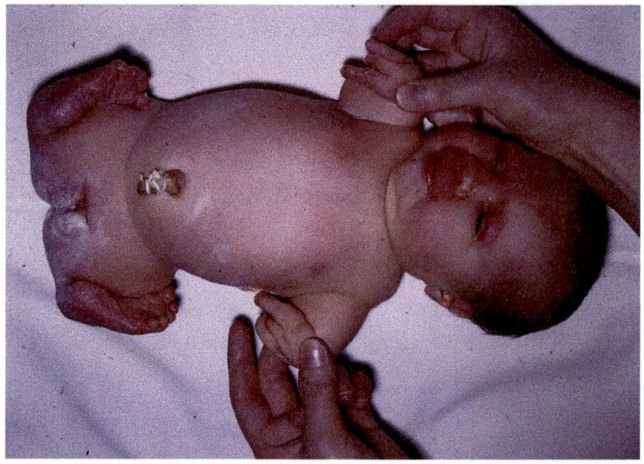

Fig. 48.2 Phocomelia thalidomide

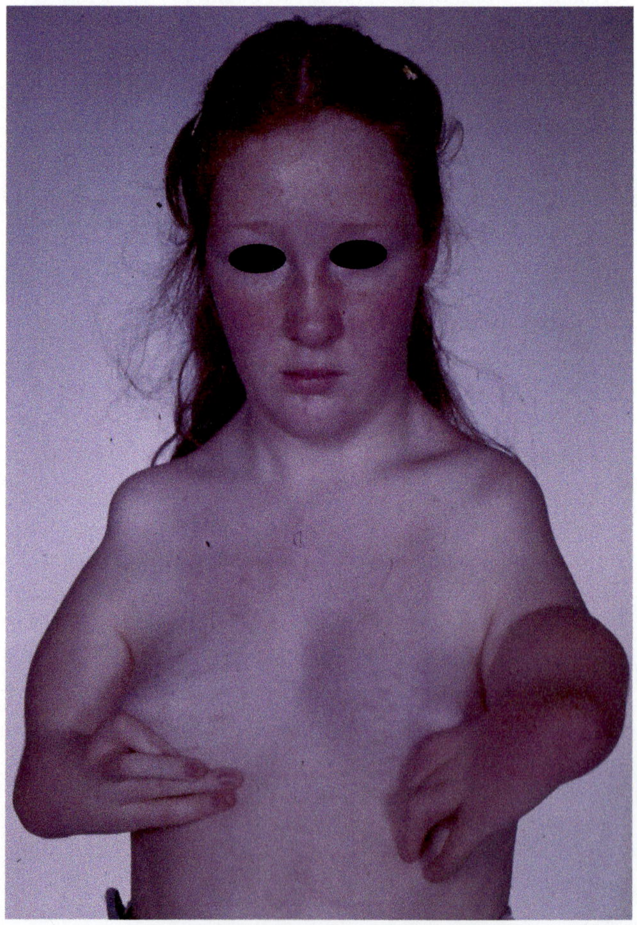

Fig. 48.3 Phocomelia thalidomide

We will now give examples of some common congenital anomalies from several of the groups listed.

In group I, failure of formation, there are conditions such as amelia, meromelia and phocomelia. Amelia is complete absence of one of more limbs [picture]. Meromelia is the partial absence of one or more limbs. Phocomelia is a type of meromelia caused by absence of long bones with rudimentary hands and feet attached directly to the trunk. In Greek, 'phoco' means seal or flipper and 'melia' means limb. Also seen in group I is split-hand/split-foot malformation (SHFM) or ectrodactyly, also known as "lobster claw hand". SHFM involves median clefts of the hands and feet with associated syndactyly, aplasia and/or hypoplasia of the phalanges, metacarpals and metatarsals (third metacarpal and phalangeal bones usually absent.). Its incidence has been reported to be about 1 in 90,000 babies affecting both sexes equally (Fig. 48.4) [3].

In group II, there are conditions such as sirenomelia or mermaid syndrome which presents with complete or partial fusion of the lower limbs due to failure of differentiation. The child may be born with one foot, no feet or both feet and may also have absence of their sacrum and coccyx. The specific cause is unknown with most cases occurring sporadically. The condition is often fatal in the newborn period and is commonly associated with other abnormalities such as spina bifida, imperforate anus, renal abnormalities and cardiac malformations. The exact incidence is unknown, but sirenomelia is estimated to occur in approximately 1 in 60,000 to 100,000 births with a male to

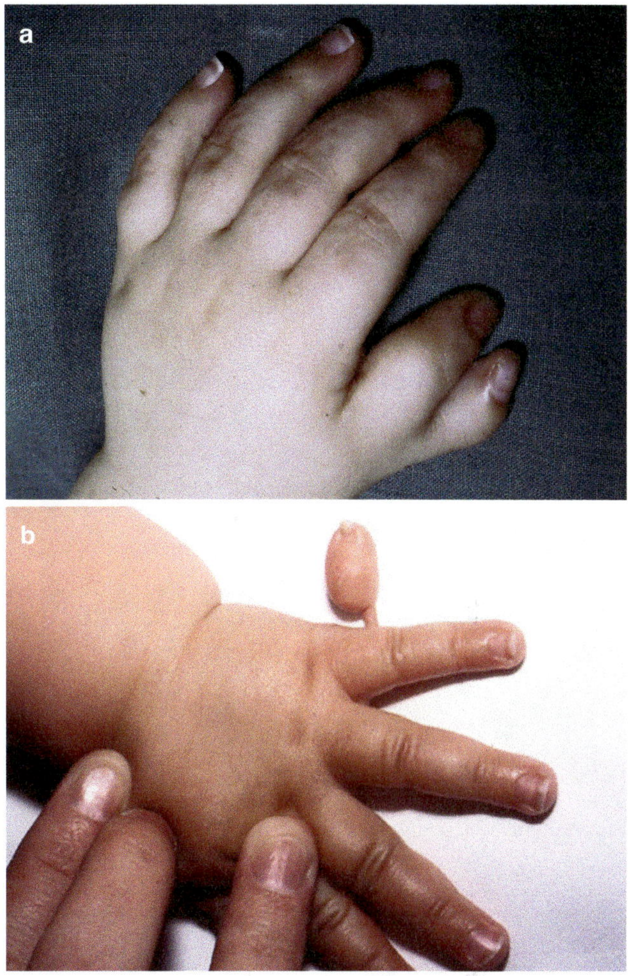

Fig. 48.5 (**a**) Polydactyly. (**b**) Accessory digit

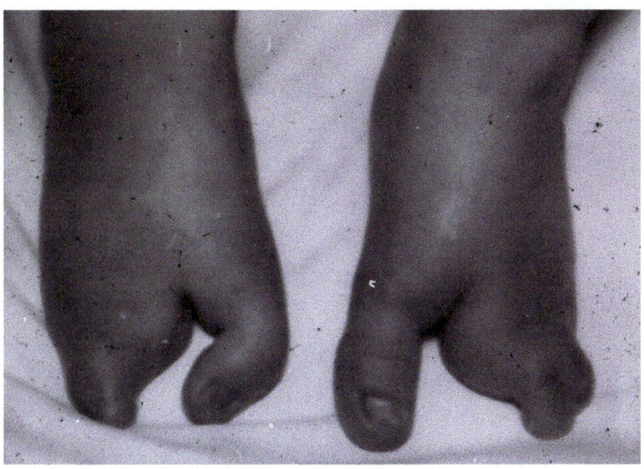

Fig. 48.4 Lobster claw

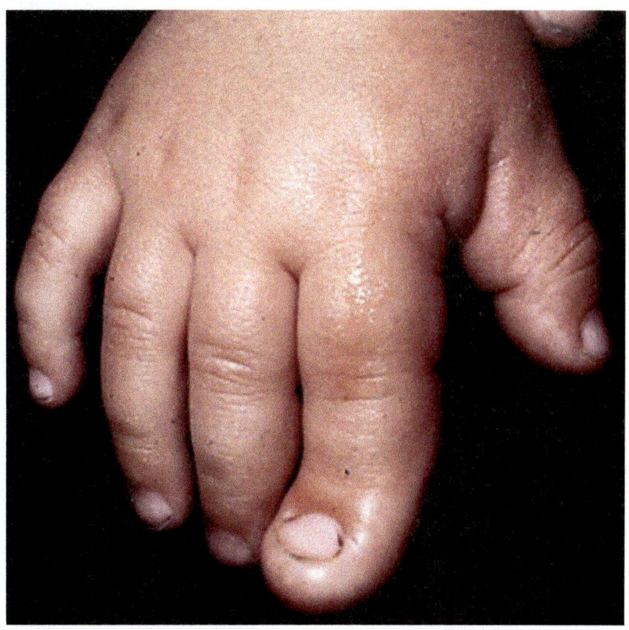

Fig. 48.6 Macrodactyly

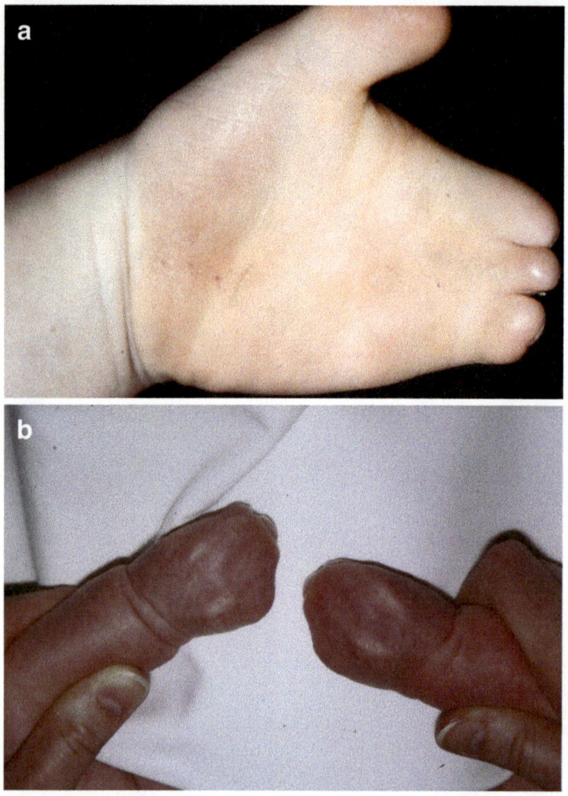

Fig. 48.7 (a) Syndactyly. (b) Apert syndrome

female bias of 2.7–1.0 [4]. The diagnosis can be made in the prenatal period by foetal ultrasound scans. Surgery has been successful in separating joined legs. In preparation for surgery, balloon-like tissue expanders are inserted under the skin; they are filled with a salt solution over a period of time expanding the balloons making the skin stretch and grow. Excess skin is then used to cover the legs once they are separated [1]. Another condition in group II is syndactyly, which is fusion of two fingers or toes due to failure of apoptosis of the AER in the hand plates and footplates. Syndactyly is seen in both Apert and Poland syndromes (Figs. 48.5a, b and 48.6).

In group III, we see duplication conditions such as polydactyly. Polydactyly is the presence of extra fingers or toes and is the most common congenital digital anomaly of the hand and foot. It may appear in isolation or in association with other birth defects [5]. Isolated polydactyly is often autosomal dominant or occasionally random, while syndromic polydactyly is commonly autosomal recessive. Surgical treatment depends on the complexity of the deformity and the functional and cosmetic implications for the patient (Fig. 48.7a and b).

Amniotic Bands

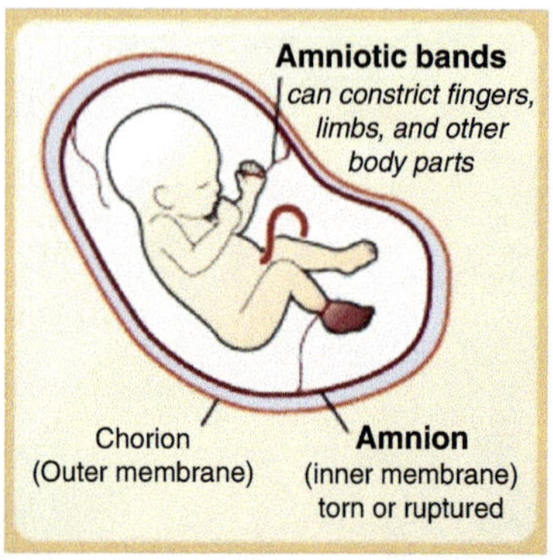

- AKA Streeter dysplasia
- Ring constriction and/or amputation of limbs (or digits).

- **Pathogenesis:**
  - Adhesions between amnion and foetal structures?
  - Tears in the amnion that separate and surrond part(s) of the foetus?

In group VI, congenital constriction band syndrome (CCBS) also known as streeter dysplasia or amniotic band syndrome can lead to amputation at any level. CCBS is a sporadic condition that may also be present in association with other congenital anomalies such as musculoskeletal, craniofacial and abdominal disorders. It has an incidence of one in 1200 to one in 15,000 live births and affects both sexes equally [6]. CBBS is most likely to affect the upper limbs and is more common at the distal extremities. CCBS is an intrauterine phenomenon probably caused by the rupture /tear of amniotic membranes which then fuse and constrict the developing foetal tissue. If severe, a child can have an in utero amputation of the limb or digit affected by the constriction band. No medical treatment exists for the condition. Avoiding certain drugs that can lead to spontaneous rupture of membranes, such as cocaine and mifepristone, may help decrease the potential risk [6]. Due to tight constrictions on the digits or extremities, urgent surgical treatment is often necessary for patients with vascular compromise soon after birth. In bands identified by 3-D ultrasound to cause neurovascular compromise, early in utero fetoscopic surgery for release can be performed. Multiple series of case reports

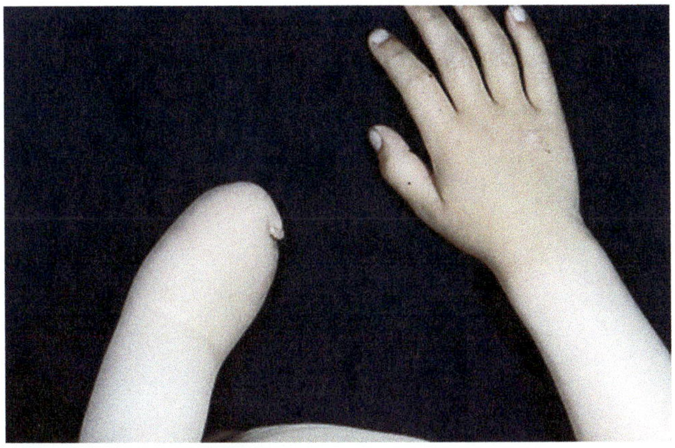

Fig. 48.8 Amputation intrauterine

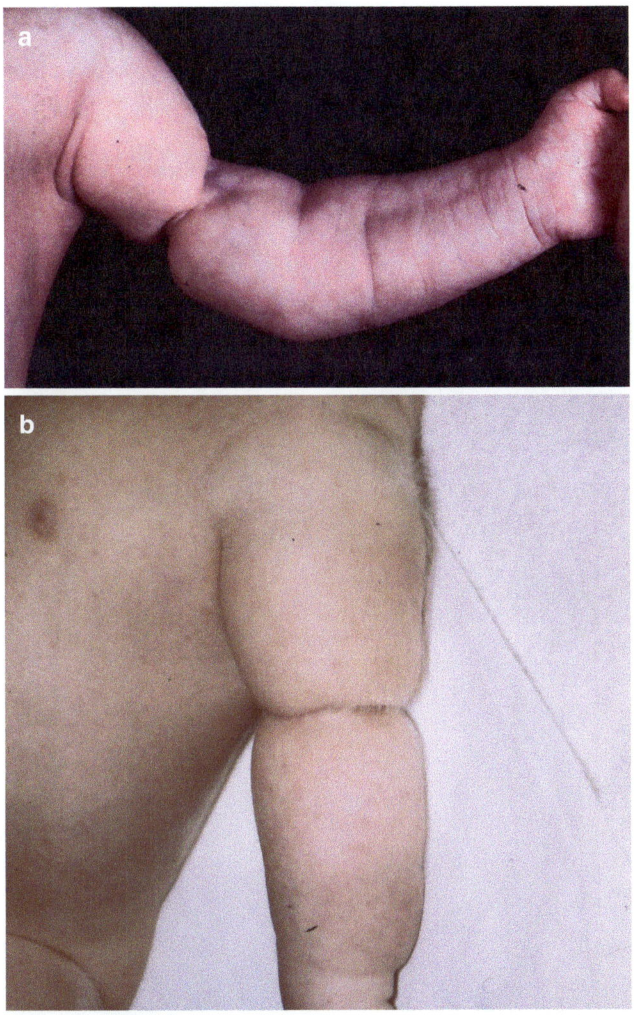

Fig. 48.9 (**a**, **b**) Constriction band of arm

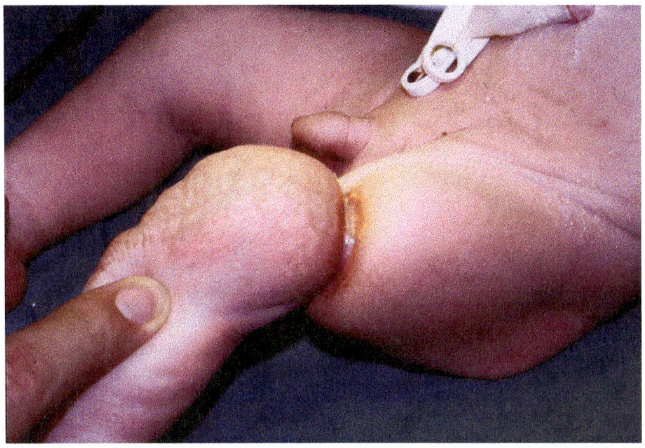

Fig. 48.10 Constriction band of leg

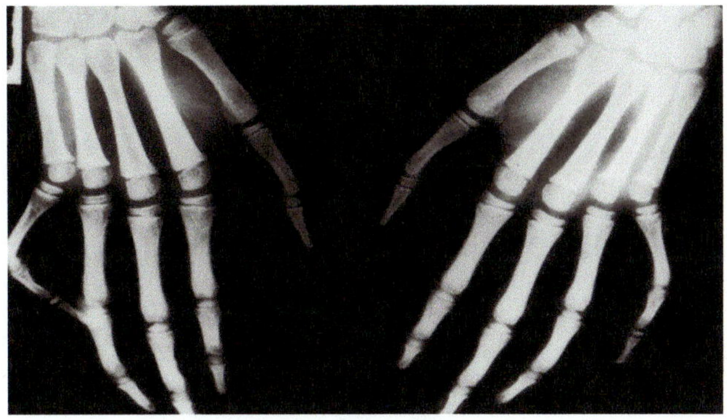

Fig. 48.11 Marfan's syndrome

have documented the effectiveness fetoscopic releases for the foetus and its relative safety of the procedure for the mothers (Figs. 48.8, 48.9a, b, and 48.10) [6].

Generalised anomalies and syndromes in group VII include achondroplasia. This is an autosomal dominant mutation in the fibroblast growth factor receptor-3 (FGF-R3) gene on chromosome 4p16.3, causing a defect in cartilage-derived bone. The condition occurs in 1 in 15,000 to 40,000 newborns [7]. Children present with macrocephaly, frontal bossing, midface hypoplasia, small chest, rhizomelic limb shortening (disproportion of the length of the proximal limb), redundant skin folds, joint laxity and short stature. Other examples include Marfan's syndrome (Fig. 48.11). This is an autosomal dominant disorder of connective tissue due to mutations in the fibrillin 1 (FBN 1) gene on chromosome 15q21.1. Patients present with tall stature, long thin digits (arachnodactyly), long thin limbs, hyperextensible joints, a high arched palate, dislocation (usually upwards) of the lenses of the eyes and severe myopia.

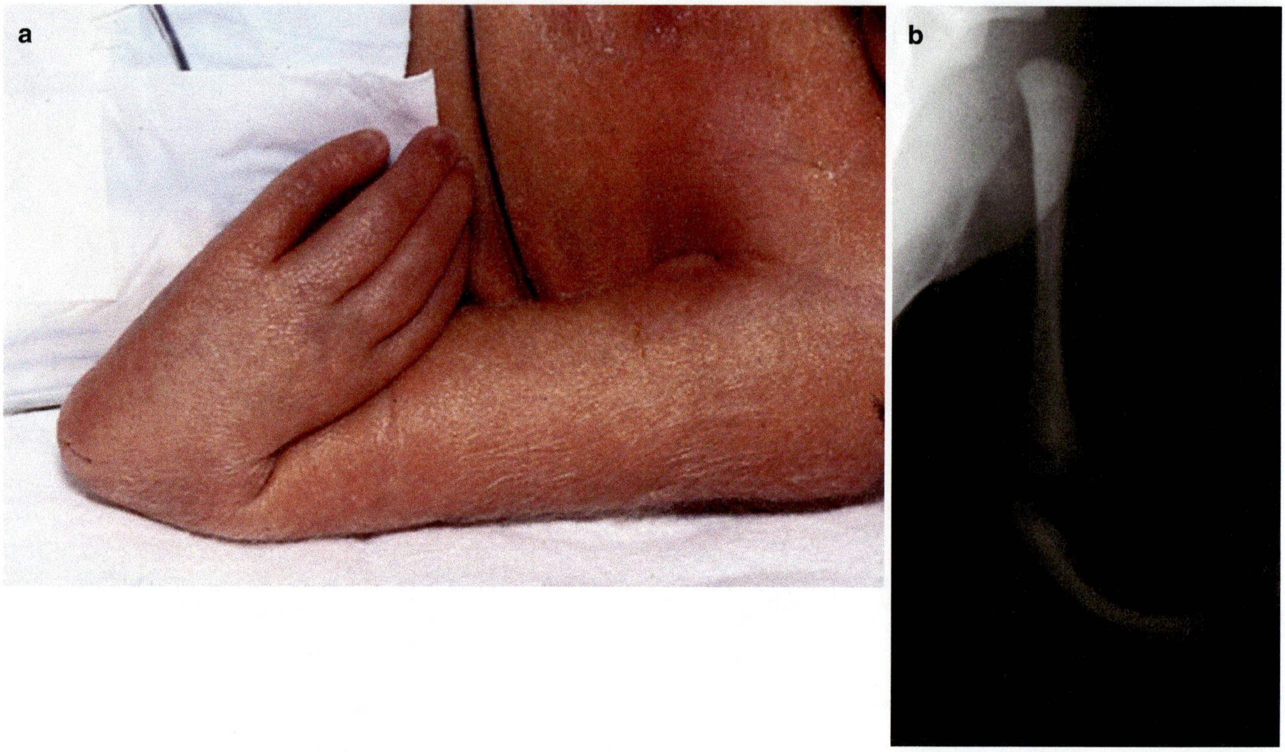

Fig. 48.12 (a) VACTERL Association Club feet. (b) VACTERL Association X-ray

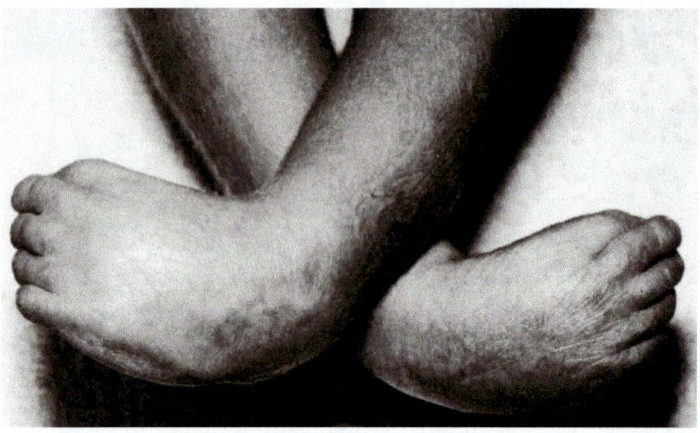

Fig. 48.13 Club feet

Club foot also known as talipes equinovarus is hypoplasia of limb muscles causing an excessively turned in foot; it is bilateral in up to 50% of cases. The patient presents at birth with hindfoot equinus and varus, forefoot adduction and a shortened foot. The incidence is approximately one in 1000 births with males more commonly affected than females [8]. Treatment follows the Ponseti method which involves serial casting and percutaneous tenotomy of Achilles tendon. Left untreated it can lead to significant long-term disability, deformity and pain (Figs. 48.12a, b and 48.13).

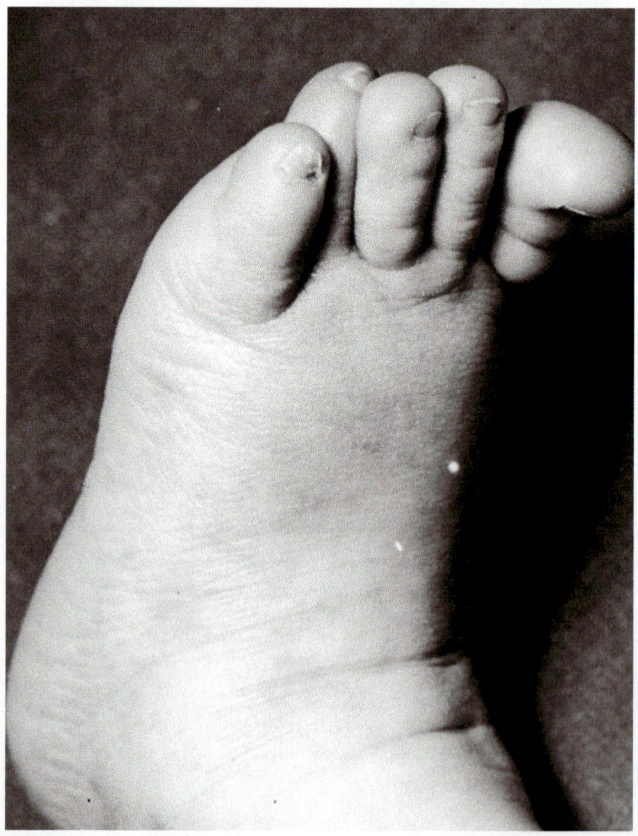

Fig. 48.14 Overriding Toe

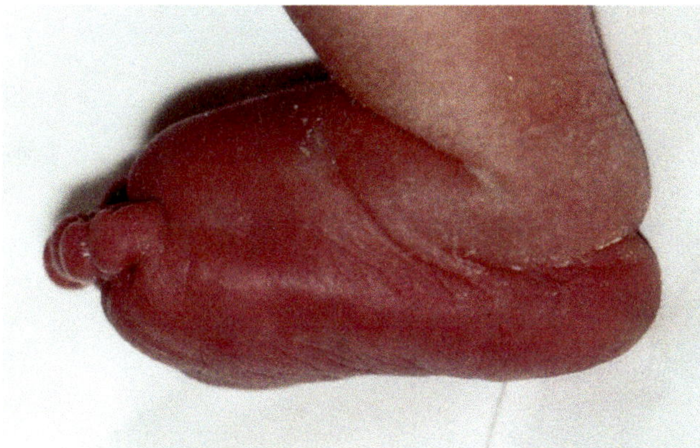

Fig. 48.15 Turner's syndrome

Developmental dysplasia of the hip is due to a poorly developed acetabulum and head of femur in utero. Dislocation commonly occurs after birth (congenital hip dislocation) and can be complicated by avascular necrosis of the femoral head. Developmental dysplasia of the hip affects 1–3% of newborns and is responsible for 29% of primary hip replacements in people up to the age of 60 years [9]. The child may present with asymmetry of skin folds at hip (Galeazzi sign) or shortening of the affected limb. The diagnosis can be further elucidated with use of Barlow's manoeuvre which checks for posterior dislocation and Ortolani's manoeuvre which checks for relocation on abduction. Treatment is with a Pavlik harness. Most cases are detected in the first few months and year of life with the manoeuvres above or with ultrasound. However, if the diagnosis is missed, extensive surgery is required in later childhood and adolescence.

Other conditions of the limbs are covered in other chapters (Figs. 48.14–48.20).

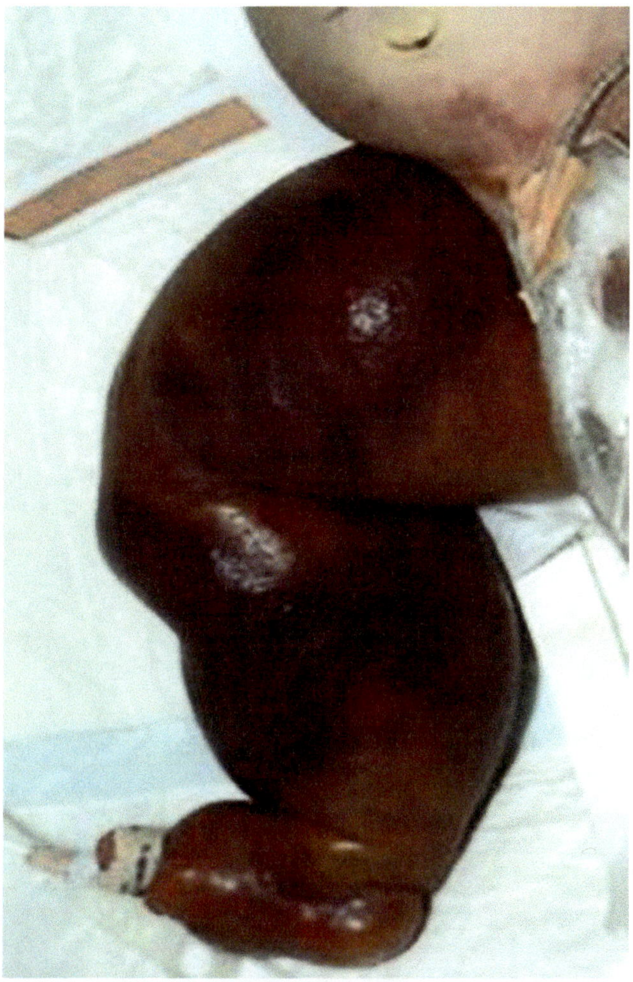

Fig. 48.16 Haemangioendothelioma of the leg

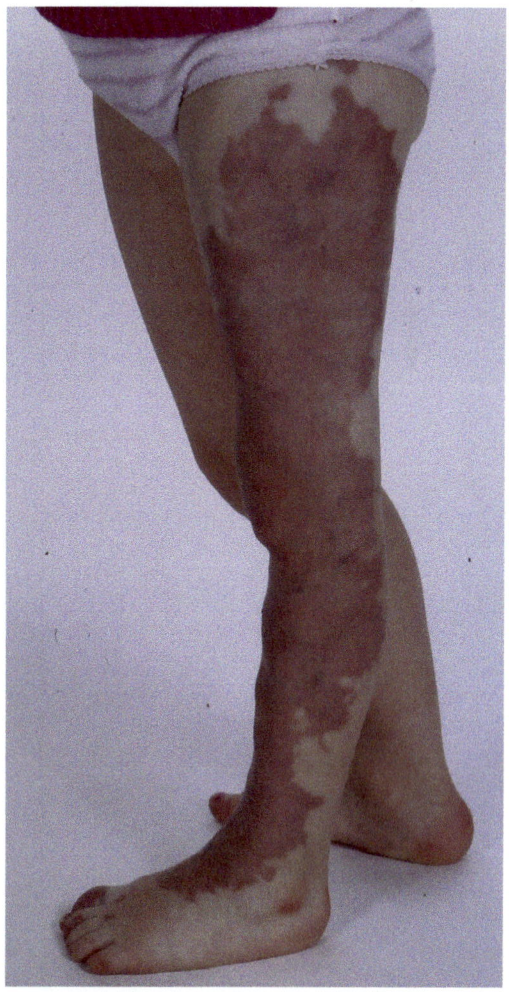

Fig. 48.17 Vascular malformation

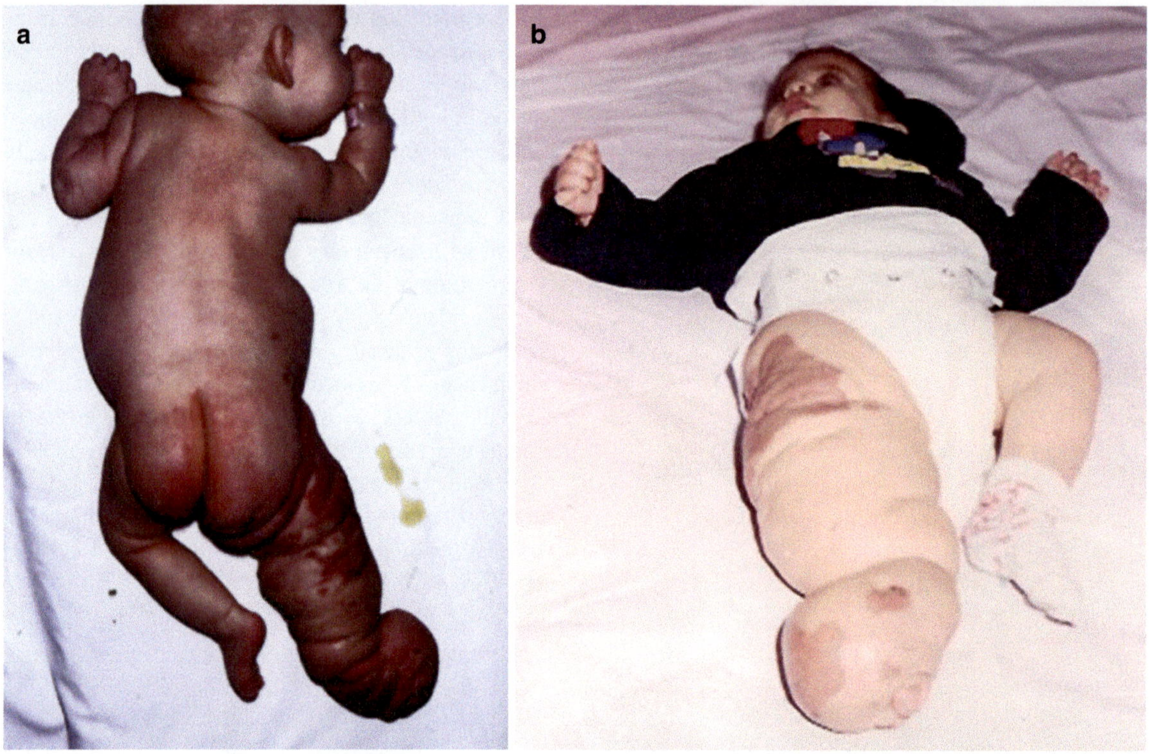

Fig. 48.18 (**a**, **b**) Klippel-Trenaunay syndrome

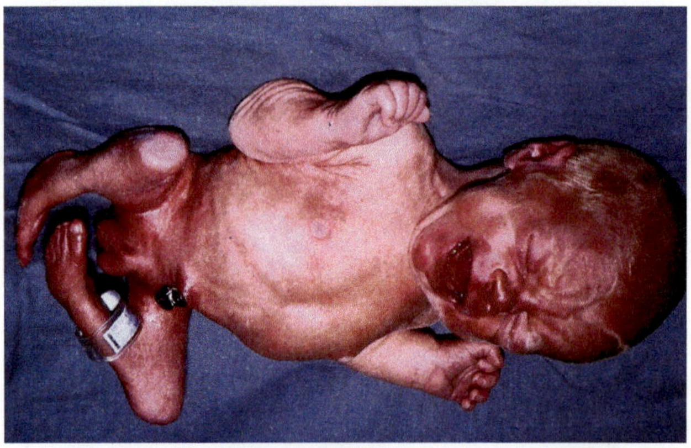

Fig. 48.19 Lumbosacral agenesis

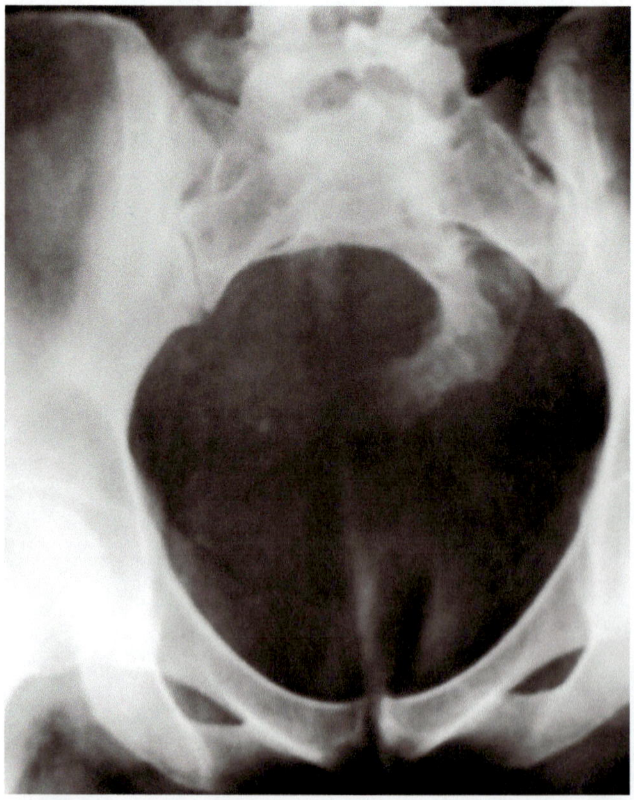

Fig. 48.20 Currarino syndrome, scimitar sacrum

References

1. Swanson AB. A classification for congenital limb malformations. J Hand Surg [Am]. 1976;1(1):8–22.
2. Swanson AB, Swanson GD, Tada K. A classification for congenital limb malformation. J Hand Surg [Am]. 1983;8(5 Pt 2):693–702.
3. Jindal G, Parmar VR, Gupta VK. Ectrodactyly/split hand feet malformation. Indian J Hum Genet. 2009;15(3):140.
4. Martinez-Frias ML, Cucalon F, Urioste M. New case of limb body-wall complex associated with sirenomelia sequence. Am J Med Genet. 1992;44:583–5.
5. Klaassen Z, Shoja MM, Tubbs RS, Loukas M. Supernumerary and absent limbs and digits of the lower limb: a review of the literature. Clin Anat. Jul 2011;24(5):570–5.
6. Choulakian MY, Bruce Williams H. Surgical correction of congenital constriction band syndrome in children: replacing z-plasty with direct closure. Can J Plast Surg. 2008;16(4):221.
7. Baitner AC, Maurer SG, Gruen MB, Di Cesare PE. The genetic basis of the osteochondrodysplasias. J Pediatr Orthop. 2000;20(5):594–605.
8. Werler MM, Yazdy MM, Mitchell AA, et al. Descriptive epidemiology of idiopathic clubfoot. Am J Med Genet A. 2013;161A(7):1569–78. https://doi.org/10.1002/ajmg.a.35955. Epub 2013 May 17
9. Sewell MD, Rosendahl K, Eastwood DM. Developmental dysplasia of the hip. BMJ. 2009;339:b4454. https://doi.org/10.1136/bmj.b4454.

Embryological Basis of Congenital Tumours

49

Philip Hammond and Srinivas Annavarapu

Introduction

During normal cellular development, there is a complex system of checks and balances to ensure regulation of the cells as they proliferate and specialize to perform their physiological functions. Various genes cooperate, concomitantly and/or sequentially with others, to activate and direct the developmental mechanisms in a developing foetus. After their target is achieved, these pathways are either kept dormant or are used elsewhere in a different context (growth, repair, etc.). Reactivation of these genes, by various mutations and/or carcinogens, can reinstate these developmental pathways. If these pathways remain active incessantly and do not obey the normal regulatory mechanisms, cell proliferation becomes independent of the growth stimulus, and this produces a mass—cancer [1–3].

Embryological Basis of Cancer

The growth and development of an embryo and that of a tumour has many parallels. Growth and development of a foetus from a single fertilized cell, the zygote, is a remarkable feat. This possibly is the best example of fast, coordinated growth where multiple complex pathways—like proliferation, migration, differentiation, apoptosis and reorganization—take place in an orchestrated manner. It is interesting to note that the growth of the embryo is faster than any of the malignant tumours [1–3].

As we continue to understand the developmental pathways operative in embryological development, startling revelations indicate that patterns of molecular pathways underpinning the tumour growth, migration, proliferation and differentiation have similar blueprints as the developmental pathways, which have either gone awry or have been arrested at a particular stage of development [2, 3]. This also sheds light of the timing of these tumours in relation to development.

Let us look at examples of the some of the many developmental/physiological pathways that are frequently deployed by tumour cells.

- The neural crest cells migrate to various locations in the body and are crucial in the development of many structures including the aortic arch, craniofacial cartilage and enteric nervous system [4]. This requires a very complex and intricate interplay of various developmental pathways. The tumour cells can use them selectively to metastasize to different locations in the body.
- Proliferation, migration and differentiation are vital in wound healing [1–3]. To enable repair of a skin wound, the epithelial cells of the skin must first transform into a mesenchymal phenotype. The epithelial cells are anchored to the basement membranes and other epithelial structures by desmosomes and hemi-desmosomes, respectively. To enable migration, they must lose their anchors and express laminins on their surfaces which help them to dissociate from their confines, interact with the extracellular connective tissue stroma and migrate to the wound area (epithelial to mesenchymal transition) [2, 3]. Angiogenesis supports this process. Once the skin wound has repaired, the mesenchymal phenotype reverts back to the epithelial type, and status quo is maintained. It is easy to see how these fundamental physiological pathways can be exploited by the various tumour cells to invade, migrate and metastasize.
- Formation of the placental bed during normal pregnancy is another fine example of many intricate pathways working in tandem. The aim of the exercise is to secure a constant and uniform blood supply from the host. The intermediate trophoblasts (X-cells) of the placenta invade deep into the maternal uterine wall to anchor the placenta and plug and remodel the uterine spiral arteriole vessels to

P. Hammond
Royal Hospital for Sick Children, Edinburgh, UK
e-mail: Philip.Hammond@nhslothian.scot.nhs.uk

S. Annavarapu (✉)
Department of Cellular Pathology, Royal Victoria Infirmary, Newcastle-Upon-Tyne, UK

R. Carachi, S. H. E. Doss (eds.), *Clinical Embryology*, https://doi.org/10.1007/978-3-319-26158-4_49

establish the placental vascular bed. All of this happens whilst the mother's immune system is temporarily suppressed lest it should reject the foetus, which harbours paternal (foreign) antigens. Many tumours adopt these developmental strategies during tissue invasion, vascular invasion and evasion of immune surveillance [2, 3].

In essence, the tumour does not need to reinvent mechanisms to grow and disseminate throughout the body. Deregulation of physiological mechanisms may serve as important primary driver for cancer. Thus, the apparently random mutations that cancer cells may accrue may not just be chance events after all, and it may be plausible that the cancer cells selectively run through the chronicles of developmental works to serve its ends [2, 3].

The Cancer Model

Cancer is a genetic disease where disordered autonomous cell proliferation is driven by a series of accumulating genetic changes influenced by hereditary factors and the somatic environment. Apparently random mutations in the DNA of the cells enable them to escape from the clutches of stringent and tightly regulated developmental pathways [1–3, 5]. As a result, cancer cells do not remain a single disease, and as additional mutations accumulate, the cells within the mass become heterogeneous, and different tumour subclones appear, each diversifying and acquiring different characteristics to give the tumour a survival advantage. This randomness may also explain why different people with cancers arising from the same cells may be different in terms of their biological progression, prognosis and response to therapy [1–3, 5].

The dysregulated growth that characterizes neoplastic cells is caused by overexpression of proto-oncogenes (which move the cell through the cell cycle) or inactivation of tumour suppressor genes (which normally restrict cell growth and proliferation). These deranged genes may occur when the DNA has point mutations, viral insertions, chromosomal or gene amplifications, deletions or rearrangements which occur through genomic instability during cell divisions and hence are more likely to be accumulated with advancing age [5]. Certain tumours, specifically retinoblastoma and Wilms' tumour, occur earlier (and more often bilaterally) when they result from germline mutations (which have been inherited) than when they result from sporadic or somatic mutations. These observations led Knudson and Strong to propose a 'two-hit' mechanism of carcinogenesis in which the first genetic defect, already present in the germ line, must be complemented by an additional spontaneous mutation before a tumour can arise. In sporadic cancer, cellular transformation only occurs when two (or more) spontaneous mutations take place in the same cell [5].

As such, most tumours represent the disordered cell division associated with the genetic degeneration of age, as evidenced by statistics indicating that a third of adults in the UK will develop a tumour at some stage in their lives. Increasingly, the genetic basis of many tumours is being discovered including some with an inherited genetic predisposition (such as breast cancer [e.g. BRCA1] or colon cancer [e.g. APC]). Less commonly, tumours present in infancy or more rarely perinatally, and these 'congenital' tumours will be outlined in this chapter.

Nephroblastoma (Wilms' Tumour)

Wilms' tumour or nephroblastoma is a common kidney tumour of childhood with a peak age at onset of 3 years [5, 6] (Fig. 49.1). Development of Wilms' tumour appears to be intricately linked to the process of foetal nephrogenesis [7]. A genetic basis has long been postulated and remains the focus of much research [8]. In a kidney with Wilms' tumour, often there are associated adjacent foci of primitive renal blastemal tissue called *nephrogenic rests* (nephroblastomatosis). These are particularly common in children with an inherited susceptibility to Wilms' tumour and may represent a premalignant lesion. Wilms' tumour genes may therefore be involved in distinct developmental pathways in the kidney, and their inactivation may interrupt normal development resulting in increased risk of malignant transformation. Various genes have been implicated in this malignant transformation including two genes on the short arm of chromosome 11, the Wilms' tumour suppressor genes—WT1 (11p13) and WT2 (11p15) [5, 8].

Embryological basis of Wilms' tumour: The kidney develops from a structure called metanephros which itself is derived from two separate mesodermal components—the ureteric bud and the metanephric mesenchyme [5]. The ureteric bud invaginates into the metanephric mesenchyme, and as this happens, both reciprocally induce sequential activation of genes that activates signalling cascades which in turn stimulates successive dichotomous branching of the ureteric bud. This complex process involves mesenchyme-to-epithelial transition (MET) and requires cooperation of diverse players such as WT-1/WT-2, Wnt1/Wnt4, c-ret receptor tyrosine kinase, BMP4, glial-derived neurotrophic factor (GDNF) and the GDNF receptor (GDNFRα) [8] (Fig. 49.2). The tips of the distal branches then sequentially transform through the comma- and S-shaped body stage until vascular endothelial cells migrate into these bodies to form glomerular tufts; the remaining distal bodies elongate to form the renal collecting duct system, together forming the functional nephron. The subcapsular zone of the foetal kidney contains a region of undifferentiated mesenchyme (nephrogenic zone) which keeps adding rows of glomeruli to the kidney. Nephrogenesis is completed by 36 gestational weeks [8].

Classical Wilms' tumours show a typical triphasic histology—renal blastemal, epithelial components and stroma.

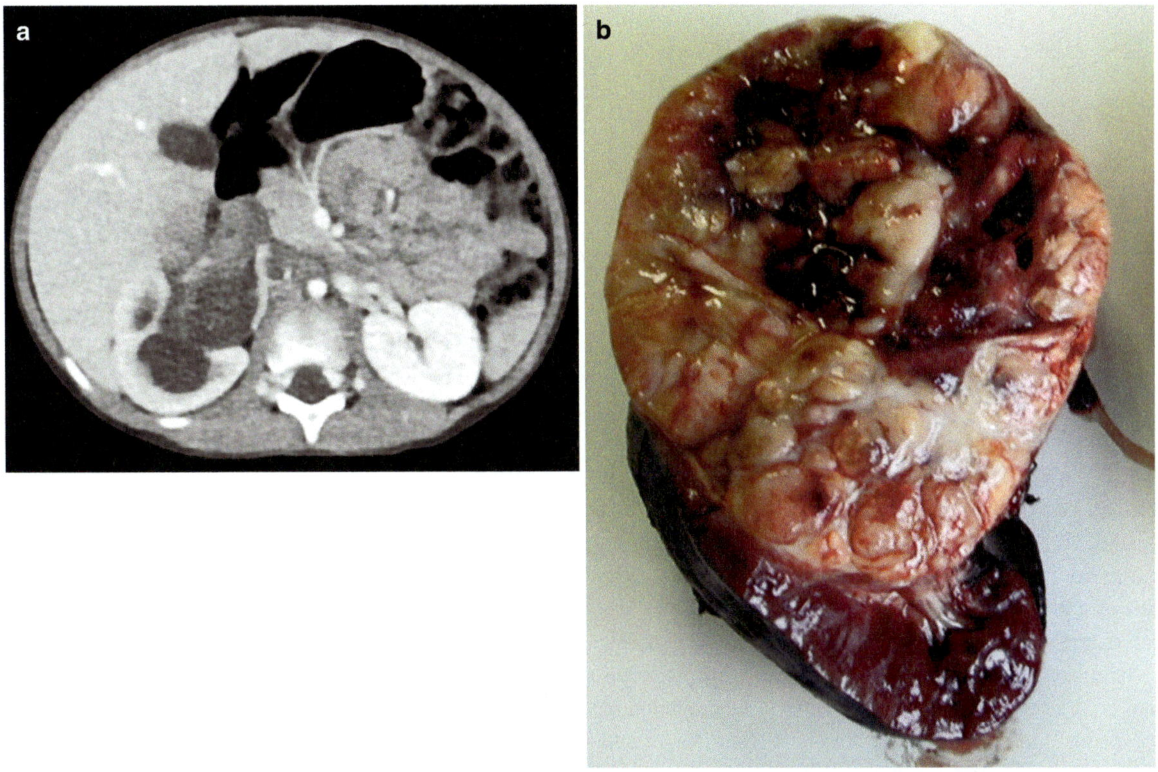

Fig. 49.1 Radiological and gross findings in Wilms' tumour. (**a**) MRI of the abdomen showing a large renal mass arising from the upper pole of the kidney. (**b**) Macroscopic appearance of Wilms' tumour arising from the upper pole of the kidney

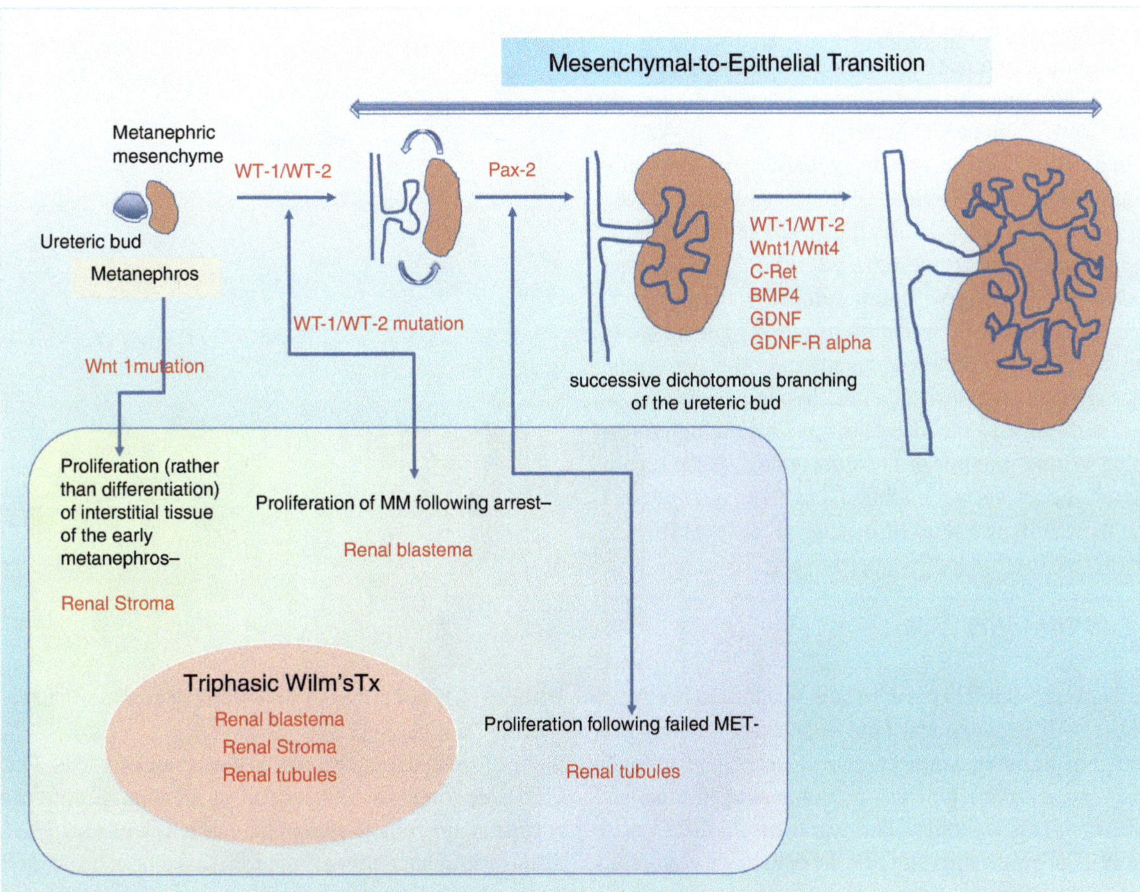

Fig. 49.2 Schematic diagram to show the development of Wilms' tumour

Each component of the tumour may reflect an arrested developmental stage of renal nephrogenesis [7–9].

Blastemal elements	• Proliferation (rather than differentiation) of undifferentiated metanephric mesenchyme • Driven by WT-1/WT-2
Epithelial elements	• Proliferation following partial differentiation of blastemal elements into renal tubules due to unsuccessful attempt at mesenchyme-to-epithelial transition (MET) • Driven by Pax-2
Stromal elements	• Proliferation (rather than differentiation) of interstitial tissue of the early metanephros • Driven by Wnt1

The histology of the Wilms' tumour shows varying proportions of these three components. Any of these elements may be the dominant component in a given tumour, though one can usually find all three elements on careful search. Rarely, Wilms' tumour can be biphasic or even monophasic. The prognosis of a given tumour appears to correlate with the histological elements. Stromal-predominant Wilms' tumour is known to show poor response to radio-/chemotherapy, whereas blastemal component shows good response. The existence of *nephrogenic rests* in normal kidneys has long been interpreted as residual embryonal elements that have failed to differentiate normally. There is a significant association between the presence of nephrogenic rests in the kidneys and Wilms' tumour (28–40%). This again suggests that Wilms' tumour may be a case of renal maldevelopment [7, 9].

A constellation of other findings may be found in children with Wilms' tumour which is further evidence of a genetic aetiology. For instance, children who develop **W**ilms' tumour in association with **a**niridia, **g**enitourinary anomalies and mental **r**etardation (termed WAGR syndrome) have a deletion at the WT1 gene locus (Fig. 49.3) [5, 6]. Children with Denys-Drash syndrome (characterized by pseudohermaphroditism, progressive glomerulopathy and Wilms' tumour) also have a point mutation of the WT1 gene [5, 6]. Children with Beckwith-Wiedemann syndrome (exomphalos, macroglossia, hyperinsulinemic hypoglycaemia) have DNA loss at the WT2 gene locus and have such a high risk of developing Wilms' tumour or hepatoblastoma that a screening programme in early childhood has been instituted to diagnose these tumours at an early stage (Fig. 49.4) [6].

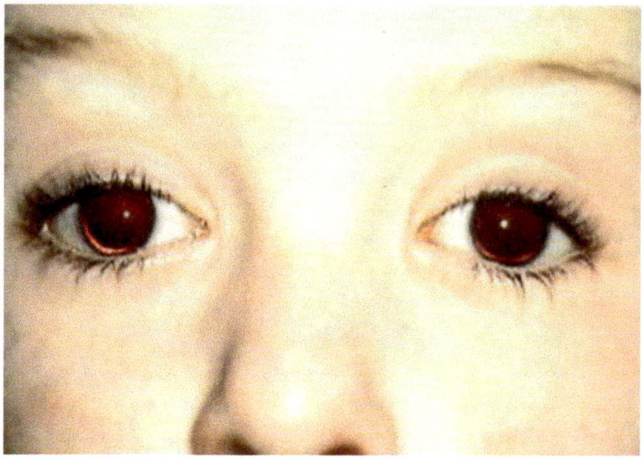

Fig. 49.3 Clinical picture of aniridia

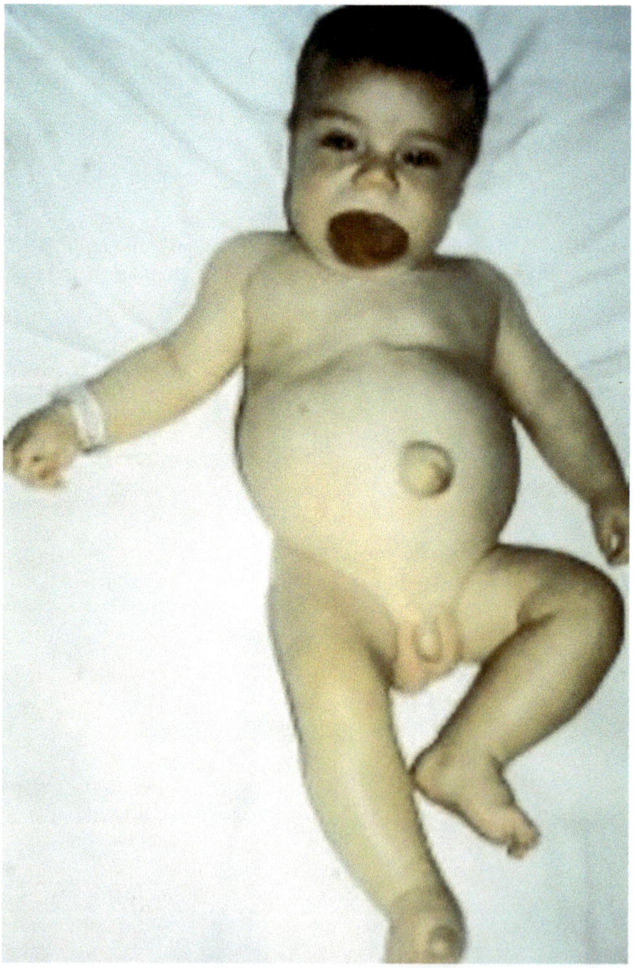

Fig. 49.4 Clinical picture of Beckwith-Wiedemann syndrome

Germ Cell Tumours

Germ cell tumours (GCTs) occur in both gonadal and extragonadal sites, with extragonadal (and testicular tumours) predominating in children younger than 3 years and gonadal (testicular and ovarian) tumours predominating during and after puberty [5]. It is thought that extragonadal GCTs arise when there is aberrant migration or deposition of germ cells along the path of migration or in abnormal locations. Therefore, GCTs are found in the sacrococcygeal area, mediastinum, retroperitoneum, pineal area of the brain as well as the ovary and testis. Malignant transformation may then occur at any of

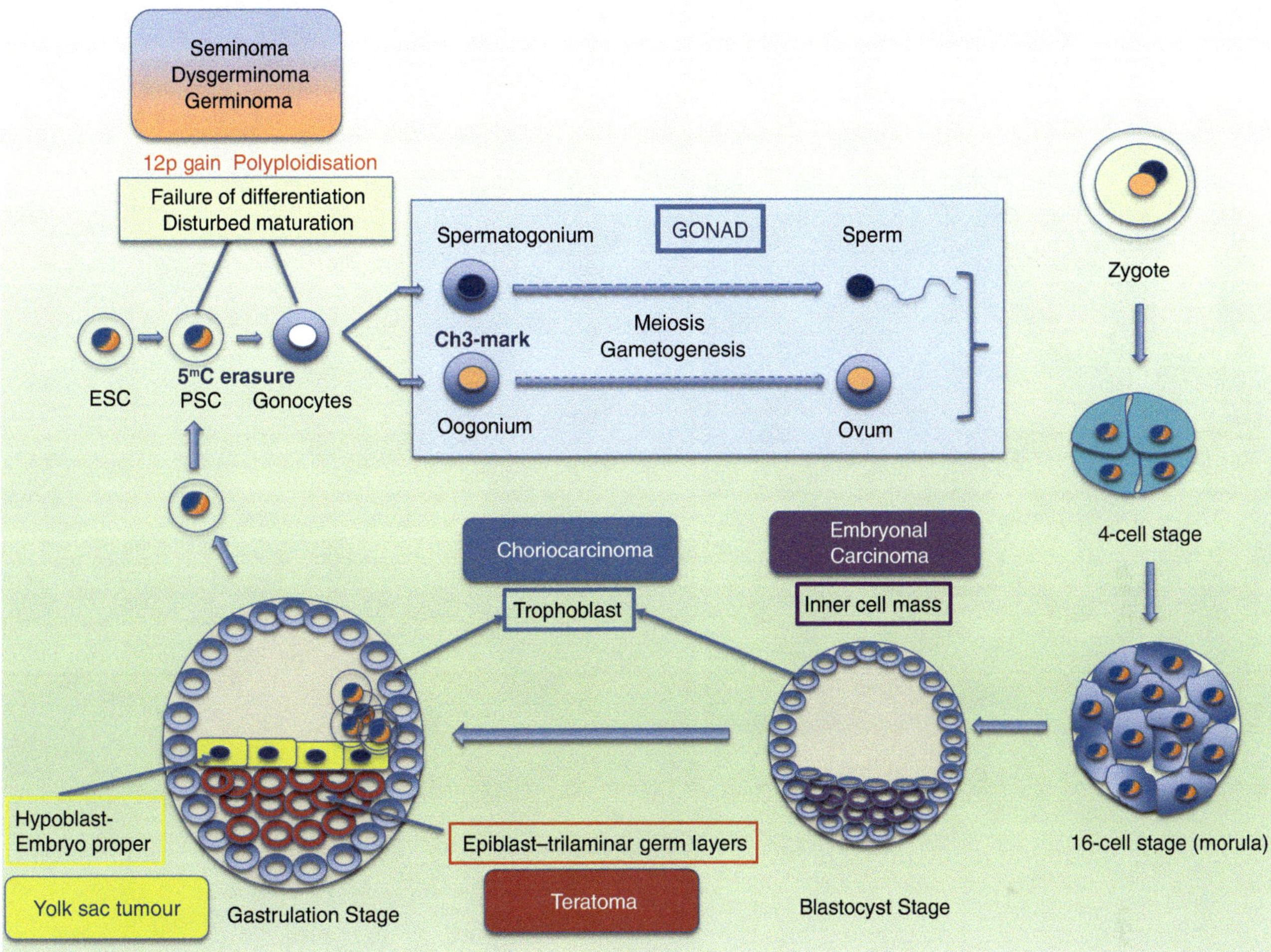

Fig. 49.5 Schematic diagram depicting development of germ cell tumours

these sites with the broad spectrum of types reflecting the totipotent/pluripotent nature of germ cells [7].

Embryological basis of germ cell tumours: Primordial germ cells (PGCs) originate from embryonal stem cells near the allantois of the embryonic yolk sac endoderm. To remove the parental genetic imprints, they undergo epigenetic erasure of $5^{m}C$ methylation. By the fifth gestational week, these PGCs with methylation erasure migrate through the mesentery to the gonadal ridge. Here the PGCs undergo genetic reprogramming by undergoing de novo methylation to acquire 'methylation mark' and become committed to gender-specific gametogenesis and eventually differentiate to form the gonads [5, 7] (Fig. 49.5).

Genetic factors (polyploidisation or 12p gain) transform these germ cells to form *intra-tubular germ cell neoplasia (ITGCN)*. If the transformed totipotent PGCs (with $5^{m}C$ methylation erasure) undergo clonal proliferation without further differentiation, it gives rise to seminoma (in males) and dysgerminoma (in females). If the transformed pluripotent PGCs (with de novo methylation) show differentiation towards inner cell mass, it gives rise to embryonal tumour. If the germ cells show intra-embryonic differentiation, it gives rise to teratoma (epiblastic differentiation/future embryo proper) or to endodermal sinus tumour (hypoblastic differentiation/future yolk sac). Extraembryonic, trophoblastic differentiation gives rise to choriocarcinoma [7, 10] (Fig. 49.5).

A teratoma is composed of representative tissue from each of the three germ layers of the embryonic disc (ectoderm, endoderm and mesoderm) [5] (Fig. 49.6). Mesodermal components such as fat, cartilage, bone and muscle are particularly common although endodermal tissues also commonly result in cystic structures lined by squamous, cuboidal or flattened epithelium [5, 7]. Most paediatric teratomas are mature (benign) with little tendency to undergo malignant degeneration [6]. Foci of malignant yolk sac tumour may be indicated by elevated serum alpha-fetoprotein (AFP) levels which may be used as a tumour marker [7]. The sacrococcygeal region is the

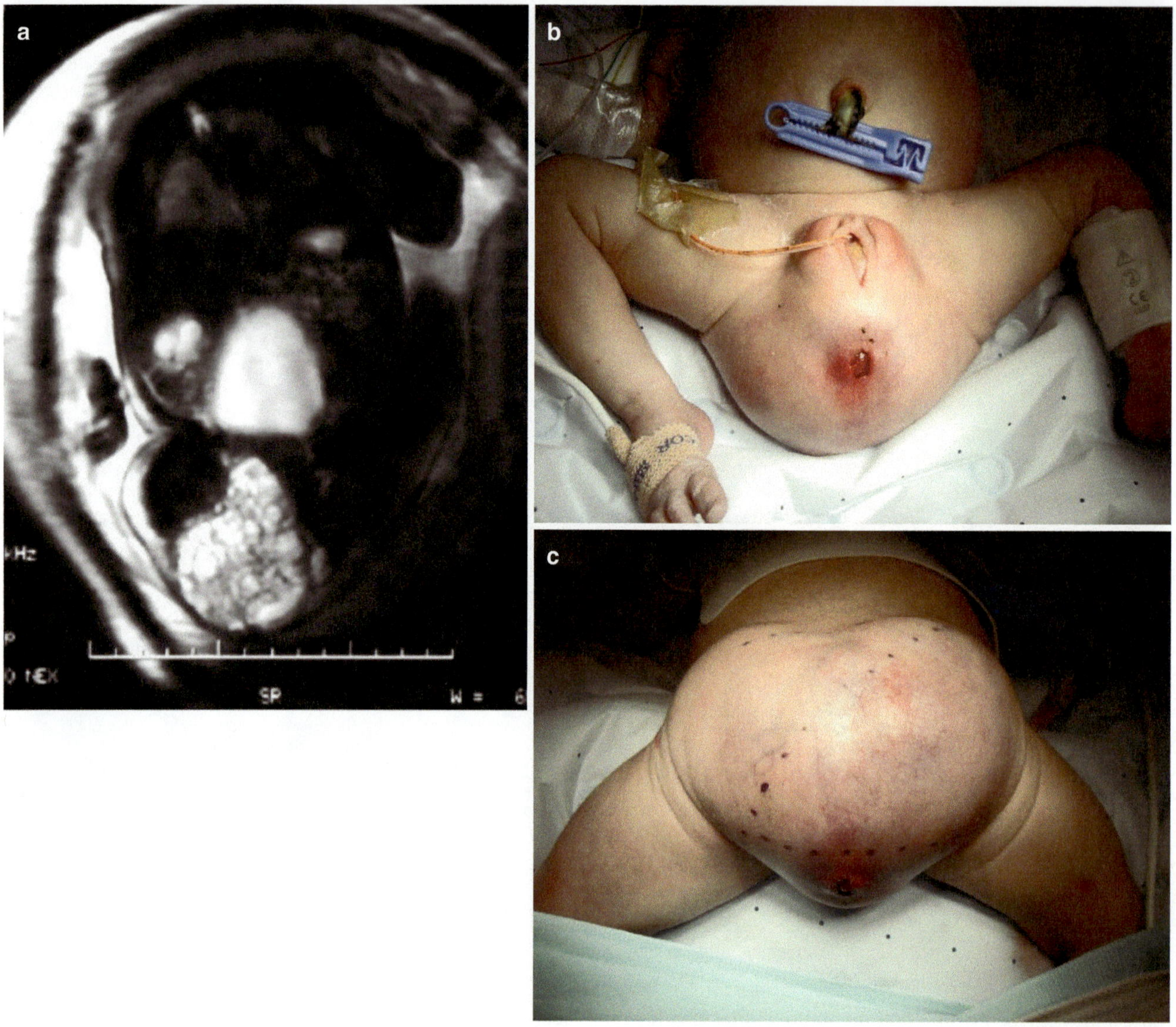

Fig. 49.6 (**a**) shows a large pelvic mass in the prenatal MRI at 32 gestational weeks. (**b**, **c**) show a sacrococcygeal teratoma

most common extragonadal location of teratomas and is usually diagnosed in the first year of life (Fig. 49.7) [6]. Over half of paediatric ovarian tumours are teratomas which often present with pain when large tumours cause torsion of the ovary. Teratomas are also the most common testicular neoplasm in childhood [6].

Germinomas (often referred to as *seminoma* when found in the testis or *dysgerminoma* when in the ovary) may secrete β-hCG (human chorionic gonadotropin) which may also be used as a serum marker of malignancy [7]. Germinoma is the predominant malignancy found in dysgenetic gonads and undescended testes. Boys with undescended testes are likely to have a threefold increased risk of later development of testicular cancer compared to other boys [11]. Gonadoblastoma is a rare tumour which is usually found in dysgenetic gonads of phenotypic females who have a fragment of the Y chromosome, and because of this risk, prophylactic gonadectomy may be recommended for these girls [5, 7].

Neuroblastoma

Neuroblastoma is the most common malignancy diagnosed in infancy and is remarkable for its broad spectrum of clinical behaviour [5, 6]. The clinical course can be variable, as spontaneous regression or maturation may occur in young infants, whilst other children develop progressive neuroblastoma with a dismal prognosis [6]. Neuroblastoma cells are found in the adrenal gland in about 1 in every 40 neonates who die of other causes; yet clinical neuroblastoma presents in only approximately 1 in 10,000 children [6].

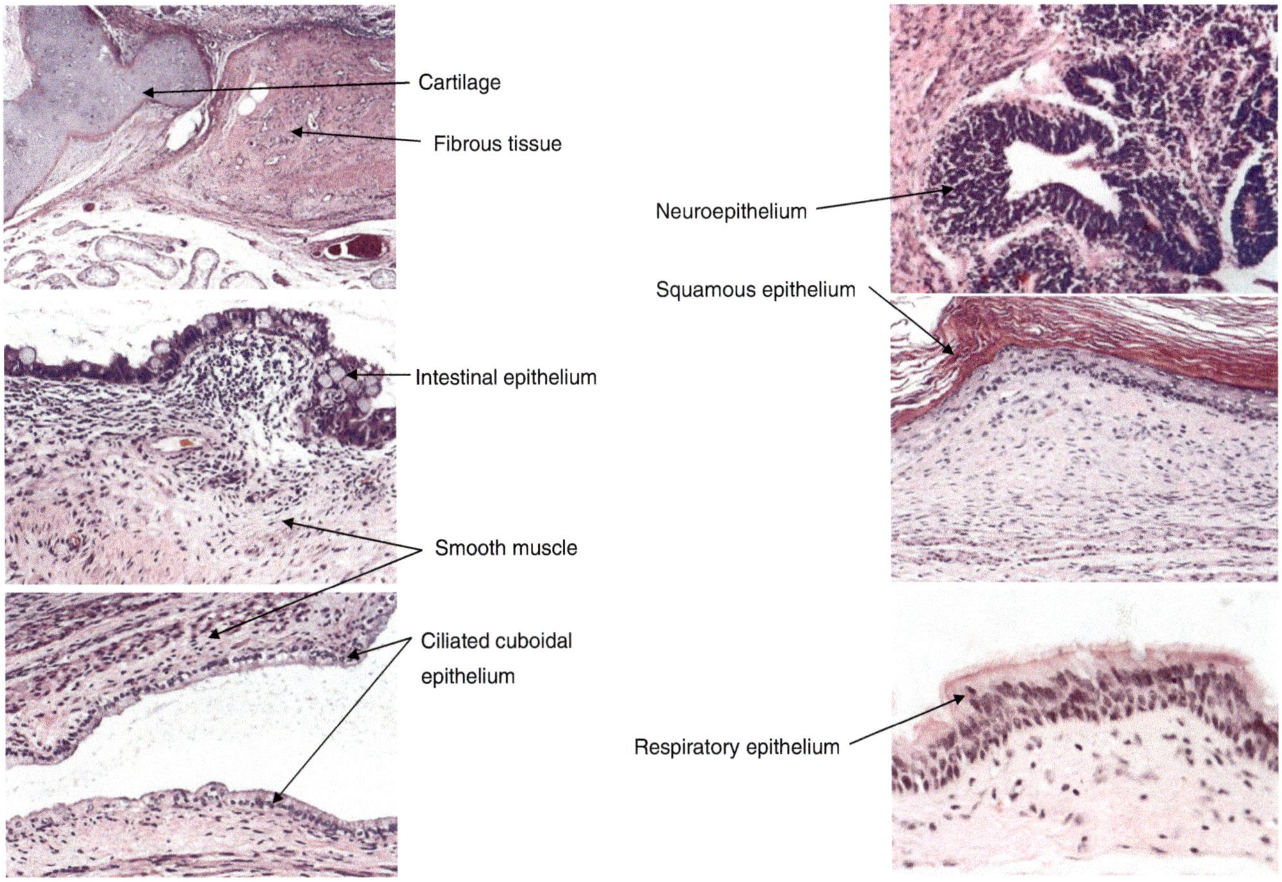

Fig. 49.7 Histology of teratoma shows divergent differentiation towards all three germ layers

In an attempt to identify early cases of neuroblastoma amenable to cure, mass screening has been attempted in several countries by evaluating urinary catecholamine metabolites in infants. Unfortunately, tumours detected by screening apparently seem to be the group which would have regressed spontaneously, whilst there is no reduction in the incidence of neuroblastoma-related deaths [6]. Clinical presentation is variable, depending on the site of primary and metastatic tumours causing a mass effect with over half having an abdominal mass. Catecholamines released by these neuroblastoma cells may be detected when excreted in urine and help with diagnosis [5, 6].

Embryological basis of neuroblastoma: Primitive neuroblasts can be identified in the foetal adrenal gland from 10 weeks of gestation. Antenatal ultrasound of the adrenal glands has identified neuroblastoma in numerous reports. Neuroblast cells are derived from primordial neural crest cells which migrate from the mantle layer of the embryonic spinal cord [4]. They populate tissues such as the adrenal medulla and sympathetic ganglion chain, and hence, neuroblastoma may occur at any site from the neck to the pelvis. The fate of these neuroblasts may be to regress spontaneously, mature by differentiation to benign ganglioneuroma or progress to malignant neuroblastoma [7]. The cell pathway is determined by complex molecular signalling mechanisms determined by genetic abnormalities (Fig. 49.8. Spectrum of neuroblastic tumours).

N-myc is a transcription factor related to a proto-oncogene found at the chromosome 2p24 locus, which has become one of the most important biological factors for the prognosis [5–7, 11]. When there are ≥10 copies of N-myc, as detected by fluorescent in situ hybridization (FISH), it is considered to represent genomic amplification which confers a poor prognosis. N-myc amplification promotes proliferation of neuroblasts and their transformation to neuroblastoma by preventing terminal differentiation and successful exit from the cell cycle [6]. Approximately 30% of all neuroblastoma cases are found to be N-myc-amplified [12].

Hepatoblastoma

Hepatoblastoma is the most common hepatic malignancy in children below 3 years of age. A number of genetic predispositions increase the risk of hepatoblastoma, and the most important ones include Beckwith-Wiedemann syndrome and familial adenomatous polyposis (FAP) [5–7].

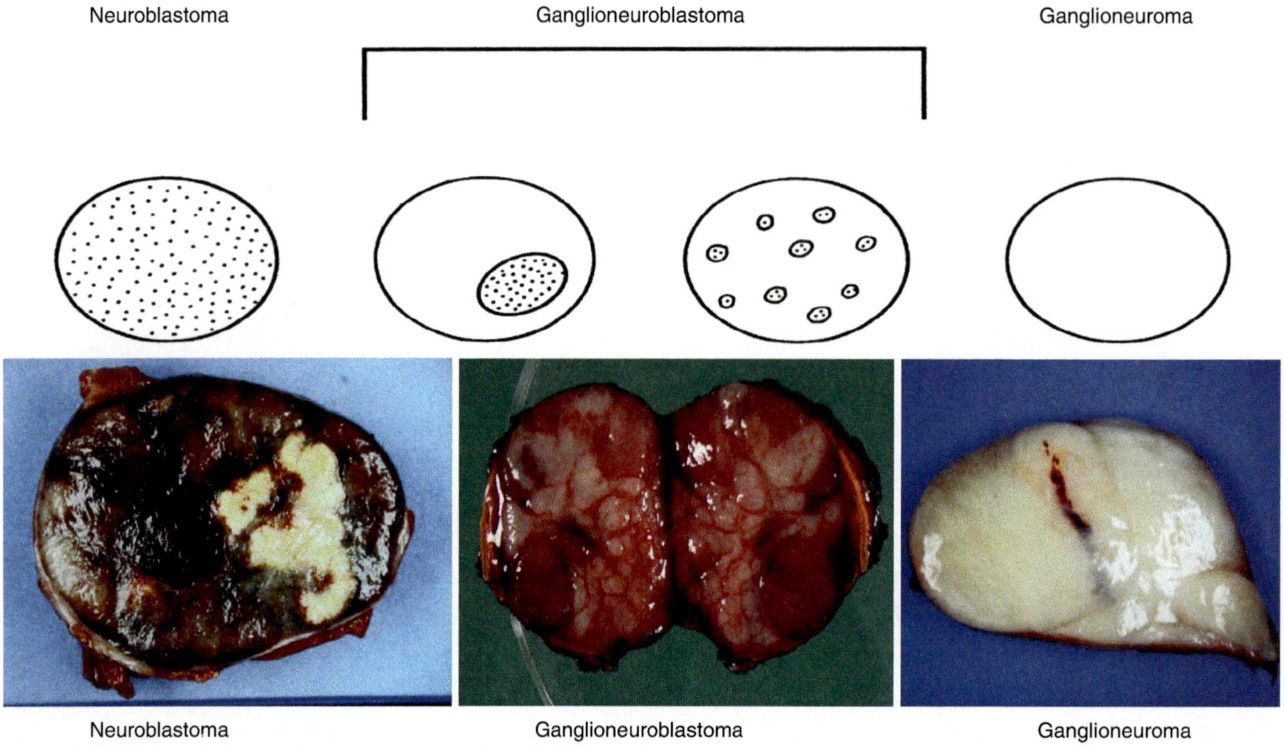

Fig. 49.8 Spectrum of neuroblastic tumours

Embryological basis of hepatoblastoma: Hepatoblastoma is an embryonal tumour that recapitulates various developmental stages of liver development. Many developmental pathways (Wnt, Notch, Shh, c-MET, etc.) play an important role in hepatic development. Wnt signalling is one such pathway that is crucial for embryonal liver development and is frequently dysregulated in hepatoblastoma [13, 14]. This is reflected by the fact that there is an increased incidence of hepatoblastoma in families with FAP, where there is a germline mutation of APC gene [14]. APC plays a central role in the proteasomal degradation of beta-catenin, a powerful nuclear transcription factor and effector of Wnt pathway, which promotes cell proliferation, migration, invasion and cell survival [14]. Thus, constitutional activation of Wnt pathway confers a proliferation advantage to tumour cells in hepatoblastoma (Fig. 49.9. Schematic diagram of development of hepatoblastoma).

Histologically, hepatoblastoma can show a diverse range of differentiation towards epithelial (foetal, embryonal, mixed foetal/embryonal, small cell undifferentiated), mesenchymal or mixed epithelial/mesenchymal components. The different subtypes possibly reflect proliferation (following developmental arrest) rather than differentiation into next stage [7]. Recent research suggests that activation of specific developmental pathways at different stages of liver development may correlate with the degree of differentiation in various histologic subtypes [13, 14]. Wnt activation is present in embryonal and in mixed epithelial/mesenchymal subtypes, whereas Notch pathway activation is seen in foetal subtype [13, 14]. Small cell undifferentiated subtype is genetically distinct from others and shows aberrations in human SMARCB1 gene, causing them to show characteristic loss of INI-1 staining on immunohistochemistry, a feature that it shares more closely with malignant renal rhabdoid tumour and the CNS atypical teratoid/rhabdoid tumours including very dismal prognosis. Apart from pure foetal subtype, where surgery alone may suffice and enjoys a very good prognosis, other histological subtypes do not have any significant bearing on the overall clinical outcome.

Rhabdomyosarcoma

Malignancies that arise from connective tissues (mesodermal origin) in the body are called sarcomas. They may arise from tissues such as muscle, adipose tissue, fascia, blood vessels, nerves, bones, synovium or blood vessels. Sarcomas that display differentiation towards the skeletal muscle are

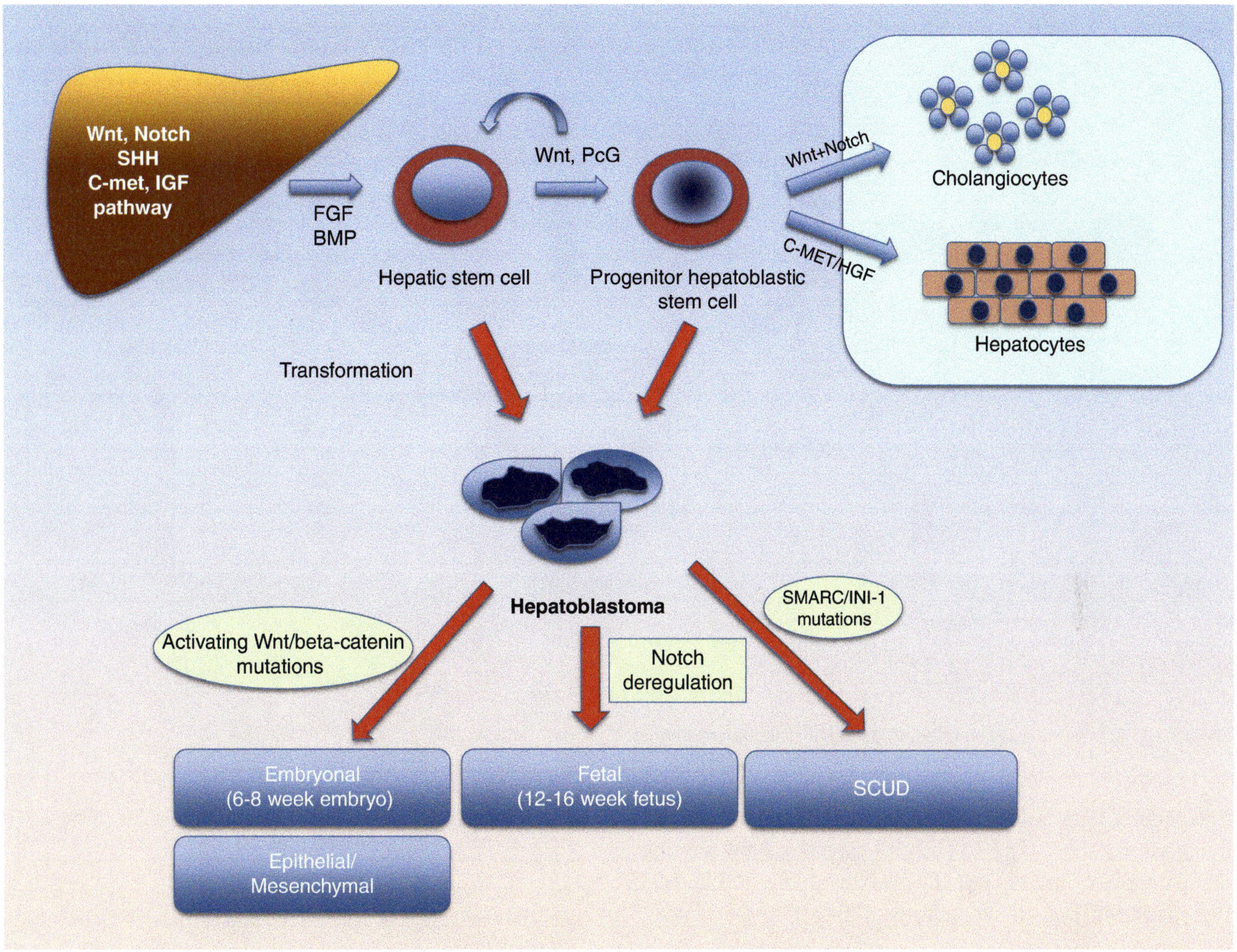

Fig. 49.9 Schematic diagram to show the development of the various subtypes of hepatoblastoma

collectively known as rhabdomyosarcoma (RMS). There principally are two types of RMS—embryonal and alveolar. These subtypes may essentially reflect the developmental stages of the skeletal muscle [7, 15].

Embryonal RMS usually affects children <5 years of age and tends to occur in the head and neck region and the genitourinary organs. Botryoid and spindle cell RMS are considered to be further subtypes of embryonal RMS that tend to have a better prognosis than the conventional embryonal RMS. Alveolar RMS is a more aggressive tumour with a poorer prognosis than embryonal RMS and occurs mostly in the trunk and extremities [6, 7, 15].

Embryological basis of RMS: The uncommitted mesodermal stem cells at around 6–8 gestational weeks under the influence of Pax3/Pax7 genes differentiate asymmetrically into cells, committed myogenic progenitor cells and primitive (multinucleated) myotubes [15]. The primitive myotubes, under the influence of MyoD and Myf5, differentiate further into rhabdomyoblasts. Myogenin and Myf6 promote differentiation of rhabdomyoblasts into maturing myotubes and finally into mature skeletal muscle fibres. The committed myogenic progenitor cells reside as 'satellite cells' in the mature skeletal muscle and serve as source of regeneration by forming more multinucleated myotubes [15] (Fig. 49.10).

Alveolar RMS is characterized by Pax3/Pax7-FOXO translocation that causes uncontrolled proliferation at the stages of uncommitted mesodermal stem cells or satellite cells within the maturing muscle tissue that may give rise to the alveolar RMS subtypes, respectively (similar to foetal muscle at 10 gestational weeks) [16]. Similarly, proliferation following arrest at the rhabdomyoblast and/or the maturing myotube stages may give rise to embryonal RMS (similar to foetal muscle at 6–7 gestational weeks) [15] (Fig. 49.10).

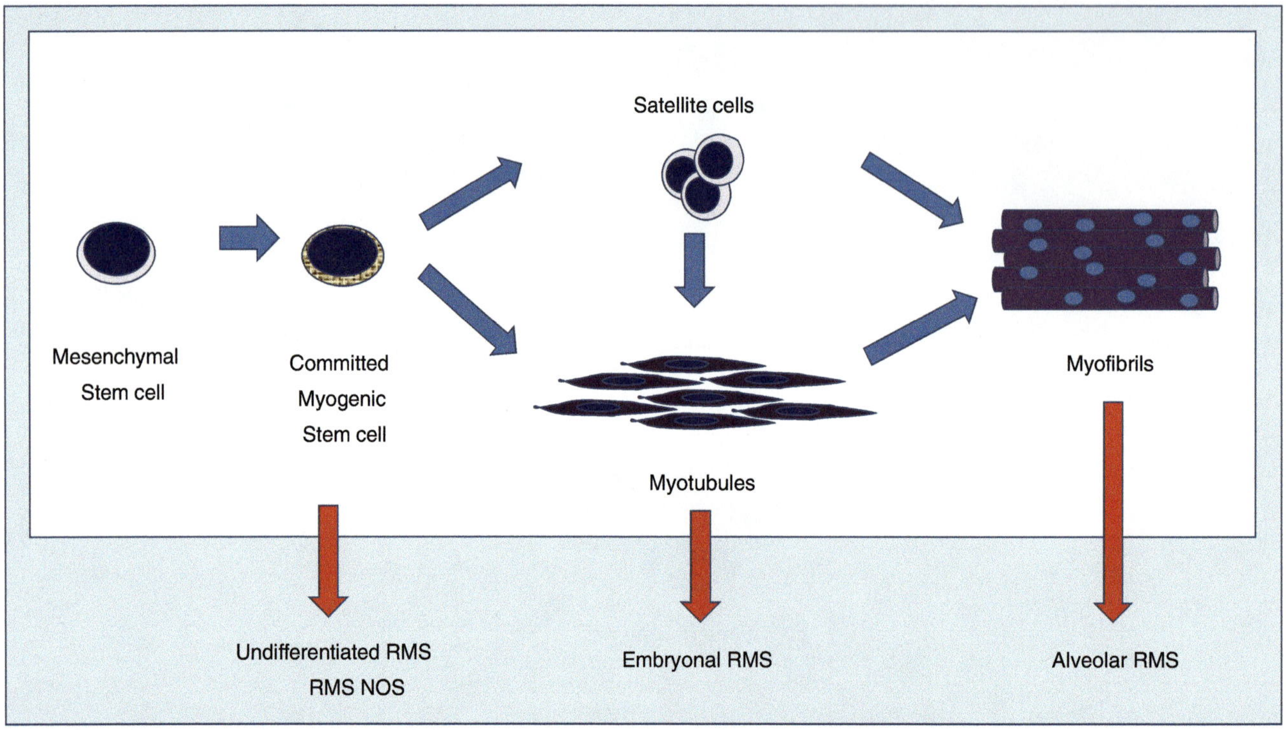

Fig. 49.10 Schematic diagram of development of rhabdomyosarcoma

The underlying genetics of embryonal RMS is not consistent though many show 11p15.5 aberrations; many show possible deregulation of multiple pathways involving IGF2, p53, Rb1, Wnt, Shh and Notch pathways [15–17].

Conclusion

The embryologic basis of tumours is demonstrated through the prism of these paediatric examples although this role can be extrapolated to the breadth of adult oncological practice. Although tumours presenting congenitally are rare, our inherited genotype has a major impact on our predisposition to the development of tumours and other conditions in later life.

References

1. Cofre J, Abdelhay E. Cancer is to embryology as mutation is to genetics: hypothesis of the cancer as embryological phenomenon. Sci World J. 2017;2017:3578090.
2. Kelleher FC, Fennelly D, Rafferty M. Common critical pathways in embryogenesis and cancer. Acta Oncol. 2006;45(4):375–88.
3. Ma Y, Zhang P, Wang F, Yang J, Yang Z, Qin H. The relationship between early embryo development and tumourigenesis. J Cell Mol Med. 2010;14(12):2697–701.
4. Jiang M, Stanke J, Lahti JM. The connections between neural crest development and neuroblastoma. Curr Top Dev Biol. 2011;94:77–127.
5. Kumar V, Abbas AK, Aster JC. Neoplasia. In: Kumar V, Abbas AK, Aster JC, editors. Robbins basic pathology. 9th ed. Philadelphia: Elsevier; 2013. p. 161–214.
6. Carachi R, Grosfeld JL. The surgery of childhood tumors. 3rd ed. Berlin Heidelberg: Springer-Verlag; 2016.
7. Gilbert-Barness E. Pathology of the fetus, infant and child. 2nd ed. Mosby Elsevier, 2007.
8. Hohenstein P, Pritchard-Jones K, Charlton J. The yin and yang of kidney development and Wilms' tumors. Genes Dev. 2015;29(5):467–82.
9. Vujanić GM, Sandstedt B. The pathology of Wilms' tumour (nephroblastoma): the International Society of Paediatric Oncology approach. J Clin Pathol. 2010;63(2):102–9.
10. Sheikine Y, Genega E, Melamed J, Lee P, Reuter VE, Ye H. Molecular genetics of testicular germ cell tumors. Am J Cancer Res. 2012;2(2):153–67.
11. Lip SZ, Murchison LE, Cullis PS, Govan L, Carachi R. A meta-analysis of the risk of boys with isolated cryptorchidism developing testicular cancer in later life. Arch Dis Child. 2013;98(1):20–6.
12. Huang M, Weiss WA. Neuroblastoma and MYCN. Cold Spring Harb Perspect Med. 2013;3(10):a014415.
13. Han ZG. Mutational landscape of hepatoblastoma goes beyond the Wnt-β-catenin pathway. Hepatology. 2014;60(5):1476–8.

14. Krawczuk-Rybak M, Jakubiuk-Tomaszuk A, Skiba E, Plawski A. Hepatoblastoma as a result of APC gene mutation. J Pediatr Gastroenterol Nutr. 2012;55(3):334–6.
15. Parham DM, Barr FG. Classification of rhabdomyosarcoma and its molecular basis. Adv Anat Pathol. 2013;20(6):387–97.
16. Rudzinski ER. Histology and fusion status in rhabdomyosarcoma. Am Soc Clin Oncol Educ Book. 2013:425–8.
17. Wachtel M, Runge T, Leuschner I, Stegmaier S, Koscielniak E, Treuner J, Odermatt B, Behnke S, Niggli FK, Schäfer BW. Subtype and prognostic classification of rhabdomyosarcoma by immunohistochemistry. J Clin Oncol. 2006;24(5):816–22.

Conjoined Twins

50

Mohamed Abdel-Latif, Mohamed Sameh Shalaby, and Sameh Abd Elhay

Introduction

Conjoined twin pregnancy is a rare embryologic event with an overall prevalence of 1.4 per 100,000 births. Its prevalence shows a marked variation among different countries, from as high as 3.22 per 100,000 births in Finland to as low as less than 0.08 per 100,000 births in northeast of Italy [1]. It occurs in about 1% of monozygotic twins. Conjoined twins are mostly premature with 3:1 female predominance. Management difficulties are primarily related to the conjoined organs and the abnormal hemodynamic circulation. Survival depends on the type, the shared organs, the associated anomalies, and the chosen time and modality of management.

Etiology

The exact etiology of conjoint twinning is unknown. There are no significant associations between conjoined twins and consanguinity, family history, maternal age, and environmental factors. There are two opposing theories (fission vs fusion) trying to identify the scenario of embryological events.

The Fission Theory

Supporters believe that in conjoined twinning only partial separation of the embryonic axes occurs rather than the complete normal separation. This could explain the higher incidence of mirror imaging in conjoined twins than other monoamniotic twins [2].

The Fusion Theory (Box 50.1)

Proponents of the "fusion theory" believe that conjoined twins result from secondary union of two originally separate but closely approximated monovular embryonic discs, usually in the third and fourth weeks after fertilization.

The Spherical Theory [3, 4]

This theory adds the three-dimensional concept to the fusion theory. The two separate monovular embryonic discs may lie adjacent to each other at variable angles and planes. Sometimes, the fusion occurs over the sphere of single yolk sac; other times, it occurs within the sphere of a single amniotic cavity. The lean of the two discs, which may be cranial, caudal, or lateral within their sphere, determines the specific type of the conjoined twins.

Box 50.1 The Fusion Theory [3, 4]

- As intact ectoderm will not fuse to intact ectoderm, union occurs at sites where the surface ectoderm is absent, is preprogrammed to break down, or fused.
- Areas of fusion are the oropharyngeal membrane, the primordia of the heart, the septum transversum, the cloacal membranes, the neural folds, and the periphery of the embryonic disc.
- By the end of the fourth week, intact ectoderm covers the entire embryo, except the umbilicus; thus, all possible sites for secondary union of the embryonic discs have closed or moved to inaccessible locations.

M. Abdel-Latif
Helwan University, Cairo, Egypt

Ain Shams University Hospitals, Cairo, Egypt

M. S. Shalaby (✉)
Ain Shams University Hospitals, Cairo, Egypt

Bristol Royal Children's Hospital, Bristol, UK
e-mail: mshalaby@doctors.org.uk

S. A. Elhay
Ain Shams University Hospitals, Cairo, Egypt

R. Carachi, S. H. E. Doss (eds.), *Clinical Embryology*, https://doi.org/10.1007/978-3-319-26158-4_50

The union is always homologous: head-to-head, front-to-front, back-to-back, and side-by-side. Conjoined twins with common yolk sac have shared portions of the gastrointestinal tract together with other duplicated portions, single umbilicus, single umbilical cord with more than three vessels, and single placenta. The duplicated structures support the theory of fusion. Conjoined twins united dorsally have a single amniotic cavity but entirely separate yolk sacs.

Classification

Conjoined twins are classified according to the prominent site of fusion followed by the suffix "pagus," which is a Greek term that means fixed or joined. They are either symmetrical or asymmetrical.

Symmetrical Conjoined Twins (Table 50.1) [3, 4]

There are two main groups of symmetrical conjoined twins, ventral and dorsal, depending on the involved aspect of embryonic disc.

Ventral union (cephalopagus, thoracopagus, omphalopagus, ischiopagus, and parapagus) occurs over a single yolk sac, while the dorsal union (craniopagus, rachipagus, and pygopagus) occurs in the neural tube.

Table 50.1 Types of symmetrical conjoined twins

Type	Incidence	Embryology	Fusion site
Cephalopagus	11%	Ventral fusion occurs in the oropharyngeal membrane and all the structures derived from and adjacent to it	Head
Thoracopagus (Fig. 50.1)	19%	Ventral fusion occurs at the cardiac primordia	Chest
Omphalopagus (Fig. 50.2)	18%	Ventral fusion takes place in the septum transversum	Umbilicus
Ischiopagus (Figs. 50.3 and 50.4)	11%	Ventral fusion is at the cloacal membrane. Twins may lie face to face with fused abdomen or caudal to caudal with straight vertebral column	Hip
Parapagus	28%	Ventral fusion occurs at the cloacal membrane. This is side-to-side fusion	Side
Craniopagus	5%	Dorsal fusion occurs in the cranial neuropore and the calvarium	Helmet
Rachipagus	2%	Dorsal fusion occurs in the midportion of the neural tube	Spine
Pygopagus (Fig. 50.5)	6%	Dorsal fusion occurs in the caudal neuropore	Rump

Ventral Union

Cephalopagus

The twins often fuse from the top of the head that may extend down to the umbilicus. The single fused head may have two faces (janiceps) facing away from each other. These twins usually have single cerebrum, separate vertebral columns, two joined "anterior" and "posterior" hearts, and shared esophagus. The brain and spinal cords are extremely abnormal. The "posterior" heart is often smaller and rudimentary. The lower abdomen and pelvis are separate. There are four upper limbs (tetrabrachius) and four lower limbs (tetrapus). These twins are nonviable.

Thoracopagus (Fig. 50.1)

The twins fuse from the upper thorax to the umbilicus. They lie face to face and have one compound or two united hearts, which possess other significant anomalies. These twins share

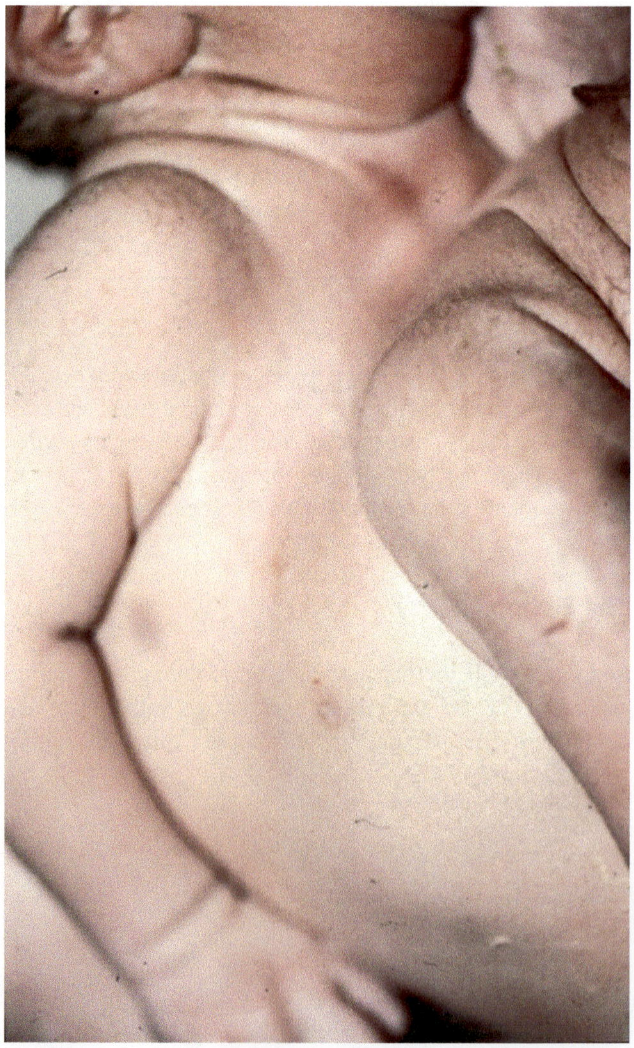

Fig. 50.1 Thoracopagus twin

the pericardium, the sternum, the diaphragm, the upper abdomen wall, and the liver, while they have two separate sets of lungs. Their foreguts are separate but rejoin at the duodenum and then become separate again at the ileum. About 25% of cases have joined biliary tree. There may be associated exomphalos. There are tetrabrachius and tetrapus. The major cardiac affection renders survival rare.

Omphalopagus (Fig. 50.2)
Twins lie face to face and join in the area of the umbilicus, often including the lower chest. The heart is never fused. Most of these twins have joined livers. While the stomach and proximal small bowel are usually separate, few cases have shared duodenum. In up to one third of these twins, the intestine usually joins at the Meckel's diverticulum; thus, they share terminal ileum and colon. Each twin has a separate pelvis, rectum, and genitourinary tract. There is associated exomphalos. There are tetrabrachius and tetrapus.

Ischiopagus (Figs. 50.3 and 50.4)
Twins may lie face to face with fused abdomen or caudal to caudal with straight vertebral column. Fusion results in a large conjoined pelvis with complex orthopedic and urogenital anatomy. Twins may share the cauda equine or the caudal part of the two spinal cords in more extreme cases. In addition, they share the lower gastrointestinal tract with anorectal agenesis and rectovesical fistula. Some of these twins share a single pericardium, fused diaphragms, fused xiphoids, and sometimes livers. Urethral duplication may be present, but the urethral meatus is always single. When two bladders are present, they lie side-by-side or anteroposterior in the midline with one bladder draining into the other. The ureters frequently drain into the contralateral bladder. The kidneys usually function normally. Urogenital sinus or cloaca may be present. In boys, there is an increased incidence of undescended testes. These twins can possess tetrapus (the most common), tripus, or bipus.

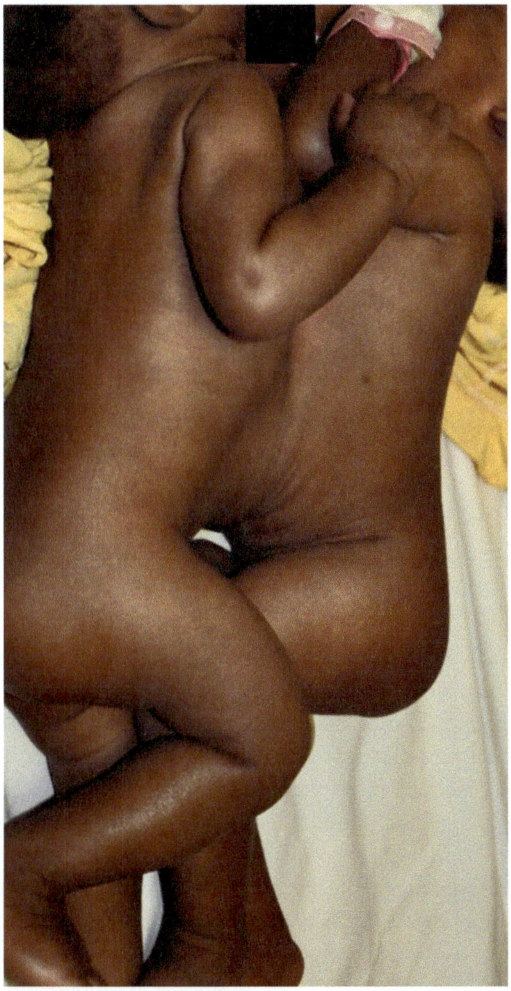

Fig. 50.2 Omphalopagus twin

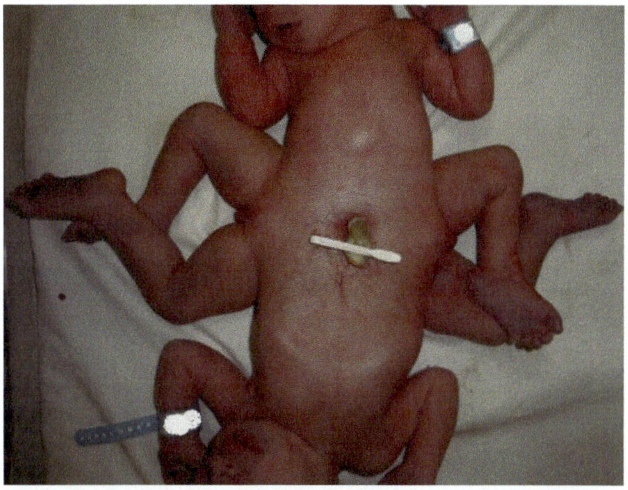

Fig. 50.3 Front view of ischiopagus twin

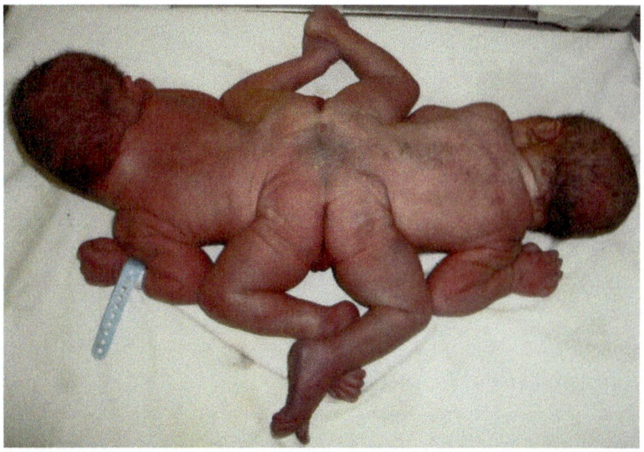

Fig. 50.4 Back view of ischiopagus twin

They have separate hearts, sacra, and symphysis pubis. The outcome is satisfactory despite the anatomical complexity.

Parapagus

The twins share the umbilicus, the lower abdomen, the ilium, the genitourinary tract, and the pelvis. There are single symphysis pubis and single or double sacra. There may be anorectal anomaly with colovesical fistula. The union may involve both trunks including the heads with two faces on the same side (diprosopic parapagus). Union may spare the heads (dicephalic parapagus) with one head may be anencephalic, while the other is normal. When the thoraces are separate, they are dithoracic parapagus. These twins have two vertebral axes, two hearts that may be fused, two sets of lungs, and two sets of foreguts. There are bibrachius, tribrachius, or tetrabrachius and bipus or tripus.

Dorsal Union

Craniopagus

The twins fuse at any portion of the skull except the face and the foramen magnum. They share the meninges and venous sinuses. The brains are separate in some cases, while others show moderate to marked degree of fusion. These twins have separate hearts, diaphragms, and foreguts. There are tetrabrachius and tetrapus. The outcome after surgical separation is extremely difficult to predict. The degree of fusion of venous structures is as important as the degree of fusion of the brain in predicting the outcome.

Rachipagus

Twins direct oppositely and fuse dorsally above the sacrum, which may extend up to the occiput. Associated vertebral anomalies and neural tube defects are common. There are tetrabrachius and tetrapus.

Pygopagus (Fig. 50.5)

Twins face away from each other and join dorsally, sharing the sacrococcygeal and perineal regions. Some twins share the urinary bladder and the anus with one or two rectums. Associated vertebral and pelvic anomalies are usually present. There are tetrabrachius and tetrapus. Sometimes during the separation, one twin has to be sacrificed in favor of the other.

Asymmetrical Conjoined Twins [5]

The estimated incidence is approximately one per million live births. They include the exoparasitic and endoparasitic twins.

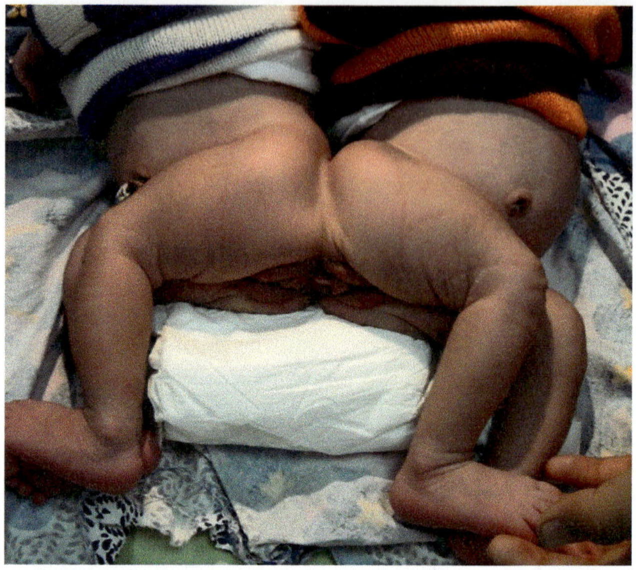

Fig. 50.5 Pygopagus twin

- *Exoparasitic twins*: They occur when one twin (parasite) is incompletely formed or wholly dependent on the cardiovascular system of the complete one (autosite).
- *Endoparasitic (fetus in fetu)*: It results from a diamniotic monozygotic twin. One twin is present inside the body of its partner. The common site is in the abdominal cavity, rarely in the cranial cavity, the neck, the posterior mediastinum, or the sacrococcygeal region. Presentation is usually in infancy or early childhood, but it may present at any age. Complete excision is curative.

Associated Congenital Anomalies

Associated congenital anomalies can either be related to the site of fusion (Fig. 50.6) or unrelated to the site of fusion. These are present in 60% of conjoined twins (Box 50.2).

Antenatal Diagnosis

Antenatal diagnosis of conjoined twins is important for further perinatal management. Prenatal diagnosis of the complex anatomy and associated anomalies has become possible with the use of ultrasonography (US), echocardiography, and MRI. Objectives of prenatal assessment are early-informed counseling of the parents about the probable outcome and the possibility of successful postnatal separation, proper obstetric management, and making decisions regarding termination of pregnancy. Prenatal suspicion of conjoined twins can be as early as 7 weeks' gestation, when differentiation between monochorionic-diamniotic and monochorionic-monoamniotic pregnancies is possible. Detection of shared yolk sac without inter-twin membrane by US raises the index

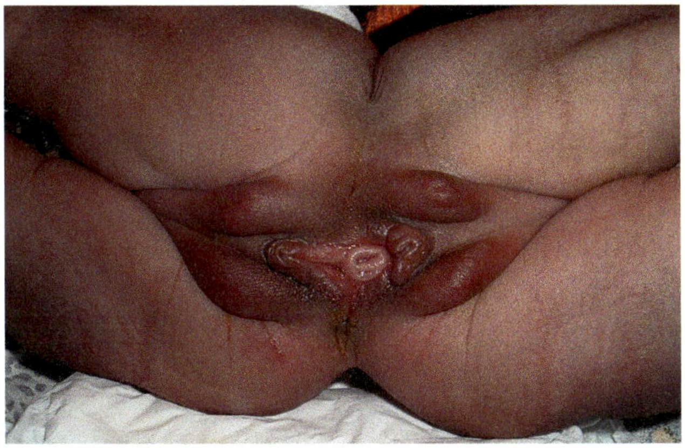

Fig. 50.6 Perineal appearance in a case of pygopagus twin

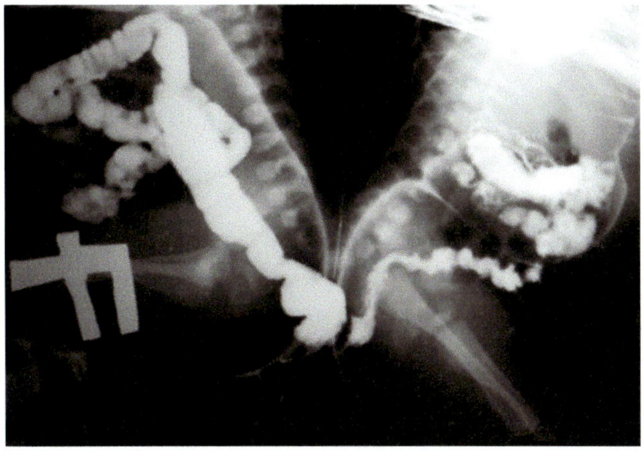

Fig. 50.7 Contrast studies are commonly used to delineate the gastrointestinal relation between the conjoined twins

Box 50.2 Associated Congenital Anomalies Unrelated to the Site of Fusion [1]

Affected systems are:

1. Genitourinary tract (19.8%)
2. Central nervous system (18.9%)
3. Musculoskeletal system (12.6%)
4. Gastrointestinal tract (9.9%)
5. Facial clefts (9.9%)
6. Congenital heart defects (6.3%)
7. Abdominal wall defects (3.6%)

Box 50.3 Sonographic Findings in Conjoined Twins [6]

1. Clear visualization of fused fetal parts
2. Twin pregnancy in a single gestational sac
3. Bifid appearance of the first trimester fetal pole
4. Inseparable fetal bodies and skin contours
5. Unchanged relative position of the fetuses overtime
6. Both fetal heads persistently at the same level and body plane
7. Fewer limbs than expected
8. A single umbilical cord with more than three vessels

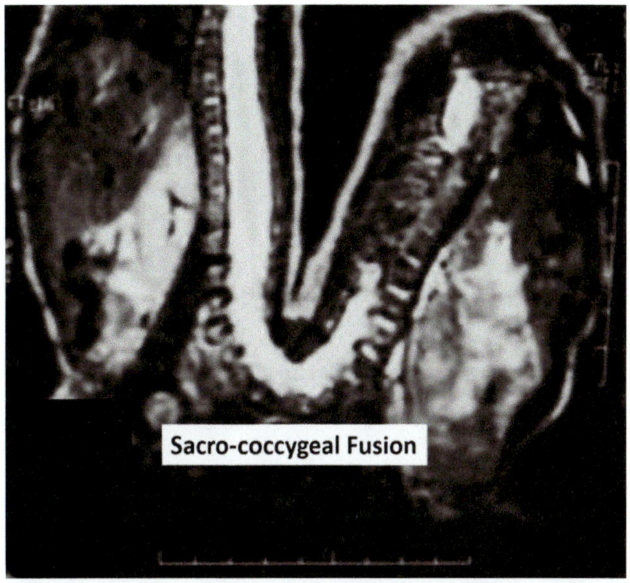

Fig. 50.8 MRI is used for the assessment of the internal anatomy

of suspicion. Better diagnosis requires an additional later scan. Sonographic findings in conjoined twins are highlighted in Box 50.3.

Management

Detailed management of different types of conjoined twins is beyond the scope of this chapter. Management should be undertaken in specialized centers following detailed assessment of the twins and their complex anatomy (Figs. 50.7 and 50.8).

Conjoined Twin Pregnancy Outcome

Live birth is present in 45.6% of cases. Stillbirth and elective termination of pregnancy cases are equal, 27.2% each [1].

Live birth cases fall into three categories depending mainly on the cardiac anatomy:

1. Death shortly after birth, thus nonsurgical management is advocated.
2. Planned separation ideally between 2 and 4 months.
3. Emergency separation at birth, which may represent the only chance of survival.

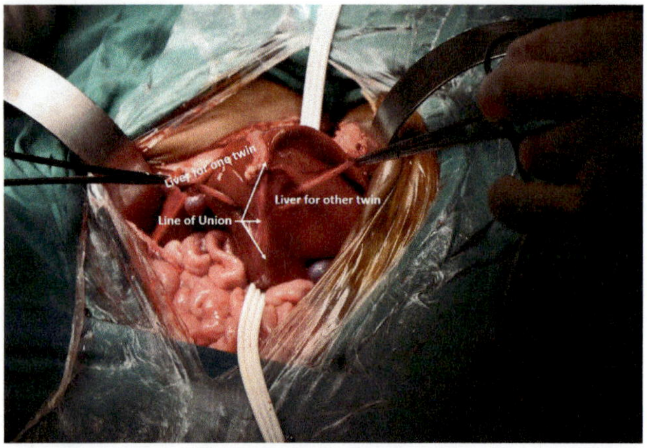

Fig. 50.9 Intraoperative photo during separation of omphalopagus twin with fused livers. Preoperative planning was crucial to help identify proper plan for separation

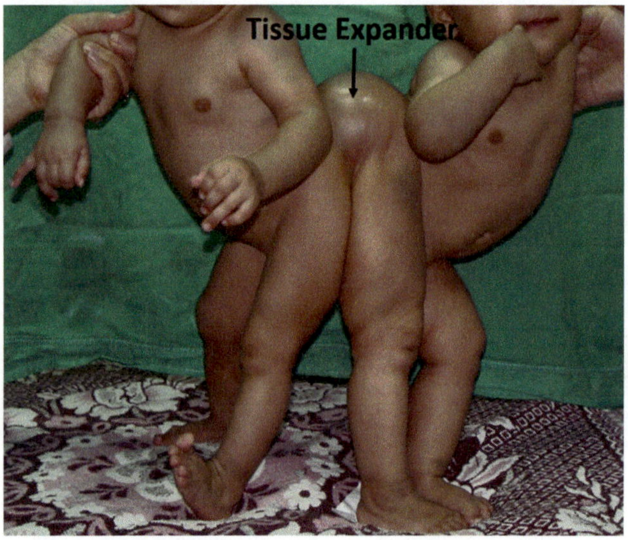

Fig. 50.10 Initial insertion of tissue expanders allows creating the space that helps in the closure at the time of separation

Survival rate is 80–90% in planned separation, compared with 30–50% survival rate in emergency separation. Planned separation has the advantages of reduced risks of anesthesia, better recognition of anatomic relationships (Fig. 50.9) and associated congenital anomalies, and usage of expanders (Figs. 50.10 and 50.11) for adequate wound coverage.

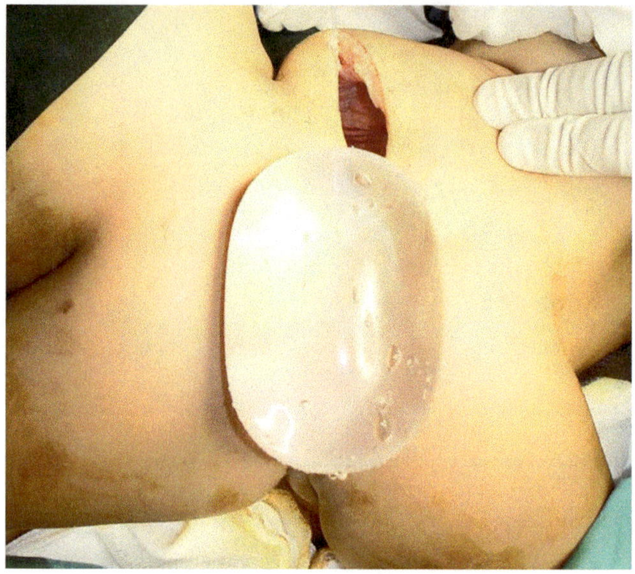

Fig. 50.11 Removal of the tissue expander at the time of separating a pygopagus twin

Acknowledgment The authors would like to dedicate this chapter to the memory of the *late Professor Alaa Fayez Hamza* on behalf of all his trainees, colleagues, and friends, for his overwhelming dedication to the care of sick children and to the pediatric surgical training.

References

1. Mutchinick OM, Luna-Muñoz L, Amar E, et al. Conjoined twins: a worldwide collaborative epidemiological study of the international clearinghouse for birth defects surveillance and research. Am J Med Genet C Semin Med Genet. 2011;157C(4):274–87.
2. Kaufman MH. The embryology of conjoined twins. Childs Nerv Syst. 2004;20:508–25.
3. Spencer R. Theoretical and analytical embryology of conjoined twins: part I: embryogenesis. Clin Anat. 2000;13(1):36–53.
4. Spencer R. Theoretical and analytical embryology of conjoined twins: part II: adjustments to union. Clin Anat. 2000;13(2):97–120.
5. Sharma G, Mobin SS, Lypka M, et al. Heteropagus (parasitic) twins: a review. J Pediatr Surg. 2010;45(12):2454–63.
6. Sharma UK, Dangol A, Chawala CD, et al. Antenatal detection of conjoined twins. J Nepal Med Assoc. 2007;46(167):133–5.

Suggested Readings

Spitz L, Kiely EM. Experience in the management of conjoined twins. Br J Surg. 2002;89(9):1188–92.

Spencer R. Conjoined twins: developmental malformations and clinical implications. Baltimore, MD: The Johns Hopkins University Press; 2003.

Index

A

Abdominal wall defects, 73, 86, 303, 304
  exomphalos, 305, 306
    antenatal/perinatal management, 306, 307
    associations, 306
    body stalk anomaly, 309
    ectopia cordis, 309
    embryology, 306
    long-term outcomes, 307
    OEIS complex, 309
    rectus muscle, divarication of, 307
  gastroschisis, 303
    antenatal/perinatal management, 305
    diagnosis, 303
    embryonic origin, 303–304
    long-term outcome, 305
    prevalence, 304
Aberrant salivary tissue, 167
Accessory auricle, 146, 147, 171
Accessory glands, 14
Accessory salivary glands, 167
Accessory spleens
  clinical Significance, 371
  diagnosis, 372
  etiology, 371
  sites, 371
Acrocephalopolysyndactyly, *see* Carpenter syndrome
Acrosome reaction, 19
Adherent labia, 387
Adrenal anomalies, 398
Adriamycin mouse-treated embryos, 87
Aicardi syndrome, 195, 197
Airway, 289
Albertus Magnus of Cologne, 3
Alimentary duplications, 331
Allantois, 32
Alobar holoprosencephaly, 232
Alpha-fetoprotein, 226
Ambiguous genitalia, 387
Amelia, 452
Ameloblastin, 103
Ameloblasts, 102
Amelogenesis imperfecta (AI)
  hypocalcification, 104
  hypomaturation, 104
  hypomaturation-hypoplasia and taurodontism, 104
  hypoplasia, 104
  incidence, 103
  proteins, 103
  Witkop classification + inheritance pattern, 103, 104
Amelogenin, 103, 104
AMELX (*Xp22.31-p22.1*), 104
American Association of Cleft Palate Rehabilitation (AACPR)
    classification, 121
Amniocentesis, 70
Amnion, 31–32
Amniotic band syndrome, 456, 458, 459
Amniotic cavity, 22
Anal membrane, 381
Anal stenosis
  anal membrane, 382
  anorectal anomalies, 382, 383
  covered anus, 382
  rectal atresia, 382
Anencephaly, 48, 225
Aneuploidy, 57
Angiogenic cells, 243
Angular dermoid cyst, 207
Anhydramnios/oligohydramnios sequence, 65
Aniridia, 198, 466
Ankyloglossia, 108
Annular pancreas, 368, 369
Anodontia, 106
Anomalies, kidney
  cardiovascular, 395
  epidemiology, 393
  gastrointestinal, 395
  long-term outcome, 395
  Mullerian, 395
  pathophysiology, 393–394
  unilateral renal agenesis, 393–395
Anophthalmia, 195, 196
Anophthalmia-oesophageal-genital syndrome, 320
Anophthalmos plus syndrome, 197
Anorchism, 444
Anorectal anomalies, 382, 383, 386
Anorectal malformations (ARMs), 87, 97
  anatomical and functional abnormalities, 381
  anomalies, 381
  congenital aberrations, 384
  continence, 383–384
  female, 385–387
    adherent labia, 387
    ambiguous genitalia, 387
    anorectal anomaly in neonate, 386
    hydrocolpos, 388
    imperforate anus, 386
    rectovaginal fistula, 386
    vaginal cyst, 387
  females, 385
  genetic factor, 381
  heterogeneous defects, 385
  level of Müller's hillock, 384
  male, 384–385

R. Carachi, S. H. E. Doss (eds.), *Clinical Embryology*, https://doi.org/10.1007/978-3-319-26158-4

Anorectal malformations (ARMs) (*cont.*)
high imperforate anus, 385
low imperforate anus, 385
meconium tracking, 386
non-genetic factor, 381
Prune belly syndrome, 384
rare/less well-known associations, 383
segmentation theories, 384
teratogenic factor, 381
voiding dysfunction, 383
Anterior abdominal wall defects, 62
Anterior neuropore, 24
Anti-Müllerian hormone (AMH), 428, 431
Aorta valves, transposition of, 251
Aortic arch anomalies
development of, 253, 276
anomalies of, 277
coarctation, 284
embryology, 284, 285
epidemiologic, genetic, morphologic and clinical correlations, 284–287
Aortic sac, 276
Aortic stenosis, 251
Aortic valves, development of, 250
Aortico-pulmonary septum, 251
APC gene, 470
Apert syndrome, 135, 137, 456
Apical ectodermal ridge (AER), 60, 451, 453
Aplasia, 166, 167
Appendages, 444, 446
Arch cartilage, 170
Arhinia, 154
Arteriovenous malformations (AVMs), 187
Ascent abnormalities, 398–399
Asplenia
clinical significance, 372–373
heterotaxia, 372
isolated asplenia, 372
syndromic asplenia, 372
vs. polysplenia, 372
Associated congenital anomalies, 478, 479
Asymmetric blastocyst cleavage, 66
Atresias, 326, 328
Atria, development of, 245, 253
Atrial septal defects (ASD), 257
Atrioventricular canal, 248, 249
Atrioventricular septal defect, 259
embryology, 259, 260
epidemiologic, genetic, morphologic and clinical correlations, 260, 261
Auriculocondylar syndrome (ARCND), 130
Autosomal dominant inheritance, 57
Autosomal recessive inheritance, 57

B
Babbitt hypothesis, 358
Baldwin and Lecco theories, 369
Bean syndrome, 190
Beckwith-Wiedemann syndrome (BWS), 109, 370, 466
Behavioural and psychiatric conditions
behavioural avoidance, 79
body image, 78
Changing Faces, 81
disfigurement, 81
evolutionary hypothesis, 79
extensive facial disfigurement, 82
facial expressions, 79
facial tumours, 82
fear, 79
helpful skills, 81
language, 81
myths, 80
neurofibromatosis type 1, 82
oppositionality, 78
predatory defence system, 79
psychopathology, 79–80
SCARED, 81
self-esteem, 78
sense of self, 77–78
social smile, 78
socio-demographic and socio-economic factors, 79
three-dimensional model, 78
victims or suffering, 81
Benjamin-Inglis classification, 182
Bianchi technique, 328
Bicoronal craniosynostosis, 137
Bilaminar cloacal membrane, 379, 381, 384
Bilaminar germ disc, 22
Bilateral ectopia, 398
Bilateral grade V reflux, 401
Bilateral hydrocele, 315
Bilateral inguinal hernia, 308, 314
Bilateral renal agenesis
associated anomalies, 397
epidemiology, 395–397
Bilateral ventriculomegaly, 74
Biliary atresia (BA)
BASM syndrome, 355
clinical forms, 373
CMV-IgM +ve-associated biliary atresia, 355–356
cystic changes, 355
incidence, 373
isolated BA, 356
occurence, 354
Biliary atresia splenic malformation (BASM) complex
associated anomalies, 373
clinical significance, 373–374
etiology, 373
Biliary atresia splenic malformation syndrome, 355
Bladder, 412, 413
duplication, 413, 414
exstrophy, 414, 415
Bladder diverticula, 350
Bleomycin, 211
Blueberry muffin rash, 50
Body stalk anomaly, 309
Boerhaave, Herman, 6
Bowel atresias, 74, 303
Bowel reduction, 74
Brain and CNS
central nervous system dysraphism, 224, 225
corpus callosum, agenesis of, 232, 233
cranial dysraphism, 227, 229
holoprosencephaly, 231, 232
hydrocephalus, 230, 231
microcephaly, 233, 234
myelomeningocele, 225–227
neurenteric cysts, 227
NTD, 223

Rathkes cleft cysts, 234
spina bifida occulta, 223, 224
spina bifida/cystica, 224
spinal dysraphism, 225
VGAM, 234, 235
Branchial arch anomalies, 171
Branchial arches, 169
Branchial cyst, 171
Branchial fistulae, 172
Branchial sinus infection, 172
Branchio-oto-renal syndrome, 147
Bronchial atresia, 292
Bronchogenic cysts, 175, 292
Bulbus cordis, 244, 249

C

Canine model, 86
Capillary malformation-arteriovenous malformation (CV-AVM) syndrome, 189
Capillary malformations, 186, 187, 189
Cardiac abnormalities
aorta and pulmonary valves, transposition of, 251
aortic and pulmonary valves, development of, 250
aortic stenosis, 251
atria, development of, 245, 253
atrial septal defects, 257–259
atrioventricular canal, anomalies of, 248, 249
atrioventricular septal defect, 259
embryology, 259, 260
epidemiologic, genetic, morphologic and clinical correlations, 260, 261
cardiac malformations, 254
embryology, 254
epidemiologic, genetic, morphologic and clinical correlations, 255, 258
common arterial trunk, 270
congenital semilunar valves malformations, 265
embryology, 265
epidemiologic, genetic, morphologic and clinical correlations, 265, 266
diagnosis, segmental approach, 254
double outlet right ventricle, 266, 267, 269
great arteries, transposition of, 269, 270
heart
anomalies of position, 252
development, 252–254
embryology, 243
internal structures of, 244, 245
heart tube, differentiation of, 243, 244
isomerism and anomalies of cardiac position, 270–272
persistent truncus arteriosus, 251
pulmonary atresia with ventricular septal defect, 266–268
semilunar valves, anomalies of, 251
septum anomalies, 247, 248
tetralogy of fallot, 266–268
total anomalous pulmonary venous drainage, 262
embryology, 262
epidemiologic, genetic, morphologic and clinical correlations, 262–264
truncus arteriosus, anomalies of, 250
univentricular hearts, 272, 273
ventricles, development of, 249
Cardiac anomalies, 300, 337
Cardiac loop formation, 252
Cardiac malformations, 254
embryology, 254
epidemiologic, genetic, morphologic and clinical correlations, 255, 258
Cardiac position, anomalies of, 270–272
Cardinal veins, 281
Cardiovascular anomalies, 395
Caroli syndrome, 356, 358
Carpenter syndrome (acrocephalopolysyndactyly), 138, 139
Caudal regression syndrome, 229
Ccorona radiata, 16
Central nervous system (CNS)
anomalies, 337
dysraphism, 224, 225
Cephalopagus, 476
Cervical thymic cysts, 173
CHARGE (coloboma, heart defects, choanal atresia, retardation of growth/development, genital anomalies, ear anomalies) syndrome, 147, 195, 197, 320
Chest wall deformities, 237
antenatal considerations, 240
causes, 239
clinical features, 239
conditions, 240
embryology, 239
epidemiology, 237
outcomes, 240
Chiari II malformation, 225
Chiari malformation, 74
Chicks model, 85
Child's self-awareness, 77
Choanal atresia, 94, 154, 155
Choanal stenosis, 154
Cholangiocytes, 356
Choledochal malformation (CM)
absorption, 358
aetiology, 358
ampulla of Vater, 358
Caroli syndrome, 356, 358
clinical features, 358
common channel, 358
distal biliary obstruction, 358
incidence, 356
types, 356
Chorion
embryo and embryonic structures, 34
primary chorionic villi, 33
secondary chorionic villi, 34
tertiary villi, 34
villous tree, 34, 35
Choroid, 193
Choroidal vascular malformations, 189
Chromosomal abnormalities, 69
Chwalle's membrane, 403, 405
Cilia, 12
Ciliary body, 193
Cleft lip (CL)
aesthetic issues, 113, 114
aetiology, 116
antenatal diagnosis, 117
antenatal management, 117
associated defects, 113
classic theory, 115
definition, 112
dental irregularities, 114

Cleft lip (CL) (*cont.*)
- disease registries and patient groups, 117
- dynamic fusion theory, 115
- environmental risk factors, 116, 117
- epidemiology, 112
- genetic factors, 117
- geographical variation, 112
- incidence, 112
- labial sounds impaired production, 114
- parents and siblings facial topography, 113
- psychosocial issues, 114
- sex variation, 113
- Streeter's merging theory, 115
- types, 113

Cleft palate
- antenatal diagnosis, 117
- antenatal management, 117
- cleft lip formation, 115
- developmental pathogenesis, 115
- disease registries and patient groups, 117
- feeding issues, 114
- formation, 116
- middle ear effusion, 115
- primary palate
  - molecular mechanisms, 115
  - normal embryogenesis, 115
- secondary palate
  - molecular mechanism, 116
  - normal embryogenesis, 116
- speech issues, 114
- types, 113
- velopharyngeal incompetence, 115

Cleft sternum, 238
Cleidocranial dysostosis, 238
Cleidocranial syndrome, 107
Cloaca, 417, 418
Cloacal membrane, 417, 418
Cloacal membrane ruptures, 415
Cloacal/bladder exstrophy, 98
CLOVES syndrome, 189
Club feet/foot, 458
CMV-IgM +ve-associated biliary atresia, 355–356
Coloboma, 192
Colonic atresia, 342–344
Common arterial trunk, 270
Common cardinal vein, 244
Common channel, 358
Common dorsal aorta, 278
Complete androgen insensitivity syndrome (CAIS), 313, 432
Complete/deformed tracheal rings
- aetiology and patho-embrology, 184
- definition, 184
- incidence, 184
- presentation, clinical features and outcome, 184

Complex microphthalmia, 196
Composite odontomes, 102
Concresence, 102
Congenital abnormalities, 72
Congenital adrenal hyperplasia (CAH), 433
Congenital anomalies, 69, 85
- aetiology, 55
- animal models, 85
- anterior abdominal wall defects, 62
- in children, 88
- chromosomal aberrations, 57, 59
- congenital left diaphragmatic hernia, 62
- congenital limb deficiency defect, 61
- definition, 55
- developmental field, 59
- epigenetic modifications, 59
- genetic defects, 55–57
- genitourinary congenital anomalies, 60
- human embryonic development, 55
- monozygotic/identical twinning, 63, 66
- neural tube defect, 59
- pathogenesis, 87
- primary congenital defect, 60
  - association, 61
  - bilateral renal agenesis, 63
  - dysplasia, 60
  - malformations, 60
  - sequence, 61
  - single/multiple defects, 60
  - syndrome, 61
  - type II thanatophoric dysplasia, 64
- secondary congenital defect, 60, 61
  - complex, 61
  - deformation, 61
  - disruption, 61
  - OEIS complex, 66
- teratogen animal models, 86

Congenital anomalies, 88
Congenital anomalies of the kidney and urinary tract (CAKUT), 337
Congenital cardiac defects, 255
Congenital cataract, 199, 200
Congenital cervical teratomas, 173, 174
Congenital dermal melanocytosis, 208
Congenital diaphragmatic hernia (CDH), 60, 85, 91, 95, 298–300
Congenital haemangiomas, 185, 187
Congenital heart defects (CHD), 71
Congenital high airway obstruction syndrome (CHAOS), 180
Congenital hydroceles, 314, 315
Congenital hypoparathyroidism, 163
Congenital hypothyroidism, 162
- aetiology, 163
- epidemiology, 162, 163
- prognosis, 163

Congenital left diaphragmatic hernia, 62
Congenital limb anomalies, 454
Congenital limb deficiency defect, 61
Congenital lobar emphysema (CLE), 293, 294
Congenital malformation
- at birth, 93
- before birth
  - abdominal malformations, 92
  - advantages and disadvantages, 91
  - defects, 92
  - head and neck malformations, 92
  - in utero *transfer*, 91
  - spontaneous or induced, and caesarean section, 92
  - thoracic malformations, 92
  - urogenital apparatus/pelvis, 92
- choanal atresia, 94
- cleft lip and palate, 94
- cloacal/bladder exstrophy, 98
- congenital diaphragmatic hernia, 95
- eye coloboma, 192, 193, 195
- gastroschisis, 96
- intestinal occlusion, 97
- myelomeningocele, 98

oesophageal atresia, 95
omphalocele, 96
sacrococcygeal teratoma, 97
severe neck lymphangioma, 95
uropathies, 97–98
Congenital malformations, 381
Congenital melanocytic naevus (CMN), 207
Congenital motility disorders
HD, developmental genetics, 336
Hirschsprung's disease, 334, 335
normal enteric nervous system development, 335, 336
Congenital muscular torticollis (CMT), 173
Congenital nasal swelling, 152, 155
Congenital oesophageal stenosis, 319, 320
Congenital portocaval shunt (CPS), 360
Congenital pulmonary airway malformation (CPAM), 294
Congenital pyloric obstruction, 322, 323
Congenital semilunar valves malformations, 265
embryology, 265
epidemiologic, genetic, morphologic and clinical correlations, 265, 266
Congenital subglottic stenosis, 180
Congenital tracheal malformations, 291
Congenital tracheomalacia, 292
Congenital tumours, 167
cancer model, 464
embryological basis, 463, 464
GCTs, 466–468
hepatoblastoma, 469, 470
neuroblastoma, 468, 469
RMS, 470, 471
Wilms' tumour, 464, 466
Congenital vascular anomalies, 181
Conjoined twin, 475
antenatal diagnosis, 478, 479
associated congenital anomalies, 478, 479
etiology, 475
fission theory, 475
fusion theory, 475
spherical theory, 475, 476
management, 479
outcome, 479, 480
symmetrical, 476
asymmetrical, 478
cephalopagus, 476
craniopagus, 478
ischiopagus, 477, 478
omphalopagus, 477
parapagus, 478
pygopagus, 478
rachipagus, 478
thoracopagus, 476
Consecutive anophthalmia, 195
Constriction band syndrome (CCBS), 452, 456
Contralateral testis, 432
Corona radiata, 19
Coronal hypospadias, 424
Coronary sinus-type defect, 258
Corpora cavernosa, 421
Corpus callosum, agenesis of, 232, 233
Corpus spongiosum, 421, 422
Cowden/PTEN hamartoma tumour syndrome, 189
CRANE database, 117
Cranial dysraphism, 227, 229
Cranial suspensory ligament (CSL), 437
Craniofacial syndrome
antenatal detection in craniosynostosis, 140
Apert syndrome, 137
carpenter syndrome, 138, 139
craniofrontonasal dysplasia, 139, 140
Crouzon syndrome, 136, 137
encephaloceles, 141, 142
epidemiology, 136
genetics
ephrin-B1 gene, 136
FGFR2 gene, 135
FGFR3 gene, 136
POLR1C and POLR1D gene, 136
TCOF1 gene, 136
TWIST1 gene, 136
incidence, 135
Muenke (Pro 250Arg) syndrome, 138
Pfeiffer syndrome, 137, 138
Saethre-Chotzen syndrome, 138, 139
Treacher Collins syndrome, 140, 141
Craniofrontonasal dysplasia, 139, 140
Craniopagus, 478
Craniosynostosis, 135–140
Criss-cross heart, 271
Crouzon syndrome, 135–137
Crouzonoid facial appearance, 135
Cryptophthalmos, 196
Cryptorchidism, 439
Cumulus oophorus, 16, 19
Currarino syndrome, 460
Cutis aplasia, 206, 207
Cyst, 444
Cystic biliary atresia, 355
Cystic hygroma, *see* Severe neck lymphangioma
Cystic hygromas, 175
Cystic lesions, 172
Cystogram reflux, 400
Cytogenetic map, 88
Cytomegalovirus (CMV), 50

D

Dandy-Walker malformation, 142
da Vinci, Leonardo, 3
Decidua basalis, 35
Decidua capsularis, 35
Decidua parietalis, 35
Definitive yolk sac, 32, 33
Dens evagination, 102
Dens invaginatus, 102, 103
Dental anomalies, *see* Tooth defects
Dental crowding, 106
Dentine defects, 104–105
Dentine dysplasia (DD), 104
Dentinogenesis imperfecta (DGI), 104, 105
Depression and anxiety disorders, 79
Dermal melanocytosis, 208
Dermoid cysts, 174, 206
Dermomyotome, 26
Dextrocardia, 252, 271
Diabetes, 52
Diaphragm
CDH, 298–300
diaphragmatic eventration, 300
DiGeorge syndrome, 59, 130

Dihydrotestosterone (DHT), 431
Direct inguinal hernias, 314
Disorders of sex development (DSD)
  abnormal embryology in, 428, 430, 432, 433
  clinical assessment of newborn with, 433–435
  normal embryology, 427
Distal ureteric segment, 403
Diverticuli, 167
Double aortic arch, 277, 286, 288
Double outlet right ventricle, 266–269
Down syndrome (DS), 70, 107, 260, 337
Doxycycline, 211
Ductal (or limiting) plate, 356
Ductal anomalies, 167
Ductal cysts, 181
Ductus venosus, 43, 360
Duplex systems
  associated anomalies, 404
  associated syndromes, 404
  ectopic ureter 2 retrogrades, 403
  epidemiology, 402
  pathophysiology, 402–403
Dynamic fusion theory, 115

E

Ectoderm, 23, 24, 453
Ectodermal clefts, 170
Ectodermal dysplasia, 106, 107
Ectopia cordis, 240, 252, 270, 309
Ectopic kidneys
  adrenal anomalies, 398
  associated syndromes, 398
  epidemiology, 398
  genital anomalies, 398
  long-term outcome, 398–399
Ectopic labium, 434
Ectopic or accessory pancreatic tissue, 369
Ectopic salivary tissue, 167
Ectopic scrotum, 434
Ectopic thyroid, 161
Ectopic thyroid gland, 160
Edelman, Gerald, 9
Egyptian sculpture, 1
Ehlers-Danlos syndrome, 152, 239
Embryogenesis, 252
ENAM (*4q13.3*), 104
Enamalin, 103, 104
Enamel defects, 103
Enamel pearls, 102
Encephalocele, 152, 153, 229
  anterior, 229
  with anencephaly, 230
Encephaloceles, 141, 142, 227, 229
Endochondral ossification, 453
Endoderm, 23, 24
  embryonic folding, 28
  foetus development, 28
  foetus, external features, 30
  germ layers, 29
Endoparasitic, 478
Endothelin receptor type B (EDNRB) gene, 336
Enteric nervous system (ENS), 335
Enterostomy, 344
Ephrin-B1 gene (EFNB1), 136
Epiblast, 22
Epididymal cyst, 444
Epididymis, 14
Epigastric hernia, 308
Epigastric hernias, 315
Epigenesis, 7
Epigenetic modifications, 59
Epignathus, 154
Epispadias, 424
Epulis, 108
Eruption cyst, 107, 108
Ethanol, 211
EUROCRAN (European Collaborations on Craniofacial Anomalies), 117
Eventration, 300, 301
Ex utero intrapartum treatment (EXIT) procedure, 154, 167, 174
EXIT (*ex utero* intrapartum treatment) procedures, 117
Exogenous agents, 86
Exomphalos, 305, 306
  antenatal/perinatal management, 306, 307
  associations, 306
  body stalk anomaly, 309
  ectopia cordis, 309
  embryology, 306
  long-term outcomes, 307
  OEIS complex, 309
  rectus muscle, divarication of, 307
Exoparasitic twins, 478
Exstrophy-epispadias complex (EEC), 414, 415
External angular dermoid, 206
External auditory canal atresia, 146
External ear malformations
  external auditory canal atresia, 146
  microtia, 145, 146
  minor deformity, 146, 147
Extrahepatic biliary atresia (EHBA), 87
Extralobar sequestration (ELS), 295
Eye
  congenital malformations of, 192, 193, 195
  disorders
    aniridia, 198
    anophthalmia, 195, 196
    congenital cataract, 199, 200
    microphthalmia, 196, 197
    nanophthalmia, 197, 198
  embryology of, 191

F

Fabricius's treatise, 4
Facial clefts
  AACPR classification, 121
  aetiology, 121
  associated anomalies
    lateral (transverse), 121
    median (lower lip), 122, 123
    median (upper lip), 121
    oblique, 121
  clefts 10–14, 125
  clefts 1–9, 124–125
  clinical features, 121
  congenital, 120
  definition, 119
  embryological classification system, 119
  epidemiology, 120
  orbito-centric descriptive classification system, 121
  patho-embryology, 124

Tessier cleft 30 (median mandibular cleft), 125
Tessier clefting system, 122
treatment, 122
FAM83H (8q24.3), 104
Familial hypocalcaemia, 163
Fanconi anaemia, 318, 320
Feingold syndrome, 320
Female genital organs
genital ducts, 11
ovary, 11
uterine tube, 11, 12
uterus, 13
Femoral hernia, 315
Fertilisation, 19–20
FGFR2 gene, 135
FGFR3 gene, 136
FGFR3 skeletal dysplasia group, 61
First branchial arch anomalies, 171
Fission theory, 475
Floyd classification, 291
Fluorescence in situ hybridization (FISH), 71, 469
Foetal alcohol spectrum disorder (FASD), 52
Foetal alcohol syndrome (FAS), 51–52
Foetal anomalies
biochemical or enzymatic assays, 71
cytogenetic analysis, 71
diagnostic tests, 70
DNA analysis, 71
first trimester screening, 70
historical perspective, 71
MRI, 72
prenatal diagnosis, 69
screening tests, 69
second trimester screening, 70
structural abnormalities, 71
three-dimensional ultrasound, 72
triple test, 70
ultrasound, 71
Foetal circulation
after birth, 43
anatomy, 42
biochemical adaptations, 42
ductus arteriosus, or arterial duct, 43
ductus venosus, 43
foramen ovale, 43
persistent pulmonary hypertension of the newborn, 42
Foetal endoluminal tracheal occlusion (FETO), 86
Foetal endoscopic tracheal occlusion (FETO), 299
Foetal intervention, 85
Foetal lamb model, 85, 86
Foetal malformation
joint consultation, 99
long-term project of care, 99
perinatal project of care, 99
prenatal diagnosis, 98
prognostic evaluation, 99
Foetus in foetu, 63
Folate deficiency, 59
Folate insufficiency, 225
Folic acid, 48
Follicular cells, 15, 16
Foramen ovale, 247
Foregut, 317
congenital oesophageal stenosis, 319, 320
congenital pyloric obstruction, 322, 323
duplication, 321
duplications, 320, 321
gastric volvulus, 321, 322
oesophageal atresia and tracheo-oesophageal fistula, 317, 318
Foregutduplication, 321
Fourth branchial arch anomalies, 172, 173
Franceschetti-Zwahlen-Klein syndrome, *see* Treacher Collins syndrome (TCS)
Frenulum, 422
Frontonasal syndrome, 153
Furstenberg's sign, 152, 153
Fusion, 102
Fusion abnormalities, 398–401
Fusion theory, 475

G

Galen of Pergamon, 2
Gallbladder, 359
Gametogenesis, 15
Ganglioneuromatosis, 340
Gardner's syndrome, 106, 107
Gastric volvulus, 321, 322
Gastrointestinal anomalies, 395
Gastro-oesophageal reflux disease (GORD), 179, 300
Gastroschisis, 73, 96, 303, 304
antenatal/perinatal management, 305
embryological origins of, 304
long-term outcome, 305
Gemination, 102
Gene silencing, 59
Genetic counselling, 338
Genetic models, 86–87
Genital anomalies, 398
Genital tubercle, 418, 421
development
corpora cavernosa, 421
glans, 420
Genital tubercle, development of, 419
Genitofemoral nerve (GFN), 439
Genitourinary congenital anomalies, 60
Germ cell tumours (GCTs), 466–468
Germ plasm theory, 8
Germinomas, 468
Ghost cells, 207
Giraldés, Joachim A. C. C., 446, 447
Glans, 420
Glanular urethra, 419, 420
Glial cell line-derived neurotrophic factor (GDNF), 336, 397
Gliomas, 152
Glomerulosclerosis, 395
Glomovenous malformations, 186
Glue ear, 115
Goitre, 162
Goldenhar syndrome, 125, 129
Goltz's syndrome, 197
Gonadoblastoma, 468
Gonioscopy, 198
Granulosa cells, 16
Great arteries, transposition of, 269, 270
Great vessels
aortic arch, 277
common dorsal aorta, 278
congenital anomalies
aorta, aortic arch anomalies coarctation and vascular rings, 284, 286, 288
patent arterial duct, 282–284

Great vessels (*cont.*)
embryology of, 276
inferior vena cava, 281
renal vein, 282
upper and lower limbs, arteries development of, 278, 279
veins, 279
cardinal, 281
vitelline and umbilical, 280
Gubernaculum, 437

H

Haemangioma/lymphangioma of tongue, 109, 173, 181, 186, 188, 208–210
Haematopoietic cells, 356
Hallermann-Streiff syndrome (HSS), 129, 130
Hamartomas, 293
Hartsoeker, Nicolaas, 2
Harvey, Willaim, 4
Heart
anomalies of position, 252
development, 252–254
internal structures of, 244, 245
Hemifacial microsomia, 127, 128
Hepatoblastoma, 469, 470
Hepatoblasts, 356
Hereditary gingival fibromatosis, 107
Hernia uteri inguinalis, 432
Hernias, 311
direct inguinal hernias, 314
femoral hernias, 314
inguinal hernia
embryological considerations and pathophysiology, 312
incidence and associations, 312–314
spigelian hernias, 315
Heterotaxia, 372
Heterotaxia syndromes/isomerisms, 271, 370
Heterotopic pancreas, *see* Ectopic or accessory pancreatic tissue
Heuser's membrane, 32
Hillocks of His, 145
Hindgut
associated anomalies, 336, 337
CAKUT, 337
cardiac anomalies, 337
CNS, 337
genetic counselling, 338
HD, syndromic associations of, 337
outcome, 339
registries, 339
therapies, 339
congenital motility disorders
HD, developmental genetics of, 336
Hirschsprung's disease, 334, 335
normal enteric nervous system development, 335, 336
Hirschsprung's disease, 339
colonic atresia, 342–344
duplications, 341, 342
histopathologica abnormalities, 340
hypoganglionosis, 340, 341
IND, 340
intestinal ganglioneuromatosis, 340
MMIHS, 341
problems, 340
ultrashort segment HD, 340
Hippocrates, 2
Hirschsprung's disease (HD), 334, 335, 339
colon, 335
colonic atresia, 342–344
developmental genetics, 336
duplications, hindgut, 341, 342
histopathologica abnormalities, 340
hypoganglionosis, 340, 341
IND, 340
intestinal ganglioneuromatosis, 340
isolated and syndromic, susceptibility genes in, 338
MMIHS, 341
problems, 340
proposed variants, 339
syndromic associations of, 337
ultrashort segment HD, 340
Holoprosencephaly, 231, 232
Homer's cyclops, 45
Horseshoe kidney
bilateral grade V reflux, 401
bilateral reflux grade V, 400
cystogram reflux, 400
epidemiology, 399
genital anomalies, 400
isotope, 399
long-term outcome, 400–401
Megacystis-megaureter reflux, 400
Hoxb gene, 239
Human congenital malformations, 86
Hunter, William, 6
Hurler syndrome, 129
Hyaline cartilage models, 452
Hydroceles, 311
congenital hydroceles, 314, 315
direct inguinal hernias, 314
epigastric and supraumbilical hernias, 315
femoral hernias, 314
inguinal hernia
embryological considerations and pathophysiology, 312
incidence and associations, 312–314
Hydrocephalus, 74, 225, 230, 231
CT scan, 231
neonatal, 231
Hydrocolpos, 388
Hydronephrosis, 398, 404
Hyoid arch, 169
Hyperdontia, 106
Hypoblast, 22
Hypocalcification amelogenesis imperfecta, 104
Hypodontia, 106
Hypoganglionosis, 340
Hypomaturation amelogenesis imperfecta, 104
Hypomaturation-hypoplasia and taurodontism amelogenesis imperfecta, 104
Hypoparathyroidism, 163
Hypoplasia, 272
Hypoplastic amelogenesis imperfecta, 104
Hypospadias, 422
Hypothalamic-pituitary-gonadal axis, 431
Hypothyroidism, 309

I

Imperforate anus, 386
Implantation, 20, 21

Indwelling silicone transanastomotic tube, 86
Infantile haemangiomas, 185, 186
Inferior vena cava, 281
Inguinal hernia, 312, 314
  development, risk factors for, 314
  embryological considerations and pathophysiology, 312
  incidence and associations, 312
    clinical presentation, 313
    development, risk factors for, 313
    ongoing debate, management and areas, 313, 314
    sex, 313
    side, 313
Inguinofemoral testis, 440
Inguinoscrotal hernia, 312
Inner ear malformations, 147–148
Interatrial communication, 258
Interatrial septum, 248
Intermediate mesoderm, 27
Intermediate trophoblasts, 463
Intersegmental arteries, 278
Interventricular septum, 249
Intestinal atresia, 327
Intestinal damage, in infants, 86
Intestinal neuronal dysplasia (IND), 340
Intestinal occlusion, 97
Intracranial calcification, 50
Intralobar sequestration (ILS), 295
Intrathoracic thyroid, 161
Intra-tracheal thyroid tissue, 161
Iris, 192
Ischiopagus, 477, 478
Isolated cleft palate (iCP), 112, 113, 116, 117
Isomerism, 270–272

K
Kaposiform haemongioendotheliomas, 189
Kasabach-Merritt phenomenon, 189
Kidney
  anomalies
    cardiovascular, 395
    epidemiology, 393
    gastrointestinal, 395
    long-term outcome, 395
    Mullerian, 395
    pathophysiology, 393–394
    unilateral renal agenesis, 393–395
  bilateral renal agenesis
    associated anomalies, 397
    epidemiology, 395–397
  ectopic kidneys
    adrenal anomalies, 398
    associated syndromes, 398
    epidemiology, 398
    genital anomalies, 398
    long-term outcome, 398–399
  horseshoe kidney
    bilateral grade V reflux, 401
    bilateral reflux grade V, 400
    cystogram reflux, 400
    epidemiology, 399
    genital anomalies, 400
    isotope, 399
    long-term outcome, 400–401
    megacystis-megaureter reflux, 400
  mesonephros, 391
  metanephros, 392
  multicystic dysplastic kidneys, 401
    associated anomalies, 402
    epidemiology, 401–402
    long-term outcome, 402
    pathophysiology, 402
  nephron, 392
  pronephros, 391
  supernumerary kidneys
    associated anomalies, 397
    associated syndromes, 397
    epidemiology, 397
    long-term outcomes, 397–398
    pathophysiology, 397
  tubules collection, 392
  upper urinary tract, 393
Kleeblattschädel deformity, 138, 139
Klippel–Trénaunay syndrome (KTS), 210, 460
Knock-in models, 87
Knock-out model, 87

L
L1 cell adhesion molecule (L1CAM), 57
Labioscrotal swellings, 418, 421
Labour suite management, 74
Lacrimo-auriculo-digital (LADD) syndrome, 166
Ladd and Gross classification, 182
Laryngeal atresia/stenosis, 290
Laryngeal cysts
  aetiology and patho-embrology, 181
  definition, 181
  incidence, 181
  presentation, clinical features and outcome, 181
Laryngeal webs and atresi
  aetiology and patho-embrology, 180
  associations, 181
  definition, 180
  incidence, 180
  presentation, clinical features and outcome, 180
Laryngeal/tracheal cleft
  aetiology and patho-embrology, 181
  associations, 182
  definitions, 181
  incidence, 181
  presentation, clinical features and outcome, 182
Laryngomalacia, 290
  aetiology and patho-embrology, 179
  associations, 179
  definition, 179
  incidence, 179
  presentation, clinical features and outcome, 179
  syndromes associated with, 179
Laryngo-tracheal groove, 177
Laryngo-tracheo-oesophageal cleft (LC), 289
Larynx, 177
Lateral plate mesoderm, 27
Lateral splanchnic branches, 278
Lateralization defect, 372
Lens, 193
Lens vesicle, 191
Lenz microphthalmia, 197
Levocardia, 271
Leydig cells, 18

Limb
abnormalities, classification of, 452
amniotic band syndrome, 456, 458, 459
clinical embryology, 453
ectoderm , differentiation of, 453
limb bud, rotation of, 454
limb bud, segmentation of, 454
limb buds, development, 453
mesoderm and ectoderm, differentiation of, 453
congenital abnormalities of, 452
congenital anomalies, 454–456
constriction band syndrome, 452
deficiencies, types of, 452
development and growth, 451
duplication, 452
limb bone and muscle development, stages of, 452
limb buds, rotation of, 451
management, principles of, 453
Lingual thyroid, 160
clinical presentation, 160
epidemiology, 160
Liver
biliary atresia
BASM syndrome, 355
CMV-IgM +ve-associated biliary atresia, 355–356
cystic changes, 355
isolated BA, 356
occurence, 354
choledochal malformation
absorption, 358
aetiology, 358
ampulla of Vater, 358
Caroli syndrome, 356
clinical features, 358
common channel, 358
distal biliary obstruction, 358
incidence, 356
types, 356
congenital vascular anomalies
congenital portocaval shunt, 360
posterior cardinal veins, 360, 362
umbilical veins, 360
vitelline veins, 360, 361
Lobar holoprosencephaly, 232
Lobster claw, 455
Lobular spleen, 371
Lop ear deformity, 146
Losee classification system, 151
Lower female genital tract, 428
Lumbosacral agenesis, 460
Lung agenesis, 293
Lungs, 289
Lymphangiomas, 175
Lymphangiomatosis, 210, 211
Lymphatic malformations (LMs), 188

M
Macrodactyly, 455
Macroglossia, 109
Macrostoma, 128
Maffucci syndrome, 190
Magnetic resonance imaging (MRI)
thyroid, 162
vascular tumours, 189
Male tract
accessory glands, 14
epididymis, 14
gametogenesis, 15
penis, 14
testes, 14
vas or ductus deferens, 14
Malpighi, Marcello, 4
Malposition, 270
Mandibular arch, 169
Mandibular congenital malformations
auriculocondylar syndrome, 130
branchial arches, 127
classification system, 130
DiGeorge syndrome/velocardiofacial syndrome, 130
Goldenhar syndrome, 129
Hallermann-Streiff syndrome, 130
hemifacial microsomia, 127, 128
Hurler syndrome, 129
Nager syndrome, 130
Pierre Robin syndrome, 128
Pruzansky classification, 130
Treacher Collins syndrome, 129
Mandibulofacial dysostosis, *see* Treacher Collins syndrome (TCS)
Marfan's syndrome, 106, 240, 457
Maternal serum alpha-fetoprotein (MSAFP), 415
Maturation stage, 103
Mature spermatozoa, 18
Meckel's and Reichert's cartilage, 147
Meckel's diverticulum, 331, 348, 350, 351
Megacystis microcolon intestinal hypoperistalsis syndrome (MMIHS), 341
Megacystis-megaureter reflux, 400
Megameatus intact prepuce (MIP), 422
Meningocele, 48, 49, 226, 227
Meningoencephaloceles, 152
Mesenchymal cap, 257
Mesenchyme-to-epithelial transition (MET), 464
Mesiodens, 106
Mesoderm, 23–26, 453
Mesonephros, 391
Metanephric cap, 392
Metanephros, 392, 404
Meyer Weigert law, 403
Microcephaly, 233, 234
Microcolon ileal atresia, 327
Microdeletion, 59
Micrognathia, 128–130
Micropenis, 431
Microphthalmia, 196, 197
Microtia, 145–146
Middle ear malformations, 147
Midgut, 326
alimentary duplications, 331
atresias, 326, 328
rotational deformities, 328, 330, 331
Midline cervical cleft (MCC), 175
Mid-shaft hypospadias, 424
Mirror/rouge test, 77
Misoprostol, 206
MMP20 (*11q22.3*), 104
Mondini deformity, 148
Motor branch, 170
Mouth duplication, 109, 128
Muenke (Pro 250Arg) syndrome, 136, 138
Mullerian anomalies, 395
Multicystic dysplastic kidneys (MCDKs), 401

associated anomalies, 402
epidemiology, 401–402
long-term outcome, 402
pathophysiology, 402
Multiple endocrine neoplasia type 2 (MEN2A), 337
Muscular element, 170
Myelomeningocele, 48, 49, 91, 98, 225–227
Myelomeningocele cystic, 227
Myelomeningocele-thoracic, 228
Myers and Cotton classification, 180
Myocardium, 254

N

Naevus, 207, 208
Nager syndrome, 130
Nanophthalmia, 197, 198
Nasal clefts, 153
Nasal dermoids, 151, 152
Nasal hypoplasia, 151
Nasal septum, 112
Nasopharyngeal teratoma, 154
Neck, 160, 161
Neonatal disorders, 85
Neoplastic cells, 464
Nephroblastoma, 464
Nephrogenic rests, 464, 466
Nephron, 392
Nervous element, 170
Neural crest cells (NCCs), 336
Neural groove, 24
Neural plate, 23, 25
Neural tube, 23–25
Neural tube defects (NTD), 71, 223
Neurenteric cysts, 227
Neuroblast cells, 469
Neuroblastic tumours, 470
Neuroblastoma, 468, 469
Neurocristopathies, 337
Neurofibromatosis, 152
Neuropathic bladder, 74
N-myc, 469
Non-involuting congenital haemangiomas (NICHs), 187
Non-syndromic clefting, 112, 117
Noonan's syndrome, 211
Norrie disease, 197
Nose and paranasal sinuses
arhinia, 154
choanal atresia, 154, 155
encephalocele, 152, 153
epignathus, 154
gliomas, 152
nasal clefts, 153
nasal dermoids, 151, 152
nasopharyngeal teratoma, 154
polyrrhinia, 154
proboscis lateralis, 153, 154
supernumerary nostrils, 154
Notochord formation, 24
Nuclear medicine, thyroid, 162
Nystagmus, 192

O

Obstructive hydrocephalus, 57
Odontomas, 102
OEIS complex, 309
Oesophageal atresia (OA), 86, 95, 317, 318
OHVIRA syndrome, 395
Oligodontia, 106
Oligohydramnios, 32
OMENS system, 127
Omphalocele, 60, 96
embryological origin of, 306
Omphalopagus, 477
Oocytes (atresia), 15
Oogenesis
cortex of the ovary, 15
female gamete, 15
fertile period, 15
follicular cells, 15
Graafian follicle, 16, 17
granulosa cells, 16
ovum, 16
primary oocyte, 15
process, 15
secondary oocyte, 15
spermatogenesis, 16
spermiogenesis, 18–19
Oppositionality, 78
Optic nerve, 193
Optic vesicle, 191
Oral cavity, 101
embryology, 102
Oral malformations
epulis, 108
eruption cyst, 107, 108
haemangioma/lymphangioma of tongue, 109
macroglossia, 109
mouth duplication, 109
ranula, 108, 109
tongue tie, 108
white sponge naevus, 107
Organoaxial gastric volvulus, 321
Orofacial cleft, 112, 113, 116, 117
Oropharyngeal teratomas, 154
Ossification, 112
Osteogenesis imperfecta, 56, 106
Ostium primum ASD, 259
Ostium primum defect, 248, 258
Ostium secundum ASD, 259
Ostium secundum defect, 248, 258
Otic vesicle, 147
Oxycephaly, 231

P

Palatal/palatine shelves, 111, 116
Pallister-Hall syndrome, 320
Pancreas
agenesis, 368
annular pancreas, 368, 369
Beckwith-Wiedemann syndrome, 370
congenital malformations, 368
development, 366, 367
ectopic/accessory pancreatic tissue, 369
heterotaxia syndromes/isomerisms, 370
hypoplasia, 368
minor duodenal papillae/accessory pancreatic duct, 370
pancreas divisum, 369, 370
pancreatic buds, 366
Pdx 1, 366

Pancreas and duodenal homeobox gene 1 (Pdx1), 366
Pancreas divisum, 369, 370
Parapagus, 478
Parathyroid gland, 158
Paraxial mesoderm, 26
Parenchymal anomalies, 400–402
Parkes Weber syndrome, 189
Parotid gland, 165
Patau's syndrome, 58
Patent arterial duct, 282
  embryology, 282
  epidemiologic, genetic, morphologic and clinical correlations, 282–284
Patent ductus arteriosus, 277
Patent foramen ovale (PFO), 247, 257
Patent processus vaginalis (PPV), 313
PAX6 gene, 198
Pectus carinatum, 237
Pectus excavatum, 237, 239, 240
Pelviureteric junction obstruction
  associated anomalies, 404
  associated syndromes, 404
  epidemiology, 404
  long-term outcome, 404–405
  pathophysiology, 404
Penile urethra, 419
Penis, 14, 423
  cloaca, 417
  cloacal membrane, development and fate of, 417, 418
  embryology of, 417
  genital tubercle, development of, 419
    corpora cavernosa, 420, 421
    corpus spongiosum, 421
    glans, 420
  male penile urethra, development
    glanular urethra, 419
    proximal penile urethra, 418
  prepuce and frenulum, development of, 422
  scrotum, development of, 421, 422
  urogenital and labioscrotal folds, development of, 418
  urogenital plate, development of, 418
Perineal hypospadias, 424
Perineal testis, 440
Periodontitis, 107
Persistent Müllerian duct syndrome (PMDS), 432
Persistent pulmonary hypertension of the newborn (PPHN), 42
Persistent truncus arteriosus, 251
Pfeiffer syndrome, 137, 138
PHACES syndrome, 181, 295
Pharyngeal arches, 169, 177
Phenylketonuria (PKU), 52
Phocomelia thalidomide, 454
Physiological hernia, 348
Pierre Robin syndrome, 128, 129
Pilomatrixoma, 206, 207
Placenta
  chorionic villi, 36
  cytotrophoblastic shell, 36
  decidua basalis, 35
  decidua capsularis, 35
  decidua parietalis, 35
  foetal and maternal parts, 35, 36
  formation, 36
  intervillous space, 36, 38
  maternal and foetal circulation, 37, 38
  placental barrier, 36
  placental septa and cotyledons, 36
Poland syndrome, 238, 240, 241
POLR1C and POLR1D gene, 136
Polycystic kidneys, 401
Polydactyly, 455
Polyhydramnios, 182
Polyorchism, 444
Polyps, 293
Polyrrhinia, 154
Polyspermy, 19
Polysplenia syndrome, 372, 373
Ponseti method, 458
Porcine (piglets) model, 86
Posterior cardinal veins, 360, 362
Posterior neuropore, 24
Posterior urethral obstruction (PUV), 423
  clinical features, 423
  management, 423, 426
Post-trematic sensory branch, 170
Potter facies, 65
Pouches
  derivatives of
    first pouch, 170
    second pouch, 171
    third pouch, 171
    fourth pouch, 171
Preauricular pits and sinuses, 146
"Preformation" drawn, 2
Pregnancy, week 2, 21
Premigratory neural crest, 86
Prepuce, 422
Pre-trematic sensory branch, 170
Primary anophthalmia, 195
Primary congenital defect
  association, 61
  bilateral renal agenesis, 63
  dysplasia, 60
  FGFR3 skeletal dysplasia group, 61
  malformations, 60
  sequence, 61
  single or multiple defects, 60
  syndrome, 61
  type II thanatophoric dysplasia, 64
Primary palate, 111, 112, 115, 116
Primary yolk sac, 32
Primitive genital tubercle, 421
Primitive gut, 32
Primitive neuroblasts, 469
Primitive ventricle, 244
Primordial germ cells (PGCs), 467
Proboscis lateralis, 153
Processus vaginalis, 312
Progress zone, 451
Pronephros, 391
Prostate gland, 14
Proteus syndrome, 190
Proximal penile urethra, 418
Prune belly syndrome, 384
Pruzansky classification, 130
Pruzansky-Kaban classification system, 130
Psychiatric disorders, 79
Pulmonary and airways
  bronchial atresia, 292
  bronchogenic cysts, 292
  congenital lobar emphysema, 293, 294

congenital tracheal malformations, 291
congenital tracheomalacia, 292
CPAM, 294
hamartomas and polyps, 293
laryngeal atresia/stenosis, 290
laryngomalacia, 290
LC, 289
lung agenesis and pulmonary hypoplasia, 293
PHACES syndrome, 295
pulmonary sequestration, 294, 295
subglottic haemangioma, 290
tracheal agenesis, 290, 291
tracheal stenosis, 291
tracheo-oesophageal fistula +/– oesophageal atresia, 290
Pulmonary atresia with ventricular septal defect, 266–269
Pulmonary hypoplasia, 293
Pulmonary sequestration, 294, 295
Pulmonary stenosis, 251
Pulmonary valves
development, 250
transposition of, 251
Pygopagus, 478–480
Pyloric atresia, 322

Q
Quadruple test, 70
Question mark ear, 130

R
Rachipagus, 478
Ranula, 108, 109, 167
Rapidly involuting congenital haemangiomas (RICHs), 187
Rathkes cleft cysts, 234
Rearranged during Transfection (RET) gene, 336
Recti, divarication of, 307
Rectovaginal fistula, 386
Renal vein, 282
Respiratory distress, 175
Ret gene, 87
Retina, 193
Rhabdomyosarcoma
Rhabdomyosarcoma (RMS), 470–472
Right aortic arch, 277, 287
Rotational deformities, 328–331
Roundabout receptor (ROBO2), 397
Rubber bleb naevus syndrome, 190
Rubella, 51
RUNX2 gene, 107

S
Saccular cysts, 181
Sacrococcygeal teratoma, 97
Saethre-Chotzen syndrome, 138, 139
Salivary glands
congenital malformations atlas
accessory salivary glands, 167
aplasia, 166, 167
congenital tumours, 167
ductal anomalies, 167
ectopic and aberrant salivary tissue, 167
parotid glands, polycystic disease of, 167
ranulae, congenital and acquired cysts of, 167
embryology
parotid gland, 165
sublingual gland, 166
submandibular gland, 166
Schleiden, Matthias, 7
Schwann, Theodor, 7
Scimitar syndrome, 262
Sclerotherapy, 211
Scrotum, development of, 421, 422
Second branchial arch anomalies, 172
Secondary anophthalmia, 195
Secondary congenital defect
anhydramnios/oligohydramnios sequence, 65
complex, 61
deformation, 61
disruption, 61
OEIS complex, 66
Secondary palate, 111, 112, 115, 116
Secondary yolk sac, 32
Secretory stage, 103
Segmental haemangiomas, 295
Segmentation theories, 384
Semialobar holoprosencephaly, 232
Semilunar valves, anomalies of, 251
Seminal vesicles, 14
Separate theory, 185
Septo-optic dysplasia holoprosencephaly, 232
Septum anomalies, 247, 248
Septum intermedium, 246
Septum primum, 245
Septum secundum, 246
Septum transversum, 356
Sertoli cells, 18
Severe microphthalmia, 196
Severe neck lymphangioma, 95
Sheep model, 85
Shh gene, 87
Shone's syndrome, 286
Simple microphthalmia, 196
Sinus venosus, 244
Sinus venosus ASD, 260
Sinus venosus defects, 258
Sirenomelia, 452
Sistrunk's procedure, 160
Skeletal element, 170
Skin and soft tissues
cutis aplasia, 206
dermoid cysts, 206
haemangiomas, 208–210
Klippel–Trénaunay syndrome, 210
lymphangiomatosis, 210, 211
naevus, 207, 208
pilomatrixoma, 206, 207
Slit-family proteins (SLIT2), 397
Small bowel atresia, 328
Small bowel duplication, 331
Small intestinal atresia, 86
Social communication disorders, 79
Somites, 26, 27
Spemann, Hans, 8
Sperm capacitation, 19
Spermatogenesis, 16, 17
Spermatogonia, 17
Spermiogenesis, 18–19
Spherical theory, 475, 476
Spigelian hernias, 315

Spina bifida, 74, 226
Spina bifida occulta, 48, 223, 224, 228
Spina bifida/cystica, 224
Spina bifida/myelomeningocoele (SBM), 75
Spleen
accessory spleens
clinical significance, 371
diagnosis, 372
etiology, 371
sites, 371
asplenia
clinical significance, 372–373
heterotaxia, 372
isolated asplenia, 372
syndromic asplenia, 372
vs. polysplenia, 372
BASM complex
associated anomalies, 373
clinical significance, 373–374
etiology, 373
lobular spleen, 371
polysplenia, 373
splenogonadal fusion
clinical significance, 374–375
continuous SGF, 374
discontinuous SGF, 374
etiology, 374
management, 375
splenogonadal fusion-limb defect syndrome, 375
wandering spleen
acquired form, 375
clinical significance, 376
congenital form, 375
definition, 375
diagnosis, 376, 377
etiology, 375, 376
management, 376–377
presentations, 375
Splenogonadal fusion (SGF)
clinical Significance, 374–375
continuous SGF, 374
discontinuous SGF, 374
etiology, 374
management, 375
splenogonadal fusion-limb defect syndrome, 375
Splenogonadal fusion-limb defect syndrome (SGFLD), 375
Splenunculi, *see* Accessory spleens
Split notochord syndrome, 229
Split-hand/split-foot malformation (SHFM), 455
Spongy urethra, 419
Stenotic semilunar valves, 266
Stereocilia, 14
Sternomastoid tumour of infancy, 173
Stigma, 80
Stomodeum, 101
Streeter dysplasia, 456
Streeter's merging theory, 115
Sturge-Weber syndrome, 189
Sub-diaphragmatic thyroid, 161
Subglottic haemangioma, 290
Sublingual gland, 166
Submandibular gland, 166
Superficial mucous retention cysts, 167
Supernumerary kidneys
associated anomalies, 397
associated syndromes, 397
epidemiology, 397
long-term outcomes, 397–398
pathophysiology, 397
Supernumerary nostrils, 154
Supernumerary teeth, 106
Supplemental teeth, 106
Supraumbilical hernias, 308, 315
Surgical animal models, 85–86
Symmetric blastocyst cleavage, 66
Syndactyly, 456
Syndromic asplenia, 372

T

Tabula rasa, 79
Tail gut, 417
Talipes equinovarus, 458
Talon cusp, 102
Taurodontism, 102
TCOF1 (Treacle) gene, 136
TCOF1/POLR1D gene, 129
TCS, *see* Treacher Collins syndrome (TCS)
Temporomandibular joint (TMJ), 106, 127
See also Mandibular congenital malformations
Teratogenesis
amniotic banding, 46
bilateral cleft lip, 45, 46
birth defects, 45, 46
foetal alcohol syndrome, 52
Homer's Cyclops, 45
infections
cytomegalovirus, 50
infectious agents, 49
rubella, 51
varicella zoster, 51
Zika virus, 51
malformed specimens, 45
Mendelian inheritance, 45
nutrients
folic acid, 48
vitamin A, 49
phocomelia/amelia, 47
PKU, 52
rubella cataract, 45, 46
teratogenic agents, 46
teratogenic drugs, 47
thalidomide, 47
type 1/type 2 diabetes, 52
Teratogenic animal models, 85–87
Teratogen-induced limb defects, 454
Teratogens, 86
Teratomas, 154, 467–469
Tessier cleft 30 (median mandibular cleft), 125
Tessier clefting system, 119, 122
Testes, 14
Testicular descent, stages of, 437
Testicular microlithiasis, 445
Testis, 17, 444
abnormal embryology, 438
appendages, 444
calcification and microcalcification of, 445, 446
cysts, 445
historical biography, 447
normal embryology, 437
position, 444
undescended testis, 438, 439, 441
vas deferens, 444
Testosterone, 437

Tetralogy of fallot, 266–269
Thalidomide, 48, 454
The School of Athens, 3
Third branchial arch anomalies, 172
Thoracopagus, 476
Thoracopagus conjoint twins, 240
Thyroglossal duct cysts, 159, 173
  classification, 159
  clinical features, 159
  diagnosis, 160
  epidemiology, 159
  management, 160
  prognosis, 160
Thyroid, 157
  computerised tomography, 162
  congenital hypothyroidism, 162
    aetiology, 163
    epidemiology, 162, 163
    prognosis, 163
  MRI, 162
  nuclear medicine, 162
  treatment and prognosis, 162
  ultrasound, 162
Thyroid hemiagenesis, 163
Tongue tie (ankyloglossia), 108
Tooth defects
  abnormal tooth shape, 102
  AI (*see* Amelogenesis imperfecta)
  autosomal dominant
    cleidocranial syndrome, 107
    ectodermal dysplasia, 106
    Gardner's syndrome, 106, 107
    Marfan's syndrome, 106
    osteogenesis imperfecta, 106
  autosomal recessive
    Down's syndrome, 107
    X-linked disease, 107
  detine, 104, 105
  enamel defects, 103
  eruption of a teeth, 106
  hyperdontia, 106
  hypodontia, 106
  oligodontia, 106
  periodontal tissue malformations
    hereditary gingival fibromatosis, 107
    periodontitis, 107
  supernumerary teeth, 106
  supplemental teeth, 106
Topobiology, 9
Total anomalous pulmonary venous drainage, 262
  embryology, 262
  epidemiologic, genetic, morphologic and clinical correlation, 262–264
Total intestinal aganglionosis (TIA), 335
Toxoplasmosis, 50
Tracheal agenesis, 290, 291
  aetiology and patho-embrology, 183
  associations, 184
  definition, 183
  incidence, 183
  presentation, clinical features and outcome, 183
Tracheal stenosis, 291
Tracheal web
  aetiology and patho-embrology, 183
  definition, 183
  incidence, 183
  presentation, clinical features and outcome, 183
Tracheoesophageal fistulae (TOF), 180, 290, 317, 318
  aetiology and patho-embrology, 182
  associations, 182
  definition, 182
  incidence, 182
  presentation, clinical features and outcome, 182
  syndromes and malformations associated with, 182
Tracheomalacia
  aetiology and patho-embrology, 183
  associations, 183
  definition, 183
  incidence, 183
  presentation, clinical features and outcome, 183
Transitional phase, 43
Treacher Collins syndrome (TCS), 121, 125, 129, 135, 136, 140, 141
Tricuspid atresia, 249
Triple test, 70
Trisomy 18, 320
Trisomy 21, 320
Trisomy 21-Down Syndrome, 70
Trophoblast, 21
Truncus arteriosus, anomalies of, 250
Tubular hypoplasia, 58
Tubular invagination, 32
Tufted angioma, 189
Tuftelin, 103
TWIST1 gene, 136
Type 2 Mayer-Rokitansky-Kuster-Hauser syndrome, 395
Type II thanatophoric dysplasia, 64

U

Uber Entwickelungsgeschichte der Thiere, 7
Ultrasound, 162
Umbilical arteries, 278
Umbilical cord, 348
Umbilical granulomas and polyps, 350
Umbilical hernia, 308, 349
Umbilical veins, 244, 280, 360
Umbilicus
  embryological considerations, 348–349
  umbilical cord, 348
  umbilical granulomas and polyps, 350
  umbilical hernia, 349–350
  urachal remnants, 350
  vitelline duct remnants, 350
Umblical hernia, 349
Undescended right testis, 439
Undescended testis (UDT), 438, 439, 441
Unilateral choanal atresia, 155
Unilateral renal agenesis (URA), 393–395
Univentricular hearts, 272, 273
Unseptatable ventricles, 272
Upper and lower limbs, arteries development of, 278, 279
Urachal remnants, 350
Urachus, 348, 351, 412, 413
Ureter
  duplex systems
    associated anomalies, 404
    associated syndromes, 404
    ectopic ureter 2 retrogrades, 403
    epidemiology, 402
    pathophysiology, 402–403
  pelviureteric junction obstruction
    associated anomalies, 404
    associated syndromes, 404
    epidemiology, 404

Ureter (*cont.*)
long-term outcome, 404–405
pathophysiology, 404
vesicoureteric junction obstruction
associated anomalies, 405
associated syndromes, 405
epidemiology, 405
long-term outcome, 405–406
pathophysiology, 405
Ureteric bud, 392
Ureteroceles, 403
Urethra, 417, 423
cloaca, 417
cloacal membrane, development and fate of, 417, 418
genital tubercle, development of, 419
corpora cavernosa, 420, 421
corpus spongiosum, 421
glans, 420
hypospadias, disorganisation hypothesis, 422
male penile urethra, development
glanular urethra, 419
proximal penile urethra, 418
prepuce and frenulum, development of, 422
posterior urethral valves, 423
clinical features, 423
management, 423, 426
scrotum, development of, 421, 422
urogenital and labioscrotal folds, development of, 418
urogenital plate, development of, 418
Urethral fistula, 434
Urogenital folds, 421
Urogenital folds, development of, 418
Urogenital plate, development of, 418
Urogenital sinus, 418
Uropathies, 97–98
Urorectal septum, 381
Uterine tube, 11, 12
Uterine wall, 13
Uterus, 13

V
VACTERL (vertebral anomalies, anal atresia, cardiac defects, tracheoesophageal fistula, renal anomalies, limb defects), 147
Vaginal cyst, 387
Valproic acid, 206
Valsalva testing/jugular vein compression, 152
Varicella zoster, 51
Vas deferens, 17, 444
Vascular anomalies, 173, 185
antenatal detection, role of, 188, 189
associations
vascular malformations, 189, 190
vascular tumours, 189
classification systems, 187, 188
clinical features
vascular malformations, 187
vascular tumours, 186, 187
epidemiology
vascular malformations, 186
vascular tumours, 186
genetics
vascular malformations, 186
vascular tumours, 186
pathoembryology
vascular malformations, 185, 186
vascular tumours, 185
Vascular element, 170
Vascular malformations, 185, 186, 459
associations, 189, 190
clinical features, 187
epidemiology, 186
genetics, 186
ISSVA classification, 188
syndromes associated with, 190
Vascular system, 275
Vascular tumours, 185
associations, 189
clinical features, 186, 187
epidemiology, 186
genetics, 186
ISSVA classification, 188
syndromes associated with, 190
Vein of galen aneurysmal malformation (VGAM), 234, 235
Veins, 279
cardinal, 281
vitelline and umbilical, 280
Velopharyngeal incompetence, 115
Venous malformations, 187
Ventral forebrain, 232
Ventral splanchnic arteries, 278
Ventricles, development of, 249
Ventricular septal defect (VSD), 254–256
Ventriculo-arterial junctions, 254
Ventriculo-peritoneal shunting, 226–227
Ventriculoseptal defects, 300
Verbal body image, 78
Vesicourachal diverticuli, 412
Vesicoureteric junction obstruction
associated anomalies, 405
associated syndromes, 405
epidemiology, 405
long-term outcome, 405–406
pathophysiology, 405
Vitamin A, 49
Vitelline intestinal remnant, 351
Vitelline veins, 244, 280, 360, 361
Vitello-intestinal duct, 32
Vocal cord paralysis
aetiology and patho-embrology, 179
associations, 180
definition, 179
incidence, 179
presentation, clinical features and outcome, 179
von Baer, Karl Ernst, 7
von Haller, Albrecht, 5, 446, 447

W
Wandering spleen
acquired form, 375
clinical significance, 376
congenital form, 375
definition, 375
diagnosis, 376, 377
etiology, 375, 376
management, 376–377
presentations, 375
Warburg's syndrome, 197
Wedge hypothesis, 415
Weismann, August, 8

Wharton's jelly, 348
Wharton's jelly umbilical cord, 352
White sponge naevus, 107
WHO International Collaborative Research on Craniofacial Anomalies Project, 117
Wiedemann-Beckwith syndrome, 146
Wilms' tumour, 464–466
Witkop classification, 104
Wolff, Caspar Friedrich, 5
Wolffian duct (WD), 431
Wyburn-Mason syndrome, 189

X
X-linked dominant inheritance, 57
X-linked genetic disease, 107
X-linked Opitz G/BBB, 320
X-linked recessive hypoparathyroidism (XLHPT), 163
X-linked recessive inheritance, 57

Y
Yolk sac, 22, 31–33
Yolk sac stalk, 32

Z
Zika virus, 51
Zona pellucida, 19